THE LIBRARY
ST. MARY'S COLLEGE OF MARYLAND
ST. MARY'S CITY, MARYLAND 20686

OXFORD MONOGRAPHS ON GEOLOGY AND GEOPHYSICS NO. 15

Series editors

H. Charnock
J. F. Dewey
S. Conway Morris
A. Navrotsky
E. R. Oxburgh
R. A. Price
B. J. Skinner

OXFORD MONOGRAPHS ON GEOLOGY AND GEOPHYSICS

1. DeVerle P. Harris: *Mineral resources appraisal: mineral endowment, resources, and potential supply: concepts, methods, and cases*
2. J. J. Veevers (ed.): *Phanerozoic earth history of Australia*
3. Yang Zunyi, Wang Hongzhen, and Cheng Yuqi (eds.): *The geology of China*
4. Lin-gun Liu and William A. Bassett: *Elements, oxides, and silicates: high-pressure phases with implications for the earth's interior*
5. Antoni Hoffman and Matthew H. Nitecki (eds.): *Problematic fossil taxa*
6. S. Mahmood Naqvi and John J. W. Rogers: *Precambrian geology of India*
7. Chih-Pei Chang and T. N. Krishnamurti (eds.): *Monsoon meteorology*
8. Zvi Ben-Avraham (ed.): *The evolution of the Pacific Ocean margins*
9. Ian McDougall and T. Mark Harrison: *Geochronology and thermochronology by the $^{40}Ar/^{39}Ar$ method*
10. Walter C. Sweet: *The conodonta: morphology, taxonomy, paleoecology, and evolutionary history of a long-extinct animal phylum*
11. H. J. Melosh: *Impact cratering: a geologic process*
12. J. W. Cowie and M. D. Brasier (eds.): *The Precambrian-Cambrian boundary*
13. C. S. Hutchison: *Geological evolution of southeast Asia*
14. Anthony J. Naldrett: *Magmatic sulfide deposits*
15. D. R. Prothero and R. M. Schoch (eds.): *The evolution of perissodactyls*

# The Evolution of Perissodactyls

*Edited by*

DONALD R. PROTHERO
*Occidental College*

ROBERT M. SCHOCH
*Boston University*

New York · Oxford
CLARENDON PRESS · OXFORD UNIVERSITY PRESS
1989

Oxford University Press

Oxford  New York  Toronto
Delhi  Bombay  Calcutta  Madras  Karachi
Petaling Jaya  Singapore  Hong Kong  Tokyo
Nairobi  Dar es Salaam  Cape Town
Melbourne  Auckland

and associated companies in
Berlin  Ibadan

Copyright © 1989 by Oxford University Press, Inc.

Published by Oxford University Press, Inc.,
200 Madison Avenue, New York, New York 10016

Oxford is a registered trademark of Oxford University Press

All rights reserved. No part of this publication may be reproduced,
stored in a retrieval system, or transmitted, in any form or by any means,
electronic, mechanical, photocopying, recording, or otherwise,
without the prior permission of Oxford University Press.

Library of Congress Cataloging-in-Publication Data
The Evolution of perissodactyls / edited by Donald R. Prothero,
Robert M. Schoch. p.  cm. — (Oxford monographs on geology and geophysics ; no. 15)
Bibliography: p. ISBN 0-19-506039-3
1. Perissodactyla, Fossil—Evolution.  2. Hyracoidea, Fossil.
I. Prothero, Donald R.  II. Schoch, Robert M.  III. Series.
QE882.U6E96  1989    569'.72—dc19   89-3052  CIP

987654321

Printed in the United States of America

## PREFACE

On August 16, 1985, a workshop on the evolution of perissodactyls, or "odd-toed" hoofed mammals, was held at the Fourth International Theriological Congress in Edmonton, Canada. Organized by Robert M. Schoch and Jens L. Franzen, it included many of the scientists from around the world who work on perissodactyls, and brought together a tremendous amount of new information about this important, but still poorly understood, group of mammals. As a consequence of that meeting, the participants agreed that there were so many significant new discoveries and ideas presented at the workshop that its results should be published. During the succeeding four years, the editors of this work solicited manuscripts from most of the original participants, as well as manuscripts from authors who were not at the original meeting, but had important information to contribute. Consequently, this volume includes much important new research on fossil perissodactyls, written by scientists from around the world who have expertise in this group. There are not only review papers of every major subordinal group, but also descriptions of important new forms, including the oldest perissodactyl-like mammal, and the oldest rhinocerotid. Most of the papers include phylogenetic analyses of their respective groups, including one of the whole order. Most surprising, this volume includes evidence that shows that hyraxes, long thought to be related to elephants, are in fact perissodactyls (as Owen thought when he created the taxon Perissodactyla in 1848). This wealth of new information is summarized in the final chapter.

In bringing these contributions together, we have also attempted to include the approaches of scientists from both the New World and the Old World wherever possible. A case in point is the differing schools of thought concerning hipparionine evolution (as represented by Chapters 12 and 13 by Woodburne and Alberdi, respectively). At the time this workshop was being organized, Leonard Radinsky, one of the world's foremost students of perissodactyls, was invited to participate. Tragically, he was unable to attend and died of cancer just days after the workshop was held. We dedicate this volume to him, and invited one of his colleagues, Jim Hopson, to write a memorial and review of his contributions to the study of perissodactyls.

At the time this book went to press, the Eocene and Oligocene timescales were undergoing radical revision, so many of these chapters could not be updated to reflect this new chronology. See p. 505 for further discussion.

## ACKNOWLEDGMENTS

This volume was prepared by the senior editor as camera-ready copy on an Apple MacIntosh computer, using Microsoft Word software and Laserwriter NT II printer. Virtually all the word-processing, and much of the editing, was cheerfully and patiently done by Steven King. Without his skill and dedication, this volume would never have been possible. Mr. Clifford Prothero helped with the page layouts, and much of the art preparation. Prothero was partially supported by a Guggenheim Fellowship, and by NSF grant EAR 87-08221 during the editing of this manuscript. Schoch thanks his wife, Cynthia, and children, Nicholas and Edward, for their patience while this volume was being prepared.

CONTENTS

Contributors                                                                                              ix

1. Leonard Burton Radinsky (1937-1985)  *James A. Hopson*                                                  3

2. A brief historical review of perissodactyl classification  *Robert M. Schoch*                          13

3. *Radinskya yupingae*, a perissodactyl-like mammal from the late Paleocene of
   southern China  *Malcolm C. McKenna, Chow Minchen, Ting Suyin, and Luo Zhexi*                          24

4. Hyracoids, the sister-group of perissodactyls  *Martin S. Fischer*                                     37

5. The evolution of the Hyracoidea: a review of the fossil evidence  *D. Tab Rasmussen*                   57

6. Character polarities in early perissodactyls and their significance for
   *Hyracotherium* and infraordinal relationships  *J. J. Hooker*                                         79

7. Origin and systematic position of the Palaeotheriidae  *Jens Lorenz Franzen*                          102

8. Phylogeny of the Family Equidae  *Robert L. Evander*                                                  109

9. Dental character variation in paleopopulations and morphospecies of fossil horses
   and extant analogues  *Bruce J. MacFadden*                                                            127

10. The evolution of Oligocene horses  *Donald R. Prothero and Neil Shubin*                              142

11. Phylogenetic interrelationships and evolution of North American late Neogene
    Equidae  *Richard C. Hulbert, Jr.*                                                                   176

12. Hipparion horses: a pattern of endemic evolution and intercontinental dispersal
    *Michael O. Woodburne*                                                                               197

13. A review of Old World hipparionine horses  *Maria-Teresa Alberdi*                                    234

14. A quantitative study of the North American fossil species of the genus *Equus*
    *Melissa C. Winans*                                                                                  262

15. A review of the tapiroids  *Robert M. Schoch*                                                        298

16. The history of the Rhinocerotoidea  *Donald R. Prothero, Claude Guérin, and Earl
    Manning*                                                                                             322

17. The phylogenetic history and adaptive radiation of the Amynodontidae
    *William P. Wall*     341

18. The allaceropine hyracodonts   *Kurt Heissig*     355

19. The systematics of indricotheres   *Spencer G. Lucas and Jay Sobus*     358

20. *Teletaceras radinskyi*, a new primitive rhinocerotid from the late Eocene Clarno Formation, Oregon   *C. Bruce Hanson*     379

21. The Rhinocerotidae   *Kurt Heissig*     399

22. *Hoploaceratherium*, n. gen., a new generic name for "*Aceratherium*" *tetradactylum*   *Leonard Ginsburg and Kurt Heissig*     418

23. Taxonomy and biochronology of *Eomoropus* and *Grangeria*, Eocene chalicotheres from the western United States and China   *Spencer G. Lucas and Robert M. Schoch*   422

24. Interrelationships and diversity in the Chalicotheriidae   *Margery C. Coombs*     438

25. The Brontotheriidae: a systematic revision and preliminary phylogeny of North American genera   *Bryn J. Mader*     458

26. European brontotheres   *Spencer G. Lucas and Robert M. Schoch*     485

27. Taxonomy of *Duchesneodus* (Brontotheriidae) from the late Eocene of North America   *Spencer G. Lucas and Robert M. Schoch*     491

28. Origin and evolution of the Perissodactyla: summary and synthesis   *Donald R. Prothero and Robert M. Schoch*     504

29. Classification of the Perissodactyla   *Donald R. Prothero and Robert M. Schoch*     530

# CONTRIBUTORS

MARIA-TERESA ALBERDI, Museo National Ciencias Naturales, C.S.I.C., Jose Gutierrez Abascal 2, 28006 Madrid, Spain

CHOW MINCHEN, Institute of Vertebrate Paleontology and Paleoanthropology, Academy of Sciences, Beijing, People's Republic of China

MARGERY C. COOMBS, Department of Zoology, 348 Morrill Science Center, University of Massachusetts, Amherst, Massachusetts 01003

ROBERT L. EVANDER, Box 725, Branchville, New Jersey 07826

MARTIN S. FISCHER, Institut für Biology III, Universität Tübingen, Auf der Morgenstelle 28, D7400 Tübingen 1, Federal Republic of Germany

JENS LORENZ FRANZEN, Forschunginstitut Senckenberg, Senckenberganlage 25, 6000 Frankfurt 1, Federal Republic of Germany

LEONARD GINSBURG, Institute de Paleontologie, Musée National d'Histoire Naturelle, 8 Rue de Buffon, 75005 Paris, France

CLAUDE GUÉRIN, Department des Sciences de la Terre, Université Claude Bernard-Lyon I, 43 Blvd. du 11 Novembre 1918, 69622 Villeurbanne Cedex, France

C. BRUCE HANSON, Department of Paleontology, University of California, Berkeley, California 94720

KURT HEISSIG, Bayerische Staatssammlung für Paläontologie und historische Geologie, Richard-Wagner-Strasse 10/2, 8000 München 2, Federal Republic of Germany

J. J. HOOKER, Department of Palaeontology, British Museum (Natural History), Cromwell Road, London SW7 5BD, England

JAMES A. HOPSON, Department of Anatomy, University of Chicago, 1025 E. 57th St., Chicago, Illinois 60637

RICHARD C. HULBERT, JR., Florida Museum of Natural History, University of Florida, Gainesville, Florida 32611

SPENCER G. LUCAS, New Mexico Museum of Natural History, P.O. Box 7010, Albuquerque, New Mexico 87194

LUO ZHEXI, Department of Paleontology, University of California, Berkeley, California 94720

BRUCE J. MacFADDEN, Florida Museum of Natural History, University of Florida, Gainesville, Florida 32611

BRYN J. MADER, Department of Zoology, 348 Morrill Science Center, University of Massachusetts, Amherst, Massachusetts 01003

EARL MANNING, Museum of Geosciences, Louisiana State University, Baton Rouge, Louisiana 70803

MALCOLM C. McKENNA, Department of Vertebrate Paleontology, American Museum of Natural History, Central Park West at 79th St., New York, New York 10024

DONALD R. PROTHERO, Department of Geology, Occidental College, Los Angeles, California 90041

D. TAB RASMUSSEN, Department of Anthropology, University of California, Los Angeles, California 90024

ROBERT M. SCHOCH, College of Basic Studies, Boston University, 871 Commonwealth Avenue, Boston, Massachusetts 02215

NEIL SHUBIN, Department of Paleontology, University of California, Berkeley, California 94720

JAY C. SOBUS, Department of Geology, University of New Mexico, Albuquerque, New Mexico 87131

TING SUYIN, Institute of Vertebrate Paleontology and Paleoanthropology, Academy of Sciences, Beijing, People's Republic of China

WILLIAM P. WALL, Department of Biology, Georgia College, Milledgeville, Georgia 31061

MELISSA C. WINANS, Vertebrate Paleontology Laboratory, Texas Memorial Museum, University of Texas, Balcones Research Center, Bldg. 6, 10100 Burnet Road, Austin, Texas 78758

MICHAEL O. WOODBURNE, Department of Earth Sciences, University of California, Riverside, California 92521

# The Evolution of Perissodactyls

# 1. LEONARD BURTON RADINSKY (1937-1985)

## JAMES A. HOPSON

Leonard B. Radinsky was a leading American vertebrate paleontologist of the second half of the twentieth century. He made major contributions on the systematics and evolution of early perissodactyls, on the evolution of the mammalian brain during the Cenozoic, and on the quantitative study of mammalian functional craniology. Len spent the greater part of his career at the University of Chicago, where he was my colleague and friend. He died on August 30, 1985, at the age of 48, after a long and brave fight against cancer.

In the following pages I shall briefly review some of the major aspects of Len's career (some of which I have published elsewhere; Hopson, 1986) and offer my assessments of his principal contributions to vertebrate paleontology and evolutionary biology. In keeping with the subject of this volume which is dedicated to his memory, I shall place special emphasis on his contributions to our understanding of the systematics and evolution of Perissodactyla. (Radinsky's publications are cited in the following text by date only and are listed separately in his complete bibliography at the end of this chapter.)

Leonard Radinsky was born on Staten Island, New York, on July 16, 1937, of Russian-Polish immigrant parents. His lifelong fascination with fossils was developed in childhood. At the age of thirteen he wrote and illustrated a short "book" entitled "Prehistoric Animals." He continued the serious pursuit of his paleontological interest at Cornell University, from which he graduated in 1958 with a bachelor's degree in geology. He entered graduate school in geology at Yale University where he began the study of vertebrate paleontology with Joseph T. Gregory.

While a graduate student at Yale, Len published his first paper, a review of tooth histology in chondrichthyans (1961). His doctoral dissertation (1963a) was a systematic revision of North American Eocene and Oligocene Tapiroidea exclusive of Tapiridae. It was begun under Gregory's supervision and completed under Elwyn L. Simons, who replaced Gregory at Yale. Following completion of his doctoral studies, Len spent the 1962-1963 academic year at the American Museum of Natural History in New York, where he described the museum's extensive collection of Eocene and Oligocene tapiroids from Asia. From this extremely productive postdoctoral year came a monographic study entitled "Early Tertiary Tapiroidea of Asia" (1965a), and short papers on the first metatarsal in fossil and living ceratomorphs (1963b), a new Eocene chalicothere (1964a), and Eocene and Oligocene fossil localities in Inner Mongolia (1964b).

In 1963, Len joined the faculty of the Biology Department of Boston University as Instructor. He became a Research Associate at the Museum of Comparative Zoology at Harvard, where he described the skeleton of the Early Eocene tapiroid *Heptodon* (1965b). The following year he moved to New York where he became Assistant Professor at Brooklyn College and Research Associate at the American Museum. During the three years (mid-1964 to mid-1967) in New York he continued research on the early evolution of the Perissodactyla, publishing a monographic review of the rhinocerotoid family Hyracodontidae

(1967b) and five shorter papers, the most significant of which dealt with the origin of perissodactyls (1966c), the families of rhinocerotoids (1966d), and the taxonomy and phylogeny of *Hyrachyus* and *Chasmotherium* (1967c). With the exception of a 1969 (1969b) paper, which was a synthesis of his work on the early adaptive radiation of the major subgroups of perissodactyls, Len did no further work on perissodactyl systematics and phylogeny after 1966.

Len began his studies of brain evolution while at Brooklyn College and the American Museum. There he developed a new measure of relative brain size in mammals, using the area of the foramen magnum as a substitute for body size (1967a). He also began the study of endocranial casts of fossil mammals as a means for understanding the evolution of the brain through geologic time. His first paper on this subject described the oldest primate endocast (1967d).

In 1967, Len came to the University of Chicago to join a developing group of comparative morphologists and paleontologists in the Department of Anatomy and the newly formed Committee on Evolutionary Biology. He was promoted to Associate Professor in 1970 and to Professor in 1976. He became a Research Associate in the Field Museum of Natural History in Chicago in 1971. After serving a year as Acting Chairman, he became Chairman of Anatomy in 1978, a position he held until 1983.

Len was a popular teacher, both of undergraduates in the College and of graduate students. In his undergraduate teaching he derived his greatest pleasure from showing non-biology majors how to appreciate both the unity and the diversity of vertebrate structure. His principal aim in his graduate teaching was to instill the ability to read the scientific literature critically. Len was a good role model to students; he derived great pleasure from his research and his enthusiasm was evident to all who heard him discuss it, whether in the classroom or at professional meetings.

In his research as well as in his training of graduate students, Len was cognizant of the need to understand the morphology of living mammals, particularly the functional significance of structure, before one could justify making statements about the adaptive significance of fossil morphology. All of Len's graduate students worked on the functional morphology of living mammals, either exclusively or in conjunction with studies on fossils, for their dissertation research.

At Chicago, Len's research until 1980 was almost entirely on the evolution of the mammalian brain as revealed by endocranial casts of fossil and living mammals. He developed a method for making latex casts of the cranial cavities of recent skulls (1968) and built up a large collection of endocasts of living species both as a basis for comparison with fossil endocasts and as material for the study of brain size and shape diversity. (His collection of fossil and recent mammalian endocasts is now housed in the Division of Mammals at the Field Museum.) From 1967 until about 1975 his research was exclusively on brain evolution in primates and carnivorans. After 1975, in addition to a few papers on primate and carnivoran brains, he published studies of the brains of mesonychids (1976a), horses (1976f), and South American ungulates (1981b), and an important review of brain size evolution in carnivores and ungulates (1978d).

From 1980 until his death in 1985, Len studied the functional craniology of mammals utilizing multivariate techniques of shape analysis. He published a series of papers on the evolution of skull shape in modern and fossil carnivorans (1981a, 1981c, 1982c, 1984a) and, with Sharon B. Emerson, a functional analysis of the cranial morphology of sabertooths (1980d). In addition, he published two papers on the evolution of skull shape in horses (1983a, 1984c) and one on jaw shape in ungulates (1985a). His last scientific paper was a review entitled "Approaches in evolutionary morphology: a search for patterns" (1985b).

At the time of his death, Len was completing revisions of a manuscript for an undergraduate textbook on the evolution of vertebrates. The revisions were completed by his wife Sharon Emerson and the book was published in 1987 as *The Evolution of Vertebrate Design* by the University of Chicago Press. As described in the Publisher's Note, the book was an outgrowth of his course on vertebrate morphology and evolution for undergraduates who were not majoring in biology. His aim in this course was to give the students an appreciation of vertebrate evolution by showing them how vertebrate body design changed through evolutionary time.

## Leonard Radinsky's contribution to perissodactyl systematics

Leonard Radinsky's published work on perissodactyl systematics falls into three main categories: (1) Systematics and morphology of Tapiroidea (1963a, 1963b, 1965a, 1965b, 1966a, 1966b, 1967c, 1983b); (2) Systematics of Eocene Chalicotherioidea (1964a); (3) Systematics and morphology of Rhinocerotoidea (1966d, 1967b, 1967c, 1983b). They are reviewed and discussed in the following paragraphs. His other publications on aspects of perissodactyl morphology and evolution, which do not include major taxonomic revisions, are briefly mentioned here. A review of the origin of perissodactyls from phenacodontids (1966c) detailed the diversity of functional adaptations in the feeding and locomotor systems of the phenacodontids and *Hyracotherium*, and suggested that perissodactyls originated from the phenacodont genus *Tetraclaenodon*. A review of the early radiations of Perissodactyla (1969b) detailed the functional changes in the molar dentition in the early members of the five superfamilies of perissodactyls; the contents of this paper were based largely on analyses in his earlier papers. A report on the brain of *Hyracotherium* (1976f) demonstrated that the earlier description of an extremely primitive brain in this genus was based on a probable "condylarth" endocast; the brain of *Hyracotherium* actually was quite large for an Early Eocene mammal.

Two papers on the allometry of fossil and modern horse skulls (1983a, 1984c) indicated that both scaling effects and reorganization were factors in the evolutionary transformation of horse skulls; reorganization occurred with the development of high-crowned teeth and involved "primarily a ventral and anterior displacement of the tooth row relative to the jaw joint and the orbit" (1984c, p. 13).

*Systematics and morphology of Tapiroidea.* In his major review of North American tapiroids exclusive of tapirids (1963a), Radinsky synonymized four of the ten named genera of isectolophids and helaletids and worked out the probable interrelationships of genera and species. He subsequently (1966b) named a new genus of helaletid, *Selenaletes*, from the Early Eocene. In his review of Asiatic tapiroids (1965a), he added two new families, Deperetellidae and Lophialetidae, to the four tapiroid families recognized by Simpson (1945), and synonymized two genera with *Deperetella* and one genus with *Colodon*. The supposed isectolophid *Indolophus*, which he had earlier removed from that family (1963a), he considered to be different from all other known tapiroids and declared it "Family *incertae sedis*." In these papers (also see 1969b), he discussed the patterns of dental evolution that characterized each of the tapiroid families. Recently, Schoch and Prins (1984), in a brief summary of the evolution of the Tapiroidea, recognized the families accepted by Radinsky and essentially repeated his conclusions on familial interrelationships and major dental trends. They also suggested that *Indolophus* may be a hippomorph.

Radinsky (1965a, 1966a) described *Rhodopagus* and *Pataecops* (originally *Pataecus*), two genera of diminutive Asiatic ceratomorphs he considered to be tapiroids and provisionally placed in the family Lophialetidae. The molars of these genera possess an unusual pattern for tapiroids, including a long ectoloph formed in part by a flattened metacone, which Radinsky considered to bear a resemblance to that of *Lophialetes*. As Radinsky (1965a, p. 189) also noted, the dentition of *Lophialetes* bears a striking resemblance to that of hyracodontid rhinocerotoids, although it is more primitive than any known rhinocerotoid in possessing a long hypoconulid on the last lower molar. *Rhodopagus* lacks a hypoconulid on $M_3$ (the lower dentition of *Pataecops* is unknown). Radinsky did not characterize *Rhodopagus* and *Pataecops* as having rhinocerotoid-like molars, but, in fact, their molars do resemble those of hyracodontids.

Lucas and Schoch (1981), on the basis of eight features of the molars of *Rhodopagus* and *Pataecops*, argued that these genera are members of the rhinocerotoid family Hyracodontidae. Prothero, Manning, and Hanson (1986) also accepted the rhinocerotoid affinities of these genera, although they hesitated to place them definitely within the Hyracodontidae. Their reason for this appears unclear to me, inasmuch as they noted that *Rhodopagus* possesses a reduced lower canine, "a hyracodont feature" (p. 361). It appears that their hesitation in placing them within the Hyracodontidae is due to (1) the possession by *Rhodopagus* of "the primitive spatulate lower incisors, rather than the conical incisors of hyracodontids," and (2) lack of knowledge on "the relative length of their metapodials (a key hyracodont synapomorphy)" (p. 361). However, Radinsky (1967b) described the incisors of the primitive hyracodontid *Triplopus* as spatulate, and described and illustrated spatulate incisors in the hyracodontids *Epitriplopus, Triplopides,* and *Ardynia*. Therefore, I fail to see why *Rhodopagus* and *Pataecops* should not be included in the Hyracodontidae on the basis of the reduced canine and the majority of molar features listed by Lucas and Schoch.

Radinsky (1965b) provided an excellent description of the skeleton of the Early Eocene tapiroid *Heptodon posticus*. He compared it with the skeleton of a living tapir and showed that "the more striking osteological changes which occurred during fifty million years of evolution from *Heptodon* to *Tapirus* are modifications of the skull correlated with development of a proboscis" (p. 101). The postcranial skeleton of *Tapirus* differed from that of *Heptodon* mainly in features correlated with its larger size. Interestingly, the modern tapir shows some cursorial specializations in its skeleton that are absent in the smaller, more gracile Eocene tapiroid, suggesting that the living tapirs passed through a more specialized running stage before acquiring their present heavy-bodied, short-legged aspect.

*Systematics of Eocene Chalicotherioidea.* Radinsky (1964a) described an Early Eocene molar dentition resembling that of a primitive tapiroid but with a distinct paraconule and paracone and a complete high metaloph. On the basis of these features he suggested that the specimen was a chalicothere, the only one known until then that was older than late Eocene in age. It was named *Paleomoropus* and placed in the primitive chalicotherioid family Eomoropidae. He also considered the European early to middle Eocene genus *Lophiaspis*, considered a lophiodontid tapiroid, to be a chalicothere. *Paleomoropus* was considered to demonstrate an early origin for the Chalicotherioidea, as ancient as that of the Equoidea and Tapiroidea. Consequently they were removed from the suborder Hippomorpha and placed in their own suborder Ancylopoda.

Fischer (1977) challenged the identification of *Lophiaspis* and *Paleomoropus* as chalicotheres, claiming, on the basis of more complete skull material, that *Lophiaspis* was a lophiodontid tapiroid. This view is accepted by Prothero, Manning, and Hanson (1986), but Hooker (1984) draws an entirely different conclusion from the resemblance of lophiodontid molars to those of chalicotheres. He accepts the chalicothere relationships of *Lophiaspis* and *Paleomoropus* (see also Schoch, 1983) and considers their similarities to *Lophiodon* as evidence of a sister group relationship of lophiodontids and chalicotheres within a paraphyletic Tapiroidea. Thus, Radinsky's allocation of *Paleomoropus* and *Lophiaspis* to the Chalicotherioidea appears at present to represent too simple an interpretation of a more complex phylogenetic problem than he envisaged.

*Systematics and morphology of the Rhinocerotoidea.* In a series of three interrelated papers on Eocene and Oligocene rhinocerotoids, Radinsky (1966d, 1967b, 1967c) greatly advanced understanding of the phylogeny of the Rhinocerotoidea. In his review of the families of Rhinocerotoidea (1966d) he clarified the monophyly of the family Rhinocerotidae by restricting it to those taxa in which the first upper incisor is chisel-shaped and the second lower incisor is a procumbent, lanceolate tusk. Forms with either stubby, conical tusks formed by the first upper and lower incisors (the *Paraceratherium* group, or indricotheres), large, erect canine tusks and reduced incisors (the *Allacerops* group), or moderately enlarged, pointed incisors and stout, stubby canines (the *Forstercooperia* group) were therefore removed from the Rhinocerotidae. This restriction of the family Rhinocerotidae has become universally accepted.

The groups that Radinsky removed from the Rhinocerotidae he placed in the Hyracodontidae, which in his view was a grade group (a paraphyletic group in current terminology) including "the primitive late Eocene non-amynodontid rhinocerotoids and the (relatively) unsuccessful lines of the late Eocene and Oligocene radiation which arose from those forms" (1966d, p. 638).

In his monographic review of the family Hyracodontidae, Radinsky (1967b) provided no uniquely diagnostic features of the family, with the possible exception of characterizing them as "basically cursorial rhinocerotoids" (p. 7). He also cites the tridactyl manus as a feature of *Triplopus cubitalis*, one of the earliest hyracodontids, and of all later hyracodontids in which the manus is known. But he cautions that the "condition of the manus should not be considered as diagnostic of the family, for some late Eocene, non-amynodontid rhinocerotoid must have retained a tetradactyl manus, since that is the condition in *Trigonias* and *Epiaceratherium*, the oldest (early Oligocene) rhinocerotids. Because the characteristic rhinocerotid incisor specialization is so primitive in *Epiaceratherium*, we may assume that its late Eocene ancestor lacked that specialization, and therefore would be considered a hyracodontid" (p. 7). With the advent of cladistic thinking in systematics, the possession of a tridactyl manus and cursorial specializations of the limbs are considered the principal (or only?) synapomorphies of the dentally-diverse Hyracodontidae (Prothero, Manning, and Hanson, 1986).

Radinsky (1966d) had removed the giant paraceratheres, or indricotheres, from the Rhinocerotidae and placed them in the Hyracodontidae. Within this family, he considered them to have been derived from some late Eocene species of *Forstercooperia* such as *F. sharamurenense*. The latter species, considered a junior synonym of *Juxia borissiaki* by Lucas, Schoch, and Manning (1981), is placed as the sister taxon to the giant paraceratheres by these authors and by Prothero, Manning, and Hanson (1986).

The Eocene Hyrachyidae had traditionally been included in the Rhinocerotoidea as "a primitive family near the structural ancestry of the whole superfamily" (Simpson, 1945, p. 257). In his review of the families of rhinocerotoids, Radinsky (1966d) removed *Hyrachyus* from the Rhinocerotoidea and reduced the family Hyrachyidae to subfamilial status within the tapiroid family Helaletidae. His reason for removing *Hyrachyus* from the Rhinocerotoidea was explicitly stated: "Since... *Hyrachyus* lacks the features diagnostic of the Rhinocerotoidea, there is no reason to assign it to that superfamily" (p. 633). He acknowledged that "some advanced individuals of one species of *Hyrachyus* show the beginnings of those characters, and may be ancestral to the primitive hyracodontid rhinocerotoid *Triplopus*" (pp. 632-633). But, as he explained in a later paper on the subject (Radinsky, 1967c), "because *Hyrachyus* has not diverged much in dentition or other skeletal features from the primitive ceratomorph tapiroid condition, and has not achieved the dental specializations that characterize the Rhinocerotoidea, I have transferred *Hyrachyus* to the Tapiroidea, an action that implies nothing about the phylogenetic position of *Hyrachyus* relative to the rhinocerotoids" (p. 19). The last part of the sentence was meant to emphasize the point (misunderstood by Savage, Russell, and Louis, 1966) that his reason for removing *Hyrachyus* from the Rhinocerotoidea was independent of his views on its relationships. In fact, he believed that different species of *Hyrachyus* may have independently given rise to the hyracodontids *Triplopus* and *Forstercooperia*. His classifying *Hyrachyus* as a tapiroid rather than as a rhinocerotoid was based on his views on the nature of higher taxa as representatives of morphologically or adaptively unified segments of an evolutionary continuum, as discussed further below.

Schoch (1982) suggested that *Hyrachyus* be returned to the Rhinocerotoidea on the basis of five derived character-states it shares with rhinocerotoids. He noted that Radinsky (1966d) had "transferred *Hyrachyus* from the Rhinocerotoidea to the Tapiroidea on the basis of admittedly primitive character-states shared between *Hyrachyus* and primitive ceratomorphs classified as tapiroids" (Schoch, 1982, p. 166). Radinsky (1983b, p. 296) took exception to this characterization of his reasoning, stating: "Contrary to Schoch's assertion, Radinsky [1966d] did not assign *Hyrachyus* to the Tapiroidea on the basis of primitive character states. He specifically noted that *Hyrachyus* lacked derived features characteristic of rhinocerotoids, and provided figures showing similarities between *Hyrachyus* and helaletid tapiroids." He disputed the features Schoch used for allying *Hyrachyus* to the Rhinocerotoidea, stating that they "are either not apparent from inspection of the teeth (e.g., crest-like molar metacones, parastyles slightly reduced), or evolved several times independently in various families of tapiroids (e.g. lingually depressed $M^3$ metacone, loss of $M_3$ hypoconulid)."

More recently, Prothero, Manning, and Hanson (1986) returned *Hyrachyus* to the Rhinocerotoidea as the primitive sister taxon to all other rhinocerotoids on the basis of the following shared derived characters that distinguish them from tapiroids: "$P^4$ protocone 'creased' lingually; $M_3$ hypoconulid lost; cylindrical odontoid process" (p. 359). They also list as diagnostic of rhinocerotoids the following features: "enlarged paralophids and metalophids; increased hypsodonty" (Table 1). As with the synapomorphies listed by Schoch, the latter two are less developed in *Hyrachyus* than in undoubted rhinocerotoids; the "crease" in the $P^4$ protocone appears to be variably developed (it is absent in the specimen of *H. modestus* figured by Radinsky; 1967c, Fig. 2a), and the loss of the $M_3$ hypoconulid occurred independently several times in tapiroids. This appears to leave the shape of the odontoid process, at least for the present, as the only clear-cut synapomorphy of *Hyrachyus* with rhinocerotoids.

It is undoubtedly the case that some of Radinsky's disagreement with Schoch on the superfamilial placement of *Hyrachyus* was due to contrasting philosophies of systematics. Radinsky practiced "evolutionary systematics" in which ancestor-descendant series representing sequential stages in evolution were recognized and given equivalent taxonomic rank. Thus, a more primitive taxon of ceratomorph perissodactyls, the Tapiroidea, was considered ancestral to a more advanced taxon, the Rhinocerotoidea. Boundaries between sequential taxa in an evolutionary continuum should be drawn, in the words of Simpson (1945, p. 19), "at points where striking or important new morphological characters or modifications become widespread or universal in the evolving population." This statement was made with reference to subdividing lineages of species and genera, but the same logic would apply to sequences of higher taxa. Although some species of *Hyrachyus* show the beginnings of rhinocerotoid features, they do not universally show the "striking or important new morphological characters" seen in even the most primitive rhinocerotoids. For this reason, Radinsky placed the Hyrachyidae in the Tapiroidea. This philosophy also underlies his reason for merging Hyrachyidae into the Helaletidae, for he states: "the family Hyrachyidae is neither diverse enough nor distinct enough to be separated on the family level from the tapiroid family Helaletidae" (Radinsky, 1966d, p. 631).

Schoch follows the cladistic tenet that in systematics only genealogical relationship is important and this is revealed by the distribution of derived features only. Grade, or paraphyletic, groups are not recognized, which includes ancestor-descendant sequences. Cladistic approaches work best when nature breaks up evolutionary continua for us and we are able to deal with morphologically discrete sets of organisms. This is usually the case when dealing with the interrelationships of higher level groups with discontinuous fossil records. But, where the fossil record is dense, as it is with perissodactyls, the pattern of distribution of derived features may be complex. This complexity may be due to high levels of variation within contemporaneous populations, to a mosaic pattern of acquisition of derived features in stratigraphically sequential populations, and to the difficulty in recognizing or characterizing derived features in their incipient stages. The temptation in cladistic analysis is to ignore such complex continua among taxa and impose discontinuities between taxa where they may not exist. "Evolutionary systematics" accepted their existence and attempted to cope with them in consciously subjective ways. Cladists, all too often, implicitly deny their existence.

I am not a specialist on Cenozoic mammals, so I am not familiar with the actual distribution of the purported synapomorphies of Rhinocerotoidea among Eocene and Oligocene ceratomorphs. Relatively brief reviews, such as those of Schoch (1982) and Prothero, Manning, and Hanson (1986) are not appropriate places to present thorough analyses of the distribution of the relevant characters, but such analyses should lie behind their systematic conclusions. In my overview of the systematic work of Leonard Radinsky, I find an admirable awareness of the complexity of the patterns of distribution of morphological features among ceratomorphs. I believe that he never fully accepted cladistics as a philosophy of systematics because, in his experience with the literature of Cenozoic mammals, it tended to yield overly simple patterns of relationship based on overly clearcut distributions of characters. Although I am an advocate of cladistics, I find that the application of its methods to the dense fossil record of Cenozoic mammals, in which differences among closely related taxa often have a complex distribution, carries with it the danger of oversimplifying such distributions in the interest of clearcut results. The controversy over the taxonomic position of *Hyrachyus* seems to me to represent such a case.

# Bibliography

Fischer, K. (1977): Neue Funde von *Rhinocerolophiodon* (n. gen.), *Lophiodon* und *Hyrachyus* (Ceratomorpha, Perissodactyla, Mammalia) aus dem Eozän des Geiseltals bei Halle (DDR). -*Zeit. Geol. Wiss.*, 5(7): 909-919.

Hooker, J. J. (1984): A primitive ceratomorph (Perissodactyla, Mammalia) from the early Tertiary of Europe.-*Zool. J. Linn. Soc.*, 82: 229-244.

Hopson, J. A. (1986): Obituary of Leonard B. Radinsky .- *Soc. Vert. Paleo. News Bull.*, 138: 63-65.

Lucas, S. G., and Schoch, R. M. (1981). The systematics of *Rhodopagus*, a late Eocene hyracodontid (Perissodactyla: Rhinocerotoidea) from China. -*Bull. Geol. Inst. Univ. Uppsala, New Series*, 9: 43-50.

Lucas, S. G., Schoch, R. M., and Manning, E. (1981): The systematics of *Forstercooperia*, a middle to late Eocene hyracodontid (Perissodactyla: Rhinocerotoidea) from Asia and western North America. -*J. Paleont.*, 55: 826-841.

Prothero, D.R., Manning, E., and Hanson, C. B. (1986): The phylogeny of the Rhinocerotoidea (Mammalia, Perissodactyla). -*Zool. J. Linn. Soc.*, 87: 341-366.

Savage, D. E., Russell, D. E., and Louis, P. (1966): Ceratomorpha and Ancylopoda (Perissodactyla) from the lower Eocene Paris Basin, France. -*Univ. Calif. Publ. Geol. Sci.*, 66: 1-38.

Schoch, R.M. (1982): *Hyrachyus*: tapiroid or rhinocerotoid? -*Evol. Theory*, 6: 166 (Abstract).

Schoch, R.M. (1983): Relationships of the earliest perissodactyls (Mammalia, Eutheria). -*Geol. Soc. Am. Abstr. Prog.*, 15(3): 144 (Abstract).

Schoch, R. M. and Prins, N. (1984): The evolution of the Tapiroidea (Mammalia, Perissodactyla). -*Geol. Soc. Am. Abstr. Prog.*, 16 (6): 647 (Abstract).

Simpson, G. G. (1945): The principles of classification and a classification of mammals. -*Bull. Amer. Mus. Nat. Hist.*, 85: 1-350.

## The publications of LEONARD B. RADINSKY

1961. Tooth histology as a taxonomic criterion for cartilaginous fishes. -*Journal of Morphology*, 109: 43-92.

1963a. Origin and early evolution of North American Tapiroidea. -*Yale Peabody Museum Bulletin*, 17: 1-106.

1963b. The perissodactyl hallux. -*American Museum Novitates*, 2145: 1-8.

1964a. *Paleomoropus*, a new early Eocene chalicothere (Mammalia, Perissodactyla), and a revision of Eocene chalicotheres. -*American Museum Novitates*, 2179: 1-28.

1964b. Notes on Eocene and Oligocene fossil localities in Inner Mongolia. -*American Museum Novitates*, 2180: 1-11.

1965a. Early Tertiary Tapiroidea of Asia. -*Bulletin of the American Museum of Natural History*, 129: 181-262.

1965b. Evolution of the tapiroid skeleton from *Heptodon* to *Tapirus*. -*Bulletin of the Museum of Comparative Zoology, Harvard*, 134: 69-106.

1966a. *Pataecops*, new name for *Pataecus*, Radinsky, 1965. -*Journal of Paleontology*, 40: 222.

1966b. A new genus of Early Eocene tapiroid (Mammalia, Perissodactyla). -*Journal of Paleontology*, 40: 740-742.

1966c. The adaptive radiation of the phenacodontid condylarths and the origin of the Perissodactyla. -*Evolution*, 20: 408-417.

1966d. The families of the Rhinocerotoidea (Mammalia, Perissodactyla). -*Journal of Mammalogy*, 47: 631-639.

1967a. Relative brain size: a new measure. -*Science*, 155: 836-837.

1967b. Review of the rhinocerotoid family Hyracodontidae. -*Bulletin of the American Museum of Natural History*, 136: 1-46.

1967c. *Hyrachyus, Chasmotherium*, and the early evolution of helaletid tapiroids. -*American Museum Novitates*, 2313: 1-23.

1967d. The oldest primate endocast. -*American Journal of Physical Anthropology*, 27: 385-388 (published 1968).

1968. A new approach to mammalian cranial analysis, illustrated by examples of prosimian primates. -*Journal of Morphology*, 124: 167-180.

1969a. Evolution of somatic sensory specialization in otter brains. -*Journal of Comparative Neurology*, 134: 495-506.
1969b. The early evolution of the Perissodactyla. -*Evolution*, 23: 308-328.
1969c. Outlines of canid and felid brain evolution. -*Annals of the New York Academy of Sciences*, 167: 277-288.
1970. The fossil evidence of prosimian brain evolution. In: Noback, C. R., and Montagna, W. (eds.): *The Primate Brain, Advances in Primatology* (vol.1), New York (Appleton-Century-Crofts), pp. 781-798.
1971. An example of parallelism in carnivore brain evolution. -*Evolution*, 25: 518- 522.
1972. Endocasts and studies of primate brain evolution. In: Tuttle, R. (ed.): *The Functional and Evolutionary Biology of Primates*. Chicago (Aldine-Atherton), pp. 175-184.
1973a. Evolution of the canid brain. -*Brain, Behavior and Evolution*, 7: 169-202.
1973b. Fossil evidence of higher primate brain evolution. -*Anatomical Record*, 175: 418 (Abstract).
1973c. Are stink badgers skunks? Implications of neuroanatomy for mustelid phylogeny. -*Journal of Mammalogy*, 54: 585-594.
1973d. *Aegyptopithecus* endocasts: oldest record of a pongid brain. -*American Journal of Physical Anthropology*, 39: 239-248.
1974a. Fossil evidence of felid brain evolution. -*Anatomical Record*, 178: 443 (Abstract).
1974b. The fossil evidence of anthropoid brain evolution. -*American Journal of Physical Anthropology*, 41: 15-28.
1974c. Prosimian brain morphology. Functional and phylogenetic implications. In: Martin, R., Doyle, G. A., and Walker, A. C. (eds.): *Prosimian Biology*. London (Duckworth), pp. 781-798.
1975a. Viverrid neuroanatomy: phylogenetic and behavioral implications. -*Journal of Mammalogy*, 56: 130-150.
1975b. Review of: Evolution of the Brain and Intelligence by H. J. Jerison. -*Evolution*, 29: 190-192.
1975c. Early primate brains: facts and fiction. -*American Journal of Physical Anthropology*, 42: 324 (Abstract).
1975d. Evolution of the felid brain. -*Brain, Behavior, and Evolution*, 11: 214-254.
1975e. Primate brain evolution. -*American Scientist*, 63: 656-663.
1976a. The brain of *Mesonyx*, a middle Eocene mesonychid condylarth. -*Fieldana: Geology*, 33: 323-337.
1976b. Cerebral clues. -*Natural History*, 85(5): 54-59.
1976c. Later mammalian radiations. In: Masterton, R. B., Hodos, W., and Jerison, H. J. (eds.):*Evolution of the Brain and Behavior in Vertebrates*. Hillsdale, N.Y. (Erlbaum Associates), pp. 27-43.
1976d. New evidence on ungulate brain evolution. -*American Zoologist*, 16: 207 (Abstract).
1976e. Review of: Phylogeny of the primates, edited by W. P. Luckett and F. S. Szalay. -*Science*, 192: 1121-1122.
1976f. Oldest horse brains; more advanced than previously realized. -*Science*, 194: 626-627.
1976g. *Thylacoleo*, marsupial lion or lamb? An exercise in behavioral paleoneurology. -*Neurosciences Abstracts*, 2: 258 (Abstract) (with J. I. Johnson).
1976h. The fossil record and neural organization. Discussion. -*Annals of the New York Academy of Sciences*, 280: 383-384. (Symposium on origin and evolution of language and speech).
1977a. Early primate brains: facts and fiction. -*Journal of Human Evolution*, 6: 79-86.

1977b. Brains of early carnivores. -*Paleobiology*, 3: 333-349.
1978a. The evolutionary history of dog brains. -*Museologia*, 10: 25-29.
1978b: Brain evolution in extinct South American ungulates. -*American Zoologist*, 18: 588 (Abstract).
1978c. Do albumin clocks run on time? -*Science*, 200: 1182-1183.
1978d. Evolution of brain size in carnivores and ungulates. -*American Naturalist*, 112: 815-831.
1979a. The evolutionary history of cat brains. -*Museologia*, 12: 35-41.
1979b. The fossil record of primate brain evolution. -Forty-ninth James Arthur lecture, American Museum of Natural History. 27 pages.
1979c. The comparative anatomy of the muscular system. In: Wake, M. H. (ed.): *Hyman's comparative vertebrate anatomy*. Chicago (University of Chicago Press), pp. 327-377.
1979d. A new look at old sabertooths. -*American Zoologist*, 19: 1011 (Abstract) (with S. Emerson).
1980a. Endocasts of amphicyonid carnivorans. -*American Museum Novitates*, 2694: 1-11.
1980b. Analysis of carnivore skull morphology. -*American Zoologist*, 20: 784 (Abstract).
1980c. Vertebrate paleontology: new approaches and new insights. -*Paleobiology*, 6: 250-270 (with J. A. Hopson).
1980d. Functional analysis of sabertooth cranial morphology. -*Paleobiology*, 6: 295-312 (with S. Emerson).
1981a. Evolution of skull shape in carnivores. 1. Representative modern carnivores. -*Biological Journal of the Linnean Society*, 15: 369-388.
1981b. Brain evolution in extinct South American ungulates. -*Brain, Behavior, and Evolution*, 18: 169-187.
1981c. Evolution of skull shape in carnivores. 2. Additional modern carnivores. -*Biological Journal of the Linnean Society*, 16: 337-355.
1982a. Some cautionary notes on making inferences about relative brain size. In: Armstrong, E., and Falk, D. (eds.): *Primate Brain Evolution: Methods and Concepts*. New York (Plenum Publishing Corp.), p. 29-37.
1982b. The late, great sabertooths. -*Natural History*, 91 (4): 51-57 (with S. Emerson).
1982c. Evolution of skull shape in carnivores. 3. The origin and early radiation of the modern carnivore families. -*Paleobiology*, 8:177-195.
1983a. Allometry and reorganization in horse skull evolution. -*Science*. 221: 1189-1191.
1983b. *Hyrachyus*, tapiroid, not rhinocerotoid. -*Evolutionary Theory*, 6: 296 (Abstract).
1984a. Basicranial length vs. skull length in analysis of carnivore skull shape. -*Biological Journal of the Linnean Society*, 22:31-41.
1984b. The skull of *Ernanodon*, an unusual fossil mammal. -*Journal of Mammalogy*, 65: 155-158 (with Ting Suyin).
1984c. Ontogeny and phylogeny in horse skull evolution. -*Evolution*, 38: 1-15.
1985a. Patterns in the evolution of ungulate jaw shape. -*American Zoologist*, 25: 303-314.
1985b. Approaches in evolutionary morphology: a search for patterns. -*Annual Review of Ecology and Systematics*, 16: 1-14.
1987. *The Evolution of Vertebrate Design*. Chicago (University of Chicago Press) (posthumous).

# 2. A BRIEF HISTORICAL REVIEW OF PERISSODACTYL CLASSIFICATION

## ROBERT M. SCHOCH

As here recognized, the Perissodactyla includes the extant Equidae, Tapiridae, Rhinocerotidae, and Hyracoidea, along with the last common ancestor of all of these living forms, and all descendants of their common ancestor. The Mesaxonia is the group composed of all perissodactyls exclusive of the hyraxes. As presently classified, within the Mesaxonia there is a basic trichotomy: the Titanotheriomorpha (extinct brontotheres), the Hippomorpha (pachynolophids, equids, and palaeotheriids), and the Moropomorpha (primitive "tapiroids," ancylopods [including chalicotheres and lophiodonts], and ceratomorphs [such as deperetellids, lophialetids, tapirids, and rhinocerotoids]). Earlier classifications of the Mesaxonia have often associated equoids and brontotheres as the Hippomorpha, and tapiroids and rhinocerotoids as the Ceratomorpha. Ancylopods (chalicotheres) have usually either been relegated to the Hippomorpha or left as a distinct grouping of uncertain relationships to the remaining mesaxonians. The relationship of the ancylopods to the ceratomorphs was first seriously proposed by Schoch (1983) and Hooker (1984).

## Introduction

It seems appropriate that a multiauthored volume of collected papers on the evolution of perissodactyls should begin with a brief review of the history of classification of perissodactyls. It is hoped that this chapter will serve as an introduction to the book, and help place the following contributions in perspective. I also believe that I should here state explicitly the scope of the Perissodactyla as encompassed in the contributions to this book.

As I currently recognize the Perissodactyla, the group includes the extant Equidae (horses, zebras, quaggas, onagers, asses, and so forth), Tapiridae (tapirs), Rhinocerotidae (rhinoceroses), and the Hyracoidea (Procaviidae: hyraxes, dassies, and conies), along with the last common ancestor of all of these extant forms and any and all descendants of that last common ancestor. It has not been customary in the present century to include the Hyracoidea within the Perissodactyla, but this move is proposed on the basis of what seems to be clear evidence of a close phylogenetic relationship between the hyracoids and all other perissodactyls (see M. Fischer, this volume, Chapter 4). As is briefly reviewed here and by M. Fischer, there is a clear historical precedent for including the hyracoids within the Perissodactyla; indeed, when the order was originally named by Owen in 1848 he included hyracoids among perissodactyls. In the comprehensive classification of the Perissodactyla at the end of this volume, Prothero and I treat the Hyracoidea as a suborder of the Perissodactyla. We hope that this concept of the Perissodactyla (i.e., including the Hyracoidea) will be generally adopted. Pending such consensus, in the interim it can be expected that various authors will use the terms *perissodactyl* and *Perissodactyla* with differing meanings; this is certainly the case among papers in this book. In some cases the Perissodactyla is considered to exclude the hyracoids, whereas other authors use it in the sense of including the hyracoids. However, in each instance the manner in which these terms are used should be clear to the sagacious reader.

*The Evolution of Perissodactyls* (ed. D.R. Prothero & R.M. Schoch) Oxford Univ. Press, New York, 1989.

## Internal composition, subordinal, and superfamilial classification of the Perissodactyla

As reviewed by Simpson (1945), the forms now allocated to the single taxon Perissodactyla were referred to a number of different groups in the eighteenth century. Linnaeus (1758) placed the horses and hippopotami (including the tapirs) in the taxon Belluae, and the rhinoceroses in the taxon Glires along with lagomorphs and rodents. Linnaeus (1758) did not deal with the hyraxes. Brisson (1762) separated the tapirs from the hippopotami and the rhinoceroses from the rodents, but placed the tapirs, rhinoceroses, and horses in three distinct orders. Storr (1780), in describing the first named hyrax genus *Procavia*, considered the hyracoids to be rodents. In the late eighteenth and early nineteenth centuries, the tapirs, rhinoceroses, and hyracoids were commonly associated with various proboscideans, artiodactyls, and perhaps other forms, as "pachyderms" (e.g., Cuvier, 1817; see Simpson, 1945). Horses were usually differentiated from all other mammals and placed in a taxon unto themselves (Simpson, 1945). I suspect that the special status afforded horses was due, in large measure, to their economic, social, and cultural importance in earlier centuries (see, e.g., Buffon, 1797; Taplin, 1797).

The concept of the Perissodactyla as used here (Prothero and Schoch, this volume, Chapters 28, 29), to include extant hyracoids, horses, rhinoceroses, and tapirs (along with all related fossil forms) has its origins in the work of de Blainville (1816) who was the first to consistently classify ungulates as having either an odd number of toes (essentially the modern Perissodactyla) on each foot or an even number of toes (essentially the modern Artiodactyla). Owen (1848) adopted de Blainville's association of hyracoids, horses, rhinoceroses, and tapirs, labelling the taxon Perissodactyla. It might be noted, however, that Owen (e.g., 1868) also variously included some forms in his Perissodactyla, such as *Coryphodon*, *Macrauchenia*, *Toxodon*, and *Nesodon*, which are now relegated to other orders.

Hyracoids were generally separated from the other perissodactyls after the middle of the nineteenth century (Huxley, 1869, 1872; Gill, 1870) and usually relegated to the distinct order Hyracoidea Huxley (1869). It was sometimes suggested that hyracoids might be in some way related to proboscideans (Simpson, 1945). Throughout the remainder of the nineteenth century and for much of the twentieth century the affinities of the hyraxes have been considered obscure. Reflecting this attitude, in his massive classification of all then known mammalian genera, Simpson (1945) united the Hyracoidea with the Pantodonta, Dinocerata, Pyrotheria, Proboscidea, Embrithopoda, and Sirenia in his new superorder Paenungulata. However, the concept that hyracoids and perissodactyls (exclusive of hyracoids) might together form a natural group persisted as a minority viewpoint (e.g., Frechkop, 1936; see also M. Fischer, this volume, Chapter 4). In a recent comprehensive classification of the mammals, McKenna (1975; adopted in its essentials by Eisenberg, 1981, and Schoch, 1984b) united the orders Condylarthra, Perissodactyla, and Hyracoidea in his mirorder Phenacodonta, and McKenna and Manning (1977) hypothesized a close relationship between perissodactyls and hyraxes. This recent trend has culminated in the work of M. Fischer (1986; this volume, Chapter 4) and the formal reinstatement of the Hyracoidea as a suborder of the Perissodactyla (Prothero and Schoch, this volume, Chapter 29).

Marsh (1884) coined the name Mesaxonia for what would be considered the orthodox Perissodactyla of the majority of the twentieth century, viz., the extant tapirs, horses, and rhinoceroses (and their extinct relations). Most authors have treated Mesaxonia as a virtual synonym of Perissodactyla (see Simpson, 1945), but with current understanding that the hyracoids should be included within the order Perissodactyla, the name Mesaxonia can be applied to the taxon of non-hyracoid perissodactyls (Prothero and Schoch, this volume, Chapter 29).

In the present century the Mesaxonia have traditionally been subdivided into five major superfamilies: the extant (each with some extinct members) Equoidea, Rhinocerotoidea, and Tapiroidea and the wholly extinct Brontotherioidea and Chalicotherioidea.

Of course, until recently there has been little, if any, concern among investigators as to whether these groups were natural (i.e., monophyletic = holophyletic; see Schoch, 1986). Indeed, at least some of them probably are not monophyletic as commonly construed. In particular, the traditional Tapiroidea is certainly not monophyletic (Hooker, this volume, Chapter 6; Schoch, this volume, Chapter 15), and among the Equoidea the genus *Hyracotherium* as currently used by many workers is blatantly paraphyletic (Hooker, this volume, Chapter 6).

MacFadden (1976) argued for the monophyly of the Equidae (= essentially Equoidea), including *Hyracotherium*, on the basis of basicranial evidence (confluence of the foramen ovale and middle lacerate foramen) and the migration of the optic foramen close to the anterior lacerate, rotundum, and anterior opening of the alisphenoid canal within the orbital region, even though *Hyracotherium* is dentally primitive relative to all other perissodactyls. Hooker (this volume, Chapter 6) agrees that the Equoidea (*sensu lato*, = Pachynolophoidea plus Equoidea of Hooker, this volume, Chapter 6) appears to be monophyletic.

Traditionally the Equoidea has been divided into two families (e.g., Simpson, 1945): the extinct and predominantly European Palaeotheriidae and the Equidae. Some investigators (e.g., Butler, 1952a, b) have suggested that palaeotheres are distinct from all other perissodactyls (or at least mesaxonians), perhaps having even arisen from a distinct ancestor (thus making the Perissodactyla nonmonophyletic if palaeotheres are included, but certain ancestral or proto-perissodactyl forms are excluded from the order). However, palaeotheres do appear to have the synapomorphic navicular facet (concave, saddle-shaped) on the astragalus that unites the Mesaxonia. MacFadden (1976), on the basis of the position of the optic foramen in *Pachynolophus*, has suggested that the palaeotheres should be included within the Equidae. Hooker (this volume, Chapter 6) has split the traditional palaeotheres into two families, the Palaeotheriidae (*Palaeotherium, Plagiolophus, Propalaeotherium, Propachynolophus,* and *Lophiotherium*) and the Pachynolophidae (*Pachynolophus, Anchilophus,* and an unnamed species that has been referred to *Hyracotherium* from Rians, France). Hooker (this volume, Chapter 6) believes that the Equidae, Palaeotheriidae, and various species of *Hyracotherium* should be united as the Equoidea, and that the Pachynolophidae (= Pachynolophoidea *sensu* Hooker) is the sister group of the Equoidea (*sensu* Hooker). In Hooker's schema, the Equoidea and Pachynolophoidea constitute the Hippomorpha (see later).

The Brontotherioidea has generally been treated as a monophyletic group, although there have been serious questions concerning the exact status of *Lambdotherium* (see Mader, this volume, Chapter 25; Hooker, this volume, Chapter 6; Prothero and Schoch, this volume, Chapter 28). Hooker (this volume, Chapter 6) has proposed a new suborder, Titanotheriomorpha, for the superfamily Brontotheriodea.

The Chalicotherioidea (= Ancylopoda of most authors, e.g., Simpson, 1945) has also generally been treated as a monophyletic group (see Coombs, this volume, Chapter 24; Lucas and Schoch, this volume, Chapter 23, and references cited therein). The status of the Chalicotheriidae (including the "eomoropids": Prothero and Schoch, this volume, Chapter 29) as a monophyletic group (although perhaps minus *Paleomoropus* and some other questionable forms) is not changed by recent suggestions that the chalicotheres are the sister group of certain "tapiroids" (see the discussion following and Hooker, this volume, Chapter 6; Schoch, this volume, Chapter 15).

At present there seems to be no question in most investigators' minds that the Rhinocerotoidea (including *Hyrachyus*) is a natural, i.e. monophyletic group (see Prothero, Guérin, and Manning, this volume, Chapter 16; Prothero, Manning, and Hanson, 1986).

Given the five traditional superfamilies of the Mesaxonia reviewed earlier, the question arises as to how to arrange these superfamilies within the order. Three of the major classificatory schemes that have been proposed, and widely adopted, in the twentieth century are as follows.

1. Wood (1934, 1937) and Simpson (1945):
Order Perissodactyla [= Mesaxonia of
    Prothero and Schoch, this volume,
    Chapter 29]
  Suborder Hippomorpha
    Superfamily Equoidea
    Superfamily Brontotherioidea
    Superfamily Chalicotherioidea
  Suborder Ceratomorpha
    Superfamily Tapiroidea
    Superfamily Rhinocerotoidea

Wood (1934) was the modern originator of the idea that the mesaxonians are divisible into two major groups, which he named in 1937 the Hippomorpha and the Ceratomorpha. In line with the above classification, Borissiak (1945) believed that the chalicotheres and brontotheres in particular were sister taxa relative to all other perissodactyls.

2. Scott (1941):
Order Perissodactyla
  Suborder Chelopoda
    Infraorder Hippomorpha
      Superfamily Brontotherioidea
      Superfamily Equoidea
        [= Hippoidea]
    Infraorder Ceratomorpha
      Superfamily Tapiroidea
      Superfamily Rhinocerotoidea
  Suborder Ancylopoda
    Superfamily Chalicotherioidea

Contrary to Wood (1934, 1937), Scott (1941) made a primary distinction between the hooved mesaxonians and the clawed mesaxonians. Thus he separated the Chalicotherioidea from all other mesaxonians as the separate suborder Ancylopoda Cope, 1889, and united the remaining mesaxonians in the suborder Chelopoda Scott, 1937.

3. Radinsky (1964):
Order Perissodactyla
  Suborder Hippomorpha
    Superfamily Equoidea
    Superfamily Brontotherioidea
  Suborder Ceratomorpha
    Superfamily Tapiroidea
    Superfamily Rhinocerotoidea
  Suborder Ancylopoda
    Superfamily Chalicotherioidea

On the assumption that the early Eocene genus *Paleomoropus* Radinsky, 1964, represents a very early eomoropid chalicothere (however, this is no longer certain: see later and Schoch, this volume, Chapter 15), Radinsky (1964, p. 6) suggested that "it would appear that chalicotheres diverged from the basic perissodactyl stock before the equoid-tapiroid split." Using a phylogenetic classification, Radinsky (1964) could have adopted Scott's (1941) classification (2, preceding). However, Radinsky (1964, p. 6) stated that Scott's (1941) classification "implies that equoids and brontotherioids are more similar to tapiroids and rhinocerotoids than either group is to chalicotherioids. Although this is suggested by the similarity between equoids and tapiroids in earliest Eocene time, I do not think that the evidence is at present definite enough to warrant formal taxonomic expression. Until more is known about the relationships between perissodactyl superfamilies in early Eocene time, I suggest that the Hippomorpha (*sensu stricto*) and Ceratomorpha be kept at subordinal rank and that the Ancylopoda be added as a third suborder." Radinsky (1964, p. 6; see also Radinsky, 1966a, 1967, 1969) believed that "rhinocerotoids were probably derived from an early Eocene tapiroid," but was unsure as to the exact ancestry of the brontotherioids. Radinsky (1964, p. 7) believed that "nothing in the known anatomy of *Hyracotherium* [early Eocene equoid] precludes its being ancestral to *Lambdotherium* [presumed by Radinsky to be a late early Eocene brontotherioid], but the morphological gap between those two genera is greater than that between ancestral tapiroid and primitive rhinocerotoid, and intermediate stages are not known." He then went on to predict that: "When details of the origin of brontotherioids become known, the classification proposed above may have to be further modified to dissociate brontotherioids from equoids. Should such separation become necessary, it would probably be best to drop the subordinal divisions of the Perissodactyla and leave the order divided into five superfamilies" (Radinsky, 1964, p. 7).

K.-H. Fischer (1964, 1977a, 1977b) questioned Radinsky's (1964) assignment of *Paleomoropus* and the European genus *Lophiaspis* (see Hooker, this volume, Chapter 6; Schoch, this volume, Chapter 15) to the chalicotheres; instead, K.-H. Fischer (1964, 1977a, b) thought that these forms might be lophiodonts (Tapiroidea). More or less picking up where Radinsky (1964) left off, and at that time still considering *Paleomoropus* to be a chalicothere, I (Schoch, 1983) noted that the chalicotheres and ceratomorphs appear to share one or more synapomorphies relative to other perissodactyls (= mesaxonians of Prothero and Schoch, this volume, Chapter 29). In March of 1984 (Schoch, 1984a, p. 17, footnote 1 to table 2) I coined the subordinal name Moropomorpha for the taxon composed of ceratomorphs (tapiroids plus rhinocerotoids) and chalicotherioids, and thus established the following classification (Schoch, 1985):

Order Perissodactyla
    Suborder Hippomorpha
        Superfamily Brontotherioidea
        Superfamily Equoidea
    Suborder Moropomorpha
        Infraorder Ancylopoda
        Infraorder Ceratomorpha
            (including "tapiroids" and "rhinocerotoids")

Independently of my work, Hooker (1984, published for September/October of that year) suggested that lophiodonts and chalicotheres are sister groups. Hooker (1984) also suggested that the Brontotherioidea are the sister group of all other mesaxonians, and that the Equoidea are the sister group of "tapiroids" plus "chalicotherioids" (= Moropomorpha of Schoch, 1984a). However, Hooker (1984) concluded that the Isectolophidae (see Schoch, this volume, Chapter 15) is the sister taxon of the remaining moropomorphs, that the Ancylopoda (*sensu* Hooker, 1984 = Lophiodontidae plus Chalicotherioidea) and the Ceratomorpha (*sensu stricto*, excluding the isectolophids) are sister taxa, and that the genus *Cymbalophus* Hooker, 1984 (see Hooker, this volume, Chapter 6) is the sister taxon of the remaining ceratomorphs. Thus Hooker (1984, p. 241, Figure 22c) classified the mesaxonians as follows:

Order Perissodactyla [i.e., Mesaxonia]
    Unnamed taxon
        Superfamily Brontotherioidea
    Unnamed taxon
        Superfamily Equoidea
        "Suborder" Tapiromorpha
            [coordinate in rank to the "Superfamily Equoidea"]
            Unnamed taxon including only the Family Isectolophidae
            Unnamed taxon
                Sub- or Infraorder Ancylopoda
                    Lophidontidae
                    Chalicotherioidea

Sub- or Infraorder
   Ceratomorpha
      *Cymbalophus*
      All other ceratomorphs

To label the same group that I had called Moropomorpha (the fact that the internal arrangement of the group differs between Schoch, 1983, 1984a, and Hooker, 1984, is immaterial here), Hooker (1984) resurrected and redefined Haeckel's (1873) name Tapiromorpha, stating (Hooker, 1984, p. 242) that "to name this entire related group, it would be possible to extend slightly the concept of Haeckel's (1873) Tapiromorpha, which has been used (as in recent times by Romer, 1945 [see also Simpson, 1945]) as an equivalent for the Ceratomorpha." However, I find this re-use of Haeckel's (1873) old term inappropriate, inadequate, and potentially confusing. As Wood (1937, p. 106) noted when considering the possibility of reviving the term Tapiromorpha for what he decided to label the Ceratomorpha, "In some ways, Haeckel's term, Tapiromorpha, would be preferable to Ceratomorpha, except that, from 1870 [it is not clear to me when the term Tapiromorpha was actually first coined; according to Simpson, 1945, it was not used until 1873] to 1892, he used it in three different senses (for Palaeotherida, Macrauchenida, Tapirida, Nasicornia, and Elasmotherida; then for Palaeotherida, Tapirida, and Macrauchenida; finally for Palaeotherida and Tapirida). Thus his progressive emendations diverged farther and farther from the concept under discussion." Note in particular that Haeckel consistently included the palaeotheres within his Tapiromorpha whereas Hooker (1984; this volume, Chapter 6; likewise Schoch, 1983, 1984a, 1985) includes the palaeotheres within the Equoidea. It was for reasons such as these that I rejected re-use of the term Tapiromorpha (although I like the "sound" of the name) and coined the new name Moropomorpha (Schoch, 1984a).

Hooker (this volume, Chapter 6) has modified and refined his earlier ideas, in particular relative to the position of *Cymbalophus*, which he now considers a primitive equoid rather than a primitive ceratomorph (see Hooker for a brief review of the development of his thinking). Using a computer-based cladistic analysis, Hooker (this volume, Chapter 6) found a first division of the mesaxonians into brontotherioids plus equoids and all other mesaxonians (i.e., moropomorphs). However, the group composed of the brontotherioids plus the equoids was united on the basis of only two very weak, poorly known, and in Hooker's opinion suspect, synapomorphies (loss of I3 distal cusp, and an upper molar cingular mesostyle), such that Hooker (this volume) chooses to acknowledge that there is currently an unresolved basal trichotomy in perissodactyl classification: Titanotheriomorpha (Hooker, this volume, Chapter 6; = Brontotherioidea), Hippomorpha, and Tapiromorpha (*sensu* Hooker, = Moropomorpha).

Within the Hippomorpha, Hooker distinguishes two sister groups, the Equoidea (including the Equidae, Palaeotheriidae, *Cymbalophus*, and various species of "*Hyracotherium*") and the Pachynolophoidea. Within what Hooker terms the Tapiromorpha, he considers the Isectolophidae to be the sister group of all other forms. The Lophiodontoidea (= Lophiodontidae) and Chalicotherioidea are sister taxa, together comprising the Ancylopoda. The Ancylopoda is the sister taxon of the remaining tapiromorphs/ moropomorphs (non-isectolophids and non-ancylopods). Hooker (this volume, Chapter 6; see also Schoch, this volume, Chapter 15) also discusses the internal arrangement of various "tapiroids" within his concept of the Ceratormorpha. Thus, based on Hooker, the outlines of a classification of the traditional Perissodactyla (i.e., Mesaxonia) plus the Hyracoidea can be formulated as follows (see also Prothero and Schoch, this volume, Chapter 29):

Order Perissodactyla
  Suborder Hyracoidea
  Suborder Mesaxonia
    Infraorder Titanotheriomorpha
      Superfamily Brontotherioidea
    Infraorder Hippomorpha
      Superfamily Pachynolophoidea
        Family Pachynolophidae
      Superfamily Equoidea
      (Including the families Equidae and Palaeotheriidae, along with various primitive forms)
    Infraorder Moropomorpha (= Tapiromorpha *sensu* Hooker)
      Unnamed taxon
        Family Isectolophidae
      Unnamed taxon [including *Kalakotia*?]
        Parvorder Ancylopoda
          Superfamily Lophiodontoidea
          Superfamily Chalicotherioidea
        Parvorder Ceratomorpha
          Breviodontidae as the sister taxon of all remaining ceratomorphs
          Unnamed taxon
            Magnafamily Deperetellidea, new rank[1]
              Family Deperetellidae
              Family Rhodopagidae
            Magnafamily Lophialetidea, new rank[1]
              Family Eoletidae, new family
              Family Lophialetidae
            Magnafamily Tapiridea, new rank[1]
              *Heptodon* as the sister taxon of the remaining Tapiridea
              Superfamily Tapiroidea
              Superfamily Rhinocerotoidea

---

[1]The magnafamily is the taxonomic category between the superfamily and the parvorder (Schoch, 1986). Even though the magnafamily would appear to be a family-group category, the naming of such taxa is not covered by the International Code of Zoological Nomenclature (Ride *et al.*, 1985), for the Code does not pertain to the nomenclature of Linnaean taxa above the level of superfamily. Here, however, I follow the spirit of the Code in attributing the magnafamilies Deperetellidea, Lophialetidea, and Tapiridea to the original describers of the family groups Deperetellidae, Lophialetinae, and Tapiridae (Radinsky, 1965; Matthew and Granger, 1925; and Burnett, 1830, respectively). In other words, when these authors established a name for any taxon at any rank in the family group they also simultaneously established names for taxa at all other ranks within the family group (see Article 36 of Ride *et al.*, 1985). Here I arbitrarily utilize the suffix -idea to denote a magnafamily.

## Relationship of perissodactyls to other mammals

Until well into the present century the exact relationship of the perissodactyls to other orders has been viewed as highly debatable (see review by Simpson, 1945). Linnaeus (1766) divided all mammals into the Unguiculata, Ungulata, and Mutica. Since then, the forms that were to be eventually united as the Perissodactyla have most often been considered typical ungulates (i.e., hoofed mammals). Indeed, the Perissodactyla and Artiodactyla have been associated as a single ungulate order (termed the Diplarthra by Cope, 1887; but see Gregory, 1910, who specifically separated the perissodactyls from the artiodactyls). Simpson (1945) separated the Perissodactyla (exclusive of the Hyracoidea) not only as a distinct order, but also as a distinct superorder (for which Simpson adopted the name Mesaxonia Marsh, 1884), of his new cohort Ferungulata (composed of the Carnivora, Condylarthra, Litopterna, Notoungulata, Astrapotheria, Paenungulata, Perissodactyla, and Artiodactyla). The conventional wisdom at the beginning of the twentieth century was that perissodactyls had risen from primitive condylarth forms. This concept was codified by Radinsky (1966b, 1969) in his certainty that some phenacodontid "condylarth," specifically a form referable to the genus *Tetraclaenodon* or a closely related form, was ancestral to the perissodactyls (mesaxonians), and this remained the orthodox view into the late 1980s (see Prothero and Schoch, this volume, Chapter 28). Radinsky felt confident enough in his views to make such statements as "the origin of the Perissodactyla is better documented than that of any other order of mammals and provides an excellent opportunity to study the emergence of a major taxon" (Radinsky, 1966b, p. 408). "Perissodactyls arose from phenacodontid condylarths in the late Paleocene" (Radinsky, 1969, p. 308). "The oldest true phenacodontid condylarth, *Tetraclaenodon*, first appears in faunas of middle Paleocene age, and by the beginning of the late Paleocene appears to have radiated into three main groups, represented respectively by *Phenacodus, Ectocion*, and an as yet unknown proto-perissodactyl" (Radinsky, 1966b, p. 408).

In recent years there have been doubts about the origin of the Perissodactyla being found among the phenacodontids. In part this is associated with the recognition that hyracoids are perissodactyls. Cladistic analyses of various ungulates (McKenna and Manning, 1977; Prothero, Manning, and Fischer, 1988; reviewed in Prothero and Schoch, this volume, Chapter 28) suggest that perissodactyls (including hyracoids), phenacolophids (which may in fact be best considered arsinoitheres; see McKenna and Manning, 1977), arsinoitheres, and tethytheres (Proboscidea, Sirenia, and Desmostylia; see McKenna, 1975) may all be more closely related to one another than any of them are to phenacodontids. The description of *Radinskya* (McKenna *et al.*, this volume, Chapter 3) as a perissodactyl-like mammal from the late Paleocene of southern China that McKenna *et al.* (this volume, Chapter 3) tentatively classify as "?Family Phenacolophidae, *incertae sedis*" very much opens up the question of the origin of perissodactyls. For those who desire paraphyletic ancestors, maybe the ancestor of the Perissodactyla will now be placed among the Phenacolophidae.

## Bibliography

Birjukov, M. D. (1974): [The new genus of the family Lophialetidae from Eocene deposits of Kazakhstan]. -*Akademiia Nauk Kasakhskoi SSR, Alma-Ara. Institut Zoologii, Materialy Po Istorii Fauny I Flory Kazakhstana,* 6:57-73.

Blainville, H. M. D. de (1816): Prodrome d'une nouvelle distribution systèmatique du règne animal. -*Bull. Sci. Soc. Philom.,* 3:105-124.

Borissiak, A. (1945): The chalicotheres as a biological type. -*Amer. J. Sci.,* 243:667-679.

Brisson, M. J. (1762): *Regnum animale in classes IX distributum sive synopsis methodica*. Edito altera auctior. Leiden (Theodorum Haak).

Buffon, G. (1797): *Buffon's Natural History*, volume 5. London (H. D. Symonds).

Butler, P. M. (1952a): The milk-molars of Perissodactyla, with remarks on molar occlusion. -*Proc. Zool. Soc. Lond.*, 121: 777-817.

Butler, P. M. (1952b): Molarization of premolars in Perissodactyla. -*Proc. Zool. Soc. Lond.*, 121: 819-843.

Burnett, G. T. (1830): Illustrations of the Quadrupeda, or quadrupeds, being the arrangement of the true four-footed beasts, indicated in outline. -*Quart. J. Sci. Lit. Arts*, 26:336-353.

Coombs, M. C. (1989): Interrelationships and diversity in the Chalicotheriidae (this volume, Chapter 24).

Cope, E. D. (1887): The Perissodactyla. -*Amer. Nat.*, 21: 985-1007, 1060-1076.

Cope, E. D. (1889): The Vertebrata of the Swift Current River, II. -*Amer. Nat.*, 23:151-155.

Cuvier, G. (1817): *Le règne animal*. Dèterville, Paris.

Eisenberg, J. F. (1981): *The Mammalian Radiations*. Chicago (University of Chicago Press).

Fischer, K.-H. (1964): Die tapiroiden Perissodactylen aus der eozänen Braunkohle des Geiseltales. *Geologie, Berlin*, 45:1-101.

Fischer, K.-H. (1977a): Neue Funde von *Rhinocerolophiodon* (n. gen.), *Lophiodon* und *Hyrachyus* (Ceratomorpha, Perissodactyla, Mammalia) aus dem Eozän des Geiseltals bei Halle (DDR). 1. Teil: *Rhinocerolophiodon. Zeit. Geol. Wiss.*, 5:909-919.

Fischer, K.-H. (1977b): Neue Funde von *Rhinocerolophiodon* (n. gen.) *Lophiodon* und *Hyrachyus* (Ceratomorpha Perissodactyla, Mammalia) aus dem Eozän des Geiseltals bei Halle (DDR). 2. Teil: *Lophiodon. Zeit. Geol. Wiss.*, 5:1129-1152.

Fischer, M. S. (1986): Die Stellung der Schliefer (Hyracoidea) im phylogenetischen System der Eutheria. Zugleich ein Beitrag zur Anpassungsgeschichte der Procaviidae. -*Cour. Forsch. Inst. Senckenberg*, 84:1-132.

Fischer, M. S. (1989): Hyracoids, the sister-group of perissodactyls (this volume, Chapter 4).

Frechkop, S. (1936): Notes sur les mammifères. Remarque sur la classification des Ongulès et sur la position systèmatique des Damans. -*Bull. Mus. Roy. Hist. Nat. Belgique.*, 12(37):1-28.

Gill, T. (1870): On the relations of the orders of mammals. -*Proc. Amer. Assoc. Adv. Sci.*, 19th meeting, pp. 267-270.

Gregory, W. K. (1910): The orders of mammals. -*Bull. Amer. Mus. Nat. Hist.*, 27:1-524.

Haeckel, E. (1873): *Natürliche Schöpfundsgeschichte* . . . Vierte auflage. Berlin (George Reimer).

Hooker, J. J. (1984): A primitive ceratomorph (Perissodactyla, Mammalia) from the early Tertiary of Europe. -*Zool. J. Linn. Soc.*, 82: 229-244.

Hooker, J. J. (1989): Character polarities in early perissodactyls and their significance for *Hyracotherium* and infraordinal relationships (this volume, Chapter 6)

Huxley, T. H. (1869): *An Introduction to the Classification of Animals*. London (John Churchill and Sons).

Huxley, T. H. (1872): *A manual of the Vertebrated Animals*. New York (D. Appleton and Co.).

Linnaeus, C. (1758): *Systema naturae per regna tria naturae, secundum classes, ordines, genera, species cum characteribus, differentiis, synonymis, locis.* Editio decima, reformata. Stockholm (Laurentii Salvii),

Linnaeus, C. (1766): *Systema naturae per regna tria naturae, secundum classes, ordines, genera, species cum characteribus, differentiis, synonymis, locis.* Editio duodecima, reformata. Stockholm (Laurentii Salvii).

Lucas, S. G. and Schoch, R. M. (1989): Taxonomy and biochronology of *Eomoropus* and *Grangeria*, Eocene chalicotheres from the western United States and

China (this volume).

MacFadden, B. J. (1976): Cladistic analysis of primitive equids, with notes on other perissodactyls. -*Syst. Zool.*, 25:1-14.

Mader, B. J. (1989): The Brontotheriidae: a systematic revision and preliminary phylogeny of North American genera (this volume, Chapter 25).

Marsh, O. C. (1884 [date uncertain, possibly 1885 or 1886]): Dinocerata. A monograph of an extinct order of gigantic mammals. -*Monogr. U. S. Geol. Surv.*, 10:1-237.

Matthew, W. D. and Granger, W. (1925): The smaller perissodactyls of the Irdin Manha Formation, Eocene of Mongolia. -*Amer. Mus. Novit.*, 199:1-9.

McKenna, M. C. (1975): Toward a phylogenetic classification of the Mammalia. -In: Luckett, W. P., and Szalay, F. S. (eds.): *Phylogeny of the Primates.* New York (Plenum Press), pp. 21-46.

McKenna, M. C. and Manning, E. (1977): Affinities and paleobiogeographic significance of the Mongolian Paleogene genus *Phenacolophus*. -*Geobios, Mem. Spec.*, 1:61-85.

McKenna, M. C., Chow, M., Ting, S., and Luo, Z. (1989): *Radinskya yupingae*, a perissodactyl-like mammal from the late Paleocene of southern China (this volume, Chapter 3).

Owen, R. (1848): Description of teeth and portions of jaws of two extinct anthracotherioid quadrupeds (*Hyopotamus vectianus* and *Hyop. bovinus*) discovered by the Marchioness of Hastings in the Eocene deposits on the N. W. coast of the Isle of Wight: with an attempt to develop Cuvier's idea of the classification of pachyderms by the number of their toes. -*Quart. J. Geol. Soc. Lond.*, 4:103-141.

Owen, R. (1868): *On the Anatomy of Vertebrates.* Vol. III, Mammals. London (Longmans, Green, and Co.).

Prothero, D. R. , Manning, E., and Fischer, M.S. (1988): The phylogeny of the ungulates.-In: Benton, M.J. (ed.): The Phylogeny and Classification of the Tetrapods.-*Syst. Assoc. Spec. Vol.* 35 (2): 201-234.

Prothero, D. R., Guérin, C., and Manning, E. (1989): The history of the Rhinocerotoidea (this volume, Chapter 16).

Prothero, D. R., Manning, E., and Hanson, C. B. (1986): The phylogeny of the Rhinocerotoidea (Mammalia, Perissodactyla). -*Zool. J. Linn. Soc.*, 87:341-366.

Prothero, D. R. and Schoch, R. M. (1989a): The origin and evolution of perissodactyls: summary and synthesis (this volume, Chapter 28).

Prothero, D. R. and Schoch, R. M. (1989b): Classification of the Perissodactyla (this volume, Chapter 29).

Radinsky, L. B. (1964): *Paleomoropus*, a new early Eocene chalicothere (Mammalia, Perissodactyla), and a revision of Eocene chalicotheres. -*Amer. Mus. Novit.*, 2179:1-28.

Radinsky, L. B. (1965): Early Tertiary Tapiroidea of Asia. -*Bull. Amer. Mus. Nat. Hist.*, 129:181-264.

Radinsky, L. B. (1966a): The families of the Rhinocerotoidea (Mammalia, Perissodactyla). -*J. Mammal.*, 47: 631-639.

Radinsky, L. B. (1966b): The adaptive radiation of the phenacodontid condylarths and the origin of the Perissodactyla. -*Evolution*, 20: 408-417.

Radinsky, L. B. (1967): *Hyrachyus, Chasmotherium*, and the early evolution of helaletid tapiroids. -*Amer. Mus. Novit.*, 2313: 1-23.

Radinsky, L. B. (1969): The early evolution of the Perissodactyla. -*Evolution*, 23: 308-328.

Reshetov, V. Yu. (1979): Early Tertiary Tapiroidea of Mongolia and the USSR [in Russian]. -*The Joint Soviet-Mongolian Expedition, Transactions*, 11: 1-144.

Ride, W. D. L., Sabrosky, C. W., Bernardi, G., and Melville, R. V., eds. (1985): *International Code of Zoological Nomenclature*, third edition. London (International Trust for Zoological Nomenclature, British Museum [Natural History]).

Romer, A. S. (1945): *Vertebrate paleontology*, second edition. Chicago (Univ. Chicago Press).

Schoch, R. M. (1983): Relationships of the earliest Perissodactyls. -*Geol. Soc. Amer., Abstr. Prog.*, 15: 144.

Schoch, R. M. (1984a): Two unusual specimens of *Helaletes* in the Yale Peabody Museum collections, and some comments on the ancestry of the Tapiridae (Perissodactyla, Mammalia). -*Postilla, Peabody Mus., Yale Univ.*, 193:1-20.

Schoch, R. M. (1984b): Introduction. -In: Schoch, R. M. (ed.): *Vertebrate Paleontology*. New York (Van Nostrand Reinhold). pp. 1-16.

Schoch, R. M. (1985): Concepts of the relationships and classification of major perissodactyl groups: notes for a workshop on fossil perissodactyls held at Fourth International Theriological Congress, Edmonton, Canada, August 1985. Privately printed and distributed.

Schoch, R. M. (1986): *Phylogeny Reconstruction in Paleontology*. New York (Van Nostrand Reinhold).

Schoch, R. M. (1989): A review of the tapiroids (this volume, Chapter 15).

Scott, W. B. (1937): *A History of the Land Mammals in the Western Hemisphere*. New York (Macmillan Co.).

Scott, W. B. (1941): Perissodactyla. The mammalian fauna of the White River Oligocene. -*Trans. Amer. Philos. Soc.*, 28: 747-980.

Simpson, G. G. (1945): The principles of classification and a classification of mammals. -*Bull. Amer. Mus. Nat. Hist.*, 85: 1-350.

Storr, G. C. C. (1780): *Prodromus methodi Mammalium . . . inaugural disputationem propositus*. Tübingen (F. Wolffer).

Taplin, W. (1797): *A compendium of practical and experimental farriery*. Philadelphia (Robert Cambell and Co.).

Wood, H. E., II (1934): Revision of the Hyrachyidae. -*Bull. Amer. Mus. Nat. Hist.*, 67:181-295.

Wood, H. E., II (1937): Perissodactyl suborders. -*J. Mammal*. 18:106.

**NOTE:** The new family Eoletidae: Here, adopting the phylogenetic hypotheses of Hooker (this volume, Chapter 6; see also Schoch, this volume, Chapter 15), the Lophialetidae is restricted to *Lophialetes, Schlosseria*, and *Simplaletes*. The Lophialetidea is considered to be presently composed of only four known genera, *Lophialetes, Schlosseria, Simplaletes*, and *Eoletes*. As *Eoletes* appears to be, according to Hooker (this volume, Chapter 6), the sister taxon of the Lophialetidae I here establish the family Eoletidae erected to be of equivalent rank with the Lophialetidae.

Type genus of the Eoletidae: *Eoletes* Birjukov, 1974.

Included genera: Only the type genus.
Distribution: Middle Eocene of Asia.

Diagnosis: Same as that for the genus *Eoletes*. For diagnoses and descriptions of *Eoletes* see Birjukov (1974), Reshetov (1979), and Hooker (this volume, Chapter 6). *Eoletes* is very similar to *Lophialetes* and *Schlosseria*, but bears only a short canine diastema and lacks the moderate to enlarged nasal incision of the latter forms.

# 3. *RADINSKYA YUPINGAE*, A PERISSODACTYL-LIKE MAMMAL FROM THE LATE PALEOCENE OF CHINA

## MALCOLM C. McKENNA, CHOW MINCHEN, TING SUYIN, and LUO ZHEXI

A new genus and species of late Paleocene herbivorous mammals from southern China, *Radinskya yupingae*, is very close to the morphotype from which the π-shaped molar pattern of the mammalian order Perissodactyla can be derived. Although *Radinskya* is more primitive in dental morphology than either known perissodactyls or known phenacolophids, it is here tentatively regarded as the most primitive phenacolophid or a sister-group of the phenacolophids, because it shares with known phenacolophids the derived enlargement of the paraconule and especially the metaconule of its upper molars. However, on $M^1$ and $M^2$, the hypocone is linked to the strong metaconule by a short metaloph, and both the protoloph and metaloph are slanted posterolingually, forming the basic π-pattern seen in Eocene and later perissodactyls. But, unlike known phenacolophids, *Anthracobune* (USNM 392235), and later perissodactyls, the metaloph neither reaches the ectoloph nor turns forward near it. $P^4$ is partially molariform in that it has a metacone, but the paracone and metacone are not widely separate. Nevertheless, $P^4$ is more molariform in *Radinskya* than in *Anthracobune*. $M^3$ is not greatly reduced in size as in condylarths, but does have an isolated hypocone as in some archaic ungulates. Phenacodonts, long considered the probable sister-group of perissodactyls, may not be as closely related to them as is *Radinskya*.

The only known specimen of *Radinskya yupingae* has a relatively well-preserved cranium and tympanic region. Its ventrally constricted occiput and the reduced triangular external exposure of the mastoid process of the petrosal bone resemble these features of perissodactyls. But otherwise, *R. yupingae* has predominantly plesiomorphous cranial are either more primitive than their homologues in Hyracoidea and Perissodactyla or are shared by a number of other primitive eutherian herbivores. The lower dentition and postcranial skeleton of *Radinskya yupingae* are unknown.

## Introduction

The origin of the mammalian order Perissodactyla (horses, tapirs, rhinoceroses, and their extinct relatives) was long sought among Paleocene phenacodonts of North America (Matthew, 1937; Morris 1966, 1968), but no transitional animals have ever been demonstrated. It was suggested that the molar morphology of *Tetraclaenodon*, a mainly Torrejonian phenacodont, was closest among mammals to the morphotype from which the π-pattern of perissodactyl molars could be derived (Radinsky, 1966; Van Valen, 1978). Phenacodonts have been regarded explicitly as the sister-group of Perissodactyla (MacFadden, 1976, 1988). Unknown Paleocene ancestors of perissodactyls in southern North America or Central America (Morris, 1966, 1968; Sloan, 1970; Gingerich, 1976) and Europe (Hooker, 1980; Godinot, 1982) were hypothesized. The discovery in the late Paleocene of southern China of *Radinskya yupingae*, a primitive eutherian herbivore whose molars possess a π-pattern, raises new questions concerning how and where the Perissodactyla originated. We give here a preliminary description of the nearly complete cranium and upper postcanine dentition of *Radinskya yupingae*. When additional preparation is completed, the ma-

*The Evolution of Perissodactyls* (ed. D.R. Prothero & R.M. Schoch) Oxford Univ. Press, New York, 1989.

terial will be described in greater detail. The lower jaw and postcranial skeleton of the animal remain unknown.

**Abbreviations**
AMNH  Department of Vertebrate Paleontology, American Museum of Natural History, New York, New York
IVPP  Institute of Vertebrate Paleontology and Paleoanthropology, Academy of Sciences, Beijing, People's Republic of China (PRC)
UCMP  Museum of Paleontology, University of California, Berkeley, California
USNM  National Museum of Natural History, Smithsonian Institution, Washington, D.C.

**Systematics**
Class Mammalia
(Order uncertain)
?Family Phenacolophidae, *incertae sedis*
*Radinskya*, new genus

*Type species: Radinskya yupingae*, new species
*Locality and diagnosis:* as for the type species.
*Age:* Late Paleocene (Li and Ting, 1983).
*Etymology:* named for the late Leonard B. Radinsky, expert student of perissodactyls.
*Diagnosis and description:* as for the type species.

*Radinskya yupingae*, new species

*Type specimen:* IVVP no. V5255 (Figs. 3.1-3.7).
*Locality:* Lat. 25° 17' 5" N., Long. 114° 30' 59" E.; about 900 meters NW (292°) of the village of Da-tang, Nan-xiong County, Guangdong Province, PRC. Lowest part of Da-tang Member of the Nung-shan Formation, late Paleocene (Li and Ting, 1983).
*Referred specimens:* only the type specimen.
*Fauna of the Da-tang Member:* Ernan- odon antelios, Petrolemur brevirostre, cf. *Huaiyangale leura, Haltictops mirabilis, Huaiyangale meilingensis, Yantanglestes datangensis*, Arctostylopidae new genus and species, *Archaeolambda* (= *Nanlinglambda*) sp., *Altilambda pactus, Altilambda minor, Minchenella grandis, Yuelophus validus*, and *Radinskya yupingae*.

*Etymology:* named for the late Zhang Yuping, who collected the type specimen.
*Diagnosis:* much smaller than known phenacolophids, phenacodonts, hyracoids, and most perissodactyls; somewhat smaller than the smallest species of *Hyracotherium* and *Rhodopagus*; braincase more expanded than in most known archaic ungulates; occiput constricted ventrally, with rectangular outline in occipital view, resembling the condition of perissodactyls and hyracoids; reduced external exposure of the mastoid part of the petrosal triangular, facing laterally instead of posteriorly, differing from the condition displayed by primitive condylarths and most perissodactyls but resembling equids; promontorium with sulci for both promontorial and stapedial arteries; pterygoid processes doubled; glenoid fossa flat and anteroposteroirly elongate, flanked medially by a crest and with a large postglenoid process; optic foramen separated by a narrow bony bridge from the foramen lacerum anterius (see MacFadden, 1976, Fig. 6A) as in hyracoids and equids; foramen ovale plesiomorphously separated from large foramen lacerum medius (piriform fenestra of MacPhee, 1981) as in hyracoids, tapirids, and some advanced rhinoceroses; short diastema between $P^1$ and alveolus of C, and between $P^1$ and $P^2$; $P^4$ with subequal, conjoined paracone and metacone, no hypocone; molars wider (linguolabially) than long (anteroposteriorly), and in pattern as similar to phenacolophids (McKenna and Manning, 1977; Tong, 1978; Zhang, 1978) and certain hyopsodonts (Gazin, 1969) as to hyracoids, perissodactyls, or *Phenacodus* and *Tetraclaenodon* (in this regard more like those of *Ectocion, Prosthecion*, and an unnamed phenacodont; Patterson and West, 1973;

Fig. 3. 1. *Radinskya yupingae*, type specimen, IVPP V5255, primitive late Paleocene perissodactyl-like mammal from southern China. Stereophotograph of dorsal view of skull. x1.

Fig. 3. 2. *Radinskya yupingae*, type specimen, IVPP 5255, primitive late Paleocene perissodactyl-like mammal from southern China. Stereophotograph of left view of skull. x1.

Fig. 3. 3. *Radinskya yupingae*, type specimen, IVPP V5255, primitive late Paleocene perissodactyl-like mammal from southern China. Stereophotograph of ventral view of skull. x1.

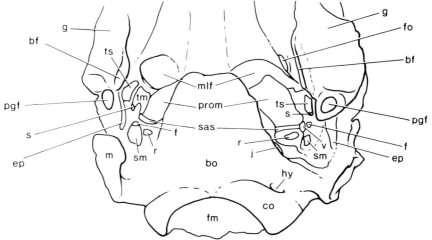

Fig. 3. 4. Crude interpretive map of the heavily damaged basicranial region of *Radinskya yupingae*. Abbreviations: bf, bony flange at medial side of squmosal glenoid fossa; bo, basiocipital; co, occipital condyle; ep, epitympanic recess; f, foramen for the facial nerve (VII); fm, foramen magnum; fo, foramen ovale; g, glenoid fossa of squamosal bone (incomplete laterally?); hy, hypoglossal foramen; j, jugular foramen (foramen lacerum posterius); m, mastoid part of petromastoid bone; mlf, foramen lacerum medius (heavily damaged on both sides); pgf, postglenoid foramen; prom, promontorium cochleae; r, cochlear femestra (fenestra "rotundra" cochleae, round window); s, stapedial foramen; sas, sulcus arteriae stapediae; sm, fossa muscularis minor for stapedial muscle; tm, fossa for tensor tympani muscle; ts, support for tympanic ring (?); v, fenestra ovalis vestibuli (fenestra ovalis, oval window for footplate of stapes).

McKenna, 1960); upper molars with incipient ectoloph formed from crested metacone and paracone, weakly linked to a low, anteriorly projecting parastyle; extremely weak molar mesostyle on narrow labial cingulum, not forming part of ectoloph; large paraconule with crest projecting anterolabially past anterolingual base of paracone toward parastyle; very large metaconule in straight line with metacone and hypocone as in phenacolophids, some phenacodonts, hyopsodonts, and primitive artiodactyls, well to rear of an imaginary line from the protocone apex to the metacone apex, lacking both pre- and post-metaconule cristae; strong metaloph of $M^1$ and $M^2$ formed from conjoined metaconule and hypocone, but not reaching metacone and separated from it by a gutter in contrast to known phenacolophids; protoloph and metaloph slanting more posterolingually to the same degree as in known phenacolophids, but less so than in phenacodonts and perissodactyls, and more so than in primitive artiodactyls that possess both metconule and true hypocone. $M^3$ slightly (not markedly) smaller than $M^2$, with large, isolated metaconule and small hypocone and metacone. Anterior and posterior cingula of the molars not hypertrophied nor connecting around the lingual side of the teeth as in known phenacolophids.

*Description*: The skull of *Radinskya* is rectangular in lateral view because of its slightly vaulted snout, a condition seen in *Meniscotherium* (Gazin, 1965) and perissodactyls (Simpson, 1952; Radinsky, 1965). The orbit is at about the middle of the total length of the skull. The premaxilla is not well preserved and the anterior part of the nasal is missing. Judged from breakage along what may be the nasal-frontal suture, which extends laterally in front of the anterior rim of the orbit, the nasal might have a broad posterior expansion, which is a plesiomorphic condition retained in *Phenacodus*, hyracoids, and various other therians. The frontal area is broad, somewhat flat, and slightly depressed at the midline. The frontal crest is moderately developed. The maxilla is short but very deep vertically. The lacrimal bone has a moderate facial expansion and bears a small tubercle. The braincase is well expanded, but constricted posteriorly near the nuchal crest, as in *Meniscotherium* (Gazin, 1965), *Hyopsodus*, and the Perissodactyla, but not as in phenacodonts. The squamosal-parietal suture extends almost horizontally and the posteriormost parietal surfaces dip ventrally before the nuchal crest is reached. A low sagittal crest is present; it also dips down and then rises to meet the nuchal crest. The palatal exposure of the palatine bone is nearly rectangular. Its anterior end extends to the level of $P^4$. The opening of the internal nares (choanae) is anterior to the level of the posterior edge of $M^2$. On either side of the long respiratory passage, each pterygoid process is elongate and doubled. On both sides of the skull, the major part of the zygoma is not preserved, although a part of the anterior base of the left zygomatic arch is preserved. Judging from what remains of the anterior and posterior ends, the zygomatic arch does not flare laterally. The anterior end of the zygoma arises opposite the metacone of $M^2$. A short crest extends forward from the anterior base of the zygoma. One cannot see whether the jugal bone extended to the glenoid fossa.

The orbital region of the type specimen of *Radinskya yupingae* is poorly preserved and the foramina here are only vaguely recognizable. The optic foramen appears to be present and its position seems similar to that of its homologue in *Phenacodus* (Kitts, 1956) and *Heptodon* (Radinsky, 1965). The sphenorbital fissure (foramen lacerum anterius) also appears to be present; it is separated anteriorly by a bony bar from the optic foramen. Because of poor preservation, it is difficult to determine how far the optic foramen is separated from the sphenorbital fissure. We cannot be certain whether it is more similar to the plesiomorphous condition (farther apart) seen in *Phenaco-*

*dus* and *Moeritherium*, or to the more derived equoid condition (close approximation). The foramen rotundum is not visible from the side and probably opens into the alisphenoid canal. The anterior opening of the alisphenoid canal is broadly open, and we interpret that, as usual, it transmitted the maxillary branch of the trigeminal nerve (V2) and an artery. The posterior opening of the alisphenoid canal is small, with a shallow groove connecting it to the foramen ovale. The glenoid fossa is flatter and somewhat more elongate anteroposteriorly than in any known perissodactyl. A bony flange of the squamosal flanks the medial side of the glenoid fossa. The postglenoid process, although damaged on both sides of the skull, appears to have possessed a sharp ventral projection. A rounded groove (fissura Glasseri) separates the posterior end of the medial flange from the postglenoid process. To the rear of the postglenoid process, the postglenoid foramen is very large. The characteristics of the glenoid fossa and postglenoid process resemble those of the condylarth *Hyopsodus* and the artiodactyl *Homacodon*, and are presumed to be plesiomorphous. In occipital view, the occiput of the skull is somewhat constricted ventrally. In other words, the occiput becomes narrow near its base in comparison with phenacodonts, a condition approaching those of some hyracoids (e.g., *Bunohyrax*) and perissodactyls. The hypoglossal foramen is distinct and lies anterior to the lateral part of the occipital condyle. The paroccipital process (= "jugular process") of the exoccipital bone is very small, tightly appressed to the mastoid process of the petrosal, and is separated from the condyle by a broad "U"-shaped gap.

The tympanic region of *Radinskya yupingae* retains many primitive eutherian characteristics. The promontorium cochleae is very large, more expanded than those of *Arctocyon* (UCMP 61456), *Pleuraspidotherium* (UCMP 61488), or *Hyopsodus* (AMNH 95496). The degree of its expansion approaches that of *Heptodon* (Radinsky, 1965). The fenestra cochlearis (fenestra rotunda) is at the posterior side of the promontorium and faces posteroventrally. The fenestra ovalis vestibuli is at the posterolateral corner of the promontorium, facing ventrolaterally. The fenestra vestibuli is smaller than the fenestra cochlearis. A rather deep groove extends from anterior of the jugular foramen toward the cochlear fenestra. This groove strikingly resembles the groove for the proximal stapedial artery of *Hyopsodus* (AMNH 95496; Cifelli, 1982). The course of the proximal stapedial artery, which appears to have extended directly over or near the cochlear fenestra, is much more posteriorly positioned than the groove for the stapedial artery in *Leptictis* (Novacek, 1986) and *Hapalodectes* (Ting and Li, 1987). A second, shallow sulcus diverges anteriorly from the proximal stapedial groove before the latter reaches the fenestra rotunda. This groove probably is for the promontorial course of the internal carotid artery, but this identification is uncertain because it is possible for the promontorium to house only the nerves of the tympanic plexus, without the promontorial artery being present (Conroy and Wible, 1978). The promontorial sulcus is separated medially by a longitudinal crest from a broad, shallow groove extending along the medial border of the promontorium. This probably represents the groove for the inferior petrosal sinus (Wible, 1983, 1986). The facial foramen, for cranial nerve VII, is anterolateral to the fenestra ovalis vestibuli. Immediately anterior to the facial foramen is a small foramen, probably for the ramus superior of the stapedial artery. Another shallow, short groove extends anteriorly from the fenestra ovalis vestibuli. It continues further anteriorly as a canal (preserved only on the right side of the skull) that eventually leads to the piriform fenestra (foramen lacerum medius). We interpret this groove and its connected canal as the course for the ramus inferior of the stapedial artery. Lateral to the canal for the ramus inferior is a large protuberance, preserved on both sides; we interpret it as the contact of the anterior crux of the ectotympanic with the petrosal and squamosal. The fossa for the stapedial muscle (M. minor fossa) is posterior to the fenestra ovalis vestibuli and facial foramen. The fossa is very large and deep and has a well-defined rim. On the left side of the skull, a small foramen lies in the posteromedial corner of the stapedial fossa. On the right side of the skull a very shallow fossa for the tensor tympani muscle (M. major fossa) is preserved. This fossa is medial to the ectotympanic contact and is formed by the ventrolateral wall of the canal for the ramus inferior.

In lateral view, the mastoid process of the petrosal has a triangular exposure. The total mastoid exposure to the exterior of the skull is very reduced compared to *Leptictis* (Novacek, 1986), *Arctocyon* (UCMP 61456), and *Pleuraspidotherium* (UCMP 61488), especially in occipital view. Posteromedially, the mastoid process is closely appressed to the small paroccipital process.

$P^1$ (?$dP^1$) is simple, blade-like, and separated from $P^2$ and the partly preserved alveolus of the canine by short diastemata. $P^2$ is double-rooted, with one major cusp (paracone) but with a lingual cingulum slightly enlarged at the posterolingual corner of the paracone base and at the anterior end (parastyle). The $P^2$ paracone has a posterior crest and lingual fluting, suggesting incipient development of a metacone. $P^3$ (missing) is three-rooted, evidently nearly the size of $P^4$.

$P^4$ is three-rooted, with a massive protocone and subequal paracone and metacone. The metacone is not so widely separated from the paracone as that of advanced phenacodonts and perissodactyls. $P^4$ has strong anterior and posterior cingula but a weaker labial cingulum and no lingual one. The anterior cingulum widens from the anterior base of the protocone to the low parastyle. The posterior cingulum lacks a hypocone; it runs from the posterior base of the protocone to the posterior base of the postmetacrista. The paraconule and metaconule are small, close to the protocone apex; the paraconule is slightly larger than the metaconule. The protocone has a central, low, transverse crest. The lingual slope of the protocone is well-developed as in primitive ungulates.

$M^1$ is four-rooted, parallelogram-shaped, wider (labiolingually) than long (anteroposteriorly), in general as similar to that of phenacolophids as to those of phenacodonts and perissodactyls. However, a perissodactyl-like π-pattern is formed from the weak ectoloph and the posterolingually slanted protoloph and metaloph. The protoloph has a strong protocone and sizable paraconule; it terminates labially as a weak crest in the gutter between the anterior cingulum and the paracone. A rather straight but low and serrate ectoloph is formed from the paracone, metacone, pre- and postparacrista, pre- and postmetacrista, and to some extent from the parastyle, without incorporation of the weak cingulum mesostyle. The paracone and metacone are embedded in the ectoloph, but are still distinct. The metaconule and hypocone are conjoined to form the metaloph, which is in a straight line with the metacone but terminates labially before reaching the lingual base of the metacone, rather than projecting anterolabially to end as a low crest anterolingual to the metacone. The metaconule is very large and very well-rounded, lacking pre- and postmetaconule cristae. It is much more posteriorly placed than in *Hyracotherium* or *Saghatherium* (Osborn, 1906), more in line with the metacone and hypocone as in arctocyonids, phenacodonts, "*Desmatoclaenus*" *mearae*, primitive artiodactyls, phenacolophids, primitive tethytheres, and some ceratomorph perissodactyls. A very weak postprotocrista still courses toward the metacone. There are shelflike anterior and posterior cingula, but these are not hypertrophied as in phenacolophids; the anterior one descends labially to the low, projecting parastyle. The labial cingulum is slightly stepped to produce a very weak

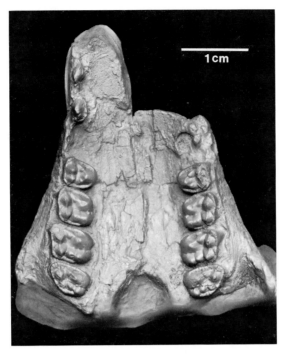

Fig. 3. 5. *Radinskya yupingae*, type specimen, IVPP V5255, primitive late Paleocene perissodactyl-like mammal from southern China, plastic cast of palate and cheek-teeth. 1 cm bar gives scale.

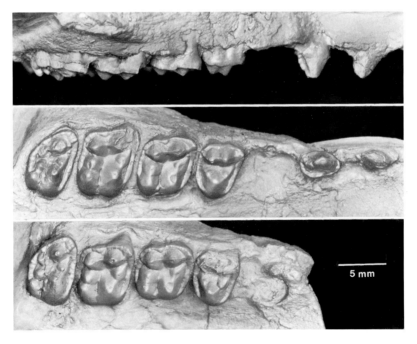

Fig. 3. 6. *Radinskya yupingae*, type specimen, IVPP V5255, primitive late Paleocene perissodactyl-like mammal from southern China, plastic casts of palate and cheek-teeth. Top: labial view of right cheek-teeth. Middle: occlusal view of right cheek-teeth. Bottom: occlusal view of left $P^4$-$M^3$, reversed for comparative purposes. 5 mm bar gives scale.

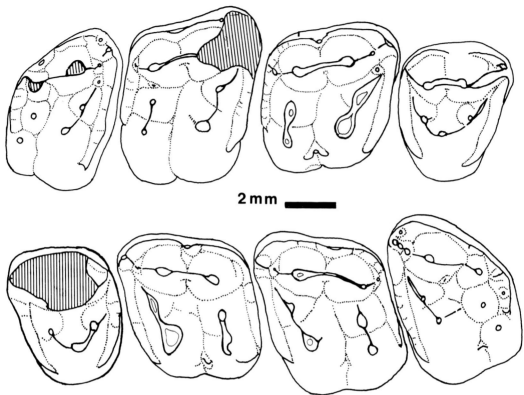

Fig. 3.7. *Radinskya yupingae*, type specimen, IVPP V5255, primitive late Paleocene perissodactyl-like mammal from southern China, dental pattern. Top: right $P^4$-$M^3$. Bottom: left $P^4$-$M^3$. 2 mm bar gives scale.

Table 3.1. Measurements (in mm) of *Radinskya yupingae*

Skull (IVPP 5255):

| | |
|---|---|
| Width across condyles | 19.6 |
| Height of occiput (top of lambdoid crest to base of occiput) | 25.9 |
| Length, from $P^1$ to posterior of occipital condyles | 82.7 |
| Width, at anterior root of zygomatic arch | 42.4 |
| Palatal width, from right to left $M^2$ parastyle | 26.3 |
| Width, between medial borders of postglenoid foramina | 30.1 |
| Diameter, postglenoid foramen | 1.5 |
| Length of glenoid fossa (approximate) | 11.0 |
| Width of glenoid fossa (approximate) | 8.1 |

| Dentition: | Length | Width |
|---|---|---|
| Right $P^1$-$M^3$ | 34.6 | -- |
| Right $P^1$ | 2.9 | 1.8 |
| Right $P^2$ | 3.9 | 2.6 |
| Right $P^3$ (alveolus) | 4.3 | -- |
| Right $P^4$ | 4.4 | 5.8 |
| Left $P^4$ | 4.3 | 5.9 |
| Right $M^1$ | 5.2 | 6.2 |
| Left $M^1$ | 5.3 | 6.2 |
| Right $M^2$ | 5.3 | 7.2 |
| Left $M^2$ | 5.1 | 7.2 |
| Right $M^3$ | 4.5 | 6.6 |
| Left $M^3$ | 4.6 | 6.7 |

mesostyle labial to and not incorporated into the ectoloph; the cingulum continues around the posterolabial corner of the tooth without bearing a metastyle. The posterior cingulum is continuous with the labial cingulum and terminates lingually on the posterior base of the crown at a point about midway between the positions of the metaconule and hypocone. A tiny stylar cusp lies at the anterior base of the hypocone.

$M^2$ is similar to $M^1$ but larger and wider. The metaconule and hypocone are not so closely conjoined as on $M^1$. $M^2$ lacks the tiny lingual stylar cusp seen on $M^1$.

$M^3$ is slightly smaller than $M^2$. In outline, $M^3$ is more rounded posteriorly than $M^2$. Its metacone is much smaller than its paracone, and its ectoloph is thus both short and weak, mostly formed by the paracone and its crests. The metaconule is distinct, larger than the metacone or hypocone, and positioned well to the rear of an imaginary line from the protocone apex to the metacone apex. A weak postprotocrista is present; it is developed to the same degree as that of phenacolophids but much weaker than that of phenacodonts.

Measurements: see Table 3.1.

## Comments

Our preliminary investigation indicates that among the cranial characteristics of eutherian herbivores, those of *Radinskya yupingae* are mostly plesiomorphous. The doubled pterygoid process is present in *Leptictis*, *Arctocyon*, *Pleuraspidotherium*, and *Procavia*. In the glenoid fossa, the medial crest and large postglenoid process separated by the fissura Glasseri are found also in *Hyopsodus* and *Homacodon*. The arrangement of the orbital foramina, so far as we can see, is probably not very different from the primitive condition seen in *Phenacodus* or *Heptodon*. The slightly elevated vault of the snout is also seen in *Meniscotherium* and *Hyopsodus*. In the ear region, the grooves, canals, and foramina for the branches of the stapedial artery are still present, in contrast to the lack of the stapedial system in hyracoids and perissodactyls (Wible, 1986). These retained primitive characters do not establish the relationships of *Radinskya*.

The paroccipito-mastoid region of *Radinskya yupingae* appears to be more similar to those of perissodactyls and hyracoids than to the plesiomorphic eutherian condition. In the primitive condition the mastoid process of the petrosal has a broad exposure on the occiput, and the external exposure of the mastoid faces posteriorly and is limited posterior to the lambdoid (nuchal) crest. This is seen in *Didelphis*, *Leptictis* (Novacek, 1986), *Arctocyon* (UCMP 61456), and *Pleuraspidotherium* (UCMP 61488). In equoids, the mastoid exposure is triangular in shape and greatly reduced in area. Moreover, it faces laterally. The lambdoid crest seems to extend to the paroccipital region along the suture of the exoccipital bone with the mastoid process of the petrosal bone. Thus the mastoid exposure is anterior to the lambdoid crest. In *Tapirus*, tethytheres (Novacek and Wyss, 1987), and *Procavia* the petrosal is enclosed into the skull by the squamosal and exoccipital bones in the paroccipito-mastoid region. The lambdoid crest extends along the suture of the squamosal with the exoccipital bone. In *Radinskya yupingae* the paroccipital process and the mastoid exposure of the petrosal bone are most similar to those of equoids, but *Radinskya* also approaches the amastoid condition in tapirids, tethytheres, and hyracoids because the total area of its mastoid exposure is reduced.

The $\pi$-pattern of $M^1$ and $M^2$ of *Radinskya yupingae* bears striking similarity to those of the perissodactyls *Hyracotherium* and *Homogalax* and, to a lesser degree to those of phenacolophids. The strong metaloph formed by a conjoined metaconule and hypocone and the loss of the postprotocrista are the key modifications in *Radinskya* that liken it to perissodactyls, hyracoids, and phenacolophids but differentiate it from all condylarths (e.g. phenacodonts). However, *Radinskya* differs from perissodactyls and hyracoids (*Saghatherium*, UCMP 86105; Rasmussen and Simons, 1988) in that its $M^1$ and $M^2$ lack a lingual cingulum around both the protocone and hypocone. Several characters suggest that *Radinskya* has affinities with phenacolophids: e.g., the large paraconule and extremely strong metaconule, and the strong but short crests of the protoloph and metaloph. The alignment of the hypocone, metaconule, and metacone in a straight line is a similarity not only to phenacolophids, but also to "*Desmatoclaenus*" *mearae* (Van Valen, 1978), *Tetraclaenodon, Ectocion*, primitive artiodactyls, and certain hyopsodonts (Gazin, 1969). Moreover, *Radinskya* differs from known phenacolophids in that its posterior cingulum does not reach the apex of the hypocone. Its $M^3$ has a moderate-sized hypocone (as in phenacolophids and advanced species of *Hyopsodus*), in contrast to such phenacodonts as *Tetraclaenodon* or *Ectocion*.

$P^4$ of *Radinskya yupingae* is closely similar to those of phenacodonts and early Eocene perissodactyls, but the metacone is closer to the paracone. However, the metacone is larger than that of hyopsodonts, in which the metacone of $P^4$ is either absent or rudimentary. The twinned paracone and metacone of $P^4$ and the weakness of the molar mesostyles, coupled with relatively large $M^3$ and possession of a moderately high $M^3$ hypocone, suggest that *Radinskya* was only collaterally related to known American or European condylarths, in which $M^3$ is reduced. In particular, *Tetraclaenodon*, a common North American Paleocene phenacodont often cited as being near the origin of the mammalian order Perissodactyla, cannot be an actual perissodactyl ancestor for this reason. The advanced phenacodonts *Phenacodus, Prosthecion*, and *Ectocion*, from Tiffanian and latter strata, likewise have a much reduced $M^3$ and have acquired large mesostyles, which "*Desmatoclaenus*" *mearae*, hyopsodonts, and the most primitive perissodactyls lack. Like paenungulates, *Radinskya* and perissodactyls appear to be derived from primitive herbivores that had not yet greatly reduced $M^3$ nor acquired more than a trace of a mesostyle on the labial cingulum.

In addition to its primitive, generally perissodactyl-like dental π-pattern, *Radinskya yupingae* is especially similar to primitive ceratomorph perissodactyls because the hypocone and metaconule of $M^1$ and $M^2$ are partly fused. However, both equoid and ceratomorph perissodactyls appear to have derived their metalophs from the fusion of the hypocone with a somewhat more anteriorly placed metaconule. Moreover, the perissodactyl metaloph primitively swings forward as the metacone is approached. Although many mammals possess diastemata and possession of such gaps is probably not very dependable evidence, the anterior diastemata of *Radinskya yupingae* suggest those of primitive perissodactyls.

The perissodactyl radiation started in the early Eocene, when their record suddenly begins in Europe, North America, and Asia. Reports of Paleocene perissodactyls in North America (Jepsen and Woodburne, 1969) have not been substantiated and probably represent an error in field data (Rose, 1981). *Hyracotherium seekinsi* from Baja California, Mexico, is early Eocene (Wasatchian) in age (Flynn and Novacek, 1984; Novacek *et al.*, 1987) rather than Paleocene as once claimed (Morris, 1966, 1968). *Radinskya yupingae* from the late Paleocene of China is thus older than any known perissodactyl. However, phenacolophids are restricted to the Paleocene of Asia and the late Paleocene locality where *Radinskya yupingae* was recovered has also produced *Yuelophus validus* (Zhang, 1978) and the phenacolophid-like primitive tethythere *Minchenella grandis* (Zhang, 1978, 1980; Domning, Ray, and McKenna, 1986). This association is indirectly suggestive but does not necessarily demand that *Radinskya yupingae* is a member of the endemic Phenacolophidae or is a sister-group of it.

Because we are fully aware of the morphological gap between *Radinskya* and other phenacolophid-like mammals, we place *Radinskya* in the Phenacolophidae with a query. If a phenacolophid, *Radinskya* is surely the most primitive one known. Moreover, it is the earliest well-documented perissodactyl-like mammal. We believe that *Radinskya* more closely qualifies as the sister-group of perissodactyls than do members of the Phenacodontidae. However, it is reasonable to argue that only minor modifications separate them from tethytheres (Domning, Ray, and McKenna, 1986) and, possibly by way of phenacolophids, from arsinoitheres as well (McKenna and Manning, 1977).

## Acknowledgments

We thank Ann R. Bleefeld, Richard L. Cifelli, John J. Flynn, Jeremy J. Hooker, Bruce J. MacFadden, Michael J. Novacek, Donald E. Savage, Robert E. Sloan, and André R. Wyss for comments and criticism. Chow Minchen thanks the American Museum of Natural History for support during a study trip to the United States. The stereophotographs depicted in Figs. 3.1-3.3 were prepared by Wang Zhefu of the photographic staff of the IVPP, Beijing, and mounted by Lisa Laumoro of the AMNH staff. Figure 3.4 was prepared by Lisa Laumoro on the basis of a sketch by Ting Suyin and stereophotographs. Chester Tarka prepared Figs. 3.5 and 3.6. Figure 3.7 was drawn by Luo Zhexi.

## Bibliography

Cifelli, R. (1982): The petrosal structure of *Hyopsodus* with respect to that of some other ungulates and its phylogenetic implications. -*J. Paleont.*, 56: 795-805.

Conroy, G.C., and Wible, J.R. (1978): Middle ear morphology of *Lemur variegatus*: some implications for primate paleontology. -*Folia Primatologica*, 29:81-85.

Domning, D.P., Ray, C.E., and McKenna, M.C. (1986): Two new Oligocene desmostylians and a discussion of tethytherian systematics. -*Smithson. Contr. Paleobiol.*, 59: 1-56.

Flynn, J.J., and Novacek, M.J. (1984): Early Eocene vertebrates from Baja California: evidence for intracontinental age correlations. -*Science*, 224: 151-153.

Gazin, C.L. (1965): A study of the early Tertiary condylarthran mammal *Meniscotherium*.-*Smithson. Misc. Coll.*, 149: 1-98.

Gazin, C.L. (1969): A new occurrence of Paleocene mammals in the Evanston Formation, southwestern Wyoming. -*Smithson. Contr. Paleobiol.*, 2: 1-17.

Gingerich, P.D. (1976): Cranial anatomy and evolution of early Tertiary Plesiadapidae (Mammalia, Primates). -*Pap. Paleont., Mus. Paleont. Univ. Mich.*, 15: 1-140.

Godinot, M. (1982): Aspects nouveaux des échanges entre les faunes mammaliennes d'Europe et d'Amérique du Nord a la base de l'Éocène. -*Geobios, Mem. Spec.*, 6: 403-412.

Hooker, J.J. (1980): The succession of *Hyracotherium* (Perissodactyla, Mammalia) in the English early Eocene. -*Bull. Brit. Mus. Nat. Hist. (Geol.)*, 33 (2): 101-114.

Jepsen, G.L., and Woodburne, M.D. (1969): Paleocene hyracothere from Polecat Bench Fm., Wyoming. -*Science*, 164: 543-547.

Kitts, D. (1956): American *Hyracotherium* (Perissodactyla, Equidae). -*Bull. Amer. Mus. Nat. Hist.*, 110 (1): 1-60.

Li, C.-K., and Ting, S. -Y. (1983): The Paleogene mammals of China. -*Bull. Carnegie Mus. Nat. Hist.*, 21: 1-93.

MacFadden, B.J. (1976): Cladistic analysis of primitive equids, with notes on other perissodactyls. -*Syst. Zool.*, 25 (1): 1-14.

MacFadden, B. J. (1988): Horses, the fossil record, and evolution, a current perspective. -*Evol. Biol.*, 22: 131-158.

MacPhee, R.D.E. (1981): Auditory regions of primates and eutherian insectivores: morphology, ontogeny and character analysis. -*Contr. Primat.*, 18: 1-282.

Matthew, W.D. (1937): Paleocene faunas of the San Juan Basin, New Mexico. -*Trans. Am. Phil. Soc., new ser.*, 30: 1-510.

McKenna, M.C. (1960): Fossil Mammalia from the early Wasatchian Four Mile Fauna, Eocene of northwest Colorado. -*Univ. Calif. Publ. Geol. Sci.*, 37 (1): 1-130.

McKenna, M.C. and Manning, E. (1977): Affinities and paleobiogeographic significance of the Mongolian Paleogene genus *Phenacolophus*. -*Géobios, Mém. Spéc.*, 1: 61-85.

Morris, W.J. (1966): Fossil mammals from Baja California: new evidence on early Tertiary migrations. -*Science*, 153: 1376-1378.

Morris, W.J. (1968): A new early Tertiary perissodactyl, *Hyracotherium seekinsi*, from Baja California. *Los Angeles County Mus. Contr. Sci.*, 151: 1-11.

Novacek, M.J. (1986): The skull of leptictid insectivorans and higher-level classification of eutherian mammals. -*Bull. Amer. Mus. Nat. Hist.*, 183: 1-112.

Novacek, M.J., Flynn, J.J., Ferrusquia-Villafranca, I., and Cipolletti, R.M. (1987): An early Eocene (Wasatchian) mammal fauna from Baja California. -*Nat. Geog. Res.*, 3 (3): 376-378.

Novacek, M.J., and Wyss, A. R. (1987): Selected features of the desmostylian skeleton and their phylogenetic implications. -*Am. Mus. Novit.*, 2870: 1-8.

Osborn, H.F. (1906): Milk dentition of the hyracoid *Saghatherium* from the Upper Eocene of Egypt. -*Bull. Amer. Mus. Nat. Hist.*, 22 (13): 263-266.

Patterson, B., and West, R.M. (1973): A new late Paleocene phenacodont (Mammalia: Condylarthra) from western Colorado. -*Breviora*, 403: 1-7.

Radinsky, L.B. (1965): Evolution of the tapiroid skeleton from *Heptodon* to *Tapirus*. -*Bull. Mus. Comp. Zool.*, 134: 69-106.

Radinsky, L.B. (1966): The adaptive radiation of the phenacodontid condylarths and the origin of the Perissodactyla. -*Evolution*, 20: 408-417.

Rasmussen, D.T., and Simons, E.L. (1988): New Oligocene hyracoids from Egypt. -*J. Vert. Paleont.*, 8: 67-83.

Rose, K.D. (1981): The Clarkforkian Land-Mammal Age and mammalian faunal composition across the Paleocene-Eocene boundary. -*Univ. Mich. Pap. Paleo.*, 26: 1-197.

Simpson, G.G. (1952): Notes on British hyracotheres. -*Zool. J. Linn. Soc.*, 42: 195-206.

Sloan, R.E. (1970): Cretaceous and Paleocene terrestrial communities of western North America. -*Proc. North Amer. Paleont. Conv.* E: 427-453.

Ting, S.Y. and Li, C.K. (1987): The skull of *Hapalodectes* (?Acreodi, Mammalia), with notes on some Chinese Paleocene mesonychids. -*Vert. PalAsiatica*, 25: 161-186.

Tong, Y.S. (1978): The late Paleocene mammals of Turfan Basin, Xinjiang. -Reports of the Xinjiang Paleontological Expeditions (III), *Mem. Inst. Vert. Paleont. Paleoanth.*, 13: 81-101 (in Chinese).

Van Valen, L.M. (1978): The beginning of the Age of Mammals. -*Evol. Theory*, 4 (2): 45-80).

Wible, J. R. (1983): The internal carotid artery in early eutherians. -*Acta Paleont. Polonica*, 28: 174-180.

Wible, J.R. (1986): Transformations in the extracranial course of the internal carotid artery in mammalian phylogeny. -*J. Vert. Paleo.*, 6: 313-325.

Zhang, Y.-P. (1978): Two new genera of condylarthran phenacolophids from the Paleocene of Nanxiong Basin, Guangdong. -*Vert. PalAsiatica*, 16 (4): 267-281.

Zhang, Y.-P. (1980): *Minchenella*, new name for *Conolophus* Zhang, 1978. -*Vert. PalAsiatica*, 18 (3): 257.

# 4. HYRACOIDS, THE SISTER-GROUP OF PERISSODACTYLS

## MARTIN S. FISCHER

The systematic position of the Hyracoidea within the Eutheria is controversial. Are they most closely related to Sirenia and Proboscidea (= Paenungulata), or are they the sister-group of the odd-toed ungulates, or perissodactyls? The test of the morphological characters (taxeopody, musculature, placenta and fetal membranes, skull morphology) used as arguments of the first assumption gives relatively weak positive evidence. The molecular data supporting the monophyly of the Paenungulata are inconsistent. On the other hand, there are apomorphic characters in the locomotory organs, the maxillare and the Eustachian sac, which suggest a grouping of hyracoids and perissodactyls.

### Introduction

Today, hyraxes rank as their own order, the Hyracoidea, in the system of the Mammalia, with the Procaviidae as the only living family. The phylogenetic history of the Hyracoidea is not satisfactorily reconstructed, although there is a reasonably good fossil record of this group from the Eocene of North Africa and from the Miocene of Eurasia. Rasmussen (this volume) reviews the fossils of the three major radiations in hyracoid evolution.

Only recently the old controversy--namely, whether hyracoids are closely related to elephants and sirenians or to perissodactyls--has been reactivated. This is partly due to the discrepant evidence from amino acid sequences and morphology, and because there is morphological evidence for both hypotheses. There is also another possible solution to this problem. Fischer (1986) and Tassy and Shoshani (1988) have argued that there might exist a more inclusive group consisting of Mesaxonia, Hyracoidea, and Tethytheria (Proboscidea and Sirenia), an idea already proposed by Gregory (1910) and Szalay (1977).

Any question concerning the origin of the Hyracoidea and their phylogenetic position within the Eutheria is confronted with the problem that the monophyly and hence the Grundplan of the Ungulata cannot be considered to be well grounded. Therefore, it is difficult to get an idea on the polarity of the characters that are relevant in the discussion on hyracoid relationships. Most of the characters used as arguments for the gigantic superorder Ungulata, e.g., Novacek (1986) or Prothero et al. (1988), are not as convincing as one would really desire. The group seems to be based more on convention than on synapomorphies. "Given the general acceptance of this concept, it is ironic that the Ungulata now seems the weakest of the higher-level groups hypothesized here" (Novacek and Wyss, 1986, p. 266).

Owen (1848) included the genus *Hyrax* (now *Procavia*) in his original definition of the Perissodactyla. However, most workers now accept a usage of the term Perissodactyla without inclusion of the hyracoids. In order to avoid any confusion, I use the term Mesaxonia Marsh, 1884, for the taxon including the Tapiridae, Rhinocerotidae, Equidae, and their extinct relatives.

### Review of the history of classification of the Hyracoidea

Cuvier (1800, 1804, 1817) placed *Hyrax* next to *Rhinoceros*, *Palaeotherium*, and *Tapirus* in the "pachydermes ordinaires qui n'ont pas le pied fourchu". Brandt (1869) reported extensively on the different approaches to the classification of hyracoids during the first seven decades of his century. Most authors have followed Cuvier's authority through most of the nineteenth century. Accordingly, de Blainville (1816) and Owen (1848) placed the hyracoids into the odd-toed animals. Even after the opinion on the systematic position of the hyracoids changed at the end of the nineteenth century, the hypothesis that they are most closely related to the mesaxonic ungulates is still found in the literature, e.g., Frechkop (1936), Grassé (1955), McKenna (1975), and Hennig (1983).

Based on the arrangement of the carpalia and tarsalia, Cope (1882) distinguished four different orders of ungulates: Taxeopoda (Hyracoidea, Condylarthra), Proboscidea, Amblypoda (Pantodonta, Dinocerata), and Diplarthra (Perissodactyla, Artiodactyla). This classification already reflects the typological principles that Osborn would later call the "laws of modification of foot structure."

> The evolution of the manus and pes of the ungulates included the following process: (1) Elevation from the plantigrade to the digitigrade position; (2) growth of certain elements and reduction of others, including the loss of lateral parts; (3) displacements of the elements of the podium and metapodium from the primitive serial arrangement; (4) coalescence of parts primitively distinct (Osborn, 1889, p. 559).

This hypothesis of ungulate phylogeny is based on the *a priori* assumed evolution of their locomotory organs. For example, since recent hyracoids are plantigrade with a serial arrangement of carpals and tarsals, they are consequently assumed to represent a primitive ungulate type. In the following, I will argue that recent hyracoids are secondarily plantigrade.

In 1891 Flower and Lydekker concentrated the first three orders of the above mentioned classification of Cope (1882) in one taxon, the Subungulata or Ungulata polydactyla in contrast to the Ungulata vera (= Diplarthra). This classification soon became established, although until recently nobody worked out the arguments for it, but only attributed what were considered to be diagnostic characters, e.g. Weber (1928). Simpson (1945) pointed out that Illiger (1811) had already used the term Subungulata for a taxon of caviamorph rodents. Therefore, he proposed the term Paenungulata for the taxon including Hyracoidea, Proboscidea, Sirenia, Embrithopoda, Pantodonta, Dinocerata, and Pyrotheria. He had some doubts concerning the inclusion of the Hyracoidea and stressed the possible polyphyletic origin of the superorder Paenungulata.

Thenius (1969) distinguished the following orders of Paenungulata: "Amblypoda" (Pantodonta, Dinocerata, Xenungulata, Pyrotheria), Desmostylia, and the superorder "Subungulata" (Sirenia, Proboscidea, Hyracoidea, Embrithopoda). McKenna and Manning (1977) departed from Simpson's definition of the Paenungulata by the exclusion of the Pantodonta and the inclusion of the Perissodactyla. McKenna and Manning (1977) placed the Tethytheria next to the Hyracoidea and Perissodactyla, returning to the ideas of Gregory (1910):

> Finally the existence of so many "cross-resemblances" between the Proboscidea and the Perissodactyla by way of *Hyrax* seems more consistent with the hypothesis that all these now very divergent orders are derived from a common protoungulate stock, than with the hypothesis that all resemblances are due to convergent evolution (Gregory, 1910, p. 366).

The assumption that very different supposed ungulate groups have descended from a common protoungulate stock is also found in papers by Matthew (1937), Whitworth (1954) and Van Valen (1971).

## Characters recently proposed as synapomorphies of Hyracoidea and Tethytheria

Different characters which have been used during the past hundred years to argue for a close relationship of hyraxes, elephants and sirenians are discussed by Fischer (1986). Most of them have been dismissed by modern proponents of this assumption, e.g., Novacek and Wyss (1986). These characters are predominantly false, symplesiomorphies, or very likely convergent. In the present paper I shall restrict myself to the discussion of those characters that have been proposed as synapomorphies for the Hyracoidea and Tethytheria in recent years. These are: the taxeopode arrangement of tarsals and carpals, various characters in skull morphology, similarities in the musculature, the zonary placenta and fetal membranes, and the molecular evidence.

### The taxeopode arrangement of the tarsalia and carpalia

The easiest way to define taxeopody is that the lunatum articulates exclusively with the capitatum in the carpus, i.e., there is no lateral shift of the two carpal rows against each other. In the tarsus it means that the talus (astragalus) has an exclusive contact distally with the naviculare.

Fischer (1986) has demonstrated that the taxeopode arrangement of carpals and tarsals is a functional result of a very special manner of rotational movements in recent hyracoids. As a unique adaptation, almost all the supination and pronation takes place in a midcarpal and midtarsal joint, i.e., between the first and second carpal row and between talus/calcaneus and naviculare/cuboid. The implications of this specialization for the adaptational history of recent hyracoids are discussed in the following.

Taxeopode arrangements in the carpus are also described in the Proboscidea, Sirenia, *Phenacodus, Meniscotherium* and according to Tassy and Shoshani (1988) in *Arsinoitherium zitteli*. Although Gawrilenko (1924) observed minor deviations from a strictly serial arrangement in the stem-group of the Elephantidae, there is sufficient reason to assume that the taxeopody is a derived character in the Grundplan of the Proboscidea (Tassy, 1985). Shoshani (1986) and Novacek and Wyss (1986) gave new, convincing arguments that the carpal structure in desmostylians is autapomorphic and can be derived from a sirenian-like carpus. Therefore, it seems plausible to accept a taxeopode carpus as a derived character for the Grundplan of the Tethytheria.

Taking into account the distribution of the character, there are quite a few possibilities for its interpretation. The following three seem to be the most plausible.

1. Tethytheria and Hyracoidea are a monophyletic group. The taxeopode arrangement is a synapomorphy of this group, and the Desmostylia have acquired an autapomorphous condition, which superficially resembles a diplarthral carpus. The taxeopody in other groups has evolved convergently. This is the position held by Novacek and Wyss (1986)

2. The taxeopody of the different groups is due to convergent evolution.

3. Tethytheria and Mesaxonia are a monophyletic group, and the Hyracoidea are related to one or the other. The taxeopod arrangement is a derived character of the phylogeny of this group. The alternate carpus of Desmostylia and Mesaxonia has evolved independently and secondarily. As phenacodontid condylarths are widely accepted to be the sister-group of the Mesaxonia (Radinsky, 1966; MacFadden, 1976), and as *Phenacodus* has a serial carpus, it may represent not only the condition of the stem-group of Mesaxonia but of Mesaxonia + Hyracoidea. This hypothesis seems very appealing.

Lacking the evidence in sirenians and, as Tassy and Shoshani (1988) stressed, accepting the discrepancy between the arrangement of tarsals and carpals, it becomes very difficult to decide between the different possible assumptions.

## Similarities in the musculature

Whereas all the seven characters mentioned by Windle and Parsons (1903) as similarities of paenungulates can be refuted (Fischer, 1983, 1986), there is another one that can be used in favor of this group. Janis (1983) described the M. styloglossus as bifurcated before its insertion in *Procavia capensis*. It is also bifurcated in *Trichechus inunguis* (Domning, 1978) and trifurcated in *Dugong dugon* (Domning, 1977). Although Mayer (1847), Miall and Greenwood (1878), and Shindo and Mori (1956) have not observed a bifurcation of this muscle in elephants, Ghetie (1944) has found it in *Elephas indicus*. There are some minor differences in the morphology of the M. styloglossus in the first-mentioned groups. Besides the paenungulates, a bifurcation is described in the artiodactyls *Cariacus mexicanus*, *Cephalophus grimmi*, and probably in *Tragulus javanicus* and *Bos taurus* (Windle and Parsons, 1901). Concerning the last, Ellenberger and Baum (1943) did not mention a bifurcation. Given its occurrence in other ungulates, the phylogenetic value of this character should not be exaggerated.

## Placenta zonaria and fetal membranes

Hyracoids have a zonary placenta like elephants, manatees, and carnivores. Perry (1974) emphasized that one of the differences between hyracoids and elephants is the early placentation. The placenta of *Procavia* extends over the whole surface of the embryo, at least until the 48-somite stage (Sturgess, 1948), thus primarily forming a placenta diffusa (Assheton and Stevens, 1905). In elephants, the placenta is restricted from the earliest ontogeny to the equatorial belt. There are no observations on the early placentation in sirenians. The only description of placentation in this group has been made on a late fetus of *Trichechus* (Wislocki, 1935). Perry (1974) remarked that the hope of Wislocki and Westhuysen (1940) that the examination of the early placentation of hyracoids may reveal their systematic position has not been fulfilled. Instead, implantation and early placentation is different in hyracoids and elephants. Amoroso and Perry (1964) found that the placenta of elephants is not of the hemochorial type, as has been widely held for a long time, but of the endotheliochoral type. *Trichechus* and *Procavia* possess the hemochorial type of placenta. Amniogenesis of *Loxodonta* and *Procavia* is different, too (Sturgess, 1948; Perry, 1974).

Although Perry (1974) ascribes no systematic value to the genesis of the allantois, it should be mentioned that *Trichechus*, *Loxodonta*, and *Procavia* have a quadrilobular allantois, a character shared with *Orycteropus* (Mossman, 1957) and the Strepsirhini (Hill, 1932). Recently this character has been used again by Novacek and Wyss (1986).

I had the chance to discuss the problem at length with W. P. Luckett. He stated that:

> The major point of significance for systematic purposes is the zonary placenta which in other mammals occurs only in Carnivora and Tubulidenta (Luckett, 1977). Carnivores differ from tubulidentates and paenungulates in the retention and modification of the distal portion of the yolk-sac in late gestation stages. In contrast tubulidentates and paenungulates exhibit the primitive eutherian condition of complete reduction of the yolk-sac in late stages. The only shared derived character which provides some support for the hypothesis of the monophyly of Paenungulata (including Tubulidentata) is therefore the common possession of a zonary chorioallantoic placenta. On the contrary, there is no evidence of shared derived placental characters to support a sister-group relationship of Hyracoidea and Mesaxonia. The quadrilobular allantois is relatively primitive and occurs in many different taxa. It appears to be related to a large al-

lantois and the necessity to support the placenta.

*Cranial morphology*
Extension of the jugal to the posterior border of the glenoid fossa

The jugal extends back to the posterior border of the glenoid fossa in hyracoids and forms an articulation facet for the condyle of the lower jaw at the interior side of the zygomatic arch. In Elephantidae, the jugal extends nearly as far back as in hyracoids, but without participation in the articulation. Sirenians possess a jugal that extends somewhat farther back than in the usual mammalian condition with a jugal ending at the level of the anterior border of the glenoid fossa. I would agree with Tassy (1981) and Tassy and Shoshani (1988) in their consideration that this character has evolved in parallel in proboscideans and hyracoids, because the condition in sirenians closely resembles the primitive eutherian status, making it difficult to assume the derived condition is a shared derived character of the Tethytheria.

Amastoidy

Proboscidea, Sirenia, and Hyracoidea are amastoid, i.e., the mastoid is not exposed at the surface of the cranium, which is a derived character within eutherians. The dorsal exposure of the mastoid in recent sirenians seems to be secondarily achieved (Novacek and Wyss, 1986). The major argument to weaken this character is its distribution within the Eutheria or even in distinct orders. Novacek and Wyss (1986) already mentioned the amastoidy in Pholidota, Cetacea, and Dermoptera. Restricting the typological notion "amastoidy" to the case where the squamosum overgrows the pars mastoidea of the periotic, we still find this condition in groups other than the paenungulates, such as the Suidae and Rhinocerotidae. Tapirs possess a somewhat transitional morphology between horses and rhinoceroses because the exposed area of the mastoid is very small.

Lamboid crest weak and occiput expanded

This character and the following two were used by Novacek (1986) as synapomorphies of the paenungulates. But a strong lamboidal crest occurs in different fossil hyracoids, e.g., *Prohyrax hendeyi*, *Megalohyrax championi*, *Pliohyrax graecus*, or *Bunohyrax fajumensis*. The character is therefore not valid for the whole group.

Fenestra rotunda with vertical orientation, partly shielded ventrally by expansion of promontorium cochleae

According to Tassy and Shoshani (1988) this character shows some homoplasy in paenungulates. They report that "it can be noticed that the fenestra rotunda of *Prorastomus* is more ventrally situated (plesiomorphic condition) than in other sirenians." The status of this character does not seem to be convincingly established.

Post-tympanic ridge of squamosal very weak or absent

Novacek (1986) used this character although he himself mentioned that the plesiomorphic condition occurs in the Sirenia.

Foramen magnum formed by basioccipital and exoccipital only

I tested this character mentioned by Shoshani (1986) in a series of neonatal to adult skulls of *Procavia capensis*. Up to the stage when $M^2$ starts to erupt, the supraoccipital nearly always forms the dorsal closure of the foramen magnum. Then the suture of the occipital bones closes, and a distinction between exoccipital and supraoccipital is almost impossible. As the character is also inconsistent in Sirenia (Tassy and Shoshani, 1988) it seems best to abandon it as a synapomorphy of the Paenungulata.

*Evidence from different molecular analyses for the relationship of Hyracoidea and Tethytheria*

Besides anatomical data, there is evidence from serological and amino acid analyses for the monophyly of the Paenungulata. Weitz (1953) tested the immunological reaction of the elephant and hyrax. He found a reaction of anti-elephant-serum to hyrax-serum and of anti-hyrax-serum to elephant-serum, but no reaction of hyrax-serum to horse-serum.

A concise summary of the results of an electrophoresis test of plasma proteins and hemoglobin is given by Buettner-Janusch et al. (1964). They observed no accordance between *Loxodonta* and *Procavia* in the plasma proteins, and the only result for the hemoglobin was that it moved faster than the human hemoglobin A. The alkali resistance test revealed no similarities between the two species.

Shoshani et al. (1978) made an immunodiffusion test of serum proteins of ungulates and paenungulates against antiserum of the chicken. They concluded that Sirenia and Proboscidea are related to Hyracoidea and Tubulidentata. Based on an analysis of albumin immunological distance, Rainey et al. (1984) stated that sirenians, elephants, hyraxes, and aardvarks form a monophyletic group. The albumins of the elephant and aardvark are the most similar.

The analysis of the primary structure of the hemoglobin of *Procavia capensis habessinica* (Kleinschmidt and Braunitzer, 1983) and its comparison to *Elephas maximus* shows that the α- and ß-chain have one substitution each which is unique to these groups. In a recent paper, Kleinschmidt et al. (1986) included *Trichechus inunguis* in their data base. The manatee shows the same unique amino acid on α110 as the elephant and hyrax. The second amino acid replacement (ß 56) claimed as an indication of a certain relationship between elephants and hyraxes in the earlier paper does not occur in the manatee. The manatee has the same amino acid at this position as some artiodactyls and the rabbit. Kleinschmidt et al. (1986) consider four more amino acid replacements as possible synapomorphies of the paenungulates. Two of them would group the manatee and hyrax closer together (α 19, ß 23), one confirms the Tethytheria (α 111) and the last occurs in all three groups, but also in the elk and the kangaroo (ß 44). Taking these data as if they were morphological data and ignoring the maximum parsimony approach, one could conclude that the evidence for the monophyly of the paenungulates comes from one unique amino acid replacement (α 110) and from another (ß 44) which is not so widely distributed in other mammals.

De Jong and coauthors (De Jong et al., 1977; De Jong and Goodman, 1981; De Jong et al., 1982) have analyzed the eye lens protein α-crystallin in 17 different mammalian orders. One of the results is that the Paenungulata and Tubulidentata show three identical nucleotide replacements. Two of them occur in one other mammalian group. *Orycteropus, Procavia* and *Trichechus* possess an additional synapomorphic nucleotide replacement placing them as the sister-group of the elephant. We can presume that there is one unique substitution (70 Lys-Glu) in Paenungulata and Tubulidentata, and that all four substitutions of the Paenungulata occur in Tubulidentata, too. This alliance is not supported by anatomical evidence (Prothero et al., 1988). Furthermore the trichotomy of *Trichechus, Procavia* and *Orycteropus* contradicts the morphologically based opinion that Sirenia and Proboscidea are sister-groups.

There are some more points in the results of De Jong and Goodman (1982) that may be criticized without entering into the discussion of their methodological principles. De Jong and Goodman (1982) made some "a priori cladistic assumptions," e.g., Mammalia are a monophyletic group, or all species under consideration are grouped in their respective traditional orders. Without these assumptions the most parsimonious phylogenetic tree has had the marsupials as the sister-group of the chicken. Cetacea, Carnivora, Rodentia, and Lagomorpha did not come out as monophyletic groups, and the pangolin became the sister-group of the bear.

Another objection can be raised. One locus (number 150) is used on various occasions as the exclusive argument for a dichotomy, despite its high variability in mammals. We find the following distribution of locus 150 in mammals (De Jong and Goodman, 1982):

150 Methionine (Met): minke whale, porpoise, horse, rhinoceros, bear, elephant, hyrax, manatus, aardvark, kangaroo, opossum, chicken;

150 Leucine (Leu): tapir, hedgehog, tupaia, rat, hamster, gerbil, guinea pig, springhaasd, pika, rabbit, lemur, galago, potto, rhesus monkey, human, frog;

150 Valine (Val): pig, camel, giraffe, hippopotamus, ox, dog, cat, mink, seal, sea lion, pangolin, bat, sloths, tamandua.

De Jong and Goodman (1982) decided that 150 Met represents the plesiomorphic conditions in mammals. The dichotomy Paenungulata/other Eutheria, the monophyly of a group including tupaia, hedgehog, rodents, primates, and lagomorphs and the sister-group Cetacea-Perissodactyla are based only on nucleotide replacements on locus 150. Further, the substitution at the last mentioned dichotomy has to be assumed as back-mutation.

At last there is a methodological argument that should be tested. De Jong and Goodman (1982, Fig. 4) present a cladogram that is most parsimonious, after the conjectures mentioned above. Their statement that ten additional nucleotide replacements are necessary to relate Hyracoidea and Mesaxonia is derived from this cladogram by counting the dichotomies separating them. Would it not be better to start with the *a priori* assumption that these two groups are sister-groups, and then compare the total number of necessary nucleotide replacements in the cladogram with the number of required substitutions in the corrected maximum parsimony cladogram?

## Apomorphic characters of the Hyracoidea and Mesaxonia

There are three complexes of characters in which I have noticed an apomorphic coincidence between Hyracoidea and Mesaxonia. These are the locomotory organs, the Eustachian sacs, the course of the internal carotid artery, and the morphology of the maxillaries.

*The locomotory organs of hyracoids and mesaxonians*

Despite the fact that similarities in the locomotory organs are supposedly due to convergent evolution, there are some characters found in hyracoids and mesaxonians that, due to their uniqueness and their very high degree of similarity, should be seriously considered as arguments for a hyracoid-mesaxonian clade.

Fischer (1986) has reconstructed the adaptational history of the locomotory organs of hyracoids, applying the constructional morphology approach (Seilacher, 1970; Reif et al., 1985). He hypothesized that recent hyracoids descended from cursorial, unguligrade ancestors. The idea that recent hyracoids are secondarily plantigrade had been expressed earlier by Thenius (1979). The arguments for this assumption are briefly reviewed here.

Analyzing the locomotory organs of hyracoids, one is very surprised to find numerous features that are not only unexpected but which seem to be in contradiction to the adaptational requirements of their present climbing way of locomotion. These include the loss of the clavicle, the tridactyl foot, the absence or advanced reduction of all muscles causing supination and pronation movements, and the possession of hoof-like structures instead of claws. Recent hyracoids have a unique way of pronation and supination in that the whole movement occurs in a midcarpal and midtarsal joint. A plausible explanation for this adaptation seems to be that the original flexibility in the lower arm and wrist joint, or in the ankle joint, has been lost once and had to be replaced by these new joints. As the loss of supination and pronation movements is characteristic of cursorial animals, Fischer (1986) concluded that recent hyracoids went

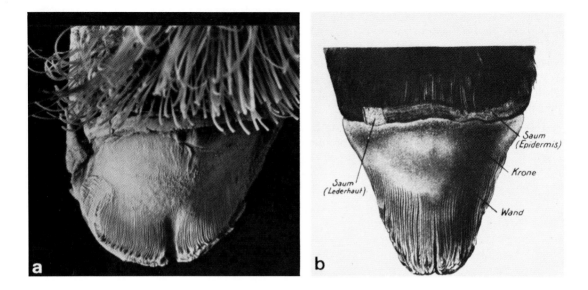
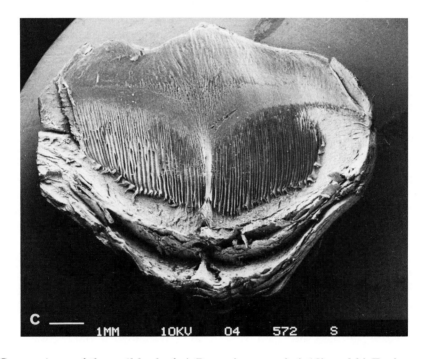

Fig. 4.1. Comparison of the nail bed of a) *Procavia capensis* (x10) and b) *Tapirus americanus* (Hildebrand, 1927) to show the deep groove in the medial plane. c) As in *Procavia capensis*, the groove corresponds with fused horny lamellae found on the lower side of the horny plate.

through a cursorial stage of which the remaining traits can be considered to be historical constraints (Reif et al., 1985). This hypothesis is supported by the reproductive ecology of hyraxes. The unusual gestation period of 7.5 months is difficult to explain as an adaption to a cave-or-tree-dwelling existence. But it can be understood as a vestige of a former plains-dwelling life. Further support comes from ontogenetic studies of Böker (1935) and Böker and Pfaff (1931), who found that the length of the pes is shortened relative to the femur and to the tibia during ontogeny.

An extensive description and comparison of the locomotory organs of hyracoids and mesaxonians with other eutherians reveals a high degree of overall similarity in these organs between the first two groups. Here, I only want to pick out those characters which seem to be unique to hyracoids and mesaxonians.

This is first the formation and course of the M. sternoscapularis. This muscle originates from the cranial end of the sternum, passes over the shoulder joint and is inserted at the superior edge (angulus superior) of the scapula. Murie (1872) clearly described the similarity between the conditions in the tapir and hyrax. An identical morphology is also found in rhinoceros (Fischer, in prep a.) and horses (Ellenberger and Baum, 1943). Only in *Sus* and *Choeropsis* has a similar muscle been described (Rübli, 1930; Campbell, 1936; Ellenberger and Baum, 1943). But in these two genera the muscle ends at the anterior border of the scapula or with an aponeurosis on the M. supraspinatus. According to Woods (1972) the M. subclavius is inserted not only on the clavicle and acromion but also on the scapular spine in *Dasyprocta* and *Thryonomys*.

The whole anatomy of the shoulder is very similar in hyracoids and mesaxonians. This is illustrated by the following examples. The acromion is lacking and thus the acromial portion of the M. deltoideus is absent. This is another unique character, because in suids, which are the only other recent mammals with no acromion, the acromial portion of the M. deltoideus is still present. After the reduction of the clavicle (a common ungulate trait), the fusion of the M. sternocleidomastoideus with the clavicular portion of the M. deltoideus, forming the M. brachiocephalicus or M. cephalohumeralis, is achieved in very much the same way in hyracoids and mesaxonians.

A closer look on the organon digitale (the organ covering the tips of the digits, or the nails) of hyracoids revealed interesting derived similarities with the hooves of mesaxonians. The terminal phalanges of fossil hyracoids and the terminal phalanx of the second toe of recent hyracoids are broadened proximally, then taper distally with a median notch at the tip. Although the terminal phalanges of the other digits are somewhat blunt, the horny plate covering these digits has preserved the characteristics of the former structure. On the lower side of the horny plate, there are numerous horny lamellae of which several in the median region are fused together and are much higher (Fig. 4.1). This structure leaves a deep groove in the "nail-bed" (Fig. 4.1), which corresponds to the median notch in the terminal phalanx. In mesaxonians, such as the tapir (Fig. 4.1), we find the same derived conditions. This not only suggests a close relationship but also that the organon digitale of recent hyracoids has evolved from hooves and has been secondarily altered. This assumption is supported by further anatomical and histological evidence (Fischer, 1986).

*The Eustachian sac (Diverticulum tubae Eustachii)*

The Eustachian sac of hyracoids was discovered by Brandt (1863) and mentioned in several publications (Brandt, 1869; George, 1875; Peter, 1894; Altmann, 1932), but it has never been adequately described. Because of its striking similarity to that of the horse, George (1875) just reproduced a description of the latter.

The Eustachian sacs (Fig. 4.2) occupy the space between the skull base, the atlas,

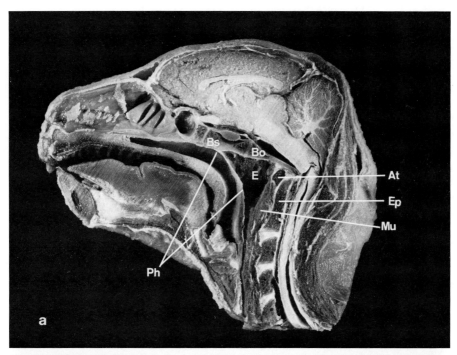
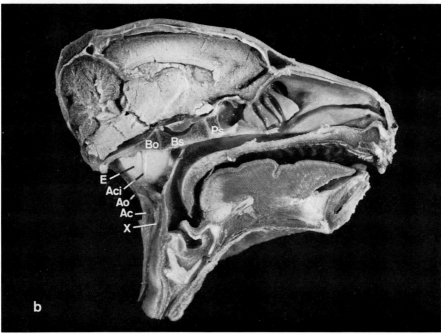

Fig. 4.2. Mediosagittal section through the head of *Procavia capensis*. a) The flexors of the head have been removed to exhibit the position of the Eustachian sac (E). b) The Eustachian sac has been prepared caudally and medially. Remark the course of the internal carotid artery (Aci) along the medial side of the Eustachian sac. Ac = A. carotis communis, Aci = A. carotis interna, Ao = A. occipitalis, At = atlas, Bo = basioccipitale, Bs = basisphenoid, E = Eustachian sac, Ep = epistropheus, Mu = prevertebral muscles, Ph = pharynx, X = N. vagus.

and the posterior wall of the pharynx. They are thin-walled, vesicle-like, ventral dilatations of the membraneous part of the Eustachian tube with a volume ranging from approximately 2 ml in hyracoids, to 400-500 ml in horses. They are not visible at first in the mediosagittal section, because they are not adjacent, but separated by the flexors of the head (M. longus capitis and M. rectus capitis ventralis). The A. carotis interna runs between these two muscles and the medial wall of the Eustachian sac. It enters the skull through the foramen lacerum medium.

Caudomedially, the Eustachian sac is adjacent to the atlas and the condyli occipitales and lateral to the M. pterygoideus medialis and the glandula parotis, while it is contiguous to the M. digastricus and the M. stylohyoideus caudolaterally. Caudally, the processus paroccipitalis pushes in its caudal wall. The region of the skull base covered by the Eustachian sac is the basioccipital, basisphenoid, bulla tympani, foramen lacerum medium, the caudal side of the processus postglenoideus, and the posterior border of the pterygoid.

The ligamentum stylohyoideum runs along the ventral side of the Eustachian sac together with the M. styloglossus, dividing the sac into two compartments. This is especially noteworthy because, even after the reduction of the stylohyal bone to a tiny ligament, it still divides the Eustachian sac. Further, although the ligamentum stylohyoideum is inserted at the processus paroccipitalis, the Eustachian sac still reflects the former conditions. In front of the processus paroccipitalis, its posterior wall has a duplication that is divisible down to the skull base, the former point of insertion of the stylohyal bone. Ontogenetic studies (Fischer, in prep. b) show that the Eustachian sac develops relatively early in the ontogeny of *Procavia*, a fact also known in horses (Daum, 1925).

A Eustachian sac (Fig. 4.3) has been described in Equidae (Ellenberger and Baum, 1943) and in Tapiridae (Turner, 1850; Zuckerkandl, 1885; Peter, 1894; Boas and Paulli, 1925; Anthony, 1920; Lechner, 1932). The situation in the Rhinocerotidae is not clear. Whereas Richter (1923) and Ellenberger and Baum (1943) mention a Eustachian sac in the rhinoceros (although most probably without their own dissections and without citation), Zuckerkandl (1885) denies its occurrence in that group. Nevertheless, it seems more plausible to accept the Eustachian sac as an apomorphic character of the Grundplan of the Mesaxonia, probably lost or altered in Rhinocerotidae, than to assume a parallel evolution of such a highly complex organ in two very closely related groups.

Murie (1872) described a structure in the manatee under the name "Eustachian sac." Recently, I had the chance to dissect the head of a manatee (Fischer, in press). It became obvious that the "Eustachian sac" of the manatee is part of the tympanic cavity. Ventrally, the tympanic cavity is bordered by a membraneous sac, which encloses the space between the basioccipital, basisphenoid, pterygoid process, and the tympanic ring. The Eustachian tube is very divergent in the manatee compared to the conditions of other mammals. For example, there is no hook-like cartilage. Elephants show the usual mammalian condition of a closed Eustachian tube. While the Eustachian sac of hyracoids and mesaxonians is a dilatation of the proximal and middle part of the Eustachian tube (the latter being closed at the pharyngeal and tympanic end), the Eustachian tube in manatees opens very early into the pterygoid sinus of the tympanic sac. In addition, the course of the internal carotid artery is different in manatees and the hyracoids and mesaxonians. In the manatee, the artery runs within a duplication of the membrane of the tympanic sac through the tympanic cavity to the foramen lacerum medium.

*The extracranial course of the A. carotis interna*

According to Wible (1986), the Hyracoidea and Mesaxonia represent an apomorphic course of the A. carotis interna.

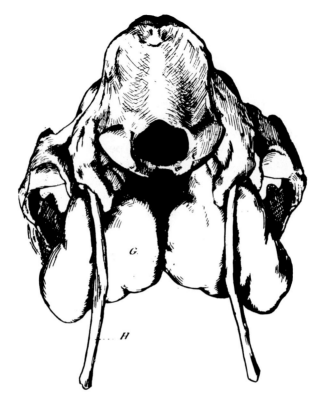

Fig. 4.3. The Eustachian sac of *Tapirus americanus*, seen from caudally. The stylohyal bone (H) divides the Eustachian sac into two compartments (From Anthony, 1920).

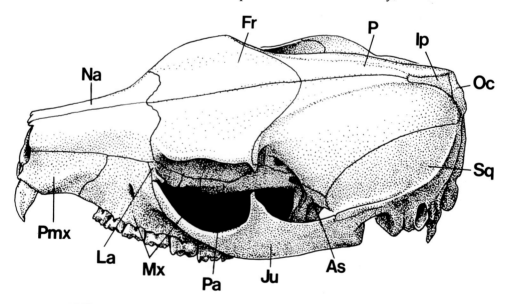

Fig. 4.4. Dorsolateral view of the skull of *Procavia capensis* to demonstrate the morphology of the tuber maxillare. Forming the floor of the orbit, the maxillare reaches far back to the alisphenoid bone. As = Alisphenoid, Fr = Frontale, Ip = Interparietale, Ju = Jugale, La = Lacrimale, Mx = Maxillare, Na = Nasale, Oc = Occipitale, P = Parietale, Pa = Palatinum, Pmx = Premaxillare, Sq = squamosum.

Wible assumed that the A. carotis interna runs through the promontory sulcus (= sulcus caroticus) in the Grundplan of the Eutheria. The sulcus caroticus is absent in Hyracoidea, Mesaxonia, and *Phenacodus* (Cifelli, 1982). *Ochotona* and "muroid rodents" (Wible, 1986) are the only other Eutheria with an extrabullar course of the A. carotis interna. Bugge (1974) has observed that in *Ochotona*, the A. carotis interna runs along the ventral side of the bulla tympani. In contrast to the observations of Wible, Tandler (1899) describes that in the rat, the A. carotis interna enters the bulla.

An examination of this region in hyracoids shows that the course of the A. carotis interna is certainly connected with the occurence of the Eustachian sac. In *Procavia* the A. carotis communis divides into the A. carotis externa, A. occipitalis and A. carotis interna immediately caudoventrally of the Eustachian sac (Fig. 4.2). As mentioned, the A. carotis interna then runs along its medial side and enters the skull through the foramen lacerum medium.

Lechner (1932) has described identical conditions in *Tapirus*. As in *Procavia*, the A. carotis interna runs first along the caudal side of the Eustachian sac, then together with the flexors of the head and finally enters the skull at the oral border of the foramen lacerum medium.

As the course of the A. carotis interna to the bulla tympani is blocked by the Eustachian sac, it is led another way. This explains the reduction of the sulcus caroticus. Cifelli (1982) interpreted the loss of the promontory sulcus in *Procavia*, mesaxonians, and *Phenacodus* as a process which has taken place three times independently. Along with the occurrence of a Eustachian sac, I think the extracranial course of the A. carotis interna is a very good synapomorphy for the Hyracoidea and Mesaxonia. The absence of a promontory sulcus in *Phenacodus* could be an argument to associate this genus with the two groups. Taking into account other characters, e.g., the tridactyl pes in hyracoids and in mesaxonians, it seems that *Phenacodus* does not belong in the stem-group of the Mesaxonia but of the Mesaxonia + Hyracoidea.

*The morphology of the tuber maxillare*

The tuber maxillare is the region of the formation of the molars (Fig. 4.4). After the eruption of the molars it usually vanishes or remains as a small rudimentary process. It persists in the adult in recent and fossil hyracoids and in Ceratomorpha (for topographical comparison, see Fischer, 1986). The same conditions are found in the stem-group of the Equidae, e.g., in *Palaeotherium* and *Propalaeotherium*. The tuber maxillare extends so far backward that it often comes in contact with the descending pterygoid process of the alisphenoid.

A plausible explanation for these apomorphic conditions is the retardation of the $M^2$ and, above all, of the $M^3$. Data vary for the tooth eruption of *Procavia*. The average eruption time may be considered to be around 2 years for the $M^2$ and 3-4 years for $M^3$.

There is no information as to the eruption time of the molars in the Tapiridae. My own observations on a skull of an 8-year-old specimen of *Tapirus indicus* (in the Landessammlung für Naturkunde in Karlsruhe) showed that $M^2$ was in use, whereas $M^3$ starts to erupt at this age.

The reason for such a delayed eruption of the molars could be a sequence of function. Mendelssohn (1965) observes in a 12-year-old female of *Procavia capensis syriaca* only the presence of the last two molars. This age is the highest recorded in the literature. Novacek and Wyss (1986) have argued that the tuber maxillare is present in ungulates, including Proboscidea and Sirenia. Tassy and Shoshani (1988) stress the point that this is clearly a parallelism. The plesiomorphic condition is seen in primitive proboscidean genera with no "horizontal" tooth succession (*Moeritherium, Numidoterium*, and deinotheres). The apomorphic condition for hyracoids and mesaxonians is not just the possession of a tuber maxillare that often forms around the

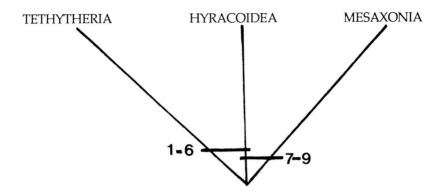

Fig. 4.5. The two cladograms summarize the arguments for the two different approaches to the systematic position of the Hyracoidea. Arguments for a sister-group- relationship between Hyracoidea and Tethytheria (Proboscidea + Sirenia) are: (1) Molecular data from amino acid sequences of α crystallin and hemoglobin, and serological data; (2) the taxeopode arrangement of carpals and tarsals; (3) the bifurcation of the styloglossus muscle just before its insertion; and with less consequence (4) the zonary placenta; (5) the amastoidy; (6) and the vertical orientation of the fenestra rotunda. Arguments for a sister-group- relationship between Hyracoidea and Mesaxonia are: (7) The similarity in the locomotory organs especially in the morphology of the hooves, or the occurence of a sternoscapular muscle; (8) the Eustachian sac and the extracranial course of the internal carotid artery; (9) the persistence of the tuber maxillare in the adult.

dental arcade, but the persistence of this structure after the last molar has erupted. Because of their "horizontal" tooth replacement, elephants and manatees must have a tuber maxillare that extends backwards, but is due to different functional requirements.

### Discussion

Evidence for the systematic position of the Hyracoidea comes from three different sources of data: the molecular evidence for the Paenungulata, the morphological evidence for the Paenungulata, and the morphological evidence for the sister-group Hyracoidea + Mesaxonia (Fig. 4.5).

As mentioned earlier, there are different molecular approaches to the problem of the systematic position of hyracoids. They all seem to produce similar results. Surprisingly enough, the data of Kleinschmidt et al. (1986) agree with those from the analysis of the eye lens protein α-crystallin (De Jong and Goodman, 1982). In both papers, we find the assumption that the paenungulates are the first clade to branch from the rest of the eutherians, and this contradicts both morphological and myoglobin data. With the exception of the position of the edentates, Miyamoto and Goodman (1986) depict the same branching pattern in their consensus approach. Without entering into the discussion of the molecular data in general, as has been done recently by Wyss et al. (1987), one should point out that the molecular approach is no real help in searching for the sister-group of hyracoids, because of significant differences in details of the alliances within the Paenungulata. There is still no support from morphological data for the inclusion of the Tubulidentata into this group. The α-crystallin sequences result in an 'aardvark-hyrax-manatee' clade which then joins the 'elephant' clade (De Jong and Goodman, 1982). In the most parsimonious tree of the analysis of α- and ß-hemoglobin, the manatee first joins the elephant clade and then the hyrax, but it comes out differently when we take only the amino acid replacements that are "likely candidates for synapomorphies grouping Sirenia, Hyracoidea, and Proboscidea into a monophyletic Paenungulata" (Kleinschmidt et al., 1986, p. 428). Two of the six "candidates" seem to be unique within mammals to the manatee and hyrax, *one* unites manatee and elephant; another, elephants and hyrax; and *two*, all these three groups. Miyamoto and Goodman (1986) present the manatee as sister-group of the aardvark, and elephants as sister-group of hyrax. Accepting the molecular data as arguments for the monophyly of Paenungulata should consequently result in the abandonment of the concept of Tethytheria (Proboscidea + Sirenia).

The morphological evidence for the Paenungulata is based on the following characters: (1) taxeopody; (2) bifurcation of the M. styloglossus; (3) placentation, and various aspects in the cranial morphology. These include: (4) extension of the jugal to the posterior border of the glenoid fossa; (5) amastoidy; (6) lambdoidal crest weak and occiput expanded; (7) fenestra rotunda with vertical orientation, partly shielded ventrally by expansion of promontorium cochleae; (8) post-tympanic ridge of squamosal very weak or absent; and (9) foramen magnum formed by basioccipital and exoccipital only.

Two of these characters can be refuted (numbers 6 and 9), because the supraoccipital always takes part in the formation of the foramen magnum in juvenile hyraxes, and because a strong lamboidal crest occurs in fossil hyracoids. Characters 4 and 8 are the plesiomorphic conditions in sirenians according to Novacek (1986) (for number 4) and Tassy and Shoshani (1988) (for number 8). They could therefore only be used as synapomorphies for the Paenungulata if we accept a reversal in sirenians, or they could be synapomorphies of elephants and hyraxes. Character 7 shows some homoplasy in paenungulates (Tassy and Shoshani, 1988). The last character of the skull, the amastoidy (number 5), is somewhat weakened by its distribution within mammals especially because the same conditions occur in some mesaxonians.

According to the personal communication of W. P. Luckett cited earlier, it is only the possession of the zonary placenta that provides some support for the hypothesis of the monophyly of Paenungulata. A zonary placenta is also found in Tubulidentata and Carnivora. Along with the molecular data, it could be used as a synapomorphy for Paenungulata including Tubulidentata. But it could as well have been evolved independently several times.

The bifurcation of the M. styloglossus before its insertion might give a little support for the monophyly of the Paenungulata despite its occurrence in other ungulates. Perhaps the best morphological character for this group is the taxeopode arrangement of carpals and tarsals, although there is, of course, no evidence for a taxeopode tarsus in sirenians. The interpretation of this character depends very much on whether the proposed assumption of a more inclusive group (Mesaxonia + Hyracoidea + Tethytheria) will be accepted or not. Anyway, it is difficult to object to this idea, and also see a serial carpus and tarsus in *Phenacodus* (a generally accepted representative of the stem-group of the Mesaxonia), and then accept it as a character of the Paenungulata.

Summing up the characters proposed as synapomorphies of the Paenungulata, they do not strongly corroborate the monophyly of the Paenungulata as argued by Wyss et al. (1987), especially if we compare them with the possible synapomorphies of the sister-group Hyracoidea + Mesaxonia. Here, the Eustachian sac seems to be the most striking character. It is a complex and unique structure of hyracoids, equids, and tapirs. Its apparent absence in rhinoceros can be due to the lack of adequate study, or its secondary loss in these ceratomorphs. The extracranial course of the internal carotid artery, whose phylogenetic importance has been recognized by Wible (1986), is a consequence of the topography of the Eustachian sac. The artery has changed its course in exactly the same way in hyracoids and mesaxonians, underlining the homogeneity of this possible synapomorphy. The loss of the promontory sulcus (Cifelli, 1982; Prothero et al., 1988) is dependent on the changed course of the artery. Supported by Tassy and Shoshani (1988), I still assume that the persistence of the tuber maxillare is a derived condition restricted to hyracoids and mesaxonians. Finally, the coincidence in many apomorphic characters of their locomotory organs is most plausible due to a common adaptational process to cursoriality in the stem-group of Hyracoidea and Mesaxonia. This assumption is strengthened by the fact that some of the apomorphies are only found in these two groups, e.g. the special morphology of their hooves, or the M. sternoscapularis.

It is very difficult to reach conclusions on the basis of the present data. If we compare the morphological arguments for the Paenungulata with those for the sister-group Mesaxonia + Hyracoidea, there seems to be a reasonable qualitative difference to decide in favor of the latter hypothesis. I think that this is still the case, even if we take into account the molecular data, but then, of course, the distance between the hypotheses becomes smaller. Perhaps, then, we are not finished, but only in the home stretch.

**Acknowledgments**
I thank J. L. Franzen, W. P. Luckett, G. Mickoleit, M. J. Novacek, D. Prothero, T. Rasmussen, G. Meyer, and P. Tassy for critically reading and commenting on the manuscript, as well as for many discussions. I especially thank D. Prothero for editing this work.

**Bibliography**
Altmann, F. (1932): Zur Kenntnis des feineren Baues des Schläfenbeines des südafrikanischen Klippschliefers (*Procavia capensis*). -*Monatsschr. Ohrenheilk. Laryngo- Rhinologie*, 66: 702-721.

Amoroso, E.C., and Perry, J.S. (1964): The fetal membranes and placenta of the African elephant (*Loxodonta africana*). -*Phil. Trans. Roy. Soc. London*, B, 248: 1-34.

Anthony, R. (1920): La poche gutturale du Tapir. -*Bull. Soc. Sci. Vétér. Lyon*, 1920: 115-131.

Assheton, R. and Stevens, T.G. (1905): Notes on the structure and development of the elephant's placenta. -*Quart. J. Micr. Sci.*, 49: 1-37.

Blainville, H.M.D. de (1816): Prodrome d'une nouvelle distribution systématique du règne animal. -*Bull. Sci. Soc. Philom. Paris*, sér. 3, 3: 105-124.

Boas, J.E.V., and Paulli, S. (1925): *The elephant's head*. 2: 81-131. -Jena (G. Fischer).

Böker, H. (1935): *Einführung in die vergleichende Anatomie der Wirbeltiere*. - Jena (G. Fischer).

Böker, H., and Pfaff, R. (1931): Die biologische Anatomie der Fortbewegung auf dem Boden und ihre phylogenetische Abhängigkeit vom primären Baumklettern bei den Säugetieren. -*Gegenbaurs morph. Jb.*, 68: 496-540.

Brandt, J.F. (1863): Bericht über eine Abhandlung, Untersuchung der Gattung *Hyrax* in anatomischer und verwandtschaftlicher Beziehung. -*Bull. Acad. Imp. Sci. St. Petersburg*, 5: 508-510.

Brandt, J.F.(1869): Untersuchung über die Gattung Klippschliefer. -*Mém. Acad. Imp. Sci. St. Petersburg*, sér. 7, 14: 1-127.

Buettner-Janusch, J., Buettner-Janusch, V., and Sale, J.B. (1964): Plasma proteins and hemoglobins of the African elephant and hyrax. -*Nature*, 201: 510-511.

Bugge, J. (1974): The cephalic arterial system in insectivores, primates, rodents, and lagomorphs, with special reference to the systematic classification. -*Acta. Anat., Suppl.*, 62: 1-159.

Campbell, B. (1936): The comparative myology of the fore-limb of the hippopotamus, pig and tapir. -*Amer. J. Anat.*, 59: 201-247.

Cifelli, R.L. (1982): The petrosal structure of *Hyopsodus* with respect to that of

some other ungulates, and its phylogenetic implications. -*J. Paleont.*, 56: 795-805.
Cope, E.D. (1882): Classification of Ungulate Mammalia. -*Proc. Amer. Phil. Soc.*, 20: 438-461.
Cuvier, G. (1800): *Lessons d'anatomie comparée.* - Paris (Baudoin).
Cuvier, G.(1804): Description ostéologique et comparative du daman. *Hyrax capensis*. -*Ann. Mus. Hist. Nat.*, 3: 171-182.
Cuvier, G.(1817): *Le régne animal.* -Paris (Déterville).
Daum, E. (1925): Beitrag zur fetalen Ausbildung der Tubenanhänge des Pferdes. -*Morph. Jb.*, 54: 322-332
De Jong, W.W., Gleaves, J.T., and Boulter, D. (1977): Evolutionary changes of a-crystallin and the phylogeny of mammalian orders. -*J. Molec. Evol.*, 10: 123-135.
De Jong, W.W., and Goodman, M. (1982): Mammalian phylogeny by sequence analysis of the eye lens protein α-crystallin. -*Z. Säugetierk.*, 47: 257-276.
De Jong, W.W., Zweers, A., and Goodman, M. (1981): Relationship of aardvark to elephants, hyraxes and sea cows from α-crystallin sequences. -*Nature*, 292: 538-540.
Domning, D.P. (1977): Observations on the myology of *Dugong dugon* (Müller). -*Smithson. Contrib. Zool.*, 226: 1-57.
Domning, D.P.(1978): The myology of the Amazonian manatee, *Trichechus inunguis* (Natterer) (Mammalia, Sirenia). -*Acta Amazon.*, 8(2), Suppl. 1: 1-81.
Ellenberger, W. and Baum, H. (1943): *Handbuch der vergleichenden Anatomie der Haustiere.* -Berlin (Springer).
Fischer, M.S. (1983): Die Extremitätenmuskulatur der Hyracoidea. Beiträge zur Fortbewegung und Anpassungsgeschichte. -M. Sc. Thesis, Univ. Tübingen.
Fischer, M.S. (1986): Die Stellung der Schliefer (Hyracoidea) im phylogenetischen System der Eutheria. Zugleich ein Beitrag zur Anpassungsgeschichte der Procaviidae. -*Cour. Forsch. Inst. Senckenberg*, 84: 1-132.
Fischer, M.S. (in press): Zur Anatomie des Gehörorganes der Seekuh (*Trichechus manatus* L.), (Mammalia, Sirenia). -*Z. Säugetierk.*
Fischer, M.S. ( in prep. a): The myology of the Rhinoceros *Diceros bicornis* and *Rhinoceros unicornis*.
Fischer, M.S. (in prep. b): Zur Ontogenese der Tympanalregion der Procaviidae (Mammalia: Hyracoidea).
Flower, W.H., and Lydekker, R. (1891): *An introduction to study of mammals living and extinct.* -London (A. and C. Black).
Frechkop, S. (1936): Notes sur les mammifères. Remarque sur la classification des Ongulés et sur la position systématique des Damans. -*Bull. Mus. Roy. Hist. Nat. Belg.*, 12(37): 1-28.
Gawrilenko, A. (1924): Die Evolution des Karpus bei den Proboscidiern. -*Anat. Anz.*, 58: 218-244.
George, H. (1875): Monographie anatomique et zoologique des mammifères du genre Daman. -*Ann. Sci. Nat. Paris*, 1 (9): 1-260.
Ghetie, V. (1944): Zungenbein, Kehlkopf und Zunge des indischen Elefanten. -*Abh. Math. Nat. Kl. Sächs. Akad. Wiss.*, 42: 1-13.
Grassé, P.-P. (1955): Ordre des hyracoides ou hyraciens. In: Grassé, P.-P. : *Traité de Zoologie* . Paris (Masson), pp. 878-898.
Gregory, W. K. (1910): Orders of mammals. -*Bull. Amer. Mus. Nat. Hist.*, 27: 1-524.
Hennig, W. (1983): Stammesgeslüchte der Chordaten. (Hennig, Wolfgang Hrsg.). -*Fortschr. Zool. Syst. Evol. Forsch.*, 2: 1-208.
Hildebrand, H. (1927): Über das Zehenendorgan des Tapirs (*Tapirus americanus*). -*Gegenbaurs Morph. Jb.*, 58: 348-366.
Hill, J.P. (1932): The developmental history of the primates.- *Phil. Trans. Roy. Soc. London B*, 221: 45-176.
Illiger, C. (1811): *Prodromus systematis mammalium et avium additus terminus*

zoographicis utriudque classis. - Berlin (Salfeld).

Janis, C. (1983): Muscles of the masticatory apparatus in two genera of hyraxes (*Procavia* and *Heterohyrax*).- *J. Morph.*, 176: 61-87.

Kleinschmidt, T., and Braunitzer, G. (1983): Die Primärstruktur des Hämoglobins vom Abessinischen Klippschliefer (*Procavia habessinica*, Hyracoidea): Insertion von Glutamin in den α-Ketten. -*Hoppe-Seyler's Z. Physiol. Chem.*, 364: 1303-1313.

Kleinschmidt, T., Czelusniak, J., Goodman, M., and Braunitzer, G. (1986): Paenungulata: A comparison of the hemoglobin sequences from elephant, hyrax, and manatee. -*Mol. Biol. Evol.*, 3: 427-435.

Lechner, W. (1932): Über die Tubendivertikel (Luftsäcke) beim Tapir.- *Anat. Anz.*, 74: 250-268.

Luckett, W. P. (1977): Ontogeny of amniote fetal membranes and their application to phylogeny. -In: Hecht, M. K., Goody, P. C. and Hecht, B. M. (eds): *Major Patterns in Vertebrate Evolution.* New York (Plenum Press), pp. 439-516.

Marsh, O.C. (1884): Dinocerata. A monograph of an extinct order of gigantic mammals.-*Monogr. U.S. Geol. Surv.*, 10: 1-237.

Matthew, W. D. (1937): Paleocene faunas of the San Juan Basin, New Mexico. - *Trans. Amer. Phil. Soc.*, 30: 1-510.

Mayer, A. F. J. C. (1847): Beiträge zu Anatomie der Elephanten und der übrigen Pachydermen. -*Nova Acta Acad. Leopold.*, 22: 1-88.

MacFadden, B. (1976): Cladistic analysis of primitive equids, with notes on other perissodactyls. -*Syst. Zool.*, 25: 1-14.

McKenna, M. C. (1975): Toward a phylogenetic classification of the Mammalia. - In: Luckett, W. P., and Szalay, F. S. (eds.): *Phylogeny of the Primates: a Multidisciplinary Approach.* -New York (Plenum Press), p. 21-46.

McKenna, M. C., and Manning, E. (1977): Affinities and palaeobiogeographic significance of the Mongolian Paleogene genus *Phenacolophus*. -*Géobios*, *Mém. spéc.*, 1977: 61-85.

Mendelssohn, H. (1965): Breeding in the Syrian hyrax, *Procavia capensis syriaca* Schreber 1784. -*Int. Zool. Yearbook*, 5: 116-125.

Miall, L. C., and Greenwood, F. (1878): Anatomy of the Indian elephant. -*J. Anat. Physiol.*, 12: 261-287, 385-400.

Miyamoto, M. M., and Goodman, M. (1986): Biomolecular systematics of eutherian mammals: phylogenetic patterns and classification. -*Syst. Zool.*, 35: 230-240.

Mossman, H. W. (1957): The foetal membranes of the aardvark. -*Mitt. naturforsch. Ges. Bern (N. F.)*, 14: 119-127.

Murie, J. (1872): On the Malayan tapir, *Rhinochoerus sumatranus*. -*J. Anat. Physiol.*, 6: 131-169.

Novacek, M. J. (1982): Information for the molecular studies from anatomical and fossil evidence on higher eutherian phylogeny.- In: Goodman, M. (ed.): *Macromolecular Sequences in Systematic and Evolutionary Biology.* New York (Plenum Press), pp. 59-81.

Novacek, M. J. (1986): The skull of leptictid insectivorans and the higher-level classification of Eutherian mammals. -*Bull. Amer. Mus. Nat. Hist.*, 183: 1-112.

Novacek, M. J., and Wyss, A. R. (1986): Higher level relationship of the recent Eutherian orders: morphological evidence. -*Cladistics*, 2: 257-287.

Osborn, H. F. (1889): The evolution of the ungulate foot. -*Trans. Amer. Phil. Soc. (N. S.)*, 16: 531-569.

Owen, R. (1848): Description of teeth and portions of jaws in two extinct anthracotheroid quadrupeds (*Hyopotamus vectianus* and *Hyop. bovinus*) discovered by the Marchioness of Hastings in the Eocene deposits on the N. W. coast of the Isle of Wight: with an attempt to develop Cuvier's idea of the classification of Pachyderms by the number of their toes. -*Quart. J. Geol. Soc. London*, 4: 103-141.

Perry, J. S. (1974): Implantation, foetal membranes and early placentation of the African elephant *Loxodonta africana*. -*Phil. Trans. Roy. Soc.*, B, 269 (897): 109-135.

Peter, B. (1894): Die Ohrtrompeten der Säugethiere und ihre Anhänge. -*Arch. Mikroskop. Anat.*, 43: 327-376.

Prothero, D. R., Manning, E. M., and Fischer, M. S. (1988): The phylogeny of ungulates.- In: Benton, M. (ed.): *The Phylogeny of the Tetrapods*. Oxford (Oxford Univ. Press), 2: 201-234.

Radinsky, L. (1966): The adaptive radiation of the phenacodontid Condylarthra and the origin of Perissodactyla. - *Evolution*, 20: 408-417.

Rainey, W. C., Lowenstein, J. M., Sarich, V.M., and Mager, D. M. (1984): Sirenian molecular systematics -including the extinct Steller's sea cow (*Hydrodamalis gigas*). -*Naturwissenschaften*, 71: 586-588.

Rasmussen, D. T. (1989): The evolution of the Hyracoidea: a review of the fossil evidence (this volume, Chapter 5).

Reif, W.-E., Thomas, R., and Fischer, M. S. (1985): Constructional morphology: the analysis of constraints in evolution.- *Acta Biotheor.*, 34: 233-248.

Richter, H. (1923): Physiologische Bedeutung und Erklärung des Luftsackes (Divertix tubae auditivae Eusachii) bei den Equiden und verwandten Tierarten (Tapiren, Rhinoceren, Hyracoiden) und bei Babirusa. -*Schweiz Arch. Tierheilk.*, 65: 61-74.

Rübli, H. (1930): Die Myologie des Wildschweines. -*Arch. Klaus Stift.*, Zürich, V (3/4): 391-431.

Seilacher, A. (1970): Arbeitskonzept zur Konstruktionsmorphologie. -*Lethaia*, 3: 39-396.

Shindo, T., and Mori, M. (1956): Musculature of Indian elephant. III: Musculature of the trunk, neck and head. -*Okajimas Fol. Anat. Jap.*, 29: 17-41.

Shoshani, J. (1986): Mammalian phylogeny: comparison of morphological and molecular results. -*Mol. Biol. Evol.*, 3: 222-242.

Shoshani, J., Goodman, M., Prychodko, W., and Morrison, K. (1978): A survey of the contemporary Paenungulata -an immunodiffusion approach. -*Abstr. Congr. Theriol. Int.*, 2: 75.

Simpson, G. G. (1945): The principles of classification and a classification of mammals. -*Bull. Amer. Mus. Nat. Hist.*, 85: 1-350.

Sturgess, I. (1948): The early placentation of *Procavia capensis*. -*Acta. Zool.*, 29: 393-479.

Szalay, F. S. (1977): Phylogenetic relationship and a classification of the eutherian mammals. -In: Hecht, M., Goody, P. C., and Hecht, B. M. (eds.): *Major Patterns in Vertebrate Evolution*. New York (Plenum Press), pp. 315-374.

Tandler, J. (1899): Zur vergleichenden Anatomie der Kopfarterien bei den Mammalia. -*Denkschr. kaiserl. Akad. Wiss., Math. Nat. Cl.*, 67: 677-784.

Tassy, P. (1981): Le crâne de *Moeritherium* (Proboscidea, Mammalia) de l'Eocène de Dor el Talha (Libye) et le problème de la classification phylogénétique du genre dans le Tethytheria McKenna 1975. -*Bull. Mus. Nat. Hist. Natur.*, 4, 3, C, 1: 87-147.

Tassy, P. (1985): La place des mastodontes miocènes de l'Ancien Monde dans la phylogénie des Proboscidea (Mammalia): hypothèses et conjectures.-*Mém. Sc. Terre Univ. Curie Paris,*: 1-861.

Tassy, P., and Shoshani, J. (1988): The Tethytheria: elephants and their relatives. -In: Benton, M. (ed.): *The Phylogeny of the Tetrapods*. Oxford (Oxford Univ. Press), 2: 283-315.

Thenius, E. (1969): Stammesgeschichte der Säugetiere (einschließlich der Hominiden). -In: Helmcke, J. G., Starck, D. and Weruth, H. (Hrsg.): *Handbuch der Zoologie*, 8, 46 and 47 Lief.: 1-722. Berlin (de Gruyter).

Thenius, E. (1979): *Die Evolution der Säugetiere*. -Stuttgart (UTB).

Turner, H. N. (1850): Contributions to the anatomy of the Tapir. -*Proc. Zool. Soc. London*, 1850: 102-106.

Van Valen, L. (1971): Adaptive zones and

the orders of mammals. -*Evolution*, 25: 420-428.
Weber, M. (1928): *Die Säugetiere*. Bd. 2. - Jena (G. Fischer).
Weitz, B. (1953): Serological relationship of hyrax and elephant. -*Nature*, 171: 261.
Whitworth, T. (1954): The Miocene hyracoids of east Africa. -*Fossil Mammals of Africa*, 7: 1-58: London (Brit. Mus. (Nat. Hist.)).
Wible, J. R. (1986): Transformations in the extracranial course of the internal carotid artery in mammalian phylogeny. -*J. Vert. Paleont.*, 6: 313-325.
Windle, B. C. A., and Parsons, F. G. (1901): On the muscles of the Ungulata. Part I: Muscles of the head, neck and forelimb. -*Proc. Zool. Soc. London*: 656-704.
Windle, B. C. A., and Parsons, F. G. (1903): On the muscles of the Ungulata: Part II: Muscles of the hind-limb and trunk. -*Proc. Zool. Soc. London*: 261-298.
Wislocki, G. B. (1935): The placentation of the manatee (*Trichechus latirostris*). -*Mem. Mus. Comp. Zool.*, 54: 159-178.
Wislocki, G. B., and Westhuysen, O. P. van der (1940): The placentation of *Procavia capensis*, with a discussion of the placental affinities of the Hyracoidea. -*Contr. Embryol.*, 28: 65-88.
Woods, C. A. (1972): Comparative myology of jaw, hyoid, and pectoral appendicular regions of New and Old World hystricomorph rodents. -*Bull. Amer. Mus. Nat. Hist.*, 147: 114-198.
Wyss, A. R., Novacek, M. J., and McKenna, M. C. (1987): Amino acid sequence versus morphological data and the interordinal relationship of mammals. -*Mol. Biol. Evol.*, 4: 99-116.
Zuckerkandl, E. (1885): Über die Ohrtrompete des Tapir und Rhinoceros. -*Arch. Ohrenheilk.*, 22: 222-223.

# 5. THE EVOLUTION OF THE HYRACOIDEA: A REVIEW OF THE FOSSIL EVIDENCE

## D. TAB RASMUSSEN

Modern hyraxes represent a small fraction of the diversity once present in the Hyracoidea. Extinct forms show convergence in body size and dental morphology towards suids, anthracotheres, chalicotheres, tapirs, and equoids. The earliest fossil hyracoids (Pliohyracidae) are from the Eocene of Algeria, where at least five species are present, including small, bunodont forms (*Seggeurius amourensis* and *Microhyrax lavocati*) and the huge, selenodont *Titanohyrax mongereaui*, thus indicating that the divergence of pliohyracid lineages occurred well before the middle Eocene. Pliohyracids belonging to at least eight genera were Africa's dominant small- to medium-sized terrestrial herbivores during the Oligocene, with numerous sympatric species ecologically differentiated from each other by dietary specializations and body size. Only two lineages are known to have survived to the early Miocene, the rest having been replaced by suiforms, ruminants and mesaxonians from Eurasia. Two later radiations of hyracoids began in the late Miocene. One gave rise to large, hypsodont, partially aquatic forms in Africa and Eurasia (Pliohyracinae). The other led to the modern family Procaviidae in Africa. During the Plio-Pleistocene, *Heterohyrax* is found in East Africa, *Procavia* in South Africa, and large, extinct *Gigantohyrax* in both areas.

## Introduction

The order Hyracoidea is represented today by only three closely related genera distributed in parts of Africa and the Middle East (Jones, 1978; Honacki et al., 1982; Olds and Shoshani, 1982). The fossil record shows that the modern species represent but a small fraction of what was once a much greater diversity. The extinct hyracoids include such varied forms that paleontologists have misidentified some species as suids (Andrews, 1904a, 1906; Hooijer, 1963), chalicotheres (Teilhard de Chardin and Piveteau, 1930; Koenigswald, 1932; Viret, 1947), and anthracotheres (Hooijer, 1963). Paleontologists have been especially impressed by the great size of some fossil taxa, giving them names such as *Megalohyrax*, *Titanohyrax*, and *Gigantohyrax* (Andrews, 1903; Matsumoto, 1922; Kitching, 1965). Altogether, known fossil hyracoids comprise as many as 19 genera distributed in Africa, Asia, and Europe, with the oldest forms dating back to the Eocene.

From the beginning, the study of fossil hyracoids was intimately tied to the study of fossil Perissodactyla, or Mesaxonia Marsh 1884, a term used here to denote the group containing Equoidea, Tapiridae, Rhinocerotidae, and Chalicotheriidae, but not Hyracoidea. Although a close phylogenetic relationship between hyracoids and mesaxonians is debatable (Fischer, this volume, Chapter 4), there can be no doubt that significant points of resemblance do exist between the two groups. In 1841, Owen named· an early Eocene equoid *Hyracotherium* because of dental resemblances to modern hyraxes. Similarly, the first fossil hyracoid that was discovered, "*Leptodon graecus*" Gaudry, 1862, was considered close to the mesaxonians *Palaeotherium* and *Rhino-ceros* until Osborn (1899) finally recognized it as a hyracoid (and placed it in the new genus *Pliohyrax*, *Leptodon* being occupied by the Cayenne kite). Soon afterwards, numerous Oligocene hyracoids that documented the unexpected morphological diversity within the order were discovered in Egypt (Andrews and Beadnell, 1902; Andrews, 1906; Schlosser, 1911; Matsumoto, 1926). Fossil hyracoids have since been recovered throughout Africa and also at Eurasian sites from Spain to China. Meyer (1978) presented a chronological history of the study of fossil hyracoids.

The fossil evidence currently available suggests that hyracoids have undergone three major evolutionary radiations. The first and most diverse radiation was restricted to Africa during the Eocene and Oligocene. During that time, hyracoids belonging to the family Pliohyracidae were the dominant small- to medium-sized terrestrial herbivores, with numerous sympatric species ecologically differentiated from each other by dietary specializations and body size. By the earliest Miocene, this early radiation was largely decimated in association with the influx of artiodactyls and mesaxonians from the northern continents (Meyer, 1978; Rasmussen and Simons, 1988). A second radiation of large-bodied, hypsodont pliohyracids occured during the late Miocene and Pliocene in Europe and Asia. The third radiation began in the late Miocene and includes the modern hyraxes and their closest extinct relatives (Family Procaviidae). The purpose of this chapter is to review each of these three radiations, to provide a current update of valid taxa and geographic distributions (Table 5.1), and to briefly discuss the phylogenetic relationships and specializations of extinct species.

## The First Radiation: African Pliohyracidae
*Fayum hyracoids*

Early Tertiary hyracoids were first discovered near the turn of the century in Oligocene deposits of the Fayum depression, Egypt. The Fayum still remains the only pre-Miocene site that has yielded a diverse and well-studied sample of fossil hyracoids. This does not mean that hyracoids were restricted to the Fayum during the Paleogene, but instead reflects the extreme scarcity of continental sedimentary deposits in Africa that date to that period. Where other terrestrial deposits have been found, mainly in Libya and Algeria, hyracoids are often present, but like the rest of the terrestrial mammalian fauna they are generally represented by small samples and poorly preserved material. With a few important exceptions, all of the Paleogene hyracoids found outside the Fayum have been referred to Fayum genera, so an understanding of Fayum hyracoids is necessary for an examination of other Paleogene sites.

The Fayum hyracoids are now all placed in the extinct family Pliohyracidae (Meyer, 1978), although previous authors divided the material into as many as three families (Matsumoto, 1926). Pliohyracids differ from members of Procaviidae in generally having larger body sizes, relatively smaller brains, complete eutherian dentitions (Fig. 5.1), and a hollow chamber inside the mandibular corpus that opens lingually through a large round or oval fenestra below the molar series. The mandibular chamber occurs in only one sex, evidently females, since individuals with chambers have relatively small tusks like those of modern female hyraxes (Meyer, 1978). In one Fayum genus, *Geniohyus,* the hollow opens broadly to the lingual side forming a large fossa. In others, the fenestra entering the chamber may be small and nearly round (e.g., *Thyrohyrax*) or relatively large and somewhat triangular (e.g., *Bunohyrax*). The internal chamber itself may be a narrow canal (as in some *Titanohyrax*) or greatly inflated, causing ballooning of the corpus and even parts of the ascending ramus (e.g., *Thyrohyrax domoricutus*). Some species lack a mandibular chamber or fossa. The function of the chamber remains unknown, but one suggestion is that it contained some kind of buccal sac that functioned in producing vocalizations (Andrews, 1907). Modern hyracoids have an inflated sac of the eustachian tube (Fischer, this volume) that is possibly analogous to the mandibular chamber of pliohyracids. Previous authors have chosen to use the occurrence or shape of this mandibular structure as an important taxonomic character. However, given the

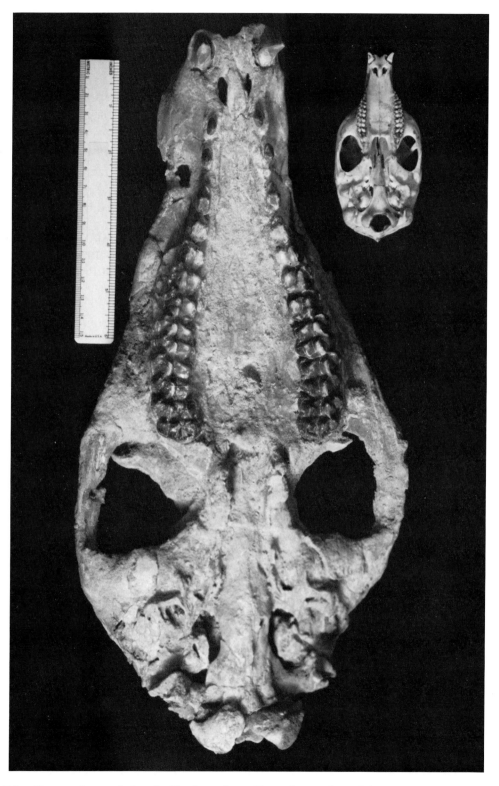

Fig. 5.1. Comparison of the skull of modern *Heterohyrax brucei* (small specimen) and of Oligocene *Megalohyrax eocaenus* from Egypt, DPC (Duke Primate Center) 6640.

observed variation in chamber structure or occurrence among otherwise similar species, and given our ignorance of the chamber's function, it is probably best to give little taxonomic weight to this feature. Pickford and Fischer (1987) revised Pliohyracidae to include only those genera with a bifurcate metastyle on $M^3$ and a premolariform lower canine. These two characters may define a clade but do not warrant separate familial rank. I retain Meyer's (1978) diagnosis of Pliohyracidae and consider Pickford and Fischer's clade to correspond with Pliohyracinae (Table 5.1).

All of the Fayum hyracoids come from the Oligocene Jebel Qatrani Formation, which comprises 340 m of fluvial mudstones and sandstones deposited as point bars and overbank deposits in a warm, freshwater, swampy tropical environment that was largely or partially forested (Bown et al., 1982; Olson and Rasmussen, 1986). Fossils are recovered at numerous quarries that can be classified for convenience into four general stratigraphic levels representing different faunal units, two each in the upper and lower sequences of the Jebel Qatrani Formation (Bown and Kraus, 1987). Successive quarries and faunal units are often separated from each other by large stratigraphic intervals that have not yet yielded mammalian fossils, and thus represent significant missing gaps in the fossil record. Faunal unit 1 is represented by quarry L-41, or the "Green Hill locality," a recently discovered site lying only 46 m above the top of the underlying Eocene Qasr el Sagha Formation, which has already produced five genera of hyracoids. Unit 2 (meter levels 58-92) contains the quarries of the "Lower Fossil Wood Zone" that were initially sampled by the early expeditions to the Fayum (e.g., Andrews, 1906; see Simons, 1968). Unit 3 (meter levels 165-209) includes a few sites near the middle of the section (e.g., quarries G, R, V). Unit 4 (meter levels 242-249) includes the rich quarries of the "Upper Fossil Wood Zone" (Simons, 1968). The Jebel Qatrani Formation is capped by a basalt radiometrically dated to 31 ± 1 million years ago (Ma) (Fleagle et al., 1986), while the base lies very near the Eocene/Oligocene boundary (about 40 Ma; Bown and Kraus, 1987).

Eight genera of hyracoids are represented in the Fayum. The rarest and in many ways the most distinctive is *Geniohyus*, which includes three species distributed in units 2 and 3. *Geniohyus mirus*, a pig-sized species, was initially identified as a suid (Andrews, 1904a, 1906) because of the inflated bunodont cusps with accessory cuspules that characterize the lower molars. It further differs from other Fayum forms in the long, narrow symphyseal region, the distinctive, simple shape of the premolars, and the very deep jaw with a broad fossa excavated on the lingual side of the corpus (Andrews, 1907; Meyer, 1978). *G. diphycus* seems to be a small version of *G. mirus*, while the smallest species, *G. magnus* (named originally as a large species of *Saghatherium*), appears to be intermediate between *Geniohyus* and *Bunohyrax* in dental morphology, and differs from both in the presence of less inflated, sharply crested molar cusps without, however, development of lophs or selenodonty (Rasmussen and Simons, 1988). In these respects, *G. magnus* resembles *Microhyrax*, a primitive Eocene hyracoid from Algeria (see later). Largely because of differences in the mandible and premolars, *Geniohyus* has been placed in a separate subfamily from other Fayum hyracoids, Geniohyinae (Meyer, 1978).

Two other Fayum genera, *Bunohyrax* and *Pachyhyrax*, resemble *Geniohyus* in the retention of a bunodont molar morphology that lacks the development of high transverse crests across the lower molars between the hypoconid and entoconid (see revision of *Pachyhyrax* by Rasmussen and Simons, 1988). Because of this cuspate molar morphology, both *Bunohyrax* and *Pachyhyrax* bear a superficial resemblance to suiforms convincing enough that they have been identified as suids and anthracotheres, respectively (Andrews, 1904a ["*Geniohyus*" *fajumensis*]; Hooijer, 1963). These two

Table 5.1. Annotated systematic list of fossil hyracoids.

| Species and Distribution | References | Synonymy |
|---|---|---|
| **PLIOHYRACIDAE OSBORN 1899**<br>Subfamily indet.<br>Gen. et sp. indet. | | |
| Dor el Talha, Libya | Savage, 1969<br>Wight, 1980 | |
| **Geniohyinae Andrews, 1906**<br>*Seggeurius amourensis* Crochet, 1986 | | |
| El Kohol, Algeria | Mahboubi *et al.*, 1984, 1986 | |
| *Geniohyus mirus* Andrews, 1904a | | |
| Fayum, Egypt (unit 2) | Andrews, 1906, 1907<br>Schlosser, 1910, 1911<br>Meyer, 1978 | see Meyer, 1978 |
| *Geniohyus diphycus* Matsumoto, 1926 | | |
| Fayum, Egypt (unit 2) | Meyer, 1978 | |
| *Geniohyus magnus* (Andrews, 1904b) | | |
| Fayum, Egypt (units 2-3) | Andrews, 1903, 1904b, 1906<br>Meyer, 1978<br>Rasmussen and Simons, 1988 | see Meyer, 1978 |
| *Geniohyus*, sp. indet. | | |
| Bugti, Pakistan | Forster Cooper, 1924<br>Pickford, 1986b | *Anthracotherium adiposum*<br>*G. adiposum* |
| **Saghatheriinae Andrews, 1906**<br>*Microhyrax lavocati* Sudre, 1979 | | |
| Gour Lazib, Algeria | Gevin *et al.*, 1975 | |
| *Bunohyrax fajumensis* (Andrews) 1904a | | |
| Fayum, Egypt (units 2-4) | Andrews, 1904a, 1906<br>Schlosser, 1910, 1911<br>Matsumoto, 1926<br>Meyer, 1978 | see Meyer, 1978<br>see Meyer, 1978<br>see Meyer, 1978 |
| *Bunohyrax* cf. *B. fajumensis* | | |
| Malembe, Angola | Hooijer, 1963<br>Pickford, 1986a | *Palaeochoerus dartevellei* |
| *Bunohyrax major* (Andrews) 1904b | | |
| Fayum, Egypt (unit 2) | Andrews, 1904b, 1906<br>Schlosser, 1910, 1911<br>Matsumoto, 1926<br>Meyer, 1978<br>Rasmussen and Simons, 1988 | see Meyer, 1978<br>see Meyer, 1978<br>see Meyer, 1978 |
| *Bunohyrax* sp. nov. | | |
| Fayum, Egypt (unit 1) | undescribed | |
| *Pachyhyrax crassidentatus* Schlosser, 1910 | | |
| Fayum, Egypt (unit 4) | Schlosser, 1910, 1911<br>Matsumoto, 1926<br>Meyer, 1978<br>Rasmussen and Simons, 1988 | |
| *Pachyhyrax* cf. *P. crassidentatus* | | |
| Malembe, Angola | Hooijer, 1963<br><br>Pickford, 1986a | Anthracotheriidarum, g. et sp. indet.<br>*Geniohyus* cf. *G. mirus* |
| *Megalohyrax eocaenus* Andrews, 1903 | | |
| Fayum, Egypt (unit 1-4) | Andrews, 1904b, 1906<br>Schlosser, 1910, 1911<br>Matsumoto, 1922, 1926<br>Meyer, 1978 | see Meyer, 1978<br>see Meyer, 1978<br>see Meyer, 1978 |
| *Megalohyrax gevini* Sudre, 1979 | | |
| Gour Lazib, Algeria | Gevin *et al.*, 1975 | |

*Megalohyrax championi* (Arambourg, 1933)
    Moruorot, Kenya      Arambourg, 1933      *Pliohyrax championi*
    other East African sites      Whitworth, 1954      *Megalohyrax* cf. *M. pygmaeus*
          Pickford, 1981
          Meyer, 1978      *Pachyhyrax championi*
          Rasmussen and Simons, 1988
    Saudi Arabia      Thomas *et al*, 1982
*Saghatherium antiquum*      Andrews and Beadnell, 1902
    Fayum, Egypt (unit 2)      Andrews, 1906      see Meyer, 1978
          Schlosser, 1910, 1911      see Meyer, 1978
          Matsumoto, 1922, 1926      see Meyer, 1978
          Meyer, 1978
*Saghatherium sobrina* Matsumoto, 1926
    Fayum, Egypt (unit 2)      Meyer, 1978
*Saghatherium, humarum* Rasmussen and Simons,1988
    Fayum, Egypt (unit 3)
*Saghatherium*, sp. nov. #2
    Fayum, Egypt (unit 1)      undescribed
*Selenohyrax chatrathi* Rasmussen and Simons, 1988
    Fayum, Egypt (unit 3)
*Thyrohyrax domorictus* Meyer, 1973
    Fayum, Egypt (units 3-4)      Meyer, 1978
          Rasmussen and Simons, 1988

*Thyrohyrax pygmaeus* (Matsumoto, 1922)
    Fayum, Egypt (unit 2)      Matsumoto, 1922, 1926      *Megalohyrax pygmaeus*
          Meyer, 1978      *Pachyhyrax pygmaeus*
          Rasmussen and Simons, 1988
*Thyrohyrax*, sp. indet.
    Fayum, Egypt (unit 2)      Rasmussen and Simons, 1988
*Thyrohyrax*, sp. indet.
    Fayum, Egypt (unit 1)      undescribed
*Titanohyrax andrewsi* Matsumoto, 1922
    Fayum, Egypt (unit 2)      Matsumoto, 1922, 1926      see Meyer, 1978
          Meyer, 1978
          Rasmussen and Simons, 1988
*Titanohyrax mongereaui* Sudre 1979
    Gour Lazib, Algeria      Gevin *et al.*, 1975
*Titanohyrax ultimus* Matsumoto, 1922
    Fayum, Egypt (unit 4)
*Titanohyrax, angustidens* Rasmussen and Simons, 1988
    Fayum, Egypt (units 3-4)      Schlosser, 1910, 1911      *Megalohyrax palaeotherioides*
          Matsumoto, 1922, 1926      *Titanohyrax palaeotherioides*
          Meyer, 1978      *Titanohyrax andrewsi* in part

*Titanohyrax*, sp. nov.#2
    Fayum, Egypt (unit 1)      undescribed
*Titanohyrax*, sp. indet.
    Zella, Libya      Arambourg and Magnier, 1961      *Megalohyrax palaeotherioides*
          Arambourg, 1963      *Megalohyrax palaeotherioides*

*Meroehyrax bateae* Whitworth, 1954
    Rusinga Island, Kenya      Whitworth, 1954
    Bukwa, Kenya      Pickford, 1981
          Meyer, 1978

**Pliohyracinae Osborn, 1899**
*Prohyrax tertiarius* Stromer, 1924
    Langenthal, S.W. Africa      Stromer, 1924, 1926
          Churcher, 1956      *Procavia tertiaria*
          Meyer, 1978

*Prohyrax* sp. nov.
    Arrisdrift, Namibia      Hendey, 1978
          Pickford and Fischer, 1987

[?] *Prohyrax tertiarius*
    Muruarot, Kenya      Madden, 1972

cf. *Prohyrax* (?=*Meroehyrax*)
    Loperot, Kenya                                  Pickford, 1981
*Parapliohyrax mirabilis* Lavocat, 1961          Pickford and Fischer, 1987
    Beni Mellal, Morocco
*Parapliohyrax ngororgensis*                     Pickford and Fischer, 1987     *Parapliohyrax* sp. nov
    Barengo District, Kenya                         Bishop and Pickford, 1987
*Sogdohyrax soriaus* Dubrovo, 1978
    Tadzhikistan, USSR
*Pliohyrax graecus* (Gaudry, 1862)
    Pikermi, Greece                                 Gaudry, 1862                   *Leptodon graecus*
    Samos, Greece                                   Major, 1899a, 1899b            *Leptodon graecus*
                                                    Osborn, 1899                   *Pliohyrax kruppii*
    Halmyropotamus, Greece                          Melentis, 1965, 1966
    Montpellier, France                             Viret and Thenius, 1952        *Pliohyrax occidentalis*
                                                    Pickford and Fischer, 1987
                                                    Melentis, 1965, 1966

*Pliohyrax rossignoli* (Viret, 1947)
    Soblay, France                                  Viret, 1947                    *Neoschizotherium rossignoli*
                                                    Viret and Mazenot, 1948        *Neoschizotherium rossignoli*
                                                    Viret, 1949a                   *Neoschizotherium rossignoli*
                                                    Viret, 1949b
                                                    Mein, 1975

*Pliohyrax* sp. indet.
    Esme-Akcakoy, Turkey                            Becker-Platen *et al.*, 1975
    Garkin, Turkey                                  Becker-Platen *et al.*, 1975
    Melambes, Crete                                 Kuss, 1976
    Sabadell, Spain                                 Golpe-Posse and Crusafont-Pairo, 1981
    Malayan, Afghanistan                            Heintz *et al.*, 1981
*Kvabebihrax kachethicus* Gabunia and Vekua, 1966
    Georgia, USSR                                   Gabunia and Vekua, 1974
*Postschizotherium chardini* Koenigswald, 1932
    Niwowan Basin, China                            Teilhard de Chardin            Chalicotheride gen. nov. indet.
                                                      and Piveteau, 1930
    Choukoutien, China                              Teilhard de Chardin            *Postschizotherium* ?
                                                      and Pei, 1934
                                                    Teilhard de Chardin,           *P.* cf. *chardini*
                                                      1938, 1939
                                                    Koenigswald, 1932, 1966
*Postschizotherium licenti* Koenigswald, 1966
    Jushe Basin, China                              Teilhard de Chardin.           *Postschizotherium* sp.
                                                      1939
*Postschizotherium intermedium* Koenigswald, 1966
    Jushe Basin, China                              Teilhard de Chardin            *P. chardini*
                                                      and Licent, 1936
                                                    Teilhard de Chardin,           *Postschizotherium* sp.
                                                      1939
                                                    Pei, 1939                      *Postschizotherium*?
*Postschizotherium* sp. indet.
    Shansi, China                                   Tung and Huang, 1974           *Pliohyrax*
                                                    Dubrovo, 1978

**PROCAVIIDAE** Thomas, 1892
*Procavia antigua* Broom, 1934
    Taungs, South Africa
    Sterkfontein, South Africa                      Broom, 1946                    *Procavia robertsi*
                                                    Broom, 1948
    other South African sites                       see Churcher, 1956
                                                    Wells, 1939                    "*Hyrax*" sp.
                                                    Meyer, 1978

*Procavia* cf. *P. antiqua*
    Langebaanweg, South Africa                      Hendey, 1976

*Procavia transvaalensis* Shaw, 1937
    Sterkfontein, South Africa
    Uitkomst, South Africa    Broom, 1937    *Procavia obermeyerae*
    other South African sites    see Churcher, 1956
        Wells, 1939
        Meyer, 1978

*Gigantohyrax maguirei* Kitching, 1965
    Makapansgat, South Africa
*Gigantohyrax*, sp. indet.
    Omo Basin, Ethiopia    Howell and Coppens, 1974
*Heterohyrax*, sp. indet.
    Omo Basin, Ethiopia    Howell and Coppens 1974

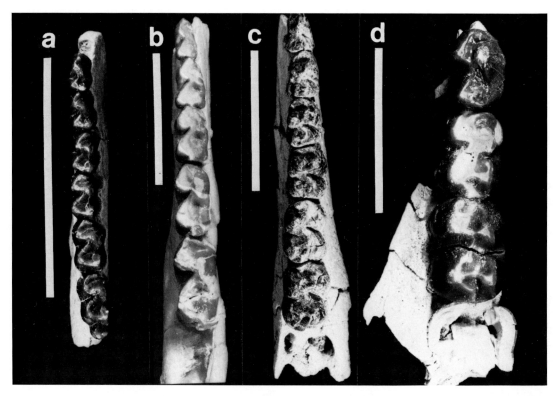

Fig. 5.2. Occlusal views of the lower teeth in four Oligocene hyracoids from Egypt illustrating diversity of dental structure: a) *Thyrohyrax domorictus*, DPC 2763, left $P_2$-$M_3$, showing lophoselenodont teeth resembling those of modern hyraxes; b) *Titanohyrax angustidens* (Rasmussen and Simons, 1988), CGM (Cairo Geological Museum) 42848, left $P_2$-$M_2$, $M_3$ erupting, showing selenodont teeth reminiscent of those in Eocene brontotheres; c) *Pachyhyrax crassidentatus*, DPC 2163, right $P_1$-$M_2$, showing bunoselenodont teeth resembling those of anthracotheres; d) *Bunohyrax major*, DPC 5291, left $P_4$-$M_2$, showing bunodont teeth resembli ng those of suids. Bar scale = 4 mm.

genera have been put in the subfamily Saghatheriinae along with the remaining Fayum taxa rather than in Geniohyinae, because of the shared occurrence of a restricted fenestra opening into the mandibular chamber and the development of a V-shaped hypocristid on the lower premolars, thus distally closing the talonid. This arrangement of subfamilies has been followed in Table 5.1. However, this system may require revision as more is learned about primitive Eocene hyracoids (see later). *Bunohyrax* includes three species: a small, undescribed form from the Green Hill locality; medium-sized *B. fajumensis* that is very similar in molar morphology to *G. mirus;* and a large species, *B. major* (Fig. 5.2d), that shows extreme specialization towards very low, broad, mound-like cusps. *Pachyhyrax* is monotypic. "*P.*" *pygmaeus* and "*P.*" *championi*, two species with highly selenodont or lophoselenodont dentitions, must be removed from *Pachyhyrax* (see Rasmussen and Simons, 1988).

Whatever the precise phyletic relationships among these three genera, *Geniohyus, Bunohyrax* and *Pachyhyrax* do share a general "suiform" molar morphology that probably reflects similar dietary adaptations to a pig-like omnivorous diet, perhaps dominated by roots, fruit, fungus, and other vegetable matter requiring crushing rather than slicing. *Pachyhyrax* (Fig. 5.2c), which may have been derived from a *Bunohyrax* -like ancestor, differs from *Geniohyus* and *Bunohyrax* in having more selenodont buccal cusps, more molariform premolars, and the presence of long spurs in the basins of the upper molars and premolars. This "suiform" group of hyracoids represents a general dietary adaptation very distinct from any Neogene members of the order, and the group apparently became extinct before the earliest Miocene in Africa, possibly due to direct competition with suids and the increasing diversity of anthracotheres that arrived from Eurasia. A slightly younger "suiform" hyracoid is known from two specimens from the Bugti beds of Pakistan (Pickford, 1986b; see below).

The remaining five genera of Fayum hyracoids differ dentally from the "suiform" group in having complete hypocristids and protocristids on the lower molars, thus incorporating the lingual cusps (metaconid, entoconid) into the buccal shearing crests. *Megalohyrax eocaenus* (Fig. 5.1) is the size of a large tapir, and has relatively simple bunoselenodont teeth and non-molariform premolars. It is the most common large hyracoid in both the upper and lower quarries, and is the only one to occur at all four faunal levels, although morphological differences are evident between lower and higher samples (Meyer, 1978). It is convenient to classify all Fayum specimens of the genus in *M. eocaenus* until evolutionary change within the lineage has been examined in detail. The relative abundance of *M. eocaenus* in most quarries suggests that it was either a successful ecological generalist, or possibly a specialized, tapir-like marsh or swamp dweller, as this was a common depositional environment in the Fayum (Olson and Rasmussen, 1986). Unfortunately, because of sympatry among many similarly sized hyracoids, postcrania have not been positively identified to species and so offer no clue yet as to locomotor specializations of the various Fayum species.

The one remaining genus of medium to large Fayum hyracoids is *Titanohyrax*, represented in the Fayum by four species. One of these, *T. ultimus*, is the largest known hyracoid, but to date only isolated teeth have been found. An upper second molar measures 41 x 46 mm, thus dwarfing the molars of the early proboscidean, *Moeritherium*, also found in the Jebel Qatrani Formation (Andrews, 1906). The remaining three Fayum species of *Titanohyrax* are smaller, especially an early, undescribed species from the Green Hill locality that falls within the size range of small *B. fajumensis*. *Titanohyrax* has a highly selenodont dentition with molariform premolars, well-developed ectolophs on the upper molars, and lunate buccal shearing crests on the lowers (Fig. 5.2b). Metastylids are prominent on the three larger species, and weak on the early Green Hill species, which is primitive in some other features as well. These dental features suggest that species of *Titanohyrax* were probably specialized folivores.

Three genera of relatively small hyracoids are known from the Fayum; all are nevertheless larger than living hyraxes. *Saghatherium* is a bunoselenodont form that is the common small hyracoid of faunal units 1-3. Some authors have suggested that *Saghatherium* may lie near the ancestry of modern hyracoids (Matsumoto, 1926; Whitworth, 1954; Churcher, 1956), in part, no doubt, because of its small size. However, most of the resemblances between *Saghatherium* and Procaviidae are primitive features also present in other Fayum genera, and *Saghatherium* does show specialized features of the dentition (such as a greatly enlarged molar series relative to the premolar series, robust crested cusps, and strong spurs on the ectolophs of the upper cheek teeth) that would seem to exclude it from any close relationship to Procaviidae. These dental features may be an adaptation to a diet requiring crushing and grinding hard objects such as nuts, pods, or seeds. A monotypic genus, *Selenohyrax* (Rasmussen and Simons, 1988) from faunal unit 3, apparently allied with *Saghatherium* and possibly descended from early representatives of that genus, has highly specialized tooth crowns dominated by delicate lunate crests nearly lacking any differentiated cusps. Like *Titanohyrax*, it was probably a specialized folivore, but smaller in size.

The final genus of Fayum hyracoids is *Thyrohyrax*, a small lophoselenodont form similar in some dental and cranial features to members of Procaviidae. *T. domorictus* (Fig. 5.2a) is the most common hyracoid of faunal unit 4, and along with the primate *Apidium phiomense*, is one of the most common mammals. *Saghatherium* is absent from the upper quarries, apparently replaced by *T. domorictus*. A few specimens of *Thyrohyrax* very similar to *T. domorictus* and possibly conspecific with it are known from faunal units 1-3, but these are relatively rare compared to the abundant species of *Saghatherium* at the lower levels. *T. domorictus* has lower cheek teeth characterized by high, nearly flat lophs forming a zigzag pattern very similar to that observed in living *Heterohyrax* and *Dendrohyrax*. (*Procavia* is more hypsodont and has a larger molar series relative to premolar series.) *Thyrohyrax* was presumably a browser like the living procaviids (Kingdon, 1974; Hoeck, 1975). Females of *T. domorictus* show greater inflation of the mandibular corpus and ascending ramus than any other Fayum species (Meyer, 1973). *T. pygmaeus* is a larger species from faunal unit 2 that is very similar to *T. domorictus* in dental morphology but has only a small mandibular chamber.

In summary, the Fayum hyracoids represent a diverse radiation of early pliohyracids, ranging in size from small species only slightly larger than modern hyraxes, to *Titanohyrax ultimus*, larger than some Oligocene Proboscidea. Their teeth suggest divergent dietary specializations: the puffy bunodont molars of a fruit or root eater such as *Bunohyrax major*; the strongly lophed teeth of a browser like *Thyrohyrax domorictus* ; and the delicate lunate crests of the folivore *Selenohyrax*. The new Green Hill locality near the bottom of the Jebel Qatrani Formation, which is still incompletely sampled, shows that this diversity was already present by the earliest Oligocene, as species of *Bunohyrax, Megalohyrax, Titanohyrax, Saghatherium,* and *Thyrohryrax* have been found there. This generic diversity was maintained throughout the time period represented by the deposition of the Jebel Qatrani Formation, although species diversity may have declined somewhat. The upper sequence of the formation has yielded eight hyracoid genera, while a single quarry in faunal unit 3 (quarry V) that measures less than 30 m long, 10 m wide, and 1 m thick, has produced six genera of hyracoids.

*Paleogene Hyracoids Outside the Fayum*

Sixty years elapsed between the discovery of the Fayum hyracoids and the first recognized discovery of Paleogene hyracoids elsewhere (Arambourg and Magnier, 1961). However, it has only been in the last decade that fossils from new localities have contributed significant new information towards an understanding of early hyracoid evolution. Paleogene hyracoids are now known from six areas outside Egypt: three Algerian sites of Eocene age that include the earliest known hyracoids, and three Oligocene localities, one in Angola and two in Libya. The Libyan material is very sparse, consisting of two specimens, one from each of two localities. Arambourg and Magnier (1961) reported "*Megalohyrax palaeotherioides*" from the Oligocene deposits at Zella (= Zallah) a taxonomic designation that was repeated by later workers (Arambourg, 1963; Savage, 1969, 1971) despite the fact that Matsumoto (1922, 1926) had long before shown that *Megalohyrax palaeotherioides* Schlosser belongs in *Titanohyrax* Matsumoto. The name *T. palaeotherioides* was identified as a *nomen nudum* by Meyer (1978), who merged material previously allocated to that species with *T. andrewsi* (Matsumoto, 1922). The "*palaeotherioides*" material has been subsequently granted status as a species distinct from *T. andrewsi*, along with a new name (*T. angustidens*) and description of new specimens (Rasmussen and Simons, 1988). The Zella specimen has never been illustrated or described, but the original identification to genus and species made by Arambourg and Magnier (1961) suggests that the specimen probably belongs in *Titanohyrax*, species indeterminate. The other isolated Libyan specimen comes from the Oligocene Dor el Talha deposits (= Dur at Talhah; Jebel Coquin), but it remains unillustrated, undescribed and unidentified to genus (Savage, 1969; Wight, 1980). Other components of the fauna recovered at Zella and Dor el Talha closely resemble the Fayum assemblage, and all three sites are probably roughly contemporaneous.

Pickford (1986a) has recently identified an Oligocene fauna showing strong affinities with the Fayum mammals from Malembe, northern Angola, based on material collected in the 1930s and described by Hooijer (1963). Two hyracoid species are represented, which in the absence of better material, are probably referable to two Fayum species, *Pachyhyrax crassidentatus* and *Bunohyrax fajumensis*. The best-preserved specimen is a complete $M_3$ set in a jaw fragment that retains on the lingual side a small remnant of a fenestra opening into a hollow mandibular chamber. The specimen was originally identified as an anthracothere (Hooijer, 1963, pl. IX, Figs. 1-2) and was later transferred to *Geniohyus* aff. *G. mirus* (Pickford, 1986a, Fig. 3) before the revision of *Pachyhyrax* by Rasmussen and Simons (1988). The main criterion for assignment to *Geniohyus* was the broad lingual opening in the jaw. However, the superior border of the opening does not appear to be flat as in *G. mirus* but rather is curved as in specimens of *P. crassidentatus* from the Fayum. The molar bears distinctive features of *P. crassidentatus* that differ from the conditions seen in *Geniohyus*, including: a high buccal side of the crown that is straight-sided with a marked lingual tilt; lingual cusps in occlusal view that are placed near the periphery of the crown; protoconid and hypoconid that have well-developed crescentic crests; and the buccal base of the tooth that bears a strong cingulum (Fig. 5.2c). The measurements of the $M_3$ from Malembe (27 x 15 mm; Hooijer, 1963) fall near the means of Fayum specimens of *P. crassidentatus* (31.8 x 16.3 mm). In contrast, *G. mirus* has lower molars that are inflated at the base, the cusps are clustered near the center of the crown, the buccal cusps are much less selenodont, and small accessory cuspules are present on the cristid obliqua and other crests. Another isolated lower molar from Malembe was assigned to *Bunohyrax* aff. *B. fajumensis* by Pickford (Hooijer, 1963, pl. IX, Fig. 3; Pickford, 1986a, Fig. 4). This molar, undoubtedly belonging to *Bunohyrax* or the very similar

*Geniohyus*, happens to be the holotype of *Palaeochoerus dartevellei* Hooijer, 1963 (a name that was then applied to an East African suid recently renamed *Kenyasus rusingensis*). The name *dartevellei* is therefore available for the Malembe species of *Bunohyrax* should it eventually prove to be distinct from Fayum species. The sub-Saharan mammal specimens from Malembe show that the typical Fayum fauna probably enjoyed a broad distribution in Africa rather than being restricted to the Tethyan coastline (Pickford, 1986a).

The most interesting and important recent additions to the hyracoid fossil record are newly described Eocene species from Algeria. A locality near Nementcha has produced undescribed dental specimens referred to the Fayum genus *Bunohyrax*, species indeterminate (Coiffait, et al., 1984). Of greater interest are two Algerian sites that have yielded hyracoids described as new genera that are not represented in the Fayum assemblage. These specimens offer for the first time a view of the Order Hyracoidea during the Eocene, and are thus of great potential importance in understanding primitive hyracoid morphology, and possibly in suggesting new clues to hyracoid origins and affinities.

The greatest diversity of Eocene forms has been found at Gour Lazib (= Gouret el Azib) in deposits formed near the flanks of the Atlas Mountains (Gevin et al., 1975; Sudre, 1979). Hyracoids have been recovered from four localities that, according to charophyte correlations, appear to be middle Eocene in age. The most unusual species represented is *Microhyrax lavocati*, a very small form that differs from Fayum taxa in ways that have been interpreted as primitive (Sudre, 1979). The type and only known specimen is a lower jaw with $P_2$-$M_3$. The premolars are relatively simple: the protoconid is placed far forward, the metaconid is distinct and not connected to the protoconid by a strong crest, and a well-defined, simple hypoconid is present. The brachydont molars have strong protoconids and metaconids and a short paracristid that drops abruptly to the mesial border of the tooth, apparently without forming a distinct paraconid. The cristid obliqua originates centrally behind the trigonid and runs to the hypoconid. The hypoconid and entoconid are connected by a low interrupted crest rather than being completely separated by a deep furrow. This structure is possibly a primitive development homologous to the stronger, higher crest that connects entoconid to hypoconid in *Megalohyrax*, *Titanohyrax*, *Saghatherium*, and *Thyrohyrax*. As in Fayum species, a distal cingulum is present that bears a slight central projection (hypoconulid). The $M_3$ bears a large lobate hypoconulid as in Oligocene species. The mandible evidently lacks a hollow internal chamber (in at least one gender). Most of the characters of *Microhyrax* stressed as primitive by Sudre (1979), such as the relatively simple premolars, are also present in some Fayum taxa, especially a new small species of *Saghatherium* from the Green Hill locality and recently described lower teeth referred to *Geniohyus magnus* from faunal unit 3 (Rasmussen and Simons, 1988). The greatest difference between *Microhyrax* and *Saghatherium* is the greater selenodonty and smaller premolar series relative to molar series of the latter (Sudre, 1979). The similarity between *Microhyrax* and *G. magnus* may be greater than that between *Microhyrax* and *Saghatherium*, and direct comparison of those two taxa should prove to be interesting.

Other specimens from Gour Lazib are referrable to two Fayum genera: a very large species, *Titanohyrax mongereaui*, and smaller *Megalohyrax gevini* (Sudre, 1979). *T. mongereaui* is about the same size as *T. ultimus* and shows all the derived specializations of the upper molars present in Fayum specimens of *Titanohyrax*, which might be considered surprising at such an early date. *T. mongereaui* is apparently older than *Microhyrax* and the other Gour Lazib hyracids (Sudre, 1979), and if also older than the Eocene El Kohol site (see following), it would therefore be the world's oldest known hyracoid. Another Gour Lazib species, *Megalohyrax gevini*, is known by one isolated upper molar. As described and illustrated by Sudre (1979), this tooth shows some characteristic features of *Titanohyrax*. These include: the anteroposteriorly compressed hypocone with pre- and posthypocristae forming a sharp V-shape in occlusal view; the high and short prehypocrista that abuts the lingual face of the metacone; the very weak metastyle; and the strongly selenodont ectoloph with sharply compressed mesostyle. Fayum specimens of *Megalohyrax* show these same features developed to a much lesser extent than in *Titanohyrax*. Future discoveries may prove *M. gevini* to be a primitive member of the *Titanohyrax* clade.

A fourth species of hyracoid is present at Gour Lazib (tentatively compared with *Megalohyrax* and *Bunohyrax* by Sudre, 1979), but it is represented only by an isolated premolar fragment. Whatever the status of this specimen, it is clear that Gour Lazib preserves a record of notable hyracoid diversity possibly dating back to the middle Eocene. *T. mongereaui* is a huge, selenodont form that differs dramatically in size and morphology from *M. lavocati*, which suggests that the Gour Lazib deposits are much younger than the earliest divergence of pliohyracid lineages.

Another Algerian site, El Kohol, is the source of a very different Eocene hyracoid, *Seggeurius amourensis* Crochet, 1986, which has been placed in Geniohyinae (Mahboubi et al., 1984, 1986). The subfamilial identification must be viewed as tentative, because *Seggeurius* shares none of the distinctive features of *Geniohyus*, but rather is placed in that taxon because it possesses simple premolars and a general "primitive" bunodont morphology. The most unique feature of *Seggeurius* is that $M^3$ bears an unbroken W-shaped crest connecting the protocone to the hypocone (= metaconule of Mahboubi et al., 1986), a feature not present in any other hyracoid genus. Like other hyracoids, *Seggeurius* has a thick mandible, a strong symphysis, and a well-developed angular region. There is a shallow depression on the lingual face of the ascending ramus bounded below by a distinct thickening of bone. It is conceivable that this feature may be functionally related to the large lingual fossa of *Geniohyus* and the internal chamber of some other pliohyracids.

Mahboubi et al. (1986) interpreted the relatively small mandibular depression of *Seggeurius* as a primitive hyracoid feature, along with the very simple premolars and the crest linking the protocone and hypocone. It is perhaps of some significance that this apparently primitive Eocene hyracoid differs markedly from the earliest known Paleocene possible mesaxonian (McKenna et al., this volume), especially in the arrangement of lingual cusps and crests, which suggests that the dental resemblances between Eocene mesaxonians and Oligocene hyracoids may be convergent. Future work at both of these Algerian localities should continue to yield new valuable specimens of the earliest hyracoids and other poorly known mammals of the African Eocene.

*Miocene Hyracoids*

The Oligocene of Africa was a world rich in hyracoids, but this situation had changed dramatically by the early Miocene. Only two pliohyracids have been found in the early Miocene, *Megalohyrax championi* and *Meroehyrax bateae* (Whitworth, 1954), despite the large, diverse collections of early Miocene mammals that have been made in Africa (e.g., Pickford, 1981). Specimens of large hyracoids have also been identified from the late Oligocene or early Miocene of Pakistan (Pickford, 1986b) and Saudi Arabia (Thomas et al., 1982). Most of the herbivorous niches occupied by hyracoids in the Oligocene were evidently usurped by artiodactyls arriving from Europe or Asia during the interval from 30 to 20 Ma. Among the families of artiodactyls that made their first appearance in Africa during the early

Miocene are Tragulidae, Suidae, Giraffidae, and Bovidae, as well as an increasing diversity of anthracotheres (Maglio and Cooke, 1978). Other ungulates arriving in Africa about the same time include the mesaxonians Chalicotheriidae and Rhinocerotidae. The hyracoid fossils from Pakistan and Saudi Arabia prove that this faunal exchange also involved movement of hyracoids from Africa to Asia, but with much less success than the movement of northern groups into Africa (Thomas et al., 1982; Pickford, 1986b).

The most common and best known of the early Miocene hyracoids is *Megalohyrax championi*, a large species found at localities in East Africa that range in age from about 19 to 16 Ma (Pickford, 1981). This species was originally described as a member of *Pliohyrax* (a Pliocene genus from Europe) by Arambourg (1933). It was transferred to *Megalohyrax* by Whitworth (1954) and then to *Pachyhyrax* by Meyer (1978). This shuffling from genus to genus in itself suggests that *M. championi* does not comfortably fit any recognized Oligocene or Miocene genus. Whitworth (1954) concluded that the Miocene species was most similar to Oligocene *Megalohyrax* and *Titanohyrax*, and suggested that the East African species could be derived from a form like Oligocene *Megalohyrax* by evolution towards molarization of premolars, greater selenolophodonty, and increasing molar height, implying that similarities to *Titanohyrax* evolved convergently, a conclusion I tentatively follow (Table 5.1). Meyer (1978) believed that *Pachyhyrax crassidentatus* from the Fayum resembled *Megalohyrax*, *Titanohyrax*, and the Miocene species in the occurrence of complete lophs or crests connecting hypoconid and entoconid. New specimens of *Pachyhyrax* show that it is a bunodont form with deep furrows separating the buccal and lingual cusps as in *Bunohyrax*, and so referral of *M. championi* to *Pachyhyrax* is inappropriate (Rasmussen and Simons, 1988).

*M. championi* was a large-bodied cursorial hyracoid (Whitworth, 1954; Fischer, 1986). The fibula is fused to the tibia at its distal end, the socket of the astragalus that receives the medial malleolus of the tibia is deeper and more restrictive than in other hyracoids, and the tarsus is narrow with the central axis and third digit strongly developed at the expense of the lateral elements. The head of the astragalus articulates only with the navicular, and the head of the calcaneum articulates solely with the cuboid. The pes has only three, or maybe four, digits. The fossil assemblages in which specimens of *M. championi* occur suggest a fairly open-country, floodplain habitat (Pickford, 1981).

Whitworth (1954) suggested that a few juvenile mandible fragments from East Africa belonged to a species distinct from *M. championi*, although they differed "only by being smaller." The size difference is not great enough to sustain a specific distinction since the dimensions of the one measurable tooth in the small sample, a $P_4$ (12.2 x 7.9 mm), are comparable to the $P_4$'s of the smallest specimens of undoubted *M. championi* (12.0 x 10.1 mm; Whitworth, 1954). East African specimens assigned to *Bunohyrax* (Whitworth, 1954) are either not hyracoid (Meyer, 1978) or possibly represent a very different kind of hyracoid (Pickford, pers. commun.). The Miocene family Myohyracidae, long considered to belong in Hyracoidea, actually belongs in Macroscelidea (Patterson, 1965).

The other legitimate early Miocene hyracoid is the rare *Meroehyrax bateae*. The heavily worn holotype (a lower jaw) suggests a small-bodied species with teeth that are generally similar to those of *Saghatherium* and *Thyrohyrax*, especially the latter (Whitworth, 1954; Meyer, 1973, 1978). However, based on the currently described material it is impossible to determine a precise taxonomic position relative to other fossil hyracoids. *Meroehyrax* differs from all Oligocene hyracoids in the occurrence of a fairly shallow fossa on the lingual side of the mandibular corpus, rather than an internal chamber as in many saghatheriines, or a large, deep fossa as in

*Geniohyus.*

*Meroehyrax* and *M. championi* are apparently the youngest known saghatheriines, the dominant group of the Paleogene. All later fossil hyracoids can be classified into the Pliohyracinae, a cohesive group of large-bodied hypsodont forms, or into the modern family Procaviidae.

**The Second Radiation: Pliohyracinae**

Two genera and four species of African hyracoids found in middle to late Miocene deposits apparently represent the earliest known members of the Pliohyracinae, a group that eventually attained broad geographic distribution and great species diversity in the Pliocene of Eurasia. The first African pliohyracine to be discovered was *Prohyrax tertiarius* from Langenthal, Namibia (Stromer, 1924, 1926), originally described as the earliest member of the Procaviidae. The known upper teeth of the holotype ($P^3$-$M^2$ and part of $M^3$) were considered to be quite close to modern *Procavia* in most morphological details, differing principally in their more bunodont hypocones, more pronounced mesostyles, and less molariform premolars (Meyer, 1978). Churcher (1956) synonymized *Prohyrax* with *Procavia*, a move not followed by subsequent authors. Tentative records of *Prohyrax* were listed for early to mid-Miocene deposits at Gebel Zelten, Libya (Savage and Hamilton, 1973), Muruarot, Kenya (Madden, 1972), and Loperot, Kenya (Pickford, 1981). However, it is possible that these referred specimens belong to *Meroehyrax*, especially since the holotype of *Meroehyrax* is a lower dentition and that of *Prohyrax* is an upper dentition, and thus they cannot be directly compared to one another.

Recent finds from Namibia and from Kenya have changed earlier views of *Prohyrax*. Hendey (1978) briefly described and illustrated a new unnamed species of hyracoid from Arrisdrift, Namibia, that he considered congeneric with *Prohyrax tertiarius*, differing from it mainly in its greater size. The new species is the most common vertebrate found in the Arrisdrift deposits and is known by excellent material including a complete cranium with intact teeth (Hendey, 1978). This new *Prohyrax* shows a premolariform $I^3$ and upper canine, a closed toothrow from $I^3$ to $M^3$, and a distally bifurcate metastyle on $M^3$. All of these are specializations shared with later members of Pliohyracinae, so Hendey (1978) included it in that subfamily. The age of the Arrisdrift *Prohyrax* is estimated to be 18-12 Ma, and Hendey (1978) suggested that it may be descended from the slightly older and smaller *P. tertiarius*.

The occurrence of pliohyracines in the Miocene of Africa is further documented by a younger genus, *Parapliohyrax*, represented by two species: *P. mirabilis* from Beni Mellal, Morocco (Lavocat, 1961), and *P. ngororaensis* from deposits about 12 Ma in age of the Ngorora Formation, Kenya (Pickford and Fischer, 1987). *Parapliohyrax* is characterized by a unique external fossa on the mandible that may be connected to the internal mandibular chamber, and by deeply excavated palatine pockets (Pickford and Fischer, 1987). The Moroccan species occurs with a fauna that is generally similar to Astaracian faunas of Europe (Savage and Russell, 1983), and *P. mirabilis* may be related to the early radiation of Eurasian pliohyracines. Pickford and Fischer (1987) presented a cladistic analysis of *Parapliohyrax* and the other pliohyracine genera.

*Eurasian Pliohyracinae*

The latest surviving members of Pliohyracinae are a group of large-bodied hyracoids known from the late Miocene and Pliocene of southern Europe and Asia. The Eurasian forms differ from other pliohyracids in the development of hypsodonty, which is carried to the greatest extent in a lineage from China, and also in some cranial features that suggest a partially aquatic mode of existence (Osborn, 1899; Dubrovo, 1978). The large, hypsodont molars differ dramatically from those of other hyracoids, but resemble chalicotheres and some other mesaxonians. The Chinese forms went unrecognized as hyracoids for 19 years (despite the publication of nine papers during that period that evaluated their systematic position), and the European forms went unrecognized as hyracoids for 37 years.

The best known genus is *Pliohyrax*, represented by two or more species distributed from Spain to Afghanistan (Golpe-Posse and Crusafont-Pairo, 1981; Heintz et al., 1981). *P. graecus*, a member of the Pontian faunas of the eastern Mediterranean, has enlarged tusk-like incisors, a complete eutherian dental formula, and double-rooted canines, as in some Oligocene pliohyracids. The skull is shorter than those of large Oligocene hyracoids, and the small orbits and external nares are situated high on the skull, which may indicate aquatic habits (Osborn, 1899; Major, 1899b). A very similar species, *P. rossignoli*, is known from the western Mediterranean (Viret and Mazenot, 1948; Viret, 1949a). Scattered finds of indeterminate species have also been reported (Viret and Thenius, 1952; Becker-Platen et al., 1975; Kuss, 1976).

Pliohyracines were represented in the southern Soviet Union by two genera closely related to *Pliohyrax*. *Kvabebihyrax kachethicus* from late Pliocene deposits of eastern Georgia (Gabunia and Vekua, 1966, 1974; Dubrovo, 1978) had eye sockets placed even higher above the plane of the forehead than in *Pliohyrax*. Another species from farther west, early Pliocene *Sogdohyrax soricus*, differs from *Kvabebihyrax* in some details of the skull and teeth (Dubrovo, 1978). Both genera have been interpreted as semi-aquatic forms that fed on coarse plant food in swampy lowlands or along watercourses (Dubrovo, 1978).

Fossils representing a distinct lineage of pliohyracines have been found in China (Teilhard de Chardin, 1939; Koenigswald, 1966). The Chinese forms, placed in the genus *Postschizotherium*, differ from *Pliohyrax* in the more projecting, tusk-like lower incisors, by the abrupt size difference between the small premolars and large molars, by the absence of $I_3$, and by the more extreme development of hypsodonty and molar cement. Three species are recognized that apparently represent stages of an evolving lineage, from oldest *P. licenti* (Pontian) which shows relatively small size, minimal formation of cement, and small rounded lower tusks, to *P. intermedium* and *P. chardini* (Nihowanian), which show changes in tusk shape and increases in body size, molar height, and cement formation. The buccal face of the upper molars in *P. chardini* is highly bowed, curving a full 90° from base to crown, as in South American toxodonts. Specimens from Shansi, China, have been described as *Pliohyrax* (Tung and Huang, 1974) but these may actually represent a species of *Postschizotherium* (Dubrovo, 1978). The extinction of *P. chardini* in the late Pliocene marks the final disappearance of Pliohyracidae, a family that appeared first in the fossil record of North Africa perhaps 50 Ma with small, brachydont *Seggeurius* and other forms, and ends in China about 2 Ma with the massive, hypsodont *Postschizotherium*.

**The Third Radiation: Procaviidae**

One pliohyracid lineage did not become extinct but gave rise to the modern family Procaviidae, although the precise origin of this family remains obscure. Poorly known *Prohyrax tertiarius* was considered to be an early procaviid (Churcher, 1956), but this species may in fact be an early pliohyracine (Hendey, 1978; see earlier). The pliohyracids with the greatest resemblance to procaviids may be Miocene *Meroehyrax* (Meyer, 1978; Pickford, 1981) and Oligocene *Thyrohyrax* (Meyer, 1973, 1978). It is conceivable that *Thyrohyrax* represents an early member of the procaviid clade. Among all Oligocene pliohyracids, *Thyrohyrax* is the only one that shows relatively high-crowned, lophodont lower molars whose cusps form a distinct Z-shaped pattern as in procaviids, and relatively simple lophoselenodont upper cheek teeth lacking specialized accessory spurs and ridges that characterize *Saghatherium*, the other small non-bunodont Oligocene hyracoid. Some details of cranial morphology may also be significant— for example, the internal nares of *Thyrohyrax* open anterior to the distal border of the $M^3$, as in Procaviidae, rather than behind the molar series as in other pliohyracids. However, given the vast temporal gap between the well-known Oligocene forms and the earliest procaviids, any suggested cladistic relationship must be viewed as extremely tenuous. Much more information about Miocene hyracoids will be required to decipher the origin of Procaviidae.

A single tooth (?$P_4$) from the latest Miocene (9-10 Ma) of Nakali, Kenya, is very similar to the $P_4$ of modern procaviids, and therefore it may be the earliest record of the family (Fischer, 1986). Undoubted procaviids are found in Plio-Pleistocene sediments of South and East Africa, where at least four species in three genera have been identified. An undescribed and indeterminate species of the modern genus *Heterohyrax* occurs in member B of the Shungura Formation, Omo Basin, Ethiopia (Howell and Coppens, 1974), which with an age of over 3 Ma (Brown *et al.*, 1985) is the oldest known procaviid except for the Nakali tooth. *Heterohyrax* is absent from the fossil record of South Africa, where fossils of *Procavia* are common, while *Procavia* is absent from East African deposits. The exclusive occurrence of these now sympatric genera in different parts of Africa during the Pliocene may suggest separate centers of evolution for the two genera.

Two species of the modern genus *Procavia* occur in the Plio-Pleistocene limestone caves of South Africa (Broom, 1934; Churcher, 1956). *P. antiqua* is similar in size and morphology to modern *P. capensis*, and may be ancestral to the living representatives of the genus, while *P. transvaalensis* is larger and differs in some other aspects of morphology. Next to a species of baboon, *P. antiqua* is the most common mammal from the Taung deposits (Broom, 1934). Many of the fossil accumulations in South African limestone caves are due in large part to the activity of leopards and other carnivores (Brain, 1981), and so the abundance of hyraxes in these deposits suggests that they were an important dietary item. *Procavia transvaalensis* is a relatively large, robust species that occurs sympatrically with *P. antiqua* at Sterkfontein and some other South Africa sites (Shaw, 1937; Broom, 1937; Churcher, 1956). In linear dimensions, *P. transvaalensis* is about 1.5 times larger than any living hyrax (i.e., similar in size to *Thyrohyrax domorictus* and *Saghatherium antiquum* of the Fayum), and in cranial morphology resembles modern species of *Procavia* more closely than *Heterohyrax* or *Dendrohyrax*. It differs from all modern hyraxes in the presence of well-developed cingula on the mesio-lingual face of the protocone, in having a more corrugated or folded buccal wall of the ectoloph (Churcher, 1956), and in a simpler convolutional pattern of the cerebral surface (Wells, 1939).

The final species of extinct procaviid is *Gigantohyrax maguirei* (Kitching, 1965) from the limeworks at Makapansgat, South Africa, where it occurs sympatrically in Member 4 breccias (Maguire, 1985) with *Procavia transvaalensis* and *P. antiqua*. *Gigantohyrax* is three times larger in linear dimensions than modern hyraxes, and it differs from *Procavia* in lacking an interparietal, in having more strongly developed temporal and sagittal crests, in the extension of the nasals forward to the anterior border of the premaxillae, and in having thick, solidly fused postorbital bars. Undescribed specimens of *Gigantohyrax* have also been recovered from member C of the East African Shungura Formation (roughly 2.5-3.0 Ma; Brown *et al.*, 1985; Howell and Coppens, 1974). If the East African specimens are correctly allocated, *Gigantohyrax* would thus have the broadest known geographic distribution of any Plio-Pleistonce procaviid. No fossil *Dendrohyrax* are known, leading one to speculate that this genus may have evolved in forested areas of West Africa that are poorly represented by fossiliferous deposits.

## Conclusions

Modern procaviids are highly specialized survivors of a morphologically diverse and geologically long-lived ungulate group. As such, the modern species offer an incomplete picture of the Order Hyracoidea, a fact that may be misleading in studying the evolutionary relationships of the order. The skeletal and dental structures of the earliest known hyracoids perhaps offer valuable clues to hyracoid affinities, but at present the fossil record dates back no earlier than the middle Eocene, when such divergent forms as *Seggeurius, Microhyrax, Megalohyrax,* and *Titanohyrax* had already evolved. Some of the most interesting questions about hyracoid evolution revolve around the earlier Paleocene and Eocene splitting and proliferation of lineages, a problem that cannot be adequately addressed without earlier African continental deposits.

The Paleogene hyracoids in Africa do offer evolutionary biologists an additional and independent example of radiation and adaptation in a group of primitive ungulates that is somewhat analogous to the diverse ungulate radiations of South America. In both cases, the endemic radiations from archaic "protoungulate" stocks occurred largely in isolation from the northern continents. These radiations were then decimated when faunal interchange with the northern continents resulted in an influx of northern ungulate (and carnivore) groups. Unlike most South American ungulates, however, some hyracoids did temporarily expand their ranges into areas outside their continent of origin, and have survived in low diversity to the present day.

## Acknowledgments

I thank Elwin L. Simons and Prithijit S. Chatrath of the Duke University Primate Center for their cooperation and assistance in the field and lab, and for access to collections and casts; the Egyptian Geological Survey and Museum for logistic support in the field and for access to collections; Patricia Holroyd-Vichodyl, Grant E. Meyer, Martin Pickford, and Martin Fischer for comments and valuable advice on the manuscript, and Asenath Bernhardt for taking the photographs.

## Bibliography

Andrews, C. W. (1903): Notes on an expedition to the Fayum, Egypt, with descriptions of some new mammals. -*Geol. Mag.*, 4: 337-343.

Andrews, C. W. (1904a): Further notes on the mammals of the Eocene of Egypt, II. -*Geol. Mag.*, 5: 157-162.

Andrews, C. W. (1904b): Further notes on the mammals of the Eocene of Egypt, III. -*Geol. Mag.*, 5: 211-215.

Andrews, C. W. (1906): *Catalogue of the Tertiary Vertebrates of the Fayum, Egypt.* -London (Brit. Mus. [Nat. Hist.]).

Andrews, C. W. (1907): Notes on some vertebrate remains collected in the Fayum, Egypt. -*Geol. Mag.*, 5: 97-100.

Andrews, C. W., and Beadnell, H. J. L. (1902): A preliminary note on some new mammals from the upper Eocene of Egypt. -*Survey Dept. Publ. Works Ministry, Cairo*: 1-9.

Arambourg, C. (1933): Mammifères miocènes du Turkana (Afrique orientale). -*Ann. Paléont.*, 22: 121-148.

Arambourg, C. (1963): Continental vertebrate faunas of the Tertiary of North Africa. -In: Howells, F. C and Bourliere, F. (eds): *African Ecology and Human Evolution.* Chicago (Alding Publ. Co.), pp. 55-64.

Arambourg, C., and Magnier, P. (1961): Gisements de vertébrés dans le bassin tertiaire de Syrte (Libye). -*C. R. Acad. Sci., Paris*, 252: 1181-1183.

Becker-Platen, J. D., Sickenbourg, O., and Tobien, H. (1975): Die Gliederung der kanozoischen Sedimente der Turkei nac Vertebraten-Faunengruppen. -*Geol. Jahrb.*, Reihe B, 15: 19-100.

Bown, T. M., and Kraus, M. J. (1987): Geology and paleoenvironments of the Oligocene Jebel Qatrani Formation and adjacent rocks, Fayum Depression, Egypt. -*U. S. Geol. Surv. Prof. Paper*, 1452: 1-114.

Bown, T. M., Kraus, M. J., Wing, S. L., Fleagle, J. G., Tiffney, B. H., Simons, E. L., and Vondra, C. F. (1982): The Fayum primate forest revisited. -*J. Hum. Evol.*, 11: 603-632.

Brain, C. K. (1981): *The hunters or the hunted?* Chicago (Univ. of Chicago Press).

Broom, R. (1934): On the fossil remains associated with *Australopithecus africanus*. -*S. Afr. J. Sci.*, 31: 471-480.

Broom, R. (1937): On some new Pleistocene mammals from limestone caves of the Transvaal. -*S. Afr. J. Sci.*, 33: 750-768.

Broom, R. (1946): The South African fossil apemen, the Australopithecinae. Part I. The occurrence and general structure of the South African ape-man. -*Mem. Transv. Mus.*, 2: 7-153.

Broom, R. (1948): Some South African Pliocene and Pleistocene mammals. -*Ann. Transv. Mus.*, 21: 1-38.

Brown, C. L., McDougall, I., Davies, T., and Maier, R. (1985): An integrated Plio-Pleistocene chronology for the Turkana Basin. -In: Delson, E. (ed.): *Ancestors: the Hard Evidence*; New York (Alan R. Liss), pp. 82-90.

Churcher, C. S. (1956): The fossil Hyracoidea of the Transvaal and Taungs deposits. -*Ann. Transv. Mus.*, 22: 477-501.

Coiffait, P. E., Coiffait, B., Jaeger, J. J., and Mahboubi, M. (1984): Un nouveau gisement à mammifères fossiles d'âge Éocène superieur sur le versant sud des Nementcha (Algérie orientale): découverte des plus anciens rongeurs d'Afrique. -*C. R. Acad. Sci. Paris*; t. 299, serie II, 13: 893-898.

Crochet, J. Y. (1986): Hyracoidea Huxley 1869. -In: Mahboubi, M., et al. (eds.): El Kohol (Saharan Atlas, Algeria): a new Eocene mammal locality in northwestern Africa; stratigraphical, phylogenetic, and paleobiogeographical data.- *Palaeontographica Abt. A*, 192: 15-49.

Dubrovo, I.A. (1978): New data on fossil Hyracoidea. -*Paleontol. J.*, 12: 375-383.

Fischer, M.S. (1986): Die Stellung der Schliefer (Hyracoidea) im phylogenetischen System der Eutheria. - *Cour. Forsch.-Inst. Senckenberg*, 84: 1-132.

Fischer, M. S. (1989) Hyracoids, the sister-group of perissodactyls. (this volume, Chapter 4).

Fleagle, J. G., Bown, T. M., Obradovich, J. M., and Simons, E. L. (1986): Age of the earliest African anthropoids. -*Science*, 234: 1247-1249.

Forster Cooper, C. (1924): The Anthracotheriidae of the Dera Bugti deposits of Baluchistan. -*Mem. Geol. Surv. India, Palaeontologica Indica*, 8 (2): 1-60.

Gabunia, L. K., and Vekua, A. K. (1966): [Peculiar representative of hyrax in the upper Pliocene in Eastern Georgia]. -*Soobsch. Akad. Nauk. gruz. SSR*, 42: 643-647 (in Georgian; summary in Russian).

Gabunia, L. K., and Vekua, A. K. (1974): [The mode of life and systematic position of the giant hyrax from Kvabebi]. -*Soobsch. Akad. Nauk. gruz. SSR*, 73: 489-493.

Gaudry, J. A. (1862-7): *Animaux fossiles et géologie de l'Attique*, vol. I, pp. 215-218; Paris.

Gevin, P., Lavocat, R., Mongereau, N., and Sudre, J. (1975): Découverte de mammifères dans la motié inférieure de l'Eocène du nord-ouest du Sahara. -*C. R. Acad. Sci. Paris.*, 280 D: 967-968.

Golpe-Posse, J.M., and Crusafont Pairo, M. (1981): Presencia de un hiracido en el Vallesiense de Can Llobateres (Sabadell), cuenca del Vallés: depresion catalan, España. -*Bol. Real Soc. Espanola de Hist. Nat. (Geol.)*, 79: 265-276.

Heintz, E., Brunet, M., and Battail, B. (1981): A cercopithecoid primate from the late Miocene of Molayan, Afghanistan, with remarks on *Mesopithecus*. -*Int. J. Primatol.*, 2: 273-284.

Hendey, Q.B. (1976): The Pliocene fossil occurrences in 'E' Quarry, Langebaanweg, South Africa. -*Annals South Afr. Mus.*, 69: 215-247.

Hendey, Q.B. (1978): Preliminary report on the Miocene vertebrates from Arrisdrift, South West Africa. -*Annals South Afr. Mus.*, 76: 1-41.

Hoeck, H. N. (1975): Differential feeding behavior of the sympatric hyrax, *Procavia johnstoni* and *Heterohyrax brucei*. -*Oecologia*, 22: 15-47.

Hooijer, D. A. (1963): Miocene Mammalia of Congo. -*Ann. Mus. Roy. Afr. Centr. Tervuren, Sci. Geol.*, 46: 1-77.

Honacki, J. H., Kinman, E., and Koeppl, J. W. (eds) (1982): *Mammal Species of the World: a Taxonomic and Geographic Reference*. Lawrence, Kansas (Allen Press, Inc. and the Association of Systematic Collections).

Howell, F. C., and Coppens, Y. (1974): Les faunes de mammifères fossiles des formations Plio-Pléistocènes de l'Omo en Ethiopie (Tubulidentata, Hyracoidea, Lagomorpha, Rodentia, Chiroptera, Insectivora, Carnivora, Primates). -*C. R. Acad. Sci. Paris*, 278 D: 2421-2424.

Jones, C. (1978): *Dendrohyrax dorsalis*. - *Mammalian Species*, 113: 1-4.

Kingdon, J. (1974): *East African Mammals: An Atlas of Evolution in Africa*, Volume I. Chicago (Univ. of Chicago Press).

Kitching, J.W. (1965): A new giant hyracoid from the Limeworks Quarry, Makapansgat, Potgieterus. -*Palaeont. Afr.*, 9: 91-96.

Koenigswald, G. H. R. von (1932): *Metaschizotherium frassi*, ein neuer Chalicotheriidae aus dem Obermiocän von Stenheim. -*Albuch. Palaeont.*, 8: 1-23.

Koenigswald, G. H. R. von (1966): Fossil Hyracoidea from China. -*Nederl. Akad. Wet., ser. B*, 9: 345-356.

Kuss, S.E. (1976): Ein erster Fund von *Pliohyrax* aus dem Vallesian von Kreta/Grichenland. -*Neues Jahrbuch Geol. Paläontol.*, 176: 157-162.

Lavocat, R. (1961): Le gisement de vertébrés miocénes de Beni Mellal (Maroc.). Part 2. Étude systematique de la fauna de mammifères. -*Serv. Géol. Maroc, Notes et Mém.*, 155: 29-92.

Madden, C. T. (1972): Miocene mammals, stratigraphy and environment of Muruarot Hill, Kenya. -*Paleobios*, 14: 1-12.

Maglio, V. J., and Cooke, H. B. S. (eds.) (1978): *Evolution of African Mammals*. Cambridge, Mass. (Harvard Univ. Press).

Maguire, J. M. (1985): Recent geological, stratigraphic and palaeontological studies at Makapansgat Limeworks. -In: Tobias, P. V. (ed.): *Hominid Evolution: Past, Present and Future*. New York (Alan R. Liss), pp. 151-164.

Mahboubi, M., Ameur, R., Crochet, J. Y., and Jaeger, J. J. (1984): Implications paléobiogéographiques de la découverte d'un nouvelle localitié Éocène à vertébrés continentaux en Afrique Nord-occidentale: el Kohol (Sud-Oranais, Algerie). -*Geobios*, 17: 625-629.

Mahboubi, M., Ameur, R., Crochet, J. Y., and Jaeger, J. J. (1986): El Kohol (Saharan Atlas, Algeria): a new Eocene mammal locality in northwestern Africa: stratigraphical, phylogenetic and paleobiogeographical data. -*Palaeontographica Abt. A.*, 192: 15-49.

Major, C. J. F. (1899a): Note upon *Pliohyrax graecus* (Gaudry) from Samos. -*Geol. Mag.*, 4: 507-508.

Major, C. J. F. (1899b): The hyracoid *Pliohyrax graecus* (Gaudry) from the upper Miocene of Samos and Pikermi. -*Geol. Mag.*, 4: 547-553.

Marsh, O.C. (1884): Dinocerata. A monograph of an extinct order of gigantic mammals.- *Monogr. U.S. Geol. Surv.*, 10: 1-237.

Matsumoto, H. (1922): *Megalohyrax* Andrews and *Titanohyrax* gen. nov.: A revision of the genera of hyracoids from the Fayum, Egypt. -*Proc. Zool. Soc. London*, 1921: 839-850.

Matsumoto, H. (1926): Contribution to the knowledge of the fossil Hyracoidea of the Fayum, Egypt, with description of several new species. -*Bull. Am. Mus. Nat. Hist.*, 56: 253-350.

McKenna, M. C., Chow M., Ting S., and Luo Z. (1989): *Radinskya yupingae*, a perissodactyl-like mammal from the late Paleocene of southern China (this volume, Chapter 3).

Mein, P. (1975): Résultats du Groupe de travail des Vértebés. -*Int. Union Géol. Sci., Rég. Comm. Médit. Néogène Stratigr., Bratislava, Czech.*, pp. 77-81.

Melentis, J. K. (1965): Neue schädel und Unterkieferfunde aus dem Pont von Pikermi (Attica) und Halmyropotamus (Euboa). -*Praklika Akad. Athenon*, 40: 424-459.

Melentis, J. K. (1966): Studien über Fossile Vertebraten Grieshenlands. 12. Neue Schädel und Unterkeferfunde von *Pliohyrax graceus* aus dem Pont von Pikermi (Attica) und Halmyropotamus (Euboa). -*Ann. Geol. Pays. Helleniques*, 17: 182-210.

Meyer, G. E. (1973): A new Oligocene hyrax from the Jebel el Qatrani Formation, Fayum, Egypt. -*Postilla*, 163: 1-11.

Meyer, G. E. (1978): Hyracoidea. -In: Maglio, V. J. and Cooke, H. B. S. (eds.): *Evolution of African Mammals*. Cambridge, Mass. (Harvard Univ. Press), pp. 284-314.

Olds, N. and Shoshani, J. (1982): *Procavia capensis*. -*Mammalian Species*, 171: 1-7.

Olson, S. L., and Rasmussen, D. T. (1986): The paleoenvironment of the earliest hominoids: new evidence from the Oligocene avifauna of Egypt. -*Science*, 23: 1202-1204.

Osborn, H. F. (1899): On *Pliohyrax kruppii* Osborn, a fossil hyracoid from Samos, lower Pliocene, in the Stuttgart Collection. A new type and the first known Tertiary hyracoid. -*Proc. 4th Int. Cong. Zool., Cambridge*, 1898: 173-174.

Owen, R. (1841): Descripton of the fossil remains of a mammal (*Hyracotherium leporinum*) and of a bird (*Lithornis vulturinus*) from the London Clay.- *Trans. Geog. Soc. London*, (2) 6: 203-208.

Patterson, B. (1965): The fossil elephant shrews (Family Macroscelididae). -*Bull. Mus. Comp. Zool., Harvard*, 133: 295-335.

Pei, W. C. (1939): New fossil material and artifacts collected from the Choukoutien region during the years 1937 to 1939. -*Bull. Geol. Soc. China*, 19: 207-234.

Pickford, M. (1981): Preliminary Miocene mammalian biostratigraphy for Western Kenya. -*J. Human Evol.*, 10: 73-97.

Pickford, M. (1986a): Première découverte d'une faune mammalienne terrestre paléogène d'Afrique sub-saharienne. -*C. R. Acad. Sci. (Paris)*, serie II, 302: 1205-1210.

Pickford, M. (1986b): Première découverte d'une hyracoide paléogène en Eurasie. -*C. R. Acad. Sci. (Paris)*, serie II, 303: 1251-1254.

Pickford, M., and Fischer, M.S. (1987): *Parapliohyrax ngororaensis*, a new hyracoid from the Miocene of Kenya, with an outline of the classification of Neogene Hyracoidea. -*Neues Jahrbuch Geol. Paläont. Abh.*, 175: 207-234.

Rasmussen, D. T., and Simons, E. L. (1988): New Oligocene hyracoids from Egypt. -*J. Vert. Paleont.*, 8: 67-83.

Savage, D. E., and Russell, D. E. (1983): *Mammalian Paleofaunas of the World*. Reading, Mass. (Addison-Wesley).

Savage, R. J. G. (1969): Early Tertiary mammal locality in southern Libya. -*Proc. Geol. Soc. London*, 1657: 167-171.

Savage, R. J. G. (1971): Review of the fossil mammals of Libya. -In: Gray, C. (ed.): *Symposium on the Geology of Libya*. Tripoli (Univ. of Tripoli), pp. 215-226.

Savage, R. J. G., and Hamilton, W. R. (1973): Introduction to the Miocene mammalian fauna of Gebel Zelton, Libya. -*Bull. Brit. Mus. (Nat. Hist.)*, *Geol.*, 22: 515-527.

Schlosser, M. (1910): Über einige fossile Säugetiere aus dem Oligocän von Ägypten. -*Zool. Anz.*, 35: 500-508.

Schlosser, M. (1911): Beiträge zur Kenntnis der oligozänen Landsaugetiere aus dem

Fayum, Ägypten. -*Beitr. Paläont. Geol. Öst.-Ung.*, 24: 51-167.
Shaw, J. C. M. (1937): Evidence concerning a large fossil hyrax. -*J. Dent. Res.*, 1: 37-40.
Simons, E. L. (1968): Early Cenozoic mammalian faunas, Fayum Province, Egypt. Part I. African Oligocene mammals: Introduction, history of study, and faunal succession. -*Bull. Peabody Mus. Nat. Hist., Yale Univ.*, 28: 1-21.
Stromer, E., von (1924): Ergebnisse der Bearbeitung mitteltertiar Wirbeltier-Reste aus Deutsch-Südwest-Afrika. -*S. B. Bayer. Akad. Wis., München*, 1923: 253-270.
Stromer, E., von (1926): Rest Land- und Süsswasser- Bewohnender Wirbeltiere aus dem Diamant-feldern Deutsch-Sudwestafrikas. -In: Kaiser, E. (ed.): *Die Diamatenwüste Südwestafrikas*, 2: 102-153. Berlin (D. Reimer).
Sudre, J. (1979): Nouveaux mammifères éocènes du Sahara occidental. -*Palaeovertebrata*, 9: 83-115.
Teilhard de Chardin, P. (1938): The fossils from Locality 12 of Choukoutien. -*Pal. Sinica*, 114: 1-50.
Teilhard de Chardin, P. (1939): New observations on the genus *Postschizotherium* von Koenigswald. -*Bull. Geol. Soc. China*, 19: 257-267.
Teilhard de Chardin, P., and Licent, E. (1936): New remains of *Postschizotherium* from S.E. Shansi. -*Bull. Geol. Soc. China*, 15: 421-427.
Teilhard de Chardin, P., and Pei, W. C. (1934): New discoveries in Choukoutien 1933-1934. -*Bull. Geol. Soc. China*, 13: 369-389.
Teilhard de Chardin, P., and Piveteau, J. (1930): Les mammifères fossiles de Nihowan (Chine). -*Ann. Paléont.*, 1930: 1-134.
Thomas, H., Sen, S., Khan, M., Battail, B., and Ligabue, G. (1982): The lower Miocene fauna of Al-Sarrar (eastern province, Saudi Arabia). -*Atlal J. Saudi Arab. Archaeol.*, 5: 109-136.
Tung Young Sheng and Huang Wan Po (1974): A new *Pliohyrax* from Shansi. -*Vertebrata Palasiatica*, 12: 212-216.
Viret, J. (1947): Découverte d'un nouvel Ancylopode dans le Pontien de Soblay (Ain). -*C. R. Acad. Sci. (Paris)*, 224: 353-354.
Viret, J .(1949a): Sur le *Pliohyrax rossignoli* du Pontien de Soblay (Ain). -*C. R. Acad. Sci. (Paris)*, 228: 1742-1744.
Viret, J. (1949b): Observations complémentaries sur quelques mammifères fossiles de Soblay. -*Eclog. Geol. Helvet.*, 42: 469-476.
Viret, J., and Mazenot, G. (1948): Nouveaux restes de mammifères dans le gisement de lignite Pontien de Soblay. -*Ann. Paléont.*, 34: 19-58.
Viret, J. and Thenius, E. (1952): Sur la présence d'une nouvelle espèce d'hyracoide dans le Pliocène de Montpellier. -*C. R. Acad. Sci. (Paris)*, 235: 1678-1680.
Wells, L. H. (1939): The endocranial cast in recent and fossil hyraces (Procaviidae). -*S. Afr. J. Sci.*, 36: 365-373.
Whitworth, T. (1954): The Miocene hyracoids of East Africa. -*Brit. Mus. (Nat. Hist.), Fossil Mammals of Africa*, 7: 1-58.
Wight, A. W. R. (1980): Paleogene vertebrate fauna and regressive sediments of Dur at Talhah, southern Sirt Basin, Libya. -In: Salem, M. J. and Busrewil, M. T. (eds.): *The Geology of Libya*, 1: 309-325; London (Academic Press).

# 6. CHARACTER POLARITIES IN EARLY PERISSODACTYLS AND THEIR SIGNIFICANCE FOR *HYRACOTHERIUM* AND INFRAORDINAL RELATIONSHIPS

## J. J. HOOKER

New ideas on the character polarities in primitive perissodactyls are advanced. These are used to hypothesize the nature of the stem perissodactyl and to analyze cladistically a variety of primitive perissodactyls. The results are an unresolved trichotomy into three suborders: Titanotheriomorpha nov., Hippomorpha, and Tapiromorpha, the last divided into the infraorders Ancylopoda and Ceratomorpha. Various other lower level rearrangements are also made.

### Introduction

Attempting to determine correctly the polarity of characters is very important for understanding the relationships of any group. This is especially true of the Order Perissodactyla, where non-recognition or faulty recognition of polarities may have formerly masked certain affinities. The genus *Hyracotherium* is central to this issue as for many years it has been interpreted by most authors as the primordial perissodactyl.

The type species of *Hyracotherium* (*H. leporinum* Owen, 1841) was first described from the marine London Clay (early Eocene) of the London Basin (England), and members of the same genus were later recovered in abundance in the continental strata of the same age in the western United States. The latter were often referred to *Eohippus* until long after they were clearly shown to be congeneric with *Hyracotherium* (Forster Cooper, 1932). Simpson (1952) revised the English species, but their straightforward classification was hampered by too much morphological variation and too few specimens covering too long a time range, apparently with little stratigraphic control. Later, additional specimens were discovered, the necessary stratigraphic control was found to have been neglected, and four morphologies were seen to succeed each other in time (Hooker, 1980). This succession was tentatively interpreted as an actual evolutionary sequence, thus providing unequivocal evidence of polarity of the characters. At this stage no possible consequences for broader perissodactyl relationships were sought.

The history of perissodactyl classification is covered by Schoch (this volume, Chapter 2), but I will briefly summarize the later stages. Relative stability had reigned for much of the earlier part of this century and was embodied in Simpson's (1945) classification. About twenty years ago some changes began to take place. Before the mid 1960s, the order was divided into two suborders, the Hippomorpha and the Ceratomorpha. The Hippomorpha comprised one modern superfamily, Equiodea (horses), and two extinct ones, the Chalicotherioidea and Brontotherioidea. The Ceratomorpha comprised the modern superfamilies Tapiroidea and Rhinocerotoidea. Radinsky

Fig. 6.1.                                Fig. 6.2.

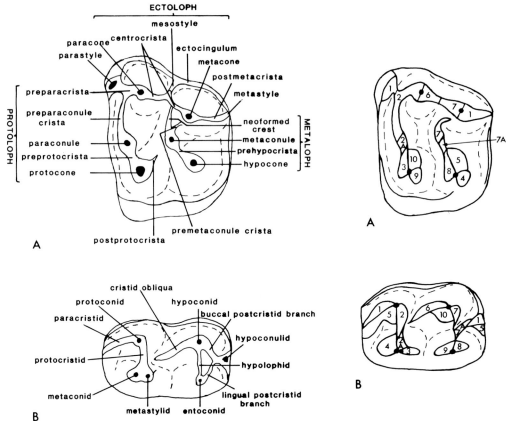

Fig. 6.1. Perissodactyl dental terminology diagram of a left upper preultimate molar (A) and a right lower preultimate molar (B) in occlusal view. Buccal is to the top of the page and mesial to the left. the premetaconule crista and neoformed crest (its analogue) of the upper molar and the lingual postcristid branch and hypolophid of the lower molar respectively do not occur together on the same tooth. The preparacrista forms part of the protoloph in bilophodont teeth and part of the ectoloph in selenolophodont teeth. Lophid developments of the paracristid, protocristid and cristid obliqua are usually termed paralophid, protolophid and metalophid respectively.

Fig. 6.2. Diagram of wear facets on a left upper (A) and a right lower (B) molar of *Cymbalophus*, representing a primitive perissodactyl. Slightly modified from Hooker (1984, Fig. 19) and essentially following the numbering system of Butler (1952a). Facets 5 and 10 function in lingual phase and the rest in buccal phase wear.

(1964) then created a third suborder for the Chalicotherioidea, as they seemed to him equally distinct from the other two. He resurrected Cope's Ancylopoda (Cope, 1889) as the name for this third suborder. MacFadden (1976) later completed the dismemberment of the Hippomorpha by showing that the Equoidea could be separated from all other perrisodactyls by the posterior orbital position of the optic foramen and from many fossil, but not modern, ceratomorphs by the confluence of the middle lacerate foramen and foramen ovale in the basicranial region. This also resulted in the removal of *Hyracotherium* from its long-held position of stem perissodactyl to stem equoid (even though it had at the same time been classified in the Equidae), although it was still considered that its teeth represented the primitive morphotype for the order. Thus he produced a cladogram of the dental characters alone, then tested and modified it to produce a more parsimonious cladogram showing *Hyracotherium* as the sister group of the Equoidea.

At this stage the Ceratomorpha appeared well defined on the basis of consistent dental characters. These include a straight upper molar ectoloph and associated lower molar metalophid extending mesially, rather than mesiolingually and joining the trigonid at a buccal position (Radinsky, 1964, p. 5). (See Fig. 6.1 for dental terminology diagram.) The apparently advanced state of this character was supported by the oblique lingually extending lower molar metalophid of phenacodontids (the nearest sister group according to Radinsky [1966b], MacFadden [1976] and most other authors) and most other condylarths. Assuming this, and also that dentally a typical *Hyracotherium* (e.g., Hooker, 1984, Fig. 20) represented the primitive perissodactyl morphotype, I (1984) interpreted the earliest English species of *Hyracotherium* as a primitive ceratomorph and erected the genus *Cymbalophus* for it. Compared to *Hyracotherium*, it showed minor but distinct trends in buccal shift of the lower molar metalophid, steeper distal slope of the unnotched protocristid, less convergence of buccal and lingual cusps, buccal tilting of the upper molar paracone, weakening of intermediate conules, and so on. Overall these characters gave it a more bilophodont aspect, and their degree of development appeared little different from that of *Homogalax*, a widely accepted primitive ceratomorph.

A consequence of this rearrangement was that chalicotheres and lophiodonts (the latter usually placed in the Ceratomorpha) were closely related, the former having reverted from a bilophodont state to a selenolophodont "hippomorph" state (see also Radinsky, 1964, p. 5). The high metaloph was taken as evidence for retention of a character evolved originally in advanced bilophodonty. The family Isectolophidae, including *Homogalax*, seemed to have split off before the lophiodont-chalicothere group and was therefore excluded from the Ceratomorpha. Although the orbit of *Cymbalophus* was unknown, it was assumed to have retained the primitive non-equoid state of the optic foramen.

Subequent study of the large North American collections of *Hyracotherium* has demonstrated in certain assemblages a dental morphotype nearly as bilophodont and ceratomorph-like as *Cymbalophus* but associated with a posterior (equoid) position of the optic foramen. The resulting incongruities appear best resolved by removing typical *Hyracotherium* from its position of primitive dental morphotype for the Perissodactyla. Thus its well-developed upper molar paraconule and metaconule, and lower molar oblique metalophid and well separated metaconid-metastylid complex would represent advanced states, and the "ceratomorph" characters of *Cymbalophus* outlined above would be primitive for perissodactyls. As has been postulated for chalicotheres, the selenolophodont condition of brontotheres and equoids would thus have been independently derived from an earlier sub-bilophodont state.

## Evidence for polarities

*Bilophodonty.* Typical bilophodont teeth in mammals (e.g., Figs. 6.3E, 6.4E) have two long transverse crests which dominate the upper and lower molars (and also the premolars if they are molariform) and the masticatory function is mainly one of shearing (see Butler, 1952a, p. 802). In the upper molar, the mesial of the two crests, the lophoid homologue of the preprotocrista plus preparaconule crista has shifted its orientation, broken its link with the parastyle, and joined the paracone via a preparacrista now made oblique (in a mesiolingual-distobuccal direction) by buccal tilting of the paracone. Thus a long protoloph is formed from these three crests. This scheme is found not only in perissodactyls but also in the orders Proboscidea, Pyrotheria, and Dinocerata. The formation of the distal loph (the metaloph) is more variable in its development in the different groups. Its development in perissodactyls is discussed below.

On the upper molars of early selenolophodont perissodactyls (i.e., brontotheres, chalicotheres, and equoids), the preparaconule crista joins the paracone as in bilophodont types, not the parastyle as in eutherian mammals generally and in phenacodontids in particular. In some of these cases, however, (e.g., *Lambdotherium*, see Fig. 6.3O), the preparacrista has reorientated itself for participation in the W-shaped ectoloph and broken away from the rest of the protoloph.

*Mesostyle.* This cusp is usually linked functionally with a W-shaped ectoloph (dilambdodonty) which provides most of the upper molar shearing component in the selenolophodont type of dentition. The mesostyle is the point of attachment of the two Vs of the W and may arise from the ectocingulum, or from a fold of the centrocrista or both. Such genera as *Lambdotherium* and *Palaeotherium* posses one but *Hyracotherium* lacks one (Figs. 6.3L, 6.3O). Most phenacodontids also possess one (e.g., *Phenacodus* and *Ectocion*, Fig. 6.3A) but usually not *Tetraclaenodon*, and mainly for this reason it has been suggested that all phenacodontids except *Tetraclaenodon* are too advanced to have given rise to perissodactyls (e.g.,Radinsky, 1966b).

Some species of *Homogalax* (e.g., *H.*, sp. 1) commonly have a small upper molar

Fig. 6.3. (next page) Upper second molars of perissodactyls (B-O) and a phenacodontid (A). A, *Ectocion osbornianus*, early Eocene, U.S.A. (composite); B, *Cymbalophus cuniculus*, early Eocene/late Palaeocene, England (composite); C, *Homogalax* sp., early Eocene, U.S.A. (UMMP 68548); D, *Lophiaspis maurettei*, early Eocene, France (holotype FSL 2084); E, *Heptodon calciculus*, early Eocene, U.S.A. (composite); F,G, "*Hyracotherium*" sp., early Eocene/late Palaeocene, Rians, France (UM RI326); H,I, "*Hyracotherium*" *seekinsi*, early Eocene/late Palaeocene, Mexico (holotype LACM 15349); J, *Pachynolophus duvali*, middle Eocene, France (MNHN unnumbered); K, *Anchilophus depereti*, middle Eocene, Switzerland (NMB Eb376); L, *Hyracotherium* aff. *vulpiceps*, early Eocene, Abbey Wood, England (M13761); M, *Propalaeotherium messelense*, middle Eocene, W. Germany (cast UCMP 65371); N, *Propalaeotherium hassiacum*, middle Eocene, W. Germany (cast UCMP 65376); O, *Lambdotherium popoagicum*, early Eocene, U.S.A. (USNM 19761). A-F, H, J-O are shown as left teeth in occlusal view. G and I are mesial views of right teeth. D, F, J-L and O are reversed from the right side. Specimen number abbreviations (also applicable to Fig. 6.4): AMNH = American Museum of Natural History; FSL = Faculté des Sciences Lyon; LACM = Los Angeles County Museum; M = British Museum (Natural History), Palaeontology Dept.; MNHN = Muséum National d'Histoire Naturelle, Paris; NMB = Naturhistorisches Museum Basel; UCMP = University of California Museum of Paleontology; UM = Université de Montpellier; UMMP = University of Michigan Museum of Paleontology; USNM = United States National Museum. A-C, E-M are x3.5; D,N and O are x2. C-K, M-O are drawn from casts.

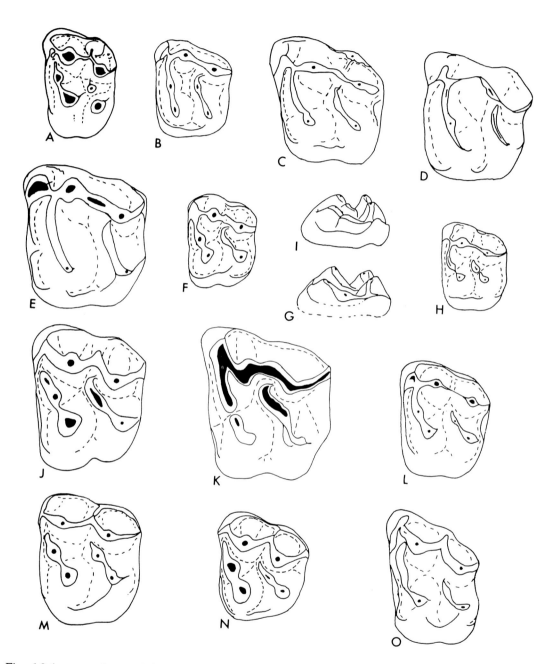

Fig. 6.3 (see caption p. 82).

84 THE EVOLUTION OF PERISSODACTYLS

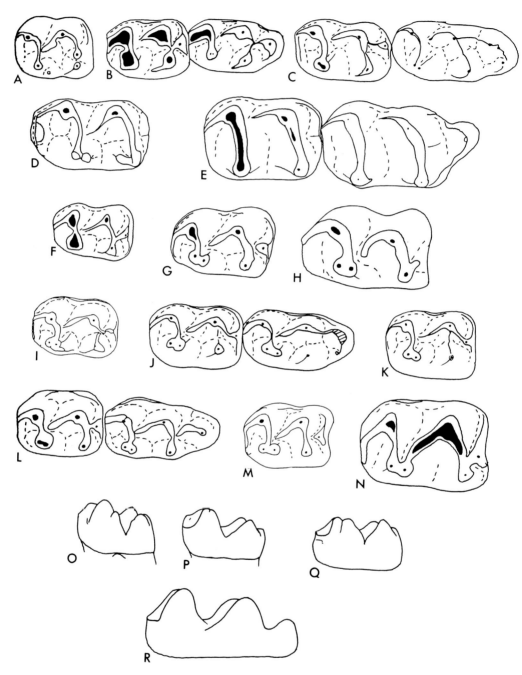

Fig. 6.4 (see caption p. 85).

mesostyle that arises from the ectocingulum but does not contact the straight, not W-shaped, centrocrista (Fig. 6.3C). As the dentition is dominantly bilophodont, functioning mainly in semitransverse shear, it is difficult to imagine this mesostyle functioning in mastication. It seems most likely that it is a vestigial feature and that perissodactyls arose from an ancestor with a dilambdodont tooth type, including a functional mesostyle. This may have had teeth somewhat like those of *Ectocion*, although *Tetraclaenodon* with a straight centrocrista sometimes has a large cingular mesostyle also.

**The primitive perissodactyl**

Thus it is possible to envisage the primitive perissodactyl dental state, which was probably rather similar to that of *Cymbalophus* (Figs. 6.3B, 6.4B, 6.4P). A consideration of the occlusal relations of the cheek teeth, especially the molars (see Butler, 1952a; Radinsky, 1969; Hooker 1984), shows how the various dental features have evolved (see Fig. 6.2). It also allows the characters discussed here to be treated as functional occluding units incorporating in most cases structures on both upper and lower teeth and sometimes including those on adjacent teeth in the same row. Wear facets are numbered according to Butler (1952a). Tooth cusps and crests are treated as homologous units, although there is not universal agreement over strict application of this procedure. Butler (1952a and pers. comm.) prefers to regard them as functional structures, the appropriate name being used according to identity of position and occlusal relations. In Van Valen's (1982) sense, they may not always be both repetitively (= serially) and historically homologous. This is particularly relevant to premolar molarization in certain species

Fig. 6.4. (previous page) Lower right molars of perissodactyls (B-N,P-R) and a phenacodontid (A,O). A,O *Ectocion osbornianus*, $M_2$, early Eocene, U.S.A. (M16471); B,P, *Cymbalophus cuniculus*, $M_{2-3}$, early Eocene/late Palaeocene, England (M36494); C, *Homogalax* sp. 1, $M_{2-3}$, early Eocene, U.S.A. (M9788); D, *Lophiaspis* cf. *maurettei*, $M_{1/2}$, early Eocene, France (MNHN Louis 194MU); E,R, *Heptodon calciculus*, $M_{2-3}$ (R, $M_3$ only), early Eocene, U.S.A. (AMNH 4858); F, "*Hyracotherium*" sp., $M_2$, early Eocene/late Palaeocene, Rians, France (UM RI136); G, *Pachynolophus duvali*, $M_{1/2}$, middle Eocene, France (MNHN unnumbered); H, *Anchilophus dumasii*, $M_2$, late Eocene, France (FSL 5854); I-K, Q, *Hyracotherium* aff. *vulpiceps*, early Eocene, Abbey Wood, England (I, $M_{1/2}$, cast M42318; J, $M_{2-3}$, M29719-21; K, $M_{1/2}$, M15143; Q, $M_{1/2}$, M32127); L, *Propachynolophus maldani*, $M_{2-3}$, early Eocene, France (MNHN GR10773); M, *Propalaeotherium hassiacum*, $M_2$, middle Eocene, W. Germany (cast UCMP 65375); N, *Lambdotherium magnum*, $M_2$, early Eocene, U.S.A. (holotype AMNH 17527). All are shown as right teeth, A-N in occlusal view, O-R in lingual view. A,D,H,I and L are reversed from the left side. For abbreviations, see Fig. 6.3. A-C, E-L, N-R are x3.5; D and M are x2. C-I,L-N and R are drawn from casts.

of *Hyracotherium*, which is in fact one of the examples that Van Valen (1982, p. 310-311) has cited. In the present paper homologies are treated as historical where information is available.

From the ancestral phenacodontid, the following changes in the molars are considered to have taken place in order to produce the primitive perissodactyl morphotype (Figs. 6.3A-B, 6.4A-B, 6.4O-P). Few cranial changes may have been effected but the evolution of the diagnostic perissodactyl astragalus, with its concave navicular facet, as well as various limb modifications is well established (Radinsky, 1966b; MacFadden, 1976).

1) The upper molar paracone has tilted slightly buccally, changing the orientation of its crests (pre- and postparacristae) from longitudinal to oblique. Corresponding changes in the lower molars are a tilting buccally and lingually respectively of protoconid plus hypoconid and metaconid-metastylid plus entoconid, as well as a buccal shift in the trigonid attachment position of the cristid obliqua to near the protoconid, making it less oblique. This was accompanied by reduction of the cingulum-derived mesostyle and its separation from the centrocrista, which may have straightened out.

2) The crests between the protocone and paracone (preprotocrista, preparaconule crista and preparacrista) strengthen and form a sublophoid structure, occluding with a more strongly crested (unnotched) protocristid and a metastylid of increased height to equal that of the metaconid.

3) Weakening of the paraconule narrows facet 2A and is associated with a close approach of the metaconid and metastylid.

4) The parastyle lengthens, occluding with more of the paralophid as the distal budding off of the hypoconulid of the more anterior lower molar lengthens this functional crest (involving facet 1).

5) The postprotocrista breaks and a new crest (prehypocrista) joins an enlarged hypocone to a mesially migrated metaconule which is in turn linked to the metacone by a strengthened distobuccally directed premetaconule crista. The new crest, prehypocrista plus premetaconule crista, is sublophoid and occludes with the postcristid, whose two halves have become a sublophoid crest, being freed by the distal "budding off" of the hypoconulid. The postcristid is weaker than the protocristid and corresponds to a notching between the metaconule and premetaconule crista on the upper tooth.

6) The distal "budding off" of the $M_3$ hypoconulid results in a distinct lobe which has buccal and lingual phase occlusal relationships with $M^3$, whose metacone and hypocone are larger and better separated from the paracone and protocone respectively.

## Cladistic analysis

Following postulation of the primitive perissodactyl state, cladistic analysis of a representative sample of early perissodactyls has been conducted, using Swofford's computer program PAUP (Swofford, 1984). This involved the global branch swapping MULPARS and reduced reversal DELTRAN options. I deduced the polarities of the various character states from a variety of "simple to complex" developmental criteria involving progressive differentiation from the primitive state, in certain cases relying on historical homology. The cladogram was rooted with a hypothetical ancestor, comprising the primitive state for all the characters, thus corresponding to the concept of the primitive perissodactyl discussed above. Explanation of the development of the relevant, often multiple, character states and how they were obtained follows. The data matrix is given in Table 6.1. The characters are mainly dental and cranial as these are the ones best known, but a few postcranial characters are discussed later. The taxa used were chosen to include one of the most primitive generally accepted fossil representatives of each of the original five main groups: extant Equoidea (*Hyracotherium vasacciense*),

Tapiroidea (*Protapirus*), Rhinocerotoidea (*Triplopus*) and extinct Chalicotherioidea (*Eomoropus*), Brontotherioidea (*Eotitanops*) and as many of their mainly more primitive relatives as possible to help elucidate relationships. Certain other less well-known taxa are discussed separately.

**Explanation of the characters**
Explanation of those characters whose meaning is obvious is omitted.

1) The separation versus confluence (1B) of the foramen ovale and middle lacerate foramen is discussed by MacFadden (1976, p. 6-8). Here, an intermediate state is recognized (1A) where the foramina are still separated but only by a narrow bridge of bone.

2) The posterior migration of the optic foramen in the orbit, to the vicinity of the posterior group of foramina, is discussed by MacFadden (1976, p. 8-9).

4, 5) Those taxa with a narial incision (4) are differentiated on details of its shape (5). State 5B involves development of a narrow groove posterodorsally on either side of the nasals. For illustration, see Franzen (1968), Radinsky (1963, p. 89-90), Reshetov (1979, p. 43-49) and Depéret (1904, pl. 16, Fig. 2).

6) For details of this sexual dimorphism, see Gingerich (1981).

7) For the coronoid process character, see Reshetov (1979, p. 23).

8) Presence of a small distal cusp on $I_3$ is well documented amongst various early Asian perissodactyls (Radinsky, 1965; Reshetov, 1979; Ranga Rao, 1972), and it occurs also in *Homogalax*. It does not occur in phenacodontids where this tooth is known, but it is here treated as primitive because of its rather random distribution amongst primitive tapiromorphs.

14-16) Diastemata. The advanced phenacodontid *Ectocion* has fairly short postcanine and post-$P_1$ diastemata, at least in the lower jaw (Rose, 1981, p. 74). The presence of both diastemata in many primitive perissodactyls suggests the primitive state. There is a trend to close the post-P1/1 diastema and thereby lengthen the postcanine diastema. This is recorded here for the upper jaw only (14), as $P_1$ is often lost. The postcanine diastema can lengthen also by loss of P1 (recorded again here only for the upper tooth (11)) or by actual extension of the mandibular symphysis. It is only the last method that is recorded under character 16. The postcanine diastema can also shorten and close (15A-B). The pattern of equoid diastematal development has been treated by Franzen (1972), and differs slightly from that envisaged here.

18) The lingual migration of the $P^3$ paraconule and distal migration of the $P_3$ metaconid eventually to become functionally the protocone and entoconid respectively has been discussed by Butler (1952b) and more recently by Van Valen (1982, p. 310-311) where it is his chosen example of historical homology.

20, 26) This strengthening (loph formation) of the postprotocrista plus premetaconule crista is on an orientation that is straight and nearly buccolingual, linking the protocone to the metacone.

21) The U-shaped crest in $P^{3-4}$ in lophialetids (*sensu stricto*) is a lophoid structure formed from the protoloph and postprotocrista plus premetaconule crista, joined at the protocone (Reshetov, 1979). It is mesiodistal lengthening of the protocone that causes the U shape. Similar lengthening of the protocone occurs on $P^4$ of *Plagiolophus*, but not on the lophoid plan (26).

29) For discussion of premolar molarization, see Butler (1952b).

30) Absence of a $P^3$ metacone and a $P_3$ hypoconid in most species of the relatively advanced equoid *Plagiolophus* is suggestive of secondary loss, in view of the presence of these cusps in phenacodontids and of other advanced fetaures of this tooth in *Plagiolophus*, rather than evolution of *Plagiolophus* independently of other perissodactyls as Butler (1952b) has suggested.

Table 6.1. Character states (left) and character distributions (right) of taxa discussed in this chapter. See text for explanation.

| Character | States | |
|---|---|---|
| | Primitive | Advanced |
| 1. Foramen ovale & middle lacerate foramen | well separated | separated by narrow bridge A, confluent B |
| 2. Optic foramen position re posterior group of orbital foramina | distant | close + |
| 3. Occiput | vertical | overhanging + |
| 4. Marked narial incision | no | yes + |
| 5. Narial incision shape | simple | lower edge stepped A, also nasals bordered by groove B |
| 6. Sexual dimorphism in size of skull, canines, anterior premolars | no | yes + |
| 7. Coronoid process of dentary | high | low + |
| 8. I3 distal cusp | present | absent + |
| 9. Incisor shape | spatulate | I3 pointed A, all pointed B |
| 10. Upper canine size | > incisors | ≤ incisors + |
| 11. P1 | present | absent + |
| 12. DP1 retained, not replaced by P1 | no | yes + |
| 13. P2 | present | absent + |
| 14. Post P1 (or DP1) diastema | open | constantly closed + |
| 15. Postcanine diastema | moderate or longer | short A, closed B |
| 16. Postcanine diastema length | moderate | long + |
| 17. P3 paraconule | absent | present + |
| 18. P3 paraconule and P3 metaconid position | normal | paraconule lingual, metaconid distal + |
| 19. P3 postprotocrista | absent | present + |
| 20. P3 postprotocrista lophoid, buccolingually orientated, joining ectoloph | no | yes + |
| 21. P3-4 lophoid loop complete and U-shaped | no | yes + |
| 22. P3 metaconule | absent | present + |
| 23. Trigonid of P3 | tapered mesially | broader, less tapered + |
| 24. P4 paraconule | present | absent + |
| 25. P4 postprotocrista | present | absent + |
| 26. P4 postprotocrista lophoid, buccolingually orientated, joining ectoloph | no | yes + |
| 27. P4 metaconule | present | absent + |
| 28. P metacone (where present) position re paracone | distal | distolingual + |
| 29. Premolar molarization trends, by hypocone and entoconid | no | yes + |
| 30. P3 metacone and P3 hypoconid | present | absent + |
| 31. Convergence of M paracone and protocone causing: buccal bending of pre- and postparacristae and obliquity of M metalophid | no | incipient A, advanced B |
| 32. M centrocrista buccal flexing causing oblique M metalophid | no | slight   A, advanced B, |
| 33. Centrocristal mesostyle constant | no | yes + |
| 34. Convergence of M metacone and hypocone causing: buccal bending of pre- and postmetacristae | no | slight A, advanced B, extreme C |
| 35. Tilting outwards of M paracone & protocone and M protoconid & metaconid and hypoconid & entoconid, with near longitudinal metaloph orientation | no | yes + |
| 36. M paraconule and M metastylid | present | absent + |
| 37. M paraconule size and M metaconid-metastylid separation | small & close | large & distant + |
| 38. M paraconule and M metastylid relatively lingual | no | yes + |
| 39. P3-4 lophoid, buccolingually orientated postprotocrista often markedly interrupted just buccal to protocone | no | yes + |
| 40. M premetaconule crista + prehypocrista & M postcristid sublophoid | no | yes + |
| 41. M preparaconule crista & preparacrista junction and M protocristid | unnotched | notched + |
| 42. M preprotocrista+preparaconule crista+preparacrista and M protocristid lophoid | no | yes + |
| 43. M metaconule and preultimate M hypoconulid | near median | lingually displaced + |
| 44. M premetaconule crista and M postcristid lingual branch | present | absent + |
| 45. M metaloph and M hypolophid | absent | present + |
| 46. M metaloph and M hypolophid lophoid | no | yes + |
| 47. Preultimate M metaloph-ectoloph joining position | well in front of metacone | at or near metacone + |
| 48. M paracone distal, M metalophid mesial end restricted | no | yes + |
| 49. M postmetacrista and M paralophid | normal | reduced + |
| 50. M preprotocrista+preparaconule crista | normal | reduced + |
| 51. M metaconule and preultimate M hypoconulid | present | reduced A, lost B |
| 52. M3 hypocone+prehypocrista | present | absent + |
| 53. M metaconule distinctly demarcated on metaloph | no | yes + |
| 54. M paracone flattened buccally | no | yes + |
| 55. M metacone flattened buccally | no | yes + |
| 56. Preultimate M antecrochet | absent | incipient A, marked B |
| 57. M parastyle tilted buccally | no | yes + |
| 58. M metastyle and adjacent ectocingulum expanded buccally | no | yes + |
| 59. M3 metacone and M3 hypoconulid | not reduced | metacone reduced A, hypoconulid reduced B, hypoconulid lost C |
| 60. M3 hypocone enlarged and M3 hypoconulid expanded | no | yes + |
| 61. M3 postmetacrista+metastyle and M3 hypoconulid extension | none | slight hypoconulid lengthening A, slight postmetacrista lengthening B, both more C |
| 62. M cingular mesostyle | present | absent + |
| 63. M metaconule distinctly mesial (even if submerged in metaloph) | no | yes + |
| 64. Trends in hypsodonty | no | yes + |
| 65. M3 parastyle buccally expanded | no | yes + |
| 66. M protocone & hypocone distally and M hypoconid & protoconid mesially recurved | no | yes + |
| 67. M's relatively broad | no | yes + |

```
                              Plagiolophus annectens
                                Plagiolophus sp. 1
                                  Propalaeotherium hassiacum
                                    Propachynolophus gaudryi
                                      Lophiotherium
                                        Propalaeotherium messelense
                                          Propachynolophus maldani
                                            Palaeotherium
                                              Hyracotherium leporinum
                                                Hyracotherium vulpiceps
                                                  Hyracotherium aff. vulpiceps
                                                    Hyracotherium vasacciense
                                                      Hyracotherium tapirinum
                                                        Cymbalophus
                                                          Anchilophus
                                                            Pachynolophus
                                                              'Hyracotherium' sp., Rians
                                                                Triplopus
                                                                  Chasmotherium
                                                                    Hyrachyus
                                                                      Protapirus
                                                                        Colodon occidentalis
                                                                          Helaletes
                                                                            Heptodon
                                                                              Lophialetes
                                                                                Boletes
                                                                                  Schlosseria
                                                                                    Rhodopagus
                                                                                      Teleolophus
                                                                                        Breviodon
                                                                                          Eomoropus
                                                                                            Lophiaspis maurettei
                                                                                              Paralophiodon buchsowillanum
                                                                                                Lophiodon
                                                                                                  Kalakotia
                                                                                                    Isectolophus
                                                                                                      Homogalax protapirinus
                                                                                                        Homogalax sp. 1
                                                                                                          Eotitanops
                                                                                                            Lambdotherium

 1.  B  ?  B  ?  ?  ?  B  ?  B  ?  B  A  ?  ?  B  ?  ?  ?  -  B  ?  ?  -  -  ?  ?  ?  ?  ?  -  ?  ?  -  ?  ?  -  ?  ?  -  ?
 2.  +  +  ?  ?  ?  ?  ?  +  +  +  ?  +  +  ?  ?  +  ?  -? -? -? -? -? -? -  -  -? -? -? -? -? -  -? -? -? -? -  -? -? ?  ?
 3.  -  -  -  -  ?  -  -  ?  -  ?  -  -  ?  ?  -  ?  -  ?  ?  -  ?  ?  ?  ?  ?  ?  ?  +  ?  -? -? ?  ?  +  +
 4.  +  +  -  ?  -  ?  +  -  -  ?  -  -  ?  -  ?  -? +  -  +  +  +  +  -  +  ?  ?  ?  ?  ?  -  ?  -  ?  -  -
 5.  -  -  0  ?  0  0  ?  -  0  0  ?  0  0  ?  0  0  ?  0? -  0  B  A  A  0  -  0  ?  ?  ?  ?  ?  0  ?  0  0  ?  0  0  ?  0  0

 6.  -  -  -  -  -  -  -  -  -  -  -  -  -  -  +  -  -  -  -  -  -  -  -  -  -  -  -  -  -  -  -  -  -  -
 7.  -  -? -  -  -? -  ?  -  ?  -  ?  -  ?  -? -  -  -? ?  ?  -  -? -? -  -? ?  ?  ?  -? +  -? ?  -? ?  -? ?  -
 8.  +  +? +? ?  +  +  ?  +  ?  +? +? +? +? +  +  +? ?  +  +? +  +? +? -  ?  -  ?  -  ?  ?  +  +  -  ?  ?  -  ?  ?  +  +
 9.  -  -? -? ?  -  -  ?  -  ?  -? -? -? -? -  -  -? ?  -  B  -  -? -  ?  -  0  ?  0  ?  0  0  ?  ?  B  B  0  ?  0  ?  A  -
10.  -  -  -  -? -  -  -  -  -  -? -  -  -? -  -  -? +  -? +  +  -  +  +  -  +  -  +? -? +? -? ?  +  -  -  -  -  -? -
11.  +  -  -  -  -  -  -? +  -  -  -? -  -  -? -  -  ?  -  -  -  -  +  -? -? +  -  -  +  -
12.  -  -  -  -  -  -  -  -  ?  +  -? -  -? -  -  -? -  -  -? -  -  -  -  -  -  -  -  -  -  -? -? -  -  -
13.  -  -  -  -  -  -  -  -  -  -  -  -  -  -  -  -  -  -  -  -  -  -  -  -  -  -  -  -  -  -  -  -  -  -
14.  0  +  +  +  +  +  ?  +  +  -? +  -  -  -? -  +  ?  +  +  +  +  +  +  +  +  +  +  +  0  -? -? 0  0  +? +  -
15.  -  -  -  -  -  -  -? A  -  -  -  -  -  -  -  -  -? -  B  -  -  -  -  -  -  -  A  -  -  -  -  -? ?  A  -  -  B  A  A  A
16.  -  -  -  -  +  -  -? 0  -  -  -? +  -  -  -  -  -  +  +  ?  -  -  -  -  -  +  0  -  -  -  -  -  -  -  0  0  0  0
17.  +  +  +  +  +  +  +? +  +  +  +  +  +  +  +  +  +  +  +  +  +  +  +  +  +  +  +  +  +
18.  -  -  -  -  -  -  -  -  -  -  -  -  -  +  +  -  -  0  0  0  0  0  0  0  0  0  0  0  0  0  0  0  0  0  0  0  0  0
19.  +  +  +  +  +  +  +  +  +  +  +  +  +  ?  ?  +  +  -  +  +  +  +  +  +  +  +  +  +  +  +  0  -? -? 0  0  +? +  -

20.  -  -  -  -  -  -  -? -  -  -  -  -  0  -  -  0  -  -  -  ?  +  +  -  +  +  +  +  +  +  +  +  0  -  -  0  0  +  +
21.  0  0  0  0  0  0  0? 0  0  0  0  0  0  0  0  0  ?  -  -  -  +  +  -  +  +  +  +  +  0  -  0  -  -  0  0  0  0
22.  -  -  -  -  -  +? ?  +  -  -  -  -  -  -  -  -  -  -  -  -  -  -  -  -  -  -  -  -  -  -  -  -  -  -  -  -  -  -  -  +
23.  +  +  +  +  +  +  +  +  +? +  +  +  +  +  +  +  +  +  +  +  +  +  +  +? +  +  +  +  +  +  +? +  +  +  +  +  +  +  +
24.  -  -  -  -  -  -  -  -  -  -  -  -  -? -  -  -  -  -  -  -  -  -  -  -  -  -  -  -  -  -  -  +  -  -  -  -  +  -
25.  -  -  -  -  -  -  -  -  -  -  -  -  -  -  ?  -  -  -  +  -  -  -  -  -  -  -  -  -  -  -  -  -  +  -  -

26.  -  -  -  -  -  -  -  -  ?  -  -  0  +  +  +  +  +  +  +  +  +  +  +  +  +  +  +  +  +  +  0  +  +  +  -  0  +
27.  -  -  -  -  -  -  -  -  ?  -  -  -  +  +  +  +  +  +  +  +  +  +  +  +  +  +  +  +  +  +  +  +  +  +  +  +  +  +
28.  -  -  -  -  -  -  -  -  -  -  -  -  -  -  -  -  -  -  -  -  -  -  -  -  -  -  -  -  -  -  -  -  -  -  -  -  -  -
29.  -  -  -  -  -  -  -  -  -  -  -  -  -  -  -  -  -  -  -  -  -  -  -  -  -  -  -  -  -  -  -  -  -  -  -  -  -  -
30.  +  -  -  -  -  -  -  -  -  -  -  -  -  -  -  -  -  -  -  -  -  -  -  -  -  -  -  -  -  -  -  -  -  -  -  -  -  -

31.  B  B  B  B  B  B  B  B  A  A  A  -  -  B  B  A  -  -  -  -  -  -  -  -  -  -  -  -  -  -  -  B  -  -  -  -  B  B
32.  B  B  B  B  B  A  A  B  A  -  -  -  -  -  A  A  -  -  -  -  -  -  -  -  -  -  -  -  -  -  B  A  -  -  -  -  B  B
33.  +  +  +  +  -  +  +  -  -  -  -  -  -  -  -  -  -  -  -  -  -  -  -  -  -  -  -  -  -  -  -  -  -  -  -  +  +
34.  A  A  A  A  A  A  A  A  -  -  -  -  -  A  A  A  A  A  A  A  A  A  A  A  A  A  A  A  B  C  A  A  A  A  A  A  -  -  A  A

35.  -  -  -  -  -  -  -  -  -  -  -  -  -  +  +  +  +  +  +  +  +  +  +  +  +  +  +  +  +  0  0  0  0  0  0  0  0  0
36.  -  -  -  -  -  -  -  -  -  -  -  -  -  +  +  +  +  +  +  +  +  +  +  +  +  +  +  +  +  -  -  +  +  +  +  -  -  -
37.  +  +  +  +  +  +  +  -  -  +  +  -  -  +  +  +  0  0  0  0  0  0  0  0  0  0  0  0  0  0  -  +  0  0  0  0  0  +
38.  -  -  -  -  -  -  -  -  -  -  -  -  -  -  -  -  0  0  0  0  0  0  0  0  0  0  0  0  0  0  -  0  0  0  +  +  -  +

39.  0  0  0  0  0  0  0  0  0  0  0  0  0  0  0  0  0  0  0  0  0  -  +  +  +  +  +  +  0  0  0  0  0  0  0  0  -  0  0  0
40.  0  0  0  0  0  0  0  0  0  0  0  0  0  0  0  0  0  -  -  +  +  +  +  +  +  +  +  +  0  0  0  0  0  0  0  0  0  +  -
41.  -  -  +  +  +  +  +  -  +  +  +  +  +  -  +  +  +  +  -  -  -  -  -  -  -  -  -  -  -  -  -  -  -  -  -  -  +  +

42.  -  -  -  -  -  -  -  -  -  -  -  -  -  -  +  +  +  +  +  +  +  +  +  +  +  +  +  +  +  +  +  +  +  +  +  +  +  +  +
43.  +  +  -  -  -  -  +  -  -  +  -  -  -  -  -  +  -  0  0  0  0  0  0  0  0  0  0  0  0  0  0  0  -  -  -  +  +
44.  +  +  +  +  +  +  +  +  -  +  +  +  +  -  +  +  -  -  -  -  -  -  -  -  -  -  -  -  -  -  -  -  -  -  -  -  +  +
45.  +  +  +  +  +  +  +  +  -  +  +  +  +  -  +  +  -  -  -  -  -  -  -  -  -  -  -  -  -  -  -  -  -  -  -  -  +  +
46.  -  -  -  -  -  0  0  0  0  0  0  0  0  -  -  -  -  -  -  -  -  -  -  -  -  -  -  -  -  -  -  -  -  -  -  0  0

47.  0  0  0  0  0  0  0  0  0  0  0  0  0  0  0  0  0  +  +  +  +  +  +  +  -  -  -  -  -  -  -  -  +  +  -  -  0  0  0  0
48.  -  -  -  -  -  -  -  -  -  -  -  -  -  -  -  -  -  +  +  +  +  +  +  +  -  -  -  -  -  -  -  -  -  -  -  -  -  -
49.  -  -  -  -  -  -  -  -  -  -  -  -  -  -  -  -  -  +  +  +  +  +  +  +  -  -  -  -  -  -  -  -  -  -  -  -  -  -
50.  -  -  -  -  -  -  -  -  -  -  -  -  -  -  -  -  -  +  +  +  +  +  +  +  -  -  -  -  -  -  -  -  -  -  -  -  -  -
51.  -  -  -  -  -  A  -  -  -  -  -  -  -  -  A  -  -  B  B  B  B  B  B  B  B  B  B  B  B  B  B  B  B  A  B  -  B  -  -  -
52.  -  -  -  -  -  -  -  -  -  -  -  -  -  -  -  -  -  -  -  -  -  -  -  -  -  -  -  -  -  -  -  -  -  -  -  -  -  -
53.  -  -  -  -  -  -  -  -  -  -  -  -  -  -  -  -  -  0  0  0  0  0  0  0  0  0  0  -  0  -  -  -  0  -  -  0
54.  -  -  -  -  -  -  -  -  -  -  -  -  -  -  -  -  -  -  -  -  -  -  -  -  -  -  -  -  -  -  -  -  -  -  -  -  -  -
55.  -  -  -  -  -  -  -  -  -  -  -  -  -  -  -  -  -  -  -  -  -  -  -  -  -  -  -  -  -  -  -  -  -  -  -  -  -  -
56.  -  -  -  -  -  -  -  -  -  -  -  -  -  -  B  -  -  -  -  -  -  -  -  -  -  -  -  A  -  A  -  -  -  -  -  -  -  -
57.  +  +  +  +  +  +  +  -  -  -  -  -  -  -  -  -  -  -  +  +  +  +  +  +  -  -  -  -  -  -  -  -  -  -  -  -  +  +
58.  -  -  -  -  -  -  -  -  -  -  -  -  -  -  -  -  -  -  -  -  -  -  -  -  -  -  -  -  -  -  -  -  -  -  -  -  -  -

59.  -  -  -  -  -  -  -  -  -  -  -  -  -  -  -  -  -  C  C  C  C  B  B  B  -  B  -  -  C  C  -  B  B  B  A  -  A  A  A  -  -
60.  0  0  0  0  0  0  0  0  0  0  0  0  0  -  -  -  -  -  -  -  -  -  -  -  -  -  -  -  -  -  -  -  -  -  -  -  0  +  +  0  0

61.  C  C  B  B  A  B  B  B  A  A  A  A  -  -  B  B  -  0  0  0  0  0  0  0  B  0  B  0  0  0  B  0  0  0  -  0  0  0  -  A
62.  +  +  +  +  +  +  +  +  +  +  +  +  +  +  +  +  +  +  +  +  +  +  +  +  +  +  +  +  +  +  +  +  +  +  -  -  +  +  +  +
63.  -  -  -  -  -  -  -  -  -  -  -  -  -  -  -  -  -  -  -  -  -  -  -  -  -  -  -  -  -  -  -  -  -  -  -  -  0  -
64.  +  +  +  +  +  +  +  -  -  -  -  -  -  -  -  -  -  -  -  -  -  -  -  -  -  -  -  -  -  -  -  -  +  +  -  -  +  +
65.  -  -  -  -  -  -  -  -  -  -  -  -  -  -  -  -  -  -  -  -  -  -  -  -  -  -  -  -  -  -  -  -  -  -  -  -  -  -

66.  +  +  -  -  -  -  -  -  -  -  -  -  -  -  -  -  -  -  -  -  -  -  -  -  -  -  -  -  -  -  -  -  -  -  -  +  +  +
67.  -  -  -  -  +  +  -  -  -  -  -  -  -  +  +  -  -  -  -  -  -  -  -  -  -  -  -  -  -  -  -  -  -  -  -  -  -
```

31-33) The convergence of upper molar paracone and protocone and of correlated lower molar protoconid and metaconid, and hypoconid and entoconid, respectively, whose initial stage occurs in *Hyracotherium* (31A), is an advancement over the primitive perissodactyl state, but a reversal to a state more like that of the phenacodontids and many other mammals (see above) (Figs. 6.3A, G, I, L). The degree of lingual shift of the lower molar trigonid attachment point of the cristid obliqua seems to be partly dependent on the upper molar paracone-protocone convergence (31A-B) and partly on buccal flexing of the centrocrista (dilambdodonty) (32A-B), as both produce a dilambdodont centrocristal trace in strictly perpendicular crown view (Figs. 6.3-6.4). Buccal centrocristal flexing may also involve the development of a cuspate swelling at the convexity of the fold (centrocristal mesostyle) (33) (Fig. 6.3M-O).

34) Convergence of upper molar metacone and hypocone is similar to that of the paracone and protocone (31) and is a contributory factor in obliquity of the lower molar paracristid (see also 58). Extreme trends cause fusion (34B) then alignment (32C) of premetacrista and metaloph (see also 47) (Figs. 6.3N, 6.4M).

35) Further reduction in convergence of upper molar paracone and protocone (from the primitive state) results in parallel sided to divergent lower molar protoconid and metaconid, and hypoconid and entoconid respectively, and in further buccal migration of the trigonid joining point of the cristid obliqua, so that this takes up an essentially mesiodistal orientation (Figs. 6.3E, 6.4E).

36-38) Enlargement of the upper molar paraconule and of facet 2A results in the metastylid moving distally away from the metaconid (37) (Figs. 6.3D, L, 6.4D, I-K, Q). Loss of the paraconule as a discrete cusp on the protoloph, as the latter becomes increasingly lophoid (42), is associated in the lower molars with a complete fusion of the metastylid with the metaconid as facet 2A is obliterated (36) (Figs. 6.3E, 6.4E, R).

Lingual migration of the upper molar paraconule on the other hand appears to be associated with a similar migration of the lower molar metastylid (38) (Figs. 6.3C, 6.4C). This is relevant even when the paraconule has disappeared within the protoloph and the metastylid has fused to the metaconid, as the long axis of the metaconid-metastylid complex remains oblique.

39) The premolar lophoid structure (20, 26) may be secondarily and distinctively broken on the postprotocrista part just buccal of the protocone on nonmolariform $P^{3-4}$ (see Wood, 1934). It is inapplicable to molariform premolars.

41, 57) Where the upper molar parastyle tilts buccally (57) following paracone-protocone convergence (31), the preparacrista becomes reorientated, breaking its link with the preparaconule crista and resuming its link with the parastyle (Fig. 6.3M-O). A break (or initially a notch) at the junction of the preparacrista and preparaconule crista is mirrored in the lower molars by a notching of the protocristid (41) (Hooker, 1984, Figs. 13, 17B). This state is reversed when trends in hypsodonty (64) are superimposed, so that the preparaconule crista once more makes uninterrupted contact with the preparacrista (Fig. 6.3K).

40, 43-47, 51) The primitive perissodactyl upper molar crest formed from the prehypocrista plus premetaconule crista and its occlusal companion, the lower molar postcristid, are modified in various ways. The premetaconule crista has a characteristic mesiolingual-distobuccal orientation as it passes up the wall of the metacone (Fig. 6.3B, F, H, J, K, O). The postcristid is slightly bowed distally (Fig. 6.4B, F-G, N). These crests may simply be strengthened into lophs (40) when they essentially retain their primitive orientation (Fig. 6.4H). Alternatively, the premetaconule crista and lingual branch of the postcristid may be lost (44) (Fig. 6.3L). In this case, a new mesiobuccally directed crest may arise from the upper molar metaconule (Fig. 6.3N) and a corresponding lower molar buccolingual one joining the hypoconid and entoconid (Fig. 6.4L). Thus are formed the metaloph and hypolophid (45). Increased strengthening into true lophoid crests is the next development (46), followed by a distal migration of the buccal attachment to the centrocrista so that it meets or nearly meets the metacone (47) (Figs. 6.3E, 6.4E). The upper molar metaconule may be weakened and lost along with the lower molar hypoconulid (often in the process of extreme loph formation) because of obliteration of facet 7A (51A-B) (Figs. 6.3E, 6.4E). Alternatively, the metaloph may be weakened, making the metaconule more distinctly demarcated (53) (Savage et al., 1965, p. 74). Like the paraconule, the metaconule may also migrate lingually where it is accompanied by the hypoconulid (43) (Figs. 6.3K, O, 6.4H, N).

The distinction on lower molars of the hypolophid from the postcristid is sometimes difficult. This is because, although the latter tends to flex distally and the former tends to form a straight line between the hypoconid and entoconid, in advanced taxa other modifications often minimize these differences. Nevertheless, it is frequently possible to glimpse the pattern of development in any one assemblage of *Hyracotherum*, provided there are enough specimens. Thus, some may have a complete postcristid (Fig. 6.4I), others simply a postcristid lacking the lingual branch (Fig. 6.4J), still others a postcristid lacking the lingual branch plus an additional low crest joining the hypoconid and entoconid, representing an incipient hypolophid (Fig. 6.4K). In *Propachynolophus* there is often the weak remnant of the buccal branch of the postcristid extending from the hypoconulid a short distance up the distal hypoconid wall (Fig. 6.4L). This state is often more clearly visible on *Homogalax* lower molars (Fig. 6.4C), athough in this genus I have not observed any individuals showing more primitive stages of these trends. Thus, like the equid $P_3$ molarization method (18), one has to rely largely on historical homology for recognition of development of the hypolophid as a crest distinct from the postcristid.

Similar recognition problems pertain with the neoformed structure replacing the premetaconule crista. Those taxa that retain the primitive crest modify its position slightly, although it nearly always retains the backwardly directed orientation (e.g., *Pachynolophus*) (Fig. 6.3J). Although its development seems to be correlated with the postcristid-hypolophid complex, this is more in terms of opposing dental pattern rather than strict occlusal relations (e.g., the premetaconule crista does not actually occlude with the lingual branch of the postcristid) (Fig. 6.2). Moreover, the two developments seem to be slightly out of phase, with the evolutionary modifications in the lower teeth occurring slightly before those in the upper. This means, for instance, that a functional lower molar hypolophid plus remains of the buccal postcristid branch, and an upper molar prehypocrista plus premetaconule crista, may dominate in a single assemblage.

It appears that the distal migration of the buccal end of the upper molar metaloph, so that it meets the centrocrista at or near the metacone, may be accomplished in two ways. The metaloph of *Rhodopagus* (and *Pataecops*) joins the centrocrista only just in front of the metacone and seems on this criterion alone similar enough to *Heptodon* (Fig. 6.3E), for example, to be a homologous development. However, the metaloph of *Rhodopagus* is oblique like *Schlosseria*, not transverse like *Heptodon*. It thus seems that the shift in position of the metaloph with respect to the metacone in *Rhodopagus* is because of advanced lingual tilting of the metacone (34B), and therefore not homologous with the actual distal migration of the metaloph buccal end as a result of change to transverse metaloph orientation (47) where primitively character state 34A pertains. When hypsodonty is superimposed on character 47 and buccal and lingual cusp tips converge (as in advanced rhinocerotoids), the metaloph in the vicinity of the occlusal edge accommodates by becoming secondarily oblique, but the broad central valley remains and the transverse deeper part of the metaloph becomes exposed by wear.

48-49) Distal migration of the upper molar paracone and associated distal restriction of facet 6 is paralleled in the lower molars by a loss of contact between the cristid obliqua (metalophid) and the trigonid, together with weakening and distal recession of the cristid obliqua (48) (Figs. 6.3D, 6.4D). Reduction of the upper molar postmetacrista and facet 1 has a similar effect on the lower molar paracristid (paralophid), which weakens (49).

54-55) Flattening out the buccal expression of the paracone (54) and metacone (55) (Fig. 6.3E) on the ectoloph follows advanced stages of lingual tilting of these cups often accompanied by dilambdodonty.

56) The upper molar antecrochet is a typical rhinocerotoid feature (56B) (Osborn, 1898), but occurs also in incipient form (56A) in more primitive ceratomorphs where there has been a trend towards hypsodonty (64).

58) Expansion of the distal corner of the upper preultimate molars takes the form of buccal migration of the metastyle and development of a strong ectocingulum buccal of the metacone. It results in a reorientation of the postmetacrista and paracristid to a mesiolingual-distobuccal direction (Figs. 6.3E, 6.4E) (see also 34).

63) The slight mesial shift of the upper molar metaconule in the transition from a phenacodontid to the primitive perissodactyl state is carried further so that it intersects a line drawn between the protocone and metacone (see Hooker, 1980, Fig. 4b).

66) The distal recurving of the upper molar protocone and hypocone and mesial recurving of the lower molar protoconid and hypoconid appear to be for slight enhancement of lingual phase occlusion in ancylopods (Figs. 6.3D, 6.4D) but are probably an accommodation to hypsodonty in palaeotheres. Those in ancylopods are not homologous with the similar functional trait of tapirids, where neoformed postprotocrista occludes with neoformed cristid obliqua (see Hooker, 1984, p. 240).

52, 59-61, 65) Modifications of M3/3. Simple cases involve a migration buccally of the $M^3$ parastyle with respect to the paracone (65) (see Radinsky, 1969, Figs. 2B, 5C) (not to be confused with character 57); or loss of the $M^3$ hypocone and prehypocrista (52) (Radinsky, 1969, Fig. 2E). From the primitive state (see above), both lengthening and shortening has taken place in a variety of ways. Lengthening first involves extension of the $M_3$ hypoconulid lobe (61A) (e.g., Hooker, 1980, Figs. 1c, 2c, 3c), then of the $M^3$ postmetacrista and enlargement of the metastyle (61B), then by further expansion of both (61C). Reduction first affects the $M^3$ metacone (59A), after which there is either a secondary expansion of the $M_3$ hypoconulid lobe mainly for lingual phase wear with an enlarged and distally projecting $M^3$ hypocone (60) (Fig. 6.4C); or progressive reduction of the $M_3$ hypoconulid (59B) (Fig. 6.4E) culminating in its loss (59C).

### Results of the computer analysis

The computer analysis of the data matrix resulted in 50 equally parsimonious trees, each with 262 steps and a fairly low consistency index (0.382) indicating much

Fig. 6.5. (next page) Cladogram of the suborder Hippomorpha. For explanation, see text; for characters up to no. 67, see Table 6.1. Unique or uniquely shared apomorphies are shown as heavy symbols, parallelisms as light symbols, normal polarities as bars and reversals as crosses. Character 68: increased crown height of upper molar protocone and hypocone; character 69: upper molar centrocristal mesostyle present in some individuals.

# PERISSODACTYL INTERRELATIONSHIPS

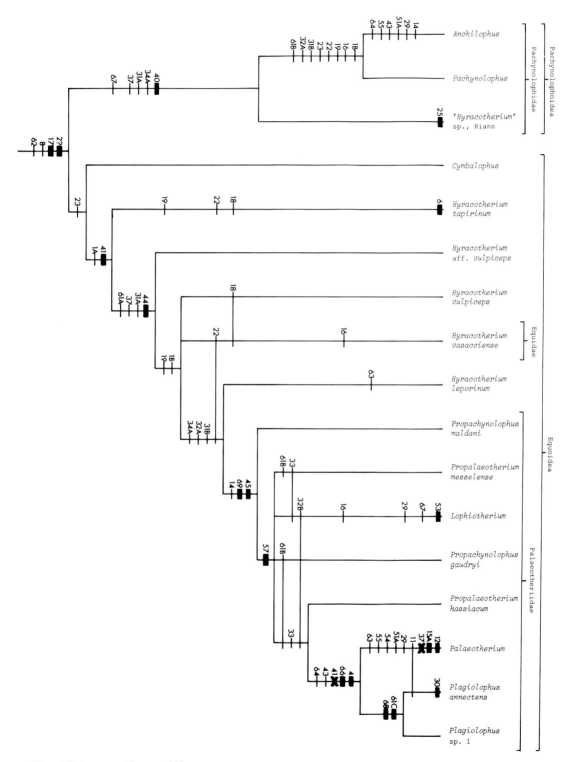

Fig. 6.5 (see caption p. 92).

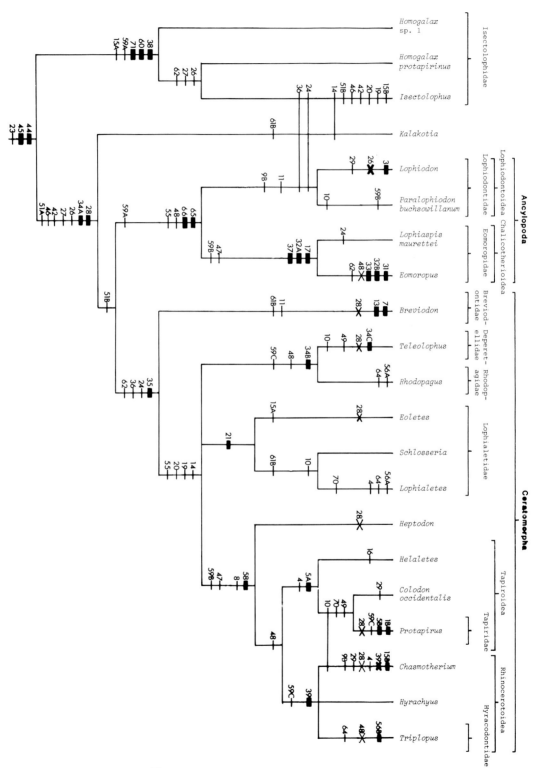

Fig. 6.6 (see caption p. 95).

homoplasy (Swofford, 1984). Some results of finding the shortest possible trees are some major character state jumps and especially reversals that, on the developmental plan advanced in the section "Explanation of characters," seem most unlikely events. They ignore, for instance, Riedl's (1978) concept of burden and canalization of evolutionary prospects. In a sequence of increasing specialization it becomes progressively difficult to revert to the primitive condition without at least affecting many other parameters. Likely reversals, on the other hand, are indicated by, for example: the selenolophodont dentition; or loss of the $P^3$ metacone by backward shift of the molarization field, but occurring anomalously with well-developed paraconule, and so on. The polarity models advanced here (above) are, moreover, in a number of cases supported by a relatively dense fossil record where the order is clear (e.g., European equoids; North American isectolophids). Also, because the data matrix covered such diverse taxa, some of the more minor characters that are easy to recognize among closely related members of small groups (e.g., *Homogalax* and *Isectolophus*) have not been included, because of the difficulty in evaluating their state or even existence in all the taxa involved.

The PAUP program nevertheless produced some major divisions that did not rely on reversals. Its first division was into Brontotherioidea plus Equoidea and the rest. "The rest" were well characterized, but the group composing the Brontotherioidea plus Equoidea was based on loss of the $I_3$ distal cusp (8) and of the upper molar cingular mesostyle. These are apparently minor characters. The former is not generally well known, and both occur in many members of the sister group. Thus, they are rather suspect. Moreover, because of lack of knowledge of the optic foramen (2) in certain key primitive forms, the computer assumed for parsimony purposes that *"Hyracotherium"* sp. from Rians (Godinot, 1981) and *Cymbalophus* both had the primitive state and linked *"Hyracotherium"* sp. from Rians with the brontotheres. The latter was dependent on reversal of character 23, considered here unlikely (see discussion by Godinot, 1981, p. 102).

The following steps were taken to produce a solution that allowed only those reversals considered likely. Firstly, it proposed as a working hypothesis that both *"Hyracotherium"* sp. from Rians and *Cymbalophus* had the advanced state for character 2 as one of the defining characters of the Equoidea. Secondly, because of the weakness of an Equoidea-Brontotherioidea link on the basis of character 8, a primary unresolved trichotomy of the Perissodactyla into Brontotherioidea, Equoidea, and the rest is considered the best reflection of our current state of knowledge of the group. Each of the three main subgroups was then taken separately and a few more characters were added to characterize better such well-accepted groups as the Family Isectolophidae and the genus *Plagiolophus*. Each subgroup was then analyzed cladistically without the aid of a computer, and the results are given in Figs. 6.5-6.6 and are discussed below.

### Infraordinal relationships

*Brontotherioidea.* The two genera considered here as representing the two main branches of this monofamilial superfamily are usually placed in separate subfamilies (Lambdotheriinae and Brontotheriinae) of the family Brontotheriidae (see Osborn, 1929). This group is defined by an overhanging occiput (3), loss of $P^4$ metaconule (27), upper molar paracone-protocone and metacone-hypocone convergence (31, 34), centrocristal flexing (32), a centrocristal mesostyle (33), notching of preparaconule crista-preparacrista and lower protocristid (41), and lingual migration of the upper

Fig. 6.6. (previous page) Cladogram of the suborder Tapiromorpha. For explanation, see text; for characters up to no. 67, see Table 6.1. Symbols as for Fig. 6.5. Character 70: nasals shortened; character 71: upper molar protoloph and metaloph oblique.

metaconule and of lower preultimate molar hypoconulid (43). All of these characters are paralleled elsewhere in the Perissodactyla, although in some cases quite distantly. The primitive brontotheriine *Eotitanops* is much more clearly defined, while *Lambdotherium* is much closer to the basal perissodactyl stock. To express the major distinction of this group from the two others, it is proposed to erect for it the monotypic suborder Titanotheriomorpha.

*Equoidea* (Fig. 6.5). This superfamily as previously defined falls into two groups: one containing the genera *Pachynolophus, Anchilophus,* and "*Hyracotherium*" sp. from Rians, the other comprising the families Equidae and Palaeotheriidae plus *Hyracotherium* and *Cymbalophus*. It is proposed that the Equoidea be restricted to the second group and the family Pachynolophidae Pavlow (1888), raised to the rank of superfamily Pachynolophoidea, be applied to the first group. The suborder Hippomorpha can thus be resurrected for the two superfamilies. Dentally, the Hippomorpha can be defined by the presence of a $P^3$ paraconule (17). This character is important as it is probably the first sign of a trend that reverses the primary bilophodonty and facilitates precocious but lasting premolar molarization, especially in the Equidae. All hippomorphs in which the orbit is known have the advanced state of the optic foramen (2). It is not known for "*Hyracotherium*" sp. from Rians or *Cymbalophus*, which are very primitive dentally, but it seems likely that the character defines the whole group as MacFadden (1976) thought, and that initially little progress was made dentally in this suborder.

The primitive equoid *Hyracotherium* is paraphyletic with respect to the Equidae and Palaeotheriidae and could be split up in such a way as to reflect this. However, species are often so difficult to distinguish and several possibilities for interrelationships are equally parsimonious that it seems that to do so would serve no useful purpose. In cases where it becomes desirable to split, a variety of generic names already exist: e.g., *Systemodon* (Cope, 1881) for the relatively primitive *H. tapirinum* (Cope, 1875) or *Pliolophus* Owen, 1858 for *H. vulpiceps* (Owen, 1858) with equid-type P3/3. The character states attributed to the various species of *Hyracotherium* were difficult to judge because of very large morphological variation in any one assemblage. The states listed reflect a qualitatively assessed mean for each. Detailed statistical study of these characters could result in a better idea of the interrelationships of this complex genus. Family Equidae would be defined on the lingual migration of the $P^3$ paraconule and distal migration of the $P_3$ metaconid after establishment of characters 19, 22, 31A, 37, 44 and 61A. Whether *Hyracotherium vasacciense* should be included here and renamed generically or whether the slightly more advanced *Orohippus* should be taken as the most primitive equid is a matter of taste. The Palaeotheriidae would be defined on upper molar metaloph and lower molar hypolophid formation (45) (paralleled by later members of the Equidae) but where the $P^3$ paraconule and $P_3$ metaconid remain in their primitive positions. Franzen (this volume), using mainly postcranial characters, restricts the family Palaeotheriidae to *Palaeotherium, Plagiolophus,* and closely related genera and relates them more distantly to other equoids than herein. The disagreement is over the polarity of certain character states and one hopes will soon be resolved by further work.

"*The rest*" (Fig. 6.6). This corresponds to and supports the recent concept of Haeckel's Tapiromorpha (Hooker, 1984), which combines as sister groups the Ceratomorpha and the Ancylopoda, deduced independently by Schoch (1983) and Hooker (1984). The latter is extended to include the formerly ceratomorph family Lophiodontidae, and the family Isectolophidae is placed as a primitive sister group to Ancylopoda plus Ceratomorpha, following Hooker (1984).

Similarity of the foot bones was demonstrated by Osborn (1913) when he compared *Pernatherium* and *Eomoropus*. Unfortunately, he did not realize that *Pernatherium* represented the foot of *Lophiodon* (see Stehlin, 1918) and the lophiodont-chalicothere relationship went unnoticed. Ancylopoda are defined by buccal expansion of the $M^3$ parastyle (65) and distal recurving of upper molar protocone and hypocone and of lower molar protoconid and hypoconid (66) (see also Schoch, 1983), although in lophiodonts the recurving of the hypocone and protoconid is not always developed.

The Ceratomorpha is defined on several advancements in "tapiroid" bilophodonty (35, 24, 36 and 62) all except the first paralled by the genus *Isectolophus*. Within the Ceratomorpha, the relationships of the various extinct Asian families differ from existing ideas but more or less reflect Reshetov's polychotomy (1975, Fig. 18) of the latter and their separation from the Helaletidae. The Lophialetidae appears to be paraphyletic, one genus (*Kalakotia*) being primitive with respect to the Ancylopoda-Ceratomorpha split. There is a possible relationship between the Rhodopaginae and Deperetellidae, in contrast to the views of Lucas and Schoch (1981). The remaining relationships are largely unresolved and not very robust, so it is proposed that subfamilies Breviodontinae and Rhodopaginae of the Lophialetidae be raised to family level and the Lophialetidae restricted to just the nominate subfamily (see Reshetov, 1979). Modern ceratomorphs (Tapiroidea and Rhinocerotoidea) are linked by characters 58, 47, and 8, the last two paralleled elsewhere. There is less evidence for cogency of these two modern superfamilies than there is for their type families. However, the Tapiroidea are tentatively defined on character 5A and the Rhinocerotoidea on characters 39, 28 and 59, the last two paralleled elsewhere. Since this manuscript was originally submitted, Prothero et al. (1986) have elucidated more characters in support of a monophyletic Rhinocerotoidea. The Helaletidae appear to be paraphyletic, and *Heptodon* has a sister relationship with the Tapiroidea plus Rhinocerotoidea. The poorly known *Dilophodon* and *Selenaletes* (see Radinsky, 1963, 1966a) may have a similar relationship and demonstrate yet another parallel loss of the $M_3$ hypoconulid. Both the family Isectolophidae and the genus *Kalakotia* (plus *Aulaxolophus*, see Ranga Rao, 1972) are left as undifferentiated members of the Tapiromorpha and do not warrant higher taxonomic status unless more characters are found to unite the Ancylopoda and Ceratomorpha independently from them. Other rank changes proposed herein are to reduce the Ancylopoda and Ceratomorpha to the level of infraorder and reinstate the Tapiromorpha (Haeckel, 1873) as a suborder, but in quite a different sense from the original. Wood (1937) has objected to use of the Tapiromorpha because it has had so many different concepts, but historically so have many other commonly used higher taxa (e.g., Lophiodontidae, Creodonta). The Tapiromorpha are defined by loss of the lower molar lingual postcristid branch (44) and development of the lower molar hypolophid and upper molar metaloph (45), while retaining the primitive position of the optic foramen. In other words, they represent a precocious development in metaloph-hypolophid bilophodonty, which becomes advanced in most members of the group, and is retained to some extent even in the chalicotheres.

### Some taxonomic modifications

Superfamily **Pachynolophoidea** Pavlow, 1888, new rank.

Includes only the nominate family, erected to be of equivalent rank to its sister group, the Equoidea.

Family **Pachynolophidae** Pavlow, 1888
Type genus: *Pachynolophus* Pomel, 1847.

Included genera: The type genus and *Anchilophus* Gervais, 1852, and new genus for "*Hyracotherium*" sp. from Rians (Godinot, 1981).

Diagnosis: Small primitive hippomorphs, $M^1$ ranging in length from 5-11 mm. Dental formula 3.1.4-3.3/3.1.4-3.3. Where known, optic foramen has posterior position in orbit close to posterior group of foramina; and foramen ovale and middle lacerate foramen are confluent. Incisors spatulate, $I_3$ without distal cusp. Post $P^1$ diastema may be open. Postcanine diastema, where known, very long. Incipient to advanced convergence of upper molar paracone and protocone, and moderate convergence of metacone and hypocone. Trigonid joining position of cristid obliqua moderately to extremely lingual. Upper molar paraconule large and metaconid and metastylid well separated. Upper molar premetaconule crista plus prehypocrista and lower molar postcristid sublophoid. Upper molars much broader than long (except in a few advanced species of *Pachynolophus* and *Anchilophus*, which are considered to have reverted to the primitive state).

Range: Early to late Eocene, Europe.

Comment: "*Hyracotherium*" *seekinsi* Morris, 1968, is known only from the upper molars of the holotype from the early Eocene of Mexico. It shares with "*Hyracotherium*" sp. from Rians the same convergent angle of buccal and lingual cusps and the same length-width proportions. However, the transverse crests are less lophoid, there being a notch between the metaconule and hypocone, and the premetaconule crista is missing (Fig. 6.3H-I). Not enough is yet known of this species to decide whether or not it belongs in the Pachynolophidae or represents a somewhat parallel development. Nevertheless, it does not seem to represent a primitive species of *Hyracotherium* as Morris thought.

Superfamily **Lophiodontoidea** Gill, 1872, new rank.

Includes only the nominate family, erected to be of equivalent rank to its sister group, the superfamily Chalicotherioidea. The Lophiodontidae currently comprises *Lophiodon*, *Paralophiodon*, and possibly *Atalonodon* (see Fischer, 1964; Dedieu, 1977).

Another potential candidate for inclusion in this family is *Schizotheriodes* Hough, 1955 (a view independently held by Prothero, Manning, and Hanson, 1986). The genus was based on associated $M^{2-3}$ from the Duchesnean (late Eocene) of the United States. Hough (1955) considered it very like *Eomoropus*, noting its large bulging $M^3$ parastyle, and placed it in the family ?Eomoropidae, but did not note the differences from *Eomoropus*, like the near absence of mesostyle and paraconule. In fact, $M^2$ has a small cingular mesostyle, but neither teeth show any sign of the W-shaped ectoloph typical of *Eomoropus*. Schiebout (1977) used the differences from eomoropids to remove *Schizotheriodes* from the Ancylopoda, but the teeth show two features in common with ancylopods: distally recurved protocone and massive buccally bulging $M^3$ parastyle. Their attachment position of the metaloph to the ectoloph is more mesial than in most chalicotherioids and their near lack of paraconule is like *Lophiodon*. A familial attribution of ?Lophiodontidae would be more appropriate than ?Eomoropidae.

Suborder **Titanotheriomorpha** nov.

Includes only the superfamily Brontotherioidea Marsh, 1873 (rank emended by Hay, 1902), erected to be of equivalent rank to the Hippomorpha and Tapiromorpha.

Diagnosis: Medium to large perissodactyls with primitive tooth and digit counts. Overhanging occiput (often reversed to the primitive state). No $P^4$ metaconule. Moderate convergence of upper molar paracone-protocone and metacone-hypocone, buccal flexing of the centrocrista plus mesostyle, with lingual trigonid joining position of cristid obliqua. Upper molar metaconule and lower preultimate molar hypoconulid (if present) lingual. Tendency to reduce or lose upper molar protoloph, prehypocrista and premetaconule crista, and $M^3$ hypocone.

## Conclusions

This attempt at cladistic analysis of primitive perissodactyls is not intended as the final answer, but is provided as the logical outcome of the new ideas advanced here on some of the character polarities. Further potential changes in polarities together with greater refinement of the data, will result in different phylogenetic arrangements, but one hopes not too different. What are really needed, however, to resolve some of the major polychotomies and weak groupings are new fossil discoveries, particularly in the vicinity of the Palaeocene-Eocene boundary.

## Acknowledgments

I would like to thank the following for access to collections in their care: Mr. J. Alexander, Drs. J. J. Flynn, M. C. McKenna, M. Novacek (American Museum of Natural History), Mmes M. Hugueney, M. Sirven (Faculté des Sciences, Lyon), Dr. M. Weidmann (Musée Géologique de Lausanne), Drs. L. Ginsburg, M. Godinot, D. E. Russell (Muséum National d'Histoire Naturelle), Drs. B. Engesser, J. Hurzeler (Naturhistorisches Museum, Basel), Drs. P. D. Gingerich, G. Gunnell (University of Michigan Museum of Paleontology), Dr. R. Emry, Mr. R. Purdy (United States National Museum), and Dr. J. Ostrom (Yale Peabody Museum). Drs. P. D. Gingerich, M. Godinot, and M. C. McKenna provided useful casts. Drs. C. Humphries, A. B. Smith and Ms. K. Shaw assisted greatly by handling the computer program. Drs. P. M. Butler, J. L. Franzen, P. D. Gingerich, D. Prothero, and R. Schoch contributed stimulating discussion and Dr. A. W. Gentry critically read the manuscript.

## Bibliography

Butler, P. M. (1952a): The milk molars of Perissodactyla, with remarks on molar occlusion. -*Proc. Zool. Soc. London*, 121: 777-817.

Butler, P. M. (1952b): Molarization of premolars in Perissodactyla. -*Proc. Zool. Soc. London*, 121: 819-843.

Cope, E. D. (1875): Systematic catalogue of Vertebrata of the Eocene of New Mexico, collected in 1874. -*U. S. Geogr. Survey West 100 Merid.*, 2:1-37.

Cope, E. D. (1881): New genus of Perissodactyl Diplarthra. -*Amer. Nat.*, 15: 1018.

Cope, E. D. (1889): The Vertebrata of the Swift Current River, II. -*Amer. Nat.*, 23: 151-155.

Dedieu, P. (1977): Sur la systématique des Tapiroidea (Mammalia) de l'Eocène européen. -*C. R. Hebd. Séanc. Acad. Sci. Paris*, (D)284: 2219-2222.

Depéret, C. (1904): Sur les caractères et les affinités du genre *Chasmotherium*. -*Bull. Soc. Géol. Fr.*, (4)4: 569-587.

Fischer, K.-H. (1964): Die tapiroiden Perissodactylen aus der eozän en Braunkohle des Geiseltales. -*Geologie*, 45: 1-101.

Forster Cooper, C. (1932): The genus *Hyracotherium*. A revision and description of new specimens found in England. -*Phil. Trans. Roy. Soc.*, (B)221: 431-448.

Franzen, J. -L. (1968): Revision der Gattung *Palaeotherium* Cuvier, 1804 (Palaeotheriidae, Perissodactyla, Mammalia). -Inaugural-Dissertation zur Erlangung der Doktorwürde der Naturwissenschaftlich- mathematischen Facultät der Albert-Ludwigs Universität zu Freiburg i. Br. 2 vols, 181 pp., 35 pls, 15 tabs, Freiburg.

Franzen, J. -L. (1972): Die taxonomische, phylogenetische, konstructive und funktionelle Bedeutung der relativen Länge des postcaninen Diastems bei den Equoidea. -*Senckenberg. Leth.*, 53: 333-352.

Franzen, J. -L. (1989): Origin and systematic position of the Palaeotheriidae (this volume).

Gervais, P. (1852): *Zoologie et Paléontolo-*

*gie françaises (animaux vertébrés) ou Nouvelles Recherches sur les Animaux Vivants et Fossiles de la France*. 1st Ed., Paris.

Gill, T. (1872): Arrangement of the families of mammals with analytical tables. - *Smithson. Misc. Colls.*, 11(230)(1): i-vi, 1-98.

Gingerich, P. D. (1981): Variation, sexual dimorphism, and social structure in the early Eocene horse *Hyracotherium* (Mammalia, Perissodactyla). -*Paleobiology*, 7: 443-455.

Godinot, M. (1981): Les mammifères de Rians (Eocène inférieur, Provence). - *Palaeovertebrata*, 10: 43-126.

Haeckel, E. (1873): *Natürliche Schöpfungsgeschichte...* Vierte Auflage. Berlin (George Reimer).

Hay, O. P. (1902): Bibliography and catalogue of the fossil Vertebrata of North America. -*Bull. U.S. Geol. Surv.*, 179: i-iii, 1-868.

Hooker, J. J. (1980): The succession of *Hyracotherium* (Perissodactyla, Mammalia) in the English early Eocene. - *Bull. Br. Mus. (Nat. Hist.), (Geol.)* 33: 101-114.

Hooker, J. J. (1984): A primitive ceratomorph (Perissodactyla, Mammalia) from the early Tertiary of Europe. - *Zool. J. Linn. Soc.*, 82: 229-244.

Hough, J. (1955): An upper Eocene fauna from the Sage Creek area, Beaverhead County, Montana. -*J. Paleont.*, 29: 22-36.

Lucas, S. G.. and Schoch, R. M. (1981): The systematics of *Rhodopagus*, a late Eocene hyracodontid (Perissodactyla: Rhinocerotoidea) from China. -*Bull. Geol. Inst. Univ. Uppsala*, (NS) 9: 43-50.

MacFadden, B. J. (1976): Cladistic analysis of primitive equids, with notes on other perissodactyls. -*Syst. Zool.*, 25: 1-14.

Marsh, O. C. (1873): Notice of new Tertiary mammals. -*Amer. J. Sci.*, (3)5: 407-410, 485-488.

Morris, W. J. (1968): A new early Tertiary perissodactyl, *Hyracotherium seekinsi*, from Baja California. -*Contr. Sci. Los Angeles*, 151: 1-11.

Osborn, H. F. (1898): The extinct rhinoceroses. -*Mem. Amer. Mus. Nat. Hist.*, 1: 75-164.

Osborn, H. F. (1913): *Eomoropus*, an American Eocene chalicothere. -*Bull. Amer. Mus. Nat. Hist.*, 32: 261-274.

Osborn, H. F. (1929): The titanotheres of ancient Wyoming, Dakota and Nebraska. 1-2. -*Monogr. U. S. Geol. Surv.*, 55: i-xxiv, 1-953.

Owen, R. (1841): Description of the fossil remains of a mammal (*Hyracotherium leporinum* ) and of a bird (*Lithornis vulturinus* ) from the London Clay. - *Trans. Geog. Soc. London*, (2)6: 203-208.

Owen, R. (1858): Description of a small lophiodont mammal (*Pliolophus vulpiceps* Owen) from the London Clay near Harwich. -*Q. J. Geol. Soc. London*, 14: 54-71.

Pavlow, M. (1888): Etudes sur l'histoire paléontologique des ongulés. II, le développement des Equidae. -*Bull. Soc. Nat. Moscou*, (NS) 2:135-182.

Pomel, A. (1847): Notes sur les mammifères et reptiles fossiles des terrains éocènes de Paris, inférieurs au dépôt gypseux. - *Archs. Sci. Phys. Nat. Genève*, 4: 326-330.

Prothero, D. R., Manning, E., and Hanson, C. B. (1986): The phylogeny of the Rhinocerotoidea (Mammalia, Perissodactyla). -*Zool. J. Linn. Soc.*, 87: 341-366.

Radinsky, L. (1963): Origin and early evolution of North American Tapiroidea.-*Bull. Peabody Mus. Nat. Hist., Yale Univ.*, 17: 1-106.

Radinsky, L. (1964): *Paleomoropus*, a new early Eocene chalicothere (Mammalia, Perissodactyla) and a revision of Eocene chalicotheres. -*Amer. Mus. Novitates*, 2179: 1-28.

Radinsky, L. (1965): Early Tertiary Tapiroidea of Asia. -*Bull. Amer. Mus. Nat. Hist.*, 129: 181-264.

Radinsky, L. ( 1966a): A new genus of early Eocene tapiroid (Mammalia, Perissodactyla). -*J. Paleont.*, 40: 740-742.

Radinsky, L. (1966b): The adaptive radiation of the phenacodontid condylarths and the origin of the Perissodactyla. - *Evolution*, 20: 408-417.

Radinsky, L. (1969): The early evolution of the Perissodactyla. -*Evolution*, 23: 308-328.

Ranga Rao, A. (1972): New mammalian genera and species from the Kalakot Zone of Himalayan foothills near Kalakat, Jammu and Kashmir State, India. -*Spec. Pap. Dir. Geol.*, 1:1-22.

Reshetov, V. Yu. (1975): Obzor rannetretichnykh tapiroobraznykh Mongolii i SSSR. -In: Kramarenko, N. N. (ed): Iskopaemaya fauna i flora Mongolii. -*Trudy Sovmest. Sov. -Mongol. Paleont. Eksped.*, 2: 19-53.

Reshetov, V. Yu. (1979): Rannetretichnye tapiroobraznye Mongolii i SSSR. - *Trudy Sovmest. Sov.-Mongol. Paleont. Eksped.*, 11: 1-141.

Riedl, R. (1978): *Order in living Organisms: a systems analysis of evolution.* (Translated by Jefferies, R. P. S.). Chichester.

Rose, K. D. (1981): The Clarkforkian land-mammal Age and mammalian faunal composition across the Paleocene-Eocene boundary. -*Univ. Mich. Mus. Paleont. Pap. Paleont.*, 26: i-ix, 1-197.

Savage, D. E., Russell, D. E., and Louis, P. (1965): European Eocene Equidae (Perissodactyla). -*Univ. Calif. Publs. Geol. Sci.*, 56:1-94.

Schiebout, J. A. (1977): *Schizotheroides* [sic] (Mammalia, Perissodactyla) from the Oligocene of Trans-Pecos Texas. -*J. Paleont.*, 51: 455-458.

Schoch, R. M. (1983): Relationships of the earliest perissodactyls (Mammalia, Eutheria). -*Abstr. Prog. Geol. Soc. Amer.*, 15(3): 144.

Schoch, R. M. (1989): A brief historical review of perissodactyl classification (this volume, Chapter 2).

Simpson, G. G. (1945): The principles of classification and a classification of mammals. -*Bull. Amer. Mus. Nat. Hist.*, 85: 1-350.

Simpson, G. G. (1952): Notes on British hyracotheres. -*J. Linn. Soc., (Zool.)*42: 195-206.

Stehlin, H. G. (1918): Le *Pernatherium rugosum* P. Gervais. -*Bull. Soc. Géol. Fr.*, (4)18: 123-138.

Swofford, D. L. (1984): PAUP -Phylogenetic analysis using parsimony. -Illinois Nat. Hist. Surv., Champaign (program and user's manual).

Van Valen, L. (1982): Homology and causes. -*J. Morph.*, 173: 305-312.

Wood, H. E. (1934): Revision of the Hyrachyidae. -*Bull. Amer. Mus. Nat. Hist.*, 67: 181-295.

Wood, H. E. (1937): Perissodactyl suborders. -*J. Mamm.*, 18: 106.

# 7. ORIGIN AND SYSTEMATIC POSITION OF THE PALAEOTHERIIDAE

## JENS LORENZ FRANZEN

Palaeotheriidae are only known from the uppermost middle Eocene until the lower Oligocene of western and central Europe. Taxonomically, the family is restricted to the genera *Palaeotherium, Paraplagiolophus, Plagiolophus, Leptolophus, Catabrotherium,* and *Pseudopalaeotherium,* whereas the true European Eocene horses comprise the genera *Hyracotherium, Propachynolophus, Pachynolophus, Propalaeotherium, Lophiotherium,* and *Anchilophus*. Characters are evaluated that clearly demonstrate that the palaeotheres, although stratigraphically younger, are still too plesiomorphic to be regarded as descendants of *Hyracotherium,* hence as equids. Except for this, they are characterized by elongated cervicals, and by metacarpals that are longer than the metatarsals. On the other hand, palaeotheres and equids show the same apomorphic position of the foramen opticum and the foramen ovale, and are therefore systematically united as Equoidea. Some evidence points to an African origin of the palaeotheres and the Equoidea as a whole, who then immigrated into Europe, probably by way of the Iberian Peninsula.

## Introduction

The Palaeotheriidae are among the first fossil mammals described by Georges Cuvier at the beginning of the last century from the famous late Eocene "gypse de Montmartre." Until recently, they were known only from the southwestern, western and central parts of Europe. The earliest palaeotheres appear at the end of the middle Eocene in Europe (Franzen and Haubold, 1986a, b). Later immigrants arrive within the Upper Eocene until the base of the Headonian (Franzen, 1968, p.144, 147; Bosma, 1974, p. 105). The last one is *Pseudopalaeotherium longirostratum* Franzen (1972), immigrating at the beginning of the Oligocene after the "Grande Coupure" of Stehlin (1909). It was accompanied by mammalian taxa of obviously Asian origin. Consequently, the whole family has been regarded as coming from central or southern Asia (Franzen, 1972a, p. 323-329), perhaps by way of southeast Europe (Heissig, 1979). A recent revision of the Paleogene mammals from Bulgaria, however, did not support this hypothesis (Nikolov and Heissig, 1985).

Since the beginning of evolutionary paleontology, the Palaeotheriidae have been tied with the phylogenetic tree of the horses. T. H. Huxley first regarded them as the earliest representatives. Later, after seeing the North American record of fossil horses with Marsh at New Haven, he considered them an early European side-branch of the horse family (Franzen, 1984). Although P. M. Butler (1952a, b), Savage, Russell, and Louis (1965), as well as myself (1968, 1972b) brought forward good reasons to consider the Palaeotheriidae as a family absolutely independent from the equids, they are still confused with European Eocene horses in most textbooks, as well as by Remy (1976), MacFadden (1976), and recently by Hooker (this volume, Chapter 6).

The decisive argument of MacFadden (1976) for including the palaeotheres in the equids was taken from the position of the optic foramen within the orbital region. Although this statement referred to a skull of *Pachynolophus,* which is in fact an equid and not a palaeothere, investigations of Brunet (1974, pp. 239-240) and Remy (1985, pp. 185, 187, 204, 217) have shown that true palaeotheres obviously display the same apomorphic position of the foramen ovale and the foramen opticum. Evidently Mac-

Fadden was right for the wrong reason, although the apomorphic position of the two foramina must not necessarily lead to an inclusion of the palaeotheres in the equids.

Remy (1976), admitting that the palaeotheres *sensu stricto* cannot be derived from any known species of *Hyracotherium*, claimed that some unknown earlier hyracothere was primitive enough to serve as a common ancestor not only for equids and palaeotheres but also for other (all?) Perissodactyla. If such a species should really turn up one day, the question would be whether it could still be called *Hyracotherium*, hence a member of the Equidae. Except for this, the Palaeotheriidae *sensu lato* [comprising European Eocene Equidae derivable from *Hyracotherium* together with true palaeotheres, which are not] is evidently a paraphyletic taxon. It should therefore be avoided.

J. J. Hooker presented a cladogram at the Edmonton congress [see Preface] showing brontotheres (*Lambdotherium*), Tapiroidea and Equidae, the last mixed with the palaeotheres *Palaeotherium* and *Plagiolophus*. But this was not a cladogram as it should be, founded on bifurcations and synapomorphies only. Instead of this, it was based mostly on characters resulting from convergent or parallel developments, as well as on at least ten reversals. There was no synapomorphy uniting *Plagiolophus* with *Propalaeotherium hassiacum* and *Propalaeotherium isselanum* as it was shown on the cladogram, and there was only one [see also Hooker, this volume, Chapter 6] associating *Palaeotherium* with *Hyracotherium leporinum*, whose synapomorphic status appears highly questionable. This character was a further migration of the metaconules of the upper molars in a mesial direction from *Hyracotherium* aff. *vulpiceps* to *H. leporinum* and *Palaeotherium*. So the argument for this kind of cladogram remains quite unclear, although its author is referring to "a combination of outgroup comparison, simple to complex morphology, intraspecific variation pattern and occlusion."

What I want to do here is to point out arguments that in my view clearly prove that the Palaeotheriidae in fact cannot be considered as descendants of *Hyracotherium*, hence as equids.

**Discussion**

Setting as premises that *Hyracotherium* is the earliest representative of the horse family, and that one species of this genus gave rise to all later equids, it is possible to figure out the evolutionary polarity of characters within the equids, although the continual transition from lower Eocene *Hyracotherium* to middle Eocene *Orohippus* as well as that from *Orohippus* to *Mesohippus* has still to be substantiated by fossil evidence. Nevertheless, viewing the trends of evolution within the horse family it seems feasible to determine the phylogenetic and systematic position of what I would call the Palaeotheriidae *sensu stricto* (Franzen, 1968). Apart from *Palaeotherium* Cuvier (1804) itself, these true palaeotheres would be comprised of the genera *Plagiolophus* Pomel (1847), *Paraplagiolophus* Depéret (1917), *Leptolophus* Remy (1965), *Pseudopalaeotherium* Franzen (1972), and *Cantabrotherium* Casanova-Cladellas and Santafé-Llopis (1987). Besides *Hyracotherium* Owen (1841), the true European Eocene equids would comprise the genera *Propachynolophus* Lemoine (1891), *Pachynolophus* Pomel (1847), *Propalaeotherium* Gervais (1849), *Lophiotherium* Gervais (1849), and *Anchilophus* Gervais (1852).

Although stratigraphically younger, ranging from the uppermost middle Eocene well up into the Oligocene, the Palaeotheriidae *sensu stricto* appear to be more primitive compared to *Hyracotherium* in the following characters:

1. The postcanine diastema: Whereas *Hyracotherium* from the early Eocene of Europe as well as North America displays an extraordinary elongation of the postcanine diastema of the upper and the lower jaws compared to the length of $P_2$-$M_3$, the diastema is still short within the earliest

Palaeotheriidae. Only the stratigraphically younger taxa of Plagiolophinae along with *Pseudopalaeotherium* show a subsequent increase of diastema width converging with that of much later anchitheriine horses. The functional explanation of the different developments of the postcanine diastema is discussed in detail by Franzen (1972b).

2. Molarization of premolars: While the buccal cusp of $P^2$ is still unsplit within the most primitive species of *Palaeotherium* (*P. castrense* and *P. ruetimeyeri*), it is already differentiated into a paracone and metacone in *Hyracotherium*, as it is with *Paraplagiolophus*, *Plagiolophus*, and *Leptolophus*.

3. Tuber coxae: While the tuber coxae is already an elongated ridgelike structure in *Hyracotherium* which provides a strong and enlarged area of attachment for the musculus tensor fasciae latae, it is still simple in *Palaeotherium* (e.g., *P. magnum*, *P. crassum*) as it is with the Phenacodontidae (e.g. *Phenacodus primaevus*).

4. Crista ilica: Whereas the crista ilica remains convex in *Palaeotherium* (e.g., *P. magnum*, *P. crassum*) as it is with the Phenacodontidae (e.g., *Phenacodus primaevus*), it is already concave in *Hyracotherium*, as in *Equus*, thus providing a direct connection between the musculus longissimus dorsi and the gluteus musculature.

5. Trochanter major: The trochanter major of the femur is already much higher than the caput femoris in *Hyracotherium*, offering a longer lever for the gluteus musculature, while it is still low in the Palaeotheriidae (e.g., *Palaeotherium magnum*, *Palaeotherium medium*, *Palaeotherium crassum*, *Plagiolophus minor*) as is the case with the Phenacodontidae (e.g., *Phenacodus primaevus*).

While these characters clearly indicate that the Palaeotheriidae are still too plesiomorphic to be regarded as descendents of *Hyracotherium*, there are other features pointing to quite another direction of evolution (autapomorphies) of the palaeotheres compared with the equids, resulting in a different type of body construction:

6. Relative length of metacarpals versus metatarsals: In contrast with equids, all of which display longer metatarsals than metacarpals, all Palaeotheriidae *sensu stricto* of which these bones are known show relatively longer metacarpals. This is demonstrated, for example, by the skeletons of *Palaeotherium magnum* from Mormoiron (Roman, 1922, Pl. 1) and of *Plagiolophus minor* from Pantin (Cuvier, 1836, Pl. 115), as well as by composite foot skeletons of *Palaeotherium medium*, *Palaeotherium curtum*, *Plagiolophus fraasi*, and *Plagiolophus minor* from Frohnstetten (Stehlin 1938, Text-figs. 5-6, 11-16), and of *Plagiolophus annectens* from Euzet-les-Bains (Depéret, 1917, Pl. 11).

7. Relative length of cervical versus thoracic vertebral series: Palaeotheres display a remarkable elongation of their cervicals relative to the series of thoracic vertebrae as shown by the skeletons of *Palaeotherium magnum* from Mormoiron (Roman, 1922, Pl. 1) and of *Plagiolophus minor* from Pantin (Cuvier, 1836, Pl. 115). Thus the series of cervicals is arriving at a length of more than 100% of that of the thoracic vertebrae in *Palaeotherium magnum* and at about 65% in *Plagiolophus minor*. In *Hyracotherium*, this relationship comes to only 42% (*H. vasacciense venticolum*), and to 40-50% in the European hyracothere *Propalaeotherium*.

Thus, palaeotheres are not descendants of *Hyracotherium*, and hence not equids. Since they appear suddenly in Europe at the turnover from middle to late Eocene (Franzen and Haubold, 1986a, 1986b), the question arises of where the palaeotheres originally came from, paleogeographically as well as phylogenetically. One possibility is that they immigrated from Asia (Franzen, 1972a; Heissig, 1979), but no related taxon has ever been discovered from that continent. During the whole Paleocene and Eocene, Asia was isolated from Europe by the Turgai Sea.

Another possibility is to derive the palaeotheres from some early brontothere, like *Lambdotherium*, as a rather late invasion from North America by way of the De Geers or the Thule land bridge. But this hypothesis is not corroborated by the facts. The facts are that the direct faunal exchange between North America and Europe across the North Atlantic was evidently restricted to the early Eocene as is demonstrated by the diverging evolution of horses and other mammals since then on both continents, and by the lack of any later immigration coming directly from North America to Europe or vice versa.

Other facts are purely morphological. They indicate that *Lambdotherium*, although it is the most primitive brontothere known so far and older than the oldest known palaeotheres, is already too advanced in some characters to be regarded as their ancestor. These features consist of: 1) strongly developed mesostyles in the upper molars; 2) a hypoconulid of the $M_3$ that is already hook-like; 3) a postcanine diastema that is already too elongated for an ancestor of the palaeotheres.

A third possibility is that the palaeotheres entered Europe from the south by way of the Iberian Peninsula. There is in fact some evidence corroborating the idea of an African origin of the palaeotheres (and of the Equoidea as a whole):

1. The African and the European continents were not widely separated during most of the Paleogene (Smith *et al.*, 1982, pp. 22, 26; Owen, 1983, pp. 62-63).

2. As Fischer has shown (this volume, Chapter 4), Hyracoidea should be included in the Perissodactyla again. This would point to an African origin of the Perissodactyla, hence of the palaeotheres.

3. A mammalian faunule recently discovered from the early Eocene of Tunisia is also suggestive of paleobiogeographic relationships between the African and the European continent at the end of the Paleocene or the beginning of the Eocene (Hartenberger *et al.*, 1985; Hartenberger, 1986; Crochet, 1986).

4. On the Iberian Peninsula, taxa of palaeotheres not known from other parts of Europe occur (Casanovas-Cladellas, 1975; Casanovas-Cladellas and Santafé-Llopis, 1980, 1987). This is either due to endemic development or to immigrations coming from outside of Europe, which would mean Africa, or to both.

5. An immigration of certain palaeotheres from the region of Spain into France was already suggested by Casanovas-Cladellas and Santafé-Llopis (1982).

6. A sequential immigration of equids and palaeotheres into Europe would be exactly parallel with those of the primates, where the Notharctidae, arriving at the base of the Eocene in Europe, are followed by the Adapidae *sensu stricto* at the turnover from the middle to the late Eocene. Both are most probably of African origin (Franzen, 1987).

Concerning the phylogenetic origin of the palaeotheres, the Phenacodontidae can only be considered a model of remote predecessors. There is no continual transition known from the fossil record connecting one family with the other. And as we have seen, the paleobiogeographic evidence is also not supporting a direct relationship of these two families.

The elongation of the metacarpals with respect to the metatarsals would rather point to a sister-group relationship of the palaeotheres with brontotheres and chalicotheres than with equids or any other group of early Perissodactyls. But one must concede that characters like length proportions, although they may suffice to disprove phylogenetic affinities, are certainly not complex enough to serve as synapomorphies in the sense of Hennig (1966). Looking at the dentition, however, there is another character linking palaeotheres with brontotheres and chalicotheres. This is the W-shaped ectolophs of the upper molars. But here again, the question arises whether this feature is really complex enough to be regarded as a synapomorphy. Interestingly, it also characterizes the Hyracoidea, and it evolved at least two times independently in the North American and in the European equids.

## Conclusion

In my view, it is therefore better to consider the similarities between palaeotheres, brontotheres, and chalicotheres as results of paraphyletic development. Together with characters 1-5, they make it impossible, however, to derive the palaeotheres from *Hyracotherium*. Hence, the Palaeotheriidae have to be considered as a family of their own, absolutely independent from all kinds of Equidae. Accepting, on the other hand, the positions of the foramen ovale and of the foramen opticum as synapomorphies, these would unite Equidae and Palaeotheriidae in a superfamily Equoidea.

## Acknowledgments

This study is a result of a research program supported by the Deutsche Forschungsgemeinschaft (Fr 396/5-1).

## Bibliography

Bosma, A. (1974): Rodent biostratigraphy of the Eocene-Oligocene transitional strata of the Isle of Wright. -*Utrecht Micropaleont. Bull.*, (spec. publ. no. 1): 1-128.

Brunet, M. (1974): Le premier crâne attribué au genre *Entelodon* (Artiodactyla) appartient en réalité au genre *Palaeotherium* (Perissodactyla). -*Ann. Paléont. (Vert.)*, 60(2): 235-242.

Butler, P. M. (1952a): The milk molars of Perissodactyla, with remarks on molar occlusion. -*Proc. Zool. Soc. London*, 121(4): 777-817.

Butler, P. M. (1952b): Molarisation of the premolars in the Perissodactyla. -*Proc. Zool. Soc. London*, 121(4): 819-843.

Casanovas-Cladellas, M. L. (1975): Datos sobre los Perisodáctilos del yacimiento de Roc de Santa; la nueva especie *Palaeotherium crusafonti.-Acta. Geol. Hispanica*, 10(3): 121-126.

Casanovas-Cladellas, M. L., and Santafé-Llopis, J. V. (1980): El *Palaeotherium* de talla grande (Palaeotheriidae, Perissodactyla) del yacimiento ludiense de Sossis (Tremp, Lérida). -*Bol. Inf. Inst. Palaeont. Sabadell*, 12(1-2): 21-29.

Casanovas-Cladellas, M. L., and Santafé-Llopis, J. V. (1982): Los Palaeotheriinae (Perissodactyla, Mammalia) de talla media del yacimiento ludiense de Sossis (Tremp, Lérida). -*Paleontologia i Evolució*, 17: 15-20.

Casanovas-Cladellas, M. L., and Santafé-Llopis, J. V. (1987): *Cantabrotherium truyolsi*, n.gen., n.sp. (Palaeotheriidae, Perissodactyla), un exemple d'endemisme dans le Paléogène Ibérique.-*Münchner Geowiss. Abh. (A)* 10: 243-252.

Crochet, J.-Y. (1986): *Kasserinotherium tunisiense* nov. gen., nov. sp., troisième marsupial découvert en Afrique (Eocène inférieur de Tunisie). -*C.R. Acad. Sci. Paris*, (2) 302 (14): 923-926.

Cuvier, G. (1804): Sur les espèces d'animaux dont proviennent les os fossiles répandus dans la pierre à plâtre des environs de Paris.-*Ann. Mus. Nat. Hist. Paris, III*: 275-303.

Cuvier, G. (1836): *Recherches sur les ossemens fossiles.* 4th ed., Paris.

Depéret, C. (1917): Monographie de la faune mammifères fossiles du Ludien inférieur d'Euzèt-les-Bains (Gard). -*Ann. Univ. Lyon, (n.s., I. Sci., Méd.)*, 40: viii, 1-290; Lyon, Paris.

Fischer, M.S. (1989): Hyracoids, the sister-group of perissodactyls (this volume, Chapter 4).

Franzen, J. L. (1968): Revision der Gattung *Palaeotherium* Cuvier (1804) (Palaeotheriidae, Perissodactyla, Mammalia). -Inauguraldiss. Naturwiss. Fak. Albert-Ludwigs-Univ., 2 vols., 1-181.

Franzen, J. L. (1972a): *Pseudopalaeotherium longirostratum* n. g., n. sp. (Perissodactyla, Mammalia) aus dem unterstampischen Kalkmergel von Ronzon (Frankreich). -*Senckenbergiana Lethaea*, 53(5): 315-331.

Franzen, J. L. (1972b): Die taxonomische,

phylogenetische, konstruktive und funktionelle Bedeutung der relativen Länge des postcaninen Diastems bei den Equoidea. -Senckenbergiana Lethaea, 53(5): 333-352.

Franzen, J. L. (1984): Die Stammesgeschichte der Pferde in ihrer wissenschaftshistorischen Entwicklung. - Natur und Museum, 114(6): 149-162.

Franzen, J. L. (1987): Ein neuer Primate aus dem Mitteleozän der Grube Messel (Deutschland, S-Hessen). -Courier Forschungsinst. Senckenberg, 91: 151-187.

Franzen, J. L. and Haubold, H. (1986a): The middle Eocene of European mammalian stratigraphy. Definition of the Geiseltalian. -Modern Geology, 10: 159-170.

Franzen, J. L., and Haubold, H. (1986b): Revision der Equoidea aus den eozänen Braunkohlen des Geiseltales bei Halle (DDR). -Palaeovertebrata, 16(1): 1-34.

Franzen, J. L., and Haubold, H. (1986c): Ein neuer Condylarthre und ein Tillodontier (Mammalia) aus dem Mitteleozän des Geiseltales. -Palaeovertebrata, 16(1): 34-53.

Gervais, P. (1849): Recherches sur les mammifères fossiles des genres Palaeotherium et Lophiodon, et sur les autres animaux de la même classe l'on a trouvés avec eux dans le midi de la France.-C. R. Acad. Sci. (Paris), 29: 381-384.

Gervais, P. (1852): Zoologie et paléontologie fran-çaise (animaux vertébrés). 3 vols., Paris.

Hartenberger, J. -L. (1986): Hypothèse paléontologique sur l'origine des Macroscelidea (Mammalia). -C. R. Acad. Sci. (Paris), (2) 302 (5): 247-249.

Hartenberger, J. -L., Martinez, C., and Said, A. Ben (1985): Découverte de Mammifères d'âge Éocène inférieur en Tunisie Centrale. -C. R. Acad. Sci. (Paris), (2) 301(9): 649-652.

Heissig, K. (1979): Die hypothetische Rolle Südosteuropas bei den Säugetierwanderungen im Eozän und Oligozän. - N. Jb. Geol. Paläontol., 2: 83-96.

Hennig, W. (1966): Phylogenetic Systematics. Urbana, Ill. (Univ. of Illinois Press).

Hooker, J.J. (1989): Character polarities in early perissodactyls and their significance for Hyracotherium and infraordinal relationships (this volume, Chapter 6).

Lemoine, V. (1891): Étude d'ensemble sur les dents des mammifères fossiles des environs des Reims. -Bull. Soc. Géol. France, (3) 19: 263-290.

MacFadden, B. (1976): Cladistic analysis of primitive equids, with notes on other perissodactyls. -Syst. Zool., 25(1): 1-14.

Nikolov, I., and Heissig, K. (1985): Fossile Säugetiere aus dem Obereozän und Unteroligozän Bulgariens und ihre Bedeutung für die Paläobiogeographie. -Mitt. Bayer. Staatsslg. Paläontol. und Histor. Geol., 25: 61-79.

Owen, H. G. (1983): Atlas of continental displacement, 200 million years to the present. Cambridge (Cambridge Univ. Press).

Owen, R. (1841): Description of fossil remains of a mammal (Hyracotherium leporinum) and of a bird (Lithornis vulturinus) from the London Clay. -Trans. Geog. Soc. London, (2) 6: 203-208.

Pomel, A. (1847): Note critique sur les caractères et les limites du genre Palaeotherium. -Arch. Sci. Phys. Nat., 5: 200-207.

Remy, J. A. (1965): Un nouveau genre de paléothéridé (Perissodactyla) de l'Éocène supérieur du Midi de la France. -C.R. Acad. Sci. Paris, 260: 4362-4364.

Remy, J. A. (1976): Étude comparative des structures dentaires chez les Palaeotheriidae et divers autres Perissodactyles fossiles. -Thèse 3ème Cycle, Univ. Louis Pasteur. -1-207; Strasbourg.

Remy, J. A. (1985): Nouveaux Gisements de Mammifères et Reptiles dans les Grès de Célas (Eocène Sup. du Gard). Étude des Palaeothériidés (Perissodactyla, Mammalia). -Palaeontographica, (A) 189: 1-225.

Roman, F. (1922): Monographie de la faune demammifères de Mormoiron (Vaucluse). Ludien supérieur. -*Mém. Soc. Géol. France, Paléont.*, 25(1): 1-39.

Savage, D. E., Russell, D. E., and Louis, P (1965): European Eocene Equidae (Perissodactyla). -*Univ. Calif. Publ. Geol. Sci.*, 56: v, 1-94.

Smith, A. G., Hurley, A.M., and Briden, J. C. (1982): *Paläokontinentale Weltkarten des Phanerozoikums*. Stuttgart (Enke).

Stehlin, H. G. (1909): Remarques sur les faunules de mammifères des couches éocènes et oligocènes du bassin de Paris. -*Bull. Soc. Géol. France*, (4)9: 488-520.

Stehlin, H. G. (1938): Zur Charakteristik einiger Palaeotheriumarten des oberen Ludien. -*Eclogae Geol. Helv.*, 31: 263-292.

# 8. PHYLOGENY OF THE FAMILY EQUIDAE

## ROBERT L. EVANDER

### Introduction

The phylogenetic progression of horses through geologic time was the first ancestor-descendant series recognized in the fossil record (Huxley, 1870), and was one of the first confirmations of evolutionary "descent with change" (Darwin, 1859). In the years since Huxley first identified fossil ancestors of the living *Equus*, the phylogeny of the Equidae has been periodically refined by leading paleontologists: Marsh (1874), Gidley (1907), Osborn (1918), Matthew (1926), Stirton (1940), and Simpson (1951). The resulting progression is decidedly more elaborate and detailed than Huxley's phylogeny, and still stands as convincing evidence for organic evolution.

A modern refinement is now due. Since the works of Hennig (1965, 1966), the field of taxonomy has undergone a major theoretical upheaval. New definitions have been offered for old terminology and new interpretations have been offered for old data. Although several recent studies have treated equids cladistically, no one has attempted a comprehensive review of the family using the criterion of derived character status to define monophyletic groups. The fossil record of horses tells a compelling story of evolutionary change. The interpretation of this record should have the advantage of lucid presentation such as cladistic analysis affords.

### Definitions and assumptions

Although I refer to the various operating taxonomic units by their generic names only, the operating taxonomic units of this phylogenetic analysis are species. The concept of a biological species lacks a geological time dimension, because testing of the species concept requires the synchroneity of living forms. A species is thus the state of a lineage at one instant in time. All of the species examined in this study approach this standard of synchroneity, as the species are based on either unusually complete individuals or on quarry collections of less complete individuals.

A monophyletic group consists of an ancestral species and all descendant species. Hennig (1965, 1966) has shown that only certain states of a character are useful in defining monophyletic groups; that only derived, or advanced, character states may properly define monophyletic groups.

I will rely principally on outgroup comparison and the age distribution of character states for determining character polarities. Arnold (1981) pointed out several methods of determining character polarities. No single method of determining character polarities is either absolutely reliable or universally applicable. The methods of outgroup comparison and age distribution of character states are both widely applicable, and the method of outgroup comparison is generally regarded as highly reliable. Where possible, I will also bring functional considerations and ontogenetic data to bear on polarity decisions.

Once character-polarity decisions have been made, and character distributions have been surveyed, species can be assembled into monophyletic groups based upon the progression of shared, derived character states. Character conflicts may be resolved unambiguously by giving more import to well-established characters with clear homologies and widely-recognized polarities. For instance, a tooth character should be accorded more importance than a particular configuration of the facial fossae, simply because the homologies of equid teeth are long established (Osborn, 1918; Stirton, 1941) and tooth character polarities correlate with a large body of knowledge on mammalian tooth evolution, beginning with Osborn (1907). The homologies of facial fossae, on the other hand, are poorly understood, and hypotheses of character polarity are only beginning to emerge.

A phylogenetic tree is a specific hypothesis of ancestry and descent between two or more taxa. A phylogenetic tree is a more specific statement than a cladogram, and is more easily falsified. Thus a phylogenetic tree is a more desirable scientific hypothesis than a cladogram (Popper, 1963).

A concise phylogenetic classification can be generated from a cladogram if the concept of a plesion is employed. Hennig (1965, 1966) suggested that only monophyletic groups be recognized as named taxa, and that sister groups be classified on the same level in a taxonomic hierarchy. The problem with Hennig's suggestions are that many monophyletic taxa are also monotypic, resulting in an inordinate number of levels in the taxonomic hierarchy. Patterson and Rosen (1975, p. 160) introduced the term plesion to refer to "groups or species, sequenced in a classification according to the convention that each group is the . . . sister-group of all those . . . that succeed it." Use of the concept of the plesion enables practical and stable, yet exhaustively phylogenetic classifications.

The following institutional abbreviations are used:

AMNH, American Museum of Natural History, New York;
F:AM, Frick American Mammals, in the AMNH;
UCR, University of California, Riverside;
UNSM, United States National Museum, Smithsonian Institution, Washington, D.C.

## Materials

This study focuses on eighteen species of fossil horses and the modern domestic horse. Many species of fossil horse are very poorly known. Descriptions are usually confined to the dentitions, and often only the upper cheek teeth are known. Therefore, I decided to accomplish a survey of fossil forms by concentrating upon the best-known species in each of several classically-recognized genera. Many of these classically-recognized genera are gradual in concept, and may not possess a rigorous phylogenetic identity. But all of these genera possess a distinct morphologic identity, so that a broad survey of the variety of forms present in the family will be reviewed.

My attempt to base several horse genera on exemplary specimens of a single species was not always successful; the specimens simply did not exist. In such cases, I was dependent upon either literature descriptions of quarry populations or upon specimens of congeneric species. A list of the genera surveyed, the particular species upon which the study focused, and the specific specimens that were studied follows. Where literature descriptions or congeneric species proved necessary, I have also listed these sources or specimens.

*Hyracotherium* was based on the species of *Hyracotherium vasacciense* (Cope, 1872), and particularly AMNH 4832. The specific identification was that of Kitts (1956). The skull and mandible of AMNH 4832 were removed from display and reprepared for this study. My knowledge of the postcranials came from Kitts (1956). AMNH 14347, a cast of the holotype of *H. leporinum* Owen (1841), which is in turn the type species of the genus, was used as a supplementary specimen.

*Orohippus* was based on the species *Orohippus pumilis* (Marsh, 1872), and particularly AMNH 12648. The specific identification was that of Kitts (1957). This mounted specimen is on display at the

American Museum of Natural History; features not visible on the mounted specimen were obtained from Kitts (1957). AMNH 12120, the holotype of *O. progressus* Granger (1908), and AMNH 96656, *Orohippus* sp., were used as supplementary specimens.

*Epihippus* was based on the species *Epihippus gracilis* (Marsh, 1871), and particularly AMNH 2042 and AMNH 2066. Specific diagnoses follow MacFadden (1980). *Mesohippus* was based on the type species *Mesohippus bairdi* (Leidy, 1850); and particularly F:AM 74001, F:AM 74002, F:AM 74003, F:AM 74004, F:AM 74006, and F:AM 74060.

*Miohippus* was based on the species *Miohippus gemmarosae* (Osborn, 1918), and particularly F:AM 109852, F:AM 109583, and the holotype, AMNH 13808. Supplementary specimens included AMNH 1196, the holotype of *M. intermedius* (Osborn and Wortman, 1895); and AMNH 1218, a referred specimen of *M. validus* (Osborn, 1904). Osborn (1918) incorrectly identified the later specimen as the holotype of the species.

*Anchitherium* was based on the type species *Anchitherium aurelianense* (Cuvier, 1812), particularly the population from Sansan that was described by both Kowalevsky (1873) and Filhol (1891). I did have available four casts of dentitions of specimens from Sansan: AMNH 10416, AMNH 111746, AMNH 111747, and AMNH 111748.

*Hypohippus* was based on the species *Hypohippus osborni* Gidley (1907), and particularly the holotype, AMNH 9407, and the paratype, AMNH 9395. Osborn (1918) fully described these two specimens.

*Megahippus* was based on the species *Megahippus mckennai* Tedford and Alf (1962), and particularly AMNH 109793 (a cast of UCR 21278), and UNSM 175375 (I had available the originals of the postcranials and AMNH 104951, a cast of the skull and mandible).

*Kalobatippus* was based on the species *Kalobatippus agatensis* Osborn (1918); and particularly AMNH 109854, and the holotype, AMNH 14211. AMNH 7269, the holotype of *K. praestans* (Cope, 1879), which is in turn the type species of the genus, was used as a supplementary specimen. *Kalobatippus* was erected by Osborn (1918). Although subsequently ignored by Stirton (1940) and Simpson (1951), *Kalobatippus* was included as an important intermediate between *Miohippus* and more advanced horses.

*Archaeohippus* was based on the species *Archaeohippus penultimus* (Matthew, 1924), and particularly F:AM 71650 and the holotype AMNH 18950.

*"Parahippus"* was based on the species *"Parahippus" pawniensis* Gidley (1907), and particularly F:AM 109857, F:AM 109858, F:AM 109859, and the holotype, AMNH 9085. I followed Matthew (1924), Stirton (1940) and Simpson (1951) in regarding *"Parahippus"* as a grade of advanced, low-crowned horses. The genus *Parahippus* is very poorly typified, and I am not at all confident that the type species is demonstrably a member of this grade of horses.

*"Merychippus"* was based on the species *"Merychippus" primus* (Osborn, 1918), and particularly the huge population of that species from Thomson Quarry (Skinner *et al.*, 1977). For the purposes of this analysis, *"Merychippus"* was considered to be a grade of mesodont horses. I have recommended elsewhere (Evander, 1986) that the usage of the taxon *Merychippus* be severely restricted.

*Protohippus* was based on the species *Protohippus simus* (Gidley, 1906), and particularly on F:AM 60353, and the holotype, AMNH 9820.

*Pliohippus* was based on the species *Pliohippus mirabilis* (Leidy, 1858), and particularly on F:AM 60801, and the holotype, AMNH 10774.

*Dinohippus* was based on the type species *Dinohippus leidyanus* (Osborn, 1918); and particularly the holotype, AMNH 17224. Osborn (1918) fully described and illustrated this specimen.

*Equus* was based on the modern domestic horse, *Equus caballus*, and particularly AMNH 16274.

*Hipparion* was based on the species *Hipparion shirleyi* MacFadden (1984); and particularly on F:AM 99384, and the holotype, F:AM 73950.

*Neohipparion* was based on the species *Neohipparion affine* (Leidy, 1969), and particularly AMNH 9815. AMNH 9815 is the holotype of *N. whitneyi*, and is mounted and on display at the American Museum of Natural History. Characters that were not visible on the mounted specimen were assemlbled using Osborn (1918).

*Pseudohipparion* was based on the species *Pseudohipparion gratum* (Leidy, 1869); and particularly F:AM 70000; F:AM 70003; F:AM 70004; F:AM 70005; and a cast of the holotype, AMNH 19788.

### Character polarity decisions
#### Foramen ovale

The specialized condition of the foramen ovale in the Family Equidae was first noted by Edinger and Kitts (1954). MacFadden (1976) has recently demonstrated the utility of this character in definition of the family. My own survey of 198 eutherian genera in 19 orders (Simpson, 1945) shows that the 'separate' condition is widely distributed in 18 mammalian orders; the foramen ovale is "confluent" within the Sirenia. The "confluent" condition is present along with the "separate" condition in four placental orders: Artiodactyla, Perissodactyla, Proboscidea, and Rodentia. Based on outgroup comparison, the "separate" condition of the foramen ovale, being present in all mammalian orders surveyed except the Sirenia, is judged primitive. It is then clear that the "confluent" condition must have arisen separately several times.

A true foramen ovale is known to exist only in eutherian mammals. The mandibular nerve of more primitive forms left the braincase via the "foramen pseudovale," located on the suture of the periotic (=petrosal) and epipterygoid (= temporal wing of sphenoid). MacIntyre (1967) traced this homologue of the foramen ovale back into the Mesozoic, where it was recognizable in both early mammals and theraspids. All Mesozoic mammals for which the basicranial anatomy is known (Kielan-Jaworowska and Trofimov, 1980; Kielan-Jaworowska, 1981) display a "separate" foramen ovale or pseudovale. Within the Cenozoic, a "confluent" foramen ovale is unknown among surveyed Paleocene forms. It is present in 6% of surveyed Eocene genera, 9% of surveyed Oligocene genera, and approximately 30% of surveyed Neogene genera. Thus, on the basis of the time distribution of character states, the "confluent" condition of the foramen ovale is judged as derived.

Ontogenetic studies of the chondocranium of *Equus caballus* provided additional evidence that the "separate" condition of the foramen ovale is primitive. The foramen ovale is well separate from the middle lacerate foramen in the chondocranium of the fetal horses (De Beer, 1937).

#### Basicranial axis

The basicranial axis of horses is uniquely specialized compared to all other living mammals: it is a "narrow," elongate bony strut flanked on either side by a wide, open space. The internal carotid artery, internal jugular vein, and the mandibular, glossopharyngeal, vagus, and accessory nerves all pass through the large "lacerate" foramen, which is located lateral to the basicranial axis of horses. Paleontologically, this specialized condition of the basicranial axis can be traced back in the Family Equidae through Oligocene time. Oligocene horses for which the crania are available demonstrate the laterally "flaring" basicranial axis typical of other mammals. Thus, both the outgroup-comparison method and the time-distribution method of character polarity determination indicate the derived nature of the "narrow" basicranial axis.

MacPhee (1981) recently reviewed the ontogeny of the basicranial region in mammals. Mammalian embryos typically demonstrate a "flaring" basicranial axis, as might be expected. De Beer (1937) showed that the chondocranium of horses also possess a "flaring" basicranial axis. Thus, ontogenetic data supports the primitive condition of the "flaring" basicranial axis. The narrow basicranial axis that was present in most horse species is derived.

*Postorbital bar*

It is widely recognized that the postorbital bar is "absent" primitively in mammals. It is present only as a specialized, secondary condition. In conjunction with my survey of the foramen ovale, I surveyed 212 genera of fossil and recent mammals for the presence of a postorbital bar. A "complete" postorbital bar is present in ochotonid lagomorphs, strepsirhine and haplorhine primates, *Calogale, Barbourofelis, Pliohyrax, Brachyhops, Daeodon,* oreodont artiodactyls, tylopod artiodactyls (except for the oromerycid *Malaquiferus* and camelid *Poebrotherium*), protoceratid artiodactyls, and all ruminants (except for the genus *Hypertragulus*). It is clear that the distribution of the postorbital bar is quite restricted. The more common absence of a postorbital bar is, on the basis of outgroup comparison, primitive. Similarly, the absence of a postorbital bar in either Mesozoic or Paleocene mammals makes the absence of a postorbital bar primitive on the basis of the time distribution of character states.

*Facial fossae*

Facial fossae are pockets, or indentations, on the rostrum of the skull. Facial fossae are rare in living mammals. They are present only in *Macropus, Rhynchocyon, Papio,* in some suids, and in some cervids. The use of facial fossae as taxonomic characters in fossil horses is a relatively recent development. M. F. Skinner and B. J. MacFadden have been the most outspoken proponents of the use of facial fossae as taxonomic characters (Skinner *et al.*, 1977; Skinner and MacFadden, 1977; MacFadden and Skinner, 1982; MacFadden, 1984). The use of facial fossae as taxonomic characters is predicated on the presumption that individual fossil horse species are characterized by a distinctive facial morphology. This presumption has been subjected to only limited testing (Skinner and MacFadden, 1977; MacFadden, 1984), but appears to be substantially correct.

The first problem facing a taxonomist wishing to use facial fossae as taxonomic characters is a precise statement of homologies. I follow Gregory (1920) in recognizing two homologous facial fossae, which I term the preorbital and the malar. The preorbital fossa (Pirlot, 1953) is the homologue of the "lacrimal" fossa of Gregory (1920), the "nasomaxillary" fossa of Skinner and MacFadden (1977), and the "dorsal preorbital" fossa of MacFadden (1984). The malar fossa remains as Gregory (1920) recognized it.

The preorbital fossa is present in four distinct conditions in the Equidae: "absent," "broad-and-open," "deep," and "forward-and-restricted." Of these four states, it is clear that the "absent" condition is primitive, on the basis of either outgroup comparison or the time distribution of character states. Because facial fossae are so rare outside the Equidae, outgroup comparison offers no further polarity resolution. By contrast, the time distribution of character states does offer some additional detail: "deep" and "forward-and-restricted" preorbital fossae made their first appearance in the middle Miocene, while the "broad-and-open" state was present well down into the Oligocene. Thus, on the basis of time distribution of character states, the "broad-and-open" condition is relatively more primitive than either the "deep" or the "forward-and-restricted" character states. No further resolution of character polarities is possible at this time.

The malar fossa occurs in either of two conditions: "absent" and "present." Because the "present" condition is restricted to Neogene equids (and, curiously, some Neogene camelids), the "absent" condition is judged primitive.

## Dental elaboration

The basic homologies of calcified tissues throughout the Vertebrata are well established on histological grounds (Peyer, 1968). Throughout primitive gnathostomes, teeth are recognized as teeth. They remain relatively monotonous, cone-shaped structures that function only to pierce and/or hold prey. By contrast, some dinosaurs and mammals elaborated this basic cone shape and utilize teeth to cut, shear, shred, or crush food effectively. It is difficult to evaluate specific changes in dental morphology without reference to this broad tendency of mammals to the elaboration of dental structures. Mammalian herbivores especially tend toward complex cheek teeth. Premolar "molarization" (Butler, 1952) occurred repeatedly throughout the history of this ecotype. In such cases a broad perspective, encompassing all gnathostomes, seems more appropriate than a narrow consideration of any specific change in dental morphology. Using this broad perspective, the following polarities are apparent:

--The absence of a hypocone in any premolar is primitive. The presence of a hypocone in any premolar is derived. A subtriangular premolar possessing a small hypocone is primitive relative to a subquadrate premolar with a large hypocone.

--The absence of a mesostyle is primitive. The presence of a mesostyle is derived.

--The absence of a pseudoparastyle on $P^2$ is primitive. The presence of a pseudoparastyle on $P^2$ is derived.

--The presence of molar ribs is primitive. The absence of molar ribs is derived. This seeming contradiction can be easily explained. The expression of molar ribs is overwhelmed only when the elaboration of labial styles proceeds to the extreme.

--The absence of a crochet is primitive. The presence of a crochet is derived. The absence of a metastylid is primitive. The presence of a metastylid is derived relative to the degree of separation from the metaconid.

--The absence of plicae is primitive. The presence of plicae is derived relative to the degree of plication and the height of the plicae.

--The absence of a protostylid is primitive. The presence of a protostylid is derived.

--The absence of cingula is primitive. The presence of cingula is derived.

None of these polarity decisions represents a radical change from traditional chronocline polarities. The evolution of the equid molar was discussed by Osborn (1907, 1918), Stirton (1941), and Quinn (1955). Premolar evolution was covered by Granger (1908), Stirton (1941), and Butler (1952).

Ontogenetic data on the development of mammalian occlusial patterns (Butler, 1956; Gaunt, 1961) support the broad tendency to the elaboration of tooth morphology. Individual tooth anlagen show progressive complication of form during development.

## Metaloph

The metaloph of horses, a joined hypocone, prehypocrista (Szalay, 1969), metaconule, and premetaconulecrista, is distinct from the metaloph of all other lophate mammals, as the metaloph typically includes the metacone. Beginning with *Hyracotherium*, the equid metaloph is directed well anterior to the metacone. In Eocene and Oligocene forms, the metaloph comes to a blind end labially. In Neogene forms the metaloph connects to the ectoloph at the mesostyles. Therefore, the time distribution method suggests that primitively the metaloph does not connect with the ectoloph. A metaloph that connects to the ectoloph is derived.

## Diastema

The archetypical gnathostome dentition consists of a series of conical teeth evenly spaced around the periphery of the jaws, each dentition being periodically replaced by successive waves of similar, conical teeth. Mammals specialize the archetypical dentition by severely limiting tooth replacement and by differentiating various tooth loci, so that different functions are performed in different areas of the mouth.

A further mammalian specialization is the creation of gaps, or diastemata, between adjacent tooth groups. Within the gnathostomes, even spacing of teeth is primitive, and the existence of any diastema must be considered derived.

All horses possess a characteristic canine-premolar diastema. The anterior dentition (incisors and canines) functions in cropping vegetation and fighting; the posterior dentition (premolars and molars) functions in mastication. *Hyracotherium* and *Orohippus* both demonstrate a short diastema between $P^1$ and $P^2$. This diastema is here considered a derived character.

*Crown height*

The origin of "hypsodont" from "brachydont" molars is an integral part of the tritubercular theory. The polarity of this character has been adequately established by Osborn (1907). In accordance with the tritubercular theory, "hypsodont" molars are more derived than "mesodont" molars, and "mesodont" molars are derived relative to primitive "brachydont" molars.

*Cementum*

Cementum is a calcified connective tissue that covers the dentine of the anatomical root of a tooth. Cementum is known from "certain reptiles" (Peyer, 1968, p. 22) and all mammals (Grue and Jensen, 1979). Thus it is not, when it appears on the crowns of the teeth of horses, a "new" tissue (Simpson, 1951, p. 134). Rather it is a phylogenetically old tissue being found in a new location on the crowns of teeth.

Cementum is present on the crowns of several taller-crowned mammals. Thus, the distribution of cement-covered crowns will be a subset of tall-crowned teeth. As hypsodont teeth are derived relative to brachydont teeth, it is apparent that cement-covered crowns must be derived relative to cement-free crowns.

*Hypoconal groove*

The hypoconal groove is a remnant of the valley between the metaloph and the postcingulum. This valley is greatly restricted by the origin of the hypoconule, and is later divided into the postfossette and the hypoconal groove by cristae stretching from the metaloph to the hypoconule and hypostyle. In brachydont forms, this valley is "deep" (relative to the height of the crown), and one would expect a homologous "deep" hypoconal groove in hypsodont forms. "Shallow" hypoconal grooves are known from some mesodont and hypsodont equids. As "shallow" hypoconal grooves are a subset of mesodont horses, the "shallow" hypoconal groove must be derived.

*Protocone*

The condition of the protocone has traditionally been used phylogenetically to subdivide mesodont and hypsodont horses (Stirton, 1940): horses with "linked" protocones belong to the genus and subgenus *"Merychippus" (Protohippus)* and its descendants, whereas "independent" protocones characterize *"Merychippus" ("Merychippus")* and its descendants.

At the level of mammals, the existence of a tall crest joining the protocone to the protoconule is decidedly rare. Thus, outgroup comparison suggests that an "independent" protocone is primitive, and a "linked" protocone derived. Similarly, a broad prespective of all mammal teeth shows that tall crests joining the protocone and protoconule are a relatively recent phylogenetic development, absent in all Mesozoic mammals. This polarity is also suggested by the ontogeny of mammalian molar patterns (Butler, 1956; Gaunt, 1961), which show that crests originate from the sulci between cusps. Crests originate by a process of cell division along a line that crosses the sulci. Clearly, the crests must grow "up" to the height of the cusps. Developmental evidence thus suggests that "independent" protocones are primitive, and that "linked" protocones are derived.

$P^1$ - In mammals, any dentition containing less than 48 permanent teeth is derived. Most horse species retain the primitive tokothere formula of 44 teeth. Some Neogene horses, however, lose $P^1$ with the eruption of $P^2$. This loss is regarded as derived, in accordance with traditional comparative practice. It is also clear in this case that ontogenetic data clearly indicate the temporary presence of a tooth at the $P^1$ locus, even in fossil species. Therefore, ontogeny supports the loss of $P^1$ as a derived character.

### Trochlea of the astragalus

The trochlea by which the perissodactyl astragalus articulates with the tibia is a broad, barrel-shaped facet, with the axis of flexion-extension running along the axis of the cylinder (Radinsky, 1966). A groove encircles the middle of the cylinder. In cursorial forms this groove tends to become deep and well pronounced. Typically, this groove is oriented in a plane that is normal to the axis of flexion-extension and "parallel" to the long axis of the foot. In many horses, however, the groove is oriented in an "oblique" plane.

A quick survey of mammalian astragali reveals that the trochlear groove is oriented in a plane normal to the axis of flexion-extension in *Didelphis, Lepus, Felis, Homo, Babirussa, Cervus, Bos, Phenacodus, Brontops, Moropus, Tapirus,* and *Diceros.* The "oblique" trochlear groove is restricted to the Equidae. On the basis of outgroup comparison, it is clear that the trochlear groove primitively lies in a plane normal to the axis of flexion-extension and "parallel" to the long axis of the foot.

Similarly, a quick survey of known Mesozoic mammal postcranials reveals that the normally oriented trochlear groove is common among earlier mammals. In *Asioryctes* and *Zalambdalestes* (Kielan-Jaworowska et al., 1979), *Protungulatum* and *Procerberus* (Szalay and Drawhorn, 1980), a trochlear groove is oriented "parallel" to the long axis of the foot. Although based on a very limited survey, the distribution of character states in time suggests that the trochlear groove is primitively oriented normal to the axis of flexion-extension.

### Number of digits

The pentadactyl limb has long been recognized as the tetrapod archetype. Loss of digits is derived. Specifically in the case of horses, the presence of metacarpal V is primitive, and the absence of this metacarpal is derived. The presence of the phalanges of digits II and IV is primitive, and the absence of these phalanges is derived.

Ontogenetic studies on the development of the carpus of horses (Ewart, 1894a, 1894b) lend support to the general thesis that loss of digits is derived, for the phalanges of digits II and IV are present in modern horse embryos, but not in modern horse adults. However, these studies do not speak specifically to the loss of metacarpal V, for no trace of metacarpal V exists in either fetal or adult modern horses.

### Accessory contacts of metatarsal III

Commensurate with the reduction of digits in the foot, a rearrangement of the tarsal elements occurs. In the typical mammal, a proximal tarsal complex (consisting of astragalus, calcaneum, and navicular) articulates with a distal tarsal row (consisting of the cuboid and three cuneiforms). The three cuneiforms are each supported upon a single metatarsal, the cuboid is supported upon two metatarsals. Metatarsal III, which becomes the principal supporting digit of modern horses, primitively articulates with only the lateral cuneiform and adjacent metatarsals. As metatarsal III becomes the principal supporting digit, its proximal articulatation expands to contact the cuboid laterally and the mesentocuneiform (homologue of the fused intermediate and medial cuneiforms) medially. The absence of accessory contacts for metatarsal III is primitive. The presence of either a mesentocuneiform-metatarsal III or a cuboid-metatarsal III contact is derived.

*Metatarsal keel*

Cursorial mammals are specialized for rapid locomotion. Cursorial mammals specialize their limbs by emphasizing fore-aft movements. This emphasis on fore-aft movement may be marked by either an exaggeration of flexion-extension, or a limitation of movement in other directions. In either case, because the joint surface determines possible joint movement, these cursorial specializations are discernible osteologically, and we can trace the development of cursorial specializations in fossil forms.

The principal action of the mammalian metapodial-phalangeal joint is flexion-extension, but in generalized forms, limited abduction-adduction also occurs. The distal metapodial surface of such generalized forms is an ellipsoid, with the long axis oriented transversely coincident with the axis of flexion-extension. The curvature around the minor axis of the ellipsoid permits abduction and adduction of the digit. In cursorially specialized forms, abduction-adduction is restricted by elimination of the curvature around this minor axis. The distal metapodial surface becomes cylindrical, and the freedom to abduct and adduct is lost. Dislocation of this specialized cylindrical joint is prevented by the development of a metapodial keel perpendicular to the axis of flexion-extension. This keel fits into a corresponding groove on the proximal phalanx (or between its tendons), thus preventing transition along the axis of flexion-extension.

Cursorial mammals also specialize the metapodial-phalangeal joints by exaggerating the range of flexion-extension (Sondaar, 1968). The extended range of flexion-extension at the metapodial-phalangeal joint is a difficult character to judge, except that it is always accompanied by the lengthening of the arc of the metapodial keel. The relationship between the range of flexion-extension and the length of the arc of the metapodial keel is quite clear: the wider the range of flexion-extension, the longer the arc of the metapodial keel.

Abbreviated metatarsal keels are present in a great variety of mammals. They are absent in monotremes, some insectivores, proboscideans, and edentates. These abbreviated keels usually separate paired sesamoid bones that are present in the tendons of the interosseus muscles. They do not articulate with the proximal phalanges. "Complete" metatarsal keels that subsume the entire arc of the metatarsals appear in few mammals. They are present only in suine suids and percoran artiodactyls, litopterns, and some equid perissodactyls. In each case, the "complete" keels are restricted to monophyletic groups within diverse monophyletic orders. Thus, outgroup comparison indicates that an "incomplete" metatarsal keel is primitive, and that a complete keel is derived.

*Radius and ulna*

In the archetypical mammal, one degree of freedom (flexion-extension) exists between the humerus and forearm; another degree of freedom is present within the forearm. A common cursorial specialization is the elimination of the freedom to pronate and supinate. In all horses, this specialization is accomplished by expansion of the proximal radius. This expansion has two effects. First, the shape of the head of the radius changes, so that it is no longer free to rotate about the bone's long axis. Second, the radius, rather than the ulna, becomes the chief weight-bearing bone at the elbow.

The ulna loses much of its function with the loss of weight bearing and pronation-supination. The distal ulna, lacking a function, is lost. But the proximal ulna, onto which the powerful triceps muscle attaches, persists. In most Neogene horses, this proximal remnant of the ulna becomes "fused" to the radius.

As might be expected, this polarity is well supported by ontogenetic data. The radius and ulna of modern horses develop separately and ossify from separate centers (Getty, 1975). The distal epiphysis of the ulna can still be identified in a newborn colt, as the proximal ulna does not fuse to the radius until the twenty-seventh to forty-second month (Getty, 1975).

**Summary of polarity decisions**

For this character analysis, I examined the condition of 33 distinct characters. I present below a summary of these characters, and give an explicit statement of the polarity decisions on the various states of each:

1. *Foramen ovale* - "separate" is primitive, "confluent" with the middle lacerate foramen is derived.
2. $P^4$ *hypocone* - "absent" is primitive, "present" is derived.
3. *Molar mesostyles* - "absent" is primitive, "present" is derived.
4. *Astragalar trochlea* - "parallel" to the axis of the foot and perpendicular to the axis of flexion-extension is primitive, "oblique" to the axis of flexion-extension is derived.
5. $P^1$ - $P^2$ *diastema* - "absent" is primitive, "present" is derived.
6. $P^2$ *hypocone* - "absent" is primitive, "present" is derived.
7. *Basicranial axis* - "flaring" is primitive, "narrow" is derived.
8. *Metacarpal V* - "present" is primitive, "absent" is derived.
9. *Hypoconules* - "absent" is primitive, "present" is derived.
10. $P^1$ *lingual cingulum* - "absent" is primitive, "present" is derived.
11. $P^2$ *shape* - "subtriangular," possessing at best a small hypocone, is primitive; fully "subquadrate," possessing hypocone subequal in size to the protocone, is derived.
12. *Cuboid-metatarsal III contact* - "absent" is primitive, "present" is derived.
13. *Metastylid* - "absent" is primitive, "present" as a distinct cusp on the apex of the metaconid is derived. A metastylid that is "distinct" from the metaconid through much of the wear of the tooth is derived relative to either "absent" or "present."
14. $P^1$ *hypocone* - "absent" is primitive, "present" is derived.
15. *Metaloph* - "independent" from the ectoloph is primitive, "connected" to the ectoloph at the mesostyle is derived.
16. *Molar ribs* - "present" is primitive, "absent" is derived.
17. *Lingual cingula on $P^2$ - $P^4$* - "absent" is primitive, "present" is derived.
18. *Mesentocuneiform-metatarsal III contact* - "absent" is primitive, "present" is derived.
19. *Metatarsal keel* - an "incomplete" keel, present only on the posterior aspect of the metatarsal, is primitive; a "complete" keel, which extends onto the anterior aspect of the distal articular surface (and can be felt there), is derived.
20. $P^2$ *pseudoparastyle* - "absent" is primitive, "present" is derived.
21. *Crochet* - "absent" is primitive, "present" is derived.
22. *Postorbital bar* - "absent" is primitive, "complete" is derived.
23. *Cementum* - "absent" from the crown of the teeth is primitive, "present" on the crown of the teeth is derived.
24. *Crown height* - "brachydont" is primitive, "mesodont" is derived. "Hypsodont" is derived relative to either "brachydont" or "mesodont."
25. *Radius and ulna* - "separate" is primitive, "fused" is derived.
26. $P_1$ - "persistent" is primitive, "lost" with the eruption of $P_2$ is derived.
27. *Plication* - "absent" is primitive, the presence of "moderate" (8-12 short plicae) plication of the upper molars is derived. "Complex" (more than 15 tall plicae) plication of the upper molars is derived relative to either "absent" or "moderate" plication.
28. *Hypoconal groove* - "deep" relative to total crown height is primitive, "shallow" or closed is derived.
29. *Protostylid* - a "low" precingulum is primitive, a tall protostylid that causes a distinct enamel column to be "present" on the anterolabial corner of

Table 8.1. Character chart. The character state of each of the 19 taxa is shown for each of the 33 characters considered in this analysis. A character that was not present on any of the specimens is represented by a blank space on the chart. A character that demonstrated the primitive condition is shown by a minus (-). A character that demonstrated a partially derived condition is shown by a slash (/). The fully derived condition is shown by a plus (+). In the case of the preorbital fossa, two separate derived conditions were found; the second derived condition is designated by an equals (=). Column headings abbreviated as follows (in sequence): *Hyracotherium, Orohippus, Epihippus, Mesohippus, Miohippus, Anchitherium, Hypohippus, Megahippus, Kalobatippus, Archaeohippus, "Parahippus," "Merychippus," Protohippus, Pliohippus, Dinohippus, Equus, Hipparion, Neohipparion, Pseudhipparion.*

| | H | O | E | M | M | A | H | M | K | A | P | M | P | P | D | E | H | N | P |
|---|---|---|---|---|---|---|---|---|---|---|---|---|---|---|---|---|---|---|---|
| 1. Foramen ovale | + | + | | + | + | + | + | + | + | + | + | + | + | + | + | + | + | + | + |
| 2. $P^4$ hypocone | - | + | + | + | + | + | + | + | + | + | + | + | + | + | + | + | + | + | + |
| 3. Molar mesostyles | - | + | + | + | + | + | + | + | + | + | + | + | + | + | + | + | + | + | + |
| 4. Astragular trochlea | - | + | | + | + | + | + | + | + | + | + | + | + | + | + | + | + | + | + |
| 5. $P^1$-$P^2$ diastema | - | - | + | + | + | + | + | + | + | + | + | + | + | + | + | + | + | + | + |
| 6. $P^2$ hypocone | - | - | + | + | + | + | + | + | + | + | + | + | + | + | + | + | + | + | + |
| 7. Basicranial axis | - | - | | + | + | + | + | + | + | + | + | + | + | + | + | + | + | + | + |
| 8. Metacarpal V | - | - | | + | + | + | + | + | + | + | + | + | + | + | + | + | + | + | + |
| 9. Hypoconules | - | - | - | + | + | + | + | + | + | + | + | + | + | + | + | + | + | + | + |
| 10. $P^1$ lingula cingula | - | - | - | + | + | + | + | + | + | + | + | + | + | + | + | + | + | + | + |
| 11. $P^2$ shape | - | - | - | + | + | + | + | + | + | + | + | + | + | + | + | + | + | + | + |
| 12. Cuboid-metatarsal III contact | - | + | | - | + | + | + | + | + | + | + | + | + | + | + | + | + | + | + |
| 13. Metastylid | - | - | - | + | + | + | + | + | + | + | + | + | + | + | + | + | + | + | + |
| 14. $P^1$ hypocone | - | - | - | + | + | + | + | + | + | + | + | + | + | + | + | + | + | + | + |
| 15. Metaloph | - | - | - | - | - | + | + | + | + | + | + | + | + | + | + | + | + | + | + |
| 16. Molar ribs | + | + | + | + | + | - | - | - | + | + | + | + | + | + | + | + | + | + | + |
| 17. Lingual cingula on $P^2$-$P^4$ | + | + | + | + | + | - | - | - | + | + | + | + | + | + | + | + | + | + | + |
| 18. Mesentocuneiform-metatarsalIII contact | + | + | + | + | + | - | - | - | + | + | + | + | + | + | + | + | + | + | + |
| 19. Metatarsal keel | - | - | - | - | - | - | - | + | + | + | + | + | + | + | + | + | + | + | + |
| 20. $P^2$ pseudoparastyle | - | - | - | - | - | - | - | - | - | + | + | + | + | + | + | + | + | + | + |
| 21. Crochet | - | - | - | - | - | - | - | - | - | - | + | + | + | + | + | + | + | + | + |
| 22. Postorbital bar | - | - | - | - | - | - | + | - | - | + | + | + | + | + | + | + | + | + | + |
| 23. Cementum | - | - | - | - | - | - | - | - | - | - | - | + | + | + | + | + | + | + | + |
| 24. Crown height | - | - | - | - | - | - | - | - | - | - | - | + | + | + | + | + | + | + | + |
| 25. Radius and ulna | - | - | - | - | - | - | - | - | - | - | - | + | + | + | + | + | + | + | + |
| 26. $P_1$ | - | - | - | - | - | - | - | - | - | - | - | + | + | + | + | + | + | + | + |
| 27. Plication | - | - | - | - | - | - | - | - | - | - | - | / | / | / | / | / | / | / | / |
| 28. Hypoconal groove | - | - | - | - | - | - | - | - | - | - | - | + | + | + | - | - | - | - | - |
| 29. Protostylid | - | - | - | - | - | - | - | - | - | - | - | - | + | - | - | - | + | + | + |
| 30. Malar fossa | - | - | - | - | - | - | - | - | - | - | - | - | - | - | + | - | - | - | - |
| 31. Lateral phalanges | - | - | - | - | - | - | - | - | - | - | - | - | - | - | + | + | - | - | - |
| 32. Preorbital fossa | - | - | - | - | - | - | - | - | - | - | - | - | - | - | = | = | + | + | + |
| 33. Protocone | + | + | + | + | + | + | + | + | + | + | + | + | + | + | + | + | + | - | - |

the tooth is derived. This character must be evaluated with extreme caution in mesodont and hypsodont forms, as an enamel flexid will appear on the crown of a worn horse tooth, regardless of whether a protostylid is "present" or not.

30. *Malar fossa* - "absent" is primitive, "present" is derived.
31. *Lateral phalanges* - phalanges of digits II and IV "present" is primitive, phalanges of digits II and IV "absent" is derived.
32. *Preorbital fossa* - "absent" is primitive, "broad-and-open" is derived. "Deep" and "forward-and-restricted" are both derived with respect to either "absent" or "broad-and-open."
33. *Protocone* - an "independent" protocone is primitive, a "linked" protocone is derived.

**Character analysis**
In Table 8.1, I present a summary of the state of these 33 characters in all taxa considered. In those cases where the material available to me did not permit a determination, the table shows a blank.

A thorough character analysis requires a complete character chart. For some taxa, it was impossible to complete the character chart based on direct observation. In such cases, the state of the missing characters was inferred during the character analysis according to the following criterion. Taxa with unknown character states were assigned to monophyletic groups based on known characters. Unknown character states were then presumed to be as primitive as in any of the other species within that monophyletic group. For instance, on the basis of dental morphology ($P^4$ hypocone is "present," molar mesostyles are "present"), *Epihippus* is judged to be a member of the monophyletic group consisting of all horses except *Hyracotherium* and *Orohippus*. As all members of this monophyletic group posses an "oblique" astragalar trochlea, *Epihippus* was presumed to possess an "oblique" astragalar trochlea.

Most of the species examined in this study were known from remarkably complete specimens, so that presumption of character states was limited to two species: *Epihippus* and *Anchitherium*. *Epihippus* was so incomplete that it was presumed to possess a "confluent" foramen ovale, an "oblique" astragalar trochlea, a "flaring" basicranial axis, a "present" metacarpal V, "absent" cuboid-metatarsal III and mesentocuneiform-metatarsal III contacts, an "incomplete" metatarsal keel, an "absent" postorbital bar, a "separate" radius and ulna, an "absent" malar fossa, an "absent" preorbital fossa, and "present" phalanges of digits II and IV. *Anchitherium* was preumed to possess a "confluent" foramen ovale, a "narrow" basicranial axis, and an "absent" postorbital bar.

A cladogram sumarizing the results of the character analysis is given in Fig. 8.1. One of the more salient features of this cladogram is the character reversals that characterize monophyletic groups of advanced horse: the secondarily independent protocone unites *Hipparion*, *Neohipparion*, and *Pseudhipparion*, whereas the secondarily "absent" preorbital fossa unites *Dinohippus* and *Equus*.

Protoconal spurs have been cited as primitive characters among horses that possess independent protocones (Stirton, 1940, p.180). Within the framework of a phylogenetic analysis that supports the breakup of the high crest between the protocone and the protoconule, I agree that the protoconal spur, as a remnant of that crest, is a patristic character among horses with independent protocones. More significant, the loss of a protoconal spur is derived.

Both Bennett (1980) and MacFadden (1984) noted as a character polarity the reduction or loss of the preorbital fossa. Such a character polarity is not supported by this cladistic analysis, in which both the outgroup and the most primitive species lack the preorbital fossa, but in which several of the terminal clades (*Hypohippus*, *Archaeohippus*, *Pliohippus*, and *Hipparion*)

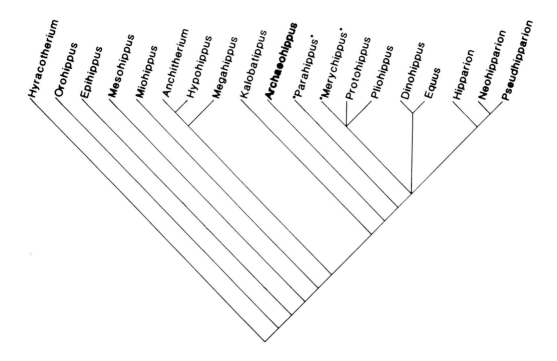

Fig. 8.1. Cladogram of the genera of the Equidae. See text for explanation of characters.

possess distinctive preorbital fossae. In this analysis, which considers a broader taxonomic base than either Bennett (1980) or MacFadden (1984), the overall tendency within the Equidae is towards the enlargement of the preorbital fossa.

### Phylogenetic analysis

A phylogenetic tree must not only be congruent with character distributions, but must also be congruent with chronostratigraphic data, as ancestral species must be older than descendant species. Using the correlation chart of Wood et al. (1941), it is possible to construct a reasonably comprehensive chronostratigraphic succession of the species being analyzed. From the youngest to the oldest, the following chronostratigraphic order is apparent: *Hyracotherium*, followed in order by *Orohippus*, *Epihippus*, *Mesohippus*, *Miohippus*, *Kalobatipus*, and "*Parahippus*," then *Archaeohippus* and "*Merychippus*" come from the same quarry (Thomson Quarry); then *Hypohippus*, *Protohippus*, *Pliohippus*, and *Pseudhipparion* come from the same formation (Valentine Formation), then *Dinohippus*, and finally *Equus*. *Anchitherium* can be placed on the same level as "*Parahippus*," based on the appearance of a congeneric species at Thomas Farm (Webb, 1981).

A phylogenetic tree, summarizing the results of this phylogenetic analysis, is given in Fig. 8.2. This tree demonstrates at once both the merits and pitfalls of phylogenetic analysis. The dangers of phylogenetic analysis are illustrated by *Epihippus*. As only dental characters of *Epihippus* were known, a huge variety of cranial and postcranial characters were presumed for this species. Most of these characters were presumed to exist in the primitive state, so that the presumptions would not bias the character analysis. But later, when testing the hypothesis that *Epihip*-

*pus* was the ancestor of all later horses, these missing characters tend, by the absence of refutation, to support any hypothesis of ancestry. Poorly known species are easily recognized as ancestors because the hypothesis of ancestry cannot be adequately tested. This concern is not trivial. Most fossil horses are known only from their dentitions. If we were to expand this analysis to include the several known species of *Hyracotherium*, *Orohippus*, and so on, the immediate result would be a tremendous increase in the number of presumptions concerning character states. At the same time, the number of unrefuted (in reality, poorly tested) hypotheses of ancestry would increase. The addition of more species to a phylogenetic analysis does not necessarily insure an increase in phylogenetic precision.

## Classification

Leidy (1869) was the first to propose a basic subdivision of the Family Equidae (his order Solidungulata) into two groups: the low-crowned Anchitheriidae and the high-crowned Equidae. Subsequently, most classifications of the family have included these two groups as subfamilies (Gidley, 1907; Simpson, 1945), although Leidy is not always given credit for their establishment (Hay, 1930; Simpson, 1945). Since Leidy first proposed the taxon, the Equinae suffered one attempted restriction: Gidley (1907) and Osborn (1918) removed the Protohippini as a primitive grade. But most authors continue to recognize the Equidae in much the same sense as Leidy did. If the Equinae are expanded only slightly to include *Kalobatippus*, *Archaeohippus*, and *Parahippus*, I find the Equinae to be a valid monophyletic group united by the possession of a complete metatarsal keel and (except for *Kalobatippus*) the presence of a pseudoparastyle on $P^2$.

Leidy's (1869) original concepts of the Anchitheriidae and Equidae were gradal in nature. It comes as no surprise that the Anchitheriinae, which shared only primitive characters, proved less stable than the Equinae, which was united by derived characters. The most primitive horses were placed in the primitive grade Hyracotheriinae (Cope, 1881), with the result that the Anchitheriinae survived as an amorphous grade of low-crowned Oligocene or Miocene horses (Simpson, 1945, p. 245). This character analysis of the family Equidae reveals that the traditional anchitheriine genera (Simpson, 1945) are a paraphyletic assemblage and are not a valid monophyletic group. However, *Anchitherium*, the nominal type genus of the Anchitheriinae, is included within a distictive clade of horses. I restrict the Anchitheriinae to this clade. An emended diagnosis of the Anchitheriinae includes the following derived character states: strongly W-shaped ectolophs in which the paraconal and metaconal ribs are not expressed labially, lingual cingula present on the upper premolars, and the presence of a mesentocuneiform-metatarsal III contact in the tarsus. In this restricted sense, the Anchitheriinae include the genera *Anchitherium*, *Hypohippus*, and *Megahippus*.

The Anchitheriinae and Equinae as here defined are together a monophyletic group united by the connection of the metaloph to the ectoloph. Five equid genera comprise a paraphyletic succession more primitive than the united anchitheriines-equids. Three of these genera had been allocated to Cope's (1881) Hyracotheriinae: *Hyracotherium*, *Orohippus*, and *Epihippus* (Simpson, 1945). This phylogenetic analysis of the Family Equidae clearly reveals that the Hyracotheriinae are a paraphyletic assemblage that no longer merits recognition. I have classified the five primitive horse genera as successive plesions.

Within the Equinae, the three most primitive genera-- *Kalobatippus*, *Archaeohippus*, and "*Parahippus*" --form a similar paraphyletic succession and have been classified as plesions. I note that *Kalobatippus* is only questionably a member of the Equinae. Although it shares a complete metatarsal keel with the remainder of the Equiinae, it also shares a mesentocuneiform-metatarsal III contact with the

# PHYLOGENY OF THE EQUIDAE

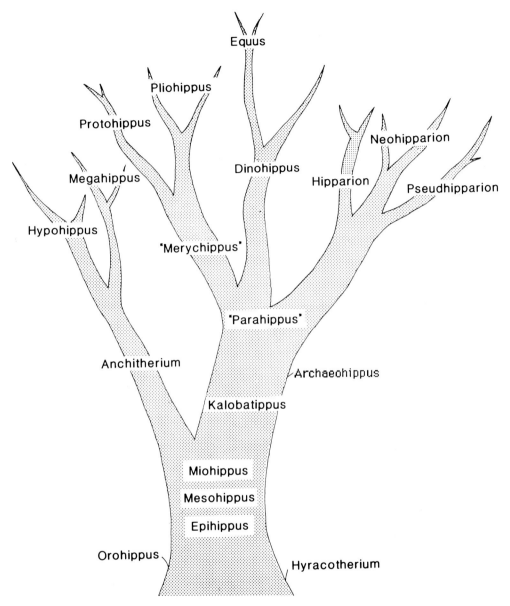

Fig. 8.2. Phylogenetic tree of the Equidae, based on the cladogram in Fig. 8.1. See text for discussion.

Anchitheriinae. *Kalobatippus* is placed with the Equinae simply because it lacks the distinctive dental specializations of the Anchitheriinae.

Mesodont and hypsodont horses are placed in an unnamed monophyletic group united not only by the presence of high crowns, but also by the presence of cementum on the molars, moderate plication, the loss of $P_1$ with the eruption of $P_2$, and the fusion of the radius and ulna. This monophyletic group can be divided into three tribes: the Hipparionini, united by the secondarily independent protocone; the Protohippini, united by the shallow hypoconal groove; and the Equini, united by the secondary loss of the preorbital fossa. The Hipparionini includes the genera *Hipparion*, *Pseudhipparion*, and *Neohipparion*. The Protohippini includes the genera *"Merychippus," Protohippus*, and *Pliohippus*, and the Equini includes the genera *Dinohippus* and *Equus*. The relationship between these three tribes remains unresolved.

Below, I present a classification of the Family Equidae based on the specimens analyzed during the study:

Family Equidae Gray, 1821
    Plesion *Hyracotherium*
    Plesion *Orohippus*
    Plesion *Epihippus*
    Plesion *Mesohippus*
    Plesion *Miohippus*
    Subfamily Anchitheriinae Leidy, 1869
        Plesion *Anchitherium*
        Genus *Hypohippus*
        Genus *Megahippus*
    Subfamily Equinae Leidy, 1869
        Plesion *Kalobatippus*
        Plesion *Archaeohippus*
        Plesion *"Parahippus"*
        Tribe Protohippini Gidley, 1907
            Plesion *"Merychippus"*
            Genus *Protohippus*
            Genus *Pliohippus*
        Tribe Equini Gidley, 1907
            Plesion *Dinohippus*
            Genus *Equus*
        Tribe Hipparionini Quinn, 1955
            Plesion *Hipparion*
            Genus *Pseudhipparion*
            Genus *Neohipparion*

## Discussion

Ever since the classic work of Marsh (1874), paleontologists have recognized a succession of North American fossil equids as belonging to the "main line" of equid evolution, or what they considered ancestors of the modern horse in the fossil record. At the same time, paleontologists have recognized a variety of fruitless branches on this phylogenetic tree, most notably the European fossil horses of the genera *Anchitherium* and *Hipparion*.

This analysis removes several taxa from direct ancestry of the modern horse, most significantly *Hyracotherium*, *"Merychippus," Protohippus*, and *Pliohippus*. The removal of *Hyracotherium* and *"Merychippus"* are an artifact of my methodology, as I examined only one species from each classically recognized genus. Other species of the classical generic grades do seem to fill an ancestral role. The removal of *Protohippus* and *Pliohippus* from the equid main line are, by contrast, significant changes not related to my method. On one hand, I am recognizing the distinctive nature of these two side branches of equid evolution, and am allocating to these two side branches the taxonomic recognition they deserve. On the other hand, I am removing from the equid main line two genera long recognized (Matthew, 1926; Stirton, 1940; Simpson, 1951) as ancestors of the living horse. These changes leave a 5-million-year gap in our knowledge of the horse. The gap is all the more perplexing because it comes during an interval when the stratigraphic record is relatively complete and fossil equids relatively abundant.

The question of the immediate ancestors to the living *Equus* has been an ever present question for mammalian paleontologists. *Dinohippus* clearly shares several derived character states with *Equus*, and the two genera belong in the same monophyletic group. However, the ancestors of this monophyletic group are not immediately apparent. The fossil record provides a lucid story of descent with change for nearly 50 million years, and we know much about the ancestors of modern horses. But during an interval immediately preceding the Pleistocene, the record becomes hazy. Both Quinn (1955) and Skinner (Skinner et al., 1977, p. 359) have warned that *Pliohippus* is unsuitable as an ancestor of the living horse. If *Pliohippus* is to be put back in the ancestral sequence, then it will be necessary to document several dramatic evolutionary reversals. Otherwise, we should begin a search for unrecognized ancestors of the modern horse. In either case, it is clear that the outstanding problem in our understanding of the ancestry of the modern horse is to be found in the identification of the Pliocene ancestors of the modern horse.

# Bibliography

Arnold, E. N. (1981): Estimating phylogenies at low taxonomic levels. -*Zeit. Zool. Syst. Evolut.-Forsch.*, 19: 1-35.

Bennett, D. K. (1980): Stripes do not a zebra make. Part 1. A cladistic analysis of *Equus*. -*Syst. Zool.*, 29: 272-287.

Butler, P. M. (1952): Molarization of the premolars in Perissodactyla. -*Proc. Zool. Soc. London*, 121: 819-843.

Butler, P. M. (1956): The ontogeny of molar pattern. -*Biol. Rev.*, 31: 30-70.

Cope, E.D. (1872): On a new genus of Pleurodira from the Eocene of Wyoming. -*Proc. Amer. Phil. Soc.*, 12: 1-6.

Cope, E.D. (1879): A new *Anchitherium*. -*Amer. Natur.*, 13: 462-463.

Cope, E. D. (1881): The systematic arrangement of the Order Perissodactyla. -*Proc. Amer. Phil. Soc.*, 19: 337-401.

Cuvier, G. (1812): *Recherches sur les ossemens fossiles*. 1st ed. 4 vols., Paris.

Darwin, C. (1859): *The origin of species by means of natural selection*. 1st. ed. London (John Murray).

De Beer, G. R. (1937): *The development of the vertebrate skull*. Oxford (Clarendon).

Edinger, T., and Kitts, D. B. (1954): The foramen ovale. -*Evolution*, 8: 389-404.

Evander, R.L. (1986): The taxonomic status of *Merychippus insignis* Leidy. -*J. Paleont.*, 60(6): 1277-1279.

Ewart, J. C. (1894a): The second and fourth digits in the horse: their development and subsequent degeneration. -*Proc. Royal Soc. Edinburgh* 1894: 185- 191.

Ewart, J. C. (1894b): Development of the skeleton of the limbs of the horse, with observations on polydactyly. -*J. Anat. Physiol.*, 28: 236-256.

Filhol, H. (1891): *Études sur les mamiferes fossiles du Sansan*. Paris (Masson).

Gaunt, W. A. (1961): The development of the molar pattern of the golden hamster (*Mesocricetus auratus* W.), together with a re-assessment of the molar pattern of the mouse (*Mus musculus*). -*Acta Anat.*, 45: 219-251.

Getty, R. (1975): *Sisson and Grossman's The anatomy of domestic animals*. Philadelphia (Saunders).

Gidley, J. W. (1906): New or little known mammals from the Miocene of South Dakota. Part N. Equidae. -*Bull. Amer. Mus. Nat. Hist.*, 22: 135-153.

Gidley, J. W. (1907): Revision of the Miocene and Pliocene Equidae of North America. -*Bull. Amer. Mus. Nat. Hist.*, 23: 865-934.

Granger, W. (1908): A revision of the American Eocene horses. -*Bull. Amer. Mus. Nat. Hist.*, 24: 221-264.

Gray, J. E. (1821): On the natural arrangement of vertebrose animals. - *London Med. Repos. Rev.*, 15: 296-310.

Gregory, W. K. (1920): Studies of comparitive myology and osteology, no. V --On the anatomy of the preorbital fossae of Equidae and other ungulates. -*Bull. Amer. Mus. Nat. Hist.*, 42: 265-283.

Grue, H., andJensen, B. (1979): Review of the formation of incremental lines in tooth cementum of terrestrial mammals. -*Danish Rev. Game Biol.*, 11(3): 1-43.

Hay, O. P. (1930): Second bibliography and catalogue of the fossil Vertebrata of

North America. - *Carnegie Inst. Wash. Publ.*, 390.

Hennig, W. (1965): Phylogenetic systematics. -*Ann. Rev. Ent.*, 10: 97-116.

Hennig, W. (1966): *Phylogenetic Systematics*. Urbana (Univ. of Illinois).

Huxley, T. H. (1870): Anniversary address of the President. -*Quart. J. Geol. Soc. London*, 26: 29-64.

Kielan-Jaworowska, Z. (1981): Evolution of the therian mammals in the late Cretaceous of Asia. Part IV. Skull structure of *Kennalestes* and *Asiorcytes*. -*Paleont. Polonica*, 42: 25-78.

Kielan-Jaworowska, Z., Bown, T. M., and Lillegraven, J. A. (1979): Eutheria. In: Lillegraven, J. A. and Clemens, W. A. (eds): *Mesozoic Mammals: the first two-thirds of mammalian history*. Berkeley (Univ. of Calif.), pp. 221-258.

Kielan-Jaworowska, Z., and Trofimov, B.A. (1980): Cranial morphology of Cretaceous eutherian mammal *Barunlestes*. - *Acta Paleont. Polonica*, 25 (2): 167-186.

Kitts, D. B. (1956): American *Hyracotherium* (Perissodactyla, Equidae) - *Bull. Amer. Mus. Nat. Hist.*, 110: 1-60.

Kitts, D. B. (1957): A revision of the genus *Orohippus* (Perissodactyla, Equidae). - *Amer. Mus. Novitates* 1864: 1-40.

Kowalevsky, W. (1873): Sur l'*Anchitherium aurelianse* Cuv. et sur l'histoire paléontologie des chevaux. -*Mém. Acad. Impér. Sci. St. Pétersburg*, sér. VII, 20(5): 1-73.

Leidy, J. (1850): [Descriptions of *Rhinoceros nebrascensis*, *Agriochoerus antiquus*, *Palaeotherium proutii*, and *Palaeotherium bairdii*]. -*Proc. Acad. Nat. Sci. Phila.*, 5: 121-122.

Leidy, J. (1858): Notice of remains of extinct Vertebrata from the valley of the Niobrara River,... -*Proc. Acad. Nat. Sci. Phila.*, 1858: 20-29.

Leidy, J. (1869): The extinct mammalian fauna of Dakota and Nebraska... - *J. Acad. Nat. Sci. Phila.*, ser. 2, 2: 1-472.

MacFadden, B. J. (1976): Cladistic analysis of primitive equids, with notes on other perissodactyls. -*Syst. Zool.*, 25: 1-14.

MacFadden, B. J. (1980): Eocene perissodactyls from the type section of the Tepee Trail Formation of northern western Wyoming. -*Univ. Wyo. Contrib. Geol.*, 18: 135-143.

MacFadden, B. J. (1984): Systematics and phylogeny of *Hipparion*, *Neohipparion*, *Nannippus*, and *Cormohipparion* (Mammalia, Equidae) from the Miocene and Pliocene of the New World. -*Bull. Amer. Mus. Nat. Hist*, 179: 1-196.

MacFadden, B. J. and Skinner, B. F. (1982): Hipparion horses and modern phylogenetic interpretation: Comments on Forsten's view of *Cormohipparion*. -*J. Paleont.*, 56: 1336-1342.

MacIntyre, G. T. (1967): Foramen pseudovale and quasi-mammals. -*Evolution*, 21: 834-841.

MacPhee, R. D. E. (1981): Auditory regions of primates and eutherian insectivores: morphology, ontogeny, and character analysis. -*Contr. Primat.*, 18: 1-228.

Marsh, O.C. (1871): Notice of some new fossil mammals from the Tertiary formation. -*Amer. J. Sci.*, 3: 35-44.

Marsh, O.C. (1872): Preliminary descriptions of new Tertiary mammals. -*Amer. J. Sci.*, 3: 202-224.

Marsh, O. C. (1874): Fossil horses in America. -*Amer. Nat.*, 8: 288-294.

Matthew, W. D. (1924): Third contribution to the Snake Creek Fauna. -*Bull. Amer. Mus. Nat. Hist.*, 38: 183-229.

Matthew, W. D.(1926): The evolution of the horse: A record and its interpretation. -*Quart. Rev. Biol.*, 1: 139-185.

Osborn, H. F. (1904): New Oligocene horses. -*Bull. Amer. Mus. Nat. Hist.*, 20 (8): 167-169.

Osborn, H. F. (1907): *Evolution of Mammalian Molar Teeth*. New York (MacMillan).

Osborn, H. F. (1918): Equidae of the Oligocene, Miocene, and Pliocene of North America: Iconographic type revision. -*Mem. Amer. Mus. Nat. Hist.*, new ser. 2: -217.

Osborn, H.F., and Wortman, J.A. (1895): Perissodactyls of the lower Miocene

White River beds. -*Bull. Amer. Mus. Nat. Hist.*, 7 (12): 343-375.

Owen, R. (1841): Description of the fossil remians of a mammal (*Hyracotherium leporinum*) and of a bird (*Lithornis vulturinus*) from the London Clay. - *Trans. Geol. Soc. London*, 2 (6): 203-208.

Patterson, C., and Rosen, D. E. (1977): Review of ichthyodectiform and other Mesozoic teleost fishes and the theory and practice of classifying fossils. -*Bull. Amer. Mus. Nat. Hist.*, 158: 81-172.

Peyer, B. (1968): *Comparative Odontology*. Chicago (Univ. of Chicago).

Piriot, P. R. (1953): The preorbital fossa of *Hipparion* . -*Amer. J. Sci.*, 251: 309-312.

Popper, K. (1963): *Conjectures and refutations: the growth of scientific knowledge*. New York (Harper and Row).

Quinn, J. H. (1955): Miocene Equidae of the Texas Gulf Coast Plain. -*Univ. Texas Publ. Bur. Econ. Geol.*, 5516: 1-102.

Radinsky, L. B. (1966): The adaptive radiation of the phenacodontid condylarths and the origin of the Perissodactyla. -*Evolution*, 20: 408-417.

Simpson, G. G. (1945): The principles of classification and a classification of the mammals. -*Bull. Amer. Mus. Nat. Hist.* , 85: 1-350.

Simpson, G. G. (1951): *Horses*. New York (Oxford Univ. Press).

Skinner, M. F., and MacFadden, B. J. (1977): *Cormohipparion* n. gen. (Mammalia, Equidae) from the North American Miocene (Barstovian - Clarendonian). -*J. Paleont.*, 51: 912-926.

Skinner, M. F., Skinner, S. M., and Gooris, R. J. (1977): Stratigraphy and biostratigraphy of late Cenozoic deposits in central Sioux County, western Nebraska. -*Bull. Amer. Mus. Nat. Hist.*, 158: 263-370.

Sondaar, P. Y. (1968): The osteology of the manus of fossil and recent Equidae. - *Verh. Koninkl. Nederl. Akad. Wetensch. afd. Natuurk.*, 25: 1-76.

Stirton, R. A. (1940): Phylogeny of North American Equidae -*Univ. Calif. Publ. Geol. Sci.*, 25: 165-198.

Stirton, R. A. (1941): Development of characters in horse teeth and the dental nomenclature. -*J. Mamm.*, 22: 434-446.

Szalay, F. S. (1969): Mixodecidae, Microsyopidae, and the insectivore-primate transition. -*Bull. Amer. Mus. Nat. Hist.*, 140: 193-330.

Szalay, F. S., and Delson, E. (1980): *Evolutionary History of the Primates*. New York (Academic).

Szalay, F. S., and Drawhorn, G. (1980): Evolution and diversification of the Archonta in an arboreal milleu. In: Luckett, W. P. (ed.): *Comparative Biology and the Evolutionary Relationships of Tree Shrews*. New York (Plenum), pp. 133-169.

Tedford, R. H., and Alf, R. M. (1962): A new *Megahippus* from the Barstow Formation, San Bernardino County, California. -*Bull. S. Calif. Acad. Sci.*, 61: 113-122.

Webb, S. D. (1981): The Thomas Farm vertebrate site. -*Plaster Jacket*, 37: 6-25.

Wood, H.E., II, Chaney, R.W., Clark, J., Colbert, E.H., Jepsen, G.L., Reeside, J.B., Jr., and Stock, C. (1941): Nomenclature and correlation of the North American continental Tertiary.-*Bull. Geol. Soc. Amer.*, 52: 1-48.

# 9. DENTAL CHARACTER VARIATION IN PALEOPOPULATIONS AND MORPHOSPECIES OF FOSSIL HORSES AND EXTANT ANALOGS

## BRUCE J. MacFADDEN

*Sampling in paleontology is very different from sampling among recent animals and constitutes a special problem, although the treatment of the samples, once obtained, is often the same in both fields.* Simpson, Roe, and Lewontin, 1960, p. 113.

Variation is analyzed for nine skull and dental characters of 12 assemblages of fossil and Recent horses, including samples interpreted to represent paleopopulations, a modern deme (*Equus burchelli*), morphospecies, and an extant species (*Equus grevyi*). The principal goal of this study was to determine if the variation observed in fossil samples is similar to that of extant *Equus*. Pairwise (i.e., fossil/extant samples) comparisons of mature individuals (juvenile and old specimens were excluded) were done using $F$-ratios, $V^2$-ratios, and the median-ratio and median-log forms of Levene's test. These tests consistently indicate that the variation observed in the fossil and recent samples is not significantly different. It is therefore proposed that for Equidae, paleopopulations and morphospecies are analogous to, respectively, demes and biological species samples.

## Introduction

Populations and species are fundamental units in systematic research. From them we can learn much about biological diversity, adaptation, and distribution. As mentioned above, a primary goal in paleontology is to interpret fossils in the same manner as neontologists study extant taxa. However, the paleontologist is confronted with a different set of problems when attempting to recognize populations or species.

In order to study extinct samples that simulate extant populations, paleontologists seek localized quarry sites that seem to represent instants in geological time. Depending upon the circumstances of taphonomy and fossilization, a particular quarry sample may approximate a biocenosis (life assemblage) from which inferences about population structure and other aspects of the paleobiology of the constituent species can be made. This type of sample has been termed a "paleopopulation" (Sylvester-Bradley, 1951, introduced the unnecessary term, "topodeme") and these are often assumed to exhibit the same biological properties as extant demes.

At the next level of biological organization, paleontologists identify extinct species based on morphology of preserved hard parts. If geographically separate paleopopulations do not differ in the morphological characters analyzed, then they are usually interpreted to represent a single species. Paleontologists are necessarily limited in their species concept; they cannot recognize morphologically indistinguishable sibling species. Given these limitations of the fossil record, terms such as "morphospecies" have been proposed (Cain, 1953; George, 1956).

With the demise of the typological species concept and the advent of the Modern Synthesis, character variation is recognized and expected in populations and species. It cannot be disputed that variation is a fundamental component of systematics and the evolutionary process.

The purpose of this paper is to test the frequent assumption, which is largely untested, that paleopopulations and morphospecies of fossil samples are analogous to, respectively, demes and species of recent samples. To state the test another way, is the character variation observed in these groups similar? If so, then they might represent similar units of biological organization. This study will use an analysis of skull and dental character variation in several quarry samples, or "paleopopulations," and geographically widespread and/or temporally distributed "morphospecies" of Tertiary Equidae from North America. These are then compared to selected samples of the extant genus *Equus*.

Studies of variation abound in the literature, and a complete discussion of these is beyond the scope of this paper. Of relevance to mammals, Yablokov's (1974) detailed survey of numerous anatomical and osteological characters (as well as exhaustive references) is an important baseline study. He found that for most osteological characters within species, coefficients of variation ($V$ s) usually fall between about 4-6%.

Of relevance to fossil horses, Forsten's (1970) study of Oligocene *Mesohippus bairdi* determined that for dental characters, $V$s are usually between 5-7%. In a very interesting series of papers, Van Valen (1963, 1964, 1965) studied dental variation in several Miocene horse quarry populations. He determined the extent of dental variation in discrete ontogenetic stages (cohorts) and then asserted that these differences related to intensities of selection (higher $V$s representing stronger selection). MacFadden (1984) studied variation in numerous skull and dental characters in both quarry and species samples of Miocene and Pliocene three-toed hipparion horses. After ontogenetic variation was removed, it was found that these horses usually exhibited $V$s between about 4-10% for measured characters.

Mayr (1969) states that $V$s of taxonomic relevance are usually less than 30%, however this figure does not filter out the various sources of variation (genetic, ontogenetic, and geographic; see Van Valen, 1965, 1969). In the context of the present paper, Simpson, Roe, and Lewontin (1960, p. 91) state that:

> Discernment of the meaning of a value of $V$ is largely a matter of experience. Its interpretation on functional zoological grounds depends on nonnumerical biological knowledge. We have compared hundreds of $V$s for linear dimensions of anatomical elements of mammals. As a matter of observation, the great majority of them lie between 4 and 10, and 5 and 6 are good average points. Much lower values usually indicate that the sample was not adequate to show the variability. Much higher values usually indicate that the sample was not pure, for instance, that it included animals of decidedly different ages or of different minor taxonomic divisions.

Despite the plethora of studies that deal with variation, there has been little research that compares variation between fossil and extant species of the same group. In many mammalian taxa (e.g., carnivores and other rarely encountered groups), the number of available fossil specimens is not sufficiently large to provide adequate statistical samples. However, this is not the case for fossil horses, which are abundant and widespread in Cenozoic terrestrial deposits of North America. They therefore provide an excellent group in which to study variation and to test the correspondence between paleopopulations and demes as well as morphospecies and extant species.

**Materials and methods**

Selected fossil and modern osteological collections were studied from the following institutions: American Museum of Natural History, New York (AMNH); Department of Biology, Midwestern State University, Wichita Falls, Texas (MSU); Department of Earth Sciences, University of California, Riverside (UCR); Florida State Museum,

University of Florida, Gainesville (UF); Museum of Comparative Zoology, Harvard University, Cambridge, Massachusetts (MCZ); Museum of Geology, South Dakota School of Mines, Rapid City (SDSM); National Museum of Natural History, Washington, D. C. (USNM); Natural History Museum of Los Angeles County, Los Angeles, California (LACM); Panhandle-Plains Museum, Canyon, Texas (PPM); Texas Memorial Museum, Austin, Texas (TMM); University of California Museum of Paleontology, Berkeley (UCMP); Yale Peabody Museum of Natural History, New Haven (YPM).

*Fossil quarry populations*

Five paleopopulations were chosen from museum collections because they included well-preserved, relatively large samples (N = 19-35). Hypsodont horses are subject to significant ontogenetic variation in dental characters. Accordingly, juvenile and old individuals were removed from the study samples; only specimens in "early maturity" and "late maturity" stages (as defined by MacFadden, 1984, p. 20) were included in this analysis. Although appropriately large samples are available for primitive (Eocene) low-crowned horses, e.g., *Hyracotherium*, these taxa can be highly dimorphic (Gingerich, 1981); they are therefore not used in this study. The five fossil quarry samples used here are as follows (abbreviations refer to collections studied in the institutions listed above):

1) *Parahippus leonensis*, three-toed primitive mixed feeder, Thomas Farm, Gilchrist County, Florida, early Miocene (Hemingfordian) age, ca. 18 Ma; N = 26 (AMNH, MCZ, SDSM, UF).

2) *Merychippus primus*, three-toed grazer, Thompson Quarry, Stonehouse Draw, Sioux County, Nebraska, middle Miocene (Hemingfordian) age, ca. 17 Ma; N = 19 (AMNH).

3) *Hipparion tehonense*, three-toed grazer, McAdams Quarry, Donley County, Texas, middle Miocene (Clarendonian) age, ca. 10 Ma; N = 23 (AMNH).

4) *Dinohippus leidyanus*, one-toed advanced grazer, Guymon Quarry, Texas County, Oklahoma, late Miocene (Hemphillian) age, ca. 6 Ma; N = 37 (AMNH).

5) *Equus simplicidens*, extinct species of the living genus of horses, Gidley Horse Quarry, Hagerman, Idaho, Pliocene (Blancan) age, ca. 3 Ma; N = 35 (AMNH, TMM, USNM, YPM).

In addition, a sixth sample represents 42 mature individuals of what is interpreted to be a deme of the extant Burchell's zebra, *Equus burchelli*, which was collected from a localized area, Guaso Nyiro, Kenya, by USNM field expeditions between 1909 and 1913.

It is surprising that there are very few relevant data or available museum collections for the dental and skull characters used here for extant populations of *Equus*. For this study, the fossil quarries actually provide a greater number of large samples.

*Fossil species samples*

These were chosen because they were represented by two or more samples (paleopopulations) from either diverse geographic locations or extensive periods of time. The temporal span represented for these species varies from a minimum of possibly several hundred thousand years to a maximum of about 6 million years. As with the paleopopulations above, in order to remove a possible component of variation caused by ontogeny, only mature individuals were included in this analysis (see MacFadden, 1984). The samples used here are as follows:

1) *Mesohippus bairdi*, a primitive, generalized three-toed horse from the early and middle Oligocene (Orellan and Whitneyan) of the High Plains (South Dakota, Wyoming, and Nebraska). This sample probably spans ca. 2-3 m. y.; it approximates the concept of a "chronospecies" (George, 1956); N = 29 (AMNH, LACM, SDSM, USNM, YPM).

2) *Cormohipparion goorisi*, a medium-sized, three-toed grazing horse from the middle Miocene (Barstovian) of the Texas

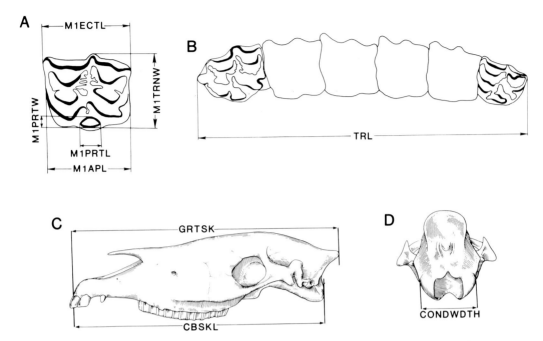

Fig. 9.1. Dental and skull characteristics of fossil and Recent horse samples measured during this study. A) Occlusal view of $M^1$. B) Tooth row length. C) Lateral view of skull. D) Condylar width. A complete description of these characters is presented in the "Materials and Methods" section.

Gulf Coastal Plain. This sample probably spans several hundred thousand to no more than 1 or 2 m. y.; N = 7 (AMNH).

3) *Protohippus simus*, a medium-sized, three-toed grazer from the middle Miocene (Barstovian) of northern Nebraska. The time span for this sample is probably the same as for *Cormohipparian goorisi*; N = 7 (AMNH).

4) *Nannippus minor*, a tiny, hypsodont, three-toed horse represented by samples from the late Miocene (Clarendonian-Hemphillian) of Florida, Mexico, and Texas. With a known temporal distribution of about 6 m.y. (from ca. 11-5 Ma), this species is the longest ranging of the taxa studied here, N = 29 (AMNH, LACM, MSU, UALP, UF).

5) *Dinohippus mexicanus*, a large, advanced one-toed extinct horse that is probably the closest sister-species of Pleistocene *Equus*. The study sample consists of individuals from the early Pliocene (Hemphillian) of Florida and Mexico, ca. 4-5 Ma; N = 25 (LACM, TMM, UF).

As with the relevant, available populations of extant *Equus* noted above, few natural (i.e., not zoological park or domesticated) species samples of *Equus* exist in museum collections. For this study, the following sample was assembled from three separate museum collections.:

6) *Equus grevyi*, Grevy's Zebra, from Kenya and Ethiopia; N = 13 (AMNH, USNM, YPM).

*Statistical analysis*

For each of these samples a suite of nine

(six dental and three cranial) characters were measured (Fig. 9.1). These were chosen because they: 1) are abundantly represented in both fossil and Recent material or 2) could be used to test Yablokov's hypothesis of drift of $V$ (the mean values of the smallest characters are in the 1-10 mm range and the largest are in the 100-1000 mm range):

1) M1ECTL; ectoloph length on $M^1$.
2) M1APL; greatest anteroposterior length of $M^1$ excluding ectoloph.
3) M1TRNW; greatest transverse width of $M^1$.
4) M1PRTL; greatest anteroposterior length of $M^1$ protocone.
5) M1PRTW; greatest transverse width of $M^1$ protocone.
6) TRL; greatest cheek tooth row length (characters 1-6 are measured excluding cement).
7) CONDWDTH; greatest width in transverse plane of occipital condyle.
8) GRTSKL; greatest skull length, distance from anteromost point on premaxilla to posteromost point on saggital-nuchal crest.
9) CBSKL; condylar-basal skull length, distance from anteromost point on premaxilla to posteromost point of occipital condyle.

All the statistical analyses were done using standard SAS programs available on the University of Florida Faculty Support Center IBM 4341 mainframe computer. The statistics of comparing variation of two or more samples has recently been the subject of much discussion in the literature. The reasoning for the methodology employed in this study is discussed in detail below.

## Results and discussion
### Patterns of variation
The coefficient of variation, $V$, is the standard unit of expressing variation in systematics. By dividing the standard deviation $s$ by the mean $x$ (and then multiplying by 100 to obtain percent), the problem of comparing, for example, variation in mouse and elephant femora is supposedly eliminated. However, this is not totally true. Workers have noted what Yablokov (1974) termed "drift," in which characters with smaller $x$s tend to exhibit larger $V$s (see also, e.g., MacFadden, 1974; Penguilly, 1974). In addition, linear characters are usually less variable than those of surface or volume (Lande, 1977, states that the ratio of these three types of characters is, respectively, 1:2:3). Therefore, although frequently used for comparisons, these limitations must be considered when comparing $V$s of different characters and taxa.

Representative variability profiles for three fossil and one Recent horse populations are shown in Figure 9.2. In these, the nine characters measured during this study (Fig. 9.1) are presented along the abscissa. As was done by Yablokov (1974), the characters are ordered in sequence of decreasing mean values, with the greatest on the left and smallest on the right. The results presented here confirm those of Yablokov (1974, and other references cited above): namely, characters with smaller mean values tend to exhibit larger $V$s. Of relevance to fossil horses, many of the smaller characters are those of the dentition, which are frequently used in taxonomy, and, as would be predicted, exhibit relatively large $V$s. In particular, measurements of the protocone (M1PRTL, M1PRTW), so commonly cited in traditional hipparion systematics, are consistently high in variation (i.e., $10 < V < 20$). This observation, along with the fact that hypsodont horse dentitions are notoriously prone to ontogenetic variation, dictates that the contributing factors of variation be carefully assessed before a particular character is used in taxonomy.

### Comparison of variation
As mentioned above, comparison of variation observed in different samples has been the subject of much discussion in the recent systematic literature. What may seem to be a routine problem to the practicing systematist is actually rather complex.

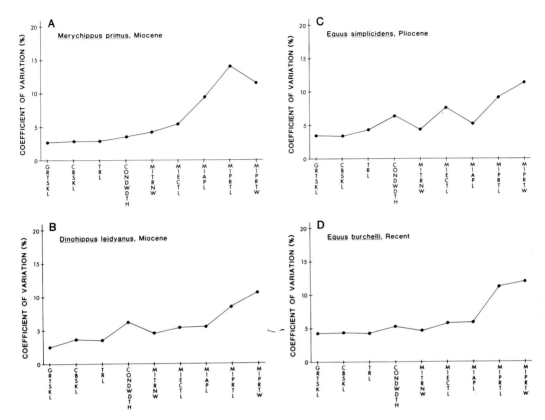

Fig. 9.2. Variability profiles of four representative paleopopulations or extant populations of horses using the nine characters measured during this study. These plots show "drift" (*sensu* YABLOKOV, 1974), i.e., characters with smaller mean values (to the right), show relatively larger coefficients of variation, $V$. A) *Merychippus primus*, middle Miocene three-toed grazer from Nebraska. B) *Dinohippus leidyanus*, late Miocene one-toed grazer from Oklahoma. C) *Equus simplicidens*, Pliocene species of modern day genus from Idaho. D) *Equus burchelli*, recent sample from Kenya. Further data for these samples are presented in "Materials and Methods."

Sokal and Braumann (1980) and Sokal and Rohlf (1981) state that, for most cases, a comparison of the variation of two samples, $V_1$ and $V_2$, should be done with an $F$-test on the sample variances, $s_1$ and $s_2$. The $F$ statistic is calculated simply by:

$$F = s_1/s_2 \text{ or } s_2/s_1 \qquad (1)$$

with the numerator as the larger of the two variances and the degrees of freedom as N-1 for each sample.

Van Valen (1978) stated that the $F$-test should not be used to compare equality of variances, because it is very sensitive to departures from normality of the samples being analyzed (i.e., it is not robust). He suggested that the Levene's test be employed instead (Levene, 1960; this is also discussed in Sokal and Braumann, 1980). Van Valen (1978, p. 34) clearly explains how to do this test: "For each sample, find the mean. Then calculate the absolute difference of each original datum from the mean. This is a new variable; the more varying the sample, the higher the values will be. Then find the mean and variance of the new variable. The means of the new variable for each population are then tested for equality by a $t$-test or an analysis of variance."

In a very detailed study, Schultz (1985) compared seven possible tests of equality of variances and coefficients of variation. He concluded that of the seven, two forms of the Levene's test are the most robust. These two use the sample median (not the mean as in Levene's original test as described above by Van Valen) as the estimate of central location. In the first method, the median-log (Levene) test, for each datum, the absolute difference is found between the log of each datum and the median of the log transformed sample:

$$Y_i = |\ln X_i - Md(\ln X_i)| \quad (2)$$

This is statistically more powerful (i.e., "the ability to reject the null hypothesis when in fact it is false," Schultz, 1985, p. 450) if the sample distributions being tested are positively skewed. On the other hand, if the sample distributions are symmetrical or negatively skewed, the median-ratio form of the Levene's test should be employed (Schultz, 1985), which is represented for each datum as:

$$Y_i = |X_i - Md(X_i)| / Md(X_i) \quad (3)$$

Given the recent discussions in the literature, four different tests were done in this study using the data from the fossil and Recent horse samples, i.e., 1) the standard $F$-test (e.g., Sokal and Rohlf, 1981), 2) the ratio of squared coefficients of variation (Lewontin, 1966), 3) the median-log form of Levene's test (Schultz, 1985), and 4) the median-ratio form of Levene's test (Schultz, 1985). Sokal and Braumann (1980) state that, when comparing $Vs$, it is necessary to correct for small sample sizes with the equation:

$$V' = V(1+(1/4N)) \quad (4)$$

As the sample size increases ($N > 5$), then this correction becomes negligible (Sokal and Braumann, 1980). All calculations of $V^2$ ratios for the samples studied here were therefore routinely corrected for sample size using equation 4.

The calculations of the statistics for the median-ratio and median-log forms of the Levene test were done as described earlier (Van Valen, 1978) but using the median instead of the mean (Schultz, 1985). Pair-wise comparisons of the fossil and recent samples were then done using a standard $t$-test.

*Results: Paleopopulations*

The nine morphological characters (Fig. 9.1) for the fossil quarry samples were compared in pair-wise ratios with those of the Recent sample of *Equus burchelli* (Table 9.3). The null hypothesis states that the variances, coefficients of variation, or Levene's statistics are the same in the fossil and recent samples. The results in Table 9.3 indicate that for the four kinds of tests previously listed, the null hypothesis is rejected in three pair-wise comparisons for the $F$-test, five for the $V^2$ ratio, two in the median-ratio Levene's test, and two in the median-log Levene's test. Given these results, the following is concluded: 1) In contrast to the results of Schultz (1985), there seems to be only a slightly greater rejection of the null hypothesis using the $F$ and $V^2$ test versus the two forms of the Levene's test. If indeed this is the case, it may be related to the fact that some of the underlying distributions are skewed (Schultz, 1985). 2) In the majority of cases (ca. 95%, depending upon the particular test), the variation observed in the nine characters of the five fossil samples is not different from that of *Equus burchelli*. Therefore, in these and similar cases it is appropriate to refer to quarry samples as paleopopulations.

*Results: Species samples*

Pair-wise comparisons of five fossil species samples and *Equus grevyi* are presented in Table 9.4. As for the populations, each of nine characters (except when not available in the fossil samples) was compared for equality of variances, squared coefficients of variation, or the Levene's statistic with the four tests discussed above. For these, the null hypothesis (no significant differences at $p = .05$) is rejected in six cases for the $F$ test, 16 cases for the $V^2$, five cases for the median-ratio and five cases for the median-log forms of the Levene's test. For the species samples there seems to be a

Table 9.1. Univariate statistics for the nine measured characters (also see Fig. 9.1 and text) of the six populations of fossil and Recent horse species studied in this report.

| CHARACTER | N | $\bar{x}$ | Median | s | MIN | MAX | V | Norm[1] |
|---|---|---|---|---|---|---|---|---|
| *Parahippus leonensis* | | | | | | | | |
| M1ECTL | 26 | 16.09 | 15.75 | 1.37 | 14.1 | 21.5 | 8.52 | 0.01* |
| M1APL | 26 | 15.81 | 15.50 | 1.39 | 14.2 | 21.2 | 8.78 | 0.01* |
| M1TRNW | 26 | 18.70 | 18.75 | 1.03 | 16.2 | 21.1 | 5.53 | 0.58 |
| M1PRTL | 25 | 5.83 | 5.80 | 0.37 | 5.2 | 6.7 | 6.35 | 0.56 |
| M1PRTW | 25 | 5.32 | 5.30 | 0.67 | 4.0 | 6.9 | 12.63 | 1.00 |
| TRL | 18 | 94.92 | 95.05 | 3.40 | 88.8 | 102.7 | 3.58 | 0.21 |
| CONDWDTH | 8 | 40.33 | 41.45 | 3.56 | 34.7 | 44.0 | 8.82 | 0.31 |
| GRTSKL | 3 | 278.43 | 272.30 | 20.21 | 262.0 | 301.0 | 7.23 | 0.43 |
| CBSKL | 3 | 261.03 | 261.03 | 15.33 | 246.7 | 277.2 | 5.87 | 0.81 |
| *Merychippus primus* | | | | | | | | |
| M1ECTL | 19 | 15.88 | 16.10 | 0.85 | 14.2 | 17.0 | 5.37 | 0.28 |
| M1APL | 19 | 15.40 | 15.20 | 1.48 | 13.1 | 19.9 | 9.63 | 0.03* |
| M1TRNW | 19 | 17.34 | 17.40 | 0.73 | 16.4 | 18.6 | 4.19 | 0.24 |
| M1PRTL | 19 | 5.65 | 5.50 | 0.81 | 4.8 | 8.1 | 14.28 | 0.01* |
| M1PRTW | 19 | 4.87 | 4.70 | 0.57 | 4.4 | 6.3 | 11.60 | 0.01* |
| TRL | 16 | 96.01 | 96.35 | 2.71 | 91.2 | 101.5 | 2.82 | 0.92 |
| CONDWDTH | 6 | 42.87 | 42.75 | 1.42 | 41.0 | 44.9 | 3.31 | 0.91 |
| GRTSKL | 4 | 282.08 | 280.30 | 7.33 | 275.6 | 292.1 | 2.60 | 0.46 |
| CBSKL | 4 | 266.10 | 264.20 | 7.10 | 260.0 | 276.0 | 2.67 | 0.42 |
| *Hipparion tehonense* | | | | | | | | |
| M1ECTL | 23 | 18.43 | 18.50 | 1.18 | 15.8 | 20.3 | 6.39 | 0.68 |
| M1APL | 23 | 18.37 | 18.30 | 1.00 | 15.5 | 19.7 | 5.42 | 0.06 |
| M1TRNW | 22 | 19.38 | 19.55 | 0.87 | 18.0 | 21.0 | 4.48 | 0.64 |
| M1PRTL | 23 | 6.37 | 6.20 | 0.68 | 5.2 | 7.7 | 10.59 | 0.32 |
| M1PRTW | 22 | 3.97 | 3.90 | 0.49 | 3.1 | 5.0 | 12.04 | 0.37 |
| TRL | 21 | 123.27 | 124.70 | 6.05 | 108.3 | 134.6 | 4.91 | 0.57 |
| CONDWDTH | 3 | 60.53 | 60.30 | 1.46 | 59.2 | 62.1 | 2.42 | 0.72 |
| GRTSKL | 4 | 384.2 | 374.80 | 24.99 | 366.6 | 420.6 | 6.51 | 0.15 |
| CBSKL | 4 | 362.63 | 357.10 | 22.21 | 342.1 | 394.2 | 6.12 | 0.31 |
| *Dinohippus leidyanus* | | | | | | | | |
| M1ECTL | 37 | 24.45 | 24.70 | 1.31 | 21.3 | 27.2 | 5.36 | 0.18 |
| M1APL | 37 | 23.65 | 23.70 | 1.31 | 20.4 | 26.4 | 5.53 | 0.73 |
| M1TRNW | 37 | 25.26 | 25.10 | 1.18 | 23.1 | 28.1 | 4.67 | 0.49 |
| M1PRTL | 38 | 8.10 | 7.85 | 0.87 | 6.2 | 9.7 | 10.70 | 0.37 |
| M1PRTW | 38 | 5.15 | 5.20 | 0.43 | 4.3 | 5.9 | 8.35 | 0.36 |
| TRL | 36 | 159.14 | 158.40 | 5.23 | 150.0 | 174.0 | 3.29 | 0.56 |
| CONDWDTH | 4 | 65.03 | 66.15 | 4.07 | 59.2 | 68.6 | 6.26 | 0.35 |
| GRTSKL | 4 | 487.00 | 486.65 | 11.95 | 476.1 | 498.6 | 2.45 | 0.14 |
| CBSKL | 4 | 453.65 | 461.65 | 17.31 | 427.8 | 463.5 | 3.82 | 0.02* |

*Equus simplicidens*

| | | | | | | | | |
|---|---|---|---|---|---|---|---|---|
| M1ECTL | 35 | 27.33 | 26.60 | 2.07 | 23.9 | 33.4 | 7.57 | 0.03 |
| M1APL | 34 | 26.59 | 26.10 | 1.33 | 23.5 | 29.4 | 5.01 | 0.31 |
| M1TRNW | 33 | 28.21 | 28.20 | 1.20 | 26.1 | 30.9 | 4.24 | 0.72 |
| M1PRTL | 30 | 11.21 | 11.35 | 1.03 | 9.0 | 13.9 | 9.20 | 0.96 |
| M1PRTW | 29 | 5.52 | 5.60 | 0.64 | 4.5 | 7.0 | 11.52 | 0.47 |
| TRL | 34 | 188.46 | 188.60 | 7.85 | 168.8 | 204.3 | 4.17 | 0.90 |
| CONDWDTH | 34 | 82.42 | 82.65 | 5.33 | 70.3 | 92.5 | 6.46 | 0.77 |
| GRTSKL | 32 | 612.63 | 610.75 | 20.24 | 581.0 | 656.8 | 3.30 | 0.40 |
| CBSKL | 34 | 570.94 | 573.25 | 19.61 | 531.7 | 614.0 | 3.44 | 0.81 |

*Equus burchelli*

| | | | | | | | | |
|---|---|---|---|---|---|---|---|---|
| M1ECTL | 42 | 22.20 | 22.35 | 1.29 | 19.0 | 25.6 | 5.80 | 0.96 |
| M1APL | 42 | 22.14 | 22.05 | 1.29 | 18.4 | 24.8 | 5.85 | 0.70 |
| M1TRNW | 42 | 23.31 | 23.40 | 1.12 | 20.7 | 25.3 | 4.82 | 0.22 |
| M1PRTL | 41 | 9.86 | 9.80 | 1.14 | 7.2 | 12.2 | 11.57 | 0.77 |
| M1PRTW | 42 | 4.83 | 4.83 | 0.58 | 3.9 | 6.9 | 12.04 | 0.01* |
| TRL | 42 | 143.80 | 144.25 | 5.81 | 126.6 | 155.0 | 4.04 | 0.54 |
| CONDWDTH | 41 | 75.00 | 74.50 | 3.90 | 66.1 | 84.9 | 5.20 | 0.55 |
| GRTSKL | 41 | 504.18 | 500.70 | 21.08 | 462.2 | 559.3 | 4.18 | 0.05 |
| CBSKL | 41 | 472.18 | 471.20 | 19.80 | 429.7 | 526.8 | 4.19 | 0.26 |

[1] Probability that sample is normally distributed; "*" indicates that null hypothesis is rejected at $p<0.05$.

Table 9.2. Univariate statistics for the nine measured characters (also see Fig. 9.1 and text) of the six species samples of fossil and Recent horses studied in this report.

| CHARACTER | N | $\bar{x}$ | Median | s | MIN | MAX | V | Norm[1] |
|---|---|---|---|---|---|---|---|---|
| *Nannippus minor* | | | | | | | | |
| M1ECTL | 26 | 14.51 | 14.5 | 0.94 | 12.9 | 17.1 | 6.44 | 0.46 |
| M1APL | 27 | 14.46 | 14.5 | 0.98 | 12.9 | 17.3 | 6.74 | 0.37 |
| M1TRNW | 25 | 14.16 | 14.3 | 0.61 | 12.9 | 15.4 | 4.34 | 0.47 |
| M1PRTL | 26 | 5.46 | 5.45 | 0.72 | 3.7 | 6.8 | 13.22 | 0.86 |
| M1PRTW | 26 | 3.27 | 3.4 | 0.42 | 2.3 | 3.6 | 12.88 | <0.01* |
| CONDWDTH | 1 | 35.4 | - | - | - | - | - | - |
| *Cormohipparion goorisi* | | | | | | | | |
| M1ECTL | 7 | 19.14 | 18.4 | 1.56 | 17.7 | 21.9 | 8.13 | 0.25 |
| M1APL | 7 | 18.2 | 17.9 | 1.01 | 17.2 | 19.8 | 5.57 | 0.24 |
| M1TRNW | 7 | 20.61 | 20.3 | 1.20 | 18.9 | 22.3 | 5.80 | 0.85 |
| M1PRTL | 7 | 6.64 | 6.4 | 0.53 | 6.1 | 7.5 | 8.05 | 0.08 |
| M1PRTW | 7 | 4.8 | 4.6 | 0.63 | 4.0 | 6.0 | 13.18 | 0.46 |
| TRL | 5 | 116.54 | 118.4 | 6.07 | 109.9 | 122.2 | 5.21 | 0.14 |
| CONDWDTH | 3 | 49.5 | 50.1 | 5.62 | 43.6 | 54.8 | 11.36 | 0.83 |
| GRTSKL | 2 | 335.65 | 335.65 | 4.74 | 332.3 | 339.0 | 1.41 | 1.0 |
| CBSKL | 2 | 308.9 | 308.9 | 7.64 | 303.5 | 314.3 | 2.47 | 1.0 |

*Protohippus simus*

| | | | | | | | | |
|---|---|---|---|---|---|---|---|---|
| M1ECTL | 7 | 22.39 | 21.9 | 1.46 | 21.2 | 25.4 | 6.51 | 0.05 |
| M1APL | 7 | 21.36 | 20.8 | 1.25 | 20.2 | 23.7 | 5.87 | 0.15 |
| M1TRNW | 7 | 22.96 | 23.0 | 1.17 | 19.4 | 24.7 | 7.54 | 0.10 |
| M1PRTL | 7 | 9.04 | 8.7 | 1.09 | 7.8 | 10.7 | 21.96 | 0.28 |
| M1PRTW | 7 | 4.94 | 4.8 | 0.64 | 3.6 | 5.5 | 12.95 | 0.04 |
| TRL | 5 | 140.92 | 139.8 | 4.51 | 135.8 | 146.6 | 3.21 | 0.61 |
| CONDWDTH | 4 | 58.98 | 57.2 | 5.26 | 55.1 | 66.4 | 8.91 | 0.25 |
| GRTSKL | 4 | 413.75 | 413.6 | 4.21 | 409.5 | 418.3 | 1.02 | 0.42 |
| CBSKL | 4 | 389.6 | 390.25 | 5.72 | 382.0 | 395.9 | 1.47 | 0.56 |

*Mesohippus bairdii*

| | | | | | | | | |
|---|---|---|---|---|---|---|---|---|
| M1ECTL | 28 | 12.43 | 12.45 | 0.97 | 9.7 | 14.7 | 7.80 | 0.47 |
| M1APL | 27 | 11.2 | 11.2 | 1.10 | 9.3 | 15.5 | 9.85 | <0.01* |
| M1TRNW | 24 | 14.04 | 14.2 | 1.16 | 10.7 | 15.8 | 8.28 | 0.01* |
| M1PRTL | 22 | 5.54 | 5.1 | 1.46 | 4.4 | 11.3 | 26.42 | <0.01* |
| M1PRTW | 19 | 6.69 | 6.6 | 0.97 | 5.0 | 8.9 | 14.49 | 0.70 |
| TRL | 29 | 68.21 | 68.0 | 2.91 | 60.9 | 74.5 | 4.27 | 0.98 |
| CONDWDTH | 18 | 32.36 | 32.8 | 1.80 | 29.1 | 35.5 | 5.58 | 0.70 |
| GRTSKL | 7 | 189.46 | 191.8 | 8.72 | 170.5 | 196.0 | 4.60 | <0.01* |
| CBSKL | 9 | 180.81 | 178.5 | 8.34 | 166.3 | 195.0 | 4.61 | 0.96 |

*Dinohippus mexicanus*

| | | | | | | | | |
|---|---|---|---|---|---|---|---|---|
| M1ECTL | 22 | 24.10 | 24.1 | 1.44 | 22.0 | 28.1 | 5.99 | 0.20 |
| M1APL | 25 | 23.34 | 23.0 | 1.37 | 21.4 | 26.9 | 5.87 | 0.23 |
| M1TRNW | 25 | 24.06 | 24.1 | 1.22 | 20.8 | 26.1 | 5.08 | 0.48 |
| M1PRTL | 26 | 9.36 | 9.2 | 1.37 | 7.5 | 12.5 | 14.69 | 0.34 |
| M1PRTW | 25 | 4.98 | 5.1 | 0.42 | 3.9 | 5.6 | 8.52 | 0.18 |
| TRL | 3 | 154.9 | 150.4 | 8.76 | 149.3 | 165.0 | 5.66 | 0.13 |

*Equus grevyi*

| | | | | | | | | |
|---|---|---|---|---|---|---|---|---|
| M1ECTL | 13 | 25.7 | 25.8 | 1.50 | 23.6 | 29.3 | 5.84 | 0.24 |
| M1APL | 13 | 26.17 | 26.3 | 1.58 | 23.6 | 29.2 | 6.03 | 0.71 |
| M1TRNW | 13 | 26.67 | 26.6 | 0.70 | 25.0 | 27.8 | 2.63 | 0.48 |
| M1PRTL | 13 | 10.98 | 11.4 | 0.98 | 8.7 | 12.0 | 8.92 | 0.04* |
| M1PRTW | 13 | 5.47 | 5.4 | 0.59 | 4.7 | 6.9 | 10.78 | 0.39 |
| TRL | 13 | 175.23 | 175.1 | 4.53 | 166.5 | 181.7 | 2.59 | 0.65 |
| CONDWDTH | 13 | 81.33 | 80.7 | 2.26 | 77.8 | 85.0 | 2.7 | 0.49 |
| GRTSKL | 13 | 611.85 | 611.3 | 15.75 | 590.5 | 639.5 | 2.57 | 0.50 |
| CBSKL | 13 | 578.59 | 580.0 | 13.56 | 558.6 | 601.5 | 2.34 | 0.65 |

[1] Probability that sample is normally distributed; "*" indicates that null hypothesis is rejected at p<0.05.

Table 9.3. Comparisons of five quarry samples of fossil horses with Recent population of *Equus burchelli*. Results of one-tailed tests are indicated for F-statistic (F), $V^2$ ratio ($V^2$), and median-ratio (MR) and median-log (ML) forms of the Levene's test. DF1 and DF2 represent degrees of freedom for, respectively, *Equus burchelli* and fossil samples for the character (CHAR) indicated (also see abbreviations in Fig. 9.1 and Materials and Methods). Probablities of rejection of the null hypothesis are given to the right of the actual statistic. The symbol "*" indicates; 1) next to the DF column that the original distribution is not normally distributed; or 2) for the PROBF, PROBV, PROBMR, and PROBML columns, the null hypothesis of no difference between the fossil sample and that of *Equus burchelli* is rejected at p<0.05.

| CHAR | DF1 | DF2 | F | PROBF | $V^2$ | PROBV | T-MR | PROBMR | T-ML | PROBML |
|---|---|---|---|---|---|---|---|---|---|---|
| *Parahippus leonensis* | | | | | | | | | | |
| M1ECTL | 41 | 25* | 1.13 | 0.36 | 2.17 | 0.01* | 0.74 | 0.46 | 0.62 | 0.54 |
| M1APL | 41 | 25* | 1.49 | 0.13 | 2.27 | 0.01* | 0.81 | 0.42 | 0.72 | 0.47 |
| M1TRNW | 41 | 25 | 0.85 | 0.66 | 1.33 | 0.20 | 0.28 | 0.78 | 0.26 | 0.79 |
| M1PRTL | 40 | 24 | 0.11 | 1.00 | 0.30 | 1.00 | 4.61 | 0.00* | 2.96 | 0.00* |
| M1PRTW | 41* | 24 | 1.32 | 0.21 | 1.11 | 0.38 | 0.29 | 0.77 | 0.61 | 0.54 |
| TRL | 41 | 17 | 0.34 | 0.99 | 0.80 | 0.68 | 0.94 | 0.35 | 0.94 | 0.35 |
| CONDWDTH | 40 | 7 | 0.83 | 0.57 | 3.02 | 0.01* | 1.55 | 0.13 | 1.56 | 0.12 |
| GRTSKL | 40* | 2 | 0.92 | 0.41 | 3.49 | 0.04* | 0.56 | 0.58 | 0.56 | 0.58 |
| CBSKL | 40 | 2 | 0.60 | 0.55 | 2.28 | 0.12 | 0.44 | 0.66 | 0.44 | 0.66 |
| *Merychippus primus* | | | | | | | | | | |
| M1ECTL | 41 | 18 | 0.44 | 0.97 | 0.87 | 0.61 | 0.21 | 0.84 | 0.09 | 0.93 |
| M1APL | 41 | 18* | 1.71 | 0.08 | 2.75 | 0.00* | 1.04 | 0.30 | 1.02 | 0.31 |
| M1TRNW | 41 | 18 | 0.42 | 0.98 | 0.76 | 0.73 | 0.36 | 0.72 | 0.40 | 0.69 |
| M1PRTL | 40 | 18* | 0.50 | 0.94 | 1.54 | 0.13 | 0.25 | 0.80 | 0.06 | 0.95 |
| M1PRTW | 41* | 18* | 0.91 | 0.57 | 0.94 | 0.54 | 0.57 | 0.57 | 0.63 | 0.53 |
| TRL | 41 | 15 | 0.22 | 1.00 | 0.50 | 0.93 | 1.51 | 0.14 | 1.50 | 0.14 |
| CONDWDTH | 40 | 5 | 0.13 | 0.98 | 0.43 | 0.82 | 1.45 | 0.15 | 1.44 | 0.16 |
| GRTSKL | 40* | 3 | 0.12 | 0.95 | 0.43 | 0.73 | 1.19 | 0.24 | 1.22 | 0.23 |
| CBSKL | 40 | 3 | 0.13 | 0.94 | 0.45 | 0.72 | 1.12 | 0.27 | 1.15 | 0.26 |
| *Hipparion tehonense* | | | | | | | | | | |
| M1ECTL | 41 | 22 | 0.84 | 0.66 | 1.23 | 0.28 | 0.44 | 0.66 | 0.42 | 0.67 |
| M1APL | 41 | 22 | 0.77 | 0.74 | 0.87 | 0.63 | 0.28 | 0.78 | 0.24 | 0.81 |
| M1TRNW | 41 | 21 | 0.60 | 0.89 | 0.87 | 0.63 | 0.03 | 0.98 | 0.03 | 0.98 |
| M1PRTL | 40 | 22 | 0.35 | 0.99 | 0.85 | 0.65 | 0.14 | 0.89 | 0.31 | 0.76 |
| M1PRTW | 41* | 21 | 0.68 | 0.83 | 1.01 | 0.47 | 0.17 | 0.86 | 0.01 | 0.99 |
| TRL | 41 | 20 | 1.08 | 0.40 | 1.49 | 0.14 | 1.12 | 0.27 | 1.12 | 0.27 |
| CONDWDTH | 40 | 2 | 0.16 | 0.85 | 0.25 | 0.78 | 2.18 | 0.03* | 2.20 | 0.03* |
| GRTSKL | 40* | 3 | 1.40 | 0.26 | 2.71 | 0.06 | 0.50 | 0.62 | 0.48 | 0.63 |
| CBSKL | 40 | 3 | 1.26 | 0.30 | 2.38 | 0.08 | 0.31 | 0.76 | 0.24 | 0.81 |
| *Dinohippus leidyanus* | | | | | | | | | | |
| M1ECTL | 41 | 36 | 1.04 | 0.45 | 0.86 | 0.68 | 0.58 | 0.57 | 0.50 | 0.62 |
| M1APL | 41 | 36 | 1.33 | 0.19 | 0.89 | 0.64 | 0.46 | 0.64 | 0.42 | 0.68 |
| M1TRNW | 41 | 36 | 1.10 | 0.38 | 0.95 | 0.56 | 0.36 | 0.72 | 0.25 | 0.81 |
| M1PRTL | 40 | 37 | 0.58 | 0.95 | 0.85 | 0.69 | 0.12 | 0.91 | 0.05 | 0.96 |
| M1PRTW | 41* | 37 | 0.53 | 0.97 | 0.48 | 0.99 | 1.47 | 0.15 | 1.18 | 0.24 |
| TRL | 41 | 35 | 0.81 | 0.74 | 0.66 | 0.89 | 1.05 | 0.30 | 1.12 | 0.27 |
| CONDWDTH | 40 | 3 | 1.09 | 0.36 | 1.62 | 0.20 | 0.01 | 0.99 | 0.08 | 0.94 |
| GRTSKL | 40* | 3 | 0.32 | 0.81 | 0.38 | 0.77 | 1.81 | 0.08 | 1.79 | 0.08 |
| CBSKL | 40 | 3* | 0.76 | 0.52 | 0.92 | 0.44 | 0.50 | 0.62 | 0.43 | 0.67 |
| *Equus simplicidens* | | | | | | | | | | |
| M1ECTL | 41 | 34* | 2.58 | 0.00* | 1.71 | 0.05 | 1.05 | 0.30 | 0.89 | 0.38 |
| M1APL | 41 | 33 | 1.37 | 0.17 | 0.74 | 0.81 | 0.51 | 0.61 | 0.61 | 0.54 |
| M1TRNW | 41 | 32 | 1.13 | 0.35 | 0.77 | 0.78 | 0.26 | 0.80 | 0.31 | 0.75 |
| M1PRTL | 40 | 29 | 0.82 | 0.71 | 0.64 | 0.89 | 1.19 | 0.24 | 1.10 | 0.28 |
| M1PRTW | 41* | 28 | 1.18 | 0.31 | 0.92 | 0.59 | 0.23 | 0.82 | 0.13 | 0.90 |
| TRL | 41 | 33 | 1.82 | 0.03* | 1.07 | 0.41 | 0.00 | 1.00 | 0.02 | 0.99 |
| CONDWDTH | 40 | 33 | 1.87 | 0.03* | 1.55 | 0.09 | 1.56 | 0.12 | 1.63 | 0.11 |
| GRTSKL | 40* | 31 | 0.92 | 0.59 | 0.53 | 0.96 | 0.54 | 0.59 | 0.52 | 0.60 |
| CBSKL | 40 | 33 | 0.98 | 0.52 | 0.67 | 0.88 | 0.63 | 0.53 | 0.56 | 0.58 |

# VARIATION IN HORSES

Table 9.4. Comparisons of five species samples of fossil horses with Recent *Equus grevyi*. Results of one-tailed tests are indicated for F-statistic (F), $V^2$ ratio ($V^2$, and median-ratio (MR) and median-log (ML) froms of Levene's test. DF1 and DF2 represent degrees of freedom for, *Equus grevyi* and fossil samples for character indicated (also see abbreviations in Fig. 9.1 and Materials and Methods). Probabilities of rejection of the null hypothesis are given to the right of the actual statistic. The symbol "*" indicates; 1) next to the DF column that the original sample is not normally distributed; or 2) for the PROBF, PROBV, PROBMR, and, PROBML columns, the null hypothesis of no difference between the fossil sample and that of *Equus grevyi* is rejected at p<0.05.

| CHAR | DF1 | DF2 | F | PROBF | $V^2$ | PROBV | T-MR | PROBMR | T-ML | PROBML |
|---|---|---|---|---|---|---|---|---|---|---|
| *Mesohippus bairdii* | | | | | | | | | | |
| M1ECTL | 12 | 27* | 0.42 | 0.97 | 1.77 | 0.15 | 1.24 | 0.22 | 1.27 | 0.21 |
| M1APL | 12 | 26* | 0.49 | 0.94 | 2.59 | 0.04* | 0.97 | 0.34 | 0.94 | 0.35 |
| M1TRNW | 12 | 23* | 2.76 | 0.04* | 9.74 | 0.00* | 2.70 | 0.01* | 2.61 | 0.01* |
| M1PRTL | 12* | 21 | 2.23 | 0.08 | 8.64 | 0.00* | 1.20 | 0.24 | 0.98 | 0.34 |
| M1PRTW | 12 | 18 | 2.69 | 0.04* | 1.79 | 0.15 | 1.05 | 0.30 | 1.13 | 0.27 |
| TRL | 12 | 28 | 0.41 | 0.97 | 2.66 | 0.04* | 1.96 | 0.06 | 1.94 | 0.06 |
| CONDWDTH12 | | 17 | 0.64 | 0.81 | 3.99 | 0.01* | 2.08 | 0.05 | 2.12 | 0.05 |
| GRTSKL | 12 | 6* | 0.31 | 0.92 | 3.31 | 0.04* | 0.37 | 0.72 | 0.40 | 0.69 |
| CBSKL | 12 | 8 | 0.38 | 0.91 | 3.95 | 0.02* | 1.51 | 0.15 | 1.49 | 0.15 |
| *Cormohipparion goorisi* | | | | | | | | | | |
| M1ECTL | 12 | 6 | 1.07 | 0.43 | 2.00 | 0.14 | 0.22 | 0.83 | 0.72 | 0.48 |
| M1APL1 | 12 | 6 | 0.41 | 0.86 | 0.88 | 0.54 | 0.22 | 0.83 | 0.31 | 0.76 |
| M1TRNW | 12 | 6 | 2.92 | 0.05 | 5.02 | 0.01* | 1.95 | 0.07 | 1.95 | 0.07 |
| M1PRTL | 12 | 6* | 0.30 | 0.93 | 0.84 | 0.56 | 0.15 | 0.88 | 0.37 | 0.72 |
| M1PRTW | 12 | 6 | 1.14 | 0.40 | 1.55 | 0.24 | 0.27 | 0.79 | 0.22 | 0.83 |
| TRL | 12 | 4 | 1.79 | 0.20 | 4.29 | 0.02* | 1.44 | 0.17 | 1.44 | 0.17 |
| CONDWDTH12 | | 2 | 6.19 | 0.01* | 18.86 | 0.00* | 1.37 | 0.19 | 1.29 | 0.22 |
| GRTSKL | 12 | 1 | 0.09 | 0.77 | 0.37 | 0.55 | 0.90 | 0.38 | 0.89 | 0.39 |
| CBSKL | 12 | 1 | 0.32 | 0.58 | 1.36 | 0.27 | 0.03 | 0.97 | 0.08 | 0.94 |
| *Protohippus simus* | | | | | | | | | | |
| M1ECTL | 12 | 6* | 0.94 | 0.50 | 1.29 | 0.33 | 0.12 | 0.90 | 0.05 | 0.96 |
| M1APL | 12 | 6 | 0.63 | 0.70 | 0.98 | 0.48 | 0.20 | 0.84 | 0.30 | 0.77 |
| M1TRNW | 12 | 6 | 6.12 | 0.00* | 8.51 | 0.00* | 1.36 | 0.19 | 1.32 | 0.20 |
| M1PRTL | 12* | 6 | 1.24 | 0.35 | 1.88 | 0.17 | 0.71 | 0.49 | 0.50 | 0.62 |
| M1PRTW | 12 | 6* | 1.17 | 0.38 | 1.49 | 0.26 | 0.10 | 0.92 | 0.12 | 0.90 |
| TRL | 12 | 4 | 0.99 | 0.45 | 1.63 | 0.23 | 0.50 | 0.62 | 0.49 | 0.63 |
| CONDWDTH12 | | 3 | 5.41 | 0.01* | 11.16 | 0.00* | 1.30 | 0.21 | 1.34 | 0.20 |
| GRTSKL | 12 | 3 | 0.07 | 0.97 | 0.17 | 0.91 | 2.42 | 0.03* | 2.42 | 0.03* |
| CBSKL | 12 | 3 | 0.18 | 0.91 | 0.43 | 0.74 | 1.30 | 0.21 | 1.30 | 0.21 |
| *Nannippus minor* | | | | | | | | | | |
| M1ECTL | 12 | 25 | 0.38 | 0.98 | 1.19 | 0.39 | 0.47 | 0.64 | 0.47 | 0.64 |
| M1APL | 12 | 26 | 0.38 | 0.98 | 1.23 | 0.36 | 0.44 | 0.66 | 0.40 | 0.69 |
| M1TRNW | 12 | 24 | 0.78 | 0.71 | 2.67 | 0.04* | 2.20 | 0.03* | 2.21 | 0.03* |
| M1PRTL | 12* | 25 | 0.54 | 0.91 | 2.16 | 0.08 | 1.81 | 0.08 | 1.52 | 0.14 |
| M1PRTW | 12 | 25* | 0.49 | 0.94 | 1.40 | 0.28 | 0.63 | 0.53 | 0.96 | 0.34 |
| *Dinohippus mexicanus* | | | | | | | | | | |
| M1ECTL | 12 | 21 | 0.92 | 0.58 | 1.04 | 0.49 | 0.41 | 0.69 | 0.39 | 0.70 |
| M1APL | 12 | 24 | 0.76 | 0.73 | 0.94 | 0.57 | 0.16 | 0.87 | 0.05 | 0.96 |
| M1TRNW | 12 | 24 | 3.04 | 0.02* | 3.66 | 0.01* | 2.48 | 0.02* | 2.45 | 0.02* |
| M1PRTL | 12* | 25 | 1.97 | 0.11 | 2.66 | 0.04* | 2.44 | 0.02* | 2.09 | 0.04* |
| M1PRTW | 12 | 24 | 0.51 | 0.92 | 0.61 | 0.85 | 0.78 | 0.44 | 0.60 | 0.55 |
| TRL | 12 | 2 | 3.74 | 0.05 | 5.40 | 0.02* | 0.46 | 0.65 | 0.44 | 0.67 |

greater frequency of rejection of the null hypothesis using the $V^2$ test. For the 152 pair-wise comparisons, only 32 (ca. 21%) were rejected. It can therefore be concluded that the variation in characters measured here is, in the great majority of cases, similar in fossil samples and that of *Equus grevyi*.

## Conclusions and Significance

A standard practice in paleontology is to assume that quarry samples, i.e., paleopopulations, and morphospecies have biological properties similar to, respectively, extant demes and species. However, this important assumption has heretofore remained little tested. The results of the present study indicate that, for horses:

1) As previously noted (e.g., Yablokov, 1974), the $V$s for the characters measured here increase with decreasing mean values.

2) The four tests employed here ($F$, $V^2$, median-log and median-ratio forms of Levene's test) indicate that in the majority of the pair-wise comparisons (95% for populations and 80% for species), there are no significant differences in the fossil and Recent samples.

3) It is therefore valid to assume that, at least in terms of variation, fossil and recent samples exhibit similar biological properties.

Within the last two decades it has become fashionable to investigate the "population dynamics" of fossil quarry samples in order to better understand the autecology of particular species. Much can be said from these kinds of studies about the age structure of a mammalian population, which in turn can lead to very interesting hypotheses about the natural history of a species. The present study suggests that once the depositional setting and taphonomy of a quarry sample is understood, then quarry samples (as exemplified by fossil horse species) probably represent similar biological units as those of modern populations and the autecology of the latter can thus be interpreted as an ancient analog.

In both populations and species there can be four principal sources of variation: ontogenetic, geographic, temporal, and evolutionary (phenotypic). As has been pointed out by previous workers ( e.g., Van Valen, 1965, 1969), if the first three sources can be controlled for, or eliminated, then the resulting observed variation is phenotypic, and it is highly important in systematics. Thus, a clear assessment of character variation and phenotypic change in both fossil and Recent samples is essential for broader interpretations of evolutionary processes.

## Ackowledgments

I thank the curators of the collections listed in Materials and Methods for permission to study relevant specimens. I am particularly indebted to Richard C. Hulbert, Jr. for his extensive assistance with the statistical analysis. Hulbert and S. David Webb provided comments on earlier versions of the manuscript. Ms. Wendy Zomlefer of UF prepared the illustrations. Computing was done using the Faculty Support Center at UF. This research was partially supported by National Science Foundation grant BSR-85 15003. This report is University of Florida Contribution to Paleobiology number 249.

## Bibliography

Cain, A. J. (1954): *Animal Species and their Evolution*. London.

Forsten, A. M. (1970): Variation in and between three populations of *Mesohippus bairdii* Leidy from the Big Badlands, South Dakota. -*Acta Zool. Fenn.*, 126: 1-16.

Gingerich, P. D. (1981): Variation, sexual dimorphism, and social structure in the early Eocene horse *Hyracotherium* (Mammalia, Perissodactyla). - *Paleobiology*, 7: 443-455.

George, T. N. (1956): Biospecies, chronospecies and morphospecies. -*Syst. Assoc. Publ.*, 2: 123-137.

Lande, R. (1977): On comparing coefficients of variation. -*Syst. Zool.*, 26: 214-217.

Levene, H. (1960): Robust tests for equality

of variances. - In: Okin, I., Ghurye, S. G., Hoeffding, W., Madow, W. G., and Mann, H. B. (eds): *Contributions to Probability and Statistics.* Stanford (Stanford Univ. Press), pp. 278-292.

Lewontin, R. C. (1966): On the measurement of relative variability. -*Syst. Zool.,* 179: 1-195.

MacFadden, B. J. (1984): Systematics and phylogeny of *Hipparion, Neohipparion, Nannippus,* and *Cormohipparion* (Mammalia, Equidae) from the Miocene and Pliocene of the New World. -*Bull. Amer. Mus. Nat. Hist.,* 179: 1-195.

Mayr, E. (1969): *Principles of Systematic Zoology.* New York (McGraw-Hill)

Penguilly, D. (1984): Developmental versus functional explanations for patterns of variability and correlation in the dentitions of foxes. -*J. Mammal.,* 65: 34-43.

Schultz, B. B. (1965): Levene's test for relative variation. -*Syst. Zool.,* 34: 449-456.

Simpson, G. G., Roe, A., and Lewontin, R. C. (1960): *Quantitative Zoology.* New York (Harcourt, Brace and Co.).

Sokal, R. R., and Braumann, C. A. (1980): Significance tests for coefficients of variation and variability profiles. - *Syst. Zool.,* 29: 50-66.

Sokal, R. R., and Rohlf, F. J. (1981): *Biometry: The principles and practice of statistics in biological research.* San Francisco (W.H. Freeman and Co.)

Sylvester-Bradley, P. C. (1951): The subspecies in paleontology. -*Geol. Mag.,* 88: 88-102.

Van Valen, L. (1963): Selection in natural populations: *Merychippus primus,* a fossil horse. -*Nature,* 197: 1181-1183.

Van Valen, L. (1964): Age in two fossil horse populations. - *Acta. Zool.,* 45: 93-106.

Van Valen, L. (1965): Selection in natural populations. III. Measurement and estimation. -*Evolution,* 19: 514-528.

Van Valen, L. (1969): Variation genetics of extinct animals. - *Amer. Nat.,* 103: 193-224.

Van Valen, L. (1978): The statistics of variation. -*Evol. Theory,* 4: 33-43.

Yablokov, A. V.(1974): *Variability of Mammals.* New Delhi, India (Amerind Pub. Co. Pvt. Ltd.).

# 10. THE EVOLUTION OF OLIGOCENE HORSES

## DONALD R. PROTHERO and NEIL SHUBIN

The popular image of Oligocene horses portrays their evolution as a single, orthogenetic lineage grading continuously from *Mesohippus* to *Miohippus*. We find that they are highly speciose, with a "bushy," branched phylogeny that includes at least four valid species of *Mesohippus* (*M. bairdi, M. westoni, M. exoletus, M. barbouri*) overlapping with several species of *Miohippus* (*M. obliquidens, M. assiniboiensis, M. intermedius, M. gidleyi, M. annectenc*, and *?M. equinanus*). The two genera are not a continuous intergrading sequence, but instead are marked by a distinctive set of changes in size, dentition, skull and foot features, and overlap by about 5 million years.

## Introduction

The evolution of horses has traditionally been among the best-documented examples of an evolutionary sequence in the fossil record. Huxley's 1870 series of European horses, and Marsh's 1879 genealogy of American horses were among the first such sequences to lend support to the theory of evolution. Early phylogenies presented the evolution of horses as a single unbranched lineage, but as more information emerged, the phylogeny began to look more branched and "bushy" (Stirton, 1940; Simpson, 1951).

Most horse phylogenies, however, were abstractions based on a few examples from each geological level. Until recently, very little work had been done on the detailed geometry of the phylogenetic pattern of horses. In the past, this may not have been so critical, since to most paleontologists (e.g., Simpson, 1953) horse evolution neatly fit into to the prevailing gradualistic neo-Darwinian view. But since the work of Eldredge and Gould (1972), detailed records have become critical in the dispute over the tempo and mode of speciation. Some authors (e.g., Stanley, 1979) have noticed the "bushiness" of the horse family tree, and the apparent stability of certain taxa (e.g., *Hyracotherium*), and have argued that the horse fossil record supports the punctuational mode of change. Others (e.g., Gingerich, 1980) have interpreted *Hyracotherium* in a gradualistic framework. Since only a small portion of the horse fossil record has been examined in detail, the dispute is far from settled.

One of the more fossiliferous portions of the horse record is the Oligocene history of *Mesohippus* and *Miohippus*. These horses are quite abundant and well preserved in localities such as the Big Badlands of South Dakota. Not only are specimens very abundant, but in the Frick Collection of the AMNH, the fossil horses have excellent stratigraphic documentation for each specimen. The stratigraphic details of the deposits bearing these horses has recently been improved (Prothero, 1982). With the help of magnetostratigraphy, greater precision of correlation and time control are also available. With this data base, it is now possible to examine the details of horse phylogeny during the Oligocene, and compare it to the two competing models of paleontological speciation.

At present, Oligocene horse phylogeny is encumbered by several sources of taxonomic confusion. The early history of horse taxonomy was characterized by extreme typological taxonomic splitting, and "horizontal" taxa that are partially defined by their stratigraphic occurrence. For example, Osborn (1918) distinguished

*The Evolution of Perissodactyls* (ed. D.R. Prothero & R.M. Schoch) Oxford Univ. Press, New York, 1989

*Mesohippus* and *Miohippus* by a stratigraphic boundary, rather than on morphological grounds. Stirton (1940) recorded 20 species of *Mesohippus* and 18 species of *Miohippus*, and did not attempt to synonymize any of them. More recent work (Forstén, 1970a, 1970b) has gone to the other extreme, lumping all species of *Mesohippus* into two species spanning an enormous amount of morphological variability across nearly eight million years. Some workers (e.g., Emry et al., 1987) are not satisfied with this arrangement, either.

The Frick Collection of the AMNH of fossil horses offers an unparalleled opportunity to resolve this dispute. Unlike the samples examined by Forstén, the Frick Collection allows us to place our phylogenetic hypothesis within a finely resolved stratigraphic context. If the diverse morphologies found at the same stratigraphic level by Forstén (and thus placed in the same species by her) have drastically different geographic and stratigraphic ranges, they may represent more than one species. If each of these morphologies has a distinctive and unique distribution in time and space, they are also useful biostratigraphically, and should, as a heuristic tool, be recognized as valid species. Thus, the systematic details of Oligocene horse evolution are important for evolutionary biology as well as for stratigraphy.

We have restricted this study to horses from the latest Chadronian, Orellan, and Whitneyan (30-33 Ma). The taxonomy of early and medial Chadronian horses is currently being investigated by Dr. R. J. Emry. We have excluded Arikareean and later horses from this study, since they are not germane to the details of horse evolution in the more fossiliferous Oligocene. Also, a satisfactory review of Arikareean horses requires a detailed study of the split between the equine and anchitheriine lineages, which is beyond the scope of the present report.

Abbreviations used in this paper are as follows: AMNH, American Museum of Natural History, New York, N.Y.; F:AM, Frick Collection, AMNH; CM, Carnegie Museum of Natural History, Pittsburgh, Penn.; l.f., local fauna; Ma, million years before present; MCZ, Museum of Comparative Zoology, Harvard University, Cambridge, Mass.; LACM, Los Angeles County Museum of Natural History, Los Angeles, Calif.; NMC, National Museum of Canada, Ottawa; SDSM, South Dakota School of Mines Museum of Geology, Rapid City, S.D.; USNM, United States National Museum, Smithsonian Institution, Washington, D.C.; YPM, Yale Peabody Museum, New Haven, Conn.

**Taxonomy and morphology**

Order Perissodactyla Owen, 1848
Family Equidae Gray, 1821
*Mesohippus* Marsh, 1875

*Palaeotherium*: Leidy, 1850
*Anchitherium*: Leidy, 1852
*Miohippus*: Hay, 1902 (in part)
*Pediohippus*: Schlaikjer, 1935 (in part)

*Revised Diagnosis.* Small equids ($M^{1-3}$ length = 27-37 mm) with molarized P2/2. $I^{1-3}$ with shallow pitted crowns. Longer face than *Epihippus*, *Orohippus*, or *Hyracotherium*, with extended premaxilla and longer diastema. Angle of jaw posteriorly rounded, without a notch. No contact between the cuboid and third metatarsal. Metacarpal five usually reduced. The genus typically has a very shallow facial fossa.

*Type species. Mesohippus bairdi* (Leidy, 1850).

*Included species. Mesohippus westoni, Mesohippus exoletus, Mesohippus barbouri,* and several Chadronian species not discussed here.

*Distribution.* Late Eocene (late Duchesnean, Porvenir l.f., Vieja Group, Texas) to mid-Oligocene (late Orellan, top of the Upper Nodules, Middle *Oreodon* beds). Known from the White River Group in South Dakota, North Dakota, Nebraska, Wyoming, Colorado, and from White River equivalents in Montana, Saskatchewan, Texas, and possibly Florida.

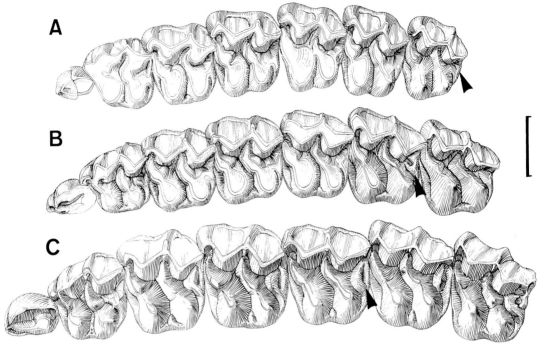

Fig. 10.1. Representative examples of hypostyles in Oligocene horses. Descriptive system modified after Clark and Beerbower (1967, Table 18 key). Type 1 hypostyles (A, *Mesohippus westoni*, F:AM 74023) are thin ridges with almost no cusp or spur projecting anteriorly. Type 2 hypostyles (B, *Mesohippus bairdi*, F:AM 115366) have a small spur projecting anteriorly from the hypostyle ridge on most of the cheek teeth (particularly apparent on $P^2$-$M^2$ in this example). Type 3 hypostyles (C, *Miohippus assiniboiensis*, F:AM 116359) have a distinct ovoid or triangular pocket developed between the posterior ridge and the anterior spur of the hypostyle (apparent on $P^3$-$M^3$ in this example). Scale bar = 1 cm.

*Discussion.* As mentioned in the introduction, the taxonomy of *Mesohippus* has gone from extreme splitting to extreme lumping. Osborn (1904, 1918) named a new species for practically every single variant he encountered in the small collections then available. Many of his "diagnostic features" are impossible to use when the variability of larger samples is considered. In addition, Osborn seldom compared specimens from different stratigraphic levels. For Osborn, each species terminated at a stratigraphic boundary, and stratigraphic position was part of species definition. Virtually no statistical techniques were employed, since biometry was in its infancy in 1918.

Since then, several authors have attempted to analyze larger samples of *Mesohippus*. Clark and Beerbower (1967) established a numerical coding system for various features of the upper teeth, and this system was adopted by Forstén (1970a, 1970b, 1971). We have adopted a similar system, shown in Figure 10.1. Each specimen was encoded for length, width, and five character states for each of the upper cheek teeth. Over 300 skulls, palates, and maxillae have been analyzed. We originally intended to analyze this data matrix by multivariate methods, but such techniques became unnecessary when obvious patterns emerged. Contrary to Forstén (1970a, 1970b), not all the tooth characters were randomly assorted or useless taxonomically. We found that the condition of the lingual cingulum, hypostyle, metaconule, and occasionally the crochet was quite useful. In addition, we considered other skeletal characters. The Frick Collection of fossil horses contains numerous skulls, and the condition of the facial fossa has proven particularly important in other horses (MacFadden, 1984) and in *Mesohippus* as well. We examined the numerous associated postcranial remains to determine if there were other useful features in the skeleton. We also examined mandibles and lower teeth, but except for size, the lower dentition showed no consistent pattern of diagnostic features. Because only skulls or maxillae with relatively unworn dentitions are identifiable to species, we do not list every *Mesohippus* specimen in the referred material below. The Frick Collection has thousands of specimens of *Mesohippus* and *Miohippus* (particularly lower jaws and postcranials) that cannot be identified to species. Only identifiable material has been cataloged, identified, and listed. We looked for evidence of sexual dimorphism, but found no sign that Oligocene horses show the type of dimorphism found in *Hyracotherium* or many Miocene horses. The completeness and quality of Frick Collection horses has made it possible to reassess the question of how many species of *Mesohippus* are distinct and recognizable.

*Mesohippus bairdi* (Leidy, 1850)
Fig. 10.1B, 10.2

*Palaeotherium bairdii* Leidy, 1850, p. 122
*Anchitherium bairdii*: Leidy, 1852, p. 572
*Anchitherium bairdi*: Leidy, 1869, p. 303
*Mesohippus bairdii*: Marsh, 1875, p. 248
*Mesohippus bairdii*: Osborn and Wortman, 1894, p. 213

*Synonym*. *Mesohippus hypostylus* Osborn, 1904, p. 170

*Revised Diagnosis.* $M^{1-3}$ length = 30-36 mm. Lingual cusp or partial lingual cingulum on upper premolars and frequently on molars. Hypostyles usually class 2 or class 3 (see Figure 10.1).

*Type*. USNM 8632, a partial skull, with left $M^{2-3}$, right $M^{1-3}$, and the posterior half of the cranium. From the head of Bear Creek, Pennington County, South Dakota. Stratigraphic level unknown, but almost certainly Scenic Member of the Brule Formation.

*Referred material.* LATE CHADRONIAN: Chadron Formation, Big Badlands, S.D.: AMNH 1180 (type specimen of *M. hypostylus*); AMNH 38836 ; F:AM 74009. Bartlett Ranch, Dawes Co., Neb: F:AM 74036, 116391. Douglas area, Converse Co., Wyo.: F:AM 116378, 116376. Seaman Hills area, Niobrara Co., Wyo.: F:AM 116330, 74060, 74025, 74034, 74029, 74026, 116370, 11636, 11636, 116366, 74033, 74028 .
EARLY ORELLAN: Lower Nodules, Lower Scenic Member, Big Badlands, S.D.: AMNH 39131, 39133, 38857, 39030, 38936, 39005, 38855, 38935, 39004, 1188, 9769, 39447, 11864; F:AM 74006, 74004, 74003, 74010, 74014, 74043, 116404, 11639, 74008. Little Badlands, Stark Co., N.D.: F:AM 74040. Geike Ranch, Sioux Co., Neb.: F:AM 7403, 74035. Munson Ranch, Sioux Co., Neb.: F:AM 116390, 116388, 116364 , 74039.
LATE ORELLAN: Top of Upper Nodules, Middle *Oreodon* Beds, Big Badlands, S.D.: F:AM 74013, 116403, 116397, 116396, 74044.

*Horizon and localities.* Late Chadronian to late Orellan, North and South Dakota, Nebraska, Wyoming, and Colorado.

*Discussion.* The most common horse in the White River Group is *Mesohippus bairdi*. Although the type specimen (Figure 10.2) is very poorly preserved, it clearly shows upper molars with slight lingual cingula and with class 2 hypostyles. Some specimens of *Mesohippus bairdi* show a few hypostyles of class 3. Occasionally (Forstén, 1970a, 1970b), crochets appear on the upper teeth, but they are extremely rare. The facial fossa is well displayed on many skulls, and shows the typical *Mesohippus* condition. Scott (1941) gave a thorough description of *Mesohippus bairdi*, so no further description is needed here. We

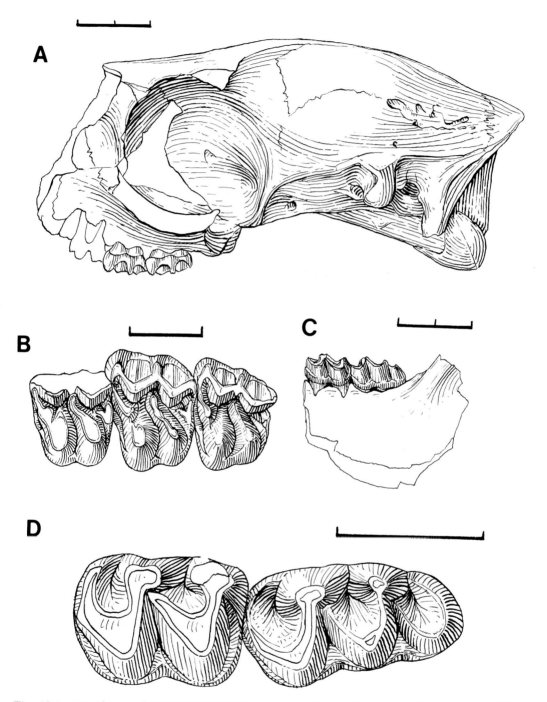

Fig. 10.2. *Mesohippus bairdi*, USNM 8632, type specimen of genus and species, in (A) left lateral view; (B) left $M^{1-3}$; (C) lower jaw fragment with $M_{2-3}$; (D) crown view of $M_{2-3}$. Scale bars in cm.

consider *Mesohippus hypostylus* to be a junior synonym of *Mesohippus bairdi*. In size (Tables 10.1, 10.2) it lies completely within the range of variation of *Mesohippus bairdi*, and the type of *Mesohippus hypostylus* (AMNH 1180) comes from deposits that also contain *Mesohippus bairdi*.

Osborn (1904, 1918) distinguished *Mesohippus hypostylus* from *Mesohippus bairdi* as follows: "Distinguished . . . from *Mesohippus bairdii* by more rudimentary hypostyle and less elevated crests. Metaloph sharp: metaconule not defined at base. $P^1$ small. Skull with pre-orbital fossa apparently deeper than in *Mesohippus bairdii*."

Of these features, none appears to be valid. The hypostyles on AMNH 1180 are all class 2, as is seen on the type specimen of *Mesohippus bairdi*. Their condition is not more "rudimentary," since both the type and most of the referred material have this kind of hypostyle. The elevation of the crests is highly variable and not taxonomically significant. The metaconules of *Mesohippus hypostylus* and *Mesohippus bairdi* are typically class 2 on $P^{2-3}$, class 3 on $P^4$ and $M^{1-3}$. The condition of the facial fossa on AMNH 1180 is difficult to judge, since the specimen has been dorsoventrally crushed. However, judging from most referred skulls, the condition of the fossa in *Mesohippus hypostylus* and *Mesohippus bairdi* is identical. Thus, *Mesohippus hypostylus* is clearly a junior subjective synonym of *Mesohippus bairdi*.

Clark and Beerbower (1967) referred most of the mid-sized Chadronian horses to *Mesohippus hypostylus*. We suspect that many of these specimens are referable to *Mesohippus bairdi*, but that others are referable to other species. Forstén (1970a, 1970b) placed *Mesohippus hypostylus* in synonymy with *Mesohippus bairdi*. We agree, although for different reasons than she gave.

*Mesohippus westoni* (Cope, 1889)
Fig. 10.1A, 10.3

*Anchitherium westoni* Cope, 1889, p. 3
*Mesohippus westoni*: Osborn 1904, p. 169
*Mesohippus westoni*: Lambe 1905a, p. 243
*Mesohippus westonii*: Russell 1975, p. 637

*Synonym.* *Mesohippus montanensis* Osborn, 1904

*Revised Diagnosis.* $M^{1-3}$ length = 30-36 mm. Lingual cingula or cingular cusps on upper cheek teeth are very faint or absent. Hypostyles usually absent or class 1, rarely class 2. No crochets.

Type. NMC 6289, a left upper molar missing the labial portion of the ectoloph.

Referred material. MIDDLE CHADRONIAN: Pipestone Springs, Jefferson Co., Montana: AMNH 9662 (type specimen of *M. montanensis*); AMNH 9663.
LATE CHADRONIAN: Chadron area, Dawes Co., Neb.: F:AM 74055, 74048, 116342, 116336. Douglas area, Converse Co., Wyo.: F:AM 116375, 116380, 116334. Seaman Hills, Niobrara Co., Wyo.: F:AM 111744, 74057, 74058, 74059, 74063, 74054, 74053, 74045, 116386, 116374, 116372, 116365, 116341, 116340, 116339, 116338, 116337, 116335, 74052.
EARLY ORELLAN: Lower Nodules, Scenic Member, Big Badlands, S.D.: F:AM 74020, 116348, 116345, 116344, 74023, 74046. Munson Ranch, Sioux Co., Neb.: F:AM 74039.
LATE ORELLAN: Upper Nodules, Middle *Oreodon* beds, Big Badlands, S.D.: F:AM 116350, 116347, 116346, 116343.

*Horizon and locality.* The type is from the early Chadronian Cypress Hills fauna, Saskatchewan. Referred material from many other Chadronian localities, and from early to late Orellan localities in Wyoming, Nebraska, and South Dakota.

*Discussion.* *Mesohippus westoni* is easily distinguished from similar-sized *Mesohippus bairdi* by the absence of lingual cingula and by its rudimentary hypostyles. It is distinguished from *Mesohippus celer* by its slightly larger size and larger $M^3$. The type of *Mesohippus westoni* is very poor, and normally the name would be considered indeterminate. However, Russell (1975) described additional

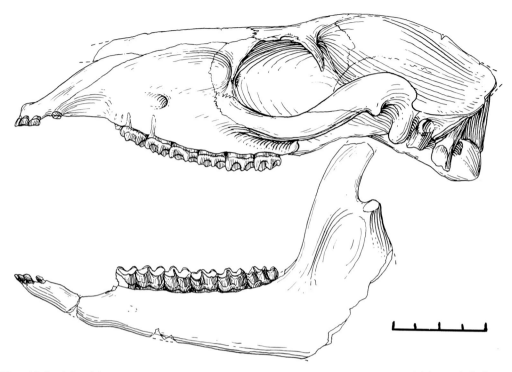

Fig. 10.3. *Mesohippus westoni*, F:AM 74023. Referred skull and mandible in left lateral view. Scale bar in cm.

topotypic material of *Mesohippus westoni* from Cypress Hills that clearly establishes the morphology of this species. As Lambe (1908) and Russell (1975) pointed out, the two diagnostic features of *Mesohippus westoni* (vestigial lingual cingulum and rudimentary hypostyle) are features that are true of Osborn's species *M. montanensis* from Pipestone Springs. The specimens referred to *M. westoni* from the Cypress Hills, Pipestone Springs, and from the localities mentioned above are identical in size (Russell, 1975; see Table 10.2). Thus, it appears that the name *westoni* is the senior synonym for Chadronian-Orellan *Mesohippus* larger than *M. celer* with rudimentary hypostyles, and faint or absent lingual cingula.

Osborn (1904, p. 40) distinguished *M. montanensis* from *M. westoni* by "sharp elevation of crests and absence of internal cingulum." Now that larger samples are known, these species cannot be distinguished by these features. The shape of molar crests is highly variable and greatly affected by wear. The lingual cingulum on the type of *Mesohippus westoni* is a little stronger than on the type of *Mesohippus montanensis*, but Russell (1975) has shown that there is some variation in this feature in topotypic samples of *Mesohippus westoni*. We have seen similar variation in the strength of the lingual cingulum in Frick Collection samples from the later Chadronian and Orellan. There appears to be no grounds for distinguishing *Mesohippus montanensis* Osborn, 1904, and it is here considered a junior synonym of *Mesohippus westoni* (Cope, 1889).

Clark and Beerbower (1967, p. 48) considered *Mesohippus montanensis* a junior synonym of *Mesohippus hypostylus* and Forstén (1970b) considered it a junior synonym of *Mesohippus bairdi*. As shown previously, *Mesohippus hypostylus* cannot be distinguished from *Mesohippus bairdi*, but *Mesohippus montanensis* (= *westoni*) can. Although the two species intergrade very slightly, there does appear to be a separation between horses with no lingual cingula and rudimentary hypostyles (*Mesohippus westoni*) and horses with stronger lingual cingula and well-developed hypostyles (*Mesohippus bairdi*). The two groups can be readily identified with well-preserved upper dentitions; they are not indistinguishable, as implied by Forstén (1970a, 1970b).

Other characters of these two species vary randomly, but the two characters discussed above are clearly associated. *M. westoni* and *M. bairdi* are diagnosable morphological clusters and deserve recognition as species. They also have different geographic and temporal ranges, which provides further evidence that they are distinct species. For example, the Pipestone Springs Main Pocket locality contains only *Mesohippus westoni*, yet coeval deposits in the High Plains contain both species. Cypress Hills contains *Mesohippus westoni* but no *Mesohippus bairdi*. Both species appear to become extinct at the end of the Orellan (top of the Middle Oreodon Beds), although Whitneyan collections are so sparse that their ranges may extend into younger deposits.

*Mesohippus exoletus* (Cope, 1874)
Fig. 10.4

*Anchitherium exoletum* Cope, 1874, p. 496
*Mesohippus exoletus*: Osborn, 1918, p. 47
*Synonyms. Mesohippus trigonostylus* Osborn, 1918, pp. 47-48; *Pedohippus trigonostylus* Schlaikjer, 1935, p. 141.

*Revised Diagnosis.* $M^{1-3}$ length = 32-36 mm. Lingual cingula on nearly all upper cheek teeth. Hypostyles all class 3 (triangular in shape). No crochets.

*Type.* AMNH 6298, a left maxilla with $M^{1-2}$ and the base of zygomatic arch, with associated fragments of a scapula, tibia, and metapodials (Fig. 10.4D). From the Cedar Creek area, Logan County, Colorado (?Orellan, according to Galbreath, 1953).

*Referred material.* LATE CHADRONIAN: 5 feet below top of Chadron Formation, Red Shirt Table, Shannon Co., S.D.: F:AM 116351. Seaman Hills, Niobrara Co., Wyo.: F:AM 116356 .
EARLY ORELLAN: Lower Nodules, Scenic Member, Big Badlands, S.D.: AMNH 674 (type of *M. trigonostylus, Metamynodon* beds); F:AM 74016, 116353, 74001; AMNH 39149. Little Badlands, Stark Co., N.D.: F:AM 116406, 116354 . Munson Ranch, Sioux Co., Neb.: F:AM 116389.
LATE ORELLAN: Upper Nodules, Middle *Oreodon* Beds, Big Badlands, S.D.: F:AM 74019, 116402, 116401, 116393, 116394, 116395, 116400, 116358, 11635, 116355, 116352. Fitterer Ranch, Stark Co., N.D.: F:AM 116405.

*Horizon and localities.* Known from the late Chadronian to late Orellan (top of Middle Oreodon beds), in Colorado, Wyoming, Nebraska, South and North Dakota.

*Discussion.* Osborn (1918, p. 47) reported the type of *Anchitherium exoletum* lost, and this was cited by Scott in 1941. Since Cope's description was inadequate and the specimen was not figured, Osborn considered the species *exoletus* to be indeterminate. Since 1941, however, the type of *M. exoletus* (AMNH 6298) has reappeared. Although the type (Fig. 10.4D) consists of only $M^{1-2}$, these teeth clearly show good examples of class 3 triangular hypostyles so characteristic of *Mesohippus trigonostylus*. In size (Table 10.2), AMNH 6298 also falls within the *M. trigonostylus* population. *M. exoletus* appears to be the a senior

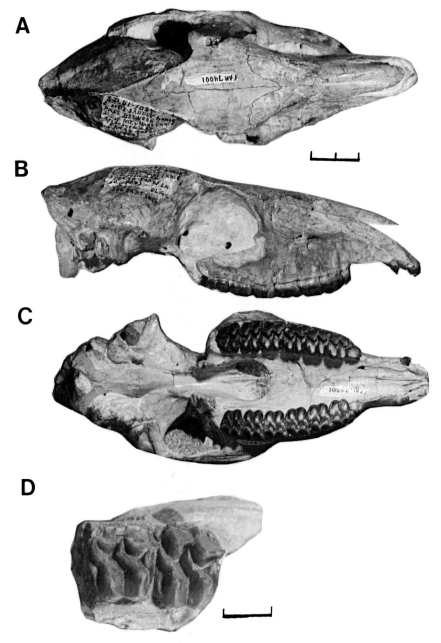

Fig. 10.4. *Mesohippus exoletus.* F:AM 74001, referred skull in (A) dorsal, (B) right lateral, and (C) ventral views. (D) AMNH 6298, type specimen. Scale bars in cm.

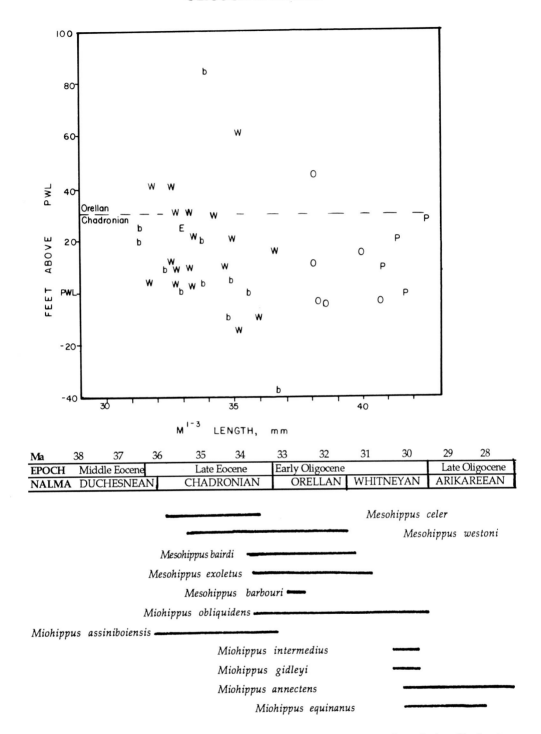

Fig. 10.5A. (top) Stratigraphic distribution versus size of horse specimens from the late Chadronian-early Orellan deposits in the Seaman Hills, near Lusk, Wyoming. Symbols as follows: b = *Mesohippus bairdi*; e = *Mesohippus exoletus*; o = *Miohippus obliquidens*; p = *Miohippus assiniboiensis*; w = *Mesohippus westoni*. All five species occur at a level about 20 feet below the Peristent White Layer (PWL). Fig. 10.5B (bottom). Temporal ranges of the equid species discussed in this chapter. NALMA = North American land mammal "age." Timescale is recalibrated to reflect new dates and correlations of the North American sections (see Chapter 28).

subjective synonym of *M. trigonostylus*.

*Mesohippus exoletus* can be distinguished from *Mesohippus bairdi* and *Mesohippus westoni* in its slightly larger size and especially by its distinctive class 3 triangular hypostyles on all upper cheek teeth. In all remaining skull, jaw, and skeletal features, *Mesohippus exoletus* and *Mesohippus bairdi* are very similar.

In the later Orellan, *Mesohippus exoletus* is generally more abundant than other species of *Mesohippus*. *M. exoletus* occurs with *M. bairdi* and *M. westoni* in deposits near the Chadronian-Orellan boundary as well, although in much smaller numbers (Fig. 10.5A).

Some may object to designating three similar-sized species of *Mesohippus* (plus two of *Miohippus*) from the same deposits (e.g., the Lusk area of Wyoming). It might be argued that under a biological species concept, these forms would be ecologically equivalent and thus would have to be the same biological species. However, we are persuaded that separate species, rather than a series of intergrading subspecies, are preferable. Horses are notoriously speciose. Many localities in the Miocene have four or five species of closely related horses, and a few quarries have over a dozen. For example, Skinner and Johnson (1984, p. 276) report eight genera of horses in Railway Quarry A of the Valentine Formation, and 12 species were reported by Schultz (1977) from the Clarendonian of Texas. Even today, two or more species of zebra can be found together in East Africa. Morphologically, these *Mesohippus* species can be unambiguously distinguished, and have distinctive stratigraphic and geographic ranges. If these taxa are to be stratigraphically useful, they cannot be lumped into a common species. It is conceivable that they are end members of a biological continuum, but our data do not support this. Instead, we find three distinct clusters of morphology. These clusters have clear biostratigraphic and biogeographic definition and utility. We cannot agree with Forstén (1970a, 1970b), who places all medium-sized *Mesohippus* from the Orellan in a single species.

*Mesohippus barbouri* Schlaikjer, 1931
Fig. 10.6

*Miohippus antiquus*: Forstén, 1974 (in part)

*Revised Diagnosis.* $M^{1-3}$ length = 34-36 mm. Metaconules reduced. No lingual cingula. Unusually short rostrum and pinched diastema, causing a flaring symphysis. Upper incisors circular and deeply cupped. Humerus and femur relatively short with respect to metapodial length.

*Type.* MCZ 17641, a complete skeleton, from the late Orellan of the Big Badlands, South Dakota.

*Referred material.* EARLY ORELLAN: Lower Nodules, Scenic Member, Big Badlands, S.D.: F:AM 116324, 116325, 74074. Harvard Fossil Reserve, Goshen Co., Wyo.: MCZ 2781, 2792.
LATE ORELLAN: Upper Nodules, Middle Oreodon Beds, Big Badlands, S.D.: F:AM 116322, 116323.

Horizon and localities. Early to late Orellan, South Dakota and Wyoming.

Description and discussion. *Mesohippus barbouri* was thoroughly described by Schlaikjer (1932). It is a very distinctive horse, with a peculiarly short face and unusually rounded, deeply cupped incisors (Fig. 10.6A-C). The upper incisors of most primitive horses are more oval in cross-section, and have only shallow pits on the crowns.

Schlaikjer (1935) referred some specimens from Harvard Fossil Reserve, Goshen County, Wyoming, to *Mesohippus barbouri*. Forstén (1974) placed these specimens in *Miohippus antiquus*. However, the two smaller specimens from Harvard Fossil Reserve (MCZ 2781 and MCZ 2792) are clearly within the size range of *Mesohippus barbouri*, and too small for *Miohippus obliquidens*, which is the most common horse from this quarry. The ratio of dental specimens of *Mesohippus barbouri* to *Miohippus obliquidens* in this quarry is 2:10. Interestingly, Forstén (1974) reported that the

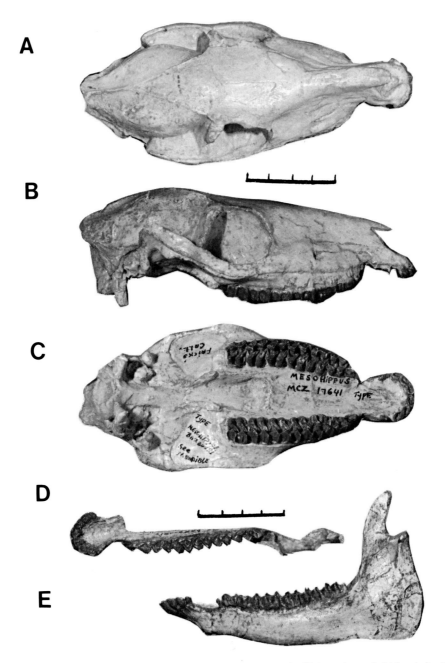

Fig. 10.6. *Mesohippus barbouri*, MCZ 1764J, type skull in dorsal (A), right lateral (B), and ventral (C) views; type mandible in crown (D) and left lateral (E) views. Scale bars in cm.

ratio of metatarsals III without a cuboid facet (the *Mesohippus* condition) to those with a cuboid facet (the *Miohippus* condition) is 2:9. Thus, we would associate these two metatarsals III without cuboid facets with the dentitions that can be referred to *Mesohippus barbouri*.

### *Miohippus* Marsh, 1874

*Anchitherium*: Leidy, 1870, p. 113
*Miohippus* Marsh, 1874, p. 249
*Mesohippus*: Osborn, 1904, p. 177 (in part)
*Pediohippus*: Schlaikjer, 1935, p. 141

*Revised Diagnosis.* Small to medium sized equids ($M^{1-3}$ length = 34-50 mm) with a slightly longer skull than *Mesohippus*, large, well developed hypostyles (usually class 3), and a deeper, more anteriorly expanded facial fossa. Facet between the third metatarsal and cuboid.

*Type species. Miohippus annectens* Marsh, 1874.

*Included species. Miohippus obliquidens, Miohippus assiniboiensis, Miohippus intermedius, Miohippus gidleyi, ?Miohippus equinanus*, and a number of exclusively Arikareean species not discussed here.

*Distribution.* Mid-Chadronian (Cypress Hills) to Arikareean of Saskatchewan, Montana, Oregon, Wyoming, Nebraska, Colorado, and North and South Dakota.

*Discussion.* The distinction between *Mesohippus* and *Miohippus* has always been controversial. Marsh (1874) created the taxon *Miohippus* for horses from the Arikareean John Day beds of Oregon. Osborn (1918) made the genus a strictly horizontal taxon for all horses from the late Whitneyan through most of the Arikareean. Osborn's diagnosis of *Miohippus* included a number of invalid, highly variable characters, but a few of these characters have stood the test of time. The articulation between the cuboid and third metatarsal, the larger hypostyles, the longer face, and the deeper facial fossa are all characters that distinguish *Miohippus* from *Mesohippus*. The cuboid-third metatarsal articulation in particular has served as a useful derived feature signaling the appearance of *Miohippus*, since this character is a result of the broadening of the third metatarsal with increased monodactyly.

Schlaikjer (1935, p. 141) created a new genus *Pediohippus* for a number of *Miohippus* specimens which had rudimentary crochets on the upper cheek teeth. Scott (1941) adopted Schlaikjer's generic taxonomy. Forstén (1974) synonymized *Pediohippus* with *Miohippus* by showing that the crochet was a highly variable feature within populations, and not taxonomically useful. Our studies confirm this conclusion. Although weak crochets occasionally occur in *Miohippus*, they are irregularly developed on adjacent teeth, and sometimes are not even symmetrically developed on both sides of the same skull. The presence of crochets alone is not diagnostic, let alone sufficient justification of a new genus.

Stirton (1940) considered *Pediohippus* to be a junior synonym of *Miohippus*. Stirton (1940) and Simpson (1951) thought that *Mesohippus* and *Miohippus* intergraded in the Orellan. Lillegraven (1970) expressed doubt that *Mesohippus* and *Miohippus* could be distinguished at all. However, in our examination of the Frick Collection of fossil horses, a clear distinction between the two genera became apparent. *Miohippus* is a significantly larger horse, with a longer face. Its facial fossa is deepest just anterior to the orbit and tapers anteriorly, becoming shallower toward the rostrum. The fossa terminates at about the level of $P^2$. There is almost no overlap or intergradation between these two morphologies, yet they overlap considerably in temporal range (Fig. 10.5B).

All *Miohippus* specimens that have associated tarsals clearly show the cuboid-metatarsal III articulation. This is never seen in *Mesohippus*. The definition of *Miohippus* is unambiguous. This genus now includes a number of derived but early horses that had previously been referred to *Mesohippus* on purely stratigraphic grounds.

Table 10.1. Skull and mandible measurements (in mm)

| | $I^1$-$M^3$ | Pmax-Occip. | W. zyg. | Pmax-Anter. Orbit | Ant.orb.-Occ. | Occip. W. | Occip. Ht. | Jaw Depth at $P_2$ | Jaw Depth at $M_3$ | $I_1$-Angle |
|---|---|---|---|---|---|---|---|---|---|---|
| *M. bairdi* (type) USNM 8632 | | | | | 111 | 59 | 47 | | | |
| *M. bairdi* AMNH 1477 | 108 | 200 | 85 | 105 | 93 | 56 | 46 | 19 | 32 | |
| *M. bairdi* AMNH 1492 | 106 | 178 | 81 | 102 | 84 | 54 | 44 | 18 | 25 | |
| *M. bairdi* F:AM 74029 | 110 | 176 | 75 | 92 | 111 | | | 21 | 31 | 152 |
| *M. westoni* F:AM 74052 | 109 | 178 | 75 | 97 | 96 | 57 | 46 | 18 | 30 | 149 |
| *M. westoni* F:AM 74063 | 114 | 184 | | 93 | 104 | | 49 | 18 | 29 | 156 |
| *M. exoletus* F:AM 7400 | 107 | 175 | 79 | 87 | 102 | 55 | 49 | | | |
| *M. barbouri* (type) MCZ 17451 | 111 | 182 | 85 | 90 | 101 | 50 | 47 | 28 | 39 | 149 |
| *M. barbouri* F:AM 116325 | 111 | 171 | 83 | 81 | 99 | 38 | 39 | 16 | 30 | 158 |
| *M. obliquidens* (type) AMNH 668 | | | | | 108 | | 46 | 21 | 27 | 148 |
| *M. obliquidens* (type of *meteulophus*) AMNH 1210 | 118 | 205 | | 110 | 117 | 59 | 54 | 20 | 28 | 163 |
| *M. assiniboiensis* F:AM 116359 | 132 | 199 | | 106 | 108 | 69 | 55 | 17 | 28 | 170 |
| *M. assiniboiensis* F:AM 116361 | 119 | 192 | 81 | 98 | 111 | | 50 | 17 | 29 | 161 |
| *M. intermedius* (type) AMNH 1196 | 140 | 220 | 98 | 114 | 124 | 68 | 55 | 17 | 28 | 170 |
| *M. gidleyi* (type) AMNH 1192 | | | | | 88 | 55 | | | | |
| *M. gidleyi* (type of *validus*) AMNH 680 | 135 | 229 | 96 | 120 | 122 | 72 | 63 | | | |
| *M. annectens* (type) YPM 12230 | 165 | 208 | 114 | 151 | 154 | 72 | 67 | | | |

Table 10.2. Measurements of upper teeth
Mean ± standard deviation in mm, sample size given in parentheses ( ).

|  | M. bairdi | M. westoni | M. exoletus | M. barbouri |
|---|---|---|---|---|
| $P^2$ Length | 11.5±0.7 (28) | 11.1±0.8 (27) | 11.3±0.8 (12) | 11.0±0.4 (5) |
| $P^3$ Length | 11.6±0.7 (31) | 11.4±0.7 (31) | 11.7±0.6 (14) | 11.6±1.0 (5) |
| $P^4$ Length | 11.9±0.6 (32) | 11.7±0.8 (32) | 11.8±0.6 (14) | 12.2±0.9 (5) |
| $M^1$ Length | 11.7±0.8 (33) | 11.6±0.8 (32) | 11.9±0.8 (15) | 12.3±0.5 (5) |
| $M^2$ Length | 11.8±0.6 (33) | 11.8±1.1 (32) | 11.8±0.9 (15) | 12.0±0.4 (5) |
| $M^3$ Length | 11.1±0.7 (32) | 11.5±0.9 (30) | 11.1±0.6 (12) | 11.4±0.6 (5) |
| $P^2$ Width | 12.8±0.1 (27) | 12.6±0.8 (32) | 12.5±0.6 (11) | 12.1±0.9 (5) |
| $P^3$ Width | 14.2±0.8 (30) | 13.8±0.7 (37) | 13.9±0.7 (12) | 13.3±1.4 (5) |
| $P^4$ Width | 14.8±0.8 (31) | 14.5±0.9 (40) | 14.7±0.8 (12) | 14.8±0.6 (5) |
| $M^1$ Width | 14.5±1.1 (31) | 14.5±0.9 (39) | 14.4±0.1 (14) | 14.7±1.0 (5) |
| $M^2$ Width | 15.2±1.0 (33) | 15.3±1.0 (40) | 15.2±1.1 (15) | 15.2±0.5 (5) |
| $M^3$ Width | 14.3±0.9 (32) | 14.4±1.0 (35) | 14.2±1.2 (13) | 14.3±1.1 (4) |
| $P^2$-$M^3$ Length | 66.5±3.1 (26) | 65.9±3.4 (30) | 66.1±1.8 (10) | 66.3±2.4 (4) |
| $P^{2-4}$ Length | 34.5±1.9 (28) | 33.8±1.7 (32) | 33.8±1.1 (11) | 33.7±2.2 (5) |
| $M^{1-3}$ Length | 33.4±1.8 (31) | 34.3±2.5 (36) | 33.7±1.8 (12) | 34.3±1.5 (5) |

|  | M. obliquidens | M. assiniboiensis | M. gidleyi | M. equinanus |
|---|---|---|---|---|
| $P^2$ Length | 11.4±1.0 (24) | 13.0±1.1 (8) | 14.8±0.5 (6) | 11.8±0.4 (2) |
| $P^3$ Length | 12.1±0.8 (27) | 14.0±0.9 (8) | 15.1±0.7 (6) | 12.1±0.1 (2) |
| $P^4$ Length | 12.5±0.9 (27) | 14.0±0.4 (7) | 15.7±0.7 (6) | 11.7±0.3 (3) |
| $M^1$ Length | 12.9±0.8 (27) | 14.0±0.8 (8) | 15.2±0.4 (6) | 11.1±0.6 (3) |
| $M^2$ Length | 12.5±0.7 (26) | 13.9±0.7 (8) | 15.0±0.9 (7) | 11.0±0.4 (2) |
| $M^3$ Length | 12.6±0.8 (26) | 13.4±0.5 (8) | 14.9±1.2 (6) | 10.4±0.5 (2) |
| $P^2$ Width | 12.9±1.1 (23) | 13.8±0.6 (7) | 16.2±0.7 (6) | 11.5±0 (2) |
| $P^3$ Width | 14.7±0.1 (26) | 15.9±0.5 (7) | 17.9±0.8 (6) | 13.2±0.8 (2) |
| $P^4$ Width | 15.5±0.9 (23) | 16.7±0.5 (7) | 18.8±1.7 (6) | 14.0±0.5 (3) |
| $M^1$ Width | 15.9±1.3 (27) | 16.3±0.9 (9) | 17.3±2.5 (6) | 13.5±0.5 (3) |
| $M^2$ Width | 16.2±1.5 (27) | 17.1±0.8 (8) | 18.8±1.2 (6) | 13.7±1.0 (2) |
| $M^3$ Width | 15.9±1.0 (26) | 16.2±0.1 (8) | 18.1±1.0 (6) | 13.0±0.9 (2) |
| $P^2$-$M^3$ Length | 71.2±3.1 (26) | 78.4±1.6 (7) | 86.6±2.3 (5) | 65.3 (1) |
| $P^{2-4}$ Length | 35.0±2.2 (26) | 39.4±0.1 (7) | 44.6±1.2 (6) | 34.2±0.7 (2) |
| $M^{1-3}$ Length | 37.3±1.6 (26) | 39.7±1.7 (8) | 43.6±1.7 (6) | 30.8±0 (2) |

Table 10.3. Measurements of lower teeth (in mm)

| | $P_2M_3$ | $P_{2-4}$ | $M_{1-3}$ | $P_2L$ | $P_2W$ | $P_3L$ | $P_3W$ | $P_4L$ | $P_4W$ | $M_1L$ | $M_1W$ | $M_2L$ | $M_2W$ | $M_3L$ | $M_3W$ |
|---|---|---|---|---|---|---|---|---|---|---|---|---|---|---|---|
| *M. bairdi* F:AM 74035 | 68.3 | 33.9 | 38.6 | 11.2 | 6.7 | 11.8 | 8.3 | 11.9 | 9.1 | 11.0 | 7.8 | 10.9 | 8.6 | 16.3 | 8.0 |
| *M. bairdi* F:AM 74029 | 76.3 | 38.6 | 38.9 | 11.7 | 7.9 | 12.3 | 8.2 | 12.4 | 9.1 | 11.4 | 7.8 | 11.6 | 8.2 | 17.2 | 8.0 |
| *M. westoni* F:AM 74052 | 70.5 | 33.1 | 37.1 | 11.6 | 7.0 | 11.4 | 7.1 | 11.2 | 9.0 | 11.8 | 8.0 | 11.9 | 7.7 | 15.6 | 7.8 |
| *M. westoni* F:AM 74063 | 71.7 | 34.8 | 37.8 | 12.9 | 7.5 | 11.8 | 8.3 | 12.4 | 8.7 | 11.6 | 7.4 | 11.8 | 7.2 | 15.6 | 7.6 |
| *M. exoletus* F:AM 116358 | 75.5 | 36.9 | 38.6 | 12.1 | 8.2 | 12.0 | 9.5 | 11.5 | 9.9 | 12.5 | 8.1 | 11.6 | 7.4 | 16.1 | 7.6 |
| *M. exoletus* F:AM 116352 | 71.4 | 35.2 | 37.7 | 12.4 | 8.1 | 12.5 | 9.2 | 12.3 | 9.0 | 11.8 | 8.7 | 11.4 | 7.3 | 15.2 | 8.1 |
| *M. barbouri* MCZ 17461 (type) | 75.4 | 35.4 | 40.0 | 11.4 | 7.0 | 11.8 | 7.4 | 11.6 | 8.2 | 12.0 | 7.3 | 12.4 | 7.0 | 16.4 | 7.3 |
| *M. barbouri* F:AM 116325 | 78.0 | 37.8 | 41.7 | 12.0 | 7.0 | 13.4 | 7.6 | 13.4 | 7.8 | 13.0 | 7.1 | 13.9 | 7.2 | 17.8 | 7.7 |
| *M. obliquidens* AMNH 668 (type) | (deciduous teeth) | | | | | | | | | 14.5 | 8.7 | 12.5 | 6.9 | | |
| *M. obliquidens* AMNH 1210 | 77.3 | 34.9 | 44.5 | 11.2 | 7.7 | 11.9 | 9.6 | 12.2 | 11.2 | 12.3 | 9.3 | 13.2 | 7.0 | 19.4 | 7.8 |
| *M. assiniboiensis* F:AM 116359 | 87.3 | 39.0 | 48.6 | 12.3 | 7.4 | 14.1 | 8.7 | 13.5 | 10.1 | 14.5 | 9.2 | 15.5 | 9.2 | 19.6 | 8.3 |
| *M. assiniboiensis* F:AM 116361 | 85.5 | 40.3 | 46.4 | 13.3 | 7.5 | 13.9 | 10.5 | 14.5 | 11.3 | 14.8 | 10.2 | 15.1 | 10.7 | 19.0 | 9.8 |
| *M. "grandis"* CMNH 9157 (type) | 89.5 | 43.3 | 47.6 | 13.8 | 8.6 | 14.0 | 11.2 | 14.8 | 12.0 | 13.6 | 9.9 | 14.7 | 9.7 | 17.9 | 9.1 |
| *M. "grandis"* CMNH 9158 | 89.0 | 42.0 | 49.0 | | | 14.5 | 11.0 | 15.0 | 11.5 | 14.0 | 10.2 | 16.8 | 10.6 | | |
| *M. intermedius* AMNH 1196 (type) | 91.1 | 43.3 | 48.7 | 16.4 | 8.4 | 13.0 | 10.5 | 14.1 | 11.7 | 15.3 | 9.6 | 14.1 | 10.3 | 19.6 | 8.4 |
| *M. gidleyi* AMNH 1195 | 100.2 | 46.3 | 53.3 | 15.5 | 10.1 | 16.3 | 13.1 | 17.7 | 13.7 | 16.9 | 12.3 | 15.3 | 11.5 | 21.4 | 9.8 |
| *M. gidleyi* AMNH 1218 | 100.4 | 47.0 | 53.6 | 13.0 | 10.0 | 16.7 | 11.5 | 16.3 | 13.3 | 16.2 | 11.6 | 16.7 | 11.8 | 22.3 | 10.0 |
| *M. equinanus* AMNH 12916 | 65.3 | 32.1 | 35.3 | 10.8 | 6.1 | 11.0 | 7.5 | 11.4 | 8.0 | 11.2 | 7.6 | 10.7 | 7.5 | 15.1 | 6.0 |

Table 10.4. Postcranial measurements (in mm)

|  | M. bairdi AMNH 1477 | M. bairdi AMNH 1492 | M. bairdi F:AM 74026 | M. westoni F:AM 74052 | M. westoni F:AM 74063 | M. exoletus F:AM 74001 |
|---|---|---|---|---|---|---|
| Atlas length | 37 | 33 | 35 | 36 |  | 35 |
| Atlas width | 36 |  | 36 | 36 |  | 36 |
| L. axis centrum | 51 | 42 |  |  |  |  |
| Axis width | 32 | 26 |  |  |  |  |
| Scapula length | 142 | 131 | 148 |  | 142 |  |
| Scapula width | 99 | 83 | 78 |  | 66 |  |
| Length humerus | 130 | 117 | 129 | 116 | 129 | 131 |
| Humerus prox W | 38 | 25 | 34 | 35 | 30 | 36 |
| Humerus dist W | 30 | 26 | 32 | 26 | 28 | 27 |
| Radius length | 146 | 128 | 133 | 119 | 134 | 32 |
| Radius prox W | 26 | 23 | 23 | 20 | 24 | 25 |
| Ulna length |  |  | 134 |  |  | 134 |
| McII length | 84 | 73 | 79 |  |  | 88 |
| McII width | 10 |  | 9 |  |  | 9 |
| McIII length | 85 | 78 | 81 | 97 |  | 91 |
| McIII width | 14 |  | 13 | 10 |  | 11 |
| McIV length | 77 | 68 | 76 |  |  | 84 |
| McIV width | 11 |  | 8 |  |  | 10 |
| Femur length |  | 157 | 196 |  | 179 | 177 |
| Femur prox W | 51 |  |  | 39 |  | 45 |
| Femur dist W | 39 | 35 | 31 |  | 34 | 39 |
| Tibia length | 197 | 170 | 205 | 177 | 189 | 187 |
| Tibia prox W | 40 | 34 | 35 | 33 | 37 | 38 |
| Tibia dist W | 28 | 24 | 26 | 21 | 23 | 25 |
| Astragalus L | 22 |  | 24 | 22 | 27 | 23 |
| Astragalus W | 21 | 21 | 19 | 18 | 18 | 20 |
| Calcaneum L | 54 | 45 | 50 | 45 | 53 | 51 |
| Calcaneum W | 21 | 18 | 23 | 23 | 22 | 24 |
| MtII length | 105 | 85 | 100 |  | 103 | 106 |
| MtII width | 12 |  | 8 |  | 10 | 11 |
| MtIII length | 109 | 94 | 109 | 99 | 111 | 112 |
| MtIII width | 12 |  | 10 | 12 | 13 | 14 |
| MtIV length | 106 | 86 |  | 94 | 105 |  |
| MtIV width | 12 |  |  | 8 | 9 |  |

Table 10.4 (continued).

| | M. barbouri MCZ 17641 | M.obliquidens F:AM 74066 | M.obliquidens F:AM 74067 | M. assiniboiensis F:AM116359 | M. assiniboiensis F:AM 74077 | M. gidleyi AMNH 1218 |
|---|---|---|---|---|---|---|
| Atlas length | | | 39 | | | |
| Atlas width | | | 41 | | | |
| L. axis centrum | | | | | | |
| Axis width | | | | | | |
| Scapula length | 134 | 135 | | | 156 | |
| Scapula width | | 92 | | | 92 | |
| Length humerus | 117 | | | | | 163 |
| Humerus prox W | | | | | | |
| Humerus dist W | 24 | | | | 32 | 44 |
| Radius length | 130 | 134 | 131 | | | |
| Radius prox W | 25 | 22 | 26 | | 27 | 32 |
| Ulna length | | | 134 | | | |
| McII length | 96 | | 90 | | 89 | 121 |
| McII width | | | 9 | | 9 | 12 |
| McIII length | 102 | | 93 | | 96 | 129 |
| McIII width | 12 | | 12 | | 12 | 18 |
| McIV length | 91 | | 82 | | 91 | 116 |
| McIV width | | | 9 | | 10 | 11 |
| Femur length | 157 | 173 | 183 | 188 | 192 | 239 |
| Femur prox W | 33 | | 42 | 54 | 44 | 47 |
| Femur dist W | | 32 | 38 | 38 | 37 | 42 |
| Tibia length | 182 | 187 | 183 | 208 | 208 | |
| Tibia prox W | | 29 | 38 | 41 | 41 | |
| Tibia dist W | | 22 | 25 | 26 | 29 | |
| Astragalus L | | | 25 | 25 | 25 | 39 |
| Astragalus W | | | 19 | 22 | 20 | 29 |
| Calcaneum L | | 46 | 53 | 59 | 53 | 76 |
| Calcaneum W | | 23 | 26 | 23 | 22 | 36 |
| MtII length | 113 | | 101 | 109 | 110 | 139 |
| MtII width | | | 9 | 6 | 9 | 14 |
| MtIII length | 117 | 102 | 110 | 115 | 118 | 146 |
| MtIII width | 15 | 12 | 15 | 6 | 14 | 23 |
| MtIV length | 109 | | 103 | 98 | 110 | 140 |
| MtIV width | | | 8 | 9 | | |

*Miohippus obliquidens* (Osborn, 1904)
new combination
Fig. 10.7

*Mesohippus obliquidens* Osborn, 1904, p. 173

*Synonyms.* *Mesohippus eulophus* Osborn, 1904, p. 173; *Mesohippus meteulophus* Osborn, 1904, p. 174 (=*Miohippus meteulophus* Osborn, 1918, p. 51); *Pediohippus antiquus* Schlaikjer 1935, p. 141 (= *Miohippus antiquus* Forstén, 1971, p. 404); *Mesohippus brachystylus* Osborn, 1918, p. 53 (= *Pediohippus brachystylus* Schlaikjer, 1935, p. 144).

*Revised Diagnosis.* $M^{1-3}$ length = 34-39 mm. Molar crests complete with metaconules weak or absent. Molar crests more obliquely oriented than in any other species (Fig. 10.7F).

*Type.* AMNH 668, a badly crushed juvenile skull, with $dP^{2-4}$ and $M^{1-2}$, and both lower rami (Fig. 10.7A). $DP^{2-4}$ have been prepared away by Morris Skinner, revealing permanent $P^{3-4}$. From the late Orellan (middle *Oreodon* beds) of the Big Badlands, South Dakota.

Referred material. LATE CHADRONIAN: Chadron Formation, Big Badlands, S.D.: F:AM 116329. Chadron area, Dawes Co., Neb.: F:AM 74073, 74047, 74067. Douglas area, Converse Co., Wyo.: F:AM 74076, 74072, 116383, 116381, 116377, 17597, 116326, 116327, 116328. Seaman Hills, Niobrara Co., Wyo.: F:AM 74071, 74064, 74066, 116412, 116387, 116384, 116368, 116333.
EARLY ORELLAN: Lower Nodules, Scenic Member, Big Badlands, S.D.: F:AM 74021; AMNH12290, 39115. Harvard Fossil Reserve, Goshen Co., Wyo.: MCZ 2790 (holotype of *P. antiquus*); MCZ 2791, MCZ 2942 (*P. antiquus* paratypes), MCZ 2789, 2785, 2788, 2783, 2786, 2779, 2782.
LATE ORELLAN: Fitterer Ranch, Stark Co., N.D.: F:AM 116408. Upper nodules, Middle Oreodon beds, Big Badlands, S.D.: F:AM 116398, 116392.
EARLY WHITNEYAN: Upper *Oreodon* beds, Big Badlands, S.D.: F:AM 116363. Vista Member, Cedar Creek area, Weld Co., Colo.: AMNH 8791 (type of *M. eulophus*).
LATE WHITNEYAN: *Leptauchenia* nodules, Big Badlands, S.D.: AMNH 1210 (type of *M. meteulophus*); AMNH 11860 (type of *M. brachystylus*); AMNH 1043; F:AM 116409. Rainey Butte, Slope Co., N.D.: AMNH 12846.

*Horizon and localities.* Late Chadronian to late Whitneyan, Wyoming, Colorado, South and North Dakota, and Nebraska.

*Discussion.* Large White River horses with slightly advanced dentitions have been given a plethora of names. In size (Table 10. 1), these horses differ very little, although they are clearly larger than typical *Mesohippus*. In just two pages, Osborn (1904, p. 173-174) named three species (*obliquidens, eulophus, meteulophus*) that are nearly identical in size and morphology (Fig. 10.8). We consider them all the same species since all of the differences Osborn cited are very slight and mainly due to intraspecific variation. It is clear athat the main reason for his separation of these species is the fact that they come from successively higher stratigraphic levels. The name *obliquidens* has already been established in the literature by Schlaikjer (1935) and Scott (1941), and has priority over *eulophus* and *meteulophus*. It is therefore the senior synonym of the three.

Schlaikjer (1935) proposed the name *Pediohippus antiquus* for part of the late Orellan horse sample from Harvard Fossil Reserve. He also reported *Mesohippus obliquidens* and *Mesohippus barbouri* from this quarry. Forstén (1974) demonstrated that most of the larger horses referred to *Pediohippus antiquus* could not be distinguished from *Miohippus obliquidens*, and she synonymized the two species. However, the species *obliquidens* has 31 years' priority over *antiquus*, so she should have utilized the name *Miohippus obliquidens* rather than the name *Miohippus antiquus*.

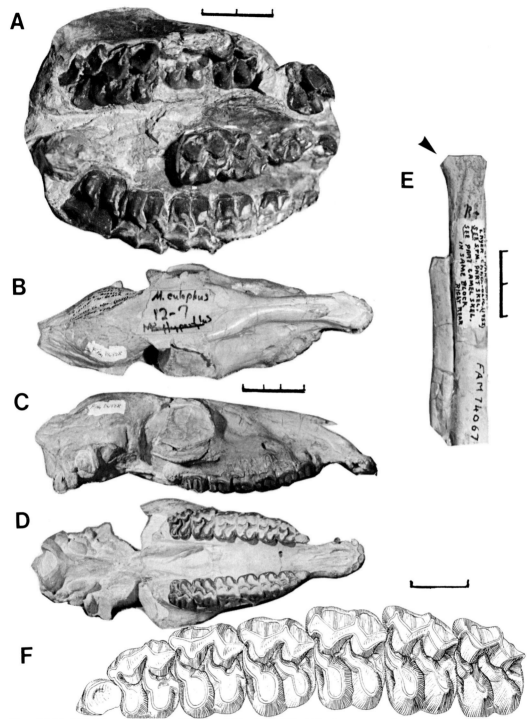

Fig. 10.7. *Miohippus obliquidens.* (A) AMNH 668, type specimen, in palatal view. Right deciduous premolars removed by Morris Skinner to show unerupted permanent premolars not previously illustrated. Referred skull, F:AM 116328, in dorsal (B), right lateral (C), and ventral (D) views. (E) Third metatarsal, F:AM 74067, showing distinct cuboid facet (arrow). (F) Referred left upper dentition, F:AM 116412. Scale bars in cm.

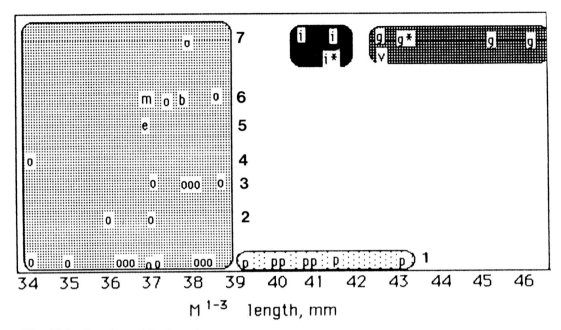

Fig. 10.8. Stratigraphic distribution versus size of specimens of *Miohippus*. Symbols as follows: b = type of *M. "brachystylus"* ; e = type of *M. "eulophus"* ; g = *M. gidleyi* ; g* = type of *M. gidleyi* ; i = *M. intermedius* ; i* = type of *M. intermedius* ; m = type of *M. "meteulophus"* ; o = *M. obliquidens* ; p = *M. assiniboiensis* ; v = type of *M. "validus."* Shaded areas indicate range of size and stratigraphic variation of *M. assiniboiensis* (light dots), *M. obliquidens* (medium dots), *M. gidleyi* (heavy dots) and *M. intermedius* (black). Numbered stratigraphic sampling levels are as follows: 1) near Persistent White Layer, Lusk area, Wyoming (late Chadronian); 2) Lower Nodules, Big Badlands, South Dakota (early Orellan); 3) Harvard Fossil Reserve, near Torrington, Wyoming (middle Orellan); 4) Upper Nodules, Big Badlands, South Dakota (late Orellan); 5) Upper *Oreodon* beds, Big Badlands, South Dakota (early Whitneyan); 6) *Leptauchenia* beds, Big Badlands, South Dakota (middle Whitneyan); 7) *Protoceras* beds, Big Badlands, South Dakota (late Whitneyan).

We do not agree with Forstén that all of the horses from this quarry are one species. At least two individuals are referable to *Mesohippus barbouri* (see earlier).

*M. obliquidens* is clearly referable to *Miohippus*, since several specimens (e.g., F:AM 74047, 74067) consist of partial skeletons that show an articulation between the cuboid and third metatarsal (Fig. 10.7E). We consider this articulation a diagnostic feature of *Miohippus*. In addition, most of the Harvard Fossil Reserve horses (here referred to *Miohippus obliquidens*) have this articulation. A number of uncrushed skulls of *M. obliquidens* (such as F:AM 116328, 116327, 116326, AMNH 1210) clearly show the typical *Miohippus* facial fossa. Osborn (1904, 1918) diagnosed *obliquidens* as lacking a facial fossa, but this is based on the type specimen, in which the facial region is crushed and impossible to interpret.

*Miohippus assiniboiensis* (Lambe, 1905b), new combination
Figs. 10.1C, 10.9

*Mesohippus assiniboiensis* Lambe, 1905b, p. 50.
*Mesohippus assiniboiensis*: Russell, 1975
 *Synonyms*. *Mesohippus planidens* Lambe, 1905b, p. 49; *Mesohippus grandis*

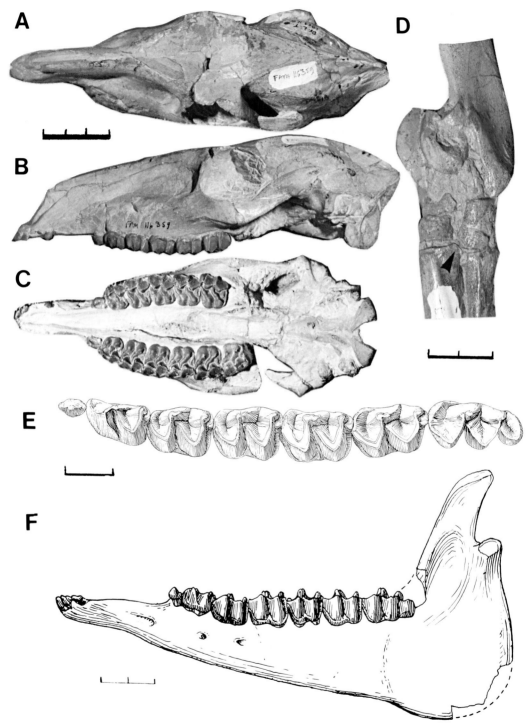

Fig. 10.9. *Miohippus assiniboiensis*. F:AM 116359, referred specimen, skull in dorsal (A), left lateral (B), and ventral (C) views, (D) articulated ankle with cuboid-Mt III contact (arrow), and mandible in dorsal (E) and left lateral (F) views. Scale bars in cm.

Clark and Beerbower, 1967

*Revised Diagnosis.* $M^{1-3}$ length = 38-42 mm. Skull longer than *Miohippus obliquidens*, but shorter than *Miohippus intermedius*. Hypostyles all class 3. Crochets frequently present.

*Type.* NMC 6305, right $P^2$ from the Chadronian of Cypress Hills, Saskatchewan.

*Referred Material.* Seaman Hills, Niobrara Co., Wyo.: F:AM 74077, 74078, 116332, 116359, 116360, 116361, 116362, 116371,11637. Douglas area, Converse Co., Wyo.: F:AM 116379. Little Badlands, Stark Co., N.D.: F:AM 116407.

*Horizon and locality.* Middle to late Chadronian, Saskatchewan, Wyoming, and North and South Dakota.

*Description and discussion.* In the Frick Collection, there is a distinctive large latest Chadronian horse that is known from a nearly complete skeleton (F:AM 116359, Fig. 10.9). The uncrushed skull shows the distinctively long facial fossa and longer, more slender rostrum that distinguishes it from *Miohippus obliquidens*. The upper incisors have deep pits in the crowns, and are oval in cross section. The cheek teeth are large relative to the size of the skull (Tables 10.1, 10.2). They have well-defined lophs with almost no metaconules, except on $P^2$. The hypostyles are mostly class 3 (Fig. 10.1C), but are much thicker and higher than the hypostyles found in *Mesohippus exoletus*. The upper cheek teeth lack a lingual cingulum, but a faint cingular cusp occurs in the valley between the lophs. On some specimens, however, a faint cingulum can be seen. The mandible (Fig. 10.9E, F) of F:AM 116359 shows a long, pinched symphysis. The incisors are spatulate, with flattened tips. In the lower cheek teeth, the metaconid and metastylid are well separated. The hypoconulids on $P_2$-$M_2$ are also very prominent.

The postcranial skeleton of F:AM 116359 includes an atlas, two cervical vertebrae, a pelvis, and most of both hind limbs. In most features, they are identical to *Mesohippus*, except that they are larger (Table 10.4). The tarsus (Fig. 10.9D) clearly shows the cuboid-metatarsal III articulation, diagnostic of *Miohippus*. This is also seen in F:AM 74077 and several other specimens referable to *Miohippus assiniboiensis*.

In all its preserved features, it is clear that this species must be referred to *Miohippus*. Yet this species is known only from the Chadronian, and it occurs in the same deposits as three species of *Mesohippus* and *Miohippus obliquidens* (Fig. 10.5A). The oldest species name that appears applicable to this horse is *Mesohippus assiniboiensis* Lambe, 1905b. Lambe's type (NMC 6305) is a very poor specimen, with few diagnostic features. But Russell (1975) referred a topotypic sample from the Cypress Hill to *Mesohippus assiniboiensis*, and this material is a good match in size and morphology (Fig. 10.9) with F:AM 116359 and other horses of this size and morphology such as F:AM 74077 (compare Russell, 1975, with Tables 10.1, 10.2).

Lambe (1905b, p. 49) described another species of horse from the same locality, which he called *Mesohippus planidens*, because it had a flattened ectoloph. We find that the only preserved ectoloph ($M^2$) on the type is no more flattened than is typical for these horses. Russell (1975, p. 648) considered *Mesohippus planidens* to be indeterminate, or possibly a synonym of *Mesohippus assiniboiensis* Lambe 1905b. The larger topotypic samples from Cypress Hills include many specimens referred to *Mesohippus assiniboiensis* that are indistinguishable from the type of *Mesohippus planidens*, except for the ectoloph. Since the slightly flattened ectoloph is not a valid character separating the two species, we suggest that the names *assiniboiensis* and *planidens* refer to the same horse. In this case, the first reviser (Russell, 1975) has established the priority of *assiniboiensis* (Lambe 1905b, p. 50) over the name *planidens* (Lambe 1905b, p. 49), even though the type specimen of *assiniboiensis* is not very diagnostic. It appears that the horses called *Mesohippus assiniboiensis* by Russell (1975) are in fact referable to *Miohippus assiniboiensis*.

Clark and Beerbower (1967, pp. 48-49) described the species *Mesohippus grandis* based on some lower jaws from the late Chadronian of the Big Badlands of South Dakota. Since the taxon is known only from lower jaws, the only feature that distinguishes it from other late Chadronian horses is its unusually large size. However, the type and referred specimens of *M. grandis* match *M. assiniboiensis* closely in size, and in morphology (Table 10.3). Both large horses are known from the latest Chadronian. Since *M. assiniboiensis* has priority over *M. grandis*, we consider Clark and Beerbower's taxon to be a junior subjective synonym.

*Miohippus intermedius*
(Osborn and Wortman, 1895)
Fig. 10.10

*Mesohippus intermedius* Osborn and Wortman, 1895, p. 354
*Miohippus intermedius*: Osborn, 1918, p. 54
*Pediohippus intermedius*: Schlaikjer, 1935, p. 145

*Revised Diagnosis.* $M^{1-3}$ length = 40-41 mm. Skull slightly larger than *Miohippus assiniboiensis*, and slightly smaller than *Miohippus gidleyi*. Distinguished from *Miohippus gidleyi* by shallower facial fossa, stronger metalophs, and by lack of metaconules.

*Type.* AMNH 1196, a nearly complete skeleton from the late Whitneyan *Protoceras* channels of the Big Badlands, South Dakota (Fig. 10.10A-C).

*Referred material.* F:AM 116418 (55 feet below Upper Whitney Ash, 66 Mountain, Phinney Ranch, Goshen Co., Wyoming) (Fig. 10.10D-E).

*Horizon and locality.* Late Whitneyan of South Dakota and Wyoming.

*Discussion.* *Miohippus intermedius* is indeed intermediate in size between *Miohippus assiniboiensis* and *Miohippus gidleyi*. Since it occurs with *Miohippus gidleyi* and falls on the small end of the size range (Fig. 10.8), it might be synonymized with *Miohippus gidleyi* (in which case *intermedius* would have priority). However, we feel that the deeper facial fossa, the distinctively primitive metaconules and metalophs of *Miohippus gidleyi*, as well as its slightly wider upper teeth (Table 10.2) warrant separation of the two species for the present.

Even with the Frick Collection, the sample of Whitneyan fossil horses is still extremely small, so it is difficult to make a final assessment of the variability of horses in the Whitneyan.

*Miohippus gidleyi* (Osborn, 1904)
10.11A-B

*Mesohippus gidleyi* Osborn, 1904, p. 178
*Miohippus gidleyi* Osborn, 1918, p. 56
*Pediohippus gidleyi* Schlaikjer, 1935, p. 145

*Synonyms.* *Miohippus validus* Osborn, 1904, p. 177; *Mesohippus grallipes* Sinclair, 1925, p. 55.

*Revised Diagnosis.* $M^{1-3}$ length = 42-46 mm. Skull slightly larger than in *Miohippus intermedius*. Facial fossa much broader and deeper than in *Miohippus intermedius*. Metaconules and lingual cingula more distinct than in *Miohippus intermedius*.

*Type.* AMNH 1192, a dorsoventrally crushed skull, from the *Protoceras* channels, Big Badlands of South Dakota (Fig. 10.11A-B).

*Referred material.* AMNH 680 (type of *M. validus*); AMNH 1193, 1218; F:AM 116413, 116414, 116415, 116416, 116417.

*Horizon and localities.* Known only from the late Whitneyan *Protoceras* channels, South Dakota.

*Discussion.* As pointed out above, *M. gidleyi* is distinguished from contemporaneous *Miohippus intermedius* by its slightly larger size, deeper facial fossa, broader cheek teeth, stronger lingual cingula, and more distinct metaconules. It is the dominant late Whitneyan horse.

*Miohippus validus* (Osborn, 1904) is

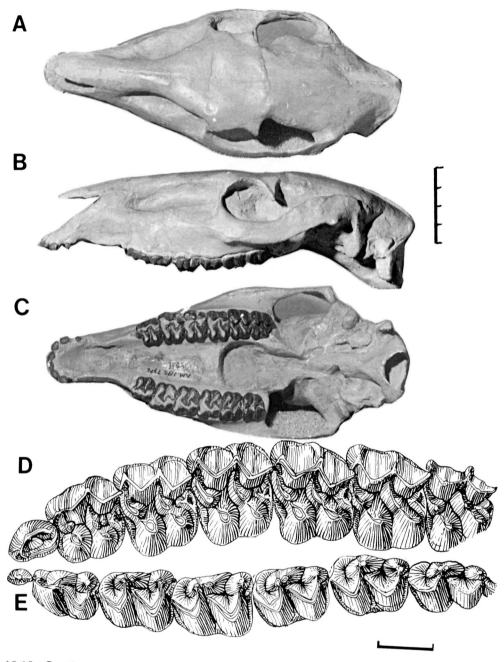

Fig. 10.10. Caption on next page.

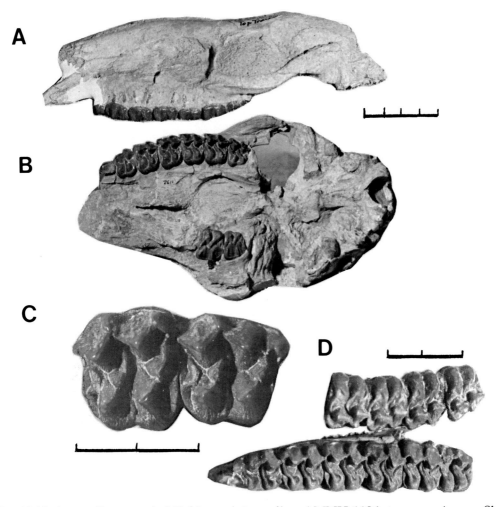

Fig. 10.10. (preceding page) *Miohippus intermedius*, AMNH 1196, type specimen. Skull in dorsal (A), left lateral (B), and ventral (C) views. Referred specimen, F:AM 116418, crown views of upper (D) and lower (E) teeth. Scale bars in cm.

Fig. 10.11. *Miohippus gidleyi*, AMNH 2611, type skull in left lateral (A) and ventral (B) views. (C) *Miohippus "crassicuspis"* (= *annectens*), AMNH 1683, type specimen. (D) *Miohippus equinanus*. F:AM 116421 (upper specimen) compared with type, AMNH 12912 (lower specimen).

here referred to *Miohippus gidleyi*. The distinctive features of *Miohippus validus* listed by Osborn serve to distinguish it only from *Miohippus intermedius*, and Osborn never compared *M. validus* with *Miohippus gidleyi* in print. The type specimen (AMNH 680, not AMNH 1218, as listed by Osborn, 1918) falls within the size range of *Miohippus gidleyi* (Fig. 10.8). The crowns of the upper teeth of the type are too worn to make detailed comparisons with *Miohippus gidleyi*, but in the preserved features they match quite well. The only possible distinguishing feature of *Miohippus validus* is a slight dorsoventral ridge subdividing the facial fossa of the type skull. However, this feature is not known in any other specimen referred to *Miohippus validus*, so it is probably due to individual variation or postmortem deformation of the type specimen. In our opinion, it does not justify the separation of *M. validus* as a distinct species.

*Miohippus annectens* Marsh, 1874
Fig. 10.11C

*Synonym.* *Miohippus crassicuspis* Osborn, 1904, p. 178.
*Revised Diagnosis.* $M^{1-3}$ length = 47-50 mm.
*Type.* YPM 12230, a skull with associated limb material, John Day Formation, Arikareean of Oregon.
*Referred material.* AMNH 683, type of *Miohippus crassicuspis*, plus much Arikareean material not considered here.
*Horizon and locality.* Late Whitneyan *Protoceras* channels, Big Badlands of South Dakota, and John Day beds, Arikareean of Oregon.
*Discussion.* *Miohippus crassicuspis* (AMNH 683) is a large *Miohippus* with remarkably distinct protoconules and metaconules (Fig. 10.11C), well-developed class 3 hypostyles, but no crochets. It is clearly much larger than, and morphologically distinct from, any other White River horse, but no previous author has reported comparing it with Arikareean horses. We find that it is a very good match in size and morphology with the type specimen of *Miohippus annectens* from the John Day beds of Oregon. The type skull of *Miohippus annectens* (YPM 12230) shows very distinct conules and well developed class 3 hypostyles. The lingual cingula are weak or expressed as intervallic cusps, as in the type of *Miohippus crassicuspis*. Reluctance to assign John Day and White River horses to the same taxa need not be a systematic barrier. The John Day and upper White River deposits are now considered so closely similar in age (Tedford *et al.*, 1987; Prothero and Rensberger, 1985) that it does not seem remarkable that a late Whitneyan High Plains horse should also continue into the Arikareean of Oregon.

?*Miohippus equinanus* Osborn, 1918
Fig. 10.11D

*Revised Diagnosis.* Small horse ($M^{1-3}$ length = 31 mm) with relatively advanced, bilophodont dentition and class 3 or 4 hypostyles.
*Type.* AMNH 12912, a palate with left and right $dP^1$-$M^3$, from the early Arikareean "Rosebud Formation" (= Sharps Formation; Macdonald, 1963), 5 miles above Manderson, Wounded Knee Creek, South Dakota.
*Referred material.* AMNH 39028 (*Leptauchenia* nodules, 6 miles east of Oelrichs, Fall River Co., S.D.); F:AM 116411 (Base of Upper *Oreodon* beds, Kodak Point, Big Badlands, S.D.). There is much additional Arikareean material from the type locality in the AMNH, LACM, and SDSM collections, as listed by Macdonald (1963, 1970).
*Horizon and localities.* Late Whitneyan and early Arikareean, South Dakota.
*Discussion.* This tiny Arikareean horse is now known from the late Whitneyan. The size and morphology of this taxon is distinctive. It is far smaller than any contemporary horse (Tables 10.2, 10.3), and far more advanced in dental morphology than

any earlier form of comparable size. The reference of this species to *Miohippus* is still questionable, since it is presently known only from upper dentitions. No skulls or postcrania have been found that could show whether it has the diagnostic features of *Miohippus*. Only the relatively derived dentition justifies its assignment to *Miohippus*, and this may change when more complete material is found.

Stirton (1940) made *M. equinanus* "ancestral" to *Archaeohippus*. Although these horses are similar in their diminutive size, they do not match in detail. *Archaeohippus* is much more derived in having the metaloph completely attached to the ectoloph by a high, continuous crest and a strong medicrista. It also has a very high-crowned triangular hypostyle. *M. equinanus* has none of these derived features of *Archaeohippus*. Until more complete skull material is available for *M. equinanus*, reference to *Archaeohippus* is based strictly on size and is therefore questionable.

A note on *"Mesohippus longipes"*
Osborn and Wortman (1894, p. 214) described a nearly complete right hind limb (AMNH 684) from the Whitneyan (Lower Poleslide Member) of South Dakota. They compared this specimen to horses and hyracodonts, and decided that it was equid. On this basis, they named the specimen *Mesohippus longipes*. In the following year, Osborn and Wortman (1895, p. 366) changed their minds and referred this specimen to the tapiroid *Colodon*. Their justification for this was the following: 1) the ectal and sustentacular facets of the astragalus were continuous, as in rhinos and hyracodonts, but not in horses; 2) the ectocuneiform had great vertical depth; 3) there was no cuboid-MT III contact.

Scott (1940, pp. 940-941) reassigned this specimen to *Mesohippus*. He claimed that the astragalus is similar to that of horses, not tapirs, but he gave no justification for this statement. Nevertheless, he admitted that the large size and the lack of cuboid-MT III articulation make it difficult to refer this specimen to the Equidae.

We examined this specimen, and it is clear that it is not equid. Although the limb is very similar in size to some specimens of *Miohippus gidleyi* (e.g., F:AM 109853), it is a much better match for *Colodon occidentalis* (e.g., AMNH 658). The sustentacular facet of the astragalus of "*M. longipes*" and *C. occidentalis* both have a distinct ventrolateral spur not found on any equid astragalus. The limb proportions are a little too slender for *Miohippus gidleyi*. As noted by previous authors, the ectocuneiform is too deep. The "connection" between the ectal and sustentacular facets is due to plastic deformation of AMNH 684. Most of the facets appear unusually rounded and indistinct. Besides, this "connection" is not found in either tapiroid or horses. It is clear that *"Mesohippus longipes"* Osborn and Wortman, 1894, is not a horse at all, but a hind limb of *Colodon occidentalis* (Leidy, 1868). The former name is here relegated to junior synonymy with *Colodon occidentalis*.

Phylogeny
A phylogenetic hypothesis for the Oligocene Equidae and their sister taxa is shown in Fig. 10.12. Most of the derived characters are dental, but characters of the skull and postcranial skeleton are used where appropriate. The temporal ranges of the species of Oligocene hores are shown in Fig. 10.5B. It is immediately apparent that Oligocene horses are characterized by a rather "bushy" branching phylogeny, and not the single "trunk" portrayed by Stirton (1940) and Simpson (1951). At any given interval of time, at least two or three species of horses existed, and frequently all can be found in the same deposits. Indeed, the pattern seen in Oligocene horses is nearly comparable in species diversity to that seen in the Neogene, except that only two genera are present. In addition, the popular myth that *Mesohippus* imperceptibly grades into a successional genus *Miohippus* is not supported. *Miohippus* appears in the mid-Chadronian, and

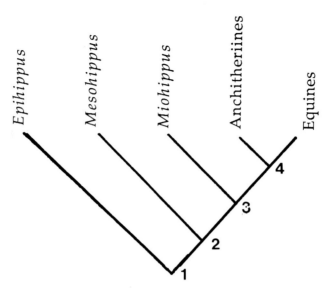

Fig. 10.12. Hypothesis of relationships of Oligocene horses and their sister-taxa. Characters at the numbered nodes as follows: 1) $P^2$ submolariform; $P_1$ single rooted; 2) $I^{1-3}$ with pitted crowns; $P^2/2$ molariform; metacarpal V reduced; extended premaxilla; longer diastema; angle of jaw posteriorly rounded without notch; 3) cuboid-metatarsal III contact; deeper, more elongate facial fossa; longer facial region; 4) metaloph connected to ectoloph.

*Mesohippus* does not become extinct until the early Whitneyan. The two genera overlap by at least four million years.

**Patterns of evolution**

Since Huxley (1870) and Kowalevsky (1873), horses have been used as one of the classic cases of evolution in the fossil record. The early European lineages connected four successive forms (*Palaeotherium-Anchitherium-Hipparion-Equus*) in the first attempt at a horse phylogeny. In 1876, Huxley visited North America and discovered that O.C. Marsh had a much more complete sequence of horses in this continent. The European "phylogeny" was in fact composed of a non-equid (*Palaeotherium*) and three successive immigration events from North America. Most subsequent horse phylogenies were strictly orthogenetic, with one continuous intergrading lineage and few or no side branches. This long-obsolete picture still appears in much of the popular literature.

Stirton (1940) and Simpson (1951) began to present a more complex, branching phylogeny of horses. They proposed cladogenic splitting events that led to the anchitherines, the hipparionines, the equines, and many other side branches. The paraphyletic nature of the genus *Merychippus* became apparent. The earlier part of horse family tree was, however, still treated as a single, unbranched "trunk," with *Mesohippus* and *Miohippus* represented as one continuous intergradational lineage. A strong gradualistic bias was evident in these phylogenies, influencing the way in which species were delineated, and their relationships interpreted. For example, Simpson (1951, p. 127) wrote, "*Mesohippus* and *Miohippus* intergrade so perfectly and the differences between them are so slight and variable that even experts find it difficult, at times nearly impossible, to distinguish them clearly." Naturally, this gradualistic bias also finds its way into the popular literature. As recently as 1980, Lewin (1980, p. 884) wrote, "A classic example of... a trend is the evolution of the modern horse, whose distant ancestor *Hydracotherium* [sic] was a three-toed [sic--it had four toes in the manus] creature no bigger than a dog. The fossil record shows an apparent steady 'progress' through time, with gradual changes in body size and form leading to the familiar *Equus*."

In recent years, however, paleontologists have been challenged to examine their biases with regard to the delineation of species, the construction of phylogenies, and the tempo of evolution (Eldredge and Gould, 1972; Gould and Eldredge, 1977). Eldredge (1979) points out that a gradualistic bias will lead to a "transformational" view of species, which are treated as arbitrary slices of a gradualistic continuum. By contrast, a "taxic" approach will recognize species as diagnostic clusters of morphology without attempting to fit them into an *a priori* bias for intergradational, stratigraphically-defined horizontal taxa. With a less arbitrary, more objective concept of species, it should be possible to assess whether they do in fact intergrade continuously, or show other patterns. One of the patterns recognized with the "taxic" approach is the sympatric persistence of closely related species through long periods of time. Stanley (1975, 1979) has used information of this kind to support the hypothesis of species selection for long-term macroevolutionary trends. The view of species as discrete, non-arbitrary units with their own intrinsic properties leads to many other fruitful hypotheses (e.g., Vrba and Eldredge, 1984).

Recent work on Neogene horses (MacFadden, 1984, 1985, 1988) has clearly shown that horses are very speciose with many genera living sympatrically in the late Miocene of North America. Skinner and Johnson (1984, p. 276) reported eight genera of horses from the late Barstovian Railway Quarry A of the Valentine Formation of Nebraska, and Schultz (1977) reported 12 species of horses from the Clarendon Beds of Texas. Unfortunately, the North American terrestrial Neogene record gives few examples of continuous fossiliferous sequences which might be used to test whether these species arise gradualistically or by splitting events (MacFadden, 1984, 1985, 1988). But the Paleogene record does have such continuous, fossiliferous sequences. For example, in the continuous sections of the Eocene of the Bighorn Basin of Wyoming, Gingerich (1980) has reported at least two lineages of *Hyracotherium* living sympatrically with the slightly larger ceratomorph *Homogalax*, which is indistinguishable from *Hyracotherium* in most features.

The new systematic analogies, combined with a robust stratigraphic context, allow us to approach several aspects of the tempo and mode of horse evolution. Three patterns in particular emerge: 1) most horse species do not intergrade morphologically; 2) advanced and primitive species lived sympatrically, and, 3) morphological stasis predominates in horse evolution. These points are elaborated below:

### 1. *Most Oligocene horse species do not intergrade*

In our studies of the White River horses, we first attempted to determine if diagnosable clusters of morphology could be recognized in *Mesohippus* and *Miohippus*. Non-arbitrary distinctions between species were indeed possible. It seems that the difficulty of distinguishing species discussed by Simpson (1951) was simply an artifact of small sample size and inadequate stratigraphic data. Taking isolated specimens from each stratigraphic level, a single orthogenetic lineage could be constructed. Larger sample sizes have clearly shown that non-arbitrary distinctions can be made between the various species of *Mesohippus* and *Miohippus*. These distinctions result in diagnosable, non-overlapping clusters of morphology that deserve to be recognized as species in the paleontological sense of the word. Admittedly, there are some specimens that appear to be intermediate or transitional between these clusters, but these are relatively few. In addition, these intermediates do not show a distribution that would be predicted by the gradualistic model. For example, the gradualistic model would predict that Chadronian faunas would be dominated by the more primitive *Mesohippus westoni* and a few intermediates between it and *M. bairdi*. In later Oligocene deposits, *M. westoni* morphologies should become rare or absent, and only intermediates and true *M. bairdi* should be found. The actual record of these taxa does not, in fact, support this model. *M. westoni* is as abundant as *M. bairdi* in the early Orellan, and quite a few are found even at the end of its range in the late Orellan. Similarly, a high proportion of the Chadronian *Mesohippus* is *M. bairdi*, even to the beginning of the range of this species.

2. Mesohippus *and* Miohippus *have overlapping temporal and geographic ranges*

Another feature of our analysis is that Oligocene horses appear almost as speciose and long-ranging as the Neogene horses discussed above. For example, in the late Chadronian-early Orellan sequences north of Lusk, Wyoming (Fig. 10.5A), the beds above the Persistent White Layer (a widespread marker bed) produce three species of *Mesohippus* (*M. bairdi, M. westoni,* and *M. exoletus*) and two of *Miohippus* (*M. obliquidens* and *M. assiniboiensis*). All of these species are extremely long-ranging, and overlap considerably (but not completely) in time (Fig. 10.5B). *Miohippus obliquidens*, for example, appears to range from the late Chadronian until the late Whitneyan, and *Mesohippus bairdi* from the mid-Chadronian until the early Whitneyan. There is no evidence of long-term changes within these species through time. Instead, they are strikingly static through millions of years. Such stasis is apparent in most Neogene horses as well (MacFadden, 1984, 1985, 1988), and in *Hyracotherium* (Gingerich, 1980; Bown, 1979, p. 105). This is contrary to the widely held myth about horse species as parts of continua, with no real distinctions between species.

Throughout the history of horses, species are well-marked and static over millions of years. At high resolution, this gradualistic picture of horse evolution becomes a complex bush of overlapping, closely related species. This pattern is predicted by the model of species selection of Stanley (1975, 1979) and Gould and Eldredge (1977).

Unfortunately, it appears that the actual transitions between species of *Mesohippus* and *Miohippus* occur in the Chadronian, where the record is relatively poor. Thus, the key question of the duration and nature of the splitting events cannot yet be answered. A handful of good specimens from the Flagstaff Rim area (currently under study by R. J. Emry) and other Chadronian localities give tantalizing hints about the separation among the species, but very few complete skulls are known. Adequate documentation of a gradualistic versus punctuated pattern would require numerous skulls and maxillae from several levels, and this kind of sample is not yet available.

**Conclusion**

The evolution of Oligocene horses is characterized by multiple speciation events, and not by orthogenetic evolution of a single continuous lineage, as it is commonly portrayed. At least four distinct species of *Mesohippus* can be recognized, each of which has a long, distinctive stratigraphic range. Although some of these species intergrade slightly, they are generally easy to distinguish, and maintain a static morphology through millions of years.

*Miohippus* is a significantly larger, more derived horse that can easily be distinguished from *Mesohippus*. Contrary to previous authors, it does not intergrade imperceptibly with *Mesohippus*, but instead is distinctive from its earliest appearance in the Chadronian. *Miohippus* and *Mesohippus* thus overlap through most of the Chadronian until the end of the Whitneyan, and are not successive genera of a continuous lineage, as is widely believed.

Two species of *Miohippus* were sympatric with at least three species of *Mesohippus* during the late Chadronian and early Orellan. After the disappearance in the Whitneyan of the last remaining species of *Mesohippus* (*M. barbouri, M. bairdi*), there is a new radiation of species of *Miohippus* (*M. obliquidens, M. intermedius, M. gidleyi, M. annectens, M. equinanus*), some of which persist into the Arikareean. Although the actual speciation events for most of these species are not preserved in the Oligocene, it is evident from the abundance of static species persisting unchanged through millions of years that horses support the punctuated, rather than gradualistic, model of speciation. *Mesohippus* and *Miohippus* are not arbitrary slices of a single continuous lineage, but genera consisting of multiple distinct species that have distinctive but overlapping temporal ranges.

## Acknowledgments

We thank R. J. Emry, R. Evander, A. Forstén, S. J. Gould, W. W. Korth, and B. J. MacFadden for helpful comments on the manuscript. Acknowledgment is made to the Donors of the Petroleum Research Fund of the American Chemical Society, and to NSF grant EAR87-08221, for partial support of the senior author during the preparation of this research. The junior author was supported by a grant from the Miller Institute of Basic Research in Science, University of California, Berkeley.

## Bibliography

Bown, T. (1979): Geology and mammalian paleontology of the Sand Creek Facies, Lower Willwood Formation (Lower Eocene), Washakie County, Wyoming. -*Mem. Geol. Surv. Wyoming*, 2: 1-151.

Clark, J., and Beerbower, J.R. (1967): Geology, paleoecology, and paleoclimatology of the Chadron Formation. -In: Clark, J., Beerbower, J.R., and Kietzke, K.K.: Oligocene sedimentation, stratigraphy, paleoecology, and paleoclimatology in the Big Badlands of South Dakota. -*Fieldiana Geol. Mem.*, 5: 21-74.

Cope, E.D. (1874): Report on the vertebrate paleontology of Colorado. -*Ann. Rept. Geol. Geogr. Surv. Terr.*, 1873: 427-533.

Cope, E.D. (1889): The Vertebrata of the Swift Current River. II. -*Amer. Nat.*, 23(267): 151-155.

Eldredge, N. (1979): Alternative approaches to evolutionary theory. -*Bull. Carn. Mus. Nat. Hist.*, 12: 7-19.

Eldredge, N., and Gould, S.J. (1972): Punctuated equilibria: an alternative to phyletic gradualism. -In: Schopf, T.J.M. (ed.): *Models in Paleobiology*. San Francisco (Freeman, Cooper), pp. 82-115.

Emry, R.J., Bjork, P.R., and Russell, L.S. (1987): The Chadronian, Orellan, and Whitneyan North American Land Mammal Ages. -In: Woodburne, M.O. (ed.): *Cenozoic Mammals of North America, Geochronology and Biostratigraphy*. Berkeley (Univ. California Press), pp. 118-152.

Forstén, A.M. (1970a): Variation in and between three populations of *Mesohippus bairdii* [sic] Leidy from the Big Badlands, South Dakota. -*Acta Zool. Fennica*, 126: 1-16.

Forstén, A.M. (1970b): *Mesohippus* from the Chadron of South Dakota, and a comparison with Brulean *Mesohippus bairdii* [sic] Leidy. -*Comm. Biol. Soc. Sci. Fennica*, 31(11): 1-22.

Forstén, A.M. (1971): Comparison of populations of *Mesohippus* from Trans-Pecos Texas and Big Badlands, South Dakota. -*Pearce-Sellards Ser., Texas Mem. Mus.*, 18: 12-15.

Forstén, A.M. (1974): The taxonomic status of the equid sample from the Harvard Fossil Reserve, Goshen County, Wyoming. -*J. Paleont.*, 48(2): 404-408.

Galbreath, E.C. (1953): A contribution to the Tertiary geology and paleontology of northeastern Colorado. -*Univ. Kans. Paleont. Contrib.*, 4: 1-120.

Gingerich, P.D. (1980): Evolutionary patterns in early Cenozoic mammals. -*Ann. Rev. Earth Planet. Sci.*, 8: 407-424.

Gould, S.J., and Eldredge, N. (1977): Punctuated equilibria: the tempo and mode of evolution reconsidered. -*Paleobiology*, 3: 115-151.

Gray, J. E. (1821): On the natural arrangement of vertebrose animals. -*London Med. Reposit.*, 15: 296-310.

Hay, O.P. (1902): Bibliography and catalogue of fossil Vertebrata of North America. -*Bull. U.S. Geol. Surv.*, 179: 1-868.

Huxley, T.H. (1870): The anniversary address of the president. -*Q. J. Geol. Soc. London*, 25: 309-310.

Kowalevsky, V. (1873): Sur l'*Anchitherium aurelianense* Cuv. et sur l'histoire paléontologiques des chevaux. -*Mém. Acad. Sci. St. Petersburg*, 20: 1-73.

Lambe, L.M. (1905a): On the tooth structure of *Mesohippus westoni* (Cope). -*Amer. Geol.*, 35: 243-245.

Lambe, L.M. (1905b): Fossil horses from the Oligocene of Cypress Hills, Assiniboia. -*Trans. Roy. Soc. Canada*, ser. 2, 11: 43-52.

Lambe, L.M. (1908): The Vertebrata of the Oligocene of the Cypress Hills, Saskatchewan. -*Geol. Surv. Br. Can. Contrib. Can. Paleont.*, 3(4): 5-64.

Leidy, J. (1850): [Description of *Rhinoceros nebraskensis, Agriochoerus antiquus, Palaeotherium proutii*, and *P. bairdii*]. *Proc. Acad. Nat. Sci. Phila.*, 5: 121-122.

Leidy, J. (1852): Description of the remains of extinct Mammalia and Chelonia from Nebraska Territory, collected during the geological survey under the direction of Dr. D.D. Owen. -In: Owen, D.D.: *Report of a geological survey of Wisconsin, Iowa, and Minnesota, and incidentally a portion of Nebraska Territory*, Philadelphia, pp. 534-572.

Leidy, J. (1868): Notice of some remains of horses. -*Proc. Acad. Nat. Sci. Phila.*, 1868: 195.

Leidy, J. (1869): The extinct mammalian fauna of Dakota and Nebraska, including an account of some allied forms from other localities, together with a synopsis of the mammalian remains of North America. -*J. Acad. Nat. Sci. Phila.*, 2: 1-472.

Leidy, J. (1870): [Remarks on a collection of fossils from Dalles City, Oregon]. -*Proc. Acad. Nat. Sci. Phila.*, 22: 111-113.

Lewin, R. (1980): Evolutionary theory under fire. -*Science*, 210: 883-887.

Lillegraven, J. A. (1970): Stratigraphy, structure, and vertebrate fossils of the Oligocene Brule Formation, Slim Buttes, northwestern South Dakota. -*Geol. Soc. Amer. Bull.*, 81: 831-850.

Macdonald, J. R. (1963): The Miocene faunas from the Wounded Knee area of western South Dakota. -*Bull. Amer. Mus. Nat. Hist.*, 125: 139-238.

Macdonald, J. R. (1970): Review of the Miocene Wounded Knee faunas of southwestern South Dakota. -*Los Angeles County Mus. Nat. Hist. Bull (Sci.)*, 8: 1-82.

MacFadden, B. J. (1984): Systematics and phylogeny of *Hipparion, Neohipparion, Nannippus*, and *Cormohipparion* (Mammalia, Equidae) from the Miocene and Pliocene of the New World. -*Bull. Amer. Mus. Nat. Hist.*, 179: 1-196.

MacFadden, B. J. (1985): Patterns of phylogeny and rates of evolution in fossil horses: hipparions frm the Miocene and Pliocene of North America. -*Paleobiology*, 11: 245-257.

MacFadden, B. J. (1988): Horses, the fossil record, and evolution, a current perspective. -*Evol. Biol.*, 22: 131-158.

Marsh, O.C. (1874): Notice of new equine mammals from the Tertiary formation. -*Amer. J. Sci.*, 7: 247-258.

Marsh, O.C. (1875): Notice of new Tertiary mammals.IV. -*Amer. J. Sci.*, 9: 239-250.

Marsh, O.C. (1879): Polydactyl horses, recent and extinct. -*Amer. J. Sci.*, 17: 449-505.

Osborn, H.F. (1904): New Oligocene horses. -*Bull. Amer. Mus. Nat. Hist.*, 20(8): 167-169.

Osborn, H.F. (1918): Equidae of the Oligocene, Miocene, and Pliocene of North America. Iconographic type revision. -*Mem. Amer. Mus. Nat. Hist.*, 2: 1-326.

Osborn, H.F., and Wortman, J.A. (1894):

Fossil mammals of the Lower Miocene White River beds. Collection of 1892. - *Bull. Amer. Mus. Nat. Hist.*, 6(7): 199-228.

Osborn, H.F., and Wortman, J.A. (1895): Perissodactyls of the Lower Miocene White River beds. -*Bull. Amer. Mus. Nat. Hist.*, 7(12): 343-375.

Owen, R. (1848): Description of teeth and portion of the jaws of two extinct anthracotheroid quadrupeds . . . -*Q. J. Geol. Soc. London*, 4: 103-141.

Prothero, D.R. (1982): Medial Oligocene magnetostratigraphy and mammalian biostratigraphy: testing the isochroneity of mammalian biostratigraphic events. -Unpubl. Doct. Dissert., New York (Columbia Univ.).

Prothero, D.R., and Rensberger, J.M. (1985): Preliminary magnetostratigraphy of the John Day Formation, Oregon, and the North American Oligocene-Miocene boundary. -*Newslett. Strat.*, 15 (2): 59-70.

Russell, L.S. (1975): Revision of the fossil horses from the Cypress Hills Formation (lower Oligocene) of Saskatchewan. -*Can. J. Earth Sci.*, 12(4): 636-648.

Schlaikjer, E.M. (1931): Description of a new *Mesohippus* from the White River Formation of South Dakota. -*Proc. New England Zool. Club*, 12: 35-36.

Schlaikjer, E.M. (1932): The osteology of *Mesohippus barbouri*. -*Bull. Mus. Comp. Zool. Harvard Univ.*, 72: 391-410.

Schlaikjer, E.M. (1935): Contributions to the stratigraphy and paleontology of the Goshen Hole area, Wyoming. IV. New vertebrates and the stratigraphy of the Oligocene and early Miocene. -*Bull. Mus. Comp. Zool. Harvard Univ.*, 76: 91-189.

Schultz, G.E. (1977): Field conference on Late Cenozoic biostratigraphy of the Texas Panhandle and adjacent Oklahoma. -*Spec. Publ. W. Texas. State Univ.*, 1: 1-160.

Scott, W.B. (1941): Perissodactyla. -In: Scott, W. B., and Jepsen, G. L. : The mammalian fauna of the White River Oligocene. -*Trans. Amer. Phil. Soc.*, 28: 747-790.

Simpson, G.G. (1951): *Horses*. -New York (Oxford Univ. Press).

Simpson, G.G. (1953): *The meaning of evolution*. -New York (Columbia Univ. Press).

Sinclair, W.J. (1925): A mounted skeleton of a new *Mesohippus* from the *Protoceras* beds. -*Proc. Amer. Phil. Soc.*, 64: 55.

Skinner, M.F., and Johnson, F.W. (1984): Tertiary stratigraphy and the Frick Collection of fossil vertebrates from north-central Nebraska. -*Bull. Amer. Mus. Nat. Hist.*, 178: 215-368.

Stanley, S.M. (1975): A theory of evolution above the species level. -*Proc. Nat. Acad. Sci.*, 72: 646-650.

Stanley, S.M. (1979): *Macroevolution*. -San Francisco (W. H. Freeman).

Stirton, R.A. (1940): Phylogeny of North American Equidae. -*Univ. Calif. Publ. Geol. Sci.*, 25: 165-198.

Tedford, R.H., Galusha, T., Skinner, M.F., Taylor, B.E., Fields, R.W., Macdonald, J.R., Rensberger, J.M., Webb, S.D., and Whistler, D.P. (1987): Faunal succession and biochronology of the Arikareean through Hemphillian interval (late Oligocene through earliest Pliocene Epochs) in North America. -In: Woodburne, M.O. (ed.): *Cenozoic Mammals of North America, Geochronology and Biostratigraphy*. Berkeley (Univ. California Press), pp. 153-210.

Vrba, E.S., and Eldredge, N. (1984): Individuals, hierarchies, and processes: towards [sic] a more complete evolutionary theory. -*Paleobiology*, 10: 146-171.

# 11. PHYLOGENETIC INTERRELATIONSHIPS AND EVOLUTION OF NORTH AMERICAN LATE NEOGENE EQUINAE

## RICHARD C. HULBERT, JR.

The phylogenetic interrelationships of thirteen late Neogene genera of advanced equids and ten additional species of "merychippine-grade" horses were investigated using computer-generated minimum length (most parsimonious) trees. Many of the 45 characters used in the analysis showed homoplasy, suggesting that further study with additional sets of characters would be useful. Two principal clades of advanced Equinae were recognized: the Hipparionini and the Equini. For both, the period from 18 Ma to 14 Ma was characterized by rapid cladogenesis and diversification. Hipparionines and protohippines numerically and taxonomically dominated North American equid faunas until about 7 Ma. From then on equines predominated, with protohippines going extinct at about 6 Ma and hipparionines at about 2 Ma. "*Merychippus*" is an artificial, paraphyletic assemblage, and merychippine species should be reassigned to strictly monophyletic groups.

## Introduction

It has long been recognized that during the latter half of the Miocene, North American equids radiated into many sympatric, hypsodont lineages. Judging by their abundance at many fossil sites, equids were a major component of later Tertiary ungulate communities of that continent. As their fossil record is seemingly so complete and overwhelmingly abundant, horses have received prominent attention in studies of evolution (e.g., Simpson, 1944; Stirton, 1947; Romer, 1949), biogeography (e.g., Colbert, 1935; Bernor *et al.*, 1980; Woodburne *et al.*, 1981), biostratigraphy and biochronology (e.g., McGrew and Meade, 1938; Stirton, 1939, 1952) and paleoecology (e.g., Shotwell, 1961; Janis, 1984). However, the phylogenetic interrelationships of advanced horses (the Equinae) have never been satisfactorily resolved. Without this basic information, evolutionary and ecologic scenarios lack proper foundation (Cracraft, 1981).

Historically, J. W. Gidley (1907) was among the first to address this problem in a modern sense. He concluded, however, that "it is moreover evident, from a study of the abundant material at hand, that although the general lines of progressive development [i.e., grades] are clearly indicated and several distinct lines of subphyla suggested, the direct lines of descent are by no means complete and the known genera cannot at present be arranged in any permanent phyletic series" (Gidley, 1907, pp. 869-870). In the subsequent eight decades, several generations of the most prominent North American mammalian paleontologists collectively arrived at such a "phyletic series." Two milestones of that effort were the reviews of Matthew (1926) and Stirton (1940). Among the major contributions of Matthew (1926) were: 1) the affirmation of the idea that horses are not a direct orthogenetic sequence from *Hyracotherium* to *Equus*; 2) the recognition of a major hipparionine-equine dichotomy among hypsodont horses; and 3) formal division of North American hipparionines into three clades (Fig. 11.1A). Stirton's (1940) well known phylogeny of the Equinae (Fig. 11.1B) is largely a derivative of Matthew's (1926), although several additional taxa were by then formally recognized (*Calippus* and *Astrohippus*). Stirton,

*The Evolution of Perissodactyls* (ed. D.R. Prothero & R.M. Schoch) Oxford Univ. Press, New York, 1989

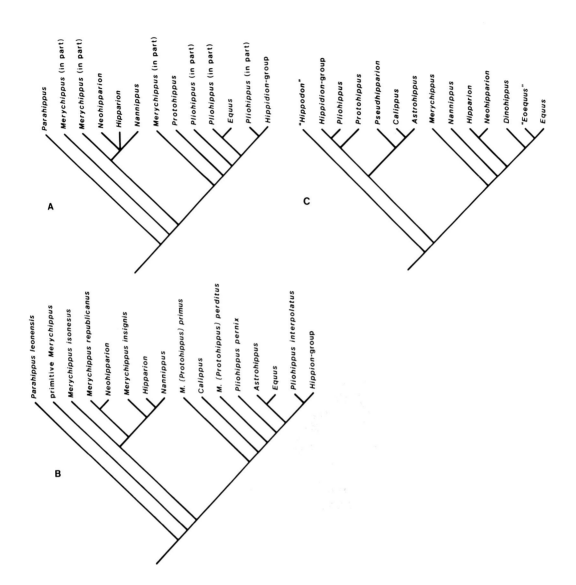

Fig. 11.1. Previous phylogenetic hypotheses of advanced equids expressed in cladistic form. A. That of Matthew (1926), based primarily on his Figure 25. B. That of Stirton (1940). C. That of Quinn (1955), based on his Figure 3. Quinn's *Hippodon* included "*Parahippus*" *leonensis* and "*Merychippus*" *gunteri*. A species or genus thought "ancestral" to another by these authors has been placed as that taxon's sister-group. Matthew and Stirton's concept of some taxa, e.g., *Merychippus* and *Pliohippus*, would today be considered paraphyletic. Quinn's taxa were not paraphyletic, but his arrangement of genera differs more from that obtained by cladistic analysis (see Figure 11.2) than do those of Matthew and Stirton.

following McGrew and Meade (1938), had an even broader concept of *Merychippus* than Matthew, and based its definition on primitive characters and temporal position. The basic pattern of Stirton's (1940) phylogeny with respect to the Equinae (i.e., a horizontal "*Merychippus*" giving rise to numerous, more vertical clades) was popularized by Simpson (1951) and has endured at the generic level with little modification to this day (e.g., MacFadden, 1985).

The difficulty in assessing the phylogenetic interrelationships of various groups, such as merychippines and advanced equids, is not unique to horses. Rather, it is the familar pattern of more primitive members of closely related clades plesiomorphically and phenetically resembling each other more than they do their derived sister-taxa. This was recognized by Morris Skinner, who proposed that non-dental cranial characters could be used to unite "merychippine-grade" taxa with their dentally more derived descendents. In a series of papers, Skinner and MacFadden (1977; MacFadden and Skinner, 1981, 1982) argued that a mesodont species from the early Barstovian of Texas, while retaining primitive "merychippine-grade" dental traits, shared derived cranial features with a particular clade of hipparionines (*Cormohipparion*) and should be classified as the oldest member of the genus. Similarly, "*Merychippus*" *sphenodus* was also assigned to *Cormohipparion* (Woodburne et al., 1981), "*M.*" *republicanus* and "*M.*" *coloradense* were allied with *Pseudhipparion* and *Neohipparion* (Webb and Hulbert, 1986), and a new species of "merychippine grade" was placed in *Hipparion* (MacFadden, 1984a). This study suggests that many of the valid species previously assigned to *Merychippus sensu lato* are either members of advanced genera, or are sister-taxa of monophyletic combinations of these genera.

Over the last 50 years, the only proposed phylogeny of advanced equids that seriously challanged that established by Matthew, Stirton, and others was that of Quinn (1955). Quinn's (1955) arrangement (expressed cladistically in Fig. 11.1C), while strictly a vertical phylogeny, did not distinguish between primitive and derived characters, did not properly consider ontogenetic and individual variation (as noted by Webb, 1969, thus oversplitting the taxa present at any particular locality), and was based almost entirely on relatively incomplete Gulf Coastal Plain material. The major differences between the phylogeny of Quinn (Fig. 11.1C) and those of Matthew (Fig. 11.1A) and Stirton (Fig. 11.1B) are: 1) Quinn's proposed sister-group relationship between hipparionines and *Equus*; 2) his placement of *Pseudhipparion* among the protohippines rather than the hipparionines; and 3) a lack of close relationship between *Pliohippus* and *Astrohippus*.

### Materials, methods, and terminology

This study is based primarily on my own work revising late Neogene equids of the Gulf Coastal Plain and the Great Plains of North America (Hulbert, 1987a, b; 1988a, b, c), a collaborative review of *Pseudhipparion* (Webb and Hulbert, 1986), and the recent work of MacFadden (1980, 1984a, 1984b, 1985, 1986). Generic allocations of most of the species considered here are substantiated in these taxonomic reviews and are summarized in Appendix 11.1. These genera, with the exception of "*Dinohippus*" (see later), have been shown to be strictly monophyletic (Hulbert, 1987a). In addition, MacFadden and I are engaged in ongoing revisions of protohippine, equine, and merychippine equids, some of whose preliminary results are incorporated in this study. Further work on these taxa may necessitate some changes in the results presented here, but these will mostly be at the species level. More limited firsthand observations of Barstovian and late Hemingfordian equids from the West Coast have been augmented by descriptions from the literature, especially the studies of Merriam (1919), Bode (1934), Dougherty

(1940), Buwalda and Lewis (1955), Munthe (1979), and Quinn (1987).

Standard equid dental terminology is used and generally follows Skinner and Taylor (1967) and MacFadden (1984a). Abbreviations employed are as follows: AMNH, American Museum of Natural History, New York; F:AM, Frick Collection of fossil vertebrates housed at the AMNH; DPOF, dorsal preorbital fossa (=nasomaxillary or lacrimal fossa, use follows that of MacFadden, 1984a); l.f., local fauna; Ma, millions of years before present on the radioisotopic timescale. Boundaries of North American land mammal "ages" and Cenozoic Epochs follow Tedford *et al.* (1987) and Berggren *et al.* (1985), respectively. These include: Hemingfordian/Barstovian, 16.5 Ma; Barstovian/Clarendonian, 11.5 Ma; Clarendonian/Hemphillian, 9.0 Ma; Miocene/Pliocene, 5.3 Ma, and Hemphillian/Blancan, 4.5 Ma. Taxonomic names in quotation marks indicate paraphyletic assemblages that are used when no alternative has been proposed. The following three terms are used to collectively refer to monophyletic groups of genera: hipparionine = *Hipparion, Cormohipparion, Nannippus, Neohipparion,* and *Pseudhipparion*; protohippine = *Protohippus* and *Calippus*; and equine = *Pliohippus, Astrohippus, "Dinohippus", Equus, Onohippidum,* and *Hippidion*. "Merychippine" informally refers to mesodont to subhypsodont equids previously included in the paraphyletic genus *Merychippus*.

Forty-five characters were selected for the phylogenetic analysis (listed in Appendix 11.2), 15 cranial characters (primarily those of the facial fossae and muzzle region), 16 of the upper dentition, ten of the lower dentition, one of the dentition in general, two post-cranial characters, and general size. As many of the characters as possible were designed so that there were only two available character states (primitive or derived, coded 0 or 1). A few characters (e.g., numbers 18, 32 and 43; numbers referring to listing of characters in Appendix 11.2) could logically be divided into a series of more than two character states, while others (e.g., 41) have two different derived states that apparently orininated from the same primitive condition. Primitive character states for the thirteen recognized late Neogene North American genera (Table 11.1) were determined from the sources listed above. In addition, ten species of "merychippine-grade" (Table 11.1) were also analyzed, and *"Parahippus" leonensis* was used as an outgroup to determine character state polarities (Hulbert, 1987a; Hulbert and MacFadden, in prep.). The most parsimonious phylogenetic arrangement(s) of these taxa were obtained with the PAUP computer program (version 2.4; Swofford, 1985).

Space limitations prevent discussion of all of the characters. Most are fairly traditional, but a few are either used here for the first time, or their inferred polarities differ from previous interpretations. The presence or absence and morphology of the DPOF has long been used in equid systematics (e.g., Gidley, 1907). The symplesiomorphic state for the DPOF of the Hipparionini and Equini is often considered to be deep, based on the condition in various anchitherines, *Archaeohippus*, and some merychippines. However, Matthew (in Osborn, 1918) long ago noted that many species of *Parahippus sensu lato* have shallow DPOFs. In *"P." leonensis*, the DPOF is shallow, suggesting the following transformation sequence: DPOF shallow (*"P." leonensis*) –> DPOF shallow to moderate (*"M."primus*) –> DPOF depth moderate (*"M." sejunctus, "M." isonesus, "M." coloradense, Calippus, Protohippus*) –> DPOF deep (independently attained by the *Hipparion* genus group and *Pliohippus*). The depth of the DPOF was later decreased independently in a number of lineages (see discussion in Hulbert, 1988b). For discussion of protocone attachment (Numbers 18 and 19) and metaconid-metastylid separation (32 and 33), see Webb and Hulbert (1986). Character

Table 11.1. Character state matrix for 13 genera of Late Neogene Equinae from North America, 10 additional species of merychippine-grade, and *"Parahippus" leonensis*, which was used as an outgroup. Characters described in Appendix 2. Taxa are: 1 = *Hippidion*; 2 = *Onohippidium*; 3 = *Equus*; 4 = *"Dinohippus;"* 5 = *Astrohippus*; 6 = *Pliohippus*; 7 = *"M." carrizoensis*; 8 = *"M." stylodontus*; 9 = *Protohippus*; 10 = *Calippus*; 11 = *"M." intermontanus*; 12 = *"Merychippus" isonesus*; 13 = *"M." sejunctus*; 14 = *"M." coloradense*; 15 = *Pseudhipparion*; 16 = *Neohipparion*; 17 = *"M. insignis;"* 18 = *"Hipparion;"* 19 = *"M.""goorisi"*; 20 = *Cormohipparion*; 21 = *Nannippus*; 22 = *"M." primus*; 23 = *"M." gunteri*; 24 = *"Parahippus" leonensis*. A "?" indicates character state unknown or cannot be determined.

```
                TAXA
     1 2 3 4 5 6 7 8 9 10 11 12 13 14 15 16 17 18 19 20 21 22 23 24

 1   1 1 1 1 0 0 ? 0 0 0  ?  0  0  0  0  0  0  1  0  0  0  0  ?  0
 2   1 1 0 1 0 0 ? 0 0 0  ?  0  0  0  0  0  0  0  0  0  0  0  ?  0
 3   0 1 0 1 1 1 1 1 1 1  1  1  1  1  0  1  1  1  1  1  1  1  ?  0
 4   0 1 0 0 1 1 1 0 1 0  1  0  0  1  0  0  1  1  1  1  1  0  ?  0
 5   ? 1 ? 1 1 1 1 1 1 1  1  0  1  1  0  1  1  1  1  1  1  1  ?  0
 6   ? 1 ? 0 ? 1 1 1 1 1  1  1  0  0  0  1  1  1  1  1  0  ?  0
 7   ? 0 ? 0 ? 0 0 0 0 0  0  0  0  0  0  0  0  1  1  1  0  ?  0
 8   1 1 1 1 1 1 0 1 1 1  1  0  0  1  1  1  1  1  1  1  1  0  0
 9   1 1 1 1 0 0 0 0 0 0  0  0  0  0  1  1  1  1  1  1  1  0  0  0
10   0 0 0 ? 1 1 1 1 0 0  0  2  2  0  0  0  0  0  0  0  0  0  0  0
11   0 0 0 0 0 0 0 0 0 0  0  0  0  0  1  0  0  0  0  0  0  0  0  0
12   0 0 0 0 0 0 0 0 0 0  0  0  0  0  0  1  0  0  0  0  0  0  1  0  0
13   0 0 0 0 0 0 0 ? 1 1  0  0  0  0  0  0  0  0  0  0  0  0  0  0
14   0 0 0 0 0 0 0 0 0 1  0  0  0  0  0  0  0  0  0  0  0  0  0  0
15   0 0 0 0 0 0 0 0 1 1  0  0  0  0  0  0  0  0  0  0  0  0  0  0
16   0 0 0 0 0 0 0 0 0 0  0  0  0  0  0  0  0  0  0  0  1  0  0  0
17   1 1 1 1 1 1 1 1 1 1  1  1  1  1  1  1  1  1  1  1  1  1  0  0
18   2 2 2 2 2 2 2 1 1 2  1  0  1  0  0  0  0  0  0  0  0  1  1  0
19   0 0 0 0 0 0 0 0 0 0  0  0  0  0  0  1  0  0  0  1  1  0  0  0
20   0 0 0 0 0 0 0 0 0 0  0  0  0  0  1  2  2  0  0  0  1  0  0  0
21   0 0 0 0 0 1 0 0 0 0  0  0  0  0  0  1  0  0  0  0  0  0  0  0
22   0 0 0 0 0 0 0 0 0 0  0  0  0  0  0  1  1  0  0  0  0  0  0  0
23   1 1 0 1 1 1 0 0 0 0  0  0  0  0  0  0  0  0  0  0  0  0  0  0
24   0 0 0 0 0 0 0 0 0 0  0  0  0  0  0  0  0  1  1  1  1  0  0  0
25   0 0 0 0 0 0 0 0 0 0  0  0  0  0  0  0  0  0  1  1  1  0  0  0
26   0 0 0 0 0 0 0 0 0 0  0  0  0  0  0  0  0  0  0  1  0  0  0  0
27   1 1 1 1 1 1 1 1 1 1  1  1  1  1  1  1  1  1  1  1  1  1  1  0
28   0 0 0 0 0 0 0 0 0 0  0  0  0  0  0  0  0  1  1  1  1  0  0  0
29   0 0 0 0 1 1 0 0 0 0  0  0  0  0  0  0  0  0  0  0  0  0  0  0
30   1 1 0 0 1 1 1 1 1 1  0  1  1  1  0  1  0  0  0  0  1  0  0  0
31   0 0 0 0 0 1 0 0 0 0  0  0  0  0  0  0  0  0  0  0  0  0  0  0
32   1 1 2 1 2 1 1 1 1 1  1  1  1  2  2  1  1  1  1  1  1  0  0
33   0 0 1 0 0 0 0 0 0 0  1  1  1  1  1  1  1  1  1  0  0  0
34   1 1 0 1 0 1 1 1 1 0  0  0  0  0  0  0  0  0  0  0  0  0  0
35   ? ? 0 0 0 2 ? 0 1 0  0  0  ?  ?  1  1  1  1  1  1  1  0  0  0
36   1 1 1 1 1 1 ? 0 0 1  0  0  ?  0  0  1  0  0  0  0  0  0  0  0
37   1 1 1 1 1 1 ? 1 1 1  1  1  1  1  1  1  1  1  1  1  1  1  1  0
38   0 0 1 1 0 0 0 0 0 0  0  0  0  0  0  1  0  0  0  1  1  0  0  0
39   0 1 1 1 1 0 0 0 1 1  0  1  0  1  1  1  0  0  0  1  1  0  0  0
40   1 1 1 1 0 1 1 1 1 1  1  1  1  1  1  1  1  1  1  1  1  1  1  0
41   2 2 2 2 2 2 ? 2 1 0  0  0  0  0  1  1  0  0  0  1  1  0  0  0
42   1 1 1 1 1 1 1 1 1 1  1  1  1  1  1  1  1  1  1  1  1  1  0  0
43   2 2 2 2 1 1 0 1 1 1  1  1  1  1  1  1  1  0  1  1  0  0  0  0
44   1 1 1 1 1 0 ? 0 0 0  0  0  0  0  0  0  0  0  0  0  0  0  0  0
45   1 1 0 0 0 0 ? 0 0 0  0  0  0  0  0  0  0  0  0  0  0  0  0  0
```

number 34, where the only apomorphy that unites the Equini occurs, is observed in the primitive state in, for example, "*P.*" *leonensis* (Hulbert, 1985, Fig. 3A), "*M.*" *sejunctus* (Osborn, 1918, Fig. 86), and "*M.*" *coloradense* (MacFadden, 1984a, Figs. 58-62). The derived state, in which the metaconid has expanded lingually much more notably than the metastylid on moderate to heavily worn lower molars, is found in *Calippus* (Quinn, 1955, Pl. 6.1, $M_1$), *Protohippus* (Quinn, 1955, Pl. 9.4 and 9.5, $M_1$), *Pliohippus* (Osborn, 1918, Fig. 89), and *Dinohippus* (Osborn, 1918, Pl. 30), for example.

## Results of phylogenetic analysis

Cladograms were generated for the 13 genera and 10 merychippine species listed in Table 11.1. The 13 genera consistently grouped into three monophyletic clades (hipparionines, protohippines, and equines) in almost all of the generated phylogenies, and the relative arrangement of the genera within each of these clades did not vary substantially from one analysis to another. With three exceptions, the relative positions of the merychippines were also stable (as shown in Fig. 11.2). The exceptions were "*M.*" *intermontanus*, "*M.*" *sejunctus*, and "*M.*" *isonesus*. The last two usually grouped together in a clade whose position could be placed with equal parsimony at a variety of locations in the cladogram. This is due to their combination of equine facial characters with a mixture of hipparionine and primitive dental characters. The situation is reversed with "*M.*" *intermontanus*. Using PAUP, there were five equally most parsimonious arrangements for all 23 taxa. However, they all differed *only* in the relative placement of the problematic species discussed earlier (i.e., the relative positions of all 13 genera and the remaining 7 merychippines are the same in all five trees). In each, the hipparionines formed the sister-group of the protohippines plus equines. "*M.*" *primus* was placed as the sister-taxon to all advanced equids. The cladogram in Fig. 11.2 is considered the most likely (evolutionarily) of the five, and the next section assumes that it accurately depicts the phylogenetic history of the Equinae.

## Discussion

As noted above, two general phylogenies of the Equinae have been proposed. One was first put forth in detail by Matthew (1926), and later refined by Stirton (1940) and Simpson (1951). A second and very different scheme was that of Quinn (1955). Comparison of the cladograms in Figs. 11.1A, 11.1B, 11.1C, and 11.2 demonstrates that the phylogeny suggested by cladistic analysis much more closely resembles those of Stirton and Matthew. None of the radical changes proposed by Quinn (e.g., a close relationship between *Calippus*, *Astrohippus*, and *Pseudhipparion*, or derivation of *Equus* from the hipparionines, thus making the hipparionines paraphyletic) are borne out in the analysis. Although the general outline of the cladogram in Fig. 11.2 is much more like the traditional one of Stirton, the interrelationships among the genera are much more explicitedly determined. Some differences with that of Stirton (1940) are that he did not recognize the close relationship between *Protohippus* and *Calippus*, and he derived *Equus* from *Astrohippus* rather than "*Dinohippus*." The results agree with the suggestions of Lance (1950), Bennett (1980), and MacFadden (1984b) that "*Dinohippus*" and *Equus* are sister-taxa.

About 50 middle Miocene equid species of merychippine-grade have been named since the mid-1800s. Many are based on inadequate material or are junior synonyms of other names. However, some can be recognized as the sister-groups of various combinations of the 13 recognized genera (Fig. 11.2) or are placed within one of these genera. The close ("ancestral") relationships of some merychippines with particular lineages of advanced equids have long been recognized (Osborn, 1918; Matthew and Stirton, 1930; Stirton, 1940;

Fig. 11.2. One of five equally most parsimonious cladograms hypothesizing the phylogenetic history of late Neogene Equinae, as determined by the PAUP computer-program. Numbers refer to derived character states shared at each node on the tree, as defined in Appendix 11.2. A complete character state matrix is presented in Table 11.1. An "R" after a number indicates a reversal (character state 1 to character state 0); a "#" indicates a transformation of the derived state 1 to a more derived state 2; and a " ' " after characters 10, 35 and 41 indicates evolution of the alternate derived character state 2 from the primitive state 0. Note the great frequency of reversal and parallelism, even in this most parsimonious arrangement.

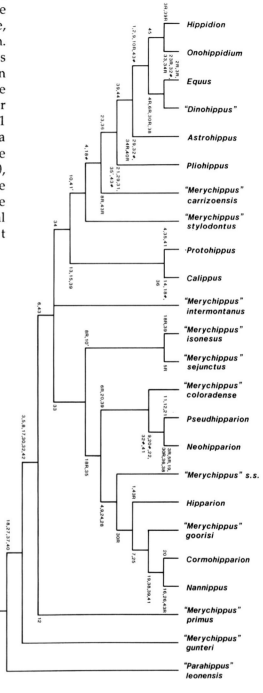

Simpson, 1951; Webb, 1969), but most workers have preferred to artificially retain them in the horizontal grade called "*Merychippus*." Efforts to reclassify some of these taxa phylogenetically have met resistance (e.g., Forstén, 1982), but the current widespread acceptance of cladistic methodology and classification should aid in breaking up this artificial taxon. A classification of the Equinae is proposed in Table 11.2 based on the cladogram from Fig. 11.2 using the conventions of Wiley (1981). Following Matthew (1932) and Sondaar (1968), *Archaeohippus* and *Parahippus sensu lato* are included in the Equinae rather than the Anchitheriinae, although they were not studied here. Taxa traditionally referred to "*Merychippus*" are placed at many levels in the hierarchy in Table 11.2, and are discussed later.

"*Merychippus*" *gunteri* is recognized as the sister-taxon of all more advanced equids (Fig. 11.2 and Table 11.2). The late Hemingfordian topotypic sample consists primarily of isolated teeth and jaw fragments, with no cranial material known. The Hipparionini, Equini, and "*M.*" *primus* share numerous apomorphies relative to "*M.*" *gunteri* and "*P.*"*leonensis*, including characters of the facial region (3, 5, and 8), the upper dentition (17 and 30), and lower cheek teeth (32). However, all the next major nodes of the equid phylogenetic tree are based on three or fewer characters (Fig. 11.2), many of which are interpreted to have evolved in parallel in other lineages. This probably reflects the numerous and rapid cladogenetic events that must have taken place between about 18 and 14 Ma. By 17 Ma (latest Hemingfordian), at least four lineages of primitive merychippines were present in North America. "*M.*" *primus* is well represented in the Sheep Creek Fauna of Nebraska, where it coexisted with "*M. isonesus*" and "*M. insignis*"-like populations of hipparionines (Skinner *et al.*, 1977). At the same time, "*M.*" *carrizoensis*, a small, low-crowned equine, is common at a number of sites in California (Munthe, 1979; Quinn, 1987). By the early Barstovian, at least nine species of merychippine-grade horses are known, representing the stock of most of the recognized subgroups of the Hipparionini and Equini. Two of the major clades, the hipparionines and protohippines, experienced further diversification in the late Barstovian, and were numerically and taxonomically dominant through the early Hemphillian (Hulbert, 1987a, 1988a). As a group the equines are poorly represented during this interval (only the genus *Pliohippus* is well known), and did not become numerous until the late Hemphillian.

Three monophyletic groups of taxa are recognized within the Hipparionini (Table 11.2). As previously noted, the one that includes "*M.*" *isonesus* and "*M.*" *sejunctus* can be placed at several locations in the cladogram with an equal degree of parsimony. The two are often considered related to the equines because they possess a malar fossa. However, they do not have the derived dental characters (34 and 41) of equines, which implies either reversals of these two traits or independent derivation of the fossa. The two also share at least one dental character (33) with hipparionines that is not known to occur in Barstovian equines. Furthermore, the malar fossa in "*M.*" *isonesus* and "*M.*" *sejunctus* is shallow and confluent with the DPOF. The malar and dorsal preorbital fossae are more distinct in equines. Some hipparionines, including species of *Pseudhipparion* and *Neohipparion*, and "*M. insignis*," and some protohippines have variably developed, shallow depressions in the malar region (Skinner and Taylor, 1967; MacFadden, 1984a; Webb and Hulbert, 1986; Hulbert, 1988a). This was probably the condition in the shared ancestor the Hipparionini and Equini, and makes an independent evolution of a malar fossa in these two species and in the equines less unlikely. The two other, more diverse subgroups of the Hipparionini are the *Neohipparion* and the *Hipparion* genus groups (informal group names of subtribal rank following the convention of Wiley, 1981).

Table 11.2. Hierarchical classification of North American Equinae, using the sequencing conventions and informal group categories of Wiley (1981).

Family Equidae
    Subfamily Equinae
        *Archaeohippus*
        "*Parahippus*" (including *Desmatippus*)
        "*Parahippus*" *leonensis*
        "*Merychippus*" *gunteri*
        "*Merychippus*" *primus*
        Tribe Hipparionini
            Genus unnamed
                "*Merychippus*" *isonesus*
                "*Merychippus*" *sejunctus*
            *Neohipparion* genus group
                "*Merychippus*" *coloradense*
                *Pseudhipparion*
                *Neohipparion*
            *Hipparion* genus group
                "*Merychippus*" sensu stricto
                *Hipparion*
                Plesion "*Merychippus*" *goorisi*
                *Cormohipparion*
                *Nannippus*
        Plesion "*Merychippus*" *intermontanus*
        Tribe Equini
            *Protohippus* genus group
                *Protohippus*
                *Calippus*
            *Equus* genus group
                Plesion "*Merychippus*" *stylodontus*
                Plesion "*Merychippus*" *carrizoensis*
                *Pliohippus*
                *Astrohippus*
                *Equus* (including "*Dinohippus*")
                *Onohippidium*
                *Hippidion*

The *Neohipparion* genus group includes at least two genera, *Neohipparion* and *Pseudhipparion*, and a Barstovian merychippine species that cannot be assigned to either genus, "*M.* "*coloradense*. "*M.*" *coloradense* has commonly been referred to *Neohipparion* (e.g., Stirton, 1940; Webb, 1969; MacFadden, 1984a), but Webb and Hulbert (1986) and Hulbert (1987b) demonstrated that *Pseudhipparion* and *Neohipparion* share numerous characters not observed in "*M.*" *coloradense*. *Neohipparion* is represented by five North American species that range in age from early Clarendonian to latest Hemphillian (ca. 11.5 to 4.5 Ma). They are moderate to large-sized (tooth row lengths of 115 to 150 mm), with only a reduced, faint, or completely absent DPOF. Trends apparent in this genus that to some extent parallel (and often exceed) those observed in younger species of *Cormohipparion* are: increases in fossette complexity, elongation of the protocone, progressively caballoid lower cheek teeth (*sensu* Forstén, 1984) with well-developed pli caballinids, and increased unworn crown height. Although described by Gidley (1903) as the typical North American hipparionine, *Neohipparion* as recognized here is generally uncommon throughout its known range relative to other equids.

*Pseudhipparion*, the other major lineage in the *Neohipparion* genus group, is well represented in the Great Plains and Gulf Coastal Plain, but is absent in more western faunas. In their recent review, Webb and Hulbert (1986) recognized six species with an overall chronologic range of late Barstovian through latest Hemphillian (ca. 14.5 to 4.5 Ma). *Pseudhipparion* consists of moderate-sized to very small (tooth row lengths of less than 80 to 115 mm) equids with shallow, anteriorly located, indistinct DPOFs. *Pseudhipparion* retained the primitive connection of the protocone to the protoselene during early to moderate wear stages, unlike other hipparionines which usually show this character state only in the $P^2$, or in late wear stages. *Pseudhipparion* is very abundant in the late Barstovian and Clarendonian of the Great Plains (e.g., Devil's Gulch, Burge, and Clarendon Faunas), but, like *Cormohipparion*, remains common through the Hemphillion only in Florida. Two interesting developments in *Pseudhipparion* are a near continental decrease in size among successive species (unique among North American hipparionine lineages--MacFadden, 1985; Webb and Hulbert, 1986), and the evolution of incipient hypselodonty (rootlessness) in the youngest species, *P. simpsoni*. In this species, the base of the crown did not close and root formation was not started just after the tooth began to wear (as is the case in all other equids), but rather these events took place much later in ontogeny. The result was a cheektooth with a potential crown height two to three times that of its sister-group *P. skinneri*, which had normal root development (Webb and Hulbert, 1986). Furthermore, the incisors of *P. simpsoni* were truly hypselodont with well-developed dentine tracts (Webb and Hulbert, 1986; Hulbert, 1987a).

Several Barstovian populations are here placed in a taxon labeled "*Merychippus*" *sensu stricto* in Fig. 11.2 and Table 11.2. Evander (1986) suggested that the genotypic material of *Merychippus insignis* is inadequate for specific and generic referral (*contra* Skinner and Taylor, 1967). Pending further study, the Lower Snake Creek merychippine population illustrated by Skinner and Taylor (1967) is therefore informally referred to as "*M. insignis*," as no other name is available. It, along with a number of described species of "*Merychippus*" (e.g., "*M.*" *calamarius*."*M.*" *californicus*), have similar dental and facial morphologies. Although "*Merychippus*" *sensu stricto* shares several derived characters with other members of the *Hipparion* genus group (Fig. 11.2), no known synapomorphies unite it. Thus, the possibility exists that this assemblage is not monophyletic. As is the case with many merychippine-grade taxa, these species are in need of a thorough modern review, and their interrelationships are currently unknown.

The remaining members of the *Hipparion* genus group include the North Amercan genera *Nannippus*, *Cormohipparion*, and "*Hipparion*," as well as all Old World hipparionines. The use of quotes around *Hipparion* when referring to North American species reflects the present uncertainty as to their exact relationship with Old World *Hipparion sensu stricto* (see MacFadden, 1980, 1984a, and Bernor and Hussain, 1985, for contrasting viewpoints). The oldest members of the group are the Barstovian "*H.*" *shirleyi* and "*M.*" *goorisi*. The latter was originally described as the oldest species of *Cormohipparion* (MacFadden and Skinner, 1981), but recent analysis of Clarendonian *Nannipus* indicates that "*M.*" *goorsi* is instead the sister-taxon to both *Cormohipparion* and *Nannipus*, and thus cannot be placed in either genus (Hulbert, 1987a, 1988c).

North American "*Hipparion*" ranges from the late Barstovian to the early Hemphillian (ca. 13.5 to 6.0 Ma) and comprises only three species (MacFadden, 1980, 1984a). They are small to moderate-sized (tooth row lengths of 100 to 135 mm) equids with deep, anteriorly located, pocketed DPOFs, and moderately retracted nasal notches. Younger species are principally recognized from West Coast faunas, and are typically rare in the Great Plains (a notable exception being the sample of "*H.*" *tehonense* from the Clarendon Fauna described by MacFadden, 1984a) and in the Gulf Coastal Plain (Hulbert, 1988c).

*Cormohipparion* forms a monophyletic assemblage of at least five species and ranges from the late Barstovian to late Blancan (ca. 14.0 to 2.0 Ma) in North America (MacFadden, 1984a; Hulbert, 1988b, 1988c). They are moderate to large-sized (tooth row lengths between 120 and 150 mm), and primitively have a deep, well pocketed, anteriorly located DPOF with a prominent anterior margin or rim.

The oldest recognized species (excluding "*M.*" *goorsi*, see above), *C. sphenodus*, is widely distributed across North America (California to Florida) and is morphologically relatively variable (Woodburne *et al.*, 1981; Bernor and Hussain, 1985). As many as three lineages may have evolved from *C. sphenodus sensu lato*, of which one probably gave rise to the initial radiation of Old World hipparionines (Skinner and MacFadden, 1977; Woodburne *et al.*, 1981; Bernor and Hussain, 1985). All three show parallel trends towards increased fossette complexity and elongation of the protocone. In North America, *Cormohipparion* was extremely abundant during the Clarendonian, but became increasingly rarer through the Hemphillian, and was apparently absent from the Great Plains and West Coast after 6.0 Ma. Advanced *Cormohipparion* were much more primitive in terms of relative unworn crown height (hypsodonty) and metapodial morphology than their Hemphillian contemporaries, and were presumably unable to successfully compete on the more arid grasslands of the terminal Miocene. Florida, however, served as a mesic refugium for such characteristic Clarendonian Chronofauna elements as protoceratids (Webb, 1981), *Pseudhipparion* (Webb and Hulbert, 1986), and *Cormohipparion* during the latest Miocene and early Pliocene. *Cormohipparian* persisted until the late Pliocene in Florida (ca. 2.0 Ma), where it was a contemporary of *Equus* and Neotropical immigrants, e.g., *Glossotherium*, *Dasypus*, and *Neochoerus* (Hulbert, 1988b).

*Nannippus* is known from the early Clarendonian to late Blancan (ca. 11.5 to 2.0 Ma) of North America and includes at least six species ranging in size from moderate to very small (tooth row lengths of 85 to 115 mm). The oldest species (as yet undescribed) is represented at several early Clarendonian localities in the Great Plains and Gulf Coastal Plain. This relatively low-crowned taxon has the facial morphology of its sister-group *Cormohipparion*, but already has such characteristic *Nannippus* features as relatively small size, reduced P$^2$ anterostyle, and a vestigial DP$^1$ (Hulbert, 1987a, in prep.). Younger species, such as *N. minor* and *N. lenticularis*, evolved greatly increased unworn crown heights, simpler enamel patterns, reduced DPOFs, and more advanced metapodials. Recently discovered early Hemphillian associated skeletons of *N. minor* from Florida demonstrate that older members of the genus retained a rudimentary fifth metacarpal and trapezium, and that the loss of these elements is not a generic character (as proposed by Matthew, 1926, and Sondaar, 1968).

The *Protohippus* and *Equus* genus groups together form the sister-group of the Hipparionini, the Equini (Fig. 11.2 and Table 11.2). They share a derived lower molar morphology (34) not found in "*M.*" *intermontanus* or the Hipparionini. The two genera of the *Protohippus* genus group are united by their muzzle morphology (13 and 15), while members of the *Equus* genus group share the loss of the protostylids (41) and the development of a true malar fossa (10, Fig. 11.2). The last is hypothesized to be secondarily lost in younger genera of the group, such as *Equus*.

*Protohippus sensu stricto* has a Gulf Coastal Plain and Great Plains distribution and ranges from at least the middle Barstovian through the early Hemphillian (ca. 14.0 to 6.0 Ma). Older specimens (i.e., those described as *P. vetus* by Quinn, 1955), are provisionally retained in *Protohippus* until recovery of more complete material. Species of *Protohippus* are of moderate size (tooth row lengths of 122 to 145 mm), show no evidence of reduction of the lateral phalanges, lack a malar fossa, and (at least in older species) have a large, broad, shallow, slightly pocketed DPOF located near the orbit (preorbital bar width is 15 to 20 mm). Recent study (Hulbert, 1988a) suggests that *Protohippus sensu stricto* consists of four sequential morphospecies, with little evidence of cladogenesis. The lineage may prove to consist merely of a single anagenetically evolving chronospecies. *Protohippus* shares with its sister-taxon *Calippus* a short, broad muzzle, although its incisors retain the normal rounded equid arcade. Trends observed within the genus are: 1) shallowing of the fossa and loss of its posterior pocket; 2) elongation and increased isolation of the protocone (ultimately isolated until mid-wear in premolars of late Clarendonian and Hemphillian *P. gidleyi*); and 3) expansion of the metaconid and metastylid of the premolars.

*Calippus*, like *Pseudhipparion* and *Protohippus*, is known only east of the Rocky Mountains, and ranges from the early Barstovian through the early Hemphillian (ca. 15.0 to 6.0 Ma). Two well-defined clades within *Calippus* form distinct subgenera (Hulbert 1988a). All species referred to *Calippus* (Appendix 11.1) have a short diastema, an extremely broad muzzle relative to tooth row length, a derived incisor morphology in which the enlarged first and second incisors are arranged in a straight or nearly straight line (not in the normal rounded incisor arcade), lack a malar fossa, and generally have a shallow, elongate, unpocketed DPOF. Recent studies indicate that this genus is more diverse than was previously thought, and that it played a prominent role in the great taxonomic abundance of equids in North America during the late Miocene. One subgenus, which is comprised of small to very small species (tooth row lengths of 75 to 105 mm), includes the traditional members of the genus like *C. placidus* and *C. regulus*. The other is made up of small to moderate-sized taxa (tooth row lengths of 100 to 135 mm) such as *C. martini*, "*Pliohippus*" *hondurensis*, and several newly described species from Florida and Nebraska. Most late Miocene faunas from the Great Plains and Gulf Coastal Plain include at least one species of each subgenus (Hulbert, 1988a).

*Pliohippus sensu stricto* is primarily known from the Great Plains, where it is continuously represented from the middle Barstovian through the early Hemphillian (ca. 14.5 to 6.0 Ma). Only scattered records are known from the Gulf Coastal Plain, and named West Coast taxa commonly referred to *Pliohippus* are either known only by dentitions and will require further study to determine their generic affinities (e.g., "*P.*" *fairbanksi*, "*P.*" *tantalus*), or seem to be more closely related to the *Dinohippus*-*Hippidion* clade (e.g., "*P.*" *spectans*, "*P.*" *edensis*). Barstovian *Pliohippus sensu stricto* is moderate-sized (tooth row lengths of 130 to 140 mm), with three well-formed but notably small phalanges on each lateral toe. Clarendonian individuals (here placed in a single species, the type species of the genus, *P. pernix*) are about 10% larger, while Hemphillian specimens (*P. nobilis*) are extremely large (tooth row lengths of up to 180 mm) and monodactyl (based on the F:AM collection). Thus, in this genus, monodactyly evolved during the Clarendonian. The DPOF is deep, elongate-oval, usually deeply pocketed, and located about 15 to 20 mm in front of the orbit. Its ventral border is well rimmed and clearly separates the DPOF from the malar fossa (Stirton and Chamberlain, 1939; MacFadden, 1984b). The variability of the malar fossa is well documented in the F:AM samples from the Devil's Gulch Horse, Burge and MacAdams quarries. It is invariably present, but can be almost as shallow as that of "*M.*" *sejunctus*, with only the posterior and ventral margins well defined. The other extreme, represented by UCMP 33481 (Stirton and Chamberlain, 1939), is a very deep, pocketed, multi-chambered malar fossa with all of its margins well defined. Dentally, *Pliohippus sensu stricto* is characterized by very strongly curved upper cheek teeth, simple enamel patterns, shallow hypoconal grooves that are lost in fairly early wear stages, and frequent connection between the protocone and hypocone in the molars. *Pliohippus* was classically thought to be ancestral to the other monodactyl equids, *Equus* and/or the South American *Hippidion* group. However, it is now clear "that a chronocline of *Pliohippus* species with dual facial fossae and simple, highly curved teeth became a dead-end phylum during early Hemphillian time" (Skinner et al., 1977, p. 359) and that monodactyly evolved at least twice within the Equidae.

*Astrohippus sensu stricto* is known only from the late Hemphillian (ca. 6.0 to 4.5 Ma) of the Great Plains, New Mexico, Florida, and Mexico (MacFadden, 1984b, 1986). The following description of the facial morphology of *Astrohippus* is summarized from MacFadden (1984b): both a DPOF and malar fossa are present; neither are as deep or well pocketed as is usual in *Pliohippus sensu stricto*; the two fossae are not well separated with poorly defined anterior margins; and the preorbital bar is relatively narrow (ca. 10 mm). Other characters of *Astrohippus* are its moderate size (tooth row lengths of 115 to 135 mm), very hypsodont teeth, simple enamel pattern, elongate protocones that do not connect to the hypocones, relatively straight crowns, widely expanded metaconids and metastylids, and shallow ectoflexids. As far as is known, all referable material is monodactyl. The decidous premolars of *Astrohippus* retain relatively well developed protostylids (Matthew and Stirton, 1930; Lance, 1950), the primitive condition. *Pliohippus sensu stricto* alone of Neogene genera reduces or entirely loses the protostylids on the $dP_3$ and $dP_4$. Thus *Astrohippus*, like "*Dinohippus*," is not easily derived directly from *Pliohippus*, and may have had a long separate ancestry of which there is no adequate fossil record.

"*Dinohippus*" first definitely appears during the early Hemphillian, and is then known from Florida (Moss Acres Site; Hulbert, 1987a), Nebraska (*Aphelops* Draw Fauna; Skinner et al., 1977), and Oregon (Rattlesnake Fauna; Merriam et al., 1925). Complete cranial material and large sample sizes are unknown until the late Hemphillian, by which time "*Dinohippus*" was perhaps the most common horse in North America. Earlier workers placed species now recognized as "*Dinohippus*" in *Pliohippus*, but the two are easily differentiated by both cranial and dental criteria (Skinner et al., 1977; MacFadden, 1984b). Of the four species usually placed in "*Dinohippus*," three ("*D.*" *spectans*, "*D.*" *interpolatus* and "*D.*" *leidyanus*) have long been recognized as being closely related and possibly conspecific (Merriam et al., 1925; Matthew and Stirton, 1930). The fourth and youngest, "*D.*" *mexicanus*, is clearly more advanced and generally regarded as the sister-taxon of *Equus* (Lance, 1950; Bennett, 1980; MacFadden, 1984b). "*Dinohippus*" seems to have no unique apomorphic characters and thus appears to be paraphyletic. Through the Hemphillian, "*Dinohippus*" shows progressive shallowing of the facial fossae, increased persistence of the hypoconal groove, increased crown height, and straightening of the upper cheek teeth. The final product of this morphocline was *Equus*. Published pre-Hemphillian records of "*Dinohippus*" are dubious at best. Only the teeth referred to "*Protohippus*" or "*Pliohippus*" *tehonensis* by Merriam (1915), Stock (1935), and Drescher (1941) from the early Clarendonian of California and *Dinohippus* sp. by Nelson et al. (1984) from Utah have a mosaic of primitive and derived characters consistent for the sister-taxon of some segment of the "*Dinohippus*"-*Astrohippus-Hippidion* clade. However, cranial material is nearly always essential for proper generic identification of late Neogene equines (Hulbert, 1988a) and is lacking in these cases.

Little can be added to the descriptions of *Onohippidium* and *Hippidion* from North America beyond those of MacFadden and Skinner (1979). They recognized *Hippidion* based on an early Hemphillian mandible from northern Texas and described a new late Hemphillian species of *Onohippidium* from Arizona. Previously these taxa had been known only from the late Pliocene and Pleistocene of South America. The mandible was referred to *Hippidion* based on its very deep premolar ectoflexids and large size. As deep ectoflexids are primitive, the mandible may represent instead a common ancestor to the *Hippidion-Onohippidium* clade, or their mutual sister-group with the "*Dinohippus*"-*Equus* clade, or even an atavistic variant of *Pliohippus nobilis* (which is common in that region). The hypodigm of *O. galushai*, however, consists of numerous skulls and mandibles, allowing unequivocal generic referral. The *Onohippidium-Hippidion* clade would appear to share a closer ancestry with "*Dinohippus*" than with *Pliohippus sensu stricto*. They have no uniquely shared derived characters with the latter, but are united with "*Dinohippus*" by a more deeply retracted nasal notch, greater preorbital bar width, large size, monodactyly, and better separated and enlarged metaconids and metastylids.

## Conclusions

It has been nearly half a century since Stirton's (1940) influential review of North American equids and over 30 years since Quinn (1955) proposed his alternate phylogeny. The subsequent years have seen great advancement in intracontinental correlation with new dating techniques, and an ever-improving fossil record. The availability of the stratigraphically-controlled Frick Collection has proven to be invaluable in the evolutionary study of many taxa, including Neogene equids. The two major competing models of equid phylogeny were tested with a computer-generated cladogram of 23 taxa using 45 characters. The results of the phylogenetic analysis were: 1) advanced Equinae form a monophyletic group characterized by numerous shared derived features of the skull and dentition; 2) several equally parsimonious arrangements of the Equinae exist, but they differ only in the relative positions of a few species; 3) many of the characters used in the study showed homoplasy. Even so, the inferred high degree of parallelism probably reflects the true evolutionary history of the group; 4) with the many homoplasous characters, and because of the relatively few ($\leq 3$) apomorphies that unite the major subgroups within the Equinae, further analysis of these taxa with additional sets of characters is desirable; 5) the advanced, late Neogene Equinae are hypothesized to consist of two monophyletic tribes, the Hipparionini and the Equini, and several "merychippine" species that do not fit into any of them; 6) the interrelationships among the major genera more closely resemble the phylogenies of Matthew (1926) and Stirton (1940) than that of Quinn (1955). Specifically, *Pseudhipparion* is a hipparionine, most closely related to *Neohipparion* and not *Calippus*. *Protohippus sensu stricto* is not ancestral to *Pliohippus* or *Equus*; rather, together with *Calippus* it constitutes the sister-group of the equines. *Equus* in the traditional sense is most closely related to "*Dinohippus*," but the latter is apparently a paraphyletic assemblage of species. In a strict cladistic framework, *Equus* is best regarded as the sister-taxon of the *Hippidion* group, and should include species previously placed in "*Dinohippus*."

## Acknowledgments

I thank Drs. Donald Prothero and Robert Schoch for asking me to contribute this paper; Bruce MacFadden, David Wright, and Robert Evander for critical reviews; and numerous museum curators, including Richard Tedford, Ernest Lundelius, Michael Voorhies, and David Whistler, for providing access to specimens. Mike Woodburne provided me with a copy of J. P. Quinn's important but as-then-unpublished Master's thesis (now published as Quinn, 1987). B. MacFadden allowed me use of his copy of the PAUP program. Support was provided by the Division of Sponsored Research, University of Florida, and by NSF Grant BSR-8515003 to Dr. MacFadden. This is University of Florida Contribution to Paleobiology No. 320.

## Bibliography

Bennett, D. K. (1980): Stripes do not a zebra make, Part 1: a cladistic analysis of *Equus* . -*Syst. Zool.*,29: 272-288.

Berggren, W. A., Kent, D. V., Flynn, J. J., and Van Couvering, J. A. (1985): Cenozoic geochronology. -*Geol. Soc. Amer. Bull.*, 96: 1407-1418.

Bernor, R. L., and Hussain, S. T. (1985): An assessment of the systematic, phylogenetic and biogeographic relationships of Siwalik hipparionine horses. -*J. Vert. Paleont.*, 5: 32-87.

Bernor, R. L., Woodburne, M. O., and Van Couvering, J. A. (1980): A contribution to the chronology of some Old World faunas based on hipparionine horses. -*Geobios*, 13: 705-739.

Bode, F. D. (1934): Tooth characters of protohippine horses with special reference to species from the *Merychippus* Zone, California. -*Contrib. Paleont., Carn. Inst. Washington Publ.*, 453: 39-64.

Buwalda, J. P., and Lewis, G. E. (1955): A new species of *Merychippus* . -*U. S. Geol. Surv., Prof. Papers*, 264G: 147-152.

Colbert, E. H. (1935): Distributional and phylogenetic studies on indian fossil mammals. II. The correlation of the Siwaliks of India as inferred by the migrations of *Hipparion* and *Equus* . -*Amer. Mus. Novit.*, 797: 1-15.

Cracraft, J. (1981): Pattern and process in paleobiology: the role of cladistic analysis in systematic paleontology. -*Paleobiol.*, 7: 456-468.

Dougherty, J. F. (1940): A new Miocene mammalian fauna from Caliente Mountain, California. -*Contrib. Paleont., Carn. Inst. Washington Publ.*, 514: 109-143.

Drescher, A. B. (1941): Later Tertiary Equidae from the Tejon Hills, California. -*Contrib. Paleont., Carn. Inst. Washington Publ.*, 530: 1-24.

Evander, A (1986): The taxonomic status of *Merychippus insignis* Leidy. -*J. Paleont.*, 60: 1277-1280.

Forstén, A. (1982): The status of the genus *Cormohipparion* Skinner and MacFadden (Mammalia, Equidae). -*J. Paleont.*, 56: 1332-1335.

Forstén, A. (1984): Supraspecific groups of

Old World hipparions (Mammalia, Equidae). -*Paläont. Zeit.*, 58: 165-171.

Gidley, J. W. (1903): A new three-toed horse. -*Bull. Amer. Mus. Nat. Hist.*, 19: 465-476.

Gidley, J. W. (1907): Revision of the Miocene and Pliocene Equidae of North America. -*Bull. Amer. Mus. Nat. Hist.*, 23: 865-934.

Hulbert, R. C. (1985): Paleoecology and population dynamics of the early Miocene (Hemingfordian) horse *Parahippus leonensis* from the Thomas Farm Site, Florida. -*J. Vert. Paleont.*, 4: 547-558.

Hulbert, R. C. (1987a): Phylogenetic systematics, biochronology, and paleobiology of late Neogene horses (Family Equidae) of the Gulf Coastal Plain and the Great Plains. -Ph.D. Dissertation, Univ. Florida, Gainesville.

Hulbert, R. C. (1987b): Late Neogene *Neohipparion* (Mammalia, Equidae) from the Gulf Coastal Plain of Florida and Texas. -*J. Paleont.*, 61: 809-830.

Hulbert, R. C. (1988a): *Calippus* and *Protohippus* (Mammalia, Perissodactyla, Equidae) from the Miocene (Barstovian-early Hemphillian) of the Gulf Coastal Plain. -*Bull. Flor. St. Mus., Biol. Sci.*, 32: 221-340.

Hulbert, R. C. (1988b): A new *Cormohipparion* (Mammalia, Equidae) from the Pliocene (latest Hemphillian and Blancan) of Florida. -*J. Vert Paleont.*, 7: 451-468.

Hulbert, R.C. (1988c): *Cormohipparion* and *Hipparion* (Mammalia, Perissodactyla, Equidae) from the late Neogene of Florida. -*Bull. Flor. St. Mus., Biol. Sci.*, 33: 229-338.

Janis, C. (1984): The use of fossil ungulate communities as indicators of climate and environment. -In Brenchley, P. (ed.): *Fossils and Climate.* (John Wiley and Sons) New York, pp. 85-104.

Lance, J. F. (1950): Paleontologia y estratigrafia del Plioceno de Yepomera, Estado de Chihuahua. 1ª Parte: equidos, excepto *Neohipparion.* -*Bol. Univ. Nac. Auto. Mexico Inst. Geol.*, 54: 1-81.

MacFadden, B. J. (1980): The Miocene horse *Hipparion* from North America and from the type locality in southern France. -*Palaeont.*, 23: 617-635.

MacFadden, B. J. (1984a): Systematics and phylogeny of *Hipparion, Neohipparion, Nannippus,* and *Cormohipparion* (Mammalia, Equidae) from the Miocene and Pliocene of the New World. -*Bull. Amer. Mus. Nat. Hist.*, 179: 1-196.

MacFadden, B. J. (1984b): *Astrohippus* and *Dinohippus* from the Yepomera local fauna (Hemphillian, Mexico) and implications for the phylogeny of one-toed horses. -*J. Vert. Paleont.*, 4: 273-283.

MacFadden, B. J. (1985): Patterns of phylogeny and rates of evolution in fossil horses: hipparions from the Miocene and Pliocene of North America. -*Paleobiol.*, 11: 245-257.

MacFadden, B. J. (1986): Late Hemphillian monodactyl horses (Mammalia, Equidae) from the Bone Valley Formation of central Florida. -*J. Paleont.*, 60: 466-475.

MacFadden, B. J., and Skinner, M. F. (1979): Diversification and biogeography of the one-toed horses *Onohippidium* and *Hippidion.* -*Postilla*, 175: 1-10.

MacFadden, B. J., and Skinner, M. F. (1981): Earliest Holarctic hipparion, *Cormohipprion goorsi* n. sp. (Mammalia, Equidae), from the Barstovian (medial Miocene) Texas Gulf Coastal Plain. -*J. Paleont.*, 55: 619-627.

MacFadden, B. J. and Skinner, M. F. (1982): Hipparion horses and modern phylogenetic interpretation: comments on Forstén's view of *Cormohipparion* . -*J. Paleont.*, 56: 1336-1342.

Matthew, W. D. (1924): Third contribution to the Snake Creek Fauna. -*Bull. Amer. Mus. Nat. Hist.*, 50: 59-210.

Matthew, W. D. (1926): The evolution of the horse. A record and its interpretation. -*Quart. Rev. Biol.*, 1: 139-185.

Matthew, W. D. (1932): New fossil

mammals from the Snake Creek quarries. -*Amer. Mus. Novit.*, 540: 1-8.

Matthew, W. D., and Stirton, R. A. (1930): Equidae from the Pliocene of Texas. - *Univ. Calif. Publ. Geol. Sci.*, 19: 349-396.

McGrew, P. O., and Meade, G. E. (1938): The bearing of the Valentine area in continental Miocene-Pliocene correlation. -*Amer. J. Sci.*, 36: 197-207.

Merriam, J. C. (1915): New horses from the Miocene and Pliocene of California. - *Univ. Calif. Publ. Geol. Sci.*, 9: 49-58.

Merriam, J. C. (1919): Tertiary mammalian faunas of the Mohave Desert. -*Univ. Calif. Publ. Geol. Sci.*, 11: 437-585.

Merriam, J. C., Stock, C., and Moody, C. L. (1925): The Pliocene Rattlesnake Formation and fauna of eastern Oregon, with notes on the geology of the Rattlesnake and Mascall deposits. - *Contrib. Paleont., Carn. Inst. Washington Publ.*, 347: 43- 92.

Munthe, J. (1979): The Hemingfordian mammal fauna of the Vedder locality, Branch Canyon Sandstone, Santa Barbara County, California. Part III: Carnivora, Perissodactyla, Artiodactyla and summary. -*Paleobios*, 29: 1-22.

Nelson, M. E., MacFadden, B. J., Madsen, J. H., and Stokes, W. L. (1984): Late Miocene horse from north central Utah and comments on the Salt Lake Group. - *Trans. Kansas Acad. Sci.*, 87: 53-58.

Osborn, H.F. (1918): Equidae of the Oligocene, Miocene, and Pliocene of North America. Iconographic type revision. -*Mem. Amer. Mus. Nat. Hist.*, 2: 1-326.

Quinn, J. H. (1955): Miocene Equidae of the Texas Gulf Coastal Plain. -*Bur. Econ. Geol., Univ. Texas Publ.*, 5516: 1-102.

Quinn, J. P. (1987): Stratigraphy of the middle Miocene Bopesta Formation, Southern Sierra Nevada, California. - *Los Angeles Co. Mus. Nat. Hist. Cont. Sci.*, 393: 1-31.

Romer, A. S. (1949): Time series and trends in animal evolution. -In: Jepsen, G. L., Simpson, G. G., and Mayr, E. (eds.): *Genetics, Paleontology, and Evolution.* Princeton, N. J. (Princeton Univ. Press), pp. 103-120.

Shotwell, J. A. (1961): Late Tertiary biogeography of horses in the northern Great Basin. -*J. Paleont.*, 35: 203-217.

Simpson, G. G. (1944): *Tempo and Mode in Evolution.* - New York (Columbia Univ. Press) .

Simpson, G. G. (1951): *Horses.* - New York (Oxford Univ. Press).

Skinner, M. F., and MacFadden, B. J. (1977): *Cormohipparion* n. gen. (Mammalia, Equidae) from the North American Miocene (Barstovian-Clarendonian). -*J. Paleont.*, 51: 912-926.

Skinner, M. F., and Taylor, B. E. (1967): A revision of the geology and paleontology of the Bijou Hills, South Dakota. -*Amer. Mus. Novit.*, 2300: 1-53.

Skinner, M. F., Skinner, S. M., and Gooris, R. J. (1977): Stratigraphy and biostratigraphy of late Cenozoic deposits in central Sioux County, western Nebraska. -*Bull. Amer. Mus. Nat. Hist.*, 158: 263-371.

Sondaar, P. Y. (1968): The osteology of the manus of fossil and recent Equidae. - *Verh. Ned. Akad. Weten.*, 25: 1-76.

Stirton, R. A. (1939): Significance of Tertiary mammalian faunas in Holarctic correlation with special reference to the Pliocene in California. - *J. Paleont.*, 13: 130-137.

Stirton, R. A. (1940): Phylogeny of North American Equidae. -*Univ. Calif. Publ. Geol. Sci.*, 25: 165-198.

Stirton, R. A. (1947): Observations on evolutionary rates in hypsodonty. - *Evolution*, 1: 32-41.

Stirton, R. A (1952): Are Petaluma horse teeth reliable in correlation? -*Bull. Amer. Assoc. Pet. Geol.*, 36: 2011-2025.

Stirton, R. A., and Chamberlain, W. (1939): A cranium of *Pliohippus fossulatus* from the Clarendon Pliocene fauna of Texas. -*J. Paleont.*, 13: 349-353.

Stock, C. (1935): Deep-well record of fossil mammal remains in California. -*Bull. Amer. Assoc. Pet. Geol.*, 19: 1064-1068.

Swofford, D. L. (1985): *PAUP, Phylogenetic Analysis Using Parsimony*. - Champaign, Illinois (published by the author).

Tedford, R. H., Galusha, T., Skinner, M. F., Taylor, B. E., Fields, R.W., MacDonald, J. R., Rensberger, J. M., Webb, S. D., and Whistler, D. P. (1987): Faunal succession and biochronology of the Arikareean through Hemphillian interval (late Oligocene through earliest Pliocene Epochs), North America. -In: Woodburne, M. O. (ed.): *Cenozoic Mammals of North America: Geochronology and Biostratigraphy*. - Berkeley, Calif. (Univ. California Press), pp. 153-210.

Webb, S. D. (1969): The Burge and Minnechaduza Clarendonian mammalian faunas of north-central Nebraska. - *Univ. Calif. Publ. Geol. Sci.*, 78: 1-191.

Webb, S. D. (1981): *Kyptoceras amatorum*, new genus and species from the Pliocene of Florida, the last protoceratid artiodactyl. -*J. Vert. Paleont.*, 1: 357-365.

Webb, S. D., and Hulbert, R. C. (1986): Systematics and evolution of *Pseudhipparion* (Mammalia, Equidae) from the late Neogene of the Gulf Coastal Plain and the Great Plains. -In: Flanagan, K. M., and Lillegraven, J. A. (eds.): *Vertebrates, Phylogeny, and Philosophy*. -Laramie, Wyoming (Univ. Wyoming), pp. 237-285.

Wiley, E. O. (1981): *Phylogenetics. The theory and practice of phylogenetic systematics*.- New York (John Wiley and Sons).

Woodburne, M. O., MacFadden, B. J., and Skinner, M. F. (1981): The North American "*Hipparion*" Datum, and implications for the Neogene of the Old World. -*Geobios*, 14: 1-32.

Appendix 11.1. List of North American species referred to the 13 recognized genera of Late Neogene Equinae. Currently undescribed species excluded. References are MacFadden(1984a, 1984b), MacFadden and Skinner (1979), Webb and Hulbert (1986), and Hulbert (1987a, 1987b, 1988a, 1988b, and 1988c).
1. *Cormohipparion*: *C. sphenodus, C. occidentale, C. plicatile, C. ingenuum, C. emsliei*.
2. *Nannippus*: *N. westoni, N. minor, N. lenticularis, N. peninsulatus* (=*N. phlegon*),
   *N. beckensis*.
3. "*Hipparion*": "*H.*" *shirleyi*, "*H.*" *tehonense*, "*H.*" *forcei*.
4. *Pseudhipparion*: *P. retrusum, P. curtivallum, P. gratum, P. hessei, P. skinneri, P. simpsoni*.
5. *Neohipparion*: *N. affine* (=*N. whitneyi*), *N. trampasense, N. leptode, N. eurystyle* (=*N. phosphorum, N. floresi, N. arellanoi, N. otomii, N. stirtoni*), *N. gidleyi*.
6. *Protohippus*: *P. vetus, P. perditus, P. supremus* (=*P. simus*), *P. gidleyi*.
7. *Calippus*: (?)*C. circulus, C. martini, C. cerasinus, C. hondurensis, C. maccartyi, C. proplacidus, C. placidus, C. regulus, C. elachistus*.
8. *Pliohippus*: *P. mirabilis* (=*P. campestris*), *P. pernix* (=*P. robustus, P. fossulatus, P. pachyops, P. lullianus*), *P. nobilis*.
9. *Astrohippus*: *A. ansae, A. stockii*.
10. *Onohippidium*: *O. galushai*.
11. *Hippidion*: no named species in North America.
12. "*Dinohippus*": "*D.*" *spectans*, "*D.*" *interpolatus*, "*D.*" *leidyanus*, "*D.*" *mexicanus*.
13. *Equus*: *E. simplicidens* (=*E. shoshonensis*), *E. cumminsii*, and various Pleistocene species.

Appendix 11.2. Description of characters and character states used in the phylogenetic analyses. Character state polarity was determined by using "*Parahippus*" *leonensis* as an outgroup (Hulbert and MacFadden, in prep.). In general, 0 = primitive character state, 1 = derived character state and 2 = more derived character state. For three characters (numbers 10, 35, and 41) that have three character states, no transformation sequence is inferred, and 1 and 2 refer to independently acquired, different derived character states.

1. depth of nasal notch, 0=posteriormost point dorsal to $P^2$-C diastema; 1=posteriormost point dorsal to $P^2$ or more posteriorly.
2. depth of nasal notch, 0=posteriormost point dorsal to $P^2$ or more anteriorly; 1=posteriormost point deeper than $P^2$.
3. dorsal preorbital fossa (DPOF), 0=absent or very shallow in depth; 1=depth moderate or deep (>5 mm).
4. DPOF, 0=not pocketed posteriorly; 1=pocketed posteriorly.
5. DPOF, 0=poorly rimmed posteriorly; 1=well rimmed posteriorly.
6. DPOF, 0=dorsal margin not well defined; 1=dosal margin well defined or rimmed.
7. DPOF, 0=anterior margin confluent with face; 1=anterior margin well defined or rimmed.
8. preorbital bar width, 0=very narrow (<5 mm); 1=moderate or broad.
9. preorbital bar width, 0=narrow or moderate (<20 mm); 1=very broad.
10. malar fossa or depression, 0=absent or very shallow and variable; 1=present, not confluent with DPOF; 2=present, confluent with DPOF.
11. zygomatic buckle (Webb, 1969), 0=present; 1=absent.
12. frontals notably domed, 0=not domed; 1=domed.
13. muzzle width relative to upper toothrow length at moderate wear-stage, 0=moderate or narrow; 1=broad (>36%).
14. incisor arcade, 0=arcuate; 1=straight.
15. relative $I^3$-$P^2$ diastema length, 0=moderate or long; 1=short.
16. $DP^1$, 0=retained with permanent dentition, relatively large; 1=not present with adult dentition, or if present is notably reduced in size.
17. relative protocone length, 0=round; 1=oval or elongate-oval.
18. protocone connenction with protoselene on $P^3$-$M^3$, 0=isolated in early wear-stages; 1=connected in early wear-stages; 2=connected at onset of wear.
19. protocone connenction with protoselene on $P^3$-$M^3$, 0=connected at middle wear-stages or earlier; 1=isolated until late wear-stages.
20. protoconal spur, 0=strongly present; 1=reduced but usually present; 2=absent.
21. protocone connection with hypocone on $M^1$-$M^2$, 0=not connected or only in latest wear-stages; 1=connected in early or moderate wear-stages.
22. orientation of long axis of the protocone on $P^2$-$P^4$, 0=approximately anteroposteriorly; 1=markedly anterolabial-posterolingually.
23. internal fossette margins, 0= moderate to complexly plicated in early wear-stages; 1=simply plicated or not plicated in early wear-stages.
24. internal fossette margins, 0=not plicated or very simple plications only in late wear-stages; 1=well plicated in late wear-stages.

25. external fossette margins, 0=not plicated or very simple plications only; 1=relatively well plicated, at least until middle wear-stages.
26. $P^2$ anterostyle, 0=large, expanded; 1=reduced.
27. pli caballin, 0=absent; 1=present, at least in early wear-stages.
28. pli caballin, 0=absent or single; 1=usually branched or multiple.
29. hypoconal groove, 0=open until at least middle wear-stages; 1=closed relatively rapidly.
30. hypoconal groove, 0=open to near base of crown; 1=closed about middle wear-stages or earlier.
31. curvature of upper cheekteeth, 0=straight or moderately curved; 1=very curved.
32. metaconid/metastylid complex, 0=small; 1=expanded; 2=elongated.
33. metaconid and metastylid, 0=poorly separated; 1=well separated.
34. metastylid positioned much less lingually than metaconid on moderate to heavily worn $M_1$-$M_3$, 0=no; 1=yes.
35. protostylid on $DP_3$ and $DP_4$, 0=present; 1=very well developed; 2=absent.
36. ectostylid on $DP_2$-$DP_4$, 0=present; 1=absent.
37. labial cingulum of $DP_2$-$DP_4$, 0=present; 1=absent.
38. pli caballinid on $P_2$-$P_4$, 0=absent, 1=present, at least in early wear-stages.
39. ectoflexid depth of $P_2$-$P_4$ in early wear-stages, 0=deep (penetrates isthmus); 1=shallow.
40. pli entoflexid, 0=usually absent, 1=generally present.
41. protostylid on $P_3$-$M_3$, 0=present; 1=very well developed; 2=absent.
42. cement on permanent cheekteeth, 0=absent or thin; 1=heavily cemented.
43. general size, 0=small (toothrow length<100 mm); 1=moderate; 2=large (toothrow length>140 mm).
44. number of digits, 0=tridactyl; 1=monodactyl.
45. metapodials, 0=moderately long to elongate; 1=short and stout.

# 12. HIPPARION HORSES: A PATTERN OF ENDEMIC EVOLUTION AND INTERCONTINENTAL DISPERSAL

## MICHAEL O. WOODBURNE

This report presents a general overview of hipparionine evolution and dispersal. Five North American genera are recognized in deposits of early Barstovian to Blancan Age. In general sequence of first appearance, these are: *Cormohipparion, Neohipparion, Hipparion sensu stricto, Pseudhipparion,* and *Nannippus.* Early species of each genus are of relatively small size, with mesodont dentitions of "merychippine" grade, but the genera can be distinguished on a combination of cranial and dental characters. All genera show increase in size and hypsodonty and reduction or loss of the dorsal preorbital fossa (DPOF), although at different times. *Cormohipparion* is unique in increasing the complexity of the enamel pattern of its upper cheek teeth. *Nannippus* is persistently of small size, develops extremely hypsodont cheek teeth, and is of gracile build. *Neohipparion* shows an early reduction of the DPOF and develops a strikingly *Equus*-like lower cheek tooth dentition. A hallmark of *Pseudhipparion* is the isolation of the hypoconal groove of the upper cheek teeth into a discrete lake.

At about 12 Ma, a species of *Cormohipparion* apparently dispersed to the Old World and gave rise to "*Hipparion*" *primigenium* and thus to other species of Cranial group 1 (characterized in early species [Vallesian age] by having a very well developed DPOF, a very wide preorbital bar, and very complex upper cheek teeth). Typical members of this hipparionine group are of circum-Mediterranean and western European distribution and range through the Turolian. Asian affinities of this group may include taxa referred to as *Sivalhippus* (Pakistan) and *Cormohipparion* ("*H.*") *weihoense*. An African radicle of this group apparently entered the continent in the Turolian and persisted into the Villafranchian.

Taxa of somewhat smaller size, retaining a large DPOF, having a very narrow preorbital bar, less complex enamel patterns, and developing an extra facial fossa between the DPOF and buccinator fossa, are referred to Cranial group 2. These species of "*Hipparion*" are largely of Asian distribution and are of Turolian age. The ancestry of this group is not known.

*Hipparion sensu stricto* apparently dispersed from North America to the Old World about 10 Ma, although an endemic origin of Old World species has been suggested. In any case, species of this group are of modest size, have a relatively small (and early reduce the size of the) DPOF, and only moderately complex upper cheek teeth. The group is typically of Turolian age and it spread widely in the Old World (it may be present in the late Turolian of sub-Saharan Africa; "*H.*" *turkanense*), and persisted into the Villafranchian of Europe and Africa to co-occur with early species of *Equus*.

A number of small *Nannippus*-like species occur in Turolian deposits of Africa and Eurasia, but their relationships are not well understood. The other three hipparionine groups exhibit trends also seen in North America: reduction and loss of the DPOF, and decrease in enamel complexity of the upper cheek teeth. Later hipparions of sub-Saharan Africa are unusual in increasing the complexity of this enamel pattern, similar to that of later species of *Cormohipparion* in North America.

### Introduction

*Hipparion* means "pony," or "small horse." As originally conceived by De Christol (1832), all fossil horses in which the protocone of the upper molars had an ovoid to circular-shaped design that was isolated from other parts of the enamel pattern (Fig. 12.1), and in which the feet were tridactyl, were included under this genus. Although their foot structure remained primitive, the dentition of hipparions was modern in being

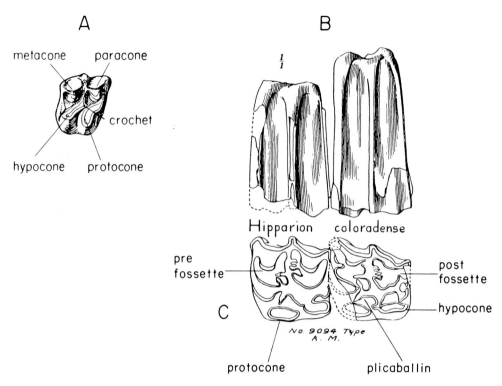

Fig. 12.1. A. Occlusal view of a right upper molar of *Archaeohippus*, showing the low-crowned and simply-crested morphology of browsing horses (including *Anchitherium* and *Sinohippus*). After Woodburne and MacFadden (1982, Fig. 2). B. Labial view and C. occlulsal view of left upper molars of *Neohipparion coloradense*, showing the taller crown height and more complex enamel pattern of hipparions. Note the isolated protocone. Other labelled structures are referred to in the text. After Osborn (1918, Fig. 146).

complexly ornamented and high-crowned (Fig. 12.1). Hipparions arrived in the Old World from North America (Fig. 12.2) about 12 Ma (e.g., Woodburne and MacFadden, 1982), where they coexisted for a short time, and in only a few places, with *Anchitherium*, an earlier Miocene immigrant from North America (e.g., Sondaar, 1971) or with derivative anchitheriine horses (*Sinohippus* Zhai, 1962; Zhegallo, 1978, Forstén, 1982). In both *Anchitherium* and *Sinohippus*, the dentition remained primitive (low-crowned and simply constructed; Fig. 12.1). Until the progressively eastward arrival in the Old World of the American genus *Equus* from about 3.0 - 2.6 Ma (Fig. 12.3), *Hipparion* was the pre-eminent equid in the Neogene record of the Palaearctic, Ethiopian, and Oriental realms (Fig. 12.2), and coexisted only for a short time and in a few places with the modern *Equus* (e.g., Eisenmann and Brunet, 1973; Lindsay et al., 1980; Churcher and Richardson, 1978). In the absence of competition from convergently modernized equids (still tridactyl, but with medium- to high-crowned cheek teeth), "*Hipparion*" flourished in its new-found home, and by some accounts (summarized in Forstén, 1968) diversified into an almost bewildering array of species, estimated by some to have reached nearly 300 forms during that span of time. By way of contrast, relatively few fossil horses have been assigned to the genus *Hipparion* in North America, and many such referrals have been shown to be attributable to other genera (e.g., MacFadden, 1984). Thus, the "time of *Hipparion*"

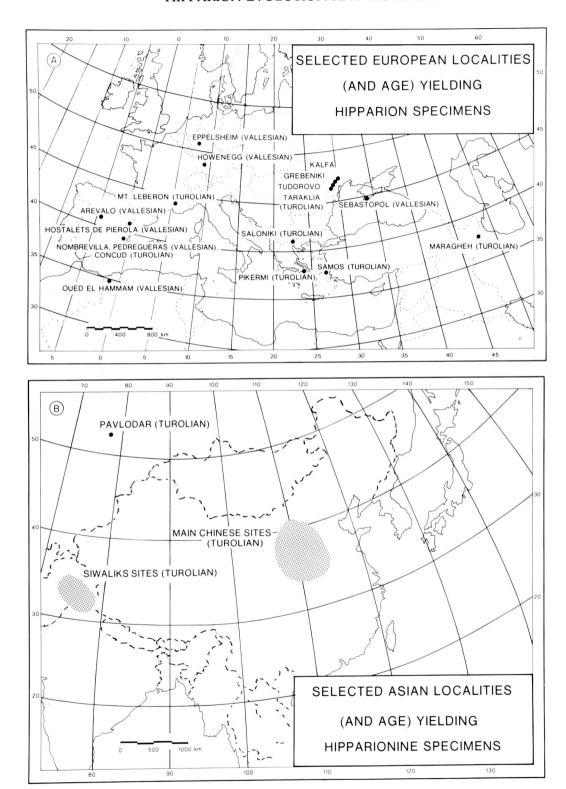

Fig. 12.2. Map of selected Old World sites (and ages) that have yielded fossil hipparionines.

is best recorded in the Old World, whereas the phyletic base for that record is the result of one or (likely) more introductions from North America (e.g., Woodburne et al., 1981; Woodburne and MacFadden, 1982; Fig. 12.3).

Most of the literature on "*Hipparion*" is of European or Eurasian origin (Von Meyer, 1829; De Christol, 1832; Kaup, 1835; Gervais, 1849; Wagner, 1850; Roth and Wagner, 1855; Hensel, 1861; Gaudry, 1862, 1867, 1873; Lydekker, 1877a, 1877b, 1882, 1884, 1885; Pilgrim, 1910; Studer, 1911; Borissiak, 1914; Sefve, 1927; Wehrli, 1941; Gromova, 1952; Pirlot, 1956; Arambourg, 1959; Tobien, 1959, 1968; Zhai, 1962; Forstén, 1968, 1978; Alberdi, 1974a, 1974b; Hussain, 1971; Sondaar, 1961, 1971, 1974; Hooijer, 1975; Sen, Sondaar, and Staesche, 1978; Liu, Li , and Zhai, 1978; Zhegallo, 1978; Eisenmann, 1976a, 1976b, 1981, 1982; Qiu, Huang, and Kuo, 1980), but various North American workers also have contributed to the discussion of this kind of horse (Cope, 1888; Osborn, 1918; Gregory, 1920; Matthew, 1929; Colbert, 1935; Stirton, 1940; Richey, 1948; MacFadden, 1980; MacFadden and Bakr, 1979; Skinner and MacFadden, 1977; Woodburne and Bernor, 1980; Bernor et al., 1980; Woodburne et al., 1981; MacFadden and Skinner, 1981, 1982; MacFadden and Woodburne, 1982; Hulbert, 1982, 1987a, 1987b, 1987c; MacFadden, 1984; Bernor and Hussain, 1985; Bernor, 1985; Webb and Hulbert, 1986).

These works have (1) contributed to the present state of knowledge of this group of horses; (2) revealed that hipparions are not a single kind of horse; (3) suggested that in the Old World even the nomen *Hipparion* may refer to more than one genus; and (4) promoted investigations into patterns of (a) endemic evolution, (b) development of provincial zoogeographic centers of diversification or hipparion biofacies, and (c) the timing and results of immigration of ancestral stocks from North America to the Old World (e.g., Fig. 12.3).

It is clear from the partial list of references cited above that study of hipparions must be international in scope, and in this context a "*Hipparion* Symposium" was convened in 1981 at the American Museum of Natural History, New York. Funded by the National Science Foundation (EAR 81-10870), about twenty workers from the Old and New Worlds were assembled for about a week to attempt the development of a standardized methodology for obtaining and recording data on hipparions so that workers could avail themselves of a consistent point of reference for further studies and interpretations. Elements of this symposium have been published in Woodburne and MacFadden (1982), Eisenmann (1982), MacFadden and Woodburne (1982), Bernor and Hussain (1985), and Bernor (1985), and the much-desired statement by designated symposium members has just appeared (Eisenmann et al., 1988).

What follows here is, (1) a definition of *Hipparion*, (2) a discussion of the evolutionary pattern of hipparions in North America, (3) comments on the apparent structure of hipparions in the Old World, (4) thoughts on the dispersal or dispersals versus endemic evolution in the Old World required by the data, and (5) concluding remarks on the degree of provincialism exhibited by Old World hipparions. General patterns of change in some aspects of hipparion morphology are then summarized.

These topics are largely based on the conventions expressed later, which include rather extensive characterizations of hipparionine generic-rank taxa. It should be pointed out here that certain parameters utilized in (2) "Morphological terms," and proposals made in (3) "Taxonomic concepts," are controversial (see later). A number of European workers (e.g., Eisenmann et al., 1986) consider the emphasis placed on the morphology of the dorsal preorbital fossa of the face in equid taxonomy (as at the moment practised by a number of North American students), to be unwarranted, limited in scope, or premature. Note however, that in previous literature certain European workers, on the one hand,

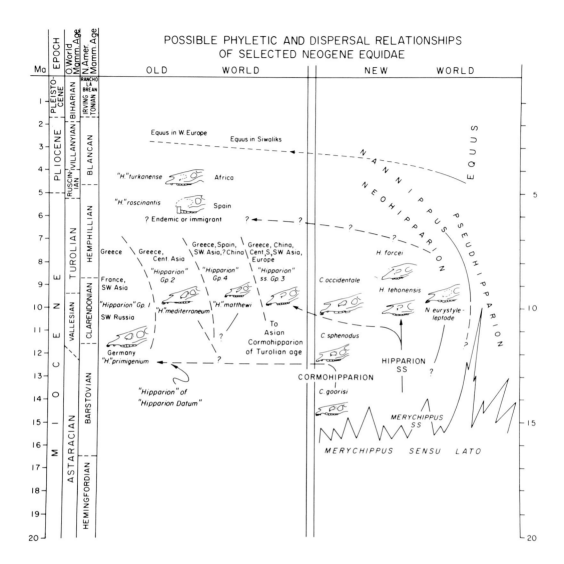

Fig. 12.3. Correlation chart and dispersal diagram for hipparions during the past 20 Ma (modified from Woodburne and MacFadden, 1982). Note that a form similar to *Cormohipparion sphenodus* is proposed as having given rise about in the early Vallesian to (1) a radiation of *Cormohipparion* that is recorded chiefly in Asian strata of Turolian age, (2) an unnamed taxon that might have been the common ancestor of Group 1 and Group 2 forms, and (3) whether or not suggestions in (2) took place, this new Old World immigrant gave rise to Group 1 horses (e.g., *Hippotherium primigenium*) and related taxa in western Europe and the Circum- Mediterranean district. A geologically younger dispersal at about 10 Ma apparently resulted in the widespread presence of *Hipparion sensu stricto* in the Old World from a North American ancestor of the same genus. Subsequent dispersals from North America to the Old World from *Neohipiparion* have been proposed but disputed (see text), whereas the *Equus* dispersal shown here is generally accepted. See also Berggren *et al.* (1985b, pp. 228-229).

advocated use of this set of characters, whereas on the other hand, certain then-contemporaneous American students suggested that the configuration of the dorsal preorbital fossa was subject to ontogenetic or sexual variation, and was to be viewed with caution as a basis of proposing phyletic relationships (examples cited in, for example, MacFadden and Woodburne, 1982, p. 191; MacFadden, 1984, pp. 29-38). MacFadden (1984) presents a recent further discussion of this question and concludes that the range of variation in the dorsal preorbital fossa and contiguous parts of the cranium was not subject to sexual variation, at least in North American hipparions, especially as demonstrated in a quarry sample of *Hipparion tehonense* (MacAdams Quarry, early Clarendonian, Texas). Skinner and Taylor (1967) arrived at a similar conclusion with respect to the dorsal preorbital fossa in specimens referred to *Merychippus insignis* from Echo Quarry, early Barstovian, of Nebraska.

While it should be clear from the following presentation that characters associated with the dorsal preorbital fossa are not the only features considered in the present morphologic analysis, it still will be important to evaluate many of the other dimensions recommended by members of the 1981 "Hipparion" Symposium when proposing phyletic relationships between hipparionine or other equid taxa (see Eisenmann et al., 1988). With regard to the taxa discussed here, attainment of this aim is an evolutionary process in itself, and awaiting its ultimate realization is not grounds for preventing the exposition of interim proposals, such as those that follow (see also MacFadden, 1986).

**Conventions and Abbreviations**
1. *Temporal terms.*

These terms are specific to the land mass in question, without repetition in the text. Some geographical divisions of the Old World are recognized (in parentheses) to aid in biogeographical discussions. New World ages follow Tedford et al. (1987); Old World is after Berggren, Kent, and Van Couvering (1985) and Berggren et al. (1985).

Barstovian - mammal age utilized as an interval corresponding to part of late Miocene time in North America, ca. 16-12 Ma (Fig. 12.3).

Clarendonian - mammal age utilized as an interval corresponding to part of late Miocene time in North America, ca. 12-9 Ma (Fig. 12.3).

Hemphillian - mammal age utilized as an interval corresponding to part of late Miocene time in North America, ca. 9-5 Ma (Fig. 12.3).

Ruscinian - mammal age utilized as an interval corresponding to early Pliocene time in the Old World, ca. 5.2-3.4 Ma (Fig. 12.3).

Turolian - mammal age utilized as an interval corresponding to part of late Miocene time in the Old World, ca. 9.5-5.2 Ma (Fig. 12.3).

Vallesian - mammal age utilized as an inteval corresponding to part of late Miocene time in the Old World, ca. 12-9.5 Ma (Fig. 12.3). Note that Berggren et al. (1985, Fig. 6; Berggren, Kent, and Van Couvering, 1985, Fig. 2) indicate that the base of the Vallesian mammal age is 12.5 Ma. Some of the oldest samples of "Hipparion" cited by Berggren, Kent, and Van Couvering (1985, p. 249; Forstén, 1972) are correlated (Wiman, 1978) to the base of planktonic foraminferal zone N14. The base of Zone N14 is now calibrated by Berggren et al. (1985) and Berggren, Kent, and Van Couvering (1985) at about 11.3 Ma. Ages for the "Hipparion" "Datum" range from 12.5 Ma to 10.8 Ma at Höwenegg, Germany, with the younger age (10.8 $\pm$ 0.4 Ma whole-rock age from an *in situ* basalt bomb) cited as being the most credible (Bernor et al., 1980). The older age (12.1-12.6 Ma; Lippolt et al., 1963) was derived from tuffs that underlie the Höwenegg sequence. All of these ages would be older by a factor of 1.0266 (Dalrymple, 1979) than originally reported, following the reassignment of calculation constants made by Steiger and Jaeger (1977), so that the Höwenegg sample should range in age from about 11 Ma to slightly less than about 13 Ma. This is consistent with the correlation (Berggren, Kent, and Van Couvering, 1986, p. 249) of an early Vallesian fauna (co-occurrence of *"Hipparion"* and *Anchitherium*; Gabunia, 1981) at Jeltokamenka (C.S.S.R.) with the upper half of magnetic chron C5A. This chron ranges in correlated age from 11.5 to 13.8 Ma, with the "upper half" being as old as ca. 12.3 Ma. See also Berggren, Kent, and Van Couvering (1985, pp. 227-228). Recognizing uncertainties in some of the data, the base of the Vallesian mammal age and the *"Hipparion"* "Datum" are here conisered to be about 12 Ma.

Berggren *et al.* (1985) and Berggren, Kent, and Van Couvering (1985) also have revised the age of the Vallesian/ Turolian boundary from about 10 Ma to 9 Ma and have indicated (Berggren *et al.*, 1985, p. 1415) that the age of the early part of Chron C5N is revised from about 9.5 Ma to about 10 Ma. Berggren, Kent, and Van Couvering (1985, Fig. 3) indicate, however, that the Vallesian/Turolian boundary is calibrated at 9.5 Ma, and Badgley *et al.* (1986) indicate that the age of the part of Chron C5N that corresponds to the local FAD ( =first appearance datum) of *"Hipparion"* in the Siwaliks, Pakistan (Fig. 12.2) is ca. 9.2 Ma. For the purposes of this report, the Vallesian/Turolian boundary is taken as ca. 9.5 Ma (as being more consistent with the discussions in Berggren, Kent, and Van Couvering, 1985; especially in Appendix 3), and the age of the *"Hipparion"* FAD in the Siwaliks is taken at about 9.2 Ma. Under this interpretation of unresolved conflicting literature statements, all hipparionine species in the Siwaliks are considered to be no older than Turolian age. As far as is known, choice of a 9.5 Ma age for the Vallesian/Turolian boundary would affect only some specimens of *Cormohipparion* cf. *C. nagriensis* that MacFadden and Woodburne (1982) indicate as reflecting the *"Hipparion"* FAD in the Siwaliks. All other Vallesian or Turolian age assignments made in this report are based on faunal correlations.

Villafranchian - mammal age utilized as an interval corresponding to part of Pliocene and early Pleistocene time in the Old World, ca. 3.4-1.7 Ma (Fig. 12.3).

Ma - Megannum in the radioisotopic time scale. Where necessary, all radioisotopic calibrations have been recalculated according to the decay constants utilized by Steiger and Jaeger (1977).

*2. Morphological terms*

Pli caballin - the loop of enamel of upper cheek teeth labial to the protocone (Fig. 12.1).

Preorbital bar - the space between the anterior rim of the orbit and the rear of the dorsal preorbital fossa, developed on, but may anteriorly exceed, the lacrimal bone (Fig 12.4).

Dorsal preorbital fossa (DPOF) - fossa located on the facial region, on the maxillary, nasal and, sometimes, lacrimal bones (Fig. 12.4).

Cheek-teeth - the major premolar and molar grinding teeth (P2/2- M3/3).

Cheek-tooth complexity: Simple = very few plications or crenulations of the enamel pattern, especially in the upper cheek teeth. Moderate = 4-5 plis on the opposing faces of the pre- and post-fossette, about 2 major plis on the anterior and posterior faces of the pre- and post-fossette, respectively, pli caballin simple on the molars (see Fig. 12.1). Complex = more than 5 plis on the opposing faces of the pre- and post- fossettes, more than 2 or 3 plis on the anterior and posterior faces of the pre- and post-fossette, respectively, 2 pli caballins on the molars. Extremely complex = more than 5 crenulations on all faces of the pre- and post- fossettes, and multiple (more than 2) pli caballins on the molars.

*3. Taxonomic concepts*

**Note:** These refer effectively to generic-rank taxa and species that comprise them. Not all species of each genus are listed, including many Chinese forms under study by Qiu Zhanxiang and R. L. Bernor. Taxa in

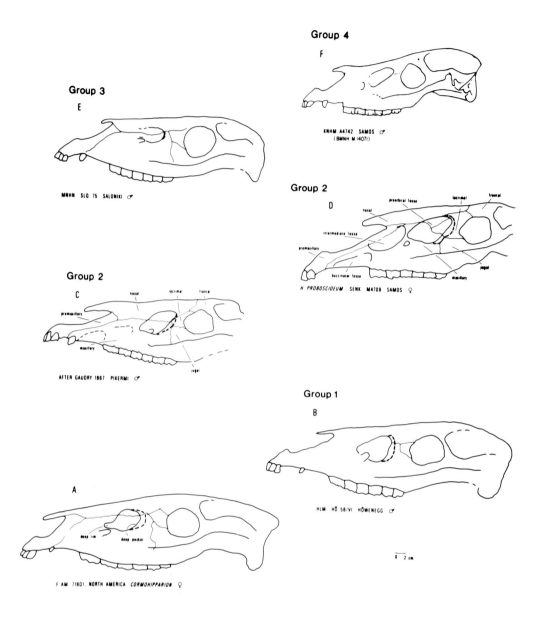

Fig. 12.4. Cranial morphology of species of "*Hipparion*" groups recognized by Woodburne and Bernor (1980), with a late Clarendonian representative of *Cormohipparion* shown for comparison. See text for discussion. After MacFadden and Woodburne (1982, Fig. 3).

**boldface** are considered to be the type species of the genus in question. Taxa designated with an asterisk (*) are considered to be the geologically oldest species of that genus. Most of the following is after Woodburne and Bernor (1980), Woodburne and MacFadden (1982), MacFadden (1984), Bernor (1985), and literature cited in those papers, modified with respect to some North American taxa after Hulbert (1982, 1987a, 1987b, 1987c) and Webb and Hulbert (1986).

hipparionine -- any horse with an isolated protocone of the upper molars and, as far as is known, tridactyl feet, including species of the following genera: *Hipparion, Neohipparion, Nannippus, Cormohipparion, Hippotherium, Proboscidipparion, Pseudhipparion* and *Stylohipparion*.

*Hipparion* (Figs. 12.4, 12.5, 12.6; Tables 12.1, 12.2) -- Hipparionine horse with a DPOF situated relatively high on the face. The DPOF is relatively short, of moderate depth and height, and is poorly pocketed posteriorly; it has no anterior rim, and is located relatively far anterior to the orbit (= preorbital bar is relatively wide). The lacrimal terminates posterior to the fossa and is dorsoventrally narrow (triangularly shaped, *H. shirleyi*, to subtriangularly shaped = dorsal and ventral borders converge toward each other) anteriorly. The nasal notch extends posteriorly to above $P^2$, or more posteriorly. The upper cheek tooth dentition is overall relatively simpler than in most other hipparions. Cranial and postcranial elements are of moderate size; metapodials are relatively slender.

This term includes species referred to Group 3 of Woodburne and Bernor (1980), including *H. prostylum* De Christol, 1832 (= *H. mesostylum* and *H. diplostylum*, Gervais (1849), Turolian (France); *H. dietrichi* (Wehrli, 1941), Turolian (Greece); *H. campbelli* Bernor, 1985, Turolian (Iran); *H. shirleyi\** MacFadden, 1984, late Barstovian; *H. tehonense* (Merriam, 1916) MacFadden, 1984, Clarendonian; *H. forcei* Richey, 1948, late Clarendonian and early Hemphillian (MacFadden, 1984).

*Cormohipparion* (Figs. 12.4, 12.5, 12.6; Tables 12.1, 12.2) -- Hipparionine horses with a DPOF that is situated relatively high on the face; it is elongate, deep posteriorly and dorsally, and usually (except geologically young species) has a well developed posterior pocket. The preorbital bar is wide. The lacrimal almost never reaches the rear of the DPOF and is always subtriangularly shaped anteriorly. The nasal notch extends posteriorly about midway between the canine and $P^2$. The upper cheek tooth dentition ranges from simple to quite complex. Cranial and postcranial elements are of small to large size.

This genus includes *C. goorisi\** Skinner and MacFadden, 1977, early Barstovian; *C. sphenodus* Woodburne, MacFadden, and Skinner, 1981, late Barstovian; *C. occidentale* (Leidy, 1856) Skinner and MacFadden, 1977, Clarendonian; *C. ingenuum* (Leidy, 1885) Hulbert, 1987c, late Clarendonian and early Hemphillian; *C. plicatile* (Leidy, 1887) Hulbert, 1987c, late Clarendonian; *Cormohipparion emsliei* Hulbert, 1987b, late Hemphillian and Blancan, and *Hipparion weihoense* Liu et al. 1978) Turolian (China). *Cormohipparion nagriensis* (Hussain, 1971) MacFadden and Woodburne, 1982, Turolian (Pakistan); *Cormohipparion theobaldi* (Lydekker, 1877a) MacFadden and Bakr, 1979, Turolian (Pakistan) may pertain to "*Hipparion*" Group 1 (see later).

*Nannippus* -- Hipparionine horses of small size, relatively very high-crowned cheek teeth with relatively simple enamel plications. The DPOF is well developed to nearly absent; it is located relatively high on the face and far from the orbit (see MacFadden, 1984; Hulbert, 1987c). The lacrimal is subtriangularly shaped anteriorly and, apparently, does not reach the rear of the DPOF. The nasal notch extends posteriorly nearly to a position above the anterior end of $P^2$. The postcranial skeketon is small, and relatively lightly built. This term includes *N. minor\** (Sellards, 1916) MacFadden, 1984, Clarendonian and

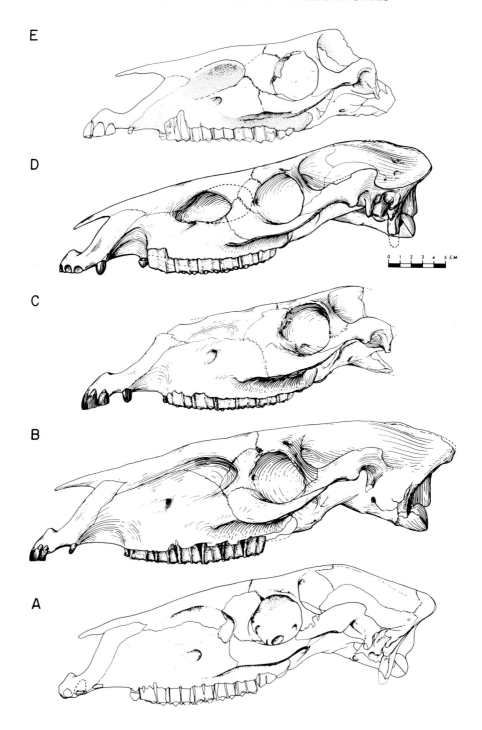

Fig. 12.5. Lateral view of merychippine and hipparionine crania from North America. A. *Merychippus insignis*, late Barstovian. B. *Neohipparion coloradense*, early Barstovian. C. *Pseudhipparion retrusum*, late Barstovian. D. *Cormohipparion goorisi*, early Barstovian. E. *Hipparion shirleyi*, late Barstovian. A and D after Woodburne and MacFadden (1982, Fig. 3). B after MacFadden (1984, Fig. 52). C after Webb and Hulbert (1986, Fig. 2). E after MacFadden (1984, Fig. 28).

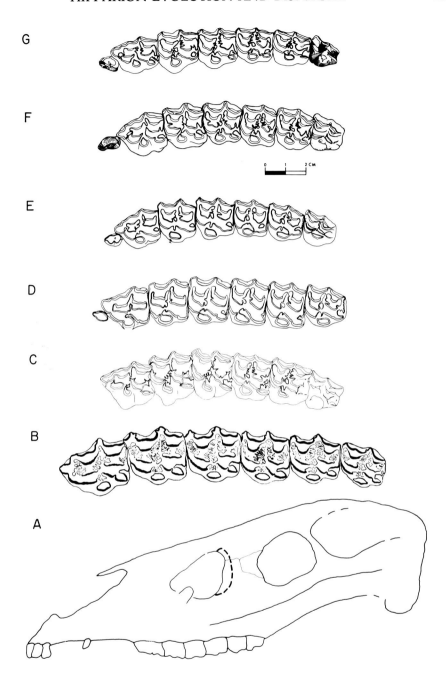

Fig. 12.6. Upper cheek tooth dentitions of merychippine and hipparionine taxa from North America, and cranium and upper cheek tooth dentition of *Hippotherium primigenium* from the Old World. A. and B. *Hippotherium primigenium*, early Vallesian, Germany. C. *Merychippus insignis*, late Barstovian. D. *Neohipparion coloradense*, early Barstovian. E. *Pseudhipparion retrusm*, late Barstovian. F. *Cormohipparion goorisi*, early Barstovian. G. *Hipparion shirleyi*, late Barstovian. A. and B. after Woodburne et al. (1981, Fig. 9). C. and F. after Woodburne and MacFadden (1982, Fig. 4). D. after MacFadden (1984, Fig. 52). E. after Webb and Hulbert (1986, Fig. 2). G. after MacFadden (1984, Fig. 28).

Table 12.1. Comparison of some cranial and upper cheek tooth parameters of some hipparionine groups.

(After Woodburne and Bernor, 1980; Bernor et al, 1980; Woodburne et al, 1981; MacFadden,1984; Bernor,1985.

| Parameter | Hipparion Group 1 | | Hipparion Group 2 | | Hipparion Group 3 | | Cormohipparion | |
|---|---|---|---|---|---|---|---|---|
| General size ($P^2$-$M^3$ length) | "H." primigenium | 154-170 | "H." moldavicum | 126-144 | H. prostylum | 136-144 | C. goorisi | 112 |
| | "H." africanum | 145-149 | "H." mediterraneum (Pikermi) | 132-139 | H. dietrichi (Pikermi) | 123-137 | C. sphenodus | 117-135 |
| | "H." catalaunicum | 151 | "H." mediterraneum (Samos) | 99 | H. dietrichi (Samos) | 137 | C. occidentale (Xmas Quarry) | 125-146 |
| | "H." gettyi | 140 | "H." proboscideum | 139-147 | H. ?prostylum (Iran) | 148-149 | C. (S). nagriensis | ---- |
| | "Hipparion" sp. (Pikermi) | 142-153 | | | H. campbelli | 155 | C. (S). theobaldi | 113 |
| | "Hipparion" sp. (Samos) | 148-161 | | | H. shirleyi | 105 | C. (S). feddeni | 154-172 |
| | "Hipparion" sp. (Mt. Lebéron) | 136 | | | H. tehonense | 123 | C. weihoense | 167 |
| | | | | | H. forcei | 133 | | |
| Lacrimal shape, anteriorly | "H." primigenium | subtriang. | "H." moldavicum | subtriang. | H. prostylum | subtraing. | C. goorisi | subtriang. |
| | "H." africanum | subtriang. | "H." mediterraneum (Pikermi) | subtr. | H. dietrichi (Pikermi) | subtriang. | C. sphenodus | subtriang. |
| | "H." catalaunicum | subtr. | "H." mediterraneum (Samos) | subtr. | H. dietrichi (Samos) | subtriang. | C. occidentale (Xmas Quarry) | subtriang |
| | "H." gettyi | rectangular | "H." proboscideum | subtriang. (modified) | H. ?prostylum (Iran) | rectangular | C. (S). nagriensis | str. |
| | "Hipparion" sp. (Pikermi) | subtriang. | | | H. campbelli | subtriang. | C. (S). theobaldi | str. |
| | "Hipparion" sp. (Samos) | rectangular | | | H. shirleyi | triang. | C. (S). feddeni | subtr. |
| | "Hipparion" sp. (Mt. Lebéron) | triangular | | | H. tehonense | subtr. | C. weihoense | subtr. |
| | H. forcei | subtr. | | | | | | |

| Parameter | Hipparion Group 1 | | Hipparion Group 2 | | Hipparion Group 3 | | Cormohipparion | |
|---|---|---|---|---|---|---|---|---|
| Preorbital fossa length | "H." primigenium | 84-88 | "H." moldavicum | 65-71 | H. prostylum | 31-52 | C. goorisi | 54 |
| | "H." africanum | ---- | "H." mediterraneum (Pikermi) | 43-49 | H. dietrichi (Pikermi) | 30-43 | C. sphenodus | 40-66 |
| | "H." catalaunicum | 70 | "H." mediterraneum (Samos) | 47-58 | H. dietrichi (Samos) | ---- | C. occidentale (Xmas Quarry) | 60-75 |
| | "H." gettyi | 67 | "H." proboscideum | 69-85 | H. ?prostylum (Iran) | 42-45 | C. (S). nagriensis | 70 |
| | "Hipparion" sp. (Pikermi) | ---- | | | H. campbelli | 52-57 | C. (S). theobaldi | 62-86 |
| | "Hipparion" sp. (Samos) | 61-67 | | | H. shirleyi | 37-51 | C. (S). feddeni | 42-55 |
| | "Hipparion" sp. (Mt. Lebéron) | 44 | H. tehonense | ca 50[3] | C. weihoense | 78 | | |
| | | | H. forcei | ca 50-70[3] | | | | |
| Snout length, $I^1$-$P^2$ | "H." primigenium | 112 | "H." moldavicum | 84 | H. prostylum | 98 | C. goorisi | 7 |
| | "H." africanum | 124 | "H." mediterraneum (Pikermi) | 104-107 | H. dietrichi (Pikermi) | 86-99 | C. sphenodus | 84 |
| | "H." catalaunicum | 103 | "H." mediterraneum (Samos) | ---- | H. dietrichi (Samos) | 98 | C. occidentale (Xmas Quarry) | 102-111 |
| | "H." gettyi | 100 | "H." proboscideum | 123-138 | H. ?prostylum (Iran) | 91-105 | C. (S). nagriensis | ---- |
| | "Hipparion" sp. (Pikermi) | ---- | H. campbelli | 111 | C. (S). theobaldi | ---- | | |
| | "Hipparion" sp. (Samos) | 116-124 | H. shirleyi | 73[3] | C. (S). feddeni | 111 | | |
| | "Hipparion" sp. (Mt. Lebéon) | ---- | H. tehonense | ca 83-108[3] | C. weihoense | 119 | | |
| | | | H. forcei | ca 88[3] | | | | |

| Parameter | Hipparion Group 1 | | Hipparion Group 2 | | Hipparion Group 3 | | Cormohipparion | |
|---|---|---|---|---|---|---|---|---|
| Preorbital bar length | "H." primigenium | 44-51 | "H." moldavicum | 17-30 | H. prostylum | 42-55 | C. goorisi | 17 |
| | "H." africanum | 44 | "H." mediterraneum (Pikermi) | 19-27 | H. dietrichi (Pikermi) | 35-44 | C. sphenodus | 19-37 |
| | "H." catalaunicum | 40 | "H." mediterraneum (Samos) | 21-22 | H. dietrichi (Samos) | 36-38 | C. occidentale (Xmas Quarry) | 40-50 |
| | "H." gettyi | 45 | "H." proboscideum | 23-24 | H. ?prostylum (Iran) | 33-40 | C. (S). nagriensis | 32 |
| | "Hipparion" sp. (Pikermi) | 31-44 | | | H. campbelli | 33-40 | C. (S). theobaldi | 55-59 |
| | "Hipparion" sp. (Samos) | 39-54 | | | H. shirleyi | 23-35 | C. (S). feddeni | 53-64 |
| | "Hipparion" sp. (Mt. Lebéron) | 46 | | | H. tehonense | 35-41 | C. weihoense | 51 |
| | | | | | H. forcei | 32-40 | | |
| Anterior rim, preorbital fossa | "H." primigenium | rel..strong | "H." moldavicum | rel. strong | H. prostylum | absent | C. goorisi | strong |
| | "H." africanum | rel. strong? | "H." mediterraneum (Pikermi) | rel. str | H. dietrichi (Pikermi) | absent | C. sphenodus | strong |
| | "H." catalaunicum | rel. str. | "H." mediterraneum (Samos) | rel. str. | H. dietrichi (Samos) | absent | C. occidentale (Xmas Quarry) | strong to weak |
| | "H." gettyi (Iran) | rel. strong | "H." proboscideum | rel. str. | H. ?prostylum | absent | C. (S). nagriensis | str. |
| | "Hipparion" sp. (Pikermi) | ? | | | H. campbelli | absent | C. (S). theobaldi | str. |
| | "Hipparion" sp. (Samos) | rel. strong | | | H. shirleyi | absent | C. (S). feddeni | strong |
| | "Hipparion" sp. (Mt. Lebéron) | rel. strong | | | H. tehonense | absent | C. weihoense | triang. |
| | | | | | H. forcei | absent | | |
| Posterior pocket of POF | "H." primigenium | moderate | "H." moldavicum | moderate | H. prostylum | moderate | C. goorsi | strong |
| | "H." africanum | ?moderate | "H." mediterraneum (Pikermi) | moderate | H. dietrichi (Pikermi) | slight | C. sphenodus | strong |
| | "H." catalaunicum | strong | "H." mediterraneum (Samos) | moderate | H. dietrichi (Samos) | weak | C. occidentale (Xmas Quarry) | moderate |
| | "H." gettyi | strong | "H." proboscideum | moderate | H. ?prostylum (Iran) | slight | C. (S). nagriensis | str. |
| | "Hipparion" sp. (Pikermi) | ?moderate | | | H. campbelli | weak | C. (S). theobaldi | str. |
| | "Hipparion" sp. (Samos) | moderate | | | H. shirleyi | weak or absent | C. (S). feddeni | mod. |
| | "Hipparion" sp. (Mt. Lebéron) | moderate | | | H. tehonense | slight | C. weihoense | strong |
| | | | | | H. forcei | weak | | |
| Complexity of upper cheek teeth | "H." primigenium | very | "H." moldavicum | moderate | H. prostylum | mod.-simple | C. goorisi | v. simple |
| | "H." africanum | very | "H." mediterraneum (Pikermi) | moderate | H. dietrichi (Pikermi) | mod.-simple | C. sphenodus | moderate |
| | "H." catalaunicum | very | "H." mediterraneum (Samos) | moderate | H. dietrichi (Samos) | mod.-simple | C. occidentale (Xmas Quarry) | complex |
| | "H." gettyi | moderate | "H." proboscideum | moderate | H. ?prostylum (Iran) | mod.-simple | C. (S). nagriensis | com. |
| | "Hipparion" sp. (Pikermi) | moderate | | | H. campbelli | mod.-simple | C. (S). theobaldi | com. |
| | "Hipparion" sp. (Samos) | moderate | | | H. shirleyi | mod.-simple | C. (S). feddeni | complex |
| | "Hipparion" sp. | moderate | | | H. tehonense | mod.-simple | C. weihoense | v. complex |
| | | | | | H. forcei | mod.-simple | | |
| Protocone shape | "H." primigenium | subovate | "H." moldavicum | subovate | H. prostylum | subrounded | C. goorisi | ovate, with spur |
| | "H." africanum | subovate | "H." mediterraneum (Pikermi) | subovate | H. dietrichi (Pikermi) | subrounded | C. sphenodus | subovate |
| | "H." catalaunicum | subovate | "H." mediterraneum (Samos) | subovate | H. dietrichi (Samos) | subrounded | C. occidental (Xmas Quarry) | subovate to long |
| | "H." gettyi | subovate | "H." proboscideum | subovate | H. ?prostylum (Iran) | subrounded to subovate | C. (S). nagriensis | sub-triangular |
| | "Hipparion" sp. (Pikermi) | subovate | | | H. campbelli | subrounded | C. (S). theobaldi | sub-triangular |
| | "Hipparion" sp. (Samos) | subovate | | | H. shirleyi | ovate, with spur | C. (S). feddeni | sub-triangular |
| | "Hipparion" sp. (Mt. Lebéron) | triangular | | | H. tehonense | subrounded | C. weihoense | — |
| | | | | | H. forcei | subrounded to subovate | | |

Notes:

|  | Group 1 |  | Group 2 | |
|---|---|---|---|---|
| Vallesian taxa: | "H." primigenium; | early Vallesian, Germany | | |
| | "H." africanum; | early Vallesian, N. Africa | | |
| | "H." catalaunicum; | early Vallesian, Spain | | |
| Turolian taxa: | "H." gettyi; | early Turolian, Iran | "H." moldavicum; | Turolian, Western Asia, Iran |
| | "Hipparion" sp., | Turolian, Greece (Pikermi) | "H." mediterraneum; | Turolian, Greece (Pikermi) |
| | "Hipparion" sp., | Turolian, Greece (Samos) | "H." mediterraneum; | Turolian, Greece (Samos) |
| | "Hipparion" sp., | Turolian, France (Mt. Lebéron) | "H." proboscideum; | Turolian, Greece (Samos) |

|  | Group 3 |  | Cormohipparion Group | |
|---|---|---|---|---|
| Barstovian taxa: (early) | H. shirleyi; Texas | | C. goorisi; Texas | |
| Barstovian taxa: (late) | | | C. goorisi; Texas | |
| | | | C. sphenodus; Great Plains | |
| Clarendonian and early Turolian taxa | H. prostylum; Turolian, France (Mt. Lebéron) | | C. occidentale; Great Plains and western US | |
| | H. dietrichi; Turolian, Greece (Pikermi, Samos) | | C. weihoense; Turolian, China | |
| | H. tehonense; early and late Clarendonian | | C. (S).nagriensis; early Turolian, Pakistan | |
| | H. forcei; late Clarendonian | | C. (S). theobaldi; Turolian, Pakistan | |
| | H. ?prostylum; Turolian, Iran | | C. (S). feddeni; Turolian, Pakistan | |
| | H. campbelli; Turolian, Iran | | C. sphenodus; Great Plains | |

Abbreviations: tri., triang. = triangular; str., subtriang. = subtriangular; rel. strong = relatively strong; mod. = moderate; v. = very; str. = strong; com. = complex

Table 12.2. Comparison of some cranial and upper cheek tooth parameters of some mesodont or primitive North American hipparions.
(After Skinner and Taylor, 1967; MacFadden and Skinner, 1981; MacFadden, 1984)

| Parameter | Merychippus insignis | Cormohipparion goorisi | Hipparion shirleyi | Neohipparion occidentale |
|---|---|---|---|---|
| General size ($P^2$-$M^3$ length) | Moderate (116.4[1,2]) | Moderate (111.2[2]) | Moderate (111.0[2]) | Moderate (127.3[3]) |
| Crown height ($M^1$ mesostyle) | Mesodont (16-20[1]) | Mesodont (22-28) | Mesodont (25-30) | Hypsodont (35-43) |
| Lacrimal shape within rear of POF | Tends broad anteriorly, | Triangular, does not reach or barely reaches POF | Triangular, just reaches rear of POF | Tends broad anteriorly, reaches within POF |
| Dorsal Preorbital Fossa (DPOF) | Relatively long (60[2]), slightly pocketed posteriorly | Relatively long (54[2]) strongly pocketed posteriorly (pocket reaches to orbit [overall most derived]) | Relatively long (55) only slightly pocketed posteriorly | Relatively long (ca.56[3]), slightly pocketed posteriorly [overall most like M. insignis] |
| Anterior rim, DPOF | Indistinct | Distinct | Indistinct | Indistinct |
| Preorbital bar | Narrow (17[1,2]) | Narrow (ca.17[1,2]) | Longer (ca.30[1,2]) | Narrow (ca.10-12[3]) |
| Protocone joins protoloph | Early wear | Late wear | Late wear | Latest wear |
| Preprotoconal groove distinct | Early wear | Late wear | Late wear | Latest wear |
| Alignment, proto- and hypocone wear stages | Anteroposteriorly premolar protocones | Obliquely, especially premolar protocones | Obliquely, especially protocones at some | Obliquely, except |
| Hypoconal facette present | Usually | Not | Not | Not |

[1] average of a number of specimens

[2] personal observation

[3] based on measurements taken from illustrations in MacFadden (1984) or MacFadden and Skinner (1981)

Hemphillian; *N. beckensis* Dalquest and Donovan, 1973, early Blancan; *N. peninsulatus* (Cope, 1885) MacFadden, 1984, Blancan.

*Neohipparion* (Figs. 12.1, 12.5, 12.6; Table 12.2) -- Hipparionine horses of moderate to large size, with moderately simple to complexly ornamented cheek teeth. The protocone ranges from subovate to very elongate and, in advanced forms, the lower cheek teeth show equine-like features, such as nearly angulate metaconids and metastylids. The DPOF is relatively weakly developed to absent. The preorbital bar (when defined) is narrow. The lacrimal is generally rectangularly shaped anteriorly in *N. coloradense*, but appears to be subtriangular anteriorly in some specimens of that species and in all later members of the genus. The nasal notch terminates about midway between $C^1$ and $P^2$. The postcranial skeleton is moderately to robustly developed. Metapodials are of moderate size to long.

This term includes *N. coloradense*\* (Osborn, 1918) MacFadden, 1984, Barstovian and early Clarendonian; *N. affine* (Leidy, 1869) MacFadden, 1984, late Barstovian and Clarendonian; *N. trampasense* (Edwards, 1982) MacFadden, 1984, late Clarendonian ?and early Hemphillian; *N. leptode* Merriam, 1915, early Hemphillian; *N. eurystyle* (Cope, 1893) MacFadden, 1984, Hemphillian; *N. gidleyi* Merriam, 1915, Hemphillian. Note that Webb and Hulbert (1986) consider *Neohipparion coloradense* to pertain to a different (unnamed) genus and to have given rise to *Neohipparion* (in their usage) and *Pseudhipparion*.

*Pseudhipparion* --Hipparionine horse of moderate to small size, develops very high crowned, nearly hypselodont (nearly rootless) cheek teeth with simple enamel plications. The protocone tends to connect to the protoloph, and the hypoconal groove tends to form an isolated enamel lake, in early wear stages. The DPOF is shallow and poorly developed to absent; when present it is located well anterior to the orbit (Webb and Hulbert, 1986). The lacrimal is subtriangularly-shaped anteriorly, and does not reach the rear of the DPOF. Some species develop a malar fossa, as well. The nasal notch extends posteriorly nearly to a positon above the anterior end of $P^2$. The postcranial skeleton tends to be small and lightly built.

This term includes an unnamed species\* Webb and Hulbert, 1986, late Barstovian; *P. retrusum* (Cope, 1889) Webb and Hulbert, 1986, latest Barstovian; *P. curtivallum* (Quinn, 1955) Webb and Hulbert, 1986, early Clarendonian; *P. hessei* Webb and Hulbert, 1986, late Clarendonian; *P. gratum* (Leidy, 1869), late Clarendonian; *P. skinneri* Webb and Hulbert, 1986, late Clarendonian and early Hemphillian; *P. simpsoni* Webb and Hulbert, 1986, late Hemphillian.

"*Hipparion*" -- hipparionine horses that differ from or cannot be placed within *Hipparion*, *Cormohipparion*, *Nannippus*, *Neohipparion*, or *Pseudhipparion*, as characterized above [includes species referred to Groups 1, 2, and 4 of Woodburne and Bernor (1980)].

Group 1 taxa include *Hippotherium primigenium*\* Von Meyer 1829, early Vallesian (Germany; Figs. 12.4, 12.5, 12.6, 12.7F, G, H; Table 12.1); *Hipparion koenigswaldi* Sondaar, 1961, late Vallesian (Spain); *H. catalaunicum* Pirlot, 1956, late Vallesian (Spain); *H. melendezi* Alberdi, 1974a, Turolian (Spain); *H. africanum* Arambourg, 1959, late Vallesian (Africa); *H. sebastopolianum* Borissiak, 1914, Vallesian (Sebastopol); *H. gettyi* Bernor, 1985, Turolian (Iran).

Members of Group 1 are of moderate to large size (Table 12.1); have a well-developed, dorsoventrally deep, DPOF that is deep posteriorly and distinctly pocketed. The anterior rim of the DPOF is distinct. The preorbital bar is wide. The lacrimal never reaches the DPOF. The nasal notch usually reaches a point above $P^2$. The cheek tooth dentition-- especially in forms of Vallesian age-- is extremely complex. In some primitive species, the lacrimal is subtriangular anteriorly. Some taxa of

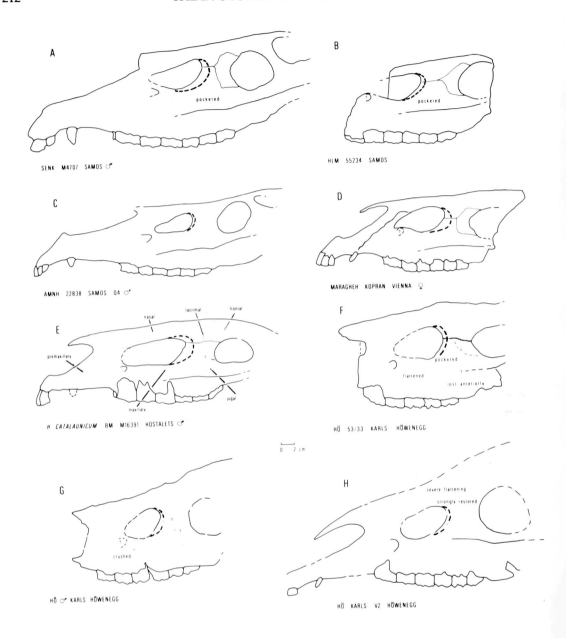

Fig. 12.7. Cranial morphology of Old World species referred to Group 1. A-C. *Hipparion* sp., late Turolian (Greece); note the anteriorly rectangular to subrectangular lacrimal bone, where present (A, B). D. "*Hipparion*" *gettyi*, Bernor 1985, early Turolian (Iran). Note the somewhat reduced anterior tip of the DPOF and the anteriorly rectangular lacrimal (*contra* the triangular shape of that bone probably present in E-G shown by the straight suture between the frontal, and lacrimal and maxillary bones, and surely present in figs. 4B and 6A). E. "*Hipparion*" *catalaunicum*, late Vallesian (Spain). F-H, *Hippotherium primigenium*, early Vallesian (Germany). After Woodburne and Bernor (1980, Fig. 5).

Turolian age referred to this group have a (?secondarily achieved) anteriorly rectangularly-shaped lacrimal (Fig. 12.7A, B, D) and a more strongly retracted nasal notch (Fig. 12.7A, C).

Group 2 taxa include *H. mediterraneum* Gaudry, 1867, Turolian (eastern Mediterranean) (Fig. 12.4C); *H. proboscideum* Studer, 1911, Turolian (eastern Mediterranean; southwest Asia) (Fig. 12.4D); *H. moldavicum* Gromova, 1952, Turolian, western Asia; *H.* aff. *moldavicum* Bernor, 1985, Turolian (Iran) (Fig. 12.8A-D).

Among other features, species of Group 2 have an extremely narrow preorbital bar (ca. 20 mm; Table 12.1), but retain a very well developed (both longitudinally and vertically) DPOF, and have developed in many cases an extra fossa between the dorsal preorbital fossa and the buccinator fossa (see Fig. 12.4C, D). The lacrimal always reaches or penetrates the rear of the DPOF, and is subtriangularly shaped anteriorly. The anterior rim of the DPOF is distinct. The nasal notch is incised to about $P^2$ or more posteriorly. The dentition is relatively complex. Postcranial elements are moderately developed.

Group 4 taxa (this may not be a "natural group", e.g., Woodburne and Bernor, 1980; Bernor and Hussain, 1985), include *H. matthewi* Abel, 1926, Turolian (Greece); *H. minus* Pavlov, 1890, Turolian (Greece) (Figs. 12.4F, 12.8E). Some of these are small-sized hipparions with a short preorbital bar (ca. 20 mm) and medially shallow (only a few mm deep) dorsal preorbital fossa. The dentition is generally not very complex.

4. *Institutional abbreviations*
AMNH American Museum of Natural History, New York
BMNH British Museum (Natural History), London, England
F:AM Frick American Mammals; fossil mammal collection in the AMNH
GIU Geologisch Instituut, Utrecht, The Netherlands
KNHM Kaiserliche Naturhistorisches Museum, Vienna, Austria
LYON Department de la Terre, Université "Claude Bernard," Lyon, France
MB Johannes Gutenberg Universität, Mainz, West Germany
MNHN Muséum National d'Histoire Naturelle, Paris, France
SENK Senckenberg Museum, Frankfurt, West Germany
HLMD Hessisches Landesmuseum, Darmstadt, West Germany
LSNK Landessammlungen für Naturkunde, Karlsruhe, West Germany
UCR Department of Earth Sciences, University of California, Riverside, California.

**Definition of** *Hipparion*
De Christol (1832) coined the term *Hipparion* for a fossil horse recovered from deposits that crop out on the flanks of Mt. Léberon in southern France (Fig. 12.2). These deposits are now known to be of Turolian age (Fig. 12.3). This horse differs from the living *Equus* in having tridactyl feet and upper teeth in which the protocone is isolated from adjacent parts of the enamel pattern (Fig. 12.1). De Christol did not designate a type species for *Hipparion*. Gervais (1849) recognized three species from the Mt. Léberon deposits, *Hipparion prostylum*, *H. mesostylum*, and *H. diplostylum*, and subsequently (Gervais, 1859) nominated *H. prostylum* as the senior synonym of the other two species. This was followed by Osborn (1918), who designated *Hipparion prostylum* as the type species (Fig. 12.9J, L, M). Forstén (1968) included *H. prostylum* as a junior synonym of *Hippotherium primigenium* Von Meyer, 1829, utilizing the nomen *Hipparion primigenius*. This brings up a consideration that will be discussed later.

Although most workers have considered only one species of *Hipparion* to be present in the Mt. Léberon deposits (e.g., Forstén, 1968; MacFadden, 1980; Pirlot, 1956), Woodburne and Bernor (1980) and Woodburne in Bernor *et al.*, (1980) suggest

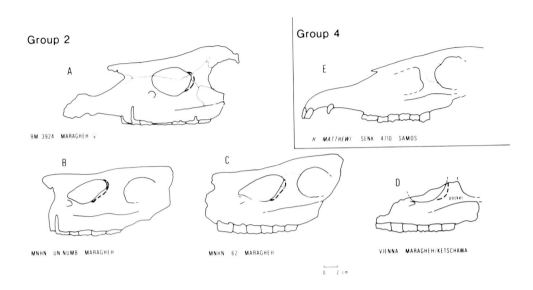

Fig. 12.8. Cranial morphology of species of "*Hipparion*" Group 2 referred (Bernor, 1985) to "*Hipparion*" aff. *moldavicum* (A-D) and to "*Hipparion*" Group 4 by Woodburne and Bernor (1980). All are of Turolian age. Note the absence of an intermediate fossa between the buccinator fossa and the DPOF in A-D (*contra* Fig. 12.4 C-D; "*H.*" *mediterraneum* and "*H*". *proboscideum*). Also note that the shape, orientation, and strength of the DPOF in E (Group 4) is different than shown in Fig. 12.4F. After Woodburne and Bernor (1980, Fig. 6).

Fig. 12.9. (opposite page) Cranial morphology of Old World species of *Hipparion sensu stricto*, all of Turolian age. A-B, *Hipparion campbelli* Bernor (1985). C-E, *Hipparion* sp., possibly similar to *H. campbelli*. F, H, I, *Hipparion dietrichi*. G, referred to *H. prostylum* by Bernor (1985) and indicated as possibly ancestral to A and B. Note that Bernor (1985) also proposed that G was derived from "*H.*" *gettyi*, a Group 1 taxon, suggesting that Group 3 horses were endemically derived in the Old World. See text for discussion as to whereas the shape of the lacrimal bone shared between these two forms is permissive of relationship, that type of lacrimal shape appears to be unusual in Group 3 species. J, L, M. *Hipparion prostylum*, from the type locality (Mt. Lebéron) in France. K, "*Hipparion*" sp., as interpreted here, also from Mt. Lebéron, is considered to be a Group 1 taxon, not *Hipparion sensu stricto* After Woodburne and Bernor (1980, Fig. 7).

# HIPPARION EVOLUTION AND DISPERSAL

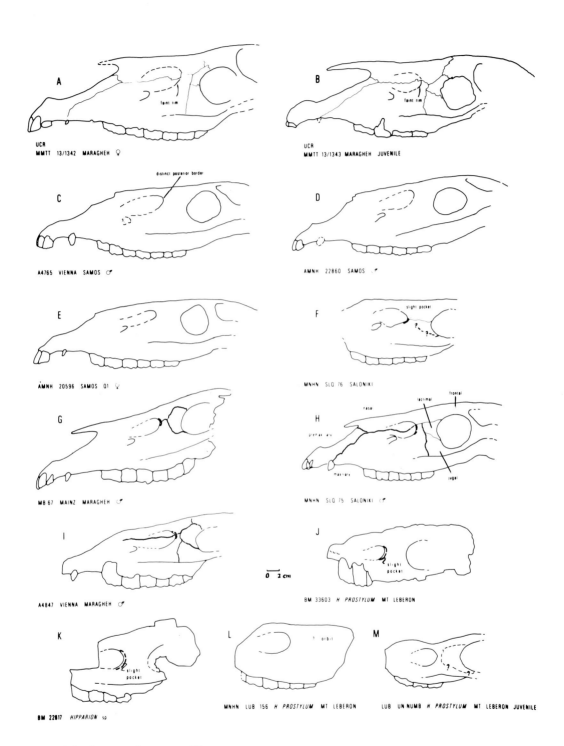

Fig. 12.9. See caption preceding page.

that there probably are two species of *Hipparion* at that locality based both on postcranial (Sondaar, 1974) and dental and cranial evidence (Bernor et al., 1980). The apparent presence of more than one species of *Hipparion* in the Mt. Léberon sample is consistent with its Turolian age. As discussed in Woodburne and Bernor (1980), the only localities in which a single species of *Hipparion* is certainly present are those of early Vallesian age. By designation, however, the type species of *Hipparion* is *Hipparion prostylum* De Christol, 1832 (Osborn, 1918); its cranium was figured by Gaudry (1867) and by Skinner and MacFadden (1977). MacFadden (1980) and Bernor et al. (1980) referred additional specimens to this species and presented additional illustrations of its cranium. For purposes of the present discussion, the specimen BMNH 26617 (Fig. 12.9K) is not considered as a member of *H. prostylum* (e.g. Bernor et al., 1980, p. 731). Based on specimens cited in the above works, *Hipparion prostylum* is a species of *Hipparion sensu stricto*, as characterized above. It is of moderate size [orbit length 46-49 mm; upper cheek tooth length ($P^2$-$M^3$) = 140- 145 mm], the dorsal preorbital fossa is located relatively high on the face (ca. 21-30 mm), relatively far anterior to the orbit (the preorbital bar is 38-47 mm wide), the dorsal preorbital fossa is relatively long (32-45 mm), only slightly pocketed posteriorly, and is effectively horizontally oriented (long axis projects to a point that essentially bisects the dorsal height of the narial aperture). The nasal notch extends posteriorly to a point above $P^2$ (see Figs. 12.4, 12.9A-J, L, M; Table 12.1). As seen below, species attributable to *Hipparion sensu stricto* are first known in the Old World from deposits of Turolian age, and comprise the geographically most widespread group of Old World hipparions during that time.

## Hipparionine evolutionary patterns in North America

Hipparions have long been considered to have been derived, ultimately, from one or more merychippine horses (e.g., Stirton, 1940). At the moment, the best-studied potential hipparionine ancestor is *Merychippus insignis* (Leidy, 1856) from the early Barstovian Olcott Formation (= lower Snake Creek beds) of Nebraska (Skinner and Taylor, 1967; Tedford et al., 1987; contra Evander, 1986). Although not necessarily the only outgroup for comparison with hipparionine taxa, it is instructive that *Merychippus insignis* is not only the type species of that genus, but also is of moderate size (Figs. 12.5, 12.6; Table 12.2), has a dorsal preorbital fossa that is relatively weakly developed (dorsal, ventral, and anterior borders are not strongly formed), and is moderately pocketed posteriorly. The preorbital bar is narrow (ca. 16 mm, and is exceeded in length by the lacrimal bone; i.e., the lacrimal penetrates the rear of the dorsal preorbital fossa). The cheek teeth are simply constructed, and the protocone connects to the protoselene in early wear. Dentally, *Merychippus insignis* generally resembles early members of other hipparionine groups, viz. *Cormohipparion goorisi*, late Barstovian, *Hipparion shirleyi*, late Barstovian, and *Neohipparion coloradense*, early Barstovian (see MacFadden, 1984, Figs. 121, 122; Figs. 28, 31; Figs. 52, 53, 54, 55, 56, 57 respectively).

Based on Skinner and Taylor (1967), MacFadden and Skinner (1981), MacFadden (1984), and Webb and Hulbert (1986), it appears possible to separate *Merychippus insignis* from *Cormohipparion goorisi* (Figs. 12.5, 12.6; Table 12.2) in that: (1) *M. insignis* possesses upper molars in which the protocone connects to the protoloph (and thus in which a preprotoconal groove is present) in earlier rather than later stages of occlusal wear. (2) The protocone and hypocone are more anteroposteriorly aligned. (3) The hypoconal groove becomes occluded by a hypoconal facette. (4) The metastylid of the lower cheek teeth is enlarged. Likewise, *M. insignis* may be distinguished from *C. goorisi* in having a lacrimal bone that is more irregularly (tends to be rectangularly) shaped and enters the dorsal preorbital fossa (the lacrimal is subtriangular in *C. goorisi* and barely reaches the DPOF), the preorbital bar is narrower in *M. insignis* (vs. wider), and the anterior rim of the DPOF is indistinct (vs. distinct), even though the two species are of comparable size, to judge from the similar cheek tooth lengths ($P^2$-$M^3$; Table 12.2). Without listing each in detail (and recognizing that the number of characters is an incomplete list), it appears (Table 12.2) that identification of lineages that are phyletically derived relative to *Merychippus insignis* is aided when character suites in addition to those of the cheek-tooth dentition are utilized. (The basal members of these phyletically derived lineages are *Cormohipparion goorisi*, *Hipparion shirleyi*, and *Neohipparion coloradense*; Figs. 12.5, 12.6; Table 12.2), and early species of *Pseudhipparion*. Thus, in addition to dental nuances, features of the dorsal preorbital fossa and adjacent parts of the facial region aid in identification of derived hipparionine lineages, as indicated earlier in the "Conventions. "

If those data can be taken as important and useful in identifying geologically early members of various hipparionine lineages (based on suites of derived characters), it appears that the earliest record of *Cormohipparion* (early Barstovian) was followed in time by, or was about synchronous with, the earliest records of *Hipparion* (late Barstovian), *Neohipparion* (early Barstovian), *Pseudhipparion* (late Barstovian) and *Nannippus* (late Clarendonian). By late Clarendonian time, members of all five North Amerian hipparionine clades were present, and presumably had existed to various degrees for some time previously.

No worker has suggested that the above-cited temporal sequence was a phyletic one. The ancient (as first-known) retraction of the nasal notch (and advanced dental features) separates species of *Hipparion* from contemporaneous or older hipparionines and merychippines (*sensu stricto*) (Table 12.2); small size and extreme dental hypsodonty separates species of *Nannippus* from all contemporary (and likely older, except for small size) hipparionines. The wide preorbital bar, a subtriangularly shaped versus more rectangularly shaped lacrimal bone, and strongly developed, pocketed, dorsal preorbital fossa (plus dental differences) separate early species of *Cormohipparion* from *M. insignis* or *Neohipparion coloradense* (Figs. 12.5, 12.6), and *N. coloradense* differs from *M. insignis* in its hypsodonty and corresponding differences in dental features (Figs. 12.1, 12.5, 12.6). Early species of *Pseudhipparion* (unnamed species and *P. retrusum*) differ from "*M.*" *insignis* and from all other hipparionines in the early tendency to reduce the expression of the DPOF, the reduced complexity of enamel plications, and the isolation of the hypoconal groove into a discrete lake. The early acquisition of elongate protocones and reduced penetration of the ectoflexid into the isthmus on $P_{2-4}$ (Webb and Hulbert, 1986) unites *Pseudhipparion* with the *Neohipparion* complex as used here. In this context, increased length of the preorbital bar (ca. 40 mm) in *P. retrusum* must have been secondarily achieved. All hipparionines in North America are derived over taxa of "merychippine grade," e.g., *M. insignis*, in the strength of the metastylid of the lower dentiton (Webb and Hulbert, 1986). In summary, all five hipparionine lineages in North America must have been established earlier than the late Clarendonian and, with the possible exception of *Nannippus*, had been separate since the early Barstovian. As developed below, this pattern is comparable to that shown by Old World hipparionine species.

It also is suggested here that the cranial configuration of *Neohipparion coloradense* is most plesiomorphic among hipparionines (= most like that of *Merychippus insignis*), recognizing the dental differences between the two species, and that early species of *Cormohipparion* may be the most derived. Early species of *Hipparion* and *Nannippus*, recognizing the extreme hypsodonty of the latter, may be more alike cranially (including a deeply incised nasal notch) than either is to *Cormohipparion*, *Neohipparion*, or *Merychippus insignis*. Thus, cladistic analysis could suggest that *M. insignis* and *N. coloradense* are sister-taxa; as could be *H. shirleyi* and *N. minor*. On the other hand, Hulbert (1987c) indicates that the cranial and dental morphology of early species of *Nannippus* (unnamed) most closely resembles that of *C. goorisi* and other species of *Cormohipparion*. The subtriangularly shaped lacrimal bone found in *Nannippus* would fit this allocation, but whether early species of *Nannippus* have the deeply pocketed DPOF characteristic of species of *Cormohipparion* has not been reported, nor has the degree to which the nasal notch is retracted, or the degree to which the lacrimal touches the rear of the DPOF.

Setting aside these still-unresolved intra-American arrangements, the overall pattern shown among hipparionines of that region resembles those portrayed by Old World hipparionines in: the progressive reduction of the dorsal preorbital fossa [late Barstovian and younger species of *Pseudhipparion* (Webb and Hulbert, 1986), late Clarendonian and Blancan species of *Cormohipparion* and *Nannippus* (Hulbert, 1987a, b) and *Neohipparion* (MacFadden, 1984)], development of taxa of small size [*Nannippus* (MacFadden, 1984); *Pseudhipparion*; (Webb and Hulbert, 1986)], deemphasis and reduction of the lateral metapodials [*Neohipparion* (MacFadden, 1984)], increase in hypsodonty [effectively all hipparionines (MacFadden, 1984; Hulbert, 1987a; 1987b, c; Webb and Hulbert, 1986)], development of an *Equus*-like lower molar morphology (distinctly angulate metaconid and metastylid), while retaining a hipparionine style upper dentition, and having lost the dorsal preorbital fossa of the face [*Neohipparion leptode* (MacFadden, 1984, Figs. 76, 77); *Hipparion houfenense* (Zhegallo, 1978, Fig. 79; MacFadden, 1984, Fig. 151)]. Contrary to most hipparionine taxa of any continent, species of *Cormohipparion* show a progressive increase in enamel complexity of the upper cheek tooth dentition, with that found in *C. emsliei* (late Hemphillian and Blancan of Florida; Hulbert, 1987b) rivalling the dental complexity seen in hipparions of Vallesian age in the Old World, but still lacking the multiple plicaballins seen in those Old World forms; Fig. 12.6B). At the same time, *C. emsliei* has a reduced DPOF, consistent with the reduction of that feature seen worldwide in hipparions of Hemphillian and late Turolian age. As suggested below, species of *Cormohipparion* appear to be present in Asian faunas of Turolian age; none are known to be older than about 9.2-9.5 Ma. Whether this Asian record resulted from a separate dispersal from North America at about 9.5 Ma remains to be determined. Finally, the development of incipient hypselodonty and a malar facial fossa appears to be unique to species of *Pseudhipparion* among all hipparionines.

## Hipparionine phyletic structure in the Old World

As suggested by Woodburne and Bernor (1980), and expanded upon by Bernor *et al.* (1980), MacFadden and Bakr (1979), MacFadden (1980), MacFadden and Woodburne (1982), Woodburne and MacFadden (1982), Bernor and Hussain (1985), Bernor (1985), Bernor *et al.* (1987), there appear to be at least three supraspecific groups of Old World hipparionine horses. In terms of the conventions expressed above, these are *Hipparion sensu stricto* (= Group 3), "*Hipparion*" Group 1, and "*Hipparion*" Group 2 of Woodburne and Bernor, 1980 (further citations above). "*Hipparion*" Group 4, *Proboscidipparion* and *Stylohipparion* may be additional separate supraspecific groups or may be referable to one of groups 1-3.

The geologically oldest member of Group 1 "*Hipparion*" appears to be *Hippotherium primigenium* Von Meyer, 1829 from deposits at Eppelsheim, Germany (Figs. 12.4, 12.6), radioisotopically dated at about 11 Ma (Berggren and Van Couvering, 1974; Bernor *et al.*, 1980; Woodburne *et* al., 1981; this report). In this instance, as in other Old World sites interpreted to be of early Vallesian age (e.g., Beglia Formation, Tunisia; Forstén, 1972; Berggren, Kent, and Van Couvering, 1985), a species of "Hipparion" appears abruptly and abundantly at a level stratigraphically above deposits in which this kind of horse is absent, but in which may be found specimens of Anchitherium and/or Leptodontomys (= Eomysops) (e.g., Berggren and Van Couvering, 1978; Engesser, 1979). This stratigraphically abrupt appearance of hipparions, which also is followed stratigraphically in many places by the rodent, Progonomys, has led to the concept of a "Hipparion Datum," as discussed further later.

As represented by cranial and dental material at Höwenegg (Tobien, 1986), Hippotherium primigenium (=Hipparion primigenium: Tobien, 1982) pertains to a horse in which the upper cheek tooth dentition is very complex [very complex fossette borders*; multiple pli caballins*, and so on (e.g., Fig. 12.6B) (Woodburne et al., 1981, Fig. 9)], the dorsal preorbital fossa is well developed and anteroventrally oriented (a line drawn from the point where the fronto-maxillary suture enters the rear of the fossa through its anterior end projects to a point posterior to $I^3$ rather than to a point anterior to $I^1$; Fig. 12.6)*, the fossa is moderately deeply pocketed posteriorly (about half the distance from the orbit to the posterior rim of the DPOF; R.L. Bernor, pers. commun.), and the nasal aperture is incised nearly to a point dorsal to the anterior end of $P^2$*. The lacrimal seems to be effectively subtriangular and strongly separated from the rear of the dorsal preorbital fossa in this species (= the preorbital bar is wide; Table 12.1). [Features designated by and asterisk (*) are considered to be derived relative to either Cormohipparion sphenodus or C. occidentale in North America; also Woodburne et al, 1981, p. 520-521]. Taxa that appear to pertain to this early hipparionine group, although differently specialized [briefly indicated by features in parentheses ( )], are H. catalaunicum, late Vallesian, Spain (very long DPOF; Fig. 12.7E); H. africanum, late Vallesian, Africa (very long DPOF and longer snout), Bernor, et al (1980, Fig. 8), Bernor et al.(1987); Hipparion gettyi, early Turolian, Iran (DPOF reduced anteriorly, lacrimal rectangular anteriorly; Fig. 12.7D), Bernor (1985); and Hipparion sp., late Turolian, Greece (nasals retracted, lacrimal rectangular anteriorly; Fig. 12.7A, C). This group apparently is also represented in deposits of Vallesian and Turolian age in sub-Saharan Africa, including forms that developed a very complex upper cheek tooth dentition and persisted into the late Pleistocene in Africa ["H". libycum (= Stylohipparion)].

On the basis of priority, typology, and morphology, the early Vallesian-aged hipparionine from Epplesheim and Höwenegg should be recognized as Hippotherium primigenium Von Meyer 1829; it is distinct from Hipparion sensu stricto (see "Conventions" earlier; Berggren, Kent, and Van Couvering, 1985, p. 228), and appears in the Old World record about 2 Ma before H. prostylum and related forms (e.g., Woodburne and Bernor, 1980; MacFadden, 1980; Bernor et al., 1980; Woodburne et al., 1981). It is also distinct from species of Cormohipparion in (at least) features designated above by an asterisk (*) (see also Woodburne et al., 1981, p. 515-521). Old World species related to H. primigenium (see above) also seem to differ from those that are more closely related to Cormohipparion [Asian taxa; see above; MacFadden and Woodburne, 1982, p. 193-195; species of C. (Sivalhippus), Bernor and Hussain, 1985]. Although analysis still is incomplete, it appears that an Old World group that stemmed from Hippotherium primigenium was distinct from members of hipparion groups 2, 3, or 4 in the Old World and from species that are either members of, or are closely related to, Cormohipparion (Fig. 12.3).

Hipparion Group 2 (based on "Hipparion" mediterraneum; Woodburne and

Bernor, 1980; see "Conventions") is of essentially Turolian age, and largely Asian in distribution (Woodburne et al., 1981: Bernor, 1985; Bernor et al., 1987); it is differently specialized than any of the Old World forms mentioned here, and almost certainly deserves recognition as a new generic-rank taxon.

Originally of relatively modest size (Table 12.1) and relatively complex cheek tooth dentition (but not as complex as seen in Vallesian members of Group 1), some Group 2 hipparions are noted for the development of an additional facial fossa between the DPOF and the buccinator fossa (found in no other Old or New World hipparionine). All members of Group 2 possess a very narrow preorbital bar (Figs. 12.4, 12.8); in some species ("*H.*" *proboscideum*) the anterior portion of the skull became hypertrophied; in others ("*H.*" *moldavicum*) more normal cranial proportions were maintained, the "extra" facial fossa (between the DPOF and buccinator fossa; Fig. 12.4C) is not yet developed (Figs. 12.8A-D), and the snout may be shorter than seen in "*H.*" *mediterraneum* or "*H.*" *proboscideum* [(Figs. 12.8A-C vs. Figs. 12.4C, D), see also Bernor, 1985, Fig. 16]. Based on specimens from Iran ("middle Maragheh," or Ketchawa-Kherajabad), thought to be between 8-9 Ma old (Bernor et al., 1980, Fig. 8; Bernor, 1985) the biochron of specimens referred to this species may include the geologically oldest record of Group 2 yet recognized. In lacking the "extra" facial fossa, "*H*," *moldavicum* is more plesiomorphic than other members of Group 2. A common ancestry of Group 2 forms and an Old World basal stock that gave rise to *H. primigenium* has been proposed (Woodburne et al., 1981; MacFadden and Woodburne, 1982), whereas Bernor (1985) and Bernor and Hussain (1985) present a contrary view, that Group 2 hipparions were derived from a Group 1 ancestor near "*Hipparion*" (=*Hippotherium*) *primigenium*. Neither proposal can be tested on the basis of direct fossil evidence, although it may be instructive that "*H.*" *moldavicum* has a simpler dentition than its plausible descendant, "*H.*" *mediterraneum*. Were the mediterranean species derived from the moldavian, the increase in dental complexity thereby recorded would represent a reversal (as one of a number of characters) that would be required should a primitive Group 2 form ("*H.*" *moldavicum*) have been derived from "*H.*" *primigenium*, and subsequently have given rise to a more derived taxon ("*H*". *mediterraneum*). In any case, no Group 1 taxon is yet known to retain a plesiomorphically large and anteroventrally oriented DPOF while concurrently showing a tendency for reduction in cranial size, dental complexity, and extreme reduction of the preorbital bar. This, along with the largely Asian distribution of Group 2 hipparionines, suggests an Asian theater of origin for these horses, but there either are no Vallesian-aged horses in that area, as currently known (*Cormohipparion* cf. *C. nagriensis* is a possible exception, see following), or their age is difficult to certify. No matter where or when these horses originated, Group 2 taxa do not show progressive loss or reduction of the DPOF, as seen in many other hipparionine clades (unless "*H.*" *matthewi* and related forms originally allocated to Group 4 are affiliated with Group 2).

Hipparion Group 3 (*Hipparion sensu stricto*) is a relatively common member of Old World Turolian faunas (see "Conventions," above). In Iran (Bernor, 1985), *Hipparion campbelli* (Figs. 12.9A, B) apparently was derived from a local, geologically older representative of *H. prostylum*, and demonstrates yet another instance of reduction of the DPOF in hipparionines (in this case coupled with increased retraction of the nasal notch). Taken together, Group 3 hipparions (including *H. dietrichi*, e.g., Woodburne and Bernor, 1980) commonly have a simpler upper cheek tooth dentition than seen in Group 1 or Group 2 forms, which is at least consistent with the proposed origin of this group from North American taxa referred to *Hipparion* with similar cranial and dental morphologies (e.g., MacFadden, 1980). This suggests that the early Turolian record of *Hipparion sensu stricto* in the Old World represents a dispersal from North America, where the group is identified in rocks as old as late Barstovian (see "Conventions"). On the other hand, Bernor (1985) proposes on the basis of dental evidence that a representative of *Hipparion prostylum* was endemically derived in Iran from a taxon ("*Hipparion*" *gettyi*) referred by him to Group 1. If this were (but isn't now) supported by intergrading cranial morphologies in the same stratigraphic sequence, it would indicate that at least some Old World and North American species identified as members of *Hipparion sensu stricto* were convergently developed. Note, however, that the lacrimal of the Iranian specimen (MB-67; Fig. 12.9G vs. Figs. 12.9A, B, *H. campbelli*, and Figs. 12.9H, I, *H. dietrichi*) is rectangular, rather than subtriangular, anteriorly. An anteriorly subtriangular lacrimal bone is typical of Old World or North American species referred to *Hipparion sensu stricto.* An anteriorly rectangular lacrimal bone is, however, also found in certain specimens referred by Woodburne and Bernor (1980; Fig. 5A, B, D) to Group 1, including the type specimen (KNHM-RLB 8401; Bernor, 1985) of "*Hipparion*" *gettyi* (see Figs. 12.7A, B, D). In the absence of stratigraphically intermediate cranial material, this lacrimal shape may strengthen the proposed relationship between "*H*". *gettyi* and the Iranian specimen referred to *H. prostylum*; it also appears to be at variance with the proposed Group 3 membership of the latter specimens. If all of these forms with an anteriorly rectangular lacrimal bone, including specimens from Samos, Greece (Fig. 12.7A, B, and ?C) can be considered a distinct lineage within Group 1, the Samos forms would be derived in being larger, having a longer snout, and in the more extreme retraction of the nasal notch. In any case, further discussion of the origin of Old World *Hipparion sensu stricto* requires identification of a synapormorphy for Old World versus American species and locating a Vallesian-aged taxon that possesses this feature. Until this takes place, it will be impossible to determine whether the characters that now unite New and Old World species of *Hipparion* were convergently derived in the two regions.

Group 4 species (see "Conventions") are probably the least well understood Old World hipparions, and may not represent a single group (Fig. 12.3F vs. Fig. 12.8E). In general these are small-sized taxa with relatively weakly developed facial features and relatively simple cheek tooth dentitions. As known, these *Nannippus*-like Old World forms lived during the Turolian. See also Bernor (1985) for a suggestion that at least "*H*." *matthewi* may be a dwarf affiliate of Group 2 hipparionines.

## Dispersals of hipparionine horses between the New and Old Worlds

As suggested above and in Fig. 12.3, the North American record appears to have been characterized by endemic evolution, with major departures being the effect the North American hipparionine radiation had on that of the Old World via a number of dispersals from North America. From the preceding it can be seen that one or more dispersals of hipparionine horses from North America to the Old World may be required by the data (e.g., Woodburne and Bernor, 1980; Bernor *et al.*, 1980; Woodburne *et al.*, 1981; MacFadden, 1984).

The geologically oldest dispersal from North America apparently resulted in the early Vallesian record of *Hippotherium primigenium* in western Europe, at about 12 Ma, the most likely ancestor of which was a species of *Cormohipparion* (e.g., *C. sphenodus*, Woodburne *et al.*, 1981) (Fig. 12.3). This, and other, stratigraphically abrupt appearances of hipparions in a number of Old World sites has led to the concept of a "*Hipparion* Datum." Berggren and Van Couvering (1978) and Berggren, Kent, and Van Couvering (1985, Appendix 3) summarize local and regional stratigraphic, paleontologic, and radioisotopic data suggesting that many of these "first" occurrences of "*Hipparion*" are about 12 Ma old. A number of these situations are discussed and graphically portrayed in Woodburne *et al.* (1981, pp. 496-497, Fig. 2), who indicate that (1) there are reasons to believe that the "first" hipparion of early Vallesian age pertains to a number of different species at different localities, rather than all being allocated to "*H.*" *primigenium*, as advocated by Forstén (1968; "*Hipparion primigenius*") and that, (2) considered regionally, the "first" hipparion is not of early Vallesian age. Thus, the "*Hipparion* Datum" ranges in age from about 12 Ma (Western Europe) to about 9.5 Ma or less in Pakistan, China, Iran, and Greece (Woodburne *et al.*, 1981, Fig. 2) (Fig. 12.3). Bernor and Hussain (1985), Berggren *et al.* (1985, note 11), Berggren, Kent, and Van Couvering (1985, Appendix 3) and Badgley *et al.* (1986) also discuss this problem.

Furthermore, the taxonomic allocation of these locally stratigraphically "first" hipparions is becoming rather complex. For the early Vallesian record (western Europe, northern Africa, eastern Paratethys) the various species that comprise this "Datum" all appear to be members of hipparion Group 1, of which "*Hippotherium primigenium*" can be claimed as the type species. It seems almost a foregone conclusion that other members of Group 1 (and especially those of Vallesian age) should be allocated to *Hippotherium*. As discussed above, species similar in cranial and dental morphology to "*Hippotherium primigenium*" differ from those allocated to *Hipparion sensu stricto* which, as indicated here, either was derived at about the beginning of the Turolian from a Group 1 taxon (Bernor, 1985) or, as is considered here to be more likely, resulted from a separate dispersal from North America at about 9.5 Ma.

Woodburne *et al.* (1981, Fig. 2), and MacFadden and Woodburne (1982, pp. 209-215) indicate that the stratigraphically oldest hipparion in the Siwalik deposits of Pakistan is *Cormohipparion* cf. *C. nagriensis*, based on dental remains referred by Hussain, 1971) to *Hipparion nagriensis*. The first occurrence of this species is (the "*Hipparion*" FAD) is calibrated at about 9.2 Ma following Badgley *et al.* (1986), contrary to an estimate of about 10 Ma (Berggren *et al.*, 1985, note 11). In the present report the age of the Vallesian/Turolian boundary is taken at 9.5 Ma (see "Conventions"), so this species is here considered to be of Turolian age. It still is the oldest species of *Cormohipparion* in the Siwaliks succession. Other species that probably are referable to *Cormohipparion* (*C. weihoense*, MacFadden and Woodburne, 1982, p. 193) are represented in China, Pakistan (*C. theobaldi*; MacFadden and Bakr, 1979), and western Russia (*C. giganteum*, Woodburne *et al.*, 1981, Fig. 2). These occurrences all appear to be of Turolian age. Referral of *C. nagriensis* and *C. theobaldi* to *Cormohipparion* is under debate in that very little is actually known of the cranial parameters of these species. It is possible that these are species of Group 1 "*Hipparion*," and had an endemic origin. On the other hand, except for a proportionately longer snout and somewhat more complex dentition (bifid plicaballins), *Cormohipparion weihoense* is virtually identical to specimens of *C. occidentale* and thus seems to have had an extrinsic origin from North America. Whether this stemmed from a *Cormohipparion* ancestor at about 9.5 Ma (e.g., the contemporaneous *C. occidentale*), or from a yet undocumented radiation that began with the earlier (ca. 12 Ma) dispersal of a form similar to *C. sphenodus* (proposed common ancestor of Group 1 and Group 2 hipparions; MacFadden and Woodburne, 1982, p. 192), is yet to be determined (see Fig. 12.3).

A further taxonomic complexity involves the Old World first occurence and origin of *Hipparion sensu stricto* Recognizing that taxonomic re-evaluations resulted in *H. antelopinum* of Woodburne *et al.* (1981, Fig 2 -- assigned to Group 1) being identified as a member of *Hipparion sensu stricto* (MacFadden and Woodburne, 1982, pp. 196-202), the stratigraphic first occurrence of Group 3 hipparions in Pakistan is calibrated at about 8 Ma (Woodburne et al., 1981, Fig. 2; MacFadden and Woodburne, 1982, p. 196). Although the Siwalik sucession is the best calibrated among Turolian-aged samples under discussion (e.g., MacFadden and Woodburne, 1982; Bernor and Hussain, 1985; Badgley et al.,, 1986), it is likely that *Hipparion sensu stricto* occurs elsewhere in geologically older sediments that locally range into the early Turolian, or ca. 9.5 Ma (Bernor et al., 1980; Woodburne et al., 1981, Fig. 2; using the Berggren, Kent, and Van Couvering, 1985, 9.5 Ma age for the Vallesian/Turolian boundary). These occurrences led those authors to propose a *Hipparion sensu stricto* dispersal from North America at about 10 Ma (= 9.5 Ma, revised, as above) from a separate (= not *Cormohipparion*) North American ancestor (taxa referred to *Hipparion sensu stricto*, e.g., MacFadden, 1980) now known to contain species as old as late Barstovian (*H. shirleyi*; MacFadden, 1984). Note the above discussion both for and against an endemic derivation of a Group 3 taxon from a member of Group 1 in Iran.

In any case, it is clear that the notion of a *Hipparion* Datum is useful only if carefully defined, and that uncritical use of such a concept, especially if extended regionally, would lead to conclusions that are very misleading as to both taxonomy and chronology. Finally, it has been suggested (Zhegallo, 1978; see MacFadden, 1984, for comment, and illustration) that a species of *Neohipparion* dispersed to the Old World about 8 Ma, as represented by certain specimens referred to the Asian species, "*Hipparion*" *houfenense*. Flynn and Bernor (1987) have suggested, however, that the putative specimens of "*H.*" *houfenense* are not actually referable to that species, and in any case are more likely an endemic derivative of an earlier Asian species, *Cormohipparion (Sivalhippus) perimense* (= "*Hipparion*" *feddeni* of MacFadden and Woodburne, 1982).

In summary, the prolific Old World record of hipparions seems best accounted for by introductions from North America at about 12 Ma (*Cormohipparion* to *Hippotherium*) and 9.5 Ma (*Hipparion* to *Hipparion*). The timing and mode of origin (endemic evolution or dispersal at about 9.5 Ma) remains too obscure to account for the Asian (including Chinese, see above) record of *Cormohipparion*. It is conceivable, if not necessarily likely, that Group 2 hipparions (with their distinctively narrow preorbital bar) represent a ca. 9.5 Ma dispersal from a form similar to *Merychippus insignis* (a species of *Neohipparion* would be too hypsodont), but the anteriorly subtriangular shape of the lacrimal in Group 2 taxa may be one (of a number of) characters that speak against this hypothesis. At the same time, the lack of a clearly stratigraphically-controlled cranial transformation series from any Group 1 taxon to forms representing Groups 2, 3, or 4 in the Old World, raises the question as to whether members of Groups 2, 3, or 4 could have resulted from additional (in taxonomy, but not necessarily in time) dispersals from North America. At the moment, no evidence has been developed to suggest the alternative, that certain North American taxa had an Old World origin.

## Provincialism of Old World hipparions

A major source of potential error that is encountered when one attempts to identify spheres of "influence" or provincialism among Old World hipparions, especially in Turolian or younger segments of geologic time, is the loss of distinctive synapomorphies. In one of the cases cited above [*C. (S.) perimense*; "*H.*" *houfenense*], the proposed ancestor has reduced, and its suggested derivative has lost, the DPOF (as well as attaining some disputed synapomorphies in their dentitions). When severe modifications are encountered that diverge from the morphology displayed by the older species of a lineage, or from species that otherwise have the morphology most typical of the lineage in question, determination of phyletic relationships and the interpretation of attendant evolutionary or dispersal patterns based on derived taxa becomes commensurately more difficult. As discussed, however, by Woodburne and MacFadden (1982), MacFadden and Woodburne (1982), Bernor and Hussain (1985), Bernor (1985) and Bernor et al. (1987), the following patterns seem to have been developed by Old World hipparions.

1. Vallesian. This time (ca. 12-9.5 Ma) witnessed the early entry and diversification of Group 1 hipparions, especially in western Europe and the circum-Mediterranean area. At the same time, it cannot be ruled out that early members of Group 2 hipparions and precursors of Turolian-aged Asian species of *Cormohipparion* were already present in Asia, at least (e.g., Woodburne and MacFadden, 1982) (Fig. 12.3). Similarly, the presence of forms ascribed to *Hippotherium* ("*Hipparion*") *primigenium* in the Ngorora Formation of East Africa (late Miocene to early Pliocene; Hooijer and Maglio, 1974; Churcher and Richardson, 1978, Fig. 20.7) indicate that this taxon or its African ancestor entered the African subcontinent in the (?late) Vallesian or early Turolian (Bernor et al., 1987; see Churcher and Richardson, 1978, pp. 388-390 who suggest that sub-Saharan African forms given this name may differ from their circum-Mediterranean representative, known either as "*H.*" *primigenium* or "*H.*" *africanum*). Whether or not accurately identified as "*H.*" *primigenium*, the persistence of taxa of this overall morphology into the early Turolian of Africa illuminates and is consistent with the notion that hipparionine evolution, at least, in sub-Saharan Africa proceeded in an autochthonous fashion for much of the later Miocene, Pliocene, and Pleistocene.

2. Early Turolian. This interval (ca. 9.5-7 Ma) saw a major increase in hipparion diversity (e.g., Bernor et al., 1980; see also Sen, Sondaar, and Staesche, 1978), as well as local endemism. At least three hipparion groups were present in western Europe and the circum-Mediterranean area. Group 4 horses may not be represented in strata of this age, with the possible exception of Iran (Bernor, 1985; "*H.*" *?matthewi*) or sub-Saharan Africa ("*H.*" *sitifense*, Churcher and Richardson, 1978). Group 2 horses were especially characteristic of western Asia at this time (e.g., Woodburne et al., 1981), while Pakistan may have represented, and China apparently did represent diversification centers for species of *Cormohipparion* (MacFadden and Woodburne, 1982; Bernor and Hussain, 1985; see also earlier). The evidence for the local endemism (above) during this time interval speaks in favor of a previously more widespread dispersal of ancestral stocks, especially for species of Group 2 and *Cormohipparion*. Group 4 taxa probably represent endemic derivations of only local significance at this time, as does the largely Mediterranean occurrence of Group 1 forms. Only *Hipparion sensu stricto* appears to have been widely represented at about 9 Ma (western Europe, Mediterranean region, Iran, Siwaliks, China; Woodburne et al., 1981, Fig. 2; MacFadden and Woodburne, 1982; Bernor and Hussain, 1985).

3. Late Turolian, ca. 7-5.2 Ma. This interval witnessed the near extinction of hipparions in western Europe and the circum-Mediterranean, but certain lineages flourished in Asia (especially China) and sub-Saharan Africa. Derivatives of "*H.*" *primigenium* persisted at Samos, Greece, as did representatives of Groups 2, 3, and 4 (e.g., Bernor and Hussain, 1985, Figs. 19-22).

Forms suggested here as being related to Group 1 taxa ("*H.*" *coelophyes*, and "*H.*" *dermatorhinum*), Group 2 ("*H.*" *richthofeni*), and Group 3 (*H. ptychodus*, and *H. fossatum*) appear to be present in China during this interval (Woodburne et al., 1981, Fig. 2). Churcher and Richardson (1978, pp. 388-402; Figs. 20.7, 20.8) indicate that local autochthones derived from "*H.*" *primigenium sensu lato* ("*H.*" *sitifense*), appear about this time, and that a possible immigrant from China (*H. turkanense*, a potential Group 3 representative; Woodburne et al., 1981) also occurs in African strata of this general age. Overall, the late Turolian saw persistent, but decreasingly abundantly represented, endemic centers of hipparionine lineages in the Old World, but with some suggestion of a dispersal connection between sub-Saharan Africa and Asia.

4. Ruscinian-Villafranchian, ca. 5.2-1.3 Ma. As seen in North America, *Nannippus* extended in time to the late Blancan, to locally coexist with *Dolichohippus simplicidens* (Skinner and Hibbard, 1972), a nearly modern equine. Similarly, in the Old World, the geologically youngest species of *Hipparion* coexisted locally in Europe and Africa from about 3.0 to 1.9 Ma with members of *Equus* (Lindsay et al., 1980; Eisenmann and Brunet, 1973; Keller et al., 1977; Opdyke et al., 1982; Berggren, Kent, and Van Couvering, 1985, p. 229). "*H.*" *afarense* and "*H.*" *baardi* apparently are derivatives of "*H.*" "*primigenium*" in Africa, having very complex cheek tooth dentitions, but lacking a DPOF (Hooijer and Maglio, 1974; Eisenmann, 1976a, 1976b; Churcher and Richardson, 1978), and "*H.*" *afarense* is suggested by Churcher and Richardson (1978) as having given rise to "*H.*" *libycum* (including *Stylohipparion*). This species, containing extremely hypsodont and complexly ornamented teeth (but lacking ectostylids in the lower cheek teeth) is represented in the Pleistocene of northern, eastern, and southern Africa, and apparently coexisted locally with *Equus* (*Dolichohippus*) *numidicus* or *Equus burchelli* as early as the latest Pliocene. In any case, hipparions persisted into the late Pleistocene of Africa (e.g., Churcher and Richardson, 1978), having become extinct in all other places in the world prior to then.

## Summary

The purpose of this summary is to distill evidence for an overview of trends in the development of dental complexity, or its simplification, and in the degree of expression of the DPOF in hipparionine horses. Individual hipparion species possess, of course, synapomorphies in individual dental complexity, DPOF morphology, cranial parameters, nasal notch retraction, overall size and (where definitely associated) limb proportions. Discussion of these features is not attempted here. An important point to keep in mind is that the world's fossil equid record does not commonly provide definite associations of crania, isolated teeth, mandibles, or postcranial elements.

This summary assumes that hipparions are a clade, and that the basic cranial configuration probably was similar to that of *Neohipparion coloradense* or, to suggest an out-group taxon, *Merychippus insignis*. In these species, the DPOF is distinctly present, has weak but nevertheless discernible dorsal, posterior, and ventral boundaries, but lacks definition anteriorly, and is weakly pocketed posteriorly. In contrast, however, to the precociously hypsodont dentition of *Neohipparion coloradense*, the mesodont cheek tooth dentition of other early (= Barstovian) hipparions and contemporaneous *Merychippus insignis* suggests that this mesodont state was also the basic condition for the ancestral hipparionine. Put another way, the cheek tooth dentition of the ancestral hipparionine was of "merychippine grade" (Figs. 12.6C, F, G).

To sketch briefly a possible scenario of descent, it is clear first of all that the ancestral hipparionine is unknown or unrecognized in the North American fossil record. Hulbert (this volume) suggests that *Parahippus leonensis* is an out-group taxon to all merychippine, and thus hipparionine, taxa. Based on the age of occurrence of at least four of the five North American hipparionine genera, their common ancestor must have existed in the early Barstovian, if not earlier. Based on the above considerations, the first hipparionine probably resembled *M. insignis* in size, overall cranial morphology, and cheek tooth crown height, length and morphology of the DPOF, length of preorbital bar, and shape of lacrimal bone (e.g., Figs. 12.5A, 12.6C; Table 12.2). In this context, the main difference between taxa of such a "merychippine grade" and the early hipparionine must have been the tendency in the latter to increase the degree to which the protocone was a separate structure, increase the distinctiveness of the pre-protoconal groove, to suppress the development of a hypoconal facette, and to enhance the size of the metastylid.

Even though plesiomorphic in many cranial characters, the line leading to *Pseudhipparion* and *Neohipparion* must have diverged early from the remainder of the hipparionine complex in its precocious hypsodonty (even though primitive species of *Pseudhipparion* are no taller than 30 mm, this still is more hypsodont than seen in *M. insignis*; Table 12.2).

The remainder of the hipparionine clade apparently can be distinguished from the "merychippine grade" and from *Neohipparion* in the early development of a subtriangularly shaped lacrimal, and more strongly developed DPOF. Early species of *Cormohipparion* are distinguished from early members of *Hipparion* in having a larger, more distinctly circumscribed, anteriorly located, and more deeply pocketed DPOF. *Nannippus* apparently at first resembled *Cormohipparion* in these features, but presumably differed from the latter in an early development of small size, greater retraction of the nasal notch, and precocious hypsodonty. If the concept of *Neohipparion* utilized by MacFadden (1984) is employed, and if the species of that group are allied with species of *Pseudhipparion* as indicated by Webb and Hulbert (1986), then species of both genera secondarily and independently achieved a subtri-angularly shaped lacrimal bone, having retained, and then further reduced, a relatively weakly developed DPOF.

In any case, by the end of the Barstovian (ca. 13 Ma), four if not all (*Nannippus*) five North American genera had evolved. Most (if one presumes the condition in *Nannippus*) Barstovian taxa were moderately hypsodont (except *Hipparion shirleyi* and unnamed species of *Pseudhipparion*). As indicated by MacFadden (1984), there was a subseqent general increase in hypsodonty in all genera, most extremely developed in species of *Nannippus* and *Pseudhipparion* (Webb and Hulbert, 1986). In almost all taxa, the upper cheek tooth dentition was moderately complex (Table 12.1, in part), or became simplified from that condition. The major exception is seen in species of *Cormohipparion*, where an early trend in increasing dental complexity is noted in *C. sphenodus* (late Barstovian-early Clarendonian). This trend culminates in *C. emsliei* (late Hemphillian and Blancan; Hulbert, 1987b) in which the dentition is nearly as complex as found in the much earlier (Vallesian) *Hippotherium primigenium*.

None of the North American hipparions undergo cranial modifications that rival those exemplified in some Old World taxa (e.g., "*Hipparion*" *proboscideum*; Fig. 12.4D; Table 12.1) but, similar to Old World taxa, early members of all North American hipparionine genera are characterized as having a DPOF that is either present or large, including *Nannippus* (Hulbert, 1987c). In contrast to North American species of *Hipparion* (in which the fossa configuration remains relatively stable) the DPOF is reduced or lost in all other genera, but at different times. The oldest species with a reduced DPOF for the various genera are: *Neohipparion affine*, late Barstovian; *Pseudhipparion retrusum*, late Barstovian; *Nannippus peninsulatus*, Blancan (although pertinent crania are not known for Hemphillian and late Clarendonian taxa); *Cormohipparion ingenuum*, late Clarendonian). Species of *Pseudhipparion* are unique among hipparionines in developing a malar fossa.

The situation in the Old World is more complex and major lineages less well understood than in North America, but many comparable patterns are seen. The Old World record of hipparions begins about 12 Ma with forms (*Hippotherium primigenium* and other members of Group 1) that are of large size, have extremely complex dentitions, and a large, very well developed DPOF (Table 12.1). Specimens of Turolian age (beginning ca. 9.5 Ma) that are referred to this group show certain reduction in fossa expression and reduced dental complexity [e.g., "*Hipparion*" *gettyi*, Maragheh, Iran (Bernor, 1985); "*Hipparion*" sp., Samos, Greece (Woodburne and Bernor, 1980, Fig. 6C); also Table 12.1 of this report]. Turolian-aged taxa from Pakistan (Siwaliks) either part of or phyletically near this group show similarly reduced dental complexity and fossa size but (with other characters) have a synapomorphically very wide preorbital bar [e.g., "*Hipparion*" *feddeni*, MacFadden and Woodburne, 1982, = "*Cormohipparion*" (*Sivalhippus*) *perimense* of Bernor and Hussain, 1985]. Other Siwalik taxa [*Cormohipparion theobaldi* and *C*. cf. *C. nagriensis*, MacFadden and Woodburne, 1982, designated as "*C*." (*S*.) *theobaldi* and "*C*." (*S*.) sp. in Bernor and Hussain, 1985], and *C. weihoense* (China) seem to be somewhat plesiomorphic in having a very large and strongly bounded DPOF, and relatively complex dentition, but it is now uncertain whether some of these species had a Group 1 ancestor or stemmed from a separate *Cormohipparion* species.

The ancestry of Group 2 hipparions, which appear abruptly in the Turolian, is similarly uncertain. The complex dentition (but not as complex as in Vallesian Group 1 taxa) is relatively unchanged throughout the known (Turolian) history of the group, except for becoming significantly longer in "*H*." *proboscideum*. Group 2 hipparionines do not follow the general trend for fossa reduction, but add an additional depression between the buccinator fossa and the DPOF (see below for potential affiliation of Group 4 taxa with Group 2). No matter whether of immigrant origin at about 9.5 Ma or endemically derived, *Hipparion sensu stricto* in the Old World resembles its North American congener in expression of the DPOF and in dental complexity. Late Turolian examples (*H. dietrichi* and *H. campbelli*) show reduction of the DPOF, in contrast to their North American and Eurasian relatives.

Group 4 hipparionines are poorly understood, but taxa included in that group (all of Turolian age, e.g., "*H*." *matthewi*, "*H*." *minus*) rival North American *Nannippus* in achieving diminutive size, DPOF reduction, and relatively great dental hypsodonty. If some or all of these taxa are allied with Group 2, then the pattern for that group would be altered accordingly.

Finally, regardless of the actual age of dispersal of hipparions into sub-Saharan Africa, the record there shows persistence of an "*H*." *primigenium*-like group of species and (later) development of increased dental complexity coupled with reduction and loss of the DPOF ("*H*." *afarense*, "*H*." *baardi*, "*H*." *libycum*). A comparable pattern is paralleled elsewhere only in *Cormohipparion* of North America.

The following points may be summarized from the foregoing discussion:

1. A large DPOF is a derived state in hipparionines, but once achieved it may be lost. In fact, the only group or genus in which some degree of fossa reduction is not seen is Old World Group 2 (*sensu stricto*), where an additional fossa is added between the DPOF and the buccinator fossa. The North American genus, *Pseudhipparion*, is unique in adding a malar fossa. The geologically earliest instance of reduction of the fossa is shown by early Clarendonian species of *Neohipparon* and late Barstovian species of *Pseudhipparion*. New World species of *Hipparion* do not reduce the fossa, but Old World species do (at about 8 Ma). It appears that at about 7-8 Ma, all hipparionines that still were extant had effectively lost the DPOF.

2. Concurrent with effectively achieving hypsodonty, all hipparionines developed at least a moderately complex cheek tooth enamel pattern. Species of *Pseudhipparion* have (?secondarily achieved) the most simple cheek tooth dentition among hipparionines. *Hippotherium primigenium* and related Old World Vallesian taxa have the greatest dental complexity, and all Old World hipparionines effectively show decreased

enamel complexity in the Turolian (excluding species in sub-Saharan Africa). Except for species of small size, most Old World hipparionines are dentally more complex than contemporaneous North American taxa. *Cormohipparion emsliei* (late Hemphillian and Blancan) appears to be an exception; in the Old World, taxa with very complex dentitions persist into the Pleistocene of sub-Saharan Africa. *Pseudhipparion simpsoni* of the North American late Hemphillian is the most hypsodont hipparionine, having achieved incipient hypselodonty. The dentition is very simple, however.

## Acknowledgments

I wish to thank Richard H. Tedford and John A. Van Couvering, American Museum of Natural History, New York, for organizing the 1981 "*Hipparion* Symposium" and for conceiving and developing the successful proposal to the National Science Foundation (EAR 81-10870) that allowed the various European and North American workers to travel to New York and supported their attendance at the symposium. The American Museum of Natural History graciously made its facilities available for this purpose and in ways large and small promoted the work that resulted. The first of what is hoped to be a series of publications is now underway (Eisenmann *et al.*, 1988). R. L. Bernor and B. J. MacFadden read an initial draft of the manuscript and offered helpful comments. Illustrations were prepared by Mrs. Linda Bobbitt, UCR. Some of the work conducted for this report was subsidized by NSF grant BSR 17396 to R.L. Bernor. This support is gratefully acknowledged.

## Bibliography

Abel, O. (1926): Geschichte der Equiden auf dem Boden Nordamerikas. - *Zool.-Bot. Gesellschaft, Wien, Verhandl.*, 74: 159-164.

Alberdi, M.T. (1974a): El genero *Hipparion* en España. Nuevas formas de Castilla y Andalucia, revision e historia evolutiva. Trabos sobre Neogeno-Cuaternario, Seccion de Paleontologia de Vertebrados y Humana. -*Inst. "Lucas Mallada" del C.S.I.S., Madrid*, 1:1-146.

Alberdi, M.T. (1974b): Descripcion del primer craneo de *Hipparion* procedente del Neogene de Teruel (España). -*Bol. R. Espan. Hist. Nat. (Geol.), Madrid*, 72:5-14.

Arambourg, C. (1959): Vertébrés continentaux du Miocenè supérieur de l'Afrique du Nord. -*Service Carte Gèol. de Algèrie Mèm.* (n.s.) 4:1-159.

Badgley, C., Tauxe, L., and Bookstein, F. 1986. Estimating the error of age interpolation in sedimentary rocks. -*Nature*, 319: 139-141.

Berggren, W.A., and Van Couvering, J.A. (1974): The late Neogene; Biostratigraphy, geochronology and paleoclimatology of the last 15 million years in marine and continental sequences. -*Palaeogeogr., Palaeoclimat., Palaeoecol.*, 16 (1- 2):1-216.

Berggren, W.A., and Van Couvering, J.A. (1978): Geochronology. - *Amer. Assoc. Petrol. Geol. Stud. Geol.*, 6: 39-55.

Berggren, W.A., Kent, D.V., Flynn, J.J., and Van Couvering, J.A. (1985): Cenozoic geochronology. -*Geol. Soc. Amer. Bull.*, 96: 1407-1418.

Berggren, W.A., D.V. Kent, and Van Couvering, J.A. (1985): Neogene geochronology and chronostratigraphy, In: N.J. Snelling, (ed.), The Chronology of the Geological Record. -*Geol. Soc. London Mem.* 10: 211-260.

Bernor, R.L. (1985): Systematics and evolutionary relationships of the hipparionine horses from Maragheh, Iran (late Miocene, Turolian Age). -*Palaeovertebrata*, 15 (4): 173-269.

Bernor, R.L., and Hussain, S.T. (1985): An assessment of the systematic, phylogenetic and biogeographic relationshps of Siwalik hipparionine horses. -*J. Vert. Paleont.*, 5(1):32-87.

Bernor, R.L., Qiu, Zh., and Tobien, H. (1987): Phylogenetic and biogeographic bases for an Old World hipparionine horse geochronology. -*Ann. Inst. Geol.*

*Publ. Hung.*, 70: 43-53.
Bernor, R.L., Woodburne, M.O., and Van Couvering, J.A. (1980): A contribution to the chronology of some Old World Miocene faunas based on hipparionine horses. -*Géobios*,13:25-47.
Borissiak, A.K. (1914): Mammiféres fossiles d'Sebastopol. I. -*Trudy geol. komiteta*, (2)87:1-153.
Christol, J. de (1832): [No title; description of *Hipparion*]. -*Ann. Sci. Indust. Midi de la France*, 1:180-181.
Churcher, C.S., and Richardson, M.L. (1978): Equidae, In: V.J. Maglio and H.B.S. Cooke, (eds.), *Evolution of African Mammals*. Cambridge (Harvard University Press), pp. 379-422.
Colbert, E.H. (1935): Siwalik mammals in the American Museum of Natural History. -*Trans. Amer. Phil. Soc.*, 26:1-401.
Cope, E.D. (1885): Pliocene horses of southwestern Texas. -*Amer. Nat.*, 19: 1208-1209.
Cope, E.D. (1888): The phylogeny of horses. -*Amer. Nat.*, 22:448-449.
Cope, E. D. (1889): A review of the North American species of *Hippotherium*. -*Proc. Amer. Phil. Soc.*, 26: 429-458.
Cope, E.D. (1893): A preliminary report on the vertebrate paleontology of the Llano Estacado. - *Ann. Rept., Geol. Surv.Texas*, 4: 1-136.
Dalquest, W.W., and Donovan, T.J. (1973): A new three-toed horse (*Nannippus*) from the late Pliocene of Scurry County, Texas. -*J. Paleont.*, 47: 34-45.
Dalrymple, G.B. (1979): Critical tables for conversion of K-Ar ages from old to new constants. -*Geology*, 7: 558-560.
Edwards, S.W. (1982): A new species of *Hipparion* (Mammalia, Equidae) from the Clarendonian (Miocene) of California. -*J. Vert. Paleont.*, 2: 173-183.
Eisenmann, V. (1976a): Equidae from the Shungura Formation. In: Coppens, Y. *et al.*, (eds.), *Earliest Man and environments in the Lake Rudolf Basin*. Chicago (Univ. Chicago Press), pp. 225-233.
Eisenmann, V. (1976b): Nouveaux crânes d'hipparions (Mammalia, Perissodactyla) Plio-Pleistocenes d'Afrique Orientale (Ethiopie et Kenya): *Hipparion* sp., *Hipparion* cf. *ethiopicum*, et *Hipparion afarense* nov. sp. -*Géobios*, 9(5): 577-605.
Eisenmann, V. (1981): Les caractères des cranes *d'Hipparion sensu lato* (Mammalia, Perissodactyla) et leur interprétation. -*C. R. Hebd. Sèances, Série D, Sci. Natur., Paris*, 293: 735-738.
Eisenmann, V. (1982): La phylogènie des *Hipparion* (Mammalia, Perissodactyla) d'Afrique d'après les caractères crâniens. -*Proc. Konink. Ned. Akad. van Wetensch.*, ser. B 85(2): 219-227.
Eisenmann, V., Alberdi, M. T., de Giuli, C., and Staesche, U. (1988): *Studying Fossil Horses*.-Leiden (E.J. Brill), pp. 1-72.
Eisenmann, V., and Brunet, M.J. (1973): Présence simultanée de cheval et d'Hipparion dans le Villafrancien moyen de France a Rocca-Neyra. -*International colloquium on the boundary between Neogene and Quaternary. Collected Papers*, 4: 104-122.
Eisenmann, V., Sondaar, P.Y., Alberdi, M.T., and De Giuli, C. (1986): Is horse phylogeny becoming a playfield in the game of theoretical evolution? Essay Review. -*J. Vert. Paleont.*, 7(2): 224-229.
Engesser, B. (1979): Relationships of some insectivores and rodents from the Miocene of North America and Europe.-*Bull. Carn. Mus. Nat. Hist.*, 14: 1-68.
Evander, R.L. (1986): The taxonomic status of *Merychippus insignis* Leidy. -*J. Paleont.*, 60(6): 1277-1280.
Flynn, L.J., and Bernor, R.L. (1987): Late Tertiary mammals from the Mongolian People's Republic.-*Amer. Mus. Novit.*, 2872: 1-16.
Forstén, A. (1968): Revision of the Palearctic *Hipparion*. -*Acta Zool. Fenn.*, 119: 1-134.
Forstén, A. (1972): *Hipparion primigenium* from southern Tunisia. -*Notes du Serv. Géol.*, 35: 7-28.

Forstén, A. (1978): *Hipparion primigenium* (v. Meyer, 1829), an early three-toed horse. -*Ann. Zool. Fenn.*, 15: 298-313.

Forstén, A. (1982): The taxonomic status of the Miocene horse genus *Sinohippus*. - *Palaeont.*, 25(3): 673-679.

Gabunia, L. (1981): Traits essentiels de l'évolution des faunes de Mammifères néogènes de la région mer Noire-Caspienne. -*Bull. Mus. Nat., d'Hist. Natur.*, 4 (3): 197-204.

Gaudry, A. (1862): *Animaux fossiles et géologie de l'Attique*. Paris.

Gaudry, A. (1867): *Animaux Fossiles et Géologie de l'Attique*. Paris (Libraire de Société géologique de France).

Gaudry, A. (1873): *Animaux fossiles du Mont Léberon (Vaucluse): Étude sur les vertébrés*. Paris (Librarie de Société géologique de France).

Gervais, P. (1849): Note sur la multiplicité des espèces d'hipparions (genre de chevaux à trois doits) qui sont enfouis à Cucuron (Vaucluse). -*C. R. Acad. Sci. Paris*, 29: 284-286.

Gervais, P. (1859): Sur une nouvelle espèce d'Hipparion (H. crassum) découverte auprès de Perpignan. -*C. R. Acad. Sci. Paris*, 48: 1117-1118.

Gregory, W.K. (1920): Studies in comparative myology and osteology, V. On the anatomy of the preorbital fossae of Equidae and other ungulates. -*Bull. Amer. Mus. Nat. Hist.*, 42(3): 265-284.

Gromova, V. (1952): Le genre *Hipparion*. - *Acad. Sci. U.S.S.R., Inst. Paleont.*, 36 (in Russian, 1952). French translation by St. Aubin, Bur. Rech. Minér., Géol., Ann. C.E.D.P. 12:1-473.

Hensel, R.F. (1861): Über *Hipparion mediterraneum*. -*Akad. Wiss. Berlin, Abh.*: 27-122.

Hooijer, D.A. (1975): Miocene to Pleistocene hipparions of Kenya, Tanzania and Ethiopia. -*Zool. Verhandl. Rijksmus. Nat. Hist. Leiden*, 142:1-80.

Hooijer, D.A., and Maglio, V.I. (1974): Hipparions from the late Miocene and Pliocene of northwestern Kenya. -*Zool Verhandl. Rijksmus. Nat. Hist. Leiden* (B), 76(4):311-315.

Hulbert, R.C. (1982): Population dynamics of the three-toed horse *Neohipparion* from the late Miocene of Florida.- *Paleobiology*, 8:159-167.

Hulbert, R.C. (1987a): Late Neogene *Neohipparion* (Mammalia, Equidae) from the Gulf Coastal Plain of Florida and Texas. -*J. Paleont.*, 61: 809-830.

Hulbert, R.C. (1987b): A new *Cormohipparion* (Mammalia, Equidae) from the Pliocene (latest Hemphillian and Blancan) of Florida. -*J. Vert. Paleont.*, 7: 451-468.

Hulbert, R.C. (1987c): *Cormohipparion* and *Hipparion* (Mammalia, Perrisodactyla, Equidae) from the late Neogene of Florida. -*Florida St. Mus., Bull. Biol. Sci.*, 33: 229-338.

Hulbert, R. C. (1989): Phylogenetic interrelationships and evolution of North American late Neogéne Equinae (this volume).

Hussain, S. T. (1971): Revision of *Hipparion* (Equidae, Mammalia) from the Siwalik Hills of Pakistan and India. - *Bayer. Akad. Wiss., Abhandl.*, 147: 1-68.

Kaup, J. J. (1835): Die zwei urweltlichen pferdeartigen Thiere, welche im Tertiären Sande bei Eppelsheim gefunden werden, bilden eine eigene Unter-Abtheilung der Gattung Pferd, in der Zahl der Fingerglieder den übergang zur Gattung *Palaeotherium* macht, und zwischen diese und Pferd zu stellen ist. -*Kaiserliche Leopoldinisch-Carolinischen Akad. Naturforscher, Verhandl.*, 9: 172-182.

Keller, H.M., Tahirkheli, R.A.K., Mirza,M.A., Johnson, G.D., Johnson, N.M., and Opdyke, N.D. (1977): Magnetic-polarity stratigraphy of the Upper Siwalik deposits, Pabbi Hills, Pakistan. -*Earth Planet. Sci. Lett.*, 36: 187-201.

Leidy, J. (1856): Notices of some remains of extinct Mammalia recently discovered by Dr. F.V. Hayden, in the badlands of Nebraska. -*Proc. Acad. Nat. Sci.*,

*Phila.*, 8: 59.
Leidy, J. (1869): The extinct mammalian fauna of Dakota and Nebraska. -*J. Acad. Nat. Sci., Phila.*, ser. 2, 7: 286.
Leidy, J. (1885): *Rhinoceros* and *Hippotherium* from Florida. -*Proc. Acad. Nat. Sci., Phila.*, 37: 32-33.
Leidy, J. (1887): Fossil bones from Florida. -*Proc. Acad. Nat. Sci., Phila.*, 39: 309-310.
Lindsay, E.H., Opdyke, N.D., and Johson, N.M. (1980): Pliocene dispersal of the horse *Equus* and late Cenozoic mammalian dispersal events. -*Nature*, 287:135-139.
Lippolt, H.-J., Gentner, W., and Wimmenauer, W. (1963): Altersbestimmungen nach der Kalium-Argon-Methode an tertiären Eruptivgesteinen Sudwestdeutschlands. -*Jahrb. Geol. Landes. Baden-Württemberg*, 6: 507-538.
Liu, T. S., Li, C. K., and Zhai, R. J. (1978): Pliocene vertebrates of Lantian, Shensi. In: Tertiary mammalian fossils of the Lantian district Shensi. Part II. -*Chin. Acad. Geol. Sci., Prof. Papers of Strat. Paleont.*, 7: 140-199.
Lydekker, R. (1877a): Notices of new and other Vertebrata from Indian Tertiary and Secondary rocks. -*Geol. Surv. India, Rec.*, 10: 30-42.
Lydekker, R. (1877b): Notices of new or rare mammals from the Siwaliks. -*Geol. Surv. India, Rec.*, 10: 76-83.
Lydekker, R. (1882): Siwalik and Narbada Equidae. -*Paleontologica Indica* (X), part 3, 2: 67-98.
Lydekker, R. (1884): Additional Siwalik Perissodactyla and Proboscidea. -*Paleontologica Indica* (X), part 1, 3: 1-34.
Lydekker, R. (1885): *Catalogue of Siwalik Vertebrata contained in the geological department of the Indian Museum. Part 1. Mammalia.* Calcutta.
MacFadden, B. J. (1980): The Miocene horse *Hipparion* from North America and from the type locality in southern France. -*Palaeont.*, 23: 617-635.
MacFadden, B. J. (1984): Systematics and phylogeny of *Hipparion, Neohipparion, Nannippus*, and *Cormohipparion* (Mammalia, Equidae) from the Miocene and Pliocene of the New World. -*Bull. Amer. Mus. Nat. Hist.*, 179 (1): 1-195.
MacFadden, B. J. (1986): Systematics, phylogeny, and evolution of fossil horses: a rational alternative to Eisenmann et al. 1987. -*J. Vert. Paleont.*, 7(2): 230-235.
MacFadden, B.J., and Bakr, A. (1979): The horse *Cormohipparion theobaldi* from the Neogene of Pakistan, with comments on Siwalik hipparions. -*Palaeont.*, 22: 439-447.
MacFadden, B.J., and Skinner, M.F. (1981): Earliest Holarctic hipparion, *Cormohipparion goorisi* n. sp., (Mammalia, Equidae), from the Barstovian (medial Miocene) Texas Gulf Coastal Plain. -*J. Paleont.*, 55: 619-627.
MacFadden, B.J., and Skinner, M.F. (1982): Hipparion horses and modern phylogenetic interpretation: comments on Forstén's view of *Cormohipparion*. -*J. Paleont.*, 56: 1336-1342.
MacFadden, B.J., and Woodburne, M.O. (1982): Systematics of the Neogene Siwalik hipparions (Mammalia, Equidae) based on cranial and dental morphology. -*J. Vert. Paleont.*, 2(2): 185-218.
Matthew, W.D. (1929): Critical observations upon Siwalik mammals (exclusive of Proboscidea). -*Bull. Amer. Mus. Nat. Hist.*, 56: 437-560.
Merriam, J.C. (1915): New species of the Hipparion group from the Pacific Coast and Great Basin provinces of North America. -*Univ. Calif. Publ. Geol. Sci.*, 9: 1-8.
Merriam, J.C. (1916): Mammalian remains from the Chanac Formation of the Tejon Hills, California. -*Univ. Calif. Publ. Geol. Sci.*, 10: 111-127.
Meyer, H. von (1829): Taschenbuch für die gesammte Mineralogie. -*Zeit. Miner.*, 23: 150-152.
Opdyke, N.S., Johnson, N.M., Johnson, G.D., Lindsay, E.H., and Tahirkheli, R.A.K. (1982): Paleomagnetism of Middle Siwalik Formations of northern Pakistan

and rotation of the Salt Range decollement. -*Palaeogeogr., Palaeoclimat., Palaeoecol.*, 37: 1-15.
Opdyke, N.S., Lindsay, E.H., Johnson, G.D., Johnson, N.M., Tahirkheli, R.A.K., and Mirza, M.A. (1979): Magnetic-polarity stratigraphy and vertebrate paleontology of the Upper Siwalik Subgroup of northern Pakistan. - *Palaeogeogr., Palaeoclimat., Palaeoecol.*, 27: 1-34.
Osborn, H.F. (1918): Equidae of the Oligocene, Miocene, and Pliocene of North America. Iconographic type revision. -*Mem. Amer. Mus. Nat. Hist.*, n.s., 2: 1-326.
Pavlov, M. (1890): Étude sur l'histoire palèontologique des ongules. IV. *Hipparion* de Russie. -*Bull Soc. Nat. Moscou*, 2(3): 683-716.
Pilgrim, G.E. (1910): Notices of new mammalian genera and species from the Tertiary of India. -*Geol. Surv. India, Rec.*, Part I,40: 63-71.
Pirlot, P.R. (1956): Les formes Européenes du genre *Hipparion*. -*Inst. Geol., Barcelona, Mem. y Commun.*, 1-123.
Quinn, J. H. (1955): Miocene Equidae of the Texas Gulf Coastal Plain.--*Bur. Econ. Geol., Univ. Texas Publ.* 5516: 1-102.
Qiu, C. S., Huang, W. L., and Kuo, Z. H. 1980. Notes on the first discovery of the skull of *Hipparion houfenense*. -*Vert. Palasiatica*, 18(2): 131-137.
Richey, K.A. (1948): Lower Pliocene horses from Black Hawk Ranch, Mount Diablo, California. -*Univ. Calif. Publ. Geol. Sci.*, 28: 1-44.
Roth, J., and Wagner, A. 1855. Die fossilen Knochenüberreste von Pikermi in Griechenland. -*Bayer. Akad. Wiss., Abh.*, 7: 361-464.
Sefve, I. (1927): Die Hipparionen Nord-Chinas. -*Paleontographica Sinica*, (C)4: 1-91.
Sellards, E.H. (1916): Fossil vertebrates from Florida. A new Miocene fauna, new species, the Pleistocene fauna. -*Ann. Rept. Flor. Geol. Surv.*, 8: 77-119.
Sen, S., Sondaar, P.Y., and Staesche, U. (1978): The biostratigraphic applications of the genus *Hipparion* with special references to the Turkish representatives. -*Proc. Konink. Ned. Akad. Wetensch., Amsterdam*, Series B. 81(3): 370-385.
Skinner, M.F., and Hibbard, C.W. (1972): Early Pleistocene pre-glacial and glacial rocks and faunas of north-central Nebraska. -*Bull. Amer. Mus. Nat. Hist.*, 146: 1-148.
Skinner, M.F., and MacFadden, B.J. (1977): *Cormohipparion* n. gen. (Mammalia, Equidae) from the North American Miocene (Barstovian-Clarendonian). -*J. Paleont.*, 51: 912- 926.
Skinner, M.F., and Taylor, B.E. (1967): A revision of the geology and paleontology of the Bijou Hills, South Dakota. -*Amer. Mus. Novitates*, 2300: 1-53.
Sondaar, P.Y. (1961): Les hipparion de l'Aragón meridional. -*Estud. Geol.*, 17: 209-305.
Sondaar, P.Y. (1971): The Samos *Hipparion*. -*Konink. Ned. Akademie Wetench. f. Natuurk.*, 25: 417-441.
Sondaar, P.Y. (1974): The *Hipparion* from the Rhone Valley. -*Géobios*, 7: 289-306.
Steiger, R.H., and Jaeger, E. (1977): Subcommission on geochronology: convention on the use of decay constants in geo- and cosmochronology. -*Earth Planet. Sci. Lett.*, 36: 359-362.
Stirton, R.A. (1940): Phylogeny of North American Equidae. -*Univ. Calif. Publ. Geol. Sci.*, 24(4): 165-198.
Studer, T. (1911): Eine neue Equidenform aus dem Obermiocän von Samos. -*Deut. Zool. Gesellsch., Abh.*, 20- 21: 192-200.
Tedford, R.H., Galusha, T., Skinner, M.F., Taylor, B.H., Fields, R.W., Macdonald, J.R., Rensberger, J.M., Webb, S.D., and Whistler, D.P. (1987): Faunal succession and biochronology of the Arikareean through Hemphillian interval (late Oligocene through earliest Pliocene Epochs), North America.--In: Woodburne, M.O. (ed.): *Cenozoic mammals of North America: geochronology and biostratigraphy*. Berkeley (Univ.

Calif. Press), pp. 153-210.
Tobien, H. (1959): Hipparion-Funde aus dem Jungtertiär des Höwenegg (Hegau). -*Aus der Heimat*, 67: 121-132.
Tobien, H. (1968): Paläontologische Ausgrabungen nach jungtertiären Wirbeltieren auf der Insel Chios (Griechenland) und bei Maragheh (NW-Iran). - *Jarhb. Ver. "Freunde der Univ. Mainz"* 17: 51-58.
Tobien, H. (1982): Preliminary report on the Equidae (Perissodactyla, Mammalia) form the Sahabi Formation (Libya). -In: Boas, N. T., Gatity, A.W., Henzelin, J, and El-Arnuti, A. (eds.): *Results from the International Sahabi Resarch Project (Geol. and Paleont.).- Garyounis Sci. Bull., Spec. Issue*, 4: 83-85.
Tobien, H. (1986): Die jungtertiäre Fossilgrabunsgsstätte Höwenegg im Hegau (Südwestdeuthchland). -*Ein Statusbericht. Carolinea*, 44: 9-34.
Wagner, A. (1850): Urweltliche Säugethier-Ueberreste aus Griechenland. - *Bayer. Akad. Wiss., Abh.*, 5: 335-378.
Webb, S.D., and Hulbert, R.C., Jr. (1986): Systematics and evolution of *Pseudhipparion* (Mammalia, Equidae) from the late Neogene of the Gulf Coastal Plain and the Great Plains. -*Univ. Wyo. Cont. Geol., Spec. Paper* 3: 237-272.
Wehrli, H. (1941): Beitrag zur Kenntnis der "Hipparionen" von Samos. -*Paläont. Zeit.*, 22: 321-386.
Wiman, S.K. (1978): Mio-Pliocene foraminiferal biostratigraphy and stratochronlogy of central and northeastern Tunisia. -*Rev. Española de Micropaleont.*, X(1): 87-143.
Woodburne, M.O., and Bernor, R.L. (1980): On superspecific groups of some Old World hipparonine horses. -*J. Paleont.*, 54: 1319-1348.
Woodburne, M. O., and MacFadden, B. J. (1982):
Woodburne, M.O., MacFadden, B.J., and Skinner, M.F. (1981): The North American "*Hipparion*" datum and implications for the Neogene of the Old World. -*Géobios*, 14: 493-524.
Zhegallo, V.I. (1978): [The hipparions of central Asia.] -*Trans. Joint Soviet-Mongolian Paleont. Exped.*, 7: 1-151. (In Russian).
Zhai, R. J. (1962): On the generic character of "*Hypohippus zitteli.*" -*Vert. Palasiatica*, 6: 48-55.

# 13. A REVIEW OF OLD WORLD HIPPARIONINE HORSES

## MARIA-TERESA ALBERDI

All Old World hipparionines are included within the genus *Hipparion sensu lato*, and among Eurasian and African forms, six morphotypes are distinguished. Five of these morphotypes follow a basic "hipparionine" pattern, whereas one morphotype exhibits "caballine" characteristics. Each of these morphotypes corresponds to a complex of nominal species. Indeed, there has been an unjustifiable proliferation of species-level names in the hipparionine literature. Hipparionines evolved from North American "*Merychippus*" and immigrated to the Old World in a major wave at about 12.5 Ma, and possibly another after 5.5 Ma. To a certain extent, parallel evolutionary trends are seen among the various lineages of hipparionines, corresponding to the six basic hipparionine morphotypes. Such evolutionary trends can be provisionally correlated with climatic trends and sea level fluctuations. Ancestor-descendant relations can also be posited among the morphotypes. For instance, morphotype 1 probably gave rise, either directly or indirectly, to all the other hipparionine morphotypes.

## Introduction

The group commonly known as "hipparionines" includes the genus *Hipparion* and the various *Hipparion*-like forms. Hipparionines evolved from North American "*Merychippus*," a tridactyl horse known from the middle and upper Miocene. Hipparionines have been classified as a tribe of the subfamily Equinae. The genera of hipparionines include:

1. *Hipparion* de Christol, 1832 (= *Hippotherium* Kaup, 1833). This genus was originally described from Europe (type locality = Mont Léberon, France). *Hipparion* fossils are abundant in continental mammal faunas of Vallesian, Turolian, Ruscinian, and lower Villafranchian ages (ca. 12.5 to 2.5 Ma, according to Berggren and Van Couvering, 1974) in Europe, Asia, and Africa. In Africa *Hipparion* seems to have survived until about 400,000 years ago (Eisenmann, 1979a: 280). In North America, *Hipparion* probably ranged from the middle Barstovian to the early Hemphillian (ca. 14 to 6.5 Ma). Three species are recognized within this genus (MacFadden, 1984, 1985).

2. *Neohipparion* Gidley, 1903, from the middle Barstovian to the latest Hemphillian (ca. 15 to 4.5 Ma) of Central and North America. Six valid species of *Neohipparion* are presently recognized (MacFadden, 1984, 1985).

3. *Nannippus* Matthew, 1926, from the latest Clarendonian to the later Blancan (ca. 10 to 2 Ma) of Central and North America, composed of four valid species (MacFadden, 1984, 1985).

4. *Cormohipparion* Skinner and MacFadden, 1977, originally named for forms from North America and later was also recognized from the Old World (MacFadden and Bakr, 1979; Woodburne and Bernor, 1980). Known from the early Barstovian to the early Hemphillian (ca. 15.5 to 8.5 Ma), three species are recognized (MacFadden, 1985). According to Woodburne *et al.* (1981) and MacFadden (1984), *C. sphenodus* appears to be phylogenetically close to the base of the dispersal and extensive radiation of *Cormohipparion* and descen-

*The Evolution of Perissodactyls* (ed. D.R. Prothero & R.M. Schoch) Oxford Univ. Press, New York, 1989

dant taxa in the Old World.

*Hipparion* fossils are abundant in continental mammal faunas of Vallesian, Turolian, Ruscinian and lower Villafranchian ages, in numerous fossil sites in Eurasia, Africa and North America. At the end of the North American Miocene, there were two groups derived from "*Merychippus*," the hipparionines and *Pliohippus*. The hipparionines included three very similar forms: *Hipparion, Neohipparion,* and *Nannippus*. In this paper, I will restrict my discussion primarily to Eurasian and African *Hipparion*. In my opinion, a detailed study will be necessary to evaluate the validity and status of these three genera and determine which differences between them correspond to generic or specific variations. Although they diversified widely in time and space, the following characters of the genus remain constant: isolated protocones on the upper cheek teeth and tridactyl and reduced limbs (with lateral digits semifunctional).

*Hipparion*, closely related to "*Merychippus*," was one of the great animal travelers. It was also progressive in the development of the skull. Its teeth had high crowns with very wrinkled enamel and isolated protocones. Conservative in the foot structure, it retained three toes on each foot. The tooth and limb structure of *Hipparion* reflect the fact that it inhabited pastures of the savanna and open plains. It expanded quickly over the Old World, where it seems to have replaced the more primitive forest horse *Anchitherium* (a form with low-crowned teeth). These two genera are found together at some localities in Spain, France, Austria and China (Forstén, 1968). The evolutionary history of *Hipparon* is well known, probably due to its habit of living in herds and thus assuring the occurrence of abundant fossils.

*Hipparion* de Christol (1832) is characterized by isolated protocones on the upper cheek teeth and many characteristic morphological and functional features on the limb bones. The most important characters of the skull are: the variation of the basal length; the general size and shape of the face; the length and variations of the nasal slit; and the shape, size, and limits of the preorbital fossa (POF). Recently, this last character has been emphasized by some workers (Skinner and MacFadden, 1977; Woodburne and Bernor, 1980; Woodburne et al., 1981; MacFadden, 1984, 1985; Bernor and Hussain, 1985). On the basis of DPOF characteristics, a new hipparionine genus from North America was established, *Cormohipparion*. Skinner and MacFadden (1977) considered *Hipparion* from the Old World to be derived from *Hipparion sensu stricto* and *Cormohipparion* of North America. On the basis of DPOF characteristics, Woodburne and Bernor (1980) established a new classification of the four Old World *Hipparion* groups.

The genus *Hipparion* (Gromova, 1952; Gabunia, 1959; Sondaar, 1961; Forstén, 1968; Alberdi, 1972, 1974) is characterized primarily on the basis of cranial, dental and metapodial features. The skulls always have the DPOF, which is well developed in the primitive forms but reduced or even lacking in later forms. It varies in shape and size for each form. The nasal slit is clearly limited, not extending further than the distal side of $P^4$. The palate is deeply concave. The occipital surface has a triangular outline. Lateral crests are well defined and the sagittal crest is well developed. It shows a supracondylar cavity. The jaws bear vertically ascending branches. The incisor crowns are low. The canines are generally placed behind and separated from the incisors. The cheek teeth have relatively high crowns, with a clear diastema between C and $P^2$ (or $P^1$). Very often, $P^1$ is called a "wolf tooth." The upper cheek teeth have isolated protocones and richly wrinkled enamel. The lower cheek teeth have a characteristic "double knot" with a deep external depression and an internal depression. The external depression is better developed on the molars than on the premolars. The posterior side of the lower M3 almost always has three lobes. The lower milk molars have ectostylids and protostylids, and the lower cheek teeth have only protostylids and sometimes reduced ectostylids (protostylids are not found in dP2 and P2). *Hipparion* was a horse with three functional digits. Some of these characters change in the most evolved forms. The lateral metapodials (II and IV) are reduced. The lateral tracks from metapodials II and IV can be found along the third metapodials. Sometimes there is a vestigial fifth metacarpal.

**Cranial characters**
The skull characters are difficult to evaluate. It is not common to find well preserved cranial remains. They are often distorted during diagenetic processes The short diastema is more primitive than the longer one. The smaller nasal slit is more primitive than the longer one. The deep and very limited dorsal preorbital fossa (DPOF) is more primitive than a lack of it.

*Dorsal preorbital fossa (DPOF)*. In regard to the dorsal preorbital fossa, more detailed discussion is necessary. Woodburne and Bernor (1980) proposed a new classification of *Hipparion* based on the skull and especially on the DPOF. Gromova (1952), in discussing the DPOF, suggested that the more primitive forms of *Hipparion* bear a well-developed DPOF and that there is a lack of them in the more derived forms. She also suggested that there are differences in shape and proportions of the DPOF among young, mature, and older individuals. Similarly, Lydekker (1884), Gaudry (1862), and Sefve (1927) have pointed out that the DPOF could have served as the location of glands that were important in the social life of the animal.

Pirlot (1953) concluded that the DPOF was probably a character of little physiological importance for the individual and most likely an unstable character. Gabunia (1959) agreed with Gromova (1952) and insisted on the exceptional stability in the length between the orbit and the DPOF. Meladze (1967) decribed five skulls found in the same strata and in the same area at Bazaleti. They probably belong to the same species of *Hipparion* since all characters (except the DPOF) are similar. He thought that differences in the DPOF could be due to sexual dimorphism, rather than to other genetic causes.

Taking into account the individual variety of the conformation (situation, shape, and depth) of the DPOF, Webb (1969) suggested that this feature could be used in postulating phylogenetic relationships at a supraspecific level. Consequently, whenever remains at a site permit it, we must determine whether such DPOF variation is intra- or interspecific. MacFadden (1980) analyzed a large sample of skulls of *H. tehonense* from the early Clarendonian Frick MacAdams Quarry, Donley County, Texas, emphasizing the variation of the DPOF. For him, the observed variations are indicative of a complex character of important taxonomic value. Sen et al. (1978) thought that the importance of the fossa is unknown.

In 1977, Skinner and MacFadden created the new genus *Cormohipparion*. In 1980, Woodburne and Bernor considered the DPOF a supraspecific character, and assembled the European and North African *Hipparion* from the Vallesian and Turolian levels, into four groups based only on the morphology of the DPOF.

Forstén (1980a) used this character in her study of the Samos *Hipparion*, along with the other characters, and in a later paper also studied Miocene *Sinohippus* (1982a). She criticized the use of the DPOF in equid systematics, alleging that the function of this character is unknown, and that the material referred to *Cormohipparion* is heterogeneous. She (1982b) also said that naming the genus *Cormohipparion* confused the nomenclature of this group, without clarifying the evolution, migration, and biostratigraphy. In 1983, Forstén studied the DPOF as a taxonomic character of some Old World *Hipparion*. In this study, she stated that until the 1970's the DPOF was erroneously considered to be a specific character. In 1984, Forstén grouped the Old World *Hipparion* into subgenera.

On the basis of her bivariate plots of the DPOF (1983, Figs. 1-6), however, Forstén thought that the relationship between the skull size and the location of the DPOF does not vary more than any other feature of the skeleton in a taxonomically homogeneous sample. A morphological study of the teeth led her to believe that the Old World *Hipparion* should remain in the genus *Hipparion*, with several species. Forstén questioned whether separation of these taxa at the supraspecific level could be practical. In this case, the definition of new taxa had to be reached by comparative analysis of all the data referring to the different parts of the body, and not only on the basis of the presence or absence of an anterior rim, or other variable data on the shape and morphology of the DPOF.

Eisenmann (1976a, 1981a, 1982) thinks that several characteristics follow the same evolutionary trend in *Hipparion* and *Equus*: lengthening of the vomer-basion distance, elongation of the muzzle and nasal slit, and shortening of the palate. In different hipparions, these characteristics do not change in a synchronous manner. They may reflect parallelisms or convergences, and they may coexist with different kinds of preorbital morphologies. Since there are contradictory opinions about the importance of the DPOF, Eisenmann thinks that its shape is not enough to define trends of evolution and obtain concrete points on which to build a chronology on a wide scale. She agrees with Gromova, Forstén, and many other authors on the fact that any attempt to understand the group must be based on the study of as many characteristics as possible, and not only on one isolated characteristic, even though it occurs on the skull. Eisenmann accepts the possibility of a polyphyletic origin of Old World *Hipparion sensu lato*, with successive immigrations. She also notes that one of the supposedly oldest forms, *H. africanum*, does not appear to be the most primitive. So, to understand the phylogeny of *Hipparion*, we must undertake complete osteological studies so that the best international correlations can be made.

MacFadden (1984) considered the DPOF to be a complex character of great taxonomic value that plays in important role in his classification of the North American hipparionines. Moreover, he considers *Hipparion* to be a polyphyletic group of horses, as they seem to have originated from at least two taxa within the merychippine complex.

In this discussion, I agree, in part, with Forstén and Eisenmann. In fact, I do not think that a classification of all *Hipparion* in the world, especially in the Old World, can be based primarily on the position, shape and morphology of the DPOF. I think that all possible characters must be analyzed in order to evaluate them as a whole. The possibility of different *Hipparion* waves from North America to the Old World may be real, but questions arise as to how and when the horses made the journey.

In my studies, at some sites with large samples of *Hipparion* remains (Piera, Spain, for example, has skulls, jaws, isolated teeth and postcranial skeletons), I found that the DPOF morphology varies in relation to the age of the animal and also among animals of the same age. It is difficult to evaluate these variations due to postdepositional deformation of the skulls. In my opinion, the DPOF is a character that must be evaluated and analyzed at a specific level. For the moment, I do not believe it is the best taxonomic character on which to base supraspecific taxa. In fact, we do not know its exact function, why it changed over time, or the relationship between variations in the DPOF and the other hipparionine characters. There is a need to clarify the taxonomy of hipparionines in relation to a larger number of characters than simply the DPOF, taking into consideration intra- and interpopulation variability and functional aspects of all characters used.

Eisenmann (1982) acknowledged that her deductions concerning the comparative craniology of tridactyl equids differ from the usual ideas on the taxonomy of these forms. However, I think that this is not a problem. In fact, this disagreement arises from different methodologies used by different authors to evaluate the variations among and within populations due to wide geographic dispersion and/or to differing environments associated with the dispersion of *Hipparion*. Unifying the methodologies is generally enough to solve this problem.

Following Eisenmann (1982), *H. africanum* and *H. catalaunicum* could not be considered closely related, in spite of the similar structure of the DPOF. However, I think that the other morphological and biometrical characters of the dentition and the skeleton in both forms clearly correspond to the same morphotype (i.e., the morphotype of "*H. primigenium*," see later). I believe that the problem lies in the skull characters, which are probably influenced by allometric phenomena. I will deal with this problem in the *Hipparion* datum discussion below. Furthermore, the craniological statements are more complicated because:

-- The samples are relatively scarce (even when they are numerous), so the evaluation of a character in relation to its age is usually doubtful.

-- It is difficult to find complete and undistorted cranial remains.

-- Due to poor taphonomic data records, it is not always clear that all of the remains in the same deposit pertain to the same epoch.

However, dental and skeletal remains are usually much more numerous and generally better preserved, as Badgley (1986) pointed out in a statistical taphonomic work. That is why I think it is necessary to observe all the remains in the different localities and then evaluate them as a whole.

**Dentition**

The taxonomy of fossil mammals is based principally on tooth morphology, because these are usually the most abundant (in many cases, the only) remains, due to their chemical composition and shape. In equids, the teeth have a functional importance reflected in their occlusal surface, which varies in relation to the environment in which they live, their diet, habits, and behavior, and so on. In this group, apart from their morphological patterns, the teeth are hypsodont (high-crowned), and the teeth vary their occlusal pattern in relation to the crown height (corresponding to the animal's age). Consequently, it is important to study the morphological characters of the teeth at the same age, and analyze their indices, percentages, and so on (Alberdi, 1972, 1974).

*Upper cheek teeth* (Fig. 13.1). *Hipparion* is a hypsodont genus of grazers, browsers, and intermediate forms, with high tooth crowns and richly developed cement. There has been considerable progressive heightening of the tooth crowns in the Equidae since *Mesohippus* in the sequence "*Merychippus*" to *Hipparion*, and "*Merychippus*" to *Equus* (Simpson, 1944). This trend towards a gradual increase of hypsodonty can be observed in genera with long geological spans and even in species like *Hipparion*. *Hipparion* is a hypsodont genus, but the early members of *H. primigenium* are relatively brachyodont.

<u>Protocone</u>- The protocone is a free and

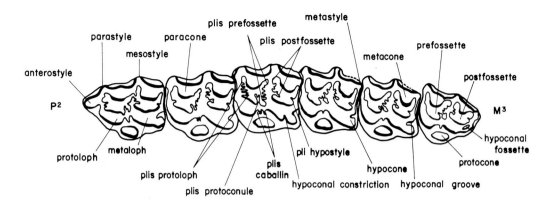

Fig. 13.1. Principal characters of *Hipparion* upper cheek tooth row (after Eisenmann et al., 1988).

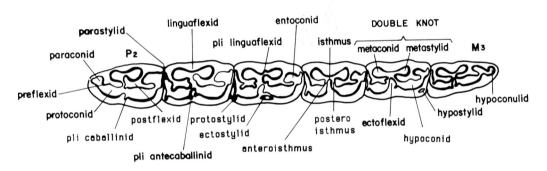

Fig. 13.2. Principal characters of *Hipparion* lower cheek tooth row (after Eisenmann et al., 1988).

isolated cone of the protoloph. It varies from oval to more or less rounded, or quadrangular (in the primitive forms), to lengthened and flattened in the most derived forms. Not only does it vary with the evolution of the genus (i.e., between different species), but also during the life of the animal (i.e., with different degrees of tooth wear). The protocone is linked to the loph in the maximum and minimum stages of wear.

It is important to make comparative studies among populations using individual of the same ontogenetic age. In this sense, Gromova (1952) utilized four levels of wear of the teeth. Musil (1969) established three levels of wear for *Equus*. Along the same lines, Alberdi (1972, 1974) established six levels for *Hipparion*, each one of ten millimeters, in populations where the average height reaches 60 mm. Alberdi did not use the first and last levels for comparative studies because they are almost uniform in all populations. Others, like Sondaar (1961), Forstén (1968), Prat (1968), and Hooijer (1975), take the measurements at a fixed point, but not at the same point for each one, to compare the teeth at the same level of wear and at the same age. Eisenmann (1980a) thinks that all methods have disadvantages. If one compares strictly the same stages of wear, the number of specimens decreases. If one does not take occlusal measurements, one is obliged to include the cement. Most of all, Eisenmann doubts whether measurements taken at the base, or near the base of the crown, i.e., on very worn teeth, are really representative of the normal shape and proportions of the teeth.

Whenever possible, I think that two measurements should be: occlusal surface measurements, as well as measurements one centimeter above the base. Eisenmann *et al.* (1988) established four levels of wear at 15 mm intervals (based on an average height of 60 mm) for *Equus* and *Hipparion*.

Caballin pli - The caballin pli is an enamel fold found on the upper cheek teeth, located in front of the most labial wall of the protocone, in the area where the protoloph and the metaloph join. It can be very complicated or absent. Generally, it is formed by several folds of enamel in primitive forms, and by a few folds in the derived forms. Gromova (1952) represents the folds on the upper cheek teeth with a tooth formula which gives the number of plis of both sides of the fossettes respectively in the numerator and pli caballin in the denominator. Forstén (1968) gives a single number which represents the mean of all tooth categories of a population. Alberdi (1972, 1974) represents the caballin pli by a diagram in which she gives the mean, the mode, and the interval for each population.

Fossettes- On the triturating surface of the teeth, the enamel folds form two fossettes, the pre- and postfossettes. These fossettes are sometimes connected to one another. This character is related to the wear level, so it is more common in $P^2$, and also more common in premolars than in molars. The fossettes can also be open at one or the other wall, and sometimes in the lingual area.

Hypocone.- The hypocone is a linked cone of the metaloph. Its form varies less than that of the protocone. It changes in relation to the degree of wear of the tooth, and also with the development of the hypoconal groove (distal) and the hypoconal constriction (lingual).

*Lower cheek teeth* (Fig. 13.2). These teeth are taxonomically very significant.

"Double knot"- The "double knot" is formed by the metaconid and metastylid cones. This character exhibits two morphological patterns in *Hipparion*: one is typically hipparionine with a round double knot, and the other is typically caballine with a sharp double knot. The last one is represented in the most derived forms.

Hipparionine lower cheek teeth are well developed and large, with an oval metaconid and somewhat angular metastylid and entoconid. Labially, they have slightly convex enamel borders on the protoconid and hypoconid. In this group, the protocone is oval.

In caballine lower cheek teeth, the metastylids and metaconids are pointedly angular and the enamel borders are straight. The more or less flattened protocone in the upper cheek teeth corresponds to this pattern.

The hipparionine double knot is found in the upper Miocene and early Pliocene (Vallesian, Turolian and Ruscinian in Eurasia). In the middle and upper Pliocene or lower Villafrachian, we find a caballine double knot pattern that disapeared with the arrival of *Equus* in Eurasia. This caballine pattern may appear in Africa as early as 6 Ma (Eisenmann, 1977), but it is not as clear nor as well documented as after 4 Ma (Eisenmann, 1979a). Later it becomes consistently associated with ectostylids (typical "*Stylohipparion*") and persists in the latest African *Hipparion* until less than 1 Ma (Eisenmann, 1979a, Fig. 1).

Pticostylid- The pticostylid, or pli caballine, is located in the labial depression between the protoconid and hypoconid. This seems to be a derived character because it is more pronounced in the caballine forms.

Ectostylid- This is an enamel stylid in the labial depression. It is well developed in the Pleistocene forms of African *Hipparion* with a caballine double knot, but these two characters are independent of one another (Eisenmann, 1977, 1983) in Eurasia, and sometimes in Africa, the caballine patterned is not associated with ectostylids. In the primitive forms of *Hipparion*, the ectostylid may be present, although not so well developed (Gromova, 1952; Sondaar, 1961).

This character never appears in $P_2$, so Eisenmann (1977) thought that the classification of such specimens was problematic when there was more than one form from the same site. The ectostylid is well developed in the milk dentition in all *Hipparion* forms.

Protostylid- This is also an external stylid which developed on the mesiolabial edges of the lower cheek teeth protoconids, except in $P_2$, in which the protostylid is always lacking. Eisenmann (1976b) evaluated the meaning of the protostylid in fossil and recent forms of *Equus* from a systematic and phyletic point of view.

The presence of this character varies in quality and quantity from one group to another. It is better developed in primitive forms, like the "*H. primigenium*" morphotype, where the height of the stylid is variable, as is the number of its subdivisions (between 1 and 6 small columns). This stylid is sometimes covered by cement.

Development of the labial depression varies. It is better developed in caballine forms than in the hipparionines. However, it can be more complicated in the primitive forms and sometimes develops complicated folds (pli linguaflexid).

The vestibular depression is important in these teeth. Eisenmann thinks that the depth of the vestibular depression of the molars decreases during evolution, both in *Equus* and *Hipparion* (Eisenmann, 1977, 1981b). Usually, however, it is deeper in molars than in premolars and may be used to distinguish between them. I am not sure whether this reduction has any relationship to the evolution of the group.

*Incisors*: These teeth do not contribute much to *Hipparion* systematics. Eisenmann (1979b) made a detailed study of the incisors of *Equus*, but she failed to obtain significant results. The characters that we can analyze in the incisors are: the degree of plication of the corresponding enamel marks at the occlusal surface, and the occurrence of longitudinal crenulation of the crown.

In general, these characters have not been significant in *Hipparion* because, in the upper and lower incisors, the infundibular marks and the occlusal surface marks depend on the degree of wear. Generally, the upper incisors are larger and have more curved crowns than the lower incisors. The infundibula in *Hipparion* is always present in the lower and upper incisors. It is only lacking when the incisors are very young or very old, and is more central in the upper incisors and more lingual in the lower ones.

In some African forms, the $I_3$ is reduced or atrophied. This is more pronounced in the lower incisors than in the upper ones (Eisenmann, 1979a). The atrophy of the third incisor appears in Pleistocene *Hipparion* (Olduvai, Cornelia) and also sometimes in some Plio-Pleistocene quarries (East Turkana, Omo; see Eisenmann, 1983, 1985). In general, this atrophy seems to be characteristic of the Pliocene forms, but not a constant feature. This atrophy produces a kind of deformation in the symphysis. I think that it could be due to a degenerative character in the latest forms.

## Hypsodonty

This is a useful character of the cheek teeth. An hypsodonty index can be calculated for each unworn or slightly worn P3-4 and M1-2 upper and lower cheek tooth by dividing the mesio-distal length x 100 by the height of the parastylid.

*Hipparion* is a hypsodont animal in which the most primitive forms are relatively less hypsodont, and this character increased with the evolution of the group. It is associated with a reduction of the characters on the occlusal cheek teeth discussed earlier (Alberdi, 1974). From a taxonomic point of view, one can calculate a set of indices from the upper and lower cheek teeth, such as the protocone index, protoflexid index, preflexid index and a double-knot index.

**Postcranial skeleton**

In general, the typical characteristics of the genus are present in every known skeletal specimen. Gromova (1952) compared the skeleton of *Hipparion* to that of *Equus*. Sondaar (1968) studied the manus, and Hussain (1975) studied the hind limbs. In all these studies, the problem was to elucidate the evolution of the functional anatomy of *Hipparion* through time and space. Advanced postcranial characters are precisely those that better reflect the adaptation of equid limbs to the environment in which they lived. If it was savannah or steppe, they showed the adaptations necessary for running. Consequently, the distal part of the limbs tended to lengthen in comparison to the proximal portion. Slender forms are more adapted to high-speed running. Because of this, the metapodials became more slender, longer, and thinner. This tendency is clear in some European and Asian species. For example, *H. mediterraneum* from Samos has exceptionally long and narrow metapodials, and *H. longipes* from Pavlodar (Gromova, 1952), was a similar large, long-legged form.

Some limb bone characters are taxonomically useful in *Hipparion*. An example is the lack of a facet against the cuneiform on the metacarpal, which has been reported in the Vallesian populations (Pirlot, 1956; Arambourg, 1959; Sondaar, 1961; Alberdi, 1972, 1974).

The most significant bones for the study of *Hipparion* functional anatomy are the metapodials, especially the third metacarpal (Mc III) and the third metatarsal (Mt III), and the carpal and tarsal bones. It is on these bones that the adaptation to running is best expressed. The carpus is formed by a set of bones that transmit the pressure that comes from the corresponding limb to the digits. In the tarsus, the astragalus has a very important function. It is the bone that takes all the pressure of the body on the corresponding limb, and transmits it to the digits, which play an important role in the forward movement of the body. The main function of the calcaneum is to contain the insertion points of the strong muscles that control the movements of the animal forward. The other bones of the tarsus play the role of transmitting the pressure that they receive (Hussain, 1975).

Mc III and Mt III are the most important bones used in a taxonomic analysis, because alterations caused by adaptation to the environment are clearly marked on them. We can make a detailed study of their maximum and minimum slenderness and of the evolving differences between *Hipparion* species.

In the osteological remains, we can evaluate the characters of a population (generally there is an abundance of remains) using different statistical methods (Alberdi, 1972, 1974; Forstén, 1968; Eisenmann, 1979c; Eisenmann and Karchoud, 1982). However, the proportions of the bones are sometimes more significant for taxonomic study, as happens in the case of the slenderness index, which can be analyzed for the generally well-preserved Mc III and Mt III.

Complete tibiae are rare in collections, as are other limb bones, so it has not been possible to compare such data for all populations. Corresponding data on the proportions of the forelimb do not show such a marked segregation into long-limbed steppe forms and short-limbed forest forms. There seems to be no difference in the length of the radius compared to the length of Mc III. However, in some Mediterranean localities, it can be observed that the radius is comparatively short and the Mc III is long (Forstén, 1968).

The study of the *Hipparion* skeleton shows an animal with a strong and massive build that represents the more primitive forms that were widespread in the Vallesian, for example, *H. primigenium*. These forms may have been represented by less massive animals in more recent levels. They survived until the Turolian.

The *Hipparion* characteristic of the Turolian has a tendency toward slenderness in the limbs accompanied by a decrease in body size. This evolution can be observed in the Spanish Turolian localities of the Teruel Basin. This kind of evolution also appears in other localities from the Mediterranean (Greece, Samos), Eastern Europe, and so on.

**The taxonomy and biostratigraphy of *Hipparion***

*Hipparion* fossils are very abundant in the Tertiary faunas of Eurasia from the Vallesian (MN 9) to the lower Villafranchian (MN 16a), and in the African Tertiary and Quaternary (from Vallesian to middle Pleistocene) (Fig. 13.3). According to some authors, *Hipparion* survived until 400,000 years ago (Eisenmann, 1979a, citing Isaac, 1977). Isaac (1978) confirmed the latter date, but emphasized that such dates should not be used if they are not well corroborated. At any rate, *Hipparion* is still present in Africa less than 1 Ma [for example, in L9 of the Shungura Formation and in Bed IV of Olduvai (Eisenmann, 1985)].

The different methodologies used in the taxonomic studies of *Hipparion* from a regional point of view (Sefve, 1927; Pirlot, 1956; Sondaar, 1961, 1974; Alberdi, 1972, 1974; Hussain, 1971; Zhegallo, 1978; Bernor, 1985; among others) or from a wider interregional Eurasian one (Gromova, 1952; Gabunia, 1959; Forstén, 1968, and others) have often resulted in differing, sometimes contradictory conclusions as to the evolutionary relationships among the different species of *Hipparion* in the Old World. There has also been an unnecessary, and often unjustified, multiplication of *Hipparion* subtaxa. Attempting to unify this group from a taxonomic point of view and build an intelligible taxonomic scheme, we find a taxonomical tangle at the generic, subgeneric and specific levels.

A comprehensive study evaluating each of the previous studies would exceed the limits of the present paper. To overcome these kinds of problems, the National Science Foundation (NSF, USA) sponsored a conference to evaluate and standardize current methods used in the study of equids (known as the "*Hipparion* Conference," New York, November, 1981). It was based on Eisenmann's methods (Eisenmann et al., 1988). We hope that its example will help solve a great number of problems that now occur in the literature.

The majority of the paleontological studies on *Hipparion* lack geological and sedimentological details of the deposits. This precludes our evaluation of the taphonomy of the deposits, and consequently we cannot judge the real value of the described associations. In other words, most studies are carried out in museums, and rarely ever in the field. Unfortunately, old museum collections are the common source for paleontological work.

In the last 30 years, the indiscriminate publication of many isotopic dates and paleomagnetic correlations has created a new problem, as such dates are used like a "magic wand" to solve everything. We have to take the limitations of such data into account if we do not want mistakes. For

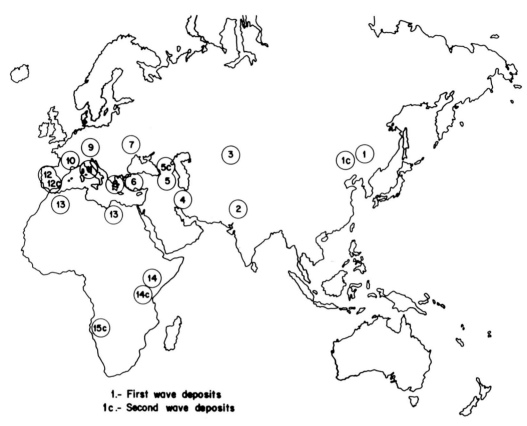

Fig. 13.3. Geographic locations of deposits where the most abundant *Hipparion* remains are found in the Old World. 1-1c) China: Shansi, Honan, Kansu, Chihli; 2) India and Pakistan: Chinji, Nagri, Dhok Pathan, Pinjor, and other localities in the Siwalik Group; 3) Central Soviet Union: Palvlodar; 4) Middle East: Maragheh; 5-5c) Soviet Georgia: Bazaleti, Oüdabno, Kvabebi; 6) Turkey: Çoban Pinar, Kavak Dere, Çalta, Camli Bayir; 7) Western Soviet Union: Sevastopol, Novo Elisabetovka, Malusteni, Kouialnik; 8) Greece: Pikermi, Samos, Saloniki, Veles, Macedonia; 9) Central Europe: Baltavar, Csakvar, Eppelsheim, Höwenegg, Gaiselberg, Polgardi, Vienna Basin; 10) France: Aubinas, Mont Leberon, Perpignan, Rhone Valley; 11) Italy: Casino and Baccinello V3; 12-12c) Spain: Nombrevilla, Los Valles de Fuentidueña, Can Llobateres, Concud, Los Mansuetos, Piera, Alcoy, Villarroya; 13) Northern Africa: Oued el Hamman, Mascara, Beglia Formation, Wadi Natrun; 14-14c) Eastern Africa: Ngorora, Npesida, Kaiso, Ekora, Afar, Shungura, Turkana, Mursi, Omo Beds, Baringo; 15c) Southern Africa: Makapansgat, Langebaanweg, and localities in the Republic of South Africa.

example, Isaac (1978, p. 194) writes: "The scatter of values suggests that the material dated may have been derived in part from reworked deposits of various ages." To visualize the taxonomic features of hipparionines in Africa and Eurasia as clearly as possible, we need to bear in mind that the African forms generally deviate from the Eurasian ones in ways that simply could be due to the different levels of geological, stratigraphic, and ecological knowledge of the African regions, and also to the fact that the African and Eurasian fossil records are not comparable.

As suggested before, I do not consider the different genera of hipparionines cited in the current literature to be valid because, in my opinion, the basic characters for the differentiation of *Hipparion* de Christol, 1832, at a generic level appear in all the different hipparionine genera cited in the literature up to now. I will not discuss the North American forms here. I understand that we need a complete revision of hipparionine forms, but such would require a more extended paper, as well as input from other researchers.

Numerous hipparionine species have been described in the literature. To avoid dealing with all the named *Hipparion* species in the different localities of Eurasia and Africa (which would obscure a view of the whole), I have decided to classify all *Hipparion* remains from Eurasia and Africa into a few morphotypes. For each of these, I will give the most important characteristics that serve to unite such forms during a given time.

The term "morphotype" is used here in the sense of a basic general scheme, with a structure similar in morphological and biometric characters as in the structure of the skeleton. This implies a parental relationship, while not excluding the existence of a possible convergence, as could be the case of the caballine forms found in Africa being dated earlier than those in Eurasia.

In Fig. 13.4, the morphotypes are shown in relation to the general stratigraphic and paleomagnetic scales (according to Pevzner and Vangengein, 1985, RCMNS Congress), and also the paleoclimatic and sea level scales (according to Vail *et al.*, 1977; Savin, 1977; Shackleton, 1984; Shackleton and Hall, 1984). With these data, a comparison between *Hipparion* morphotype distributions and the general climatic conditions prevalent during those times can be made.

These morphotypes are denoted by widely used names that represent large groups, and also evolutionary trends that appeared during particular periods (for example, *H. c. concudense* - *H. c. aguirrei* - *H. gromovae* - *H. periafricanum* of the Teruel Basin, Spain). We will call this the evolutionary trend of "*H. concudense*." This form can be found in Spain and some parts of Greece and the Soviet Union. I distinguish a large *Hipparion* group with five hipparionine morphotypes, and one caballine morphotype, as follows:

*Morphotype 1.* The morphotype of "*H. primigenium*," a group of horses with a strong build, large upper and lower cheek teeth, characteristic enamel "wrinkles" that decrease through time. They have strong metapodials that are slender in the more derived forms, and a well developed DPOF. They lived from about 12.5 Ma to 6.5 Ma. During the Vallesian of Eurasia, they were widely distributed (Spain, Europe, Soviet Union, China, Pakistan, India, Greece, Turkey, and so on) and were also found in parts of North and East Africa. They were also present in the Turolian, but to a lesser degree. In Spain, they are known from a few teeth from the upper Turolian of Venta del Moro and Arenas del Rey. In this morphotype all the forms cited as species and subspecies of *H. primigenium* are included, as well as many others that have descriptions similar to *H. primigenium* and lived during this time.

Morphotype 1 is widespread in Eurasia and Africa. It includes different types of *H. primigenium* and perhaps *H. plocodus* of

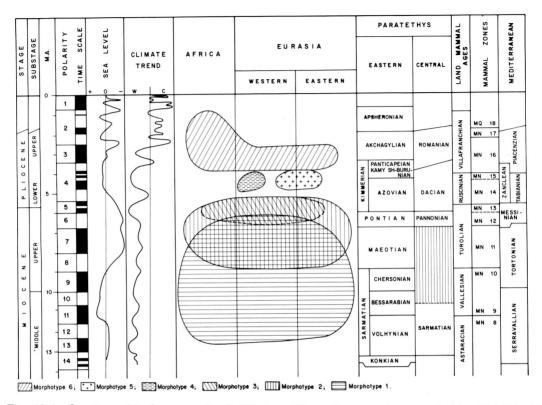

Fig. 13.4. Stratigraphic framework of different *Hipparion* morphotypes in the Old World. Stratigraphy and paleomagnetic scale from Pevzener and Vangengeim (1985). Paleoclimatic diagram redrawn from Shackleton (1984) and Shackleton and Hall (1984). Sea level oscillations redrawn from Savin (1977) and Vail et al. (1977)

China sensu Forstén (1968; 1978a c, d; 1981; 1982c; 1985a, b); *H. primigenium*, of Alberdi (1972, 1974, 1986), and Alberdi et al., (1978); *H. theniusi, H. primigenium* and *H. macedonicum*, of Koufos (1980, 1984, 1986); *H. depereti* and *H. primigenium* of Sondaar (1974); *Hipparion* from Kayadibi and Esme Akça, Turkey (Sen et al., 1978; Staesche and Sondaar, 1979); *H. theobaldi* and *H. nagriensi* of Hussain (1971); the majority of Group 1 of *Cormohipparion* as cited by Bernor et al. (1980) and other papers of Bernor, Woodburne, MacFadden, and their co-workers after 1980; *H. theobaldi, H. feddni, Cormohipparion theobaldi* and *C. nagriensi* of MacFadden and Woodburne (1982); *Hipparion* from Hungary, cited as *H.* sp. "A," *H.* sp. "B," *H.* sp. "C" and *H.* sp. "D" by Kretzoi (1983); *H. gettyi* of Bernor (1985) from Maragheh; *Cormohipparion (Sivalhippus)* sp. of Bernor et al. (1985) in part; and *Hipparion* from Oüdabno and Elda Georgia (Alberdi and Gabunia, 1985). From African localities, morphotype 1 is represented by *H. primigenium* in North Africa: Bou Hanifia and Marceau, Algeria (Arambourg, 1959; Forstén, 1968; Alberdi, 1974; Eisenmann, 1981d); in the Beglia formation, Tunisia (Forstén, 1972); and in Sahabi, Libya (Tobien, 1982). Both Arambourg and Eisenmann identify it as *H. africanum*. In East Africa, it is found in the Ngorora formation (Nakali), Kanapoi, Ekora, and Chemeron and the Aterir beds, Kenya (Aguirre and Alberdi, 1974; Hooijer, 1975; Eisenmann, 1981d). Literature about the African localities is poor.

*Morphotype 2.* This is the morphotype of *H. concudense-mediterraneum*. The distinctive skeletal build, upper and lower cheek teeth, and wrinkled dentition are found in a medium-sized *Hipparion*. It has a marked DPOF. Through time, this morphotype exhibits a decrease in size, a simplification of the occlusal surfaces, and an increase of hypsodonty. For example, this evolutionary trend appears clearly in the Teruel basin from *H. concudense* to *H. periafricanum* (Alberdi, 1972, 1974, 1978) and allows us to include small *Hipparion* species, such as *H. periafricanum*, *H. matthewi*, and so on, within this morphotype. This group is found in Eurasian (western Europe, Greece, Soviet Union, China, and so on) and in African (North Africa, such as *H. sitifense*) deposits between 9 and 5.5 Ma. They coexist in some localities with late appearances of morphotype 1, which is poorly represented.

In morphotype 2, I include the following different forms cited in literature: *H. sitifense*, *H. mediterraneum*, *H. matthewi*, and *H. periafricanum* of Forstén (1968, 1978c: only *H. mediterraneum* and *H. matthewi* of Forstén, 1980b; 1981); *Hipparion* from Samos (Sondaar, 1971; Forstén, 1980a); the evolutionary lineage of Teruel *Hipparion* and *H. mediterraneum* of Alberdi (1972, 1974, 1978); *H. prostylum* of Sondaar (1974); *Hipparion* from Kinik, Turkey, of Sen et al. (1978) and of Staesche and Sondaar (1979); *H. antelopinum* of Hussain (1971); all *Hipparion* forms from the lower and middle Turolian of Bernor et al. (1980); *H. theniusi*, *H. concudense*, *H. dietrichi*, *H. mediterraneum*, *H. nagriense*, *H. gromovae*, *H. matthewi*, and *H. periafricanum* from Greece (Koufos, 1980), and *H. mediterraneum* and *H. brachypus* from Pikermi (Koufos, 1987); "*H.*" *antelopinum* and *H.* sp. of MacFadden and Woodburne (1982); *H. antelopinum* from the Siwaliks (Hussain and Bernor, 1984); and *H. prostylum*, *H. campbelli*, *H.* aff. *moldavicum*, and *H. matthewi* from Maragheh (Bernor, 1985). *Hipparion* sp., (cf. *sitifense*) from Sahabi (Tobien, 1982) could possibly be included here.

*Morphotype 3*. This is a medium-sized *Hipparion*, with a smaller dentition, a simpler triturating surface, longer mesiodistal length of the cheek teeth (more pronounced in the lowers), and more slender legs than *H. concudense*. The DPOF is unknown. It corresponds to *Hipparion* morphotype I of Alberdi and Morales (1981). It probably evolved from morphotype 2. This group occurs only around the Mio-Pliocene boundary (6.5 to 5.5 Ma).

Within morphotype 3, I include some forms from the Soviet Union (perhaps the Oüdabno site of Alberdi and Gabunia, 1985); Spanish *Hipparion* from Orrios and Venta del Moro (Alberdi, 1986; Alberdi and Morales, 1981); some new localities from the south of France (Sondaar, 1974) and possibly from Baccinello V3 in Italy (Alberdi, 1986). The *Hipparion* from Sahabi cited as *H.* sp. cf. *sitifense* by Tobien (1982) is close to this morphotype, but there is only one tooth, so placing it in morphotype 2 or 3 is difficult.

*Morphotype 4*. This morphotype is similar to *H. crassum*, a large and massive form with conservative teeth. It has a short, robust build, strong teeth, a wrinkled occlusal surface that is reminiscent of, but more brachydont than, morphotype 1, and massive, short, and wide metapodials, which give morphotype 4 a characteristic look. It is found in the lower Ruscinian (5 to 4 Ma), and its presence seems to be restricted to the Tethys and Paratethys. It has not been found outside of the Mediterranean area, including Africa. It is poorly represented by only a few teeth and skeletal bones, but no skulls.

This *Hipparion* species is cited as *H. crassum* in Forstén (1968) from southern France, Alcoy in Spain, some remains from Hungary, and possibly the material from Malusteni and Beresti in Rumania. This last-named material has been called *H. malustenense* by Radulesco and Samson (1967) and Samson (1975). Also included here is *H. ankyranum* of Ozansoy (1965) from Turkey. In this morphotype, I put: *H.* sp form B from the Rhone Valley (Sondaar,

1974); *H. crassum* from Alcoy and Zeneta in Spain, Montpellier and Perpignan in France (Alberdi, 1974, 1986). In the latter paper, I suggested that *Hipparion* from Casino, Italy, is a possible *H. crassum*. Heintz *et al.* (1975) cited *H. crassum* from Çalta, Turkey, as did Sen *et al.* (1974), Sen *et al.* (1978), and Staesche and Sondaar (1979). *H.* sp. "G" from Gödöllo, Hungary, of Kretzoi (1983) and *H. crassum* from the lignites of Ptolomais (Macedonia, Greece) of Koufos (1982) are also included here.

*Morphotype 5-* This morphotype is similar to morphotype 3, but it appears later. It has stronger, more slender legs and smaller dentitions than morphotype 4. It corresponds to *Hipparion* sp. morphotype II of Alberdi and Morales (1981). It had a wide geographic distribution, but there are few known remains. In Eurasia, it is known from La Calera, Villalba Alta and Layna in Spain, and Bazaleti and Urmia in the Soviet Union (Alberdi, 1986; Alberdi and Morales, 1981; Alberdi and Gabunia, 1985). I saw the complete skulls of the Bazaleti *Hipparion* kept in the Paleobiology Institute and State Museum of Tbilisi. These skulls have a shallow and marked DPOF in specimens B-51 and B-50, and a well-developed DPOF in the specimen from the State Museum of Tbilisi. The dimensions of the DPOF in each one of these skulls are different. The differences among the specimens have already been discussed by Meladze (1967) in a more representative sample (see below). *H. fissurae* from Spain (Crusafont and Sondaar, 1971) could belong to this morphotype, as well as *H. garedzicum* from the Soviet Union, *H. longipes* from the Soviet Union (Gromova, 1952) and Turkey (Heintz *et al.*, 1975). At present, I do not know of any specimens from Africa that belong in this morphotype. As in morphotype 4, its geographical distribution occurs in Eurasia from about 5 to 4 Ma.

*Morphotype 6.* This is the only caballine morphotype and the best representative form is *H. rocinantis* from the Spanish localities (Villarroya, Las Higueruelas, and La Puebla de Almoradier) (Alberdi, 1972, 1974, 1986b; Alberdi *et al.* 1984). It was described by Hernandez-Pacheco (1921); cited as *H. crusafonti* by Villalta (1948, 1952) from the Vilarroya site; and also described as *H. crusafonti* from Kvabebi (Georgia, Soviet Union) by Vekua (1972).

At the American Museum of Natural History in New York, there is a specimen from China that corresponds to morphotype 6. It was described by Qiu *et al.* (1980) as *H. houfenense* Teilhard and Young, and they thought these remains to be similar to *H. turkanenses*, and that both were derived from *H. hippidiodum*. Eisenmann (1982a), however, thinks that *H. turkanense* is very similar to *H. antelopinum*, and could be its descendant after migration. It is also possible that *Proboscidihipparion* belongs to this morphotype. According to Qiu *et al.* (1980), it is possible that caballine *Hipparion* first appeared in China, before appearing in Europe. If this is true, it is possible that caballine *Hipparion* of Africa is derived from caballine *Hipparion* of China.

This morphotype can be characterized as a large horse with a caballine pattern on the occlusal surfaces of the lower cheek teeth, long metapodials, and a skull without a DPOF. This form has a wide geographic distribution in Eurasia: it is known from deposits in Spain, Georgia and China, as well as from eastern and southern Africa. It appears in the Pliocene or lower Villafranchian (MN 16a) about 4 Ma, and it disappears from Eurasia at about the same time that *Equus* appears, with the appearance of *E. stenonis livenzovensis* in the middle Villafranchian (MN 16 b) (Eisenmann, 1981c; Alberdi and Bonadonna, 1983; Bonadonna and Alberdi, 1987). However, in Africa, this morphotype with many species may have coexisted with *Equus*. These African hipparions develop original characters and strong ectostylids so that they may be grouped in the subgenus *Stylohipparion*. They may persist until 0.4 Ma (Isaac, 1978). These species are included in *Neohipparion* by Zhegallo (1978) and Forstén (1984).

These six morphotypes, well represented in Eurasia, have a wide geographic distribution. There are abundant remains of morphotypes 1, 2, and 6; remains of 3, 4, and 5, on the other hand, are scarce. A reduction in *Hipparion* species occurs at the end of the Miocene due to a general climatic cooling and a lowering of sea level. This sea-level drop may have allowed morphotypes 4 and 5 to migrate, even though remains are still scarce. The caballine forms seem to have appeared during a relatively cooler period. The Pliocene cooling trend which reached its maximum between 3 and 2.5 Ma could be responsible for *Equus* replacing *Hipparion* in Eurasian localities. All of these data are well recorded in Eurasia between 12.5 and 2.5 Ma. The persistence of *Hipparion* with *Equus* in Africa until recent times should have a climatic explanation. However, such an explanation is difficult to determine, given our current knowledge.

Some of these morphotypes were present in Africa. Morphotype 1 is found in north and east Africa (see previously), and some remains of morphotype 2 are found in the north (*H. sitifense* from Mascara) (Arambourg, 1959 and Eisenmann, 1981d). In the Sahabi locality, there are two poorly represented morphotypes. One corresponds to morphotype 1, and the other, based on a single tooth, could be close to morphotype 2 or 3 (Tobien, 1982). In Egypt, Wadi Al-Natrum has so few remains (El-Khashab, 1977) that we can only say that *Hipparion* is present, although it is impossible to identify to the specific level.

Morphotype 4 has never been described from African localities. Morphotype 5 is also unknown from Africa at present. Morphotype 6 is widespread in Africa, composed of several different species. It coexisted with *Equus* after disappearing from Eurasia at this time. According to Eisenmann (1977), the caballine forms in Africa appeared at about 6.5 Ma, while the hipparionine forms persisted until 9 Ma. However, the Mpesida locality (about 7 Ma) seems to include a hipparionine form (Aguirre and Alberdi, 1974).

Eisenmann (1976a) described a caballine *Hipparion* from Afar (*H. afarense*). Based on Churcher and Richardson's revision of the African horses (1978), the forms they described could be grouped in three of the morphotypes described above. Morphotype 1 includes the species that they assigned to *H. primigenium*. Morphotype 2 includes the species attributed to *H. sitifense*. Morphotype 6 includes all the species they included in *H. albertense, H. bairdi, H. namaquense, H. turkanense, H. afarense,* and *H. libycum*.

The majority of the forms of *Hipparion* described by Eisenmann at Koobi Fora (1983) have the caballine pattern of morphotype 6. Included here are *H. hasumense, H.* sp. A, *H. ethiopicum, H.* cf. *etiopicum,* and *H.* sp. B. At the same time, independently from the forms I included in this morphotype, Eisenmann described *H. cornelianum*, characterized by the atrophy of its incisors, that in my opinion could correspond to morphotype 6 on the basis of its lower cheek teeth. In this morphotype, I would also include the Omo Valley *Hipparion* chronologically situated between 7 and 6 Ma (Eisenmann, 1985).

Summarizing the previous discussion and the accompanying chart (Fig. 13.4), we can see that the first arrival of *Hipparion* to Eurasia and its widespread expansion is synchronous in the sense that it appears at the same time *sensu lato* in very distant localities, as explained by Berggren and Van Couvering's (1974) discussion of the *Hipparion* datum. I doubt that the first appearance of *Hipparion* in the Old World is diachronic, as suggested by Bernor *et al.* (1985), because morphotype 1 (*H. primigenium* group) is widely distributed in time and demonstrates a clear evolutionary trend through time. If we make an exhaustive comparative study through Vallesian time and covering all of Eurasia and Africa, we find differences between the different forms that can be linked to the span of time. For example, if I now compare the *Cormohipparion* (*Sivalhippus* ) group from the Indo-Pakistan region (Hussain and Bernor, 1984; Bernor and Hussain, 1985), belonging to the *H. primigenium* group, with other species of this morphotype of *H. primigenium* from Höwenegg, I find that these two samples correspond to the same morphotype 1, but they should show variations between them. These variations should correspond to the evolution of this morphotype through time, because the last locality is older than the first one. Höwenegg is about 12.5 Ma, according to Berggren and Van Couvering (1974), but the Siwalik sample is around 9.5 Ma, according to Hussain and Bernor (1984). Comparing the central European locality of Höwenegg with Can Llobaters in Spain; I find that the differences between them are slight because of the small difference in time. Therefore, I think that the *Hipparion* datum event is correct and that the appearance of *Hipparion* in Eurasia and Africa is synchronous *sensu lato*.

It is important to observe that the *Hipparion* species decrease in Eurasia at a time shortly after the Mediterranean "salinity crisis" when the climate was warmer and the sea level lower. In Africa, this is not so easy to detect. In general, there are fewer African localities, and these are less productive than those in the Soviet Union, Greece, Spain, and the other Eurasian basins. The persistence of species of *Hipparion* in Africa could result from the climatic conditions, but we have very little data on African paleoclimates. Therefore, we cannot exactly evaluate the specific influence of paleoclimate on the persistence of the *Hipparion* species in Africa.

J. A. H. Van Couvering (1980), studying the evolution of an ecological community in East Africa during the late Cenozoic, concluded that from the early Miocene to Plio-Pleistocene times, the African environment changed from a non-seasonal rainforest community to a seasonal weather pattern with a savanna-mosaic dominant community. Her results are directly related to the climatic trend shown by isotopic and faunal results (Fig. 13.4; Shackleton, 1984; Shackleton and Hall, 1984). The most important changes that took place during the middle and late Miocene and the Pliocene are explained by the African fossil record (Van Couvering, 1980), and these changes occurred gradually. This could be a reason for the persistence of these animals in Africa. Following these hypotheses, Bernor (1983, 1984) suggested that some climatic bioprovinces related to the climatic changes. I think that we are only at the beginning of this kind of climatic analysis. Therefore, we will have to be very cautious about this topic.

**Evolution and phylogeny**
The evolution of the different forms of *Hipparion* through time seems to follow a clear evolutionary trend within each morphotype. There is a simplification of dental morphology, a reduction in the size of the animal, an increase in the dental hypsodonty, and an increased slenderness of the limbs (which can be observed in the metapodials). This phenomenon can clearly be described in those morphotypes, or groups of species corresponding to the morphotypes, which have a wide representation in time and space, as is the case with morphotypes 1 and 2. In different geographical areas, the same evolutionary trends can be seen, and the same phylogenetic trends can be followed. These latter trends are not always easy to follow, because many of the species are not always represented.

The first wave of *Hipparion* in Eurasia evolved from the North American "*Merychippus*" and spread quickly into

Eurasia and Africa in the form of *H. primigenium* and similar species (grouped here as morphotype 1). They persisted until the end of the Miocene, but became less numerous through time.

Morphotype 2 dispersed widely in Eurasia and even in North Africa. It does not seem possible that it corresponded to a new wave of *Hipparion*. I think that this type could have derived from some evolutionary stage of morphotype 1, as might be the case for the "intermediate *primigenium - concudense* " form *sensu* Alberdi (1972, 1974). This "intermediate" form may have evolved into morphotype 2 with a clear and pronounced phylogenetic trend that was favored in closed basins, such as the Teruel Basin of Spain. This was probably due to the geological characteristics of a specific basin at a given time. This happened at the same time in Greece and in other geographical areas (Koufos, 1980). The geological characteristics could have favored evolutionary and phylogenetic lines, such as those which developed from *H. c. concudense* to *H. periafricanum* Alberdi (1972, 1974, 1978). I have found isolated stages of this evolution in other areas within basins that were more or less closed.

Morphotype 3 is restricted to between 6.5 and 5.5 Ma; it is scarce in Eurasia and even scarcer in Africa. It could have evolved from morphotype 2 (perhaps from *H. c. aguirrei* or *H. gromovae*; the most probable ancestral form is the latter one). Selection pressure seems to have emphasized particular characteristics of the species. This is apparent in the fact that the evolutionary trend of morphotype 2 followed a somewhat similar gradual rhythm. Morphotype 3 differs in that there was a sudden reduction in the size of the dentition, an increase in the mesio-distal length of the lower teeth, and a decrease in the metapodials, which became thinner and longer, producing a sudden increase in the slenderness.

This change of the evolutionary trend in morphotype 2 would also explain the appearance of morphotype 5 from morphotype 2. In this case, the general scheme of morphotyope 2 varied substantially, which could be explained by the selective pressure of the environment. This event could produce the morphotype 5 with long, slender metapodials. According to Forstén (1968), this change in limb proportions can be explained by the change from a typically forest to a steppe environment. This can be illustrated by the *H. longipes* of Pavlodar (Gromova, 1952), which was recently found by Heintz *et al.* (1975) in Çalta, Turkey. *H. fissurae* of Layna, Spain, can be included in this group, as well as other new finds (Alberdi and Aliatá, in prep.). The dentition of the latter is not as hypsodont as it should be for an animal with metapodials so well adapted to run on the steppe.

The origin of morphotype 4 is not so clear. It could easily be explained as a new immigrant. I considered it to be an immigrant in Spain (Alberdi, 1972, 1974). From a circum-Mediterranean point of view, it could be considered as one of the residual groups of morphotype 1 (based on its conservative molars). It could have survived sporadically around the Mediterranean where the ecological pressures affected the length of the metapodials; the metapodials became more reduced and massive, while the dental characteristics were conserved. The same thing could have occurred with *E. stehlini*, which was a horse of the *E. stenonis* type. It had very reduced metapodials in comparison with the original species; this was probably due to ecological characteristics of the environment (Bonadonna and Alberdi, 1987).

Forstén (1968) suggested that *H. ankyranum* (which she considered a synonym of *H. crassum*) evolved directly from *H. primigenium*, replacing the latter as a forest form. From Turkey, this *H. crassum* morphotype could have migrated into western Europe (southern France and eastern Spain). Taking into account the geological

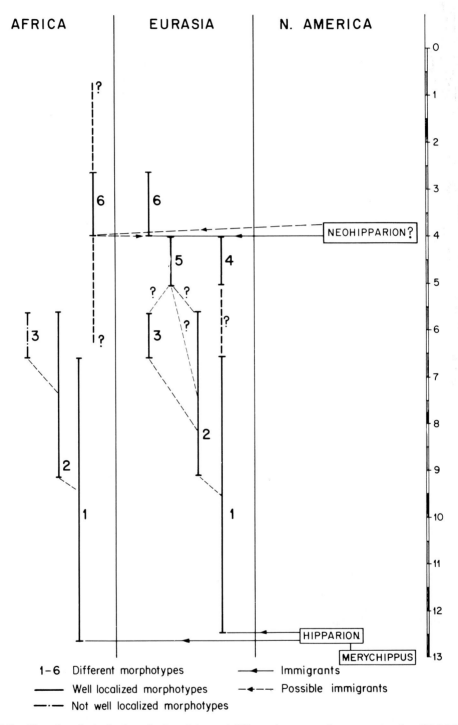

Fig. 13.5. Sketch of phyletic relationships of *Hipparion* morphotypes in the Old World Neogene.

characteristics (climate, tectonic, and so on) of this epoch in the Mediterranean, the increase of continental sediment, as well as the fact that the majority of the localities with *H. crassum* are in lignite mines, I think that this species originated in the place where it first appeared. It migrated quickly because travel was much easier at that time. Since morphotype 5 seems to be a consequence of a strong steppe adaptation, it should be in theory very difficult to find it in the same localities as morphotype 4, such as Çalta, Turkey (Ginsburg et al., 1974; Sen et al., 1974). The beds containing *H. crassum* are so geographically restricted that it is necessary to analyze the deposits in which *H. crassum* appears, together with a clearly steppic fauna (Heintz et al., 1975). On the other hand, these two morphotypes are adapted to a period of low sea level. One could have lived in grasslands, and the other could have existed in the marshes. Both probably looked for food in the same areas.

Morphotypes 4 and 5 survived the climatic changes of around 5.5 Ma (see Fig. 13.4) during the salinity crisis of the Mediterranean. This allowed a wider dispersion of the fauna. Steininger et al. (1985, Fig. 26.5) wrote of "the Messinian 'continentalization' effect in relation to the Turolian chronofauna," that could have favored the passage of the Euroasiatic fauna to Africa. This could be when *Hipparion* crossed Spain on its way from Europe to North Africa (Alberdi, 1974; Forstén, 1978b) as represented by *H. gromovae* from Spain and *H. sitifense* from North Africa (which I believe are the same species; Alberdi, 1974, 1986), but the demonstration of this theory will be the subject of another work (Alberdi, in prep.)

The origin of morphotype 6 is more problematic because, up to now, we have not found any form from which it could have evolved. Eisenmann (1977) documented the possible presence of the caballine pattern in African *Hipparion* since 6.5 Ma. She also said (1979a) that it is not as clear nor as well documented after 4 Ma. As far as I am concerned, I think that morphotype 6 could be the result of convergent evolution of the climatic and ecological conditions that emphasized its distinctive characteristics. It is unlikely that the caballine-patterned *Hipparion* in Eurasia is from Africa.

Zhegallo (1978), and lately Forstén (1982b), following his idea, grouped the caballine *Hipparion* within the subgenus *Neohipparion*. I think that these forms could be the result of a new migratory wave from North America (*Neohipparion*) that would have produced morphotype 6 in Eurasia, e.g., *H. rocinantis* in China, Georgia, and Spain. This crossing could have happened during the Ruscinian (Steininger et al., 1985).

On the other hand, Qiu et al. (1980) stated that the Chinese form (*H. houfenense*) and *H. turkanense* in Africa evolved from *H. hippidiodum* of China. Qiu and Chow (1985) said that *H. houfenense* was close to *H. crusafonti* (synonymous with *H. rocinantis*) and placed it within the Ruscinian *Hipparion* of China, where they are poorly represented. But Eisenmann (1981a, 1982) said that *H. hippidiodum* and *H. turkanense* skulls don't resemble each other.

I think that the first caballine forms had already reached Asia during the Ruscinian, and from there spread all over Eurasia. Their dispersal in Africa could have developed in a way parallel to the caballine forms already living there. In any case, the fact that the African stratigraphy is still poorly known should be taken into account; the existence of caballine forms in Africa before the possible second wave of caballine *Hipparion* in Eurasia is not yet clearly established. Moreover, I think that the caballine *Hipparion* was much more successful in Africa than in Europe, where it appeared in more restricted areas.

If we could correlate the stratigraphy of Qiu et al. (1980) with that of Eisenmann

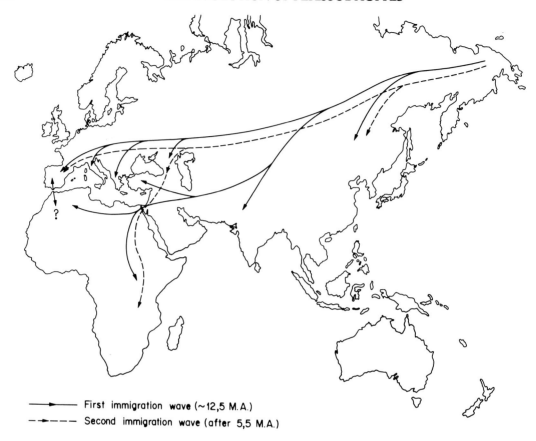

Fig. 13.6. Possible *Hipparion* migration waves into Eurasia and Africa during the Neogene.

(1977) their data would probably be complementary, and could explain the immigration of *Hipparion* from North America as the origin of the caballine forms.

## Paleobiological and paleoecological deductions

The discussion above raises several questions. It is interesting to analyze the kind of adaptation that permitted the persistence of *Hipparion* for a long period of time, lasting around 10 million years (between 12.5 to 2.5 Ma), and in which type of environment this persistence developed.

We can deduce from the climatic data (Fig. 13.4), that these animals preferred a warm and humid environment. In fact, the first cold episode around 3 to 2.5 Ma produced extinctions in Eurasia where, due to the latitude, the influence of this cold must have been more pronounced. Actually, it seems that this animal could have survived in regions such as Africa, where the arrival of aridity, linked to the formation of the first Arctic ice-sheet, could have been mitigated by the temperature that was higher in Africa than in the Eurasian regions. The fossil record of *Hipparion* in Africa, subsequent to the extinction of the Eurasian *Hipparion* (2.5 Ma), is found in savanna and fluvial environments of eastern and southern Africa. This phenomenon is more obvious in the east than the south. We also observe a sudden reduction of *Hipparion* species in the Ruscinian (fewer localities and scarcer remains) after the first cold episode of the Messinian; however, it did not reach the intensity of the middle

Villafranchian. This hypothesis agrees with that of Van Couvering (1980) concerning the slow progressive climatic change in Africa relative to the sudden changes in Eurasia.

The co-occurrence of morphotype 4, with its massive build and large dentition, and of morphotype 5, with more reduced dental structure and elongated limbs, at a time of lower sea level, could be explained by environmental considerations discussed above. Morphotype 4 always appears in lignite deposits typical of flooded areas, marshes, and pools, which are due to the sea-level decrease. These animals, living in a marsh environment, could have become more massive. Meanwhile, the higher lands of the same area, covered by grassland or savanna, could represent a more suitable habitat for a runner, such as morphotype 5.

## Conclusion

The first wave of immigrant horses in the middle upper Miocene produced a wide expansion of *Hipparion* in the Old World and the differentiation of various species. Between the Messinian and the formation of the first Arctic ice sheet, the passage of a second wave from North America could have taken place and resulted in morphotype 6, independent of previous forms of *Hipparion* of Eurasia and Africa (Fig. 13.6). From this point of view, the presence of caballine forms in Africa before the Messinian could be explained by convergent evolution *in situ* of *Hipparion* forms that evolved from the Old World forms and that could have mixed with the new post-Messinian wave. The validity of this explanation could be tested if the African Neogene chronostratigraphy were more certain, but today, this information cannot yet be confirmed (see Fig. 13.5).

On the other hand, the presence of a *Hipparion* caballine form (*H. houfenense* Qiu, 1985) in China during the Ruscinian could test the hypothesis of the new post-Messinian wave quoted above. The existence of this second wave of *Hipparion* coming from North America will have to be tested by the study of the phyletic relations between *Neohipparion* and *H. rocinantis* of the Old World.

Finally, I have discussed a scheme of possible migration paths of the two waves of *Hipparion* in the Old World during the Neogene (Fig. 13.6).

## Acknowledgments

I am very grateful to Drs. V. Eisenmann and F. P. Bonadonna for reading the manuscript and providing me with documentation, as well as for their many useful comments. I especially thank Drs. D.R. Prothero and R.M. Schoch for their corrections of the English text. I also thank Ms. M. T. Montero for typing the manuscript, and Mr. J. Arroyo for drawing the figures.

## Bibliography

Aguirre, A., and Alberdi, M. T. (1974): *Hipparion* remains from the northern part of the Rift Valley (Kenya). - *Koninkl. Nederl. Academie Wetensch., Amsterdam*, 77: 146-157.

Alberdi, M. T. (1972): El género *Hipparion* en España, nuevas formas de Castilla y Andaluca, revisión e historia evolutiva. -Ph.D. Dissertation, Univ. Complutense de Madrid: 1-368.

Alberdi, M. T. (1974): El género *Hipparion* en España. Nuevas formas de Castilla y Andaluca, revisión e historia evolutiva. -*Trabajos sobre Neógeno-Cuaternario, Madrid*, 1: 1-146.

Alberdi, M. T. (1978): El género *Hipparion* en la provincia de Teruel (España). - *Teruel*, 60: 1-16.

Alberdi, M. T. (1986): The Pliocene species of *Hipparion* and their biostratigraphical meanings. -*Geobios*, 19: 517-522.

Alberdi, M. T. (in press): Los Perissodactyla del yacimiento de las Higueruelas y su entorno. -*Cuadernos Estudios Manche-*

gos, Ciudad Real.
Alberdi, M. T., Antunes, M. T., Sondaar, P. Y., and Zbyszewski, G. (1978): Les *Hipparion* du Portugal. -*Ciências da Terra (UNL)*, 4: 129-156.
Alberdi, M. T., and Bonadonna, F. P. (1983): El *Equus stenonis* Cocchi como indicador biostratigráfico del Plio-pleistoceno en Italia y España. -*Cuadernos de Laboratorio Xeoloxico de Laxe*, 5: 167-187.
Alberdi, M. T., and Gabunia, L. K. (1985): Comparison of Georgian and Spanish hipparions. -*Bull. Acad. Sci. Georgian SSR*, 118: 641-644.
Alberdi, M. T., Jimenez, E., Mazo, A. V., Morales, J., Sese, C., and Soria, D. (1986): Paleontología y Biostratigrafía de los yacinmientos Villafranquinses de Las Higueruelas y Valverde de Calatrava II (Campo de Calatrava, Ciudad Real). - *I Reunión Regional de Castilla -La Mancha*. Albacete, mayo 1984 , 5: 249-277.
Alberdi, M. T., and Morales, J. (1981): Significado biostratigráfico del género *Hipparion* en España. -*Teruel*, 66: 61-66.
Arambourg, C. (1959) Vertébrés continentaux du Miocène supérieur de l'Afrique du Nord. -*Mém. Serv. Carte Géol. Algérie*, 4: 1-161.
Badgley, C. (1986): Taphonomy of mammalian fossil remains from Siwalik rocks of Pakistan. -*Paleobiology*, 12: 119-142.
Berggren, W. A., and Van Couvering, J. A. (1974): The Late Neogene. Biostratigraphy, geochronology and paleoclimatology of the last 15 million years in marine and continental sequences. - *Palaeogeog., Palaeoclim., Palaeoecol.*, 16: 1-216.
Bernor, R. L. (1983): Geochronology and zoogeographic relationships of Miocene Hominoidea. -In: Ciochon, R. L. and Corruccini, R. S. (eds.): *New Interpretations of Ape and Human Ancestry*: New York (Plenum), pp. 21-64.
Bernor, R. L. (1984): A zoogeographic theater and biochronologic play: the time / biofacies phenomena of Eurasian and African Miocene mammal provinces. - *Paléobiologie continentale, Montpellier*, 14: 121-142.
Bernor, R. L. (1985): Systematic and evolutionary relationships of the hipparionine horses of Maragheh, Iran (Late Miocene, Turolian Age). -*Palaeovert.*, 15: 173-269.
Bernor, R. L., and Hussain, S. T. (1985): An assessment of the systematic, phylogenetic and biogeographic relationships of Siwalik Hiparionine horses. -*J. Vert. Paleont.*, 5: 32-87.
Bernor, R. L., Woodburne, M. O., and Van Couvering, J. A. (1980): A contribution to the chronology of some Old World Miocene faunas based on hipparionine horses. -*Geobios*, 13: 705-739.
Bernor, R. L., Qiu, Z. X., and Tobien, H. (1985): Phylogenetic and biostratigraphic bases for an Old World hipparionine horse geochronology. - *Abst. VIIIth Congress Inter. RCMNS*, 15-22 September, 1985.
Bonadonna, F. P., and Alberdi, M. T. (1987): *Equus stenonis* Cocchi as a biostratigraphical marker in the Neogene-Quartenary of the Western Mediterranean Basin: consequence on Galerian-Villafranchian chronostratigraphy. - *Quat. Sci. Rev.*, 6: 55-66.
Christol, J. de (1832): Description of *Hipparion*. -*Ann. Sci. l'Industrie du Midi de la France*, 1: 180-181.
Churcher, C. S., and Richardson, M. L. (1978): Equidae- In: Maglio, V. J. and Cooke, H.B.S. (eds.): *Evolution of African Mammals*: Cambridge, Mass. (Harvard Univ. Press), pp. 379-422.
Crusafont, M., and Sondaar, P. Y. (1971): Une nouvelle espèce d'*Hipparion* du Pliocène terminal d'Espagne.- *Palaeovert.*, 4: 59-66.
Eisenmann, V. (1976a): Nouveaux cranes d'hipparions (Mammalia, Perisso-

dactyla) Plio-Pléistocenes d'Afrique orientale (Ethiopie et Kenya) : *Hipparion* sp., *Hipparion* cf. *ethiopicum* et *Hipparion afarense* nov. sp. -*Geobios*, 9: 577-605.

Eisenmann, V. (1976b): Le protostylids: valuer systématique et signification phylétique chez les espèces actuelles et fossiles du genre Equus (Perissodactyla, Mammalia).- *Zeit. Saugetierk.*, 41: 349-365.

Eisenmann, V. (1977): Les hipparions africains: valeur et signification de quelques caracterès des jugales inférieures. -*Bull. Mus. Nat. Hist. Naturelle*, 438: 69-86.

Eisenmann, V. (1979a): Le genre *Hipparion* (Mammalia, Perissodactyla) et son interêt biostratigraphique en Afrique. -*Bull. Soc. Géol. France, Paris*, 21: 277-281.

Eisenmann, V. (1979b): Étude des cornets des dents incisives inférieures des *Equus* (Mammalia, Perissodactyla) actuels et fosiles.-*Palaeont. Italica*, 71: 55-75.

Eisenmann, V. (1979c): Les métapodes d'*Equus* sensu lato (Mammalia, Perissodactyla). -*Geobios*, 12: 863-886.

Eisenmann, V. (1980a): Les Chevaux *Equus* (*sensu lato*) fossiles et actuels: crânes et dents jugales supérieures.- *Cahiers de Paléontologie CNRS, Paris*: 1-186.

Eisenmann, V. (1980b): Caractères spécifiques et problémes taxonomiques relatifs a certains hipparions africains. -*Actes du VIII$^e$ Congrés panafricain de Prehistoire*, 1: 77-81.

Eisenmann, V. (1981a): Les caractères évolutifs des crânes d'*Hipparion* sensu lato (Mammalia, Perissodactyla) et leur interpretation. -*C. R. Acad. Sci. Paris*, 293: 735-738.

Eisenmann, V. (1981b): Étude des dents jugales inférieures des *Equus* (Mammalia, Perissodactyla). Actuels et fossiles, -*Palaeovert.*, 10: 127-226.

Eisenmann, V. (1981c): The arrival of *Equus*.- *Nature*, 298: 865.

Eisenmann, V. (1982): La phylogénie des *Hipparion* (Mammalia, Perissodactyla) d'Afrique d'aprés les caractères craniens. -*Proc. Konink. Nederl. Akad. Wetensch. Palaeont.*, 85: 219-227.

Eisenmann, V. (1983): 5. Family Equidae. - In: Harris, J M. (ed.): The fossil ungulates: Proboscidea, Perissodactyla and Suidae. -*Koobi Fora Research Project*, 2: 156-214.

Eisenmann, V. (1985): Les Équidés des gisements de la Vallée de l'Omo en Éthiopie.- In: Coppens, Y. and Howell, F. C. (eds.): Les Faunes Plio-Pléistocènes de la Basse Vallée de l'Omo (Éthiopie), 1: Périssodactyles. Artiodactyles (Bovidae). -*Cahiers de Paléontologie. CNRS, Paris*: 1-55.

Eisenmann, V., Alberdi, M. T., de Giuli, C., and Staesche, U. (1988): *Studying Fossil Horses*.- Leiden (E.J. Brill), pp. 1-72.

Eisenmann, V., and Karchoud, A. (1982): Analyses multidimensionnelles des métapodes d'Equus. -*Bull. Mus. Nat. Hist. Natur.*, 4: 75-103.

El-Khashab, B. (1977): Some studies on Egyptian Vertebrates Fossils. -*Geol. Surv. Egypt Min. Auth. Paper*, 62: 1-40.

Forstén, A. M. (1968): Revision of the Palearctic *Hipparion*. - *Acta Zool. Fenn*, 119: 1-134.

Forstén, A. M. (1972): *Hipparion primigenium* from southern Tunisia. -*Notes du Service Géol.*, 35: 7-28.

Forstén, A. M. (1978a): *Hipparion primigenium* (v. Meyer, 1829), an early three-toed horse. -*Ann. Zool. Fenn.*, 15: 298-313.

Forstén, A. M. (1978b): *Hipparion* and possible Iberian-North Africa Neogene connections. -*Ann. Zool. Fenn.*, 15: 294-297.

Forstén, A. M. (1978c): A review of Bulgarian *Hipparion* (Mammalia, Perissodatyla). -*Geobios*, 11: 31-41.

Forstén, A. M. (1978d): *Hipparion*

(Mammalia, Equidae ) from Luan Fu, S. E. Shansi, China. -*Publ. Natuurhist. Genootsch. Limburg NLD*, 28: 8-17.

Forstén, A. M. (1980a): How many *Hipparion* species at Samos? -*N. Jb. Geol. Paläont. Mh.*, 7: 391-396.

Forstén, A. M. (1980b): Hipparions of the *Hipparion mediterraneum* group from South-Western Soviet Union. -*Ann. Zool. Fenn.*, 17: 27-38.

Forstén, A. M. (1981): Correlation of *Hipparion* faunal components in the Eastern Mediterranean and Paratethys areas. -*Paläont. Zeit.*, 5: 141-155.

Forstén, A. M. (1982a): The taxonomic status of the Miocene horse genus *Sinohippus*.-*Palaeont.*, 25: 673-679.

Forstén, A. M. (1982b): The status of the genus *Cormohipparion* Skinner and MacFadden (Mammalia, Equidae). -*J. Paleont.*, 56: 1332-1335.

Forstén, A. M. (1982c): Temporal differentiation of Central European *Hipparion* teeth. -*N. Jb. Geol. Paläont. Mh.*, 6: 336-346.

Forstén, A. M. (1983): The preorbital fossa as a taxonomic character in some Old World *Hipparion*. -*J. Paleont.*, 57: 686-704.

Forstén, A. M. (1984): Supraspecific grouping of Old World *Hipparion* (Mammalia, Equidae). -*Paläont. Zeit.*, 58: 165-171.

Forstén, A. M. (1985a): Chinese Turolian *Hipparion* in the Lagrelius Collection. -*Bull. Geol. Inst. Univ. Uppsala*, 11: 113-124.

Forstén, A. M. (1985b): *Hipparion primigenium* from Höwnegg / Hegan, FRG. -*Ann. Zool. Fenn.*, 22: 417-422.

Gabunia, L. K. (1959): Histoire de genre *Hipparion*. -*Bur. Rech. géol. Min., Traduct.* n. 2696, 1: 515.

Gaudry, A. (1862): *Animaux fossiles et géologie de l'Attique*. -Paris.

Gidley, J.W. (1903): A new three-toed horse.-*Bull. Amer. Mus. Nat. Hist.*, 19: 465-476.

Ginsburg, L., Heintz, E., and Sen, S. (1974): Le gisement pliocène à mammifères de Çalta (Ankara, Turquie). -*C. R. Acad. Sci. Paris*, 278: 2739-2742.

Gromova, V. (1952): Le genre *Hipparion*. - *Bur. Rech. Géol. Min. Traduc.* n. 12: 1

Heintz, E., Ginsburg, L., and Sen, S. (1975): *Hipparion longipes* Gromova du Pliocène de Çalta (Ankara, Tuquie), le plus dolichopodial des hipparions. - *Konink. Ned. Akad. Wetensch. Amsterdam*, 78: 77-82.

Hernandez-Pacheco, E. (1921): La llanura manchega y sus mamíferos fósiles (yacimiento de La Puebla de Almoradier). -*Com. Invest. Paleont. Prehistorica, Madrid*, 28: 1-42.

Hooijer, D. A. (1975): Miocene to Pleistocene hipparions of Kenya, Tanzania and Ethiopia. -*Zool. Verhandl.*, 142: 3-80.

Hussain, S. T. (1971): Revision of *Hipparion* (Mammalia, Equidae) from the Siwalik Hills of Pakistan and India. -*Verlag Bayer. Akad. Wissensch. München*, 147: 1-68.

Hussain, S. T. (1975): Evolutionary and functional anatomy of the pelvic limb in fossil and recent Equidae (Mammalia, Perissodactyla). -*Anat. Histol. Embryol.*, 4: 179-222.

Hussain, S. T., and Bernor, R. L. (1984): Evolutionary history of Siwalik hipparions. -*Cour. Forsch. -Inst. Senckenberg*, 69: 181-187.

Isaac, G. (1977): Olorgesailie. Archeological studies of a middle Pleistocene Lake basin in Kenya. -In: Butzer, K. W., and Freeman, L. G. (eds.): *Prehistoric Archeology and Ecology Series*. - Chicago (Univ. Chicago Press), pp. 1-271.

Isaac, G. (1978): 13. The Olorgesailie Formation: stratigraphy, tectonics and the palaeogeographic context of the Middle Pleistocene archaeological sites. - In: Bishop, W. W. (ed.): *Geological Background to Fossil Man*. -London (The Geological Society), pp. 173-206.

Kaup, J. J. (1833): Über die Gattung Dinotherium, Zusätze und Verbesserungen zum ersten heft der description d'ossemens fossiles. -*Neues Jahrb. Min. Geol. Paläont.*, 509-517.

Koufos, G. D. (1980): Palaeontological and stratigraphical study of the continental Neogene deposits of Axios basin. -*Sci. Ann. Fac. Phys. Math., Univ. Thessaloniki*, 19: 1-322 (Doctoral thesis).

Koufos, G. D. (1982): *Hipparion crassum* Gervais, 1859 from the lignites of Ptolemais (Macedonia, Greece). - *Konink. Nederl. Akad. Wetensch. Palaeont.*, 85: 229-239.

Koufos, G. D. (1984): A new hipparion (Mammalia, Perissodactyla) from the Vallesian (late Miocene) of Greece. -*Paläont. Zeit.*, 58: 307-317.

Koufos, G. D. (1986): Study of the Vallesian hipparions of the Lower Axios Valley (Macedonia, Greece). -*Geobios*, 19: 61-79.

Koufos, G. D. (1987): Study of the Pikermi hipparions. Part I: Generalities and taxonomy. Part II: Comparisons and odontograms.- *Bull. Mus. Nat. Hist. Natur.*, 9: 197-252, 327-363.

Kretzoi, M. (1983): Gerinces indexfajok felsö-neozoi refegtaneinkban *Hipparion*.- *M. All Földtain Intézet. évi Jelentése* A2 1981: 513-521.

Lydekker, R. (1884): Indian Tertiary and post-Tertiary Vertebrata.- *Palaeontologica Indica*, ser. 10, 3: 1-8.

MacFadden, B. J. (1980): The Miocene horse *Hipparion* from North America and from the type locality in southern France. -*Palaeontol.*, 23: 617-635.

MacFadden, B. J. (1984): Systematics and phylogeny of *Hipparion, Neohipparion, Nannippus,* and*Cormohipparion* (Mammalia, Equidae) from the Miocene and Pliocene of the New World. - *Bull. Amer. Mus. Nat. Hist.*, 179: 1-196.

MacFadden, B. J. (1985): Patterns of phylogeny and rates of evolution in fossil horses: hipparions from the Miocene and Pliocene of North America. - *Paleobiology*, 11: 245-257.

MacFadden, B. J., and Bakr, A. (1979): The horse *Cormohipparion theobaldi* from the Neogene of Pakistan, with comments on Siwalik hipparions. - *Palaeontol.*, 22: 439-447.

MacFadden, B. J., and Skinner, M. F. (1981): Earliest Holarctic hipparion, *Cormohipparion goorisi* n. sp. (Mammalia, Equidae), from the Barstovian (Medial Miocene) Texas Gulf Coastal Plain. -*J. Paleont.*, 55: 619-627.

MacFadden, B. J., and Skinner, M. F.(1982): hipparion horses and modern phylogenetic interpretation: comments on Forsten's view of *Cormohipparion* . -*J. Paleont.*, 56: 1336-1342.

MacFadden, B. J., and Woodburne, M. O. (1982): Systematics of the Neogene Siwalik *Hipparion* (Mammalia, Equidae) with special reference to the Yale-CSP collection from the Potwar Plateau. -*J. Vert. Paleont.*, 2: 185-218.

Matthew, W.D. (1926): The evolution of the horse; a record and its interpretation. -*Q. Rev. Biol.*, 1: 129-185.

Meladze, G. K. (1967): *Hipparionovaia fauna Arkneti i Bazaleti.* Izdat. "Metsnierabl," Tbilissi: 1-168

Musil, R. (1969): Die Pferde de Pekarna-Höhle. -*Tierzücht Züchtungsbiol.*, 86: 147-193.

Ozansoy, F. (1965): Étude des gisements continentaux et des mammifères du cenozoique de Turquie. -*Mém. Soc. Géol. France*, 44: 1-92.

Pirlot, P. L. (1953): The preorbital fossa of *Hipparion*. -*Comm. Am. J. Sci.*, 251: 309-312.

Pirlot, P. L. (1956): Les formes européennes du genre *Hipparion*. -*Mem. Com. Inst. Geol. Dip. Prov. Barcelona CSIC*, 14: 1-121.

Prat, F. (1968): Recherches sur les equidés pléistocènes en France. -These Coct. Et. Sc. Facult. Sc. Univ. Bordeaux, 226: 1-662.

Pevzner, M. A., and Vangengein, E. A. (1985): Magnetostratigraphy and correlation of biostratigraphic subdivisions of the Paratethyan and Mediterranean Neogene. -*Abstr. VIIIth Congress RCMNS*, 15-22 September 1985, Budapest, pp. 461-462.

Qiu Z. X., and Chow M. Z. (1985): Three Ruscinian *Hipparion* species of China. -*Abstr. of Papers and Posters IVth ITC Edmonton, Alberta, Canada 13-20 Aug.1985:* abstract number 0508.

Qiu Z. X., Huang W. L., and Kuo Z. H. (1980): Notes on the first discovery of the skull of *Hipparion houfenense.* -*Vert. PalAsiatica*, 18: 131-137.

Radulesco, C., and Samson, P. (1967): Sur la significaton de certains Equides du Pléistocène inferieur et moyen de Roumanie. -*N. J.b Geol. Paläont. Abh.*, 127: 157-178.

Samson, P. (1975): Les Equides fossiles de Roumanie (Pliocène Moyen- Pléistocene supérieur). -*Geol. Romania*, 14: 165-352.

Savin, S. M. (1977): The history of the Earth surface temperature during the past 100 million years. -*Ann. Rev. Earth Planet. Sci.*, 5: 319-355.

Sefve, I. (1927): Die Hipparionen Nord Chinas. -*Paleont. Sinica*, 4: 1-93.

Sen, S., Heintz, E. and Ginsburg, L. (1974): Premiers résultats des foulles effectués à Çalta, Ankara, Turquie. -*Bull. Min. Research Explorat. Inst, Turk.*, 83: 112-118.

Sen, S., Sondaar, P. Y., and Staesche, U. (1978): The biostratigraphical applications of the genus *Hipparion* with special references to the Turkish representatives. -*Konink. Nederl. Akad. Wet. Amsterdam, Palaeont.*, 81: 370-385.

Shackleton, N. J. (1984): Oxygen isotope evidence for Cenozoic climatic change. In: Brenchley, P. (ed.): *Fossils and Climate*, New York (John Wiley), pp. 27-34.

Shackleton, N. J., and Hall, M. A. (1984): Oxygen and carbon isotope stratigraphy of Deep Sea Drilling Project-Hole 552A: Plio-Pleistocene glacial History. -In: Roberts, D. G., Schnitker, D. *et al.* (eds.): *Initial Report of Deep Sea Drilling Project*, 71: 599-609.

Simpson, G. G. (1944): *Tempo and Mode in Evolution.* New York (Columbia Univ. Press.).

Skinner, M. F. and MacFadden, B. J. (1977): *Cormohipparion* n. gen. (Mammalia, Equidae) from the North American Miocene (Barstovian-Clarendonian). -*J. Paleont.*, 51: 912-926.

Sondaar, P. Y. (1961): Les *Hipparion* de l'Aragon meridional. -*Estudios Geol., Madrid*, 17: 209-305.

Sondaar, P. Y. (1968): The osteology of the manus of fossil and recent Equidae with special reference to phylogeny and function. -*Verhandl. Konink. Nederl. Akad., Westesch. Nat., Amsterdam*, 25: 1-76.

Sondaar, P. Y. (1971): The Samos *Hipparion* I and II. -*Koninkl. Neder. Akad., Westensch., Amsterdam*, 74: 417-441.

Sondaar, P. Y. (1974): The *Hipparion* of the Rhone Valley. -*Geobios*, 7: 289-306.

Staesche, U., and Sondaar, P. Y. (1979): *Hipparion* aus dem Vallesium und Turolium (Jungtertiar) der Turkei. -*Geol. Jb.*, 33: 35-79.

Steininger, F. F., Rabeder, G., and Rögl, F. (1985): Chapter 26. Land Mammal distribution in the Mediterranean Neogene: A consequence of geokinematic and climatic events. -In: Stanley, D. J., and Wezel, F. C. (eds.): *Geological Evolution of the Mediterranean Basin.* -Raimondo Selli Commem. Vol. (Springer Verlag), pp. 559-571.

Tobien, H. (1982): Preliminary report on the Equidae (Perissodactyla, Mammalia) from the Sahabi Formation (Libya). -In: Boaz N. T., Gatity, A. W., Heinzelin, J., and El-Arnuti, A. (eds.): Results from the International Sahabi Research Project (Geology and Paleontology).- *Garyounis Scientific Bull., Special Issue*, 4: 83-85.

Vail, P. R., Mitchum, R. M., Jr., and Thompson, S. III (1977): Seismic stratigraphy and global changes of sea level, Part 4: Global cycles of relative changes of sea level. -*Amer. Assoc. Pet. Geol. Mem.*, 26: 83-97.

Van Couvering, J. A. H. (1980): Community evolution in East Africa during the late Cenozoic.- In: Behrensmeyer, A. K., and Hill, A. P. (eds.): *Fossils in the Making. Vertebrate Taphonomy and Paleoecology*. Chicago (Univ. Chicago Press), pp. 212-298.

Vekua, A. K. (1972): Fauna Akchequiliense de Kwabebi (Georgia). -*Akad Nauk. Moskva*, 556: 1-350.

Villalta, J. F. de (1948): Una nueva especie de *Hipparion* del Villafranquiense. - *Ext. Museo de la Ciudad de Sabadell*, 4: 5-10.

Villalta, J. F. de (1952): Contribución al conocimiento de la fauna de mamíferos fósiles del Plioceno de Villarroya (Logroño). -*Bol. Inst. Geol. Min. España*, 64: 1-204.

Webb, S. D. (1969): The Burge and Minnechaduza Clarendonian mammalian faunas of north-central Nebraska. - *Univ. Calif. Publ. Geol. Sci.*, 78: 1-191.

Woodburne, M. O. and Bernor, R. L. (1980): On superspecific groups of some Old World hipparionine horses. -*J. Paleont.*, 54: 1319-1348.

Woodburne, M. O., MacFadden, B. J. and Skinner, M. F. (1981): The North American *Hipparion* datum and implications for the Neogene of the Old World. -*Geobios*, 14: 493-524.

Zhegallo, V. I. (1978): The *Hipparion* of Central Asia. -*The joint Soviet-Mongolian Paleont. Exp. Trudy*, 7: 1-152.

# 14. A QUANTITATIVE STUDY OF NORTH AMERICAN FOSSIL SPECIES OF THE GENUS *EQUUS*

## MELISSA C. WINANS

The relationships of North American fossil species of *Equus* are reviewed, using multivariate quantitative characters of skull, mandible, and metapodials. The genus can be divided into no more than five groups: 1) *E. simplicidens* group (early Blancan); 2) *E. scotti* group (late Blancan to Rancholabrean); 3) *E. laurentius* group (Rancholabrean); 4) *E. francisci* group (Irvingtonian to Rancholabrean); 5) *E. alaskae* group (Irvingtonian to Rancholabrean). A number of qualitative characters appear to support this division, but until a companion study of qualitative characters is completed, it is unclear whether these groups represent species or some higher-level grouping. The small number of Blancan and Pleistocene horse groups is in sharp contrast to the higher diversity of the family during the late Miocene and early Pliocene, and is matched by reductions in the diversity of several other groups of large herbivores. I attribute this reduction to the adverse effects of increasing aridity and cooling of the climate, and to the increase in areas of grassland. These changes had a profound effect on the composition of plant communities, and presumably on the herbivores that depended on them.

## Introduction

The genus *Equus* in North America is generally considered to include all Blancan to Recent members of the family Equidae except for members of the Blancan genus *Nannippus* and the Pleistocene South and Central American genera *Hippidion* and *Onohippidium*. From 1842 to the present, 59 species and 5 subspecies of *Equus* have been named from North American fossil material. Of this number, 13 have subsequently been demonstrated to belong to genera other than *Equus*, and 3 are invalid because they were preoccupied names (for details, see Winans, 1985). To date, all methods proposed to differentiate these species have employed qualitative or univariate or bivariate quantitative comparisons, with greatest emphasis upon characters observed in the teeth and metapodials (the skeletal elements most often preserved). As with many other taxa, a substantial percentage of the species have been based on individual, isolated, often fragmentary type specimens. While not necessarily bad *per se*, the selection of such specimens as types always raises the possibility that some of the differentiating characters that have been chosen may be ones with large ranges of ontogenetic or individual variation, and thus of doubtful taxonomic significance. A survey of publications on the subject of horse taxonomy, coupled with numerous discussions with other paleontologists who have worked with *Equus*, suggests that many of the commonly used characters do suffer from this problem, and that a majority of the species and subspecies named to date are of doubtful validity. However, no two lists of which characters and species are valid agree completely, and attempts to arrive at a satisfactory revision have not been successful. Even among the extant species, where the taxonomist usually has the advantages of complete specimens and advance knowledge of each specimen's identity, the magnitude of differences in skeletal morphology between species frequently is not significantly greater than the magnitude of the normal range of variation within any single species such that it is difficult to find consistent and easily applied qualitative or quantitative differences. For the paleontologist, often limited to isolated and fragmentary skeletal parts, reliable differentiation becomes even more difficult.

These problems have long been recognized. For example, Gidley (1901), in the first revision of the genus, commented extensively, and for the most part negatively, on the reliability of characters used by previous taxonomists. However, he was unsuccessful in defining a completely reliable alternative set of characters, and his revision includes a number of the same characters which his opening discussion had condemned. Later extensive revisions by Savage (1951), Hibbard (1955), and Dalquest (1978, 1979) and smaller-scale discussions, synonymies, and rejections of synonymies too numerous to list here (for an extensive, but still incomplete, list, see Winans, 1985) have each added their increment to our understanding of the genus. Nonetheless, the classification schemes proposed to date are frustratingly difficult to apply to the identification of new specimens, for in many cases identifications based on one scheme are contradicted by those based on other schemes.

This paper is one more round in the ongoing effort. The multivariate quantitative analyses reported here are the first part of a larger study of both qualitative and quantitative characters, the full results of which will be reported at a later time. I hope that the relationships which these analyses suggest will form the basis for further discussion and exploration, by others as well as by myself.

**Abbreviations used:**

| | |
|---|---|
| AMNH | American Museum of Natural History |
| IGM | Instituto Geologico de Mexico |
| KUMNH | Kansas University Museum of Natural History |
| LACM | Los Angeles County Museum of Natural History |
| Ma | Millions of years before present |
| SMP | Shuler Museum of Paleontology, Southern Methodist University |
| TAMU | Texas A & M University |
| TMM | Texas Memorial Museum, University of Texas at Austin |
| UA | University of Arizona |
| UCMP | University of California Museum of Paleontology |
| USNM | United States National Museum, Smithsonian Inst. |
| YPM | Yale Peabody Museum |

**Materials and methods**

I believe that one major obstacle to a satisfactory solution has been a tendency (common to earlier work on nearly all taxa, not just *Equus*) to base studies on very limited numbers of specimens. Such an approach makes it difficult to gauge whether differences found among individual specimens are consistent within the population as a whole, and thus potentially of taxonomic significance, or are so variable within the population as to have no diagnostic value. This study is aimed at amassing a large enough body of data to make it possible to obtain a clear picture of the ranges and patterns of variation in skeletal morphology.

Ideally, a study of this sort should concentrate exclusively on large site samples that appear on stratigraphic and faunal evidence to represent accumulations of material over a geologically brief period of time. These should provide information about how the characters vary within local populations. However, few such "ideal" samples actually exist. Thus all reasonably complete specimens have been included (as summarized in Figs. 14.1 and 14.2), but in interpreting the results, greatest weight has been given to assemblages of 10 or more individuals from a single site. Where specimens from a site show a strong bimodal distribution in size or proportions, these subgroups have been treated separately. With the exception of a few type specimens, only adult specimens were included in the study, because the developing skeleton undergoes large changes in size and proportions, which might blur the distinctions between taxonomic groups. No effort was

Fig. 14.1. Sites included in the study. The count of minimum number of individuals is based on complete skulls, rami, or metapodials, or on number of type specimens. A star (*) after the site name indicates that the sample was divided into large and small subgroups. A "T" after the site name indicates a site from which type specimens came. Not all sites could be located precisely enough to be mapped. Some map numbers encompass more than one site.

| Site Name and Location | | Number of Individuals (minimum) | Map Number |
|---|---|---|---|
| American Falls, Gooding Co., Idaho | | 12 | 14 |
| Arkalon Gravel Pit, Seward Co., Kansas | * | 7 | 26 |
| Arredondo I, Alachua Co., Florida | | 1 | |
| Arroyo Cedazo, Aguascalientes, Mexico | T | 5 | 15 |
| Bautista Creek, Riverside Co., California | T | 1 | 2 |
| Blanco Creek, Goliad Co., Texas | T | 1 | |
| Cameron, Milam Co., Texas | T | 1 | |
| Channing, Hartley Co., Texas | | 38 | 25 |
| Charleston, South Carolina | T | 1 | |
| Cita Canyon, Randall Co., Texas | | 10 | 5 |
| Coleman II, Sumter Co., Florida | | 1 | |
| Comosi Wash, Santa Cruz Co., Arizona | T | 1 | |
| Cueva Quebrada, Val Verde Co., Texas | * | 10 | 27 |
| Dam Site, Gooding Co., Idaho | | 9 | 20 |
| Dry Cave, Eddy Co., New Mexico | * | 10 | 22 |
| El Golfo, Baja California | | 1 | 28 |
| Froman Ferry, Canyon Co., Idaho | T | 1 | |
| Gilliland, Baylor & Knox Cos., Texas | | 8 | 9 |
| Grandview, Owyhee Co., Idaho | | 5 | 3 |
| Hagerman, Gooding Co., Idaho | | 19 | 4 |
| Hay Springs, Sheridan Co., Nebraska | *T | 21 | 11 |
| Holloman Gravel Pit, Tillman Co., Oklahoma | T | 1 | 10 |
| Hornsby Springs, Alachua Co., Florida | | 1 | |
| Ichtucknee River, Columbia Co., Florida | | 1 | |
| Ingleside, San Patricio Co., Texas | | 1 | 17 |
| Inglis IA, Citrus Co., Florida | | 1 | |
| Irvington 2, Alameda Co., California | | 2 | 6 |
| La Carreta Ravine, Guanajuato, Mexico | T | 1 | |
| Manhattan, Nye Co., Texas | T | 1 | |
| Martinez, Contra Costa Co., California | T | 1 | |
| McKittrick, Santa Barbara Co., California | | 29 | |
| Mississippi River near Natchez, Mississippi | T | 1 | |
| Moore Pit, Dallas Co., Texas | T | 1 | 24 |
| Mount Blanco, Crosby Co., Texas | T | 1 | |
| Natural Trap Cave, Bighorn Co., Wyoming | | 12 | 21 |
| Pawnee Loop Branch, Niobrara River, Nebraska | T | 1 | |
| Petite Anse Parish, Louisiana | T | 1 | |
| Pool Branch, Polk Co., Florida | | 2 | 12 |
| Port Charlotte, Charlotte Co., Florida | | 2 | |
| Port Kennedy, Pennsylvania | T | 3 | 13 |
| Rainbow Beach, Gooding Co., Idaho | | 29 | 20 |

| | | | |
|---|---|---|---|
| Rancheria Parga, Aguascalientes, Mexico | T | 1 | |
| Rancho La Brea, Los Angeles Co., California | | 23 | 19 |
| Rock Creek, Briscoe Co., Texas | T | 10 | 8 |
| San Josecito Cave, Tamaulipas, Mexico | | 10 | 23 |
| Scharbauer Ranch, Midland Co., Texas | T | 1 | |
| Slaton Quarry, Lubbock Co., Texas | | 2 | 7 |
| Sulphur Springs, Oklahoma | T | 1 | |
| Tajo de Tequixquiac, Mexico | T | 1 | 16 |
| Tolumne Co., California | T | 1 | |
| Vallecito, San Diego Co., California | | 10 | 1 |
| Valley of Mexico | T | 4 | |
| Vero, Alachua Co., Florida | | 3 | |
| Wharton Co., Texas | T | 1 | |
| Yukon River near Tofty, Alaska | T | 18 | |

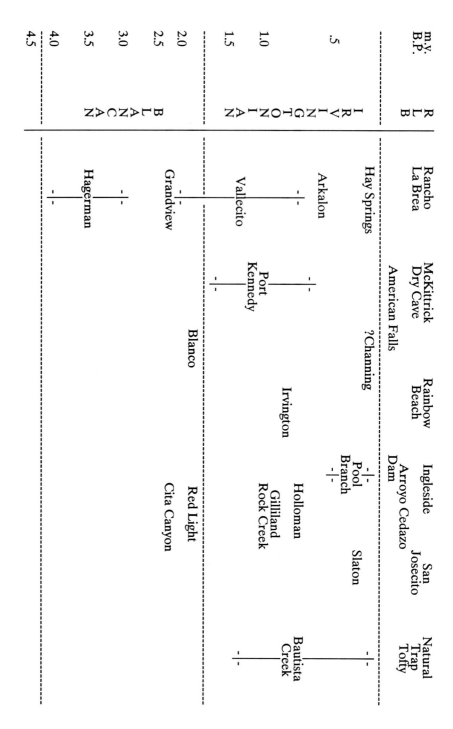

Fig. 14.2. Age ranges of sites from which type specimens or large samples came. Sources: Lundelius *et al.*, 1986, Kurten and Anderson, 1980.

made to separate males and females, because F-tests of measurements taken on samples of the extant species and of the large groups of fossils from Rancho La Brea and Hay Springs indicate that, except for measurements involving the area around the canines, differences in size and proportions between the two sexes are lost within the overall intraspecific variation.

Skulls, mandibles, and metapodials have been studied so far. Measurements taken are detailed in Figs. 14.3-14.5. Space limitations forbid publication of more than the briefest summaries of the data, but people interested in more detail can find more extensive summaries in Winans (1985), or can arrange to obtain a copy of the original database from the author.

The techniques used require that no specimens have missing measurements. However, so few fossil specimens are complete that it would have been virtually impossible to assemble a large enough body of data if only complete specimens had been used. Therefore, if a specimen lacked only a few measurements, values were estimated for each missing measurement by regressing values for the same measurement from other specimens within the same site group or subgroup on the non-missing measurement with which it was most highly correlated. The effect of such estimates is to shift these specimens slightly toward the mean value for each missing measurement. This somewhat diminishes the degree of separation between groups. In the case of type specimens, larger numbers of missing values were permitted. However, 31 types, consisting of palates, tooth rows, or individual teeth, were judged to be too incomplete to give any reasonable chance of accurate estimates. Thus the fitting of these species into the groups proposed here must await the elucidation of tooth characters.

As reflected in the summary statistics (Table 14.2), no single measurement or pair of measurements can be used to delineate unambiguously separated groups. This is a common problem, which sometimes can be solved by using a technique that considers all measurements at once, rather than one or two at a time. A number of different methods can be used to do this. The one used for this study is principal components analysis, as described by Davis (1987), Reyment, Blackith, and Campbell (1984), and Morrison (1976), among others.

In geometric terms, the set of N different measurements taken on each of M specimens can be treated as a cloud of M points in a space of N dimensions. Each point represents one specimen, with each of its coordinates along one of the N orthogonal axes representing the magnitude of one of the N measurements. Principal components analysis performs a rotation of these coordinate axes, keeping them orthogonal, to a position such that one axis (called the "first axis" or "first principal component") coincides with the direction of maximum variation in the cloud of points, and each successive axis coincides with the direction of the next largest variation. Thus each principal component represents a smaller percentage of the total variation than the previous one. Because the component axes have been rotated away from the original axis positions, each component usually reflects the influence of more than one of the original measurements. The first principal component is generally considered to be strongly affected by differences in the relative sizes of the objects under study, unless the variables are standardized by some method such as log transformation. However, other aspects of the pattern of variation also contribute to the distribution along the first axis.

Principal components analysis does not unambiguously identify subgroups within the data, but may make it easier for the researcher to perceive patterns of distribution within the data. In the jargon of multivariate analysis, principal components analysis "reduces the dimensionality" of a data set. This refers to the fact that although the total number of principal components is the same as the total number of original variables, so much of the total variation usually is accounted for by

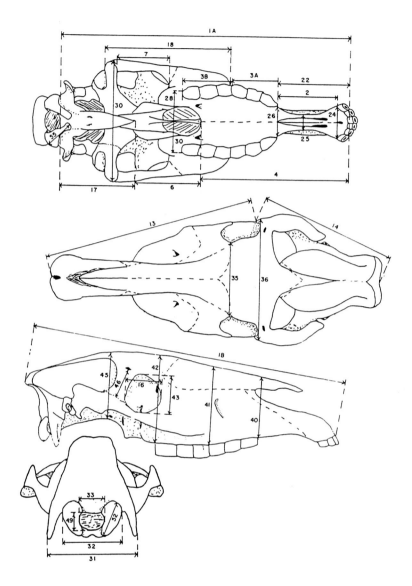

Fig. 14.3. Diagram of measurements taken on skulls. Numbers along measurement lines are those used in later figures and tables. Numbers do not run consecutively. 1A = Basal length: occipital notch - post. $I^1$; 1B = Dorsal length: nuchal crest - alv. $I^1$; 2 = Diastema: ant $P^2$- post. $I^1$; 3A = Total length $P^2$-$P^4$; 3B = Total length $M^1$-$M^3$; 4 = Length post. $I^1$ - post. border palate; 6 = Length post. border palate - post. edge vomer; 7 = Anteroposterior diameter dorsal opening zygomatic arch; 13 = Diagonal distance center of nuchal to postorbital bar; 14 = Diagonal distance ant. med. premax. to postorbital bar; 16 = Anteroposterior diameter of orbit; 17 = Length post. border vomer to occipital condyle notch; 18 = Ant. end crista facialis to post edge glenoid fossa; 22 = Midline length ant. $P^2$ to post. $I^1$; 24 = Width rostrum at post. $I^3$; 25 = Least width rostrum in diastema; 26 = Width rostrum at ant. $P^2$; 28 = Width at metastyle of $M^3$; 30 = Width across glenoid fossae; 31 = Greatest width across mastoid processes; 32 = Width across occipital condyles; 33 = Horizontal diameter foramen magnum; 35 = Least width of top of skull across orbits; 36 = Width across postorbital bars; 40 = Height of skull at ant. end $P^2$; 41 = Height of skull at junction $P^4$ and $M^1$; 42 = Height of skull at post. end $M^3$; 43 = Vertical diameter of orbit; 45 = Height of skull behind pterygoids; 46 = Height of postorbital bar; 49 = Vertical diameter foramen magnum; 52 = Greatest width posterior face occipital condyle; 55 = Greatest width ventral face occipital condyle.

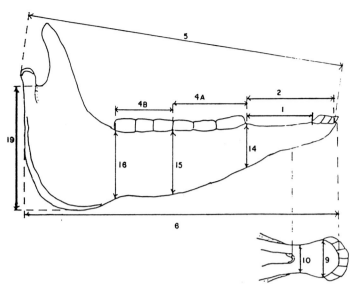

Fig. 14.4. Diagram of measurements taken on mandibles. Numbers along measurement lines are those used in later figures and tables. 1 = Length diastema; 2 = Horizontal distance post. $I_1$ to ant. $P_2$; 4A = Total length $P_2$-$P_4$; 4B = Total length $M_1$-$M_3$; 5 = Diagonal length post. $I_1$ to post. side condyle; 6 = Greatest length parallel to cheek tooth row; 9 = Width (both rami) at post. $I_3$; 10 = Least width (both rami) in diastema; 14 = Ramus height at ant. $P_2$; 15 = Ramus height at junction $P_4$ and $M_1$; 16 = Ramus height at post. $M_3$; 19 = Vertical distance condylar notch to bottom angular process.

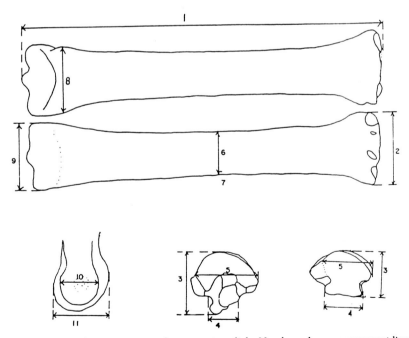

Fig. 14.5. Diagram of measurements taken on metapodials. Numbers along measurement lines are those used in later figures and tables. 1 = Greatest length; 2 = Lateral diameter top articular face; 3 = Anteroposterior diameter top articular face; 4 = Width plantar process; 5 = Greatest lateral width principal prox. articular facet; 6 = Lateral diameter at midpoint of shaft; 7 = Anteroposterior diameter at midpoint of shaft; 8 = Greatest lateral diameter above trochlea; 9 = Greatest lateral width trochlea; 10 = Anteroposterior diameter median ridge of trochlea; 11 = Anteroposterior diameter larger side of trochlea.

the first few components that later components can sometimes be ignored without unduly biasing one's view of the pattern of variation. In the case of the skulls, for example, there were 33 original measurements. It requires between 2 and 10 principal components (depending on the group of specimens studied) to represent more than 95% of the original variation, and it appears that the first 2 or 3 of these components contain most of the useful information. The original measurements of each specimen can be converted into a set of scores with respect to the new principal components axes, which can be plotted against each other (or otherwise analyzed) just as the original measurements could be. The results can then be examined to determine the frequency with which specimens were found grouped together or apart. Since the degree to which each measurement contributes to the formation of the factors can be determined, it may be possible to draw conclusions about what actual morphological differences or similarities are implied by the positions of the data points on the factor plots.

After the groups had been delineated, based on examination of the principal components plots, discriminant analysis was used to try to establish a rule for differentiating among these groups. Like principal components analysis, discriminant analysis reduces the dimensionality of the data set. Geometrically, it also performs a rotation of the coordinate axes, but in this case the aim is to arrive at an orientation that maximizes, not the total variation, but the separation between the groups one is attempting to differentiate. Because of this, one must already have decided what groups exist within the data set before the method can be applied. The goal of this method is to arrive at a linear combination of the original variables that maximizes the separation between the groups. Especially when more than two groups are involved, this method may fail to completely discriminate between the groups, because even if the original clouds of points are completely non-overlapping it may not be possible to find any single baseline along which none of the groups overlap.

Discriminant analysis can be and often is used to identify specimens not included in the original analysis which established the groups. However, two things must be kept in mind when doing so: 1) The odds that the identification is correct will be no better than, and usually worse than, the percentage of correct identifications which the procedure is able to make in the original data set (see Table 14.1 for examples). 2) This method will place the specimen in one of the existing groups, even if it is a representative of some new group which was not included in the original analysis.

**Results**

For the methods used here, each skeletal element must be analyzed separately. In addition, useful insights may sometimes result from subdividing the specimens in ways based on criteria other than those being analyzed. In this study, for example, after the preliminary analyses of all specimens together, the data set was arbitrarily subdivided into Blancan, Irvingtonian, and Rancholabrean age groups, and each age group analyzed separately. I do not wish to imply by this that any given species group was restricted to a single time period, and in fact the results seem to show that several of them spanned more than one time period. Nevertheless, this served to reduce the probable amount of drift in the morphology of a given species by confining the analysis to a shorter period of time.

Analysis of all fossil skulls, mandibles, or metapodials together (Fig. 14.6) produces groupings in which the clearest separation is between large and small specimens. Specimens from each group or subgroup tend to cluster together consistently, but there is a good deal of overlap between groups. The skull analysis (Fig. 14.6A) produces four distinct groups: small Irvingtonian and Rancholabrean horses; Blancan horses; large Irvingtonian horses together with two

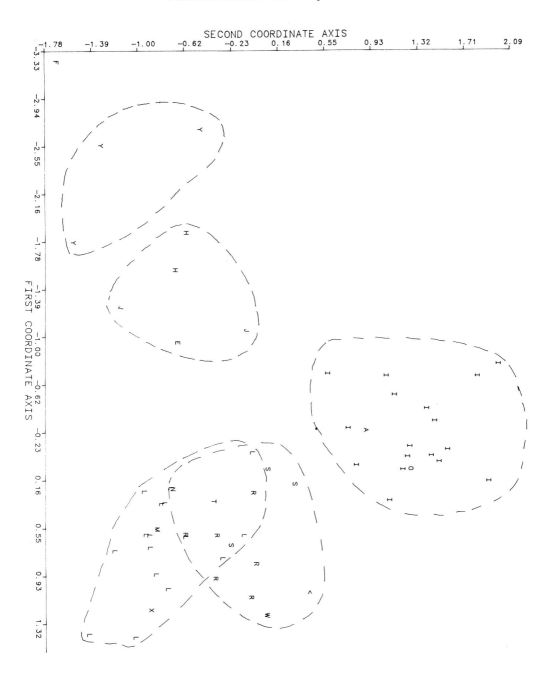

Fig. 14.6A. Principal components plots of the distribution of all skulls. Dashed lines delineate species groups discussed in text. Symbols for samples as follows: A = Crawfish Draw; E = *Equus niobrarensis alaskae* type; H = Hay Springs small horses; I = Hagerman; J = San Josecito; L = Rancho La Brea; M = McKittrick; N = *Equus niobrarensis* type; O = *Equus shoshoniensis*; R = Rock Creek; S = Hay Springs large horses; T = *Equus hatcheri* type; X = *Equus mexicanus* type; < = Ingleside.

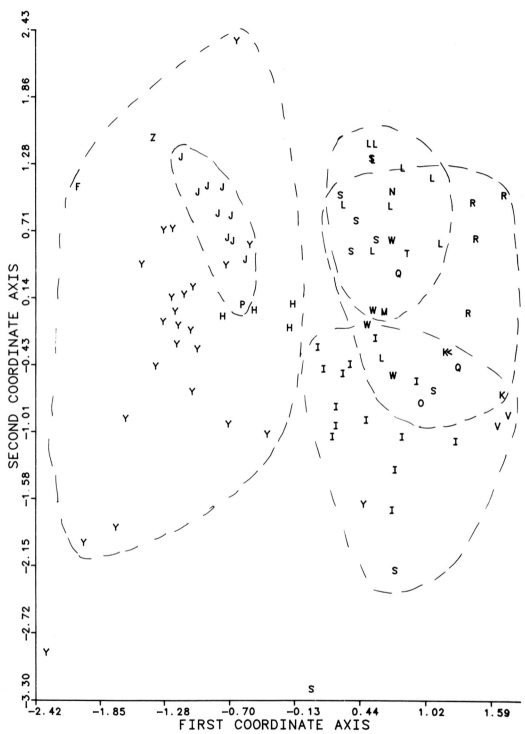

Fig. 14.6B. Principal components plots of the distribution of all mandibles. Symbols: F = *Equus francisi* type; H = Hay Springs small horses; I = Hagerman; J = San Josecito; K = Arkalon large horses; L = Rancho La Brea; M = McKittrick; N = *Equus niobrarensis* type; O = *Equus shoshoniensis*; P = Pool Branch; Q = Gilliland; R = Rock Creek; S = Hay Springs large horses; V = Vallecito; W = American Falls; Y = Channing; Z = *Equus zoytalis* type; < = Ingleside.

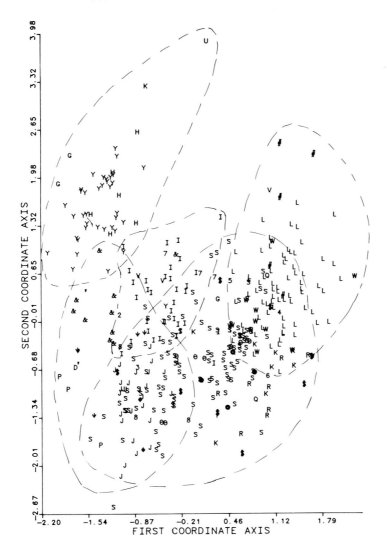

Fig. 14.6C. Principal components plots of the distribution of all metacarpals. Dashed lines delineate species groups discussed in the text. Symbols for samples as follows: 1 = Port Charlotte; 2 = Inglis; 3 = Coleman; 4 = Arredondo; 5 = Hornsby Springs; 6 = Ichtucknee; 7 = Vero; 8 = Irvington; 9 = Silverton; A = Crawfish Draw; C = Cita Canyon; D = *Equus nevadanus* type; E = *Equus niobrarensis alaskae* type; F = *Equus francisci* type; G = Quarry G; H = Hay Springs small horses; I = Hagerman; J = San Josecito; K = Arkalon large horses; L = Rancho La Brea; M = McKittrick; N = *Equus niobrarensis* type; O = *Equus shoshoniensis*; P = Pool Branch; Q = Gilliland; R = Rock Creek; S = Hay Springs large horses; T = *Equus hatcheri* type; U = *Equus calobatus* type; V = Vallecito; W = American Falls; X = *Equus mexicanus* type; Y = Channing; Z = *Equus zoytalis* type; * = Rainbow Beach; $ = Dam; # = Grandview; & = Natural Trap Cave; @ = Cueva Quebrada large horses; + = Loup Fork; Δ = Cueva Quebrada small horses; Σ = *Equus quinni* type; φ = Slaton; θ = Dry Cave large horses; ψ = Dry Cave small horses; ∞ = *Equus midlandensis* type; ¢ = *Equus altidens* type; < = Ingleside; > = Arkalon small horses.

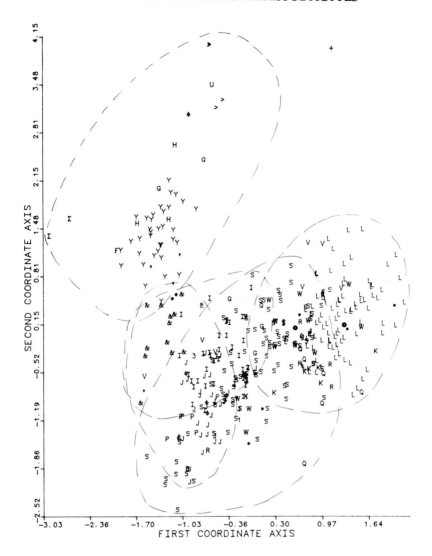

Fig. 14.6D. Principal components plots of the distribution of all metatarsals. Dashed lines delineate species groups discussed in the text. Symbols for samples as follows: 1 = Port Charlotte; 2 = Inglis; 3 = Coleman; 4 = Arredondo; 5 = Hornsby Springs; 6 = Ichtucknee; 7 = Vero; 8 = Irvington; 9 = Silverton; A = Crawfish Draw; C = Cita Canyon; D = *Equus nevadanus* type; E = *Equus niobrarensis alaskae* type; F = *Equus francisci* type; G = Quarry G; H = Hay Springs small horses; I = Hagerman; J = San Josecito; K = Arkalon large horses; L = Rancho La Brea; M = McKittrick; N = *Equus niobrarensis* type; O = *Equus shoshoniensis*; P = Pool Branch; Q = Gilliland; R = Rock Creek; S = Hay Springs large horses; T = *Equus hatcheri* type; U = *Equus calobatus* type; V = Vallecito; W = American Falls; X = *Equus mexicanus* type; Y = Channing; Z = *Equus zoytalis* type; * = Rainbow Beach; $ = Dam; # = Grandview; & = Natural Trap Cave; @ = Cueva Quebrada large horses; + = Loup Fork; Δ = Cueva Quebrada small horses; Σ = *Equus quinni* type; φ = Slaton; θ = Dry Cave large horses; ψ = Dry Cave small horses; ∞ = *Equus midlandensis* type; ¢ = *Equus altidens* type; < = Ingleside; > = Arkalon small horses.

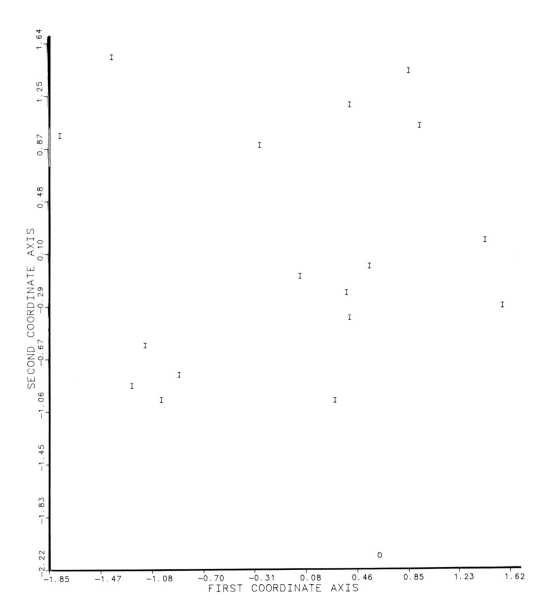

Fig. 14.7A. Principal components plots of the distribution of all Blancan skulls. Symbols: I = Hagerman; O = *Equus shoshoniensis*.

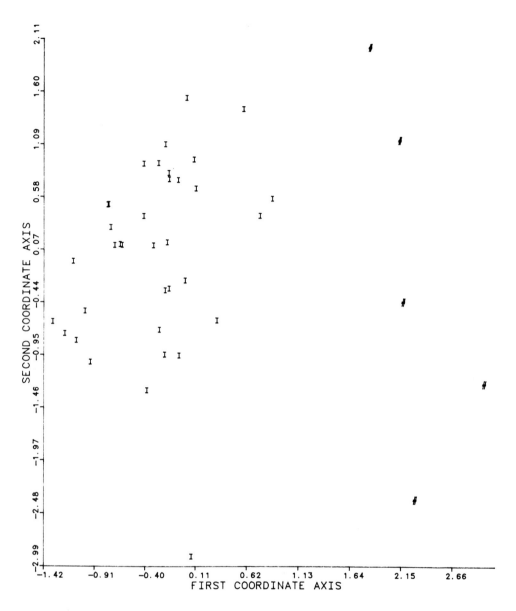

Fig. 14.7B. Principal components plots of the distribution of all Blancan metacarpals. Symbols: I = Hagerman; # = Grandview.

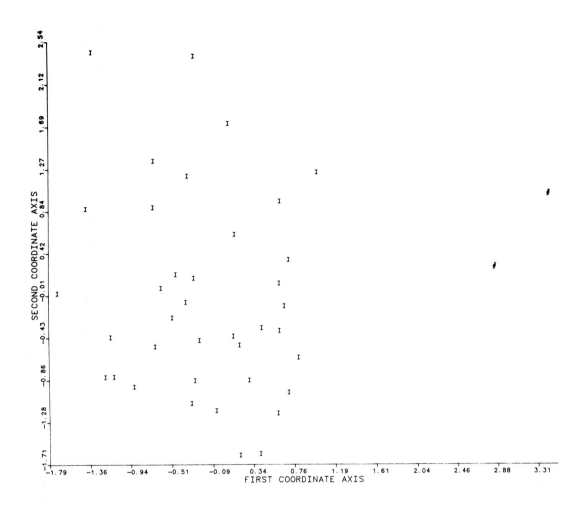

Fig. 14.7C. Principal components plots of the distribution of all Blancan metatarsals. Symbols: I = Hagerman; # = Grandview.

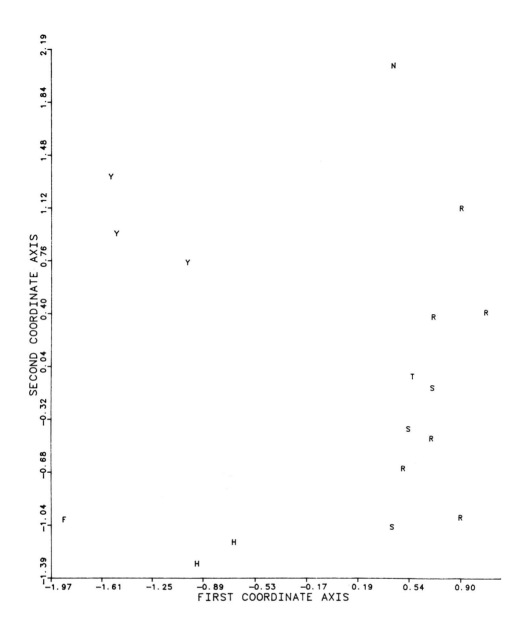

Fig. 14.8A. Principal components plots of the distribution of all Irvingtonian skulls. Symbols: H = Hay Springs small horses; N = *Equus niobrarensis* type; R = Rock Creek; S = Hay Springs large horses; T = *Equus hatcheri* type; Y = Channing.

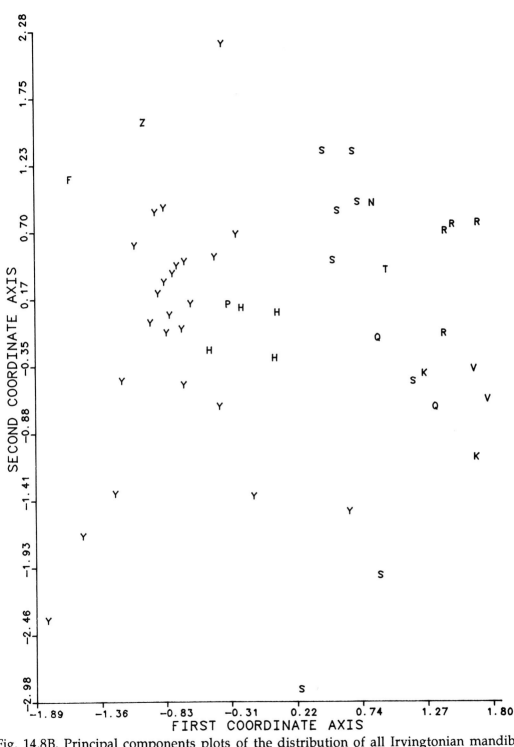

Fig. 14.8B. Principal components plots of the distribution of all Irvingtonian mandibles. Symbols: F = *Equus francisi* type; H = Hay Springs small horses; N = *Equus niobrarensis* type; P = Pool Branch; Q = Gilliland; R = Rock Creek; S = Hay Springs large horses; T = *Equus hatcheri* type; V = Vallecito; Y = Channing; Z = *Equus zoytalis* type.

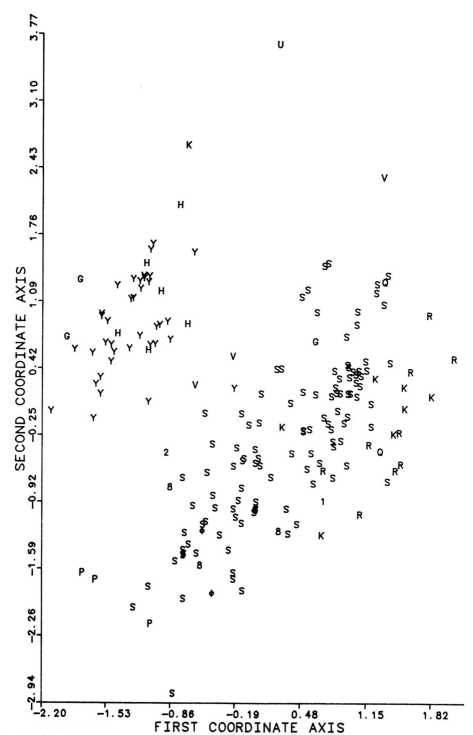

Fig. 14.8C. Principal components plots of the distribution of all Irvingtonian metacarpals. Symbols: G = Quarry G; H = Hay Springs small horses; K = Arkalon large horses; P = Pool Branch; Q = Gilliland; R = Rock Creek; S = Hay Springs large horses; U = *Equus calobatus* type; V = Vallecito; Y = Channing; 1 = Port Charlotte; 2 = Inglis; 8 = Irvington; ϕ = Slaton.

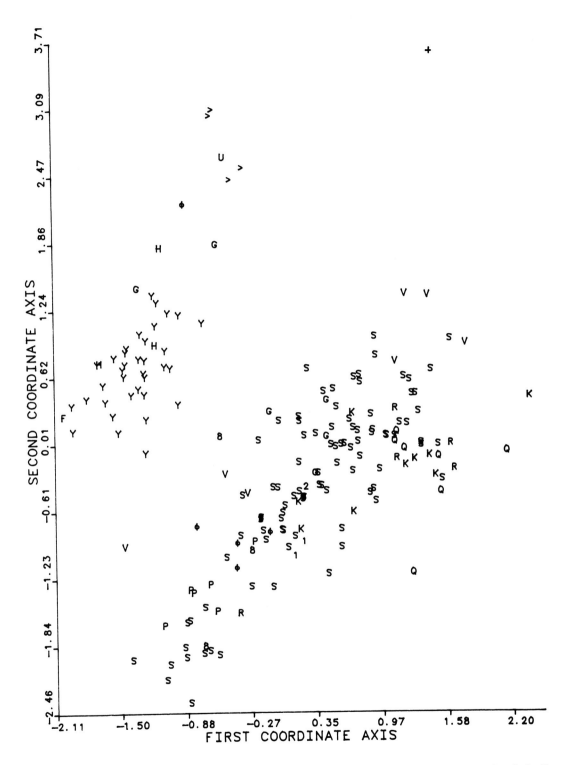

Fig. 14.8D. Principal components plots of the distribution of all Irvingtonian metatarsals. Symbols: F = *Equus francisi* type; G = Quarry G; H = Hay Springs small horses; K = Arkalon large horses; Q = Gilliland; R = Rock Creek; S = Hay Springs large horses; V = Vallecito; Y = Channing; 1 = Port Charlotte; 2 = Inglis; 8 = Irvington; + = Loup Fork; > = Arkalon small horses; φ = Slaton.

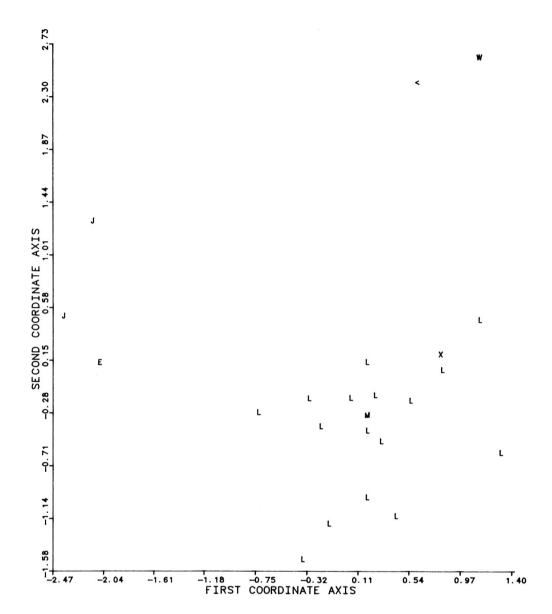

Fig. 14.9A. Principal components plots of the distribution of all Rancholabrean skulls. Symbols: E = *Equus niobrarensis alaskae* type; J = San Josecito; L = Rancho La Brea; M = McKittrick; W = American Falls; X = *Equus mexicanus* type; < = Ingleside.

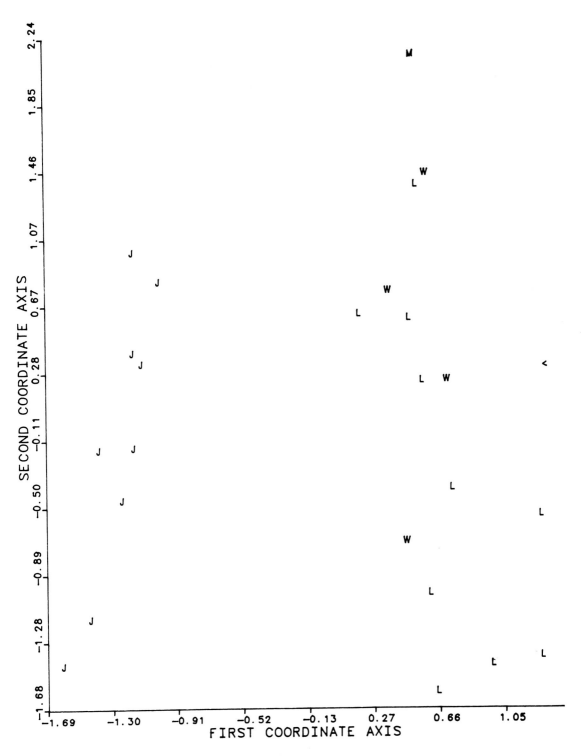

Fig. 14.9B. Principal components plots of the distribution of all Rancholabrean mandibles. Symbols: J = San Josecito; L = Rancho La Brea; M = McKittrick; W = American Falls; < = Ingleside.

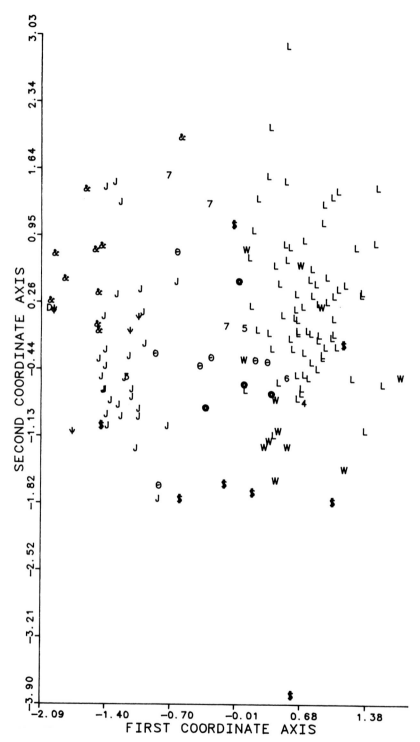

Fig. 14.9C. Principal components plots of the distribution of all Rancholabrean metacarpals. Symbols: D = *Equus nevadanus* type; J = San Josecito; L = Rancho La Brea; W = American Falls; 4 = Arredondo; 5 = Hornsby Springs; 6 = Ichtucknee; 7 = Vero; $ = Dam; & = Natural Trap Cave; @ = Cueva Quebrada large horses; θ = Dry Cave large horses; ψ = Dry Cave small horses.

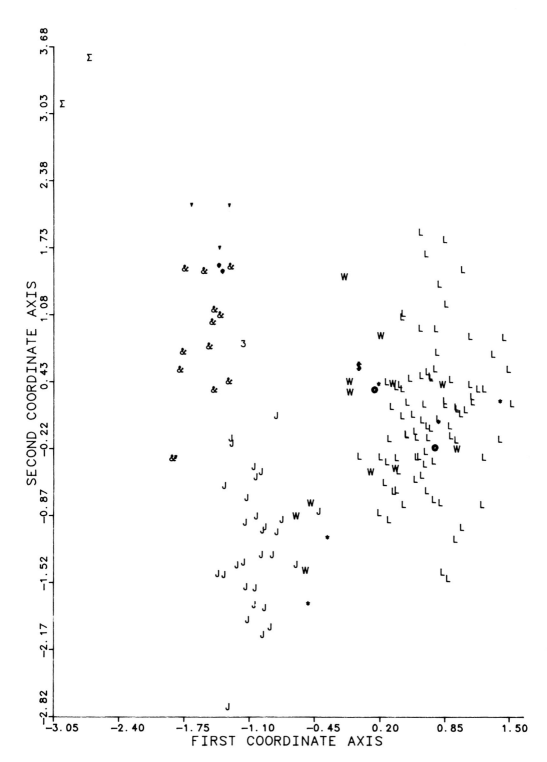

Fig. 14.9D. Principal components plots of the distribution of all Rancholabrean metatarsals. Symbols: J = San Josecito; L = Rancho La Brea; W = American Falls; * = Rainbow Beach; & = Natural Trap Cave; @ = Cueva Quebrada large horses; Δ = Cueva Quebrada small horses; Σ = *Equus quinni* type; ɸ = Slaton.

Table 14.1. Discriminant functions for the five groups of *Equus*. To use the Classification Functions to identify new specimens, multiply each of the listed measurements by the value listed for *E. shoshonensis*. Total all of these products, then add the constant listed at the bottom of this column. Repeat this process for each of the other species. The species with the highest total is the most probable identification for the specimen. The results of testing

| | | | Skulls | | |
|---|---|---|---|---|---|
| | *E. shoshonensis* | *E. scotti* | *E. laurentius* | *E. alaskae* | *E. francisci* |
| Variable | | | | | |
| 1A | 1.40984 | 1.74821 | 1.57525 | 1.66280 | 1.21677 |
| 1B | 1.51663 | 1.36098 | 1.34497 | .96548 | 1.35196 |
| 4 | .46696 | .19518 | -.16491 | -.28920 | -.08025 |
| 6 | .31151 | -.19440 | -.02769 | .04115 | -.19815 |
| 24 | -5.46297 | -4.96119 | -4.31898 | -2.87948 | -4.32974 |
| 43 | 6.48798 | 6.06482 | 5.81969 | 4.88995 | 5.35300 |
| 45 | -.11781 | .32405 | .64891 | .37005 | .42011 |
| 49 | 6.06214 | 6.01390 | 5.69115 | 3.98772 | 5.07599 |
| 52 | -2.98252 | -2.36243 | -1.11715 | -1.37628 | -1.26798 |
| Const. | -944.84212 | -975.10750 | -915.13212 | -685.81705 | -665.66956 |
| Percent classified correctly | 100.0 | 100.0 | 94.7 | 100.0 | 100.0 |

| | | | Mandibles | | |
|---|---|---|---|---|---|
| | *E. shoshonensis* | *E. scotti* | *E. laurentius* | *E. alaskae* | *E. francisci* |
| Variable | | | | | |
| 2 | 1.33614 | 1.22472 | 1.23294 | 1.08217 | .97553 |
| 9 | .41512 | .49266 | .57585 | .48762 | .36950 |
| 15 | 2.30462 | 2.71074 | 2.46784 | 2.16029 | 2.22328 |
| Const. | -202.91693 | -232.45318 | -215.46321 | -164.75940 | -152.09676 |
| Percent classified correctly | 93.3 | 60.0 | 75.0 | 100.0 | 66.7 |

| | | | Metacarpals | | |
|---|---|---|---|---|---|
| | *E. shoshonensis* | *E. scotti* | *E. laurentius* | *E. alaskae* | *E. francisci* |
| Variable | | | | | |
| 1 | 1.91566 | 1.77734 | 1.84774 | 1.79413 | 2.16980 |
| 2 | -.80266 | -.98313 | -.67604 | -.71224 | -1.02142 |
| 4 | 1.06244 | .87730 | 1.20808 | 1.27109 | .81840 |
| 7 | -.76693 | -.31710 | -.40203 | -.35261 | -.11151 |
| 8 | 3.46903 | 2.78398 | 2.68840 | 2.35760 | 1.86753 |
| 9 | -1.38275 | -.11375 | -.00543 | -.68658 | -.76083 |
| 10 | .78346 | .07500 | -.21310 | -.19666 | -.35816 |
| 11 | .73597 | 1.46649 | 1.45912 | 1.17232 | 1.28358 |
| Const. | -291.70041 | -287.14780 | -320.24085 | -251.23860 | -295.47733 |
| Percent classified correctly | 88.9 | 63.9 | 75.7 | 80.0 | 93.9 |

|  | | | Metatarsala | | |
|---|---|---|---|---|---|
| | E. shoshonensis | E. scotti | E. laurentius | E. alaskae | E. francisci |
| Variable | | | | | |
| 1 | 1.68267 | 1.62134 | 1.65248 | 1.60431 | 1.96879 |
| 2 | 1.24648 | .61031 | .89460 | .91074 | .85486 |
| 3 | .95357 | 1.60362 | 1.32693 | 1.16222 | 1.53522 |
| 4 | -.29589 | -.11977 | -.26739 | -.32199 | -.74128 |
| 5 | -1.23410 | -1.50167 | -1.08443 | -1.28293 | -1.39643 |
| 7 | .40053 | 1.05467 | .91022 | 1.13094 | .57506 |
| 8 | 1.78901 | 1.19854 | 1.22853 | 1.15149 | .90646 |
| 9 | -1.26770 | -.26916 | -.07830 | -.84206 | -1.33765 |
| Const. | -270.54361 | -282.27695 | -316.13859 | -251.90844 | -305.19829 |
| Percent classified correctly | 76.3 | 61.2 | 86.7 | 81.3 | 91.5 |

Table 14.2. (next page) Standard statistics for all specimens referred to each of the five valid species. Sample sizes may vary from one measurement to another because not all specimens were complete. Abbreviations: N, number of specimens; S.D., standard deviation; V, coefficient of variation.

Table 14.2  **Skulls**

### *Equus alaskae* group

| VAR. | N | MEAN | S.D. | V | OBS. RANGE |
|---|---|---|---|---|---|
| 1A | 4 | 89.50 | 10.02 | .0205 | 477.00-499.00 |
| 1B | 4 | 544.50 | 15.26 | .0280 | 526.00-562.00 |
| 2 | 4 | 95.36 | 2.63 | .0276 | 92.70-98.60 |
| 3A | 4 | 90.76 | 7.08 | .0780 | 83.35-98.75 |
| 3B | 4 | 78.15 | 3.76 | .0481 | 73.75-81.35 |
| 4 | 4 | 253.00 | 13.19 | .0521 | 238.00-268.00 |
| 6 | 4 | 118.55 | 4.73 | .0399 | 113.20-123.10 |
| 7 | 4 | 75.55 | 8.59 | .1137 | 65.70-82.90 |
| 13 | 4 | 378.00 | 9.63 | .0255 | 368.00-391.00 |
| 14 | 4 | 211.00 | 8.29 | .0393 | 201.00-220.00 |
| 16 | 4 | 60.56 | 6.42 | .1060 | 51.00-64.80 |
| 17 | 4 | 123.36 | 4.56 | .0369 | 119.80-129.80 |
| 18 | 4 | 212.75 | 14.03 | .0660 | 196.00-230.00 |
| 22 | 4 | 122.89 | 5.47 | .0445 | 117.60-129.85 |
| 24 | 4 | 75.44 | 12.46 | .1651 | 58.40-88.20 |
| 25 | 4 | 46.35 | 4.24 | .0914 | 40.00-48.70 |
| 26 | 4 | 79.04 | 10.18 | .1288 | 70.80-91.85 |
| 28 | 4 | 110.89 | 8.99 | .0811 | 103.10-123.80 |
| 30 | 4 | 202.74 | 5.17 | .0255 | 197.00-209.20 |
| 31 | 3 | 103.53 | 5.66 | .0546 | 97.00-106.80 |
| 32 | 4 | 80.36 | 2.24 | .0279 | 78.20-83.10 |
| 33 | 4 | 35.01 | 3.21 | .0918 | 30.35-37.40 |
| 35 | 3 | 153.73 | 4.98 | .0324 | 149.70-159.30 |
| 36 | 4 | 216.31 | 12.35 | .0571 | 202.25-232.00 |
| 40 | 1 | 101.00 | | | |
| 41 | 4 | 119.19 | 14.40 | .1208 | 98.55-131.40 |
| 42 | 3 | 132.92 | 9.50 | .0715 | 126.00-143.75 |
| 43 | 4 | 50.36 | 3.16 | .0627 | 46.90-54.50 |
| 45 | 4 | 93.06 | 7.53 | .0810 | 82.55-99.00 |
| 46 | 4 | 46.90 | 5.03 | .1073 | 41.50-52.90 |
| 49 | 4 | 31.84 | 7.58 | .2381 | 22.85-41.40 |
| 52 | 4 | 45.13 | 0.42 | .0093 | 44.85-45.75 |
| 55 | 4 | 41.60 | 1.65 | .0397 | 39.30-43.10 |

### *Equus francisci* group

| VAR. | N | MEAN | S.D. | V | OBS. RANGE |
|---|---|---|---|---|---|
| 1A | 6 | 444.67 | 27.95 | .0629 | 422.00-490.00 |
| 1B | 2 | 506.00 | 41.01 | .0811 | 477.00-535.00 |
| 2 | 9 | 91.44 | 7.18 | .0786 | 78.80-100.90 |
| 3A | 8 | 85.24 | 6.61 | .0776 | 75.20-93.65 |
| 3B | 6 | 71.53 | 6.46 | .0904 | 61.10-78.80 |
| 4 | 8 | 233.63 | 20.04 | .0858 | 206.00-260.00 |
| 6 | 7 | 103.10 | 5.23 | .0507 | 95.60-110.25 |
| 7 | 7 | 69.52 | 4.43 | .0637 | 64.60-77.50 |
| 13 | 9 | 326.00 | 80.51 | .2470 | 116.00-376.00 |
| 14 | 3 | 180.00 | 16.09 | .0894 | 163.00-195.00 |
| 16 | 10 | 59.74 | 4.96 | .0830 | 49.70-70.00 |
| 17 | 5 | 118.37 | 9.13 | .0771 | 110.15-129.35 |
| 18 | 7 | 192.29 | 15.51 | .0807 | 173.00-215.00 |
| 22 | 9 | 113.47 | 8.71 | .0767 | 101.55-123.30 |
| 24 | 8 | 64.21 | 5.54 | .0864 | 54.60-72.65 |
| 25 | 8 | 41.92 | 3.99 | .0951 | 34.20-46.00 |
| 26 | 8 | 63.71 | 6.40 | .1004 | 51.00-71.15 |
| 28 | 7 | 99.86 | 5.32 | .0532 | 91.40-106.00 |
| 30 | 7 | 186.96 | 17.51 | .0937 | 165.55-203.85 |
| 31 | 5 | 101.15 | 11.72 | .1159 | 89.70-119.40 |
| 32 | 7 | 74.94 | 9.76 | .1302 | 63.75-88.85 |
| 33 | 7 | 32.71 | 3.85 | .1178 | 26.45-37.10 |
| 35 | 7 | 135.30 | 26.13 | .1931 | 102.00-186.00 |
| 36 | 7 | 196.86 | 19.34 | .0983 | 171.00-227.00 |
| 40 | 8 | 87.99 | 3.25 | .0369 | 83.60-94.95 |
| 41 | 8 | 106.79 | 11.42 | .1069 | 93.00-125.25 |
| 42 | 4 | 118.66 | 12.45 | .1049 | 104.00-134.45 |
| 43 | 9 | 47.39 | 5.55 | .1172 | 36.40-53.75 |
| 45 | 3 | 99.13 | 20.30 | .2048 | 85.00-122.40 |
| 46 | 7 | 51.26 | 5.81 | .1133 | 44.30-61.50 |
| 49 | 7 | 32.03 | 2.33 | .0726 | 29.50-35.15 |
| 52 | 7 | 44.26 | 3.49 | .0789 | 38.10-48.55 |
| 55 | 6 | 39.45 | 2.44 | .0619 | 36.80-42.60 |

### *Equus laurentius* group

| VAR. | N | MEAN | S.D. | V | OBS. RANGE |
|---|---|---|---|---|---|
| 1A | 25 | 542.32 | 21.56 | .0398 | 498.00-594.00 |
| 1B | 24 | 612.42 | 21.46 | .0350 | 556.00-656.00 |
| 2 | 29 | 105.48 | 13.82 | .1310 | 85.70-150.45 |
| 3A | 29 | 108.76 | 6.80 | .0625 | 95.20-122.35 |
| 3B | 20 | 90.04 | 4.83 | .0537 | 82.60-99.60 |
| 4 | 24 | 281.58 | 18.06 | .0641 | 245.00-332.00 |
| 6 | 25 | 134.38 | 7.39 | .0550 | 121.60-146.70 |
| 7 | 26 | 80.81 | 6.68 | .0827 | 65.60-97.65 |
| 13 | 24 | 424.83 | 15.55 | .0366 | 392.00-462.00 |
| 14 | 27 | 228.67 | 10.35 | .0453 | 194.00-242.00 |
| 16 | 26 | 63.91 | 5.27 | .0825 | 49.45-77.40 |
| 17 | 26 | 129.97 | 8.35 | .0642 | 114.15-144.10 |
| 18 | 26 | 229.00 | 13.15 | .0574 | 203.00-260.00 |
| 22 | 29 | 139.22 | 12.46 | .0895 | 118.90-176.25 |
| 24 | 28 | 82.42 | 5.55 | .0673 | 67.60-92.65 |
| 25 | 29 | 50.39 | 7.87 | .1563 | 37.60-76.40 |
| 26 | 30 | 75.79 | 9.12 | .1203 | 51.45-88.20 |
| 28 | 22 | 122.06 | 15.08 | .1235 | 68.55-147.55 |
| 30 | 30 | 218.81 | 19.39 | .0886 | 132.00-244.00 |
| 31 | 32 | 121.33 | 6.01 | .0495 | 112.60-135.50 |
| 32 | 35 | 95.05 | 4.68 | .0492 | 86.20-104.30 |
| 33 | 36 | 39.02 | 3.71 | .0951 | 32.55-50.90 |

### *Equus scotti* group

| VAR. | N | MEAN | S.D. | V | OBS. RANGE |
|---|---|---|---|---|---|
| 1A | 9 | 556.44 | 15.66 | .0281 | 532.00-586.00 |
| 1B | 8 | 611.75 | 19.04 | .0311 | 581.00-641.00 |
| 2 | 12 | 112.95 | 7.17 | .0635 | 99.20-120.20 |
| 3A | 15 | 108.94 | 7.41 | .0680 | 98.00-121.70 |
| 3B | 14 | 91.29 | 5.93 | .0649 | 81.40-104.70 |
| 4 | 12 | 307.92 | 28.11 | .0913 | 260.00-370.00 |
| 6 | 8 | 122.02 | 9.27 | .0760 | 110.60-142.00 |
| 7 | 12 | 76.29 | 7.72 | .1012 | 59.50-91.00 |
| 13 | 9 | 433.33 | 7.11 | .0164 | 425.00-445.00 |
| 14 | 8 | 218.89 | 9.99 | .0457 | 209.10-235.00 |
| 16 | 13 | 67.24 | 7.71 | .1147 | 52.60-86.90 |
| 17 | 7 | 130.39 | 4.93 | .0378 | 126.10-137.50 |
| 18 | 12 | 239.83 | 22.51 | .0938 | 221.00-302.00 |
| 22 | 12 | 144.08 | 8.22 | .0570 | 124.70-155.30 |
| 24 | 14 | 81.28 | 4.35 | .0536 | 73.10-89.35 |
| 25 | 11 | 49.39 | 6.15 | .1245 | 41.40-60.30 |
| 26 | 14 | 81.86 | 5.89 | .0719 | 71.80-93.50 |
| 28 | 12 | 116.38 | 8.82 | .0758 | 103.90-135.45 |
| 30 | 9 | 220.65 | 12.24 | .0555 | 200.25-238.00 |
| 31 | 9 | 118.38 | 5.50 | .0464 | 110.00-128.70 |
| 32 | 10 | 91.80 | 4.17 | .0454 | 85.30-98.50 |
| 33 | 10 | 36.64 | 5.58 | .1524 | 32.60-51.80 |

| VAR. | N | MEAN | S.D. | V | OBS. RANGE | VAR. | N | MEAN | S.D. | V | OBS. RANGE |
|---|---|---|---|---|---|---|---|---|---|---|---|
| 35 | 26 | 163.66 | 15.41 | .0941 | 135.00-194.80 | 35 | 11 | 159.70 | 12.29 | .0770 | 137.00-177.20 |
| 36 | 26 | 230.65 | 13.30 | .0577 | 206.00-259.00 | 36 | 10 | 225.50 | 10.30 | .0457 | 208.00-239.00 |
| 40 | 22 | 114.28 | 7.98 | .0698 | 98.60-132.00 | 40 | 9 | 122.77 | 8.16 | .0664 | 105.00-131.00 |
| 41 | 24 | 136.22 | 7.79 | .0572 | 116.90-149.40 | 41 | 10 | 131.44 | 17.25 | .1312 | 89.00-151.00 |
| 42 | 21 | 148.82 | 10.37 | .0697 | 126.00-171.10 | 42 | 9 | 146.52 | 13.00 | .0887 | 122.00-172.00 |
| 43 | 25 | 54.71 | 3.97 | .0725 | 42.00-59.65 | 43 | 13 | 55.37 | 5.54 | .1001 | 43.50-65.10 |
| 45 | 27 | 116.07 | 4.81 | .0415 | 107.00-125.35 | 45 | 7 | 112.37 | 9.24 | .0823 | 102.00-128.00 |
| 46 | 29 | 52.52 | 4.05 | .0771 | 42.70-58.10 | 46 | 12 | 54.86 | 4.66 | .0849 | 49.10-66.30 |
| 49 | 36 | 39.05 | 3.64 | .0933 | 33.50-55.10 | 49 | 9 | 37.51 | 2.51 | .0669 | 35.20-42.40 |
| 52 | 36 | 54.94 | 3.26 | .0594 | 49.60-62.35 | 52 | 10 | 50.07 | 2.42 | .0484 | 47.70-54.70 |
| 55 | 36 | 48.95 | 3.00 | .0612 | 43.40-54.15 | 55 | 10 | 46.35 | 1.66 | .0357 | 43.20-48.80 |

*Equus simplicidens* group

| VAR. | N | MEAN | S.D. | V | OBS. RANGE | VAR. | N | MEAN | S.D. | V | OBS. RANGE |
|---|---|---|---|---|---|---|---|---|---|---|---|
| 1A | 19 | 537.89 | 14.33 | .0266 | 507.00-562.00 | 28 | 14 | 116.69 | 7.19 | .0616 | 108.50-132.50 |
| 1B | 19 | 604.83 | 26.36 | .0436 | 528.00-637.00 | 30 | 19 | 206.94 | 21.28 | .1028 | 141.20-230.00 |
| 2 | 23 | 106.75 | 13.18 | .1234 | 85.80-125.20 | 31 | 18 | 111.96 | 5.76 | .0514 | 102.05-125.00 |
| 3A | 21 | 105.71 | 7.06 | .0668 | 92.80-116.90 | 32 | 20 | 82.82 | 16.29 | .1967 | 18.10-94.90 |
| 3B | 19 | 88.26 | 3.63 | .0411 | 81.60-96.00 | 33 | 20 | 39.77 | 12.54 | .3152 | 31.75-89.35 |
| 4 | 20 | 287.32 | 17.56 | .0611 | 236.00-307.00 | 35 | 18 | 148.22 | 15.32 | .1034 | 110.70-171.00 |
| 6 | 16 | 141.43 | 11.75 | .0830 | 119.30-159.20 | 36 | 18 | 217.29 | 13.50 | .0621 | 186.00-241.00 |
| 7 | 20 | 75.77 | 6.63 | .0875 | 65.10-87.40 | 40 | 11 | 98.99 | 11.04 | .1116 | 87.00-123.70 |
| 13 | 19 | 436.00 | 10.70 | .0245 | 422.00-459.00 | 41 | 14 | 116.63 | 16.16 | .1385 | 90.00-140.00 |
| 14 | 18 | 214.36 | 7.20 | .0336 | 202.45-229.00 | 42 | 11 | 124.81 | 11.55 | .0925 | 110.00-149.00 |
| 16 | 22 | 64.38 | 4.83 | .0750 | 56.30-77.95 | 43 | 17 | 55.39 | 5.73 | .1034 | 44.10-65.70 |
| 17 | 14 | 106.72 | 6.85 | .0642 | 95.70-116.00 | 45 | 12 | 95.18 | 4.70 | .0494 | 89.00-104.00 |
| 18 | 20 | 235.05 | 15.88 | .0675 | 192.00-257.00 | 46 | 21 | 49.43 | 6.02 | .1217 | 35.40-59.50 |
| 22 | 20 | 137.63 | 10.51 | .0764 | 115.10-154.60 | 49 | 20 | 35.34 | 1.91 | .0541 | 32.75-39.20 |
| 24 | 21 | 67.96 | 11.30 | .1664 | 31.85-79.30 | 52 | 20 | 47.19 | 2.42 | .0512 | 44.00-53.80 |
| 25 | 19 | 45.86 | 8.50 | .1853 | 35.60-63.80 | 55 | 18 | 45.27 | 2.83 | .0625 | 40.50-49.55 |
| 26 | 19 | 76.59 | 11.45 | .1495 | 48.55-90.25 | | | | | | |

**Mandibles**

*Equus alaskae* group                    *Equus francisci* group

| VAR. | N | MEAN | S.D. | V | OBS. RANGE | VAR. | N | MEAN | S.D. | V | OBS. RANGE |
|---|---|---|---|---|---|---|---|---|---|---|---|
| 1 | 7 | 84.52 | 8.04 | .0951 | 71.95-99.20 | 1 | 26.79 | 15.21 | .1981 | 33.65-98.10 | |
| 2 | 6 | 108.63 | 5.91 | .0544 | 100.60-114.70 | 2 | 17 | 100.86 | 10.96 | .1087 | 82.50-119.70 |
| 4A | 11 | 81.57 | 6.61 | .0811 | 69.70-93.25 | 4A | 33 | 85.10 | 4.22 | .0496 | 78.30-99.05 |
| 4B | 11 | 74.86 | 4.14 | .0553 | 69.40-82.45 | 4B | 37.03 | 5.74 | .0745 | 60.15-91.55 | |
| 5 | 3 | 422.67 | 8.96 | .0212 | 417.00-433.00 | 5 | 17 | 402.53 | 29.50 | .0733 | 344.00-467.00 |
| 6 | 3 | 393.67 | 13.58 | .0345 | 381.00-408.00 | 6 | 18 | 381.61 | 29.67 | .0777 | 326.00-455.00 |
| 9 | 4 | 62.86 | 1.62 | .0258 | 61.50-64.80 | 9 | 14 | 52.32 | 8.53 | .1631 | 28.55-59.25 |
| 10 | 8 | 40.16 | 3.18 | .0792 | 36.30-45.00 | 10 | 21 | 37.04 | 7.54 | .2036 | 17.15-57.65 |
| 14 | 11 | 62.01 | 3.69 | .0594 | 56.40-69.40 | 14 | 43 | 58.86 | 5.16 | .0877 | 49.60-74.70 |
| 15 | 11 | 82.25 | 4.85 | .0590 | 76.10-89.00 | 15 | 40 | 80.89 | 7.61 | .0941 | 55.20-100.75 |
| 16 | 10 | 109.76 | 7.16 | .0653 | 93.30-118.05 | 16 | 37 | 112.51 | 9.11 | .0809 | 81.80-136.70 |
| 19 | 7 | 205.50 | 14.95 | .0728 | 175.50-220.00 | 35 | 192.81 | 19.11 | .0991 | 151.00-235.00 | |

*Equus laurentius* group                    *Equus scotti* group

| VAR. | N | MEAN | S.D. | V | OBS. RANGE | VAR. | N | MEAN | S.D. | V | OBS. RANGE |
|---|---|---|---|---|---|---|---|---|---|---|---|
| 1 | 25 | 96.60 | 14.82 | .1534 | 59.30-118.50 | 1 | 25 | 97.15 | 17.08 | .1758 | 58.55-124.00 |
| 2 | 23 | 120.50 | 13.63 | .1131 | 84.35-153.70 | 2 | 23 | 122.38 | 15.87 | .1297 | 86.55-148.80 |
| 4A | 29 | 103.13 | 9.29 | .0901 | 86.05-120.85 | 4A | 39 | 103.03 | 7.18 | .0697 | 87.75-119.40 |
| 4B | 17 | 94.05 | 4.12 | .0438 | 84.80-99.65 | 4B | 25 | 98.12 | 8.41 | .0857 | 89.00-119.60 |
| 5 | 16 | 471.56 | 33.56 | .0712 | 389.00-524.00 | 5 | 14 | 471.21 | 54.77 | .1162 | 357.00-564.00 |
| 6 | 17 | 449.94 | 33.69 | .0749 | 369.00-497.00 | 6 | 15 | 454.33 | 52.48 | .1155 | 347.00-545.00 |
| 9 | 18 | 70.17 | 8.48 | .1208 | 46.15-81.50 | 9 | 25 | 65.05 | 11.60 | .1783 | 31.45-81.25 |
| 10 | 28 | 44.28 | 4.27 | .0965 | 33.75-51.40 | 10 | 26 | 43.67 | 7.45 | .1707 | 22.20-55.05 |

| VAR. | N | MEAN | S.D. | V | OBS. RANGE | VAR. | N | MEAN | S.D. | V | OBS. RANGE |
|---|---|---|---|---|---|---|---|---|---|---|---|
| 14 | 29 | 66.74 | 7.78 | .1166 | 47.00-80.40 | 14 | 35 | 69.29 | 10.02 | .1447 | 43.70-87.10 |
| 15 | 29 | 90.63 | 8.53 | .0942 | 64.40-103.00 | 15 | 37 | 95.67 | 12.23 | .1278 | 71.10-117.55 |
| 16 | 17 | 123.62 | 9.09 | .0735 | 110.90-137.70 | 16 | 22 | 129.34 | 7.11 | .0550 | 119.50-144.30 |
| 19 | 21 | 204.43 | 20.32 | .0994 | 161.00-228.00 | 19 | 26 | 211.08 | 29.78 | .1411 | 143.00-268.00 |

*Equus simplicidens* group

| VAR. | N | MEAN | S.D. | V | OBS. RANGE | VAR. | N | MEAN | S.D. | V | OBS. RANGE |
|---|---|---|---|---|---|---|---|---|---|---|---|
| 1 | 22 | 98.32 | 19.26 | .1959 | 52.90-145.60 | 9 | 19 | 56.76 | 5.55 | .0978 | 42.90-69.95 |
| 2 | 20 | 119.85 | 18.49 | .1542 | 88.20-169.40 | 10 | 24 | 39.25 | 3.53 | .0900 | 30.80-45.55 |
| 4A | 20 | 100.55 | 6.87 | .0683 | 87.20-109.95 | 14 | 26 | 63.63 | 5.87 | .0922 | 53.10-74.30 |
| 4B | 16 | 93.14 | 4.12 | .0442 | 86.20-101.15 | 15 | 22 | 89.02 | 5.40 | .0607 | 74.70-96.25 |
| 5 | 16 | 480.06 | 34.29 | .0714 | 414.00-544.00 | 16 | 16 | 127.94 | 10.48 | .0819 | 106.70-144.15 |
| 6 | 16 | 454.38 | 32.29 | .0711 | 387.00-515.00 | 19 | 19 | 216.89 | 19.56 | .0902 | 186.00-255.00 |

## Metacarpals

*Equus alaskae* group — *Equus francisci* group

| VAR. | N | MEAN | S.D. | V | OBS. RANGE | VAR. | N | MEAN | S.D. | V | OBS. RANGE |
|---|---|---|---|---|---|---|---|---|---|---|---|
| 1 | 51 | 225.25 | 10.25 | .0455 | 207.00-263.00 | 1 | 54 | 251.29 | 13.31 | .0529 | 223.00-313.00 |
| 2 | 51 | 47.77 | 2.50 | .0524 | 42.70-52.60 | 2 | 52 | 45.30 | 3.08 | .0681 | 31.35-53.60 |
| 3 | 51 | 31.30 | 2.19 | .0702 | 28.05-41.80 | 3 | 53 | 30.68 | 2.56 | .0836 | 24.00-41.25 |
| 4 | 50 | 28.97 | 3.48 | .1201 | 21.70-46.75 | 4 | 53 | 25.93 | 2.83 | .1094 | 17.75-35.80 |
| 5 | 51 | 38.55 | 3.04 | .0789 | 31.20-43.65 | 5 | 52 | 36.57 | 3.73 | .1021 | 24.60-44.60 |
| 6 | 51 | 31.46 | 2.62 | .0835 | 22.70-36.20 | 6 | 54 | 30.14 | 2.32 | .0769 | 24.70-39.40 |
| 7 | 51 | 25.13 | 1.70 | .0679 | 22.60-31.05 | 7 | 54 | 25.45 | 1.85 | .0727 | 21.40-32.45 |
| 8 | 51 | 41.91 | 1.80 | .0431 | 38.00-45.80 | 8 | 54 | 39.39 | 2.28 | .0579 | 32.05-46.10 |
| 9 | 51 | 42.38 | 1.79 | .0424 | 38.90-47.10 | 9 | 49 | 40.32 | 2.12 | .0527 | 37.40-47.60 |
| 10 | 48 | 28.46 | 1.70 | .0599 | 21.90-32.50 | 10 | 53 | 28.36 | 1.45 | .0513 | 26.30-33.10 |
| 11 | 50 | 32.58 | 1.56 | .0479 | 28.00-36.25 | 11 | 52 | 32.22 | 1.71 | .0532 | 27.90-39.25 |

*Equus laurentius* group — *Equus scotti* group

| VAR. | N | MEAN | S.D. | V | OBS. RANGE | VAR. | N | MEAN | S.D. | V | OBS. RANGE |
|---|---|---|---|---|---|---|---|---|---|---|---|
| 1 | 109 | 248.20 | 10.08 | .0406 | 221.00-269.00 | 1 | 139 | 233.86 | 10.87 | .0465 | 204.00-275.00 |
| 2 | 110 | 57.90 | 4.39 | .0759 | 44.90-67.90 | 2 | 138 | 51.94 | 3.61 | .0695 | 38.90-59.00 |
| 3 | 110 | 36.33 | 2.13 | .0588 | 30.60-41.90 | 3 | 139 | 34.25 | 2.36 | .0690 | 28.55-40.35 |
| 4 | 111 | 34.46 | 2.34 | .0681 | 25.50-38.65 | 4 | 139 | 30.67 | 2.66 | .0867 | 23.60-38.10 |
| 5 | 111 | 46.43 | 3.59 | .0775 | 37.10-53.95 | 5 | 139 | 41.62 | 2.76 | .0664 | 35.50-49.55 |
| 6 | 111 | 39.00 | 3.12 | .0800 | 27.70-48.35 | 6 | 139 | 35.85 | 3.11 | .0868 | 28.10-45.45 |
| 7 | 111 | 29.99 | 2.10 | .0702 | 24.45-39.70 | 7 | 139 | 28.18 | 2.72 | .0965 | 23.40-39.70 |
| 8 | 109 | 51.40 | 2.73 | .0531 | 42.45-57.70 | 8 | 139 | 48.09 | 3.38 | .0703 | 39.65-58.05 |
| 9 | 103 | 53.09 | 3.23 | .0608 | 42.60-58.65 | 9 | 132 | 49.12 | 3.56 | .0725 | 40.30-56.50 |
| 10 | 107 | 34.15 | 1.75 | .0513 | 27.55-38.00 | 10 | 138 | 32.44 | 2.48 | .0764 | 26.35-39.60 |
| 11 | 104 | 39.80 | 2.08 | .0524 | 31.85-44.20 | 11 | 139 | 37.77 | 3.14 | .0830 | 31.25-49.45 |

*Equus simplicidens* group

| VAR. | N | MEAN | S.D. | V | OBS. RANGE | VAR. | N | MEAN | S.D. | V | OBS. RANGE |
|---|---|---|---|---|---|---|---|---|---|---|---|
| 1 | 41 | 243.61 | 10.86 | .0446 | 226.00-278.00 | 7 | 41 | 26.51 | 2.21 | .0832 | 23.10-32.25 |
| 2 | 41 | 51.62 | 3.60 | .0697 | 46.20-60.40 | 8 | 40 | 46.84 | 3.06 | .0653 | 42.50-56.20 |
| 3 | 41 | 33.44 | 2.33 | .0696 | 29.30-38.65 | 9 | 39 | 45.55 | 3.34 | .0733 | 41.50-56.45 |
| 4 | 41 | 30.23 | 2.75 | .0910 | 24.80-36.50 | 10 | 39 | 31.79 | 1.91 | .0601 | 28.70-38.10 |
| 5 | 41 | 41.91 | 2.95 | .0705 | 36.35-48.20 | 11 | 40 | 35.07 | 2.29 | .0653 | 30.25-43.20 |
| 6 | 41 | 33.89 | 3.08 | .0910 | 29.10-42.75 | | | | | | |

## Metatarsals

### Equus alaskae group

| VAR. | N | MEAN | S.D. | V | OBS. RANGE |
|---|---|---|---|---|---|
| 1 | 51 | 263.13 | 8.15 | .0309 | 242.00-283.00 |
| 2 | 52 | 46.90 | 2.72 | .0580 | 41.45-52.00 |
| 3 | 51 | 41.33 | 2.23 | .0541 | 38.10-49.30 |
| 4 | 50 | 21.70 | 1.75 | .0807 | 18.60-25.35 |
| 5 | 48 | 42.12 | 2.48 | .0590 | 37.05-45.85 |
| 6 | 52 | 31.42 | 2.85 | .0908 | 25.90-36.65 |
| 7 | 51 | 31.01 | 1.86 | .0601 | 27.10-35.45 |
| 8 | 50 | 43.13 | 2.25 | .0522 | 38.60-48.80 |
| 9 | 50 | 42.88 | 1.91 | .0445 | 38.20-46.65 |
| 10 | 51 | 29.26 | 1.95 | .0667 | 24.10-33.95 |
| 11 | 51 | 33.61 | 1.45 | .0431 | 30.25-37.25 |

### Equus francisci group

| VAR. | N | MEAN | S.D. | V | OBS. RANGE |
|---|---|---|---|---|---|
| 1 | 58 | 294.03 | 15.65 | .0532 | 258.00-346.00 |
| 2 | 56 | 43.80 | 3.91 | .0894 | 33.40-58.25 |
| 3 | 54 | 40.38 | 3.87 | .0960 | 31.90-52.75 |
| 4 | 53 | 19.97 | 2.82 | .1413 | 14.05-29.55 |
| 5 | 53 | 40.19 | 3.16 | .0786 | 31.30-48.45 |
| 6 | 59 | 29.61 | 2.38 | .0806 | 23.15-37.75 |
| 7 | 59 | 29.67 | 2.65 | .0894 | 24.00-39.05 |
| 8 | 59 | 39.53 | 3.47 | .0877 | 30.85-50.60 |
| 9 | 58 | 39.40 | 3.72 | .0945 | 30.10-53.50 |
| 10 | 59 | 28.83 | 2.27 | .0790 | 23.30-36.60 |
| 11 | 57 | 32.46 | 2.64 | .0814 | 24.50-41.35 |

### Equus laurentius group

| VAR. | N | MEAN | S.D. | V | OBS. RANGE |
|---|---|---|---|---|---|
| 1 | 111 | 289.82 | 10.45 | .0360 | 259.00-311.00 |
| 2 | 112 | 57.57 | 3.50 | .0609 | 47.30-64.30 |
| 3 | 111 | 50.29 | 2.60 | .0517 | 44.55-59.35 |
| 4 | 112 | 26.80 | 2.12 | .0790 | 20.00-31.90 |
| 5 | 107 | 52.01 | 3.45 | .0664 | 41.70-59.55 |
| 6 | 113 | 38.37 | 2.81 | .0732 | 27.40-47.15 |
| 7 | 113 | 35.93 | 2.31 | .0643 | 28.15-40.60 |
| 8 | 112 | 53.35 | 3.15 | .0591 | 45.90-59.70 |
| 9 | 107 | 54.40 | 2.87 | .0529 | 44.00-60.50 |
| 10 | 108 | 35.51 | 2.89 | .0815 | 24.20-39.70 |
| 11 | 109 | 41.61 | 2.11 | .0508 | 34.85-45.85 |

### Equus scotti group

| VAR. | N | MEAN | S.D. | V | OBS. RANGE |
|---|---|---|---|---|---|
| 1 | 130 | 274.82 | 13.97 | .0508 | 239.00-317.00 |
| 2 | 5l321 | | 3.92 | .0766 | 41.75-62.60 |
| 3 | 127 | 46.88 | 3.68 | .0786 | 37.15-56.90 |
| 4 | 129 | 24.76 | 2.53 | .1022 | 18.50-32.60 |
| 5 | 132 | 45.70 | 3.59 | .0784 | 38.60-58.00 |
| 6 | 131 | 34.09 | 2.92 | .0856 | 26.40-42.70 |
| 7 | 131 | 33.48 | 2.36 | .0704 | 28.05-41.05 |
| 8 | 132 | 48.30 | 3.98 | .0825 | 38.25-58.70 |
| 9 | 126 | 48.92 | 4.13 | .0845 | 39.15-57.70 |
| 10 | 129 | 32.50 | 2.39 | .0735 | 26.20-37.70 |
| 11 | 125 | 37.74 | 2.86 | .0757 | 31.00-44.40 |

### Equus simplicidens group

| VAR. | N | MEAN | S.D. | V | OBS. RANGE |
|---|---|---|---|---|---|
| 1 | 40 | 273.53 | 7.72 | .0282 | 259.00-296.00 |
| 2 | 40 | 50.17 | 2.33 | .0465 | 46.10-57.00 |
| 3 | 40 | 42.43 | 2.61 | .0615 | 38.30-49.45 |
| 4 | 40 | 23.29 | 1.92 | .0826 | 19.20-28.20 |
| 5 | 40 | 44.77 | 2.34 | .0522 | 40.35-49.35 |
| 6 | 40 | 32.40 | 2.52 | .0778 | 26.40-38.70 |
| 7 | 40 | 30.82 | 2.20 | .0714 | 26.10-34.90 |
| 8 | 40 | 46.58 | 2.51 | .0538 | 42.05-52.85 |
| 9 | 39 | 44.65 | 2.46 | .0551 | 38.40-53.50 |
| 10 | 40 | 31.65 | 1.42 | .0449 | 28.75-36.20 |
| 11 | 40 | 35.86 | 1.87 | .0521 | 31.20-42.40 |

early Rancholabrean specimens; all other large Rancholabrean horses. In the mandible analysis (Fig. 14.6B), the small and Blancan groups separate from the rest of the points (though not as clearly), but the other two groups do not separate. Analysis of the metapodials (Figs. 14.6C and D) produces a more or less continuous V-shaped distribution in the plane of the first and second coordinate axes which closely parallels the distribution of a scatter diagram of greatest length versus any arbitrarily chosen width measurement. One wing of this distribution contains a group of specimens with elongated metapodials (usually referred to as "stilt-legged"), and the other contains all "stout-legged" specimens, regardless of size, with the smallest ones down near the vertex of the V.

Analysis of separate geological age groups helps clarify some of the relationships. For example, analysis of the Blancan samples of metatarsals and metacarpals shows a clear division into early (Hagerman and Texas) and late (Grandview, Arizona, and Nebraska) Blancan groups (Figs. 14.7B and C). Similarly, analysis of Irvingtonian metapodials (Figs. 14.8C and D) yields division into stilt-legged and stout-legged groups.

The pattern of distribution of Irvingtonian skulls (Fig. 14.8A) is more difficult to interpret. Since sample sizes are very small, some of the apparent gulfs between groups may not be significant. Analysis of the mandibles (Fig. 14.8B) suggests a division into two groups, as did the metapodials. The group that is correlated with the stout-legged metapodials plots together with the late Blancan sample, suggesting that these specimens form a unified taxonomic group.

Separate analysis of the Rancholabrean sample (Fig. 14.9) produces a division into three groups with two outlying specimens. The small horses group together in the skull and mandible analyses, but separate into stilt- and stout-legged in the metapodial plots. The significance of the outlying specimens from Ingleside and American Falls specimens is uncertain since each of these site samples had only one specimen complete enough to be included in the analysis. However, in the overall analysis these specimens group with the Late Blancan and Irvingtonian specimens, and, as will be discussed in the next section, other lines of evidence also suggest that these specimens should be grouped separately from other Rancholabrean large horses and together with the late Blancan and Irvingtonian large horses.

Taking all of the analyses together, it appears that the only easily justifiable division would yield five groups, each of which is named for the most senior type specimen included in the group: 1) Early Blancan large horses, as represented by the Hagerman and Texas samples (*E. simplicidens* group); 2) Late Blancan to early Rancholabrean large horses (*E. scotti* group); 3) later Rancholabrean large horses (*E. laurentius* group); 4) Small, non-stilt-legged horses, of Rancholabrean age (*E. alaskae* group); 5) Irvingtonian to Rancholabrean stilt-legged horses (*E. francisci* group). Each of these groups is briefly described below.

## *Equus simplicidens* group

Included types: *E. pons* Quinn, 1958, UA 9, right maxilla with $dP^2$-$M^2$, Comosi Wash, Santa Cruz Co., Arizona; *E. shoshonensis* Gidley, 1930, USNM 11986, skull, mandible, and other skeletal elements of a male, Hagerman, Twin Falls Co., Idaho, Blancan; *E. simplicidens* Cope, 1892, TMM 40282-6, fragmentary upper cheek tooth, Mount Blanco, Crosby Co., Texas, early Blancan.

Site samples referred to this group: All specimens from Comosi Wash, Hagerman, and Blancan sites in Crosby Co., Texas.

Age and geographic range: Early Blancan, known from the western part of the United States.

General quantitative description: This is a large horse. The basicranial length (dimension 1A of the skull) is greater than 500 mm in all specimens so far referred to this group. The rostrum is elongated and narrow, but not substantially wider at the anterior end than across the diastema (Table 14.2). The metapodials are stout, with a ratio of length to proximal width in the metacarpals generally less than 5.0, and in the metatarsals generally less than 6.0. Members of this group average slightly larger than members of the *E. scotti* group.

*Equus scotti* group

Included types: *E. hatcheri* Hay, 1915, USNM 7868, skull and mandible of a female, Hay Springs, Sheridan Co., Nebraska, Irvingtonian, about 0.5 Ma; *E. niobrarensis* Hay, 1913 (1913a), USNM 4999, nearly complete skull and mandible of a female, Hay Springs, Sheridan Co., Nebraska, Irvingtonian, about 0.5 Ma; *E. scotti* Gidley, 1900, AMNH 10606, partial skeleton of a young adult, Mayfield Ranch, Rock Creek, Briscoe Co., Texas, Irvingtonian.

Site samples referred to this group: All specimens from Rock Creek, Bautista Creek, Grandview, Irvington, Gilliland, Pool Branch, Port Charlotte, Vallecito, American Falls, and Ingleside; large horses from Hay Springs, Arkalon.

Age and geographic range: Late Blancan to early Rancholabrean, known from the middle west of the United States and from a few sites in southernmost California.

General quantitative description: Like members of the *E. simplicidens* group, members of this group are large, with a basicranial length (dimension 1A of the skull) greater than 500 mm in all skulls so far referred, but the rostrum is wider in proportion to skull length. The metapodials are stout, with the average ratio of length to proximal width of the metacarpals generally less than 5.0, and that of the metatarsals generally less than 6.0. Members of this species average slightly smaller than members of the *E. simplicidens* group (Table 14.2).

*E. laurentius* group

Included types: ?*E. laurentius* Hay, 1913 (1913a), KUMNH 347, skull and mandible, both male, but not the same individual, north side of the Kaw River, Douglas Co., Kansas, probably Holocene; *E. mexicanus* Hibbard, 1955, IGM 48, skull of an old male, Tajo de Tequixquiac, Estado de Mexico, Mexico, Rancholabrean; *E. midlandensis* Quinn, 1957, TMM 998-3, fragmentary right and left mandibles, left $P^2$, right $P^3$-$M^1$, limb bones, Scharbauer Ranch, Midland Co., Texas, Late Rancholabrean.

Site samples referred to this group: All specimens from Rancho La Brea, McKittrick; Rainbow Beach, Dam; large horses from Cueva Quebrada, Dry Cave.

Age and geographic range: Rancholabrean, known from all over the United States and from Mexico, but not so far, from Canada.

General quantitative description: This is the most recent group of large horses without stilt-legged metapodials. The basicranial length (dimension 1A) of all skulls so far referred is greater than 500 mm. The ratio of length to proximal width is generally less that 5.0 for metacarpals and 6.0 for metatarsals. The skull and rostrum are wider than average proportionate to length (Table 14.2).

*Equus francisci* group

Included types: *E. (Onager) altidens* Quinn, 1957, TMM 31186-3, -4, -10, -35, -36, partial skeleton, Blanco Creek, Powers Ranch, east of Berclair, Goliad Co., Texas, Rancholabrean, about 0.3 Ma; *E. (Onager) arellanoi* Mooser, 1958, TMM 42428-12, left ramus of mandible with all cheek teeth, Arroyo Cedazo, Aguascalientes, Mexico, Irvingtonian, about 0.5 Ma; *E. (Asinus) calobatus* Troxell, 1915, YPM 13470, possibly associated left tibia, metatarsal, astragalus, right metacarpal, first, second, and third phalanges, and atlas, Rock Creek, Mayfield Ranch, Briscoe Co., Texas, Irvingtonian, about 0.9 Ma. (Hibbard, 1953, chose the metatarsal as the lectotype); *E. francisci* Hay, 1915, TAMU 2518, incomplete skull and mandible, right metatarsal, other fragmentary limb elements, all belonging to a single individual, Wharton Co., Texas, Rancholabrean; *E. quinni* Slaughter *et al.*, 1962, SMP 60578, right metatarsal, Moore Pit, Trinity River, Dallas Co., Texas, Rancholabrean, less than 0.1 Ma; *E. zoytalis* Mooser, 1958, IGM 56-4, right mandible and symphysis of a juvenile male with all canines and incisors, right $P_2$-$M_3$, Arroyo Cedazo, Aguascalientes, Mexico, Irvingtonian, about 0.5 Ma.

Site samples referred to this group: All specimens from Channing, Silverton, Coleman, Quarry G, Loup Fork; small horses from Hay Springs, Arkalon.

Age and geographic range: Irvingtonian to Rancholabrean, known from the middle west of the United States and from Mexico, but not, so far, from Canada.

General quantitative description: This group encompasses all horses with stilt-legged metapodials, having a ratio of length to proximal width generally greater that 5.0 for metacarpals and 6.0 for metatarsals. The size range from largest to smallest metapodial is greater than that of most other groups, raising the possibility that the group should be subdivided. However, there is no clear break in the distribution of metapodial size, and skulls which can be reliably associated with the larger metapodials are so far unknown. Until more information becomes available, I prefer to maintain these specimens as a single group. All skulls so far referred are somewhat more elongated and narrower than average, especially in the area of the rostrum (Table 14.2).

## *Equus alaskae* group

Included types: *E. conversidens leoni* Stock, 1950, 1953, no type ever named (Stock seems to have regarded all specimens from San Josecito Cave as constituting a hypodigm), Nuevo Leon, Mexico, Rancholabrean, about 25,000 years B.P. These specimens are deposited at LACM; *E. lambei* Hay 1917, USNM 8426, skull and mandible of a female, Gold Run Creek, Klondike, Yukon, Alaska, Pleistocene; *E. niobrarensis alaskae* Hay 1913 (1913b), USNM 7700, skull of a male, Yukon River near Tofty, Alaska, Rancholabrean.

Site samples referred to this group: All specimens from San Josecito Cave, Natural Trap Cave; small horses from Dry Cave.

Age and geographic range: Late Irvingtonian to Rancholabrean, known from Alaska, the middle west of the United States, and Mexico.

General quantitative description: This is a group of small but stout-legged horses, with basicranial length (dimension 1A) of all skulls so far referred less than 500 mm. The ratio of length to proximal width is generally less that 5.0 for metacarpals and 6.0 for metatarsals. The skull and rostrum are substantially wider than average proportionate to length (Table 14.2).

Discriminant analysis based on these five groups (Table 14.1) yields functions that can correctly identify nearly all skulls, but which misidentify on the order of 25% of other skeletal elements. This is a very common outcome for analyses of closely related taxa. The characters selected by the skull analysis reflect a number of characters which have been proposed by many previous workers, such as length of the rostrum (dimension 4) relative to the skull as a whole (dimensions 1A and 1B) and breadth of the rostrum (dimension 24), as well as several smaller features which have seldom been mentioned by previous workers (dimensions 6, 43, 45, 49, 55).

Given the very low error rate, discriminant analysis probably could be used as a tool for identification of any reasonably complete skull. However, the higher error rates on mandibles and metapodials do not make this a particularly attractive tool for their identification. Furthermore, as discussed above, this method makes no provision for the possibility that a specimen might belong to a new group not provided for by the discriminant function. A more desirable alternative for those who possess the necessary computer resources would be to repeat the principal components analysis using the same database with the new specimens added. However, this will work only for reasonably complete skulls, mandibles, and metapodials. Very fragmentary specimens and isolated teeth or tooth rows cannot be classified on the basis of presently available information.

Discussion

The taxonomic rank of these groups is problematic. Some qualitative characters appear to lend support to the five-fold division implied by the multivariate quantitative tests described here. For example, the *E. simplicidens* group is clearly differentiated from other groups by a number of primitive *Pliohippus*-like characters, including nearly circular protocones, distinct facial fossae on the skull, a very deep ectoflexid on all lower molars, and a broadly U-shaped linguaflexid. However, the distribution of qualitative characters within *Equus* as a whole still is not well enough understood to allow any definitive assessment to be made of the significance of such similarities. It is possible that some of the groups which I have defined encompass more than one species, which the numerical methods used here are unable to separate. Resolution of this uncertainty will have to await improved understanding of tooth proportions and the distribution of qualitative characters. In spite of the questions which remain, this study demonstrates that the use of large samples gives one a much clearer view of the patterns and ranges of variation within the genus *Equus*, and that multivariate quantitative analysis can assist in clarifying our understanding of this group of horses. The data and analyses which I have described strongly suggest that the taxonomy of *Equus* in North America may be less complex than previously thought.

Due to the gaps in the distribution of sites with useful material, the full geographic range of these groups is uncertain. However, all appear to have been widely distributed. The geologic age ranges of the specimens referred to each of the five groups suggest that no more than four of them were extant at the same time. The fact that several sites were found which contained both stilt-legged and stout-legged individuals makes it probable that the *E. francisci* group was sympatric with the *E. scotti* and *E. laurentius* groups. It is uncertain at present whether any of the other groups were sympatric.

If one accepts for the sake of discussion that these groups may represent species, then the maximum diversity of *Equus* on this continent at any one time (four species) would be in sharp contrast to the maximum Miocene and Pliocene diversity of eight (D. A. Winkler, pers. comm.; Woodburne and MacFadden, 1982; MacFadden, 1984b) to sixteen (Webb, 1984) species in two or three genera. This difference in diversity may be partly due to differing philosophies about what constitutes sufficient morphological difference to justify erecting a new species, for I have tended to place more emphasis on similarities than differences, which led to a very conservative classification scheme. A less conservative interpretation might have produced as many as two or three times as many groups. This would still be a lower level of diversity, but not by as great a factor. Which point of view is correct can be endlessly debated, but never conclusively settled. However, the fact that the diversity of camels and antelopes also is believed to have declined around the beginning of the Blancan lends weight to the possibility that there actually was a decrease in horse diversity around this time.

A major factor in the decline in horse diversity was the loss of the three-toed horses. Since three-toed horses are generally considered to have eaten a larger amount of plant material other than grass than did the one-toed horses (Shotwell, 1961), their disappearance may have been a consequence of the shift toward increasing areas of grassland and decreasing areas of savannah and woodland (reported by

Shotwell, 1961, for the northern Great Basin) during the Pliocene. The apparently lower diversity of the one-toed horses may reflect the fact that the range of food sources available to *Equus* was much narrower than that available to the grass- and herb-eating three-toed horses. However, it must also be remembered that the fossil record of *Equus* is much shorter than that of earlier horses, about 3.5 million years in contrast to nearly 15 million years for some other genera such as "*Merychippus*." In that relatively short period of time, the genus increased from a single species to up to four contemporary species. This implies that *Equus* may have been undergoing the first stages of a radiation which eventually would have led to a considerably higher level of diversity. If this were the case, it could account for the rather confused patterns of variation within the group, for this is a common characteristic of groups which are undergoing rapid diversification.

### Acknowledgments

This study has been greatly enhanced by discussions with Ernest Lundelius, John Wilson, Wann Langston, Dave Webb, Bruce MacFadden, William Akersten, Ted Downs, Donald Savage, Morris Skinner, Dick Tedford, Clayton Ray, Bill Turnbull, John White, George Miller, Les Marcus, Nancy Neff, and my father, Ross C. Winans, who was instrumental in the fabrication of the special measuring and photographic equipment. Financial assistance was provided by the Geological Society of America, Sigma Xi, the Geology Foundation of the Department of Geological Sciences and the Graduate School of the University of Texas at Austin, the Dee Fund of the Field Museum of Natural History, and by grant number EAR-8018799 of the National Science Foundation. This paper is a contribution of the Vertebrate Paleontology Laboratory of the Texas Memorial Museum.

### Bibliography

Cope, E.D. (1892): A contribution to the vertebrate paleontology of Texas.-*Proc. Amer. Phil. Soc.*, 30: 123-131.

Dalquest, W.W. (1978): Phylogeny of American horses of Blancan and Pleistocene age. -*Ann. Zool. Fennica*, 15: 191-199.

Dalquest, W.W. (1979): The little horses (genus *Equus*) of the Pleistocene of North America. -*Amer. Midl. Nat.*, 101(1): 241-244.

Davis, J.C. (1987): *Statistics and Data Analysis in Geology*, Second Edition. - New York (John Wiley & Sons, Inc.).

Gidley, J.W. (1900): A new species of Pleistocene horse from the Staked Plains of Texas. -*Bull. Amer. Mus. Nat. Hist.*, 13(13): 111-116.

Gidley, J.W. (1901):Tooth characters and revision of the North American species of the genus *Equus*. -*Bull. Amer. Mus. Nat. Hist.*, 14: 91-141.

Gidley, J.W. (1930): A new Pliocene horse from Idaho. - *J. Mammal.*, 11(3): 300-303.

Hay, O.P. (1913a): Notes on some fossil horses, with descriptions of four new species. -*Proc. U. S. Natl. Mus.*, 44(1969): 569-594.

Hay, O.P. (1913b): Description of the skull of an extinct horse found in central Alaska. -*Smithsonian Misc. Coll.*, 61(2): 1-18.

Hay, O.P. (1915): Contributions to the knowledge of the mammals of the Pleistocene of North America. -*Proc. U.S. Natl. Mus.*, 48(2086): 515-575.

Hay, O.P. (1917): Description of a new species of extinct horse, *Equus lambei*, from the Pleistocene of Yukon Territory. -*Proc. U.S. Natl. Mus.*, 53(2212): 435-443.

Hibbard, C.W. (1953): *Equus (Asinus) calobatus* Troxell and associated vertebrates from the Pleistocene of Kansas. -*Trans. Kansas Acad. Sci.*, 56(1): 111-126.

Hibbard, C.W. (1955): Pleistocene vertebrates from the Upper Becerra (Becerra Superior) Formation, Valley of Tequixquiac, Mexico, with notes on other Pleistocene forms. -*Contrib. Mus. Paleont. Univ. Mich.*, 12: 47-96.

MacFadden, B.J. (1984): *Astrohippus* and

*Dinohipppus* from the Yepomera local fauna (Hemphillian, Mexico) and implications for the phylogeny of one-toed horses. - *J. Vert. Paleont.*, 4(2): 273-283.

Mooser, O. (1958): La fauna "Cedazo" del Pleistoceno en Aguascalientes. -*Ann. Inst. Biol. Mex.*, 29: 409-452.

Morrison, D.F. (1976): *Multivariate Statistical Methods*, Second Edition. -New York (McGraw-Hill).

Quinn, J.H. (1957): Pleistocene Equidae of Texas. -*Bur. Econ. Geol. Rept. Invest.*, 33: 1-51.

Quinn, J.H. (1958): New Pleistocene *Asinus* from southwestern Arizona. -*J. Paleont.*, 32(3): 603-610.

Reyment, R.A., Blackith, R.E., and Campbell, N.A. (1984): *Multivariate morphometrics*. - New York (Academic Press).

Savage, D.E. (1951): Late Cenozoic vertebrates from the San Francisco Bay region. -*Univ Calif. Publ. Geol. Sci.*, 28: 215-314.

Shotwell, J.A. (1961): Late Tertiary biogeography of horses in the northern Great Basin. -*J. Paleont.*, 35(1): 203-217.

Slaughter, B.H., Crook, W.W., Jr., Harris, R.K., Allen, D.C., and Seifert, M. (1962): The Hill-Shuler local faunas of the upper Trinity River, Dallas and Denton Counties, Texas. -*Bur. Econ. Geol. Rept. Invest.*, 48: 1-75.

Stock, C. (1950): 25,000-year-old horse; the skeleton of an Ice Age horse makes a return trip to Mexico. -*Eng. Sci. Monthly*, 14: 16-17.

Stock, C. (1953): El caballo pleistoceno (*Equus conversidens leoni* sp. nov.) de la cueva de San Josecito, Aramberri, Nuevo Leon. -*Mem. Congr. Cient. Mex.*, 3: 170-171.

Troxell, E.L. (1915): The vertebrate fossils of Rock Creek, Texas. -*Amer. J. Sci.*, Fourth Series, 39(234): 613-638.

Webb, S.D. (1984): Ten million years of mammal extinctions in North America. - *In* Martin, P.S., and Klein, R.G. (eds.), *Pleistocene Extinctions - A Prehistoric Revolution*: 189-210.

Winans, M.C. (1985): Revision of North American fossil species of the genus *Equus* (Mammalia: Perissodactyla: Equidae). -Ph.D. dissertation, University of Texas at Austin. 265 pp.

Woodburne, M.O., and MacFadden, B.J. (1982): A reappraisal of the systematics, biogeography, and evolution of fossil horses. -*Paleobiology*, 8(4): 315-327.

# 15. A REVIEW OF THE TAPIROIDS

## ROBERT M. SCHOCH

The traditional Tapiroidea (*sensu lato*) is a paraphyletic grouping that has included lophiodont ancylopods (e.g., *Paleomoropus*), plesiomorphous ceratomorphs, tapirids and closely related forms, and even plesiomorphous rhinocerotoids (e.g., *Hyrachyus*). Within the context of a review of some of the better-known "tapiroid" genera (concentrating on North American forms), the new genus *Plesiocolopirus* is erected for the species *Colodon? hancocki* Radinsky, 1963. Tapiroids were most common during the Eocene, at which time they underwent a prolific radiation. In North America, tapiroids (exclusive of hyrachyids and their relatives) were low in both diversity and absolute numbers after the late Eocene, but persisted until the close of the Pleistocene.

### Introduction

Traditionally the Tapiroidea (*sensu lato*) has been considered to be the group composed of all perissodactyls possessing complete cross lophs on the upper and lower molars, but bearing short ectolophs and brachydont teeth (Radinsky, 1963); i.e., the Tapiroidea is a grade taxon and blatantly paraphyletic (see Schoch, 1986, for a discussion of the concept of paraphyly as used here). Thus, even though Radinsky (e.g., 1969, 1983, and earlier papers) suggested that various tapiroids gave rise to the rhinocerotoid families, he excluded the derived rhinocerotoids from the Tapiroidea. Furthermore, many of the nominal families, and even genera, traditionally included within the Tapiroidea are also probably paraphyletic. The Tapiroidea are in need of a thorough revision, but such a revision will be a difficult task and is beyond the scope of the present paper. Here I review the major tapiroid genera and put forth some of my ideas concerning these forms primarily with the idea of stimulating further investigation; the present discussion is not intended as a taxonomic revision. More detailed information on North American tapiroids can be found in Schoch (in prep.).

The tapiroids have generally been allocated to six to eight distinct families. 1) The Isectolophidae were small, relatively generalized, early and middle Eocene perissodactyls known from North America and Asia. 2) The Helaletidae were small to medium-sized, cursorial forms that developed greatly enlarged nasal incisions and may have given rise to the true Tapiridae. Known primarily from the early Eocene to late Oligocene of North America, helaletids have also been recovered in Asia. 3) The true tapirs (family Tapiridae, probably a monophyletic family) were and are medium-sized, short-legged, folivorous forms that developed short, mobile proboscises. In North America tapirids are known from the Oligocene through Pleistocene; tapirids are also known from fossil forms in Europe and Asia and are currently extant in Asia and South America. 4) The Lophiodontidae are early to late Eocene forms known best from Europe, but are also known probably from North America in the form of the early Eocene genus *Paleomoropus*, and possibly the late Eocene-?early Oligocene *Schizotheriodes* Hough, 1955 (=

?*Toxotherium* Wood, 1961). The enigmatic genus *Paleomoropus*, known only from the type specimen, was originally assigned to the Eomoropidae (Chalicotherioidea), but in fact may represent a tapiroid in the loose sense and has been assigned to the Lophiodontidae by Fischer (1964, 1977). *Schizotheriodes*, originally considered a possible eomoropid, of the late Eocene of North America may also be a lophiodont (Hooker, this volume). 5) The Lophialetidae (including the Rhodopaginae, but see Lucas and Schoch, 1981; Prothero *et al.*, 1986, and Hooker, this volume, Chapter 6) known only from the middle and late Eocene of Asia. 6) The Deperetellidae known from the middle Eocene to early Oligocene of Asia. 7) Also sometimes included under the aegis of the Tapiroidea are the Hyrachyidae, medium-sized, cursorial forms that are perhaps the sister-taxon of all other members of the Rhinocerotoidea (Radinsky, 1967b, considered *Hyrachyus* the ancestor of the Hyracodontidae in particular). Hyrachids are known from the early and middle Eocene of North America and Europe and the middle to late Eocene of Asia. Here I consider *Hyrachyus* to be the most plesiomorphous known rhinocerotoid and the sister taxon of all other rhinocerotoids. 8) The Indolophidae of the late Eocene of Asia (known only from a single specimen described by Pilgrim, 1925) are sometimes considered a divergent tapiroid lineage (Radinsky, 1965b, 1969); it has also been suggested that this family has affinities with hippomorphs (Schoch, 1984a and below).

Major references on tapiroid evolution include Fischer (1964, 1977), Hatcher (1896), Radinsky (1963, 1965a, 1965b), Reshetov (1979), Schaub (1928), Schoch, (1984a), Simpson (1945b), Stehlin (1903), and Wortman and Earle (1893). For descriptions of the osteology and muscular anatomy of fossil and recent tapirs, see especially Cuvier (1822), Murie (1872), Radinsky (1965a, b), Reshetov (1977, 1979), Scott (1941), and Stjernman (1932).

### Abbreviations

AMNH   American Museum of Natural History, New York, New York
MCZ   Museum of Comparative Zoology, Harvard University, Cambridge, Massachusetts
NMC   National Museum of Canada, Ottawa
PU   Princeton University Geological Museum (now housed in YPM)
USNM   National Museum of Natural History, Washington, D. C.
UOMNH   University of Oregon Museum of Natural History, Eugene, Oregon
YPM   Peabody Museum of Natural History, Yale University, New Haven, Connecticut

### Diagnostic features of the tapiroids

As a paraphyletic grouping, tapiroids are generally primitive ceratomorphs that share only plesiomorphies in common. Radinsky, to whom more than to any other single worker we owe the current concept of the Tapiroidea (*sensu lato*), diagnosed the group as such (1963, p. 9): "Very small to large perissodactyls with complete cross lophs on upper and lower molars but ectoloph short and metalophid incomplete or absent. Teeth brachydont." In general, the tapiroid skeleton is relatively complete and, compared to other perissodactyls, close to the primitive eutherian morphotype. Extant tapirs are distinguished by their possession of a short, mobile proboscis; however, this is not a trait seen in a majority of the fossil forms currently allocated to the Tapiroidea (*sensu lato*).

### Relationship of the Tapiroidea to other perissodactyls

Wood (1937; adopted by Simpson, 1945a) united the superfamilies Tapiroidea and Rhinocerotoidea in the suborder Ceratomorpha and aligned the Ceratomorpha with the suborder Hippomorpha (composed of the Equoidea, Brontotherioidea, and Chalicotherioidea) within the Perissodactyla (reviewed by Schoch, 1985a, b, and this volume). Scott (1941) united the Tapiroidea and the Rhinocerotoidea in the infraorder Ceratomorpha, restricted the infraorder Hippomorpha to the Brontotherioidea and Equoidea, and recognized the distinct suborder Ancylopoda (Cope, 1889) for the Chalicotherioidea. Scott (1941) united his Ceratomorpha and Hippomorpha in the suborder Chelopoda. Radinsky (1964) adopted essentially the same classification as Scott (1941), but considered the Ceratomorpha and Hippomorpha to be of subordinal rank coordinate with the Ancylopoda. Radinsky (1964) did not resolve the trichotomy among Ceratomorpha, Hippomorpha, and Ancylopoda.

A major theme that the three classificatory schemes reviewed above share is that the tapiroids and rhinocerotoids are regarded as being closely related to one another relative to all other perissodactyls. Another way in which this concept is expressed is repeated suggestions in the literature that rhinocerotoids were derived from tapiroids.

Also of particular note is that, in the preceding schemes, the chalicotheres are not regarded as particularly closely related to tapiroids or rhinocerotoids. In the last few years several workers have independently suggested that perhaps chalicotheres and tapirs share a close relationship. In 1964 Radinsky described *Paleomoropus* and classified this taxon as an early eomoropid chalicothere. Simultaneously, Radinsky (1964) transferred the European genus and species *Lophiaspis maurettei* Depéret 1910 from the Lophiodontidae (Tapiroidea) to the Eomoropidae. Fischer (1964, 1977), however, regarded both *Lophiaspis* and *Paleomoropus* as lophiodonts; specifically Fischer (1977) stated that *Lophiaspis* and *Lophiodon* can be derived from *Paleomoropus*. Coombs (1982) appears to have tentatively accepted Fischer's suggestion that *Paleomoropus* and *Lophiaspis* are not chalicotheres. In 1983 I suggested that the Ceratomorpha and chalicotheres might be sister-groups (Schoch, 1983c). In March of 1984 (Schoch, 1984a), I coined the subordinal name Moropomorpha to contain the infraorders Ceratomorpha (Tapiroidea and Rhinocerotoidea) and Ancylopoda (Chalicotherioidea); in this scheme the Moropomorpha and suborder Hippomorpha (composed of the Brontotherioidea and the Equoidea) were regarded as sister-groups (Schoch, 1985a, b). Independently, Hooker (1984) came to view the chalicotheres and lophiodonts as sister-groups and used the term Ancylopoda for the group composed of the Lophiodontidae and Chalicotherioidea. According to Hooker (1984) the Ancylopoda (*sensu* Hooker, 1984) is the sister-group of all other ceratomorphs except the Isectolophidae, and the Isectolophidae is the sister-group of the Ancylopoda plus the Ceratomorpha (Ceratomorpha in the traditional sense minus the Lophiodontidae and Isectolophidae). To name his concept of ceratomorphs plus ancylopods plus isectolophids Hooker (1984, published for September/October of 1984) resurrected and redefined the old name Tapiromorpha (Haeckel, 1873). Thus the composition, if not internal arrangement, of Hooker's (1984) Tapiromorpha was essentially identical to my concept of the Moropomorpha [I prefer the new name Moropomorpha rather than resurrecting Haeckel's Tapiromorpha as the name for this taxon, since the name Tapiromorpha has previously been established in use as essentially synonymous with the Tapiroidea or the Ceratomorpha (see, for example, Simpson, 1945a; see also Schoch, this volume, Chapter 2)]. In his most recent work, Hooker (this volume, Chapter 6) continues to regard the chalicotheres (specifically as represented by *Eomoropus* and *Lophiaspis*; he does not treat in detail *Paleomoropus*) as the sister-group of the Lophiodontidae.

Recently, and independently of the work of Fischer, Hooker, or Schoch, Bakker *et al.* (in prep.) have suggested that chalicotherioids and ceratomorphs are closely related to each other relative to hippomorphs. In fact, these authors classify chalicotheres as ceratomorphs. Bakker et al. (in prep.) suggest, however, that *Paleomoropus* itself is not a chalicothere but a tapiroid. Whereas *Paleomoropus* has derived features uniting it with tapiroids, or ceratomorphs in general, according to Bakker et al. (in prep.) this genus lacks any derived characters uniquely shared with chalicotheres. Although I initially considered *Paleomoropus* a chalicothere (e.g., Schoch 1983c, 1984a), I now have my doubts as to its affinities. Here I provisionally regard it as a primitive "tapiroid" (specifically, an ancylopod lophiodont). I continue to be convinced, however, that chalicotheres and ceratomorphs share a close common ancestry.

## Relationships among the tapiroids

As it is North American forms that I am most familiar with, I will discuss the North American genera and their possible phylogenetic relationships first; then I will comment on how some of the European and Asian forms may fit into tapiroid phylogeny. A hypothesis as to the phylogeny and interrelationships of the North American tapiroids, along with the Rhinocerotoidea, is shown in Figure 15.1. When comparing my cladogram to that of Hooker (his Figure 6.6, this volume) some basic similarities are immediately obvious (it should be noted the Schoch and Hooker cladograms were derived independently). Both hypotheses unite the "helaletids" (viz. *Heptodon, Helaletes, Colodon* and related forms), the tapirids, hyrachyids, and the rhinocerotoids as a monophyletic group relative to all other perissodactyls. Both unite the lophidontoids-chalicotherioids (represented by *Lophiodon, Paralophiodon, Lophiaspis,* and *Eomoropus* in the Hooker cladogram and by *Paleomoropus* in the Schoch cladogram) with the tapiroids and rhinocerotoids as a monophyletic group relative to all other perissodactyls.

Following Radinsky (1963), I have continued to consider *Homogalax* the primitive sister taxon of all other ceratomorphs. *Homogalax*, however, may bear a number of autapomorphies (Bakker et al., in prep.) and would not be directly ancestral to any later ceratomorphs. Node 2 on the cladogram marks my general concept of the Ceratomorpha. Hooker (this volume, Chapter 6) suggests that *Homogalax* and *Isectolophus* share a number of synapomorphies, and that the monophyletic Isectolophidae (*Homogalax* and *Isectolophus*) is the sister group of all other moropomorphs (ancylopods and ceratomorphs). Hooker (this volume, Chapter 6) would thus exclude the Isectolophidae from the Ceratomorpha.

Radinsky (1963) believed that *Heptodon* may have been directly ancestral to the Hyrachyidae, Rhinocerotoidea, and advanced tapiroids. Such a suggestion is not incongruous with the unresolved trichotomy at Node 6 of the cladogram. *Hyrachyus* is clearly the sister-group of at least some rhinocerotoids (Node 7); for this reason I have excluded it from the Tapiroidea (see Schoch, 1982). Initially Radinsky (1963) also excluded *Hyrachyus* from the Tapiroidea, but he later (Radinsky, 1965b, 1966b) allocated the genus to the Tapiroidea on the basis "that *Hyrachyus* lacked derived features characteristic of rhinocerotoids" (Radinsky, 1983, p. 296). Recent studies (Hooker, this volume, Chapter 6; Prothero et al., 1986) confirm that *Hyrachyus* is a probable sister-group of all other rhinocerotoids and is best referred to the Rhinocerotoidea. Given the relationships shown in the cladogram of Figure 15.1, if the Tapiroidea is restricted to Node 8, then the tapiroids *sensu stricto* would be a natural (i.e., monophyletic) group (see also Hooker, this volume, Chapter 6).

The relationships of *Desmatotherium, Selenaletes,* and *Dilophodon* are ambiguous, in part because these taxa are very poorly known (see Radinsky, 1963, 1966a; Schoch, 1984a).

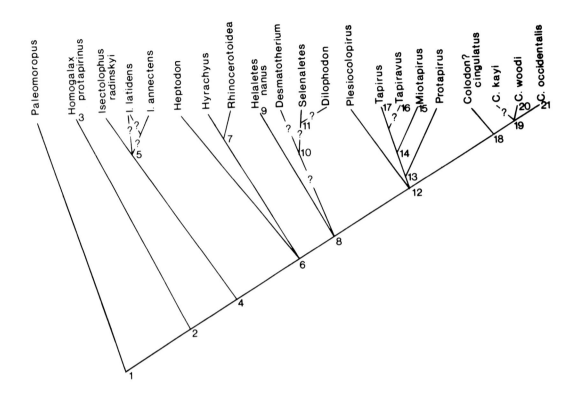

Figure 15.1. Hypothesis of the interrelationships of North American tapiroids. Character-states corresponding to the node points include: 1. $M^{1-3}$ metaconules usually not distinct, or only barely distinct, on the metalophs; metalophs form distinct, transverse shearing crests; metalophs relatively high where they join the ectolophs. 2. Complete cross lophs, both anteriorly (protolophs and protolophids) and posteriorly (metalophids and hypolophids) on the upper and lower molars; metalophid incomplete or absent. 3. Dog-like incisors (see Bakker, Cooke, and Schain, MS). 4. $M^{1-3}$ protoconules and metaconules never distinct; metastylids absent on $M_{1-3}$. 5. $P^3$ variably submolariform with one or two internal cusps. 6. $M_{1-3}$ metalophids greatly reduced; $M_3$ hypoconulid short and narrow or absent; postcanine diastema present; lingual cusp(s) on $P^2$ (variable). 7. $P^4$ protocone "creased" lingually; development of relatively crest-like metacones on $M^{1-3}$, especially posteriorly; $M^3$ metacone lingually depressed and $M^3$ relatively triangular (amynodontids are derived in bearing labially deflected $M^3$ metastyles); $M_3$ hypoconulid lost (see Radinsky, 1983; Savage et al., 1966; Schoch, 1982; Prothero et al., 1986). 8. $P_1$ absent; nasal incision enlarged (nasal incision unknown in *Desmatotherium*, *Selenaletes*, and *Dilophodon*; anterior premolars unknown in *Selenaletes*). 9. $P^{2-4}$ submolariform; modification of the hindlimb toward a more cursorial condition than seen in *Heptodon* or tapirids (see Schoch, 1984a). 10. Metaloph bypasses hypocone on $P^{3-4}$ ($P^{3-4}$ not known in *Selenaletes*); $M_3$ hypoconulid extremely reduced (*Desmatotherium*) or absent (*Dilophodon* and *Selenaletes*). 11. $M_3$ hypoconulid totally absent; very small size. 12. Nasal incision further enlarged (relative to condition of Node 8); shortened nasals. 13. Molar metacones convex and labially place; $M_3$ hypoconulid absent; atrophy of upper canines. 14. Upper and lower premolars relatively molariform (posterior cross crest developed on $P_{3-4}$; development of hypocone on $P^1$; upper molar crests better developed than in *Protapirus*; upper premolar series longer than upper molar series. 15. $M^{1-2}$ relatively square in occlusal view. 16. $M_{2-3}$ trigonids and talonids separated by a deep groove or constriction; lower cheek teeth elongated anteroposteriorly. 17. Premolars highly molariform; teeth relatively high-crowned. 18. $P^1$ with posterolingual cusp; $P^{2-4}$ hypocones variably differentiated from protocones. 19. $P^{3-4}$ molariform. 20. Small size. 21. $M^2$ metacone flat to concave and reduced; $P^1$ submolariform; canines vestigial to absent; tridactyl manus.

*Plesiocolopirus hancocki, Colodon*, and the Tapiridae form an unresolved trichotomy at Node 12. This is compatible with the traditional view that *Colodon* is closely related to the Tapiridae and with Radinsky's (1963) opinion that in terms of known morphology *Plesiocolopirus hancocki* is intermediate between *Helaletes* and *Colodon*, and that *P. hancocki* is also intermediate between *Helaletes* and *Protapirus*. *Colodon* and the Tapiridae are each monophyletic, and the group composed of *Plesiocolopirus, Colodon*, and the Tapiridae is also monophyletic. In order to express these relationships *Plesiocolopirus* and *Colodon* should perhaps be transferred from the Helaletidae to the Tapiridae, and the monophyletic group containing *Protapirus, Miotapirus, Tapiravus*, and *Tapirus* could be given the rank of subfamily Tapirinae. The reason for not making these changes here is that the family Tapiridae is very well known as it is presently constituted; I feel that any further changes in the suprageneric systematics of the tapiroids should wait until a more thorough analysis of tapiroids is undertaken (including all known forms, not just North American tapiroids).

**North American tapiroid genera**

In this section I review the North American tapiroid genera that I consider valid. In the following section I will comment on selected non-North American tapiroid forms.

Family Uncertain (perhaps Lophiodontidae Gill, 1872)

*Paleomoropus* Radinsky, 1964

Type species: *Paleomoropus jepseni* Radinsky, 1964.

Type specimen of the type species: PU 13254, left $M^1$, right $M^2$, left $M^3$.

Distribution: early Eocene of North America.

Characteristics: Perissodactyls with a molar cusp pattern as follows (quoted from Radinsky, 1964, p. 3): "parastyle large and anterolabially displaced; paracone sharply conical, with thin anterior and posterior ridges (to protoloph and metacone, respectively); no mesostyle; metacone lingually displaced and slightly convex labially; protoloph with symmetrical medial bulge, followed lingually by a sharp constriction (on both anterior and posterior sides), which delimits a lophid protoconule; lingual half of protoloph straight, protocone not deflected posteriorly; metaloph high and uninterrupted (no metaconule), terminating high at the ectoloph; basal cingula on anterior, lingual, posterior, and posterolabial sides. $M^3$ differs from $M^1$ and $M^2$ in being larger and having a larger and more labially displaced parastyle and a slightly shorter metacone. A large wear facet on the posterior side of the $M^3$ hypocone (lingual end of the metaloph) indicates that $M_3$ had a large hypoconulid."

At present *Paleomoropus* is only known from PU 13254.

*Schizotheriodes* Hough, 1955

Type species: *Schizotheriodes parvus* Hough, 1955.

Type specimen: USNM 20205, $M^{2-3}$.

Distribution: late Eocene of North America.

Discussion: Hough (1955) named *Schizotheriodes parvus* solely on the basis of an associated $M^{2-3}$. Primarily on the basis of the prominent parastyle seen on $M^3$, Hough referred it to the ?Eomoropinae. Radinsky (1964, p. 24) suggested that the "absence of a distinct protoconule is a strong argument against placing *Schizotheriodes* in the Chalicotherioidea" and further noted that "the molar cusp pattern of *Schizotheriodes* resembles that of the Tapiroidea [*sensu lato*]." Most recently Hooker (this volume, Chapter 6) has argued persuasively that at present, *Schizotheriodes* is best referred questionably to the Lophiodontidae (see also Schiebout, 1977; Wilson and Schiebout,

1984; Emry, 1979; Prothero et al., 1986).

In the literature on *Schizotheriodes*, there has been considerable confusion as to the correct spelling of this generic name. When Hough (1955, p. 34) named the genus, she consistently spelled it as *Schizotheriodes*, but in the abstract to the same paper (Hough, 1955, p. 22), the genus is spelled *Schizotherioides*. Furthermore, in her original paper, Hough (1955, p. 34) misspelled *Chasmotheroides* Wood, 1934, as "*Chasmotheriodes*" and this may have led some authors to believe that Hough's intention was to spell her new genus *Schizotheroides*. Schiebout (1977) and Emry (1979) both adopted the spelling *Schizotheroides*.

However, according to the *International Code of Zoological Nomenclature* (Ride et al., 1985, Article 32), the original spelling of a name "is to be preserved unaltered unless it is demonstrably incorrect as provided under Section c of this Article." Article 32, Section c of the *Code* states that an "original spelling is an 'incorrect original spelling' if . . . there is in the original publication itself, without recourse to any external source of information, clear evidence of an inadvertent error, such as a *lapsus calami* or a copyist's or printer's error (incorrect transliteration or latinization and use of an inappropriate connecting vowel are not to be considered inadvertent errors)."

There is no clear evidence that Hough (1955) did not intend to spell the name of her new genus as *Schizotheriodes* (even if she misspelled *Chasmotheroides*). In fact, the ending *-odes* is even recognized as an ending of a genus-group name in an example presented in the *Code* under Article 30, Section b. As for the two different spellings in the abstract and main text of Hough (1955), one might suggest that either spelling can be used as both are original spellings. However, under Article 32 of the *Code*, the first reviser of the genus would then determine which spelling should be used subsequently, and which spelling should be considered an "incorrect original spelling." As the first reviser of the genus, Radinsky (1964, p. 23-24) adopted the spelling *Schizotheriodes*.

*Toxotherium* Wood, 1961

Type species: *Toxotherium hunteri* Wood, 1961.

Type specimen: NMC 8918, right ramus with $P_{2-4}$.

Distribution: early Oligocene of North America.

Discussion: Wood (1961, p. 2) diagnosed his distinctive new taxon, based only on the type specimen, as follows: "Single, disproportionately large, bulbous-rooted lower front tooth (incisor or canine); premolar paraconid regions, especially of $P_3$ usually distinct; $P_{2-4}$ carry asymmetric trigonid and talonid crescents; premolar metaconids, especially of $P_4$, relatively isolated cusps for stage of premolar metamorphosis; teeth incipiently high-crowned." Wood (1961) suggested that *Toxotherium* might be a rhinocerotoid or possibly close to the lophialetids. On the basis of additional specimens, and considering both *Toxotherium woodi* (Skinner and Gooris, 1966) and *Schizotheriodes jackwilsoni* (Schiebout, 1977) to be synonyms of *Toxotherium hunteri*, Emry (1979) thought *Toxotherium* was possibly an amynodont. It is also possible that *Toxotherium* is a lophiodont (see Prothero et al., 1986), perhaps synonymous with *Schizotheriodes* at the generic level.

Family Isectolophidae Peterson, 1919

Characteristics: The isectolophids are a probable paraphyletic grade of small to medium-sized cursorial tapiroids (i.e., the molars have cross lophs combined with short ectolophs) that bear a full placental dentition, lack any major diastemata, and lack enlarged nasal incisions. The premolars are relatively nonmolariform. The paracones and metacones on $M^{1-3}$ tend to be subequal and convex. The metalophids on

$M_{1-3}$ tend to be well-developed.

*Homogalax* Hay, 1899

Type species: *Systemodon primaevus* Wortman, 1896 (junior subjective synonym of *Homogalax protapirinus*).

Type specimen: AMNH 144, upper dentition [AMNH 4460, maxilla with $P^1$-$M^3$, mandible with $P_2$-$M_3$, (listed by Radinsky, 1963, p. 13) is the type of *H. protapirinus* (Wortman, 1896)].

Distribution: early Eocene of North America and Asia.

Characteristics: *Homogalax* is a genus of small, cursorial forms with completely nonmolariform premolars ($P^3$ always has only one lingual cusp), $M^{1-3}$ sometimes bear visible (but poorly-developed) protoconules and metaconules; the cusps on $P^{1/}{}_1$, $P_{2-3}$ are high and sharp, and metastylids are present on $M_{1-3}$. Recent fossil finds demonstrate that *Homogalax* possessed distinctive, autapomorphous, "dog-like" incisors (described by Bakker et al., in prep.).

*Isectolophus* Scott and Osborn, 1887

Type species: *Isectolophus annectens* Scott and Osborn, 1887.

Type specimen: PU 10400, $P^4$, $M^3$, $M_3$.

Distribution: middle Eocene of North America and Asia.

Characteristics: Species of *Isectolophus* overlap in size with those of *Homogalax*, but some *Isectolophus* are larger. In *Isectolophus* $P^1$ is relatively small, and $P^3$ may be submolariform (it bears one or two lingual cusps). $M^{1-3}$ are relatively longer and narrower than in *Homogalax*, and their protoconules and metaconules are never distinct. Metastylids are lacking on $M_{1-3}$ (see Schoch, 1983a).

Family Helaletidae Osborn, 1892, in Osborn and Wortman, 1892

Characteristics: The helaletids, another paraphyletic grouping, were small to medium-sized cursorial tapiroids distinguished by the following characteristics (based on Radinsky, 1963): tendency to lose the canines and first lower premolars; long post-canine diastema; tendency for $P^{1-4}$ to become molariform via a hypocone separating posteriorly from the protocone; $M^{1-2}$ relatively flat, short, and lingually displaced, but with a labial cingulum; $M^3$ relatively long and narrow with a reduced metacone and metaloph shorter than protoloph; tendency for $P_{2-4}$ to become molariform, but without developing hypolophids; $M_{1-3}$ metalophids greatly reduced or absent; $M_3$ hypoconulid small or lost; enlarged nasal incisions developed in derived forms.

*Heptodon* Cope, 1882

Type species: *Lophiodon ventorum* Cope, 1880 (junior subjective synonym of *Heptodon calciculus*).

Type specimen: AMNH 4850, left ramus with $P_4$-$M_3$ [AMNH 4858, left and right rami with $P_3$-$M_3$, (listed by Radinsky, 1963) is the type of *Heptodon calciculus* (Cope, 1880)].

Distribution: early Eocene of North America and Asia.

Characteristics: Small to medium-sized helaletids with a complete placental dentition; $P^{2-4}$ nonmolariform with only one lingual cusp, protoloph better developed than metaloph, and protocone steeply conical; $M^{1-2}$ metacones slightly convex to flat and relatively unshortened and not as lingually displaced as in other helaletids; $M^3$ metacone relatively unreduced; $P_2$ is a labiolingually compressed blade; $P_{3-4}$ trigonids relatively high; nasal incision not enlarged.

*Helaletes* Marsh, 1872

Type species: *Helaletes boops* Marsh, 1872 (junior subjective synonym of *Helaletes nanus*).

Type specimen: YPM 11807, calvarium, mandible, and partial skeleton [YPM 11080, fragments of both maxillae with right $P^2$-$M^3$ and left $P^3$-$M^2$, (listed by Radinsky, 1963), is the type of *Helaletes nanus* (Marsh, 1871a)].

Distribution: middle Eocene of North America; ?early to ?late Eocene of Asia. (*Helaletes* may be known from the early or middle Eocene of the Canadian High Arctic; see West and Dawson, 1978, fig. 10, a right maxilla, which these authors refer to "*Hyrachyus*.")

Characteristics (based on Radinsky, 1963): Small helaletids with small canines; $P^{2-4}$ submolariform with a tendency for the hypocone to separate from the protocone, but metalophs do not bypass the hypocones as in *Desmatotherium*; $M^{1-2}$ metacones slightly convex to flat, and slightly shortened; $M^3$ metaloph relatively shorter and metacone not as reduced as in *Colodon*; $P_1$ absent; $P_{2-4}$ trigonids low, paralophids relatively long, talonids relatively narrower than in *Colodon*; small entoconids present on $P_{3-4}$; nasal incision greatly enlarged. Relative to *Heptodon* and *Tapirus*, *Helaletes* bears modifications of the hindlimb (e.g., a longer ilium and a higher greater trochanter of the femur) toward a more cursorial condition (see Schoch, 1984a).

*Desmatotherium* Scott, 1883

Type species: *Desmatotherium guyotii* Scott, 1883 (junior subjective synonym of *Desmatotherium intermedius*).

Type specimen: PU 10166, right maxilla fragment bearing $P^2$-$M^3$ and root of $P^1$ and an isolated right upper canine (PU 10095, a right $M^{1-3}$, is the type specimen of *Desmatotherium intermedius* Osborn, Scott, and Speir, 1878).

Distribution: middle Eocene of North America.

Characteristics: *Desmatotherium* is superficially similar to *Helaletes* (Radinsky, 1963, synonymized *Desmatotherium* with *Helaletes*), but differs in being larger, possessing metalophs that bypass the hypocones on $P^{3-4}$, and having an extremely reduced hypoconulid on $M_3$. *Desmatotherium* is known only from four fragmentary specimens that are only doubtfully referred to a single species (see Schoch, 1984a).

*Dilophodon* Scott, 1883

Type species: *Dilophodon minusculus* Scott, 1883.

Type specimen: PU 10019, right ramus with $P_2$-$M_3$ and symphysis.

Distribution: middle Eocene of North America.

Characteristics: Very small to small helaletids; premolars relatively nonmolariform to submolariform; metalophs bypass the hypocones on $P^{3-4}$; $M^{1-2}$ metacones lingually displaced, slightly convex labially, and bear small labial cingula; $P_1$ absent; $M_3$ lacks a hypoconulid.

Emry (in manuscript at the time of this writing) is describing a new genus and species of diminutive tapiroid that appears to be very close to, or synonymous with, *Dilophodon*.

*Selenaletes* Radinsky, 1966a

Type species: *Selenaletes scopaeus* Radinsky, 1966a.

Type specimen: AMNH 8320, lower jaw fragment with $P_4$, $M_1$.

Distribution: early Eocene of North America.

Characteristics: Very small helaletids with relatively nonmolariform premolars; lower molars similar to those of *Heptodon*, but much smaller; differs from *Dilophodon* in having a narrower $P_4$ talonid with a smaller entoconid and less reduced molar metalophids; $M_3$ is smaller than $M_2$ and lacks a hypoconulid.

*Selenaletes* is known only from three lower jaw fragments.

*Plesiocolopirus* new genus

Type species: *Colodon? hancocki* Radinsky, 1963, p. 67.

Type specimen: UOMNH 20377, anterior part of the skull with left $P^1$-$M^3$ (described and illustrated by Radinsky, 1963).

Distribution: late Eocene or early Oligocene of North America.

Diagnosis: Genus of medium-sized (known $M^{1-3}$ length 44-45 mm) tapiroids (*sensu* Radinsky, 1963, and this chapter) most similar phenetically to *Colodon*; nasal incision enlarged and nasals shortened as in *Colodon*; small canines present; $M^{1-2}$ metacones convex labially, shortened, and lingually displaced relative to *Colodon* (*C. kayi*, *C. woodi*, *C. occidentalis*, and *C.? cingulatus*) and *Protapirus*; differs from *Colodon* in that $P^1$ lacks a posteriolingual cusp and only bears a cingulum lingually; differs from both *Helaletes* and *Colodon* in that $P^{2-4}$ are relatively short, wide, and nonmolariform to barely submolariform (at most a very shallow groove suggests incipient separation of a protocone and a hypocone); small hypoconulid present on $M_3$.

Etymology: *Plesio-* from plesiomorphous in reference to the presumed overall primitive morphology of the genus relative to *Colodon* and the Tapiridae (see Figure 15.1); *-colo-* from *Colodon*, the taxon that the new genus phenetically most closely resembles; and *-pirus* from *Tapirus* in recognition that this genus may share a close common ancestry with the Tapiridae.

Included species: Only the type species, *Plesiocolopirus hancocki* (Radinsky, 1963), new combination.

Discussion: Radinsky (1963) recognized that his new species, *Colodon? hancocki* represents a new genus (compare the diagnosis of *Plesiocolopirus* to Radinsky's own diagnosis of *Colodon* quoted below), but did not name the genus, since the only known species is phenetically very similar to *Colodon*, and Radinsky considered it expedient to relegate his new species to *Colodon?*. Given the hypothesis of relationships for tapiroid genera here espoused (Fig. 15.1), including *Plesiocolopirus hancocki* in the genus *Colodon* renders *Colodon* a paraphyletic genus whereas excluding it from *Colodon* renders *Colodon* a monophyletic genus. Whenever possible, I believe that genera should be constituted such that they are natural, monophyletic (*sensu stricto*, = holophyletic: see Schoch, 1986), groupings; therefore I choose to separate this species as a distinct genus.

*Colodon* Marsh, 1890

Type species: *Colodon luxatus* Marsh, 1890 (junior subjective synonym of *Colodon occidentalis*).

Type specimen: YPM 11830, an almost complete lower dentition [an isolated $M_3$, Academy of Natural Sciences, Philadelphia (presently missing, see Radinsky, 1963, p. 63), (listed by Radinsky, 1963) is the type of *Colodon occidentalis* (Leidy, 1868)].

Distribution: late Eocene to late Oligocene of North America; middle Eocene to late Oligocene or early Miocene of Asia.

Characteristics (quoted from Radinsky, 1963, p. 57): "Medium-sized to large helaletids with relatively short, wide cheek teeth. Canines small or absent. $P^1$ with lingual loph or cusp; $P^{2-4}$ essentially molariform, protocone and hypocone more or less well separated, metaloph as prominent as protoloph and extending to hypocone. $M^{1-2}$ metacone flat to concave, more lingually displaced and shortened than in *Helaletes*. $P_1$ absent. $P_{2-4}$ with wide talonids and relatively large entoconids. Greatly enlarged nasal incisions with reduced nasals."

Family Tapiridae Burnett, 1830

Characteristics: This family includes relatively primitive, transitional, and derived forms; all of the included members, however, are united in bearing relatively convex and labially placed metacones on the upper molars and by the lack of a hypoconulid on $M_3$. Scott (1941, p. 749) defined the family as follows: "The upper molars have the outer cusps convex externally and from them two straight parallel crests cross the crown and are fused with the inner cusps. The lower molars are extremely simple and have two straight transverse crests, between which the valley is open internally and externally." Along with *Plesiocolopirus* and *Colodon*, the nasal incision in tapirids is enlarged and the nasals are shortened; derived tapirids possess short proboscises. Another trend among derived tapirids is increasing molarization of the premolars.

*Protapirus* Filhol, 1877

Type species: *Protapirus priscus* (Filhol, 1874) (from the Phosphorites du Quercy, France).

Type specimen: Upper and lower jaw fragments with $P^4$-$M^3$, $M_1$-$M_3$ described and illustrated by Filhol (1874; see also Cerdeno and Ginsburg, 1988).

Distribution: late Oligocene of North America; early to late Oligocene of Europe.

Characteristics: Relatively small- to medium-sized tapirids with nonmolariform to submolariform premolars and at most an incipient proboscis (see Sinclair, 1901; Scott, 1941; Wortman and Earle, 1893; Schoch, 1983b).

*Miotapirus* Schlaikjer, 1937

Type species: *Miotapirus harrisonensis* Schlaikjer, 1937.

Type specimen: MCZ 2949, anterior portion of the skull and jaws with complete dentition and a considerable portion of the skeleton.

Distribution: early Miocene of North America.

Characteristics: Medium-sized tapirids; in many respects intermediate in general morphology between *Protapirus* and *Tapirus* (and *Tapiravus*). The upper and lower premolars are relatively molariform. The molar crests are higher and better developed than in *Protapirus*, but not as well-developed as in *Tapirus*. $M^{1-2}$ are relatively square in occlusal view. *Miotapirus* apparently possessed a short proboscis.

Schlaikjer (1937, p. 232) diagnosed the genus as follows: "Superior premolar length greater than superior molar length. Protocone and hypocone on the premolars distinctly separated. $P^4$ larger than $M^1$. Internal cingula on premolars diminished. Nasals reduced and nasal excavation about as in *Tapirus roulini* [= *Tapirus roulinii* = *Tapirus pinchaque* according to Hershkovitz, 1954]. Inferior premolars with well developed hypolophid. $I_1$ larger than $I_2$, and $I_3$ diminutive. $P_2$ proportionately short (anteroposteriorly). $M_1$ smaller than $P_4$. Symphysis constricted but massive. Lunar with anterior magnum facet. Astragalus with small anterior cuboid facet. Ectocuneiform in contact with metatarsal IV."

*Miotapirus* is known from only a very few specimens (see Macdonald, 1970; Schlaikjer, 1937; Schoch, 1984b).

*Tapiravus* Marsh, 1877

Type species: *Tapiravus validus* (Marsh, 1871b).

Type specimen: YPM 13474, left $P^4$ (type cannot presently be located, but AMNH 15592 is a cast of the original).

Distribution: ?middle to ?late Miocene of North America.

Characteristics: Small- to medium-sized tapirids; dentition similar to that of *Tapirus*, but teeth lower-crowned (relatively brachyodont); lower cheek teeth elongated anteroposteriorly; $M_{2-3}$ trigonids and talonids separated by a deep groove or constriction; maxilla relatively deep compared to *Tapirus*; lachrymal processes less prominent than in *Tapirus*; sulcus on the lateral face of the ascending portion of the maxilla shallower and more constricted than in *Tapirus*; mandibular symphysis narrow relative to *Tapirus*.

For discussions of this relatively poorly known genus, see Gazin and Collins (1950),

Olsen (1960), Schoch (1984b), and Schultz *et al.* (1975).

*Tapirus* Brünnich, 1772

Type species: *Tapirus terrestris* (Linnaeus, 1758).

Type specimen: None in existence (see Hershkovitz, 1954; Honacki *et al.*, 1982. Type area is Pernambuco, Brazil).

Distribution: ?middle Miocene to Pleistocene of North America; ?early Miocene to Pleistocene of Europe; late Miocene to Recent of Asia; ?Pliocene to Recent of South America.

Characteristics: Medium- to large-sized tapirids; premolars highly molariform; teeth brachydont, but relatively high-crowned compared to other tapirids; upper canine vestigial and functionally replaced by $I^3$. A small proboscis is present in species of *Tapirus*.

There is no current revision of all species, recent and fossil, that have been referred to *Tapirus*. Dentally this may be a very uniform genus, but cranial differences (and in the case of extant species, differences in pelage) between the known forms have been considered to warrant full generic rank for at least each of the living species (Hershkovitz, 1954; Simpson, 1945b). Most paleontologists have merely found it convenient (a poor criterion for making scientific judgments) to lump all these forms in the single genus *Tapirus* (e.g., Kurtén and Anderson, 1980; Lundelius and Slaughter, 1976; Merriam, 1913; Simpson, 1945b; Ray and Sanders, 1984). If more than one genus were generally recognized among Pleistocene and recent tapirs, then most isolated tapir teeth would be generically indeterminant. As it is, such teeth are now referrable to *Tapirus*, but we know no more nor less about their affinities. Hershkovitz (1954), the last reviser of extant forms, was influenced by Simpson (1945b) and chose to recognize a distinct subgenus for each of the four extant species.

Family Hyrachyidae Wood, 1927

Characteristics: The hyrachyids (here considered to include only the genus *Hyrachyus*) were a relatively conservative lineage of primitive rhinocerotoid perissodactyls that Wood suggested ranged in size from that of a Russian wolfhound to a mustang (Wood, 1927, 1934). In many ways, hyrachyids preserved general dental and skeletal features observed in helaletids such as *Heptodon* (Radinsky, 1967a). Most hyrachyids have a full placental dentition, the incisors are spatulate, and the premolars are nonmolariform to submolariform with no tendency toward bilophodonty of the premolars. Hyrachyids are distinguished by the development of relatively crest-like upper molar metacones, especially posteriorly, and the loss of cingula labial of the metacones. The $M^3$ metacone is lingually depressed and the $M^3$ is relatively triangular. $M_3$ lacks a hypoconulid. The nasal incision is slight. Hyrachyids were moderately cursorial; they bore a tetradactyl manus and a tridactyl pes.

It has been suggested that the European Eocene genus *Chasmotherium* is a hyrachyid (Radinsky, 1967a; Schoch, 1984a; see also Savage *et al.*, 1966) or at least closely related to hyrachyids and rhinocerotoids generally (Hooker, this volume, Chapter 6).

*Hyrachyus* Leidy, 1871a

Type species: *Hyrachyus agrestis* Leidy, 1871a (junior subjective synonym of *Lophiodon modestus;* see Hay, 1902, p. 638; Leidy, 1871b; Wood, 1934).

Type specimen: USNM 660 left lower jaw with $dP_{1-4}$, $M_1$ [USNM 661, isolated $dP^3$ or $dP^4$, (listed by Radinsky, 1967a) is the type of *Hyrachyus modestus* (Leidy, 1870)].

Distribution: early to middle Eocene of North America and Europe; middle to late Eocene of Asia.

Characteristics: See above under characteristics of the family. As I have pointed out elsewhere (Schoch, 1984a), the genus *Hyrachyus* is in need of a thorough revision. Whereas Wood's (1934) revision may have been "oversplit," Radinsky's (1967a) revision probably suffers from an extreme case of taxonomic "lumping" (Radinsky replaced Wood's four genera and over a dozen species with one genus and two species).

## Comments on some tapiroids from outside of North America

Family Tapiridae Burnett, 1830

Discussion: The tapirids appear to be a close-knit, monophyletic group. The following genera have been named solely on the basis of European and/or Asian material and are not currently known from North America: 1) *Palaeotapirus* Filhol, 1888, and/or *Paratapirus* Depéret, 1902, from the early to ?middle Miocene of Europe and the early to middle Miocene of Asia, 2) *Tapiriscus* Kretzoi, 1951, from the late Miocene or early Pliocene of Europe; 3) *Megatapirus* Matthew and Granger, 1923a, from the Pleistocene of Asia; and 4) *Eotapirus* Cerdeno and Ginsburg, 1988, from the Oligocene and early Miocene of Europe. All of these genera appear to be phenetically close to *Tapirus*, *Tapiravus*, and/or *Miotapirus*.

It cannot be overstressed that the original material for many tapiroid (*sensu lato*) taxa needs to be restudied. To give one example, Kretzoi's sole description and diagnosis of his new genus and species *Tapiriscus pannonicus* (Kretzoi, 1951, p. 411, his English translation, which is not an abstract; the taxon is first named on p. 393, and the paper includes no illustrations) reads as follows: "Our Miocene and Pliocene Tapirids are accompanied from time to time by an unnamed Tapirid-branch, little, in molar structure brachydont, narrow and lowly-specialized. The form differs very much from the great *Tapirus priscus* of our *Hipparion*-faunae. Dimensions: $P_2$ 16.8 X 10.8 mm, $P_3$ 15.8 X 12.8 mm, $M_3$ 19.6 X 12.3 mm." Recently, Cerdeno and Ginsburg (1988) have undertaken a study of *Protapirus*, *Paratapirus*, and *Eotapirus* from the Oligocene and early Miocene of Europe.

Family Deperetellidae Radinsky, 1965b

Discussion: The Deperetellidae, composed of only the genera *Deperetella* Matthew and Granger, 1925a, and *Teleolophus* Matthew and Granger, 1925b, was originally diagnosed by Radinsky (1965b, p. 214) as follows: "Medium-sized to large tapiroids with full placental dentition and post-canine diastemata. Premolars submolariform to molariform. Molars high-crowned for a tapiroid and bilophodont. Upper molar protoloph, paracone, and metaloph forming a slightly oblique inverted U, with the metacone (posterior ectoloph extension) greatly reduced and lingually displaced in $M^{1-2}$, and lost completely in $M^3$; lower molars with two parallel cross lophs, and paralophid and metalophid extremely reduced. $M_3$ without a hypoconulid. Manus tridactyl in advanced forms."

Originally Matthew and Granger (1925a, b) considered *Deperetella* and *Teleolophus* to be closely related to *Colodon* (Helaletidae), a view adopted by Simpson (1945a). Radinsky (1965b) considered the similarities between the molars of *Deperetella* and *Teleolophus* and those of *Colodon* to be the result of convergence, and therefore established the family Deperetellidae. Hooker (this volume, Chapter 6) suggests that the Deperetellidae and Rhodopaginae (see following) may constitute a monophyletic group.

On the basis of what appears to be a damaged $M^1$ or $M^2$ and a few other tooth fragments recovered from a middle Oligocene fissure filling near Treuchtlingen, southern West Germany, Heissig (1978) named the new genus and species of tapiroids (*sensu lato*) *Haagella pilgrimi*. Heissig (1978) suggested that *Haagella* is a European representative of the Deperetellidae. Heissig's published illustrations and descriptions do not convince me that *Haagella* is definitely a deperetellid; however, I do not feel competent to further assess the affinities of this taxon without having studied the original material upon which it is based. Heissig (written commun., 1989) currently believes that *Haagella* is closely related to, or possibly a subgenus of, *Colodon*.

Family Lophialetidae Matthew and Granger, 1925b (Radinsky, 1965b)

Discussion: The Lophialetidae of the middle to late Eocene of Asia were diagnosed by Radinsky (1965b, p. 188) as follows: "Small to medium-sized, lightly built tapiroids with postcanine diastemata. Premolars non-molariform. Molars of advanced forms having oblique cross crests and a rhinocerotoid-like cusp pattern, with long flat metacones and relatively high paralophids and metalophids. M3 usually with a relatively long and narrow hypoconulid. Nasal incision enlarged, and manus tridactyl." Within this family Radinsky (1965b) included *Lophialetes* Matthew and Granger, 1925b, *Schlosseria* Matthew and Granger, 1926, and *Breviodon* Radinsky, 1965b. Radinsky (1965b) also questionably assigned the genera *Rhodopagus* Radinsky, 1965b, and *Pataecops* Radinsky, 1966d (replacement name for *Pataecus* Radinsky, 1965b) to the Lophialetidae. Genera that have since been assigned to the family by various authors include *Parabreviodon* Reshetov, 1975, *Eoletes* Birjukov, 1974, *Simplaletes* Qi, 1980, *Kalakotia* Ranga Rao, 1972, and *Aulaxolophus* Ranga Rao, 1972.

Lucas and Schoch (1981; see also Prothero *et al.* 1986) suggested that *Rhodopagus* and *Pataecops* (which together constitute the Rhodopaginae Reshetov, 1975) are actually hyracodontid rhinocerotoids. Hooker (this volume, Chapter 6), however, suggests that the Rhodopaginae may be the sister group of the Deperetellidae.

According to Qi (1980), *Simplaletes*, based on several worn lower dentitions, is very similar to *Schlosseria* and thus *Simplaletes* is readily considered a lophialetid. Hooker (this volume, Chapter 6) suggests that *Eoletes, Schlosseria,* and *Lophialetes* do form a monophyletic group, the Lophialetidae, which in turn is one member of an unresolved trichotomy with the Deperetellidae (including the Rhodopaginae) and the helaletid-tapirid-rhinocerotoid clade. *Breviodon* Radinsky, 1965b, was based exclusively on lower dentitions. Reshetov (1975) based his new genus and species, *Parabreviodon dubius*, on AMNH 81751, a skull and upper cheek teeth that Radinsky (1965b) had assigned to cf. *Breviodon acares*. At present the validity of *Parabreviodon* as a distinct genus is unclear. *Breviodon*, according to Hooker (this volume, Chapter 6), is the sister taxon of the previously noted unresolved trichotomy (Lophialetidae, Deperetellidae, and the helaletid-tapirid-rhinocerotoid clade).

*Kalakotia* Ranga Rao, 1972, and *Aulaxolophus* Ranga Rao, 1972, are relatively plesiomorphous moropomorphs that share some characters with the isectolophids (Hooker, this volume, Chapter 6); they must be removed from the Lophialetidae.

Family Lophiodontidae Gill, 1872

Discussion: The type genus of the family Lophiodontidae is *Lophiodon* Cuvier, 1822, known from the early to late Eocene of Europe. Species of *Lophiodon* were medium- to large-sized mammals, lacking enlarged nasal incisions. Radinsky (1963, p. 91) described the significant dental features of *Lophiodon* as follows (see also Depéret, 1903, 1910; Filhol, 1888; Stehlin, 1903): "Large canines, long prominent post-canine diastemata; $P^1_1$ lost; $P^{2-4}$ nonmolariform to submolariform, with prominent protoloph, short, usually incomplete metaloph, small hypocone distinct in posterolingual corner in advanced forms; $M^{1-2}$ with prominent parastyle, convex paracone, metacone as long as paracone but flattened and slightly lingually displaced; $M^3$ similar to $M^{1-2}$ except metacone shortened; $P_2$ a labiolingually compressed central cusp with small talonid; $P_{3-4}$ with prominent paralophid, short, high protolophid, low metalophid to the hypoconid, small entoconid variably developed; $M_{1-3}$ with low metalophid variably developed; $M_3$ with narrow hypoconulid."

Besides *Lophiodon*, such genera as *Lophiodocherus* Lemoine, 1880, from the early Eocene of Europe; *Atalonodon* Dal Piaz, 1929, of the middle Eocene of Europe; *Lophiaspis* Depéret, 1910, of the early to middle Eocene of Europe; *Chasmotherium* Rütimeyer, 1862, of the middle Eocene of Europe; and *Paralophidon* Dedieu, 1977 (= *Rhinocerolophiodon* Fischer, 1977; see Schoch, 1984a, footnote 2 to table 2) of the middle and ?late Eocene of Europe have been assigned to the Lophidontidae (see, for instance, Simpson, 1945a; Radinsky, 1963).

I have not had the opportunity to study any original material of *Lophiodochoerus*, and, as Radinsky (1963) noted, the published illustrations of this genus are inadequate for taxonomic assessment. As the name implies, Lemoine (1880) originally considered this taxon to represent an artiodactyl.

*Lophiaspis*, long considered a lophiodont, was reinterpreted as an eomoropid chalicothere by Radinsky (1964). Fischer (1964, 1977) suggested that both *Lophiaspis* and *Paleomoropus* are lophidonts. Hooker (this volume, Chapter 6) retains *Lophiaspis* as an eomoropid and furthermore has demonstrated what appears to be a close relationship between lophidontids and eomoropids.

*Atalonodon*, based on a single lower jaw, apparently lacks an M3 hypoconulid but is said to otherwise closely resemble *Lophiodon* (Radinsky, 1963). I have not studied the original material and I cannot make an accurate assessment of the systematic position of this genus on the basis of the published descriptions and illustrations. Hooker (this volume, Chapter 6) considers *Atalonodon* to be possibly a lophiodontid.

*Chasmotherium* was transferred to the Hyrachyidae by Radinsky (1967a), an assignment accepted by Schoch (1984a). Hooker (this volume, Chapter 6) also tentatively associates *Chasmotherium* with *Hyrachyus* and the rhinocerotoids.

*Paralophiodon*, based by Dedieu (1977) on the classical species *Lophiodon buchsowillanum*, was assigned by Dedieu (1977) to the Isectolophidae, whereas Fischer (1977) retained it within the Lophiodontidae. In the most recent study of the taxon, Hooker (this volume, Chapter 6) retains *Paralophiodon* as a lophidontid.

Family Isectolophidae Peterson, 1919
  Discussion: Sahni and Khare (1971) referred their middle Eocene Asian genus, *Sastrilophus*, to the Isectolophidae. I provisionally retain it within this family pending further study (see below, under "Indolophidae," for further discussion of *Sastrilophus*).

Family Helaletidae Osborn, 1892, in Osborn and Wortman, 1892
  Discussion: The genus *Veragromovia* Gabunia, 1961, was based on an isolated $M^3$ that Radinsky (1965b, p. 234) suggested is possibly referable to *Helaletes*.

Family Indolophidae Schoch, 1984a
  Discussion: *Indolophus* Pilgrim, 1925, from the late Eocene of Burma was long considered to show close affinities to the isectolophids (Simpson, 1945a). Radinsky (1963, 1965b) thoroughly redescribed this genus and removed it from the Isectolophidae, relegating it to family *incertae sedis* within the Tapiroidea. Among important features of *Indolophus* are the following (see Radinsky, 1965b): convex metacones; $P^2$ relatively molariform; $P^{3-4}$ protocones labiolingually flattened and

elongated posteriorly; $P^{2-4}$ anterior lophs high and acute; $P^2$-$M^1$ parastyles extremely small; and $M^1$ metaloph almost terminates before it reaches the ectoloph. In the last character-state *Indolophus* differs from all other tapiroids (as Radinsky, 1963, 1965b, noted), and indeed differs from all other moropomorphs. On this basis I suggested that *Indolophus*, sole known genus of the Indolophidae, be removed from the Moropomorpha and assigned to the Hippomorpha (Schoch, 1984a; Schoch and Prins, 1984).

In their original description of *Sastrilophus* from the Eocene of India, Sahni and Khare (1971) suggested that *Sastrilophus* is closely allied to *Indolophus* and that both genera might be referable to the Isectolophidae. From Sahni and Khare's (1971) description, drawing, and photographs, however, it appears that the molar metalophs of *Sastrilophus* exhibit an important derived character-state of the Moropomorpha, namely the metalophs terminate relatively high on the ectolophs. Therefore, I separate *Sastrilophus* from the Indolophidae and retain it within the Moropomorpha.

Family ?Hyracodontidae Cope, 1879 (Rhinocerotoidea)
*Ergilia* Gromova, 1952

Discussion: Gromova (1952) based a new genus and species of perissodactyl, *Ergilia pachypterna*, on some foot and limb bones from the Oligocene of the Mongolian People's Republic and referred this taxon to the Helaletidae. Radinsky (1965b) pointed out that the known remains of *Ergilia* are virtually identical to the corresponding remains of the hyracodontid rhinocerotoid *Ardynia* Matthew and Granger, 1923b, and thus synonymized the two genera.

**Paleobiology and evolutionary patterns among tapiroids**

All extant tapirs are usually included in a single genus, *Tapirus*, composed of four species: *T. terrestris*, *T. bairdii*, and *T. pinchaque* of Central and South America, and *T. indicus* of southeast Asia (see Eisenberg, 1981; Hershkovitz, 1954; and Walker, 1975, for discussions of living tapirs and references to more detailed works). In general body form, living tapirs present the appearance of being slightly plump and short-legged (Janis, 1984); they are rounded in the back with short tails, and in the front their heads and bodies are relatively tapered with the head low to the ground. The forefoot bears four digits (the digit homologous to the fifth digit of other mammals is smaller than the rest), and the hindfoot bears three digits. Overall, the postcranial skeleton of *Tapirus* is very similar to that of *Heptodon*; the main differences in the skeleton of *Tapirus* relative to that of *Heptodon* seem to be due to the fact that *Tapirus* is about 40% larger and more graviportal than *Heptodon* and has a relatively larger head (Radinsky, 1965a). The snout and upper lips of tapirs form a short, fleshy, mobile proboscis. The head and body length of species of extant *Tapirus* range from 180 to 250 cm; weight varies from approximately 225 to 300 kg.

Extant tapirs live in tropical forests, and woody and grassy habitats. They range from sea level to 4,500 m. At least low-land tapirs seem to be dependent on permanent water supplies. Tapirs apparently selectively browse on leaves, sprouts, and small branches, including aquatic plants, and also ingest a good deal of fruit and seeds (Eisenberg, 1981; Janis, 1984). Tapirs tend to be solitary, nonterritorial, and have no distinct breeding season. All of these characteristics can be considered generally primitive for ungulates (Janis, 1984; Walker, 1975).

The earliest generally accepted tapiroid is *Homogalax* of the early Eocene. *Homogalax* was a relatively small and generalized, primitive perissodactyl with a skull length of approximately 16 cm (PU 16168) and a body weight of approximately 10 kg (Janis, 1984). Compared to *Hyracotherium* (which in some respects approximates an ancestral perissodactyl morphotype), *Homogalax* and all later tapiroids (and ceratomorphs more generally) were characterized by the development of higher and unbroken crests on the molars, which increased the amount of shear along the fronts of the protolophs and metalophs of the upper molars with a concomitant reduction in the crushing function of the molars (Radinsky, 1969). *Homogalax* was relatively folivorous compared to contemporaneous Eocene perissodactyls (Janis, 1984).

Tapiroids were most common during the Eocene, at which time they underwent a prolific radiation. In North American at this time, several lineages, described above, were present: the conservative isectolophids; the *Desmatotherium, Selenaletes, Dilophodon* group; the remaining helaletids; and the hyrachyids. General trends among the helaletids in particular involved improvement of shear along the transverse crests of the molars, the molarization of the premolars, and enlargement of the nasal incisions, culminating in the development of the tapiroid proboscis. In Asia the Deperetellidae also progressively molarized the premolars and formed a more efficient bilophodont dentition emphasizing shear along the paralophs and metalophs (Radinsky, 1965b; Reshetov, 1979). In contrast, the Asian Lophialetidae and the European Lophiodontidae generally reduced the premolars, elongated the molar rows, and emphasized ectoloph shear--in some respects approximating the rhinocerotoid dentition (Depéret, 1903; Radinsky, 1965b, 1969). *Lophialetes* also possessed a distinct *Tapirus*-like proboscis (Reshetov, 1979).

Among tapiroids in North America, aside from the hyrachyid-rhinocerotoid clade, only the *Plesiocolopirus-Colodon*-tapirid clade survived after the end of the Eocene. These forms were generally extremely rare elements of their faunas. General trends were toward molarization of the premolars; increasing bilophodonty of the molars and premolars (emphasizing transverse shear along the protolophs and metalophs); enlargement of the nasal incision, shortening of the nasals, and the development of the proboscis; and a general increase in size. As compared to other ungulates, however, tapir evolution was relatively slow. Modern tapirs appear to retain many plesiomorphous characters, and thus extant tapirs have been considered "living fossils" (Janis, 1984).

The ancestor-descendant relationships among the post-Eocene North American tapirs are uncertain (as is to be expected; see Schoch, 1986, for my views on such matters). Radinsky (1963) suggested that *Plesiocolopirus* might have been ancestral to *Protapirus* and the Tapiridae. Later, Radinsky (1965a) suggested that Oligocene *Protapirus* appears too late to be directly ancestral to *Tapirus*, because *Tapirus* has been reported from the middle or late Oligocene of Europe (Schaub, 1928). Radinsky (1969, p. 317) suggested that "the line that gave rise to the modern family Tapiridae, which first appears in the Oligocene, probably diverged from *Helaletes* in the middle Eocene." Based on the work of Radinsky, Janis (1984, fig. 2) suggested that *Plesiocolopirus* directly gave rise to both *Protapirus* and *Tapirus* (in Europe). The dating and identification of *Tapirus* from the Oligocene of Europe appears to be questionable. The specimens referred to *Tapirus* in Schaub (1928) may pertain to *Palaeotapirus*, and furthermore, neither *Palaeotapirus* nor *Tapirus* are listed as occurring before the early Miocene in the recent compendium by Savage and Russell (1983). The fossil tapirids are in desperate need of a thorough systematic and stratigraphic revision. It is also not inconceivable that an ancestor can persist after giving rise to its sister-group and thus occur contemporaneously with, or later than, its presumed descendant.

Tapiroids (exclusive of hyrachyids and their derivatives) were low both in diversity and in absolute numbers after the late

Eocene, but persisted until the close of the Pleistocene in North America. Simpson (1945b) demonstrated that all post-Pliocene North American tapir occurrences are south of the limit of continental glaciation and are in what are presumed to have been humid mesothermal areas. Tapirs appear to have been present when man first entered the North American continent, but went extinct in these latitudes at the end of the Pleistocene when the mesic woodlands disappeared (Janis, 1984).

The general decline in diversity and abundance of tapiroids since the Eocene (exclusive of hyrachyids and their derivatives) has generally been attributed to climatic changes and the rise of the ruminant artiodactyls and rhinocerotoids (Radinsky, 1965b; Reshetov, 1979). Janis (1984) has suggested that the change at the end of the Eocene from relatively nonseasonal tropical forest habitats (with fiber distributed more or less uniformly throughout plants) to more temperate seasonal habitats (with fiber content differing between leaf and stem) favored the ruminant artiodactyl foraging strategy (involving the selective consumption of low-fiber portions of plants) over that of the tapiroids (specialists in consuming a high amount of foliage with a moderate amount of fiber content). However, Janis (1984, p. 85) suggests that tapirs "may have been able to maintain themselves at low numbers in seasonal woodland, despite ruminant competition, by adopting the strategy of selecting only nonfibrous herbage." To this end, the development of the mobile proboscis of tapirs would have been extremely useful in carrying out such a very selective foraging strategy. Closely related to the tapiroids, the rhinocerotoids generally dealt with the changing environmental conditions by developing higher-crowned teeth with increased ectoloph shear (paralleled in the Lophialetidae and Lophiodontidae, as mentioned above), increasing in overall body size, and adopting various grazing strategies.

**Bibliography**

Bakker, R. T., Cooke, J. C., and Schain, J. (in prep.): The dawn horses revisited and the basal bushiness of the Perissodactyla. -*Hunteria* (in prep.).

Birjukov, M. D. (1974): [The new genus of the family Lophialetidae from Eocene deposits of Kazakhstan]. -*Akad, Nauk Kasakhskoi SSR, Alma-Ara. Inst. Zool., Materialy Po Istorii Fauny I Flory Kazakhstana* 6: 57-73.

Brünnich (1772): Zool. Fund. [Not seen; referenced in Hershkovitz, 1954.]

Burnett, G. T. (1830): Illustrations of the Quadrupeda, or quadrupeds, being the arrangement of the true four-footed beasts, indicated in outline. -*Quart. J., Sci., Lit., Arts* 26: 336-353.

Cerdeno, E., and Ginsburg, L. (1988): Les Tapiridae (Perissodactyla, Mammalia) de l'Oligocène et du Miocène inférieur Européens. -*Ann. Paleont., Paris*, 74: 71-96.

Coombs, M. C. (1982): Chalicotheres (Perissodactyla) as large terrestrial mammals. -*Proc. Third N. Amer. Paleo. Conv.* 1: 99-103.

Cope, E. D. (1879): On the extinct species of the Rhinoceridae of North America and their allies. -*Bull. U. S. Geol. and Geog. Surv. Terr.*, 5: 227-237.

Cope, E. D. (1880): The bad lands of the Wind River and their fauna. -*Amer. Nat.* 14: 745-748.

Cope, E. D. (1882): Two new genera of Mammalia from the Wasatch Eocene. -*Amer. Nat.* 16, 1029.

Cope, E. D. (1889): The Vertebrata of the Swift Current River, II. -*Amer. Nat.* 23: 151-155.

Cuvier, G. (1822): *Recherches sur les Ossemens Fossiles* (Nouvelle Édition). Paris (Chez G. Dufour et E. D'Ocagne), vol. 2, part 1, pp. 145-222.

Dal Piaz, G. (1929): *Atalonodon*, nuovo genera di Perissodattilo dell'eocene di Gonnesa (Sardegna). -*Mem. Inst. di Geol. e Mineral., Univ. Padua*, 8, mem. 5: 1-9.

Dedieu, P. (1977): Sur la systématique des Tapiroidea (Mammalia) de l'Eocene européen. -*C. R. Hebd. Séances de l'Acad. Sci., Paris* (D)284: 2219-2222.

Depéret, C. (1902): Les Vertébrés oligocènes de Pyrimont-Challonges (Savoie).- *Mém. Soc. Paleont. Suisse*, 29: 3-91.

Depéret, C. (1903): Études paléontologique sur les *Lophiodon* du Minervois. - *Arch., Mus. Sci. Nat. Lyon*, vol. 9, mem. 1: 1-48.

Depéret, C. (1910): Études sur la famille des lophiodontidés. -*Bull. Soc. Geol. France*, 10: 558-577.

Eisenberg, J. F. (1981): *The Mammalian Radiations*. Chicago (The University of Chicago Press).

Emry, R.J. (1979): Review of *Toxotherium* (Perissodactyla, Rhinocerotoidea) with new material from the early Oligocene of Wyoming. -*Proc. Biol. Soc. Wash*. 92: 28-41.

Emry, R. J. (MS): A diminutive new ceratomorph perissodactyl from the middle Eocene of Nevada, with comments on "tapiroid" systematics.

Filhol, H. (1874): Note relative à la découverte d'un animal appartenant au genre des Tapirs dans les gisements de phosphates de chaux de Quercy. - *Mém. Soc. Phil. Nat.*, 1-3.

Filhol, H. (1877): Recherches sur les Phosphorites du Quercy. -*Bibliothèque de l'Ecole des hautes Etudes, Paris*, 16: 1-338.

Filhol, H. (1888): Etude sur les vertebres fossiles de Issel (Aude). -*Mem. Soc. Géol. France*, 5: 1-188.

Fischer, K.-H. (1964): Die tapiroiden Perissodactylen aus der eozänen Braunkohle des Geiseltales. -*Geologie, Berlin*, 45: 1-101.

Fischer, K.-H. (1977): Neue Funde von *Rhinocerolophiodon* (n. gen.), *Lophiodon* und *Hyrachyus* (Ceratomorpha, Perissodactyla, Mammalia) aus dem Eozän des Geiseltals bei Halle (DDR). 1. Teil: *Rhinocerolophiodon*. -*Zeit. Geol. Wiss.*, 5: 909-919.

Gabunia, L. K. (1961): Obailinskaya fauna - drevneishii kompleks iskopaemykh mlekopitayushikh SSSR. -*Soob. Akad. Nauk. Gruzinskoi SSSR*, 27: 711-713.

Gazin, C. L., and Collins, R. L. (1950): Remains of land mammals from the Miocene of the Chesapeake Bay region. -*Smith. Misc. Coll*. 116: 1-21.

Gill, T. (1872): Arrangement of the families of mammals with analytical tables. - *Smith. Misc. Coll*. 11:1-98.

Gromova, V. (1952): Primitivnye tapiroobraznyeiz Paleogena Mongolii. -*Trudy Paleont. Inst. Akad. Naur SSSR*, 41: 99-119.

Haeckel, E. (1873). *Natüliche Schöpfundsgeschichte* . . . Vierte auflage.: Berlin (George Reimer).

Hatcher, J. B. (1896). Recent and fossil tapirs. -*Amer. J. Sci*. (4th series), 1: 161-180, pls. 2-4.

Hay, O. P. (1899): On the names of certain North American fossil vertebrates. - *Science*, 9: 593-594.

Hay, O. P. (1902): Bibliography and catalogue of the fossil Vertebrata of North America. -*Bull. U. S. Geol. Surv*., 179: 1-868.

Heissig, K. (1978): Fossil führende Spaltenfüllungen Süddeutschlands und die Ökologie ihrer oligozänen Huftiere.- *Mitt. Bayer. Staatsslg. Paläont. Hist. Geol.*, 18: 237-288.

Hershkovitz, P. (1954): Mammals of Northern Colombia, Preliminary Report No. 7: Tapirs (Genus *Tapirus*), with a systematic review of American species. -*Proc. U. S. Nat. Mus., Smith. Inst.*, 103: 465-496.

Honacki, J. H., Kinman, K. E., and Koeppl, J. W.,(eds.) (1982): *Mammal Species of the World*. -Lawrence, Kansas (Assoc. Systematics Coll).

Hooker, J. J. (1984): A primitive ceratomorph (Perissodactyla, Mammalia) from the early Tertiary of Europe. -*J. Linn. Soc.*, 82: 229-244.

Hooker, J. J. (1989): Character polarities in early perissodactyls and their significance for *Hyracotherium* and infraordinal relationships (this volume, Chapter 6).

Hough, J. (1955): An upper Eocene fauna from the Sage Creek area, Beaverhead County, Montana. -*J. Paleont.*, 29:22-36.

Janis, C. (1984): Tapirs as living fossils. In: Eldredge, N., and Stanley, S. M. (eds.): *Living Fossils*. New York (Springer-Verlag), pp. 80-86.

Kretzoi, M. (1951): A Csakvari Hipparion-fauna. -*Fodtani Kozlony, Budapest*, 81: 384-417.

Kurtén, B., and Anderson, E. (1980): *Pleistocene Mammals of North America*. New York (Columbia University Press).

Leidy, J. (1868): Notice of some Eocene remains of extinct pachyderms. -*Proc. Acad. Nat. Sci., Phila.*, 20: 230-233.

Leidy, J. (1870): Remarks on a collection of fossils from the western territories. -*Proc. Acad. Nat. Sci. Phila.*, 22: 109.

Leidy, J. (1871a): Report on vertebrate fossils from the Tertiary formations of the West. -*Ann. Rept. U. S. Geol. Surv. Wyoming and portions of contiguous territories*, pp. 340-370.

Leidy, J. (1871b): Remarks on fossil vertebrates from Wyoming. -*Proc. Acad. Nat. Sci. Phila.*, 23: 229.

Lemoine, V. (1880): Communication sur les ossements fossiles des terrains tertiaires inférieurs des environs de Rheims. -*Assoc. Fr. pour l'avance. Sci. Congres de Montpellier*, 1-17.

Linnaeus, C. (1758): *Systema naturae per regna tria naturae, secundum classes, ordines, genera, species cum characteribus, differentiis, synonymis, locis*. Stockholm (Laurentii Salvii).

Lucas, S. G., and Schoch, R. M. (1981): The systematics of *Rhodopagus*, a late Eocene hyracodontid (Perissodactyla: Rhinocerotoidea) from China. -*Bull. Geol. Inst. Univ. Uppsala*, 9: 43-50.

Lundelius, E. L., Jr., and Slaughter, B. H. (1976). Notes on American Pleistocene tapirs. In: Churcher, C. S. (ed.) *ATHLON. Essays on Palaeontology in Honour of Loris Shano Russell*, Life Sci. Misc. Publ., Royal Ontario Museum (Canada), pp. 226-243.

Macdonald, J. R. (1970): Review of the Miocene Wounded Knee faunas of southwestern South Dakota. -*Bull. L.A.C.Mus. Nat. Hist.* 8: 1-82.

Marsh, O. C. (1871a): Notice of some new fossil mammals from the Tertiary formation. -*Amer. J. Sci.* (3rd series), 2: 35-44.

Marsh, O. C. (1871b). [Untitled]. -*Proc. Acad. Nat. Sci., Phila.*, 23: 9-10.

Marsh, O. C. (1872): Preliminary description of new Tertiary mammals. -*Amer. J. Sci.* (3rd series), 4:202-224.

Marsh, O. C. (1877): Notice of some new vertebrate fossils. -*Amer. J. Sci.* (3rd series), 14: 249-256.

Marsh, O. C. (1890): Notice of new Tertiary mammals. -*Amer. J. Sci.* (3rd series), 39: 523-525.

Matthew, W. D., and Granger, W. (1923a): New fossil mammals from the Pliocene of Sze-Chuan, China. -*Bull. Amer. Mus. Nat. Hist.*, 48:563-598.

Matthew, W. D., and Granger, W. (1923b): The fauna of the Ardyn Obo Formation. -*Amer. Mus. Novit.*, 98, 1-5.

Matthew, W. D., and Granger, W. (1925a): New mammals of the Shara Murun Eocene of Mongolia. -*Amer. Mus. Novit.*, 196: 1-11.

Matthew, W. D., and Granger, W. (1925b): The smaller perissodactyls of the Irdin Manha Formation, Eocene of Mongolia. -*Amer. Mus. Novit.*, 199: 1-9.

Matthew, W. D., and Granger, W. (1926): Two new perissodactyls from the Arshanto Eocene of Mongolia. -*Amer. Mus. Novit.*, 208: 1-5.

Merriam, J. C. (1913): Tapir remains from Late Cenozoic beds of the Pacific coast region.- *Univ. Calif. Publ., Bull. Dept. Geol.* 7: 169-175.

Murie, J. (1872): On the Malayan tapir, *Rhinochoerus sumatranus* (Gray). -*J. Anat. Physiol.* 6,: 131-169.

Olsen, S. J. (1960): Age and faunal relationship of *Tapiravus* remains from Florida. -*J. Paleont.*, 34: 164-167.

Osborn, H. F., and Wortman, J. L. (1892): Fossil mammals of the Wasatch and Wind River beds, Collection of 1891. - *Bull. Amer. Mus. Nat. Hist.*, 4: 81-148.

Osborn, H. F., Scott, W. B., and Speir, F. (1878): Paleontological report of the Princeton Scientific Expedition of 1877. -*Contrib. E. M. Mus. Geol. Arch. Princeton College*, 1: 1-107.

Peterson, O. A. (1919): Report upon the material discovered in the upper Eocene of the Uinta basin by Earl Douglas [sic] in the years 1908-1909, and by O. A. Peterson in 1912. -*Ann. Carnegie Mus.*, 12: 40-168.

Pilgrim, G. E. (1925): The Perissodactyla of the Eocene of Burma. -*Palaeont. Indica*, new series, vol. 8, mem. 3: 1-28.

Prothero, D. R., Manning, E., and Hanson, C. B. (1986): The phylogeny of the Rhinocerotoidea (Mammalia, Perissodactyla). -*Zool. J. Linn. Soc.*, 87:341-366.

Qi, T. (1980): A new lophialetid genus of Inner Mongolia. -*Vert. PalAsiatica*, 18: 215-219.

Radinsky, L. B. (1963): Origin and early evolution of North American Tapiroidea. -*Yale Univ. Peabody Mus. Nat. Hist. Bull.*, 17: 1-115.

Radinsky, L. B. (1964): *Paleomoropus*, a new early Eocene chalicothere (Mammalia, Perissodactyla), and a revision of Eocene chalicotheres. -*Amer. Mus. Novit.*, 2179: 1-28.

Radinsky, L. B. (1965a): Evolution of the tapiroid skeleton from *Heptodon* to *Tapirus*. -*Bull. Mus. Comp. Zool., Harvard Univ.*, 134: 69-106.

Radinsky, L. B. (1965b): Early Tertiary Tapiroidea of Asia. -*Bull. Amer. Mus. Nat. Hist.*, 129, 181-264.

Radinsky, L. B. (1966a): A new genus of Eocene tapiroid (Mammalia, Perissodactyla). -*J. Paleont.*, 40: 740-742.

Radinsky, L. B. (1966b): The families of the Rhinocerotoidea (Mammalia, Perissodactyla). -*J. Mamm.*, 47: 631-639.

Radinsky, L. B. (1966c): The adaptive radiation of the phenacodontid condylarths and the origin of the Perissodactyla. -*Evolution*, 20: 408-417.

Radinsky, L.B. (1966d): *Pataecops*, new name for *Pataecus* Radinsky, 1965. -*J. Paleont.*, 40: 222.

Radinsky, L. B. (1967a): *Hyrachyus*, *Chasmotherium*, and the early evolution of helaletid tapiroids. -*Amer. Mus. Novit.*, 2313: 1-23.

Radinsky, L. B. (1967b): A review of the rhinocerotoid family Hyracodontidae (Perissodactyla). -*Bull. Amer. Mus. Nat. Hist.*, 136: 1-45.

Radinsky, L. B. (1969): The early evolution of the Perissodactyla. -*Evolution*, 23: 308-328.

Radinsky, L. B. (1983): *Hyrachyus*: tapiroid not rhinocerotoid. -*Evol. Theory*, 6: 296 (abstract).

Ranga Rao, A. (1972): New mammalian genera and species from the Kalakot Zone of Himalayan foot hills near Kalakot, Jammu and Kashmir State, India. -*Special Paper of the Directorate of Geology, Oil, and Natural Gas Commission, Dehra Dun, India*. 1: 1-22.

Ray, C. E., and Sanders, A. E. (1984): Pleistocene tapirs in the Eastern United States. In: H. H. Genoways and M. R. Dawson (eds.): Contributions in Quaternary Vertebrate Paleontology: A Volume in Memorial to John E. Guilday. -*Spec. Publ. Carnegie Mus. Nat. Hist.*, 8: 283-315.

Reshetov, V. Yu.: (1975): Obzor rannetretichnykh tapiroobraznych Mongolii i SSSR. -*The Joint Soviet-Mongolian Expedition, Transactions* 2: 19-53.

Reshetov, V. Yu. (1977): Morphology of skull of Asiatic Eocene tapiroid (*Lophialetes expeditus* Matthew et Granger, 1929). -*J. Palaeont. Soc. India*, 20: 41-47.

Reshetov, V. Yu. (1979): Early Tertiary Tapiroidea of Mongolia and the USSR [in Russian]. -*The Joint Soviet-Mongo-*

lian *Expedition, Transactions* 11: 1-144. [Summarized by R. M. Schoch (1982) *J. Vert. Paleont.*, 2: 386-387.]

Ride, W. D. L., Sabrosky, C. W., Bernardi, G., and Melville, R. V. (1985): *International Code of Zoological Nomenclature*, third ed. London (International Trust for Zoological Nomenclature).

Rütimeyer, L. (1862): Eocän Säugetieren aus dem Gebiet des Schweizerischen Jura. - *Neue Denkschr. Schweiz. Gesell. Naturw.* 19.

Sahni, A. and Khare, S. K. (1971): Three new Eocene mammals from Rajauri District, Jammu and Kashmir. -*J. Palaeont. Soc. India*, 16: 41-53.

Savage, D. E. and Russell, D. E. (1983): *Mammalian Paleofaunas of the World*: Reading, Mass. (Addison-Wesley Publishing Co.).

Savage, D. E., Russell, D. E., and Louis, P. (1966): Ceratomorpha and Ancylopoda (Perissodactyla) from the lower Eocene Paris Basin, France. -*Univ. Calif. Publ. Geol. Sci.*, 66: 1-38.

Schaub, S. (1928): Der Tapirschädel von Haslen. Ein Beitrag zur Revision der oligocänen Tapiriden Europas. -*Abh. Schw. Paläont. Gesell.*, 47, 1-28.

Schiebout, J.A. (1977): *Schizotheroides* [sic] (Mammalia, Perissodactyla) from the Oligocene of Trans-Pecos Texas. -*J. Paleont.* 51: 455-458.

Schlaikjer, E. M. (1937): A new tapir from the Lower Miocene of Wyoming. -*Bull. Mus. Comp. Zool. Harvard College*, 80: 231-251.

Schoch, R. M. (1982): *Hyrachyus*: tapiroid or rhinocerotoid? -*Evol. Theory*, 6:166 (abstract).

Schoch, R. M. (1983a): A new species of *Isectolophus* (Mammalia, Tapiroidea) from the Middle Eocene of Wyoming. - *Postilla, Peabody Mus., Yale Univ.*, 188: 1-4.

Schoch, R. M. (1983b): *Tanyops undans* Marsh, 1894: a junior subjective synonym of *Protapirus obliquidens* Wortman and Earle, 1893 (Mammalia, Perissodactyla). -*Postilla, Peabody Mus., Yale Univ.*, 190: 1-7.

Schoch, R. M. (1983c): Relationships of the earliest Perissodactyls. -*Geol. Soc. Amer., Abstr. with Prog.*, 15: 144 (abstract).

Schoch, R. M. (1984a): Two unusual specimens of *Helaletes* in the Yale Peabody Museum collections, and some comments on the ancestry of the Tapiridae (Perissodactyla, Mammalia). -*Postilla, Peabody Mus., Yale Univ.*, 193: 1-20.

Schoch, R. M. (1984b): The type specimens of *Tapiravus validus* and ?*Tapiravus rarus* (Mammalia, Perissodactyla), with a review of the genus, and a new report of *Miotapirus* (*Miotapirus marslandensis* Schoch and Prins new species) from Nebraska. -*Postilla, Peabody Mus., Yale Univ.*, 195: 1-12.

Schoch, R. M. (1985a): Historical review of subordinal classifications of the Perissodactyla. In: Fuller, W.A., Nietfeld, M.T., and Harris, M. A. (eds.): *Abstracts of Papers and Posters, Fourth International Theriological Congress, Edmonton, 13-20 August 1985*, abstract 0557. University of Alberta, Edmonton (Canada).

Schoch, R. M. (1985b): Concepts of the relationships and classification of major perissodactyl groups: notes for a workshop on fossil perissodactyls held at IV ITC, Edmonton, Canada, August 1985. -Privately printed and distributed. [available from the author]

Schoch, R. M. (1986): *Phylogeny Reconstruction in Paleontology*. New York (Van Nostrand Reinhold).

Schoch, R. M. (1989): A brief historical review of perissodactyl classification (this volume, Chapter 2).

Schoch, R. M. (in prep.): Tapiroidea. In: Janis, C., Jacobs, L., and Scott, K., (eds.): *The Tertiary Mammals of North America*. Cambridge (Cambridge Univ. Press).

Schoch, R. M., and Prins, N. (1984): The

evolution of the Tapiroidea (Mammalia, Perissodactyla). -*Geol. Soc. Amer., Abstr. with Prog.*, 16, 647 (abstract).

Schultz, C. B., Martin, L. D., and Corner, R. G. (1975): Cenozoic mammals from the central Great Plains. -*Bull. Univ. Neb. State Mus.*, 10, 1-21.

Scott, W. B. (1883): On *Desmatotherium* and *Dilophodon*, two new Eocene lophiodonts. -*Contrib. E. M. Mus. Geol. Arch. Princeton College*, 3: 46-53.

Scott, W. B. (1941): Perissodactyla. The mammalian fauna of the White River Oligocene. -*Trans. Amer. Phil. Soc.*, 28, 747-980.

Scott, W. B., and Osborn, H. F. (1887): Preliminary report on the vertebrate fossils of the Uinta Formation. -*Proc. Amer. Phil. Soc.*, 24: 255-264.

Simpson, G. G. (1945a): The principles of classification and a classification of mammals. -*Bull. Amer. Mus. Nat. Hist.*, 85: 1-350.

Simpson, G. G. (1945b): Notes on Pleistocene and Recent tapirs. -*Bull. Amer. Mus. Nat. Hist.*, 86: 33-81.

Sinclair, W. J. (1901): The discovery of a new fossil tapir in Oregon. -*J. Geol.* 9: 702-707.

Skinner, S. M., and Gooris, R. J. (1966): A note on *Toxotherium* (Mammalia, Rhinocerotoidea) from Natrona County, Wyoming. - *Amer. Mus. Novit.*, 2261: 1-12.

Stehlin, H. G. (1903): Die Saugetiere des Schweizerischen Eocaens. Die Fundorte, Die Sammlungen. *Chasmotherium. Lophiodon.* -*Abh. Schw. Paläont. Gesell.*, 30, 1-153.

Stjernman, R. O. G. (1932): Vergleichend-Anatomische Studien über die Extremitäten-Muskulatur (Vorder- und Hintergliedmassen) bei *Tapirus indicus.* - Hakan Ohlssons Buchdruckerei, Lund, pp. 1-154.

Walker, E. P. (1975): *Mammals of the World* (3rd. ed.). Baltimore, Maryland (The Johns Hopkins University Press).

West, R. M., and Dawson, M. R. (1978): Vertebrate paleontology and the Cenozoic history of the North Atlantic region. -*Polarforshung (German Society of Polar Research)* 48: 103-119.

Wilson, J.A., and Schiebout, J.A. (1984): Early Tertiary vertebrate faunas, Trans-Pecos Texas: Ceratomorpha less Amynodontidae. -*Pearce-Sellards Series, Texas Mem. Mus.*, 39: 1-47.

Wood, H. E., II (1927): Some early Tertiary Rhinoceroses and hyracodonts. -*Bull. Amer. Paleont.*, 13 (50): 165-265.

Wood, H. E., II (1934): Revision of the Hyrachyidae. -*Bull. Amer. Mus. Nat. Hist.*, 67: 181-295.

Wood, H. E., II (1937): Perissodactyl suborders. -*J. Mamm.*, 18: 106.

Wood, H. E., II (1961): *Toxotherium hunteri*, a peculiar new Oligocene mammal from Saskatchewan. -*Nat. Hist. Papers, Nat. Mus. Canada*, 13: 1-4.

Wortman, J. L. (1896): Species of *Hyracotherium* and allied perissodactyls from the Wasatch and Wind River beds of North America. -*Bull. Amer. Mus. Nat. Hist.*, 8: 81-110.

Wortman, J. L. and Earle, C. (1893): Ancestors of the tapir from the Lower Miocene of Dakota. -*Bull. Amer. Mus. Nat. Hist.*, 5: 159-180.

# 16. THE HISTORY OF THE RHINOCEROTOIDEA

## DONALD R. PROTHERO, CLAUDE GUÉRIN, and EARL MANNING

The largest and most ecologically diverse group of perissodactyls is the Superfamily Rhinocerotoidea. This group includes the amynodonts (some of which were hippo-like or tapir-like aquatic forms), the hyracodonts (which included dog-sized cursorial forms, and gigantic forms which browsed treetops), and the true rhinoceroses of the Family Rhinocerotidae. All three groups diverged in the later Eocene from a form like *Hyrachyus*, and spread over the northern hemisphere Both the amynodonts and hyracodonts were reduced to a few surviving genera by the early Oligocene, but the rhinocerotids began to diversify. Most early rhinos were hornless, but the first horned rhinos had paired horns on the tip of their noses, a feature that evolved independently in two different groups, the Diceratheriinae and Menoceratinae. By the late Oligocene, rhinos began to diverge into the major subfamilies and tribes that dominated the northern hemisphere and Africa during the Miocene: the hippo-like grazing single-horned Teleoceratinae, the prehensile-lipped browsing hornless Aceratheriinae, and the Rhinocerotinae, which includes all the five living species. These first two groups were almost completely wiped out by the extinctions at the end of the Miocene, leaving North America without rhinos, and only certain rhinocerotines surviving in Eurasia and Africa. During the Plio-Pleistocene in Eurasia, the dominant rhinos were several species of large derived *Dicerorhinus*, and the wide-ranging woolly rhino. Iranotheres and giant frontal-horned elasmotheres were also present, but all of these groups were extinct at the end of the Pleistocene. Today, only two genera of dicerotine rhinos survive in Africa, and three species of rhinos of the genera *Rhinoceros* and *Dicerorhinus* barely survive in Asia. All five of these species are heavily poached and on the brink of extinction, a sad remnant of one of the most diverse and successful groups of mammals in the entire Cenozoic.

### Introduction

One of the most taxonomically and ecologically diverse, widespread, and successful perissodactyl groups is the Rhinocerotoidea. Rhinocerotoids have adapted to many herbivorous modes of life, from sheep-sized runners (*Hyracodon*), to ecological vicars of hippos (*Teleoceras, Brachypotherium, Metamynodon*), to tapir-like animals with a proboscis (*Cadurcodon, Aceratherium*), to the largest land mammal that ever lived (*Paraceratherium*, formerly known as *Indricotherium* or *Baluchitherium*). Rhinocerotoids far outnumber horses, hyraxes, tapirs, chalicotheres, or titanotheres in terms of valid fossil genera or species. In Eurasia, Africa, and North America, rhinos occurred in great numbers in the past; in a few localities they outnumbered all other mammals. This is true despite the fact that rhinos are often one of the largest herbivorous mammals in most Tertiary faunas. One would think that such a diverse and dominant group of large mammals would be well studied and understood.

Yet the opposite is the case. Of all the perissodactyls, rhinos have been among the least studied in the last few decades. There was some early phylogenetic work on the group (e.g., Cope, 1880; Gaudry, 1888;

*The Evolution of Perissodactyls* (ed. D.R. Prothero & R.M. Schoch) Oxford Univ. Press, New York, 1989

Pavlowa, 1892; Scott and Osborn, 1898), but the last general reviews or phylogenies of rhinos prior to the work of Prothero, Manning, and Hanson (1986) were those of Wood (1927), Matthew (1931), Viret (1958), and Heissig (1973). Due to the complexity of the group, misconceptions or erroneous ideas about rhinos are common among both scientists and the public. In the popular image, the diagnostic character of the rhino is its horn. Yet many rhinos were hornless, and the first horn combination was not a single horn, but paired horns at the tip of the nose, which was evolved twice independently. Similarly, nearly every rhino illustrated in the popular books or the textbooks, such as Romer (1966), is identified by the wrong generic name, such as "*Caenopus*" (= *Subhyracodon*), "*Baluchitherium*" (= *Paraceratherium*) or "*Diceratherium*" (= *Menoceras*, if it is the small Agate Springs Quarry rhino).

Much of the information that has been published in the last thirty years has not been synthesized, compelling the paleontologist to undertake extra bibliographic work. The great complexity of the superfamily also confuses those who have not worked with the group in detail. Many paleontologists have tried to identify the rhino material in their faunas. Some have succeeded, but most have failed due to the difficult literature and the complexity of the task. Yet even fragmentary rhino material can now be correctly identified to species in many cases, yielding much interesting faunal information. In this chapter, we present our view of the general pattern of rhino evolution.

## Phylogeny and classification

To unscramble the morass of misconceptions about rhinos, their systematics and phylogeny must be brought up to date. Several phylogenetic hypotheses have been presented in recent years for certain members of the Rhinocerotidae (e.g., Heissig, 1981; Guérin, 1982; Groves, 1983) and for the whole Rhinocerotoidea by Prothero, Manning, and Hanson (1986). As a consequence, rhino classifications can differ greatly: what Heissig (1973, this volume, Chapter 21) considers tribes are considered subfamilies by Guérin (1980b, 1982). Heissig (1973) synonymized the Tribes Elasmotherini and Iranotherini, while Antunes *et al.* (1972) considered them independent subfamilies. One of the commonest problems in earlier rhino phylogenies, such as those of Wood (1927), was that the crests of premolars undergoing molarization are highly variable. This can be shown by examining a number of quarry samples of rhinos such as *Trigonias* or *Subhyracodon*. Because of premolar variations, Gregory and Cook (1928) named seven species and two genera for a single, uniform-sized quarry sample of *Trigonias osborni*. This variability is also seen in a number of quarry samples of *Subhyracodon* (Prothero, in prep.). In some cases, the premolars differ on either side of the same skull. As a result, primitive rhinos are tremendously oversplit, and older phylogenies are often based on variable differences in premolars. Once premolars have become fully molarized, they are no more or less reliable than any other anatomical feature. In taxa such as *Hyrachyus*, *Triplopus*, *Trigonias*, *Subhyracodon*, and *Hyracodon*, however, they must be used with caution.

## The evolution of the rhinocerotoids

The last general discussion of the history of rhinos was by Viret (1958), but much has happened in the last 30 years. The following discussion is summarized in the diagrams of the summary chapter of this volume (Prothero and Schoch, this volume, Fig. 28.2), which show the distribution of the major rhinocerotoid genera in time and space. In this chapter, we have incorporated the new argon-argon dates (see Chapter 28) that place the Eocene-Oligocene boundary at about 33.7 Ma, and place the Chadronian in the late Eocene.

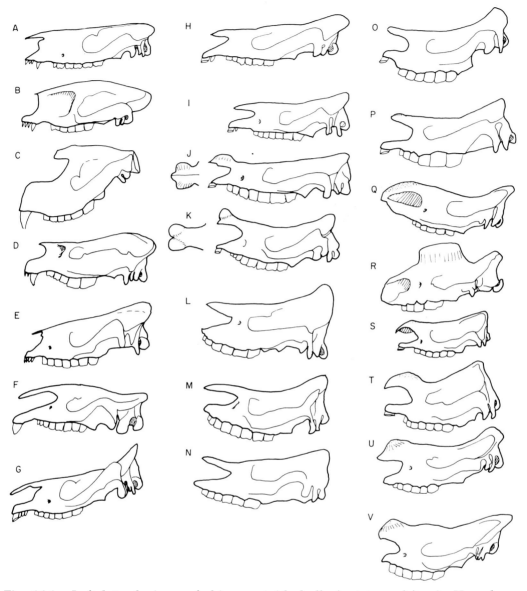

Fig. 16.1. Left lateral views of rhinocerotoid skulls (not to scale). A, *Hyrachyus*. B, *Sharamynodon*. C, *Cadurcodon*. D, *Metamynodon*. E, *Hyracodon*. F, *Paraceratherium* (= *Indricotherium*). G, *Trigonias*. H, *Amphicaenopus*. I, *Subhyracodon*. J, *Diceratherium* (showing dorsal view of paired nasal ridges). K, *Menoceras* (showing dorsal view of paired nasal knobs). L, *Aceratherium*. M, *Aphelops*. N, *Chilotherium*. O, *Brachypotherium*. P, *Teleoceras*. Q, *Coelodonta*. R, *Elasmotherium*. S, *Dicerorhinus*. T, *Rhinoceros*. U, *Diceros*. V, *Ceratotherium*. From Prothero et al. (1986, Fig. 5).

However, Fig. 28.2 was drafted before these revised concepts became available, so they may not always match the following discussion. From these diagrams, it is apparent that there have been a great variety of genera and family-level taxa of rhinos diversifying and diminishing throughout the Tertiary.

The oldest known rhinocerotoid is *Hyrachyus* (Fig. 16.1A) from the late Wasatchian and Bridgerian of North America. Radinsky (1966, 1967) placed *Hyrachyus* in the tapiroids based on shared primitive characters, but Prothero et al. (1986) gave evidence to show that it was a very primitive rhinocerotoid. It was a very cosmopolitan animal, occurring not only in the United States, but also on Ellesmere Island (West et al., 1977), Europe, and possibly Asia (Radinsky, 1967). *Hyrachyus* was probably the only rhinocerotoid to cross the European-American land bridge over the North Atlantic, which was severed by the middle Eocene (McKenna, 1975). Once this route was disconnected, there was still periodic migration between Asia (east of the Turgai Straits) and North America across Beringia, but Europe became isolated from the rest of the world in the late Eocene. Instead of rhinocerotoids, European Eocene faunas were dominated by a number of endemic perissodactyls, such as palaeotheres and lophiodonts. In the early Oligocene, this endemic fauna was wiped out by a new wave of immigrants at the "Grande Coupure." Among the immigrants were the rhinocerotoids.

Meanwhile, the three major rhinocerotoid families began to diversify in Asia and North America in the late Eocene. Each of these three groups can be most easily recognized by the condition of the $M^3$. Primitively, $M^3$ has a strong parastyle and metaloph, with a subquadrate shape, as seen in *Hyrachyus* (Fig. 16.2C). Amynodonts elongated the tooth anteroposteriorly, resulting in a quadrangular tooth with a strong metastyle (Fig. 16.2A). In hyracodonts and rhinocerotids, the metastyle is shortened and inflected lingually (Fig. 16.2B). In some hyracodonts and all rhinocerotids, the metastyle is lost completely, producing a triangular tooth. In some populations of the last European rhino, *Coelodonta antiquitatis*, the $M^3$ reverted to the classical quadrangular shape.

## Amynodonts and hyracodonts

The first of these three families, the Amynodontidae, are known from the early Uintan (middle Eocene) of North America, although the most primitive form (*Caenolophus*) is known from the late Eocene of Asia. During the late Eocene and early Oligocene, amynodonts reached their maximum diversity, particularly in Asia (Wall, 1982, this volume, Chapter 17). Two Duchesnean North American amynodonts (*Amynodontopsis* from the Sespe of California, *Procadurcodon* from the Clarno of Oregon) were short-ranging immigrants from Asia. *Metamynodon*, a very hippo-like form, ranged through almost ten million years in North America and three distinct species are currently recognized. *Metamynodon* occurs not only on the High Plains, but also from the late early Oligocene of Mississippi (Manning et al., 1985). It may have been a coastal browser as well as a river dweller in the Oligocene. By the early Oligocene, amynodonts had declined considerably, and were extinct in North America by the Whitneyan. One lingering form, *Cadurcotherium*, managed to persist until the middle Miocene (Dera Bugti beds) of Pakistan, almost 15 million years after the rest of its family was extinct. *Cadurcotherium* was not only long-lived, but very mobile, since it also occurred in the early Oligocene of Europe. Wall (this volume, Chapter 17) reviews the phylogeny, paleobiology, and paleogeography of amynodonts in much greater detail.

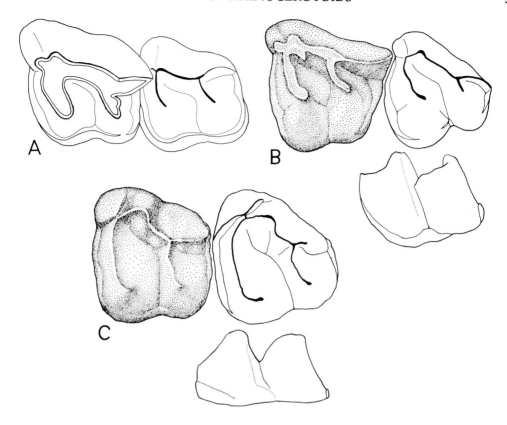

Fig. 16.2. Second and third left upper molars of: A, *Amynodon*. B, *Hyracodon*. C, *Hyrachyus*. From Radinsky (1966).

Like the amynodonts, the hyracodonts first appeared in the middle Eocene and flourished until the late Oligocene. All hyracodonts have long, laterally compressed metapodials, despite their enormous range in body size. More advanced hyracodonts also have distinctive conical incisors not found in any other rhino group. *Triplopus* (Radinsky, 1967) and *Forstercooperia* (Lucas et al., 1981) are the oldest known taxa, occurring in the middle Eocene of both Asia and North America. In Europe, hyracodonts were represented during the entire Oligocene by the genus *Eggysodon*.

The Hyracodontidae are composed of two subfamilies: the small, cursorial Hyracodontinae and the large to gigantic indricotheres, the Indricotheriinae. The latter subfamily is discussed in detail by Lucas and Sobus (this volume, Chapter 19). From the primitive *Forstercooperia*, they grew to enormous proportions in Asia in the Oligocene, producing *Paraceratherium* (= *Indricotherium*, *Baluchitherium*), the largest land mammal that ever lived. By the middle Miocene, indricotheres had vanished from Asia. Despite their gigantic proportions, their limbs did not become graviportal. Instead, their metapodials are still very long, an indication of their cursorial ancestry. Heissig (this volume, Chapter 21) suggests that the indricotheres were actually rhinocerotids, since the most primitive *Forstercooperia* has a four-digit manus. However, this is merely a primitive feature, retained in all the primitive members of the rhinocerotoid families, and sometimes secondarily regained in advanced forms. The long metapodials, however, are diagnostic of hyracodonts, and establish that indricotheres belong in this family. In addition, the enlarged upper and lower incisors of indricotheres do not resemble the

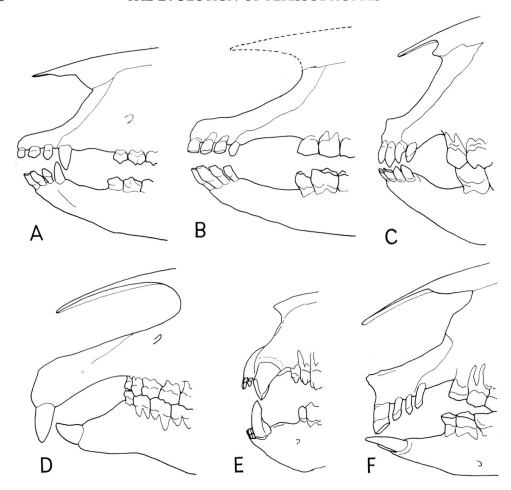

Fig. 16.3. Anterior dentitions of rhinocerotoids. A, *Hyrachyus*. B, *Ardynia*. C, *Hyracodon*. D, *Paraceratherium* (= *Indricotherium*). E, *Metamynodon*. F, *Trigonias*. From Radinsky (1966).

chisel/tusk combination of rhinocerotids (compare Fig. 16.3D with 16.3F).

The small, cursorial Hyracodontinae includes a number of lesser known forms from the late Eocene and early Oligocene of Asia and North America. Only one genus, *Hyracodon*, survived into the later Oligocene of North America. It persisted remarkably unchanged except for differences in size and in the molarization of the premolars. Since molarizing premolars are highly variable, there has been excessive oversplitting of the genus. After all other characters are taken into account, Prothero (in prep.) recognized only five valid species. The earliest species, *H. primus* and *H. petersoni* of the Duchesnean and early Chadronian, are followed by the slightly larger Chadronian form, *H. priscidens*, which changed little in four million years. In the latest Chadronian, the type species, *H. nebraskensis*, appears and persists unchanged through the Orellan, Whitneyan, and possibly earliest Arikareean. Most of the invalid species of *Hyracodon* have been synonymized with *H. nebraskensis*. In the late Whitneyan, a larger, more advanced species, *H. leidyanus*, appeared and persisted sympatrically with *H. nebraskensis*. They last co-occur in the lower Sharps Formation of

South Dakota, which conventionally was considered earliest Arikareean. However, Tedford et al. (1985) have shown that the lowest part of the Sharps may be Whitneyan. When *Hyracodon* disappeared, it was the last of its subfamily. It lasted almost ten million years after the rest of the Hyracodontinae were extinct.

**Late Eocene and Oligocene Rhinocerotidae**
While the Amynodontidae and Hyracodontidae were successful in the late Eocene and Oligocene and then became extinct, the third family, the Rhinocerotidae, have become increasingly diverse and successful since the Oligocene. Like the hyracodonts and amynodonts, they appear to have arisen in the late Eocene of Eurasia. The oldest well-known rhinocerotid is the newly described form from the Clarno Formation (Duchesnean) of Oregon (Hanson, this volume, Chapter 20). Although it resembles a primitive hyracodont in size and most features, it has already begun to develop the diagnostic features of the Rhinocerotidae. Not only is the $M^3$ metacone nearly lost, but the $I^1$ and $I_2$ have begun to develop into the characteristic chisel/tusk combination (Fig. 16.3F).

In Europe, rhinocerotids first appear in the early Oligocene (upper Sannoisian) with *Ronzotherium*, a small-sized form which is the most common and best-known European Oligocene rhino (Heissig, 1969; Brunet, 1979). Three successive species are known: *R. velaunum* from the Upper Sannoisian, *R. filholi* from the Lower Stampian, and *R. romani* from the Upper Stampian. *R. brevirostre* also occurs in the Oligocene of Mongolia. Another small European Oligocene form is *Protaceratherium albigense* from the Middle and Upper Stampian (Hugueney and Guérin, 1981). It occurs with "*Aceratherium*" (or *Protaceratherium?*) *minutum*, a small aceratheriine (de Bonis, 1973) whose generic affinities are controversial (Antunes and Ginsburg, 1983).

Other European rhinos from the Lower Oligocene are poorly known, like *Epiaceratherium bolcense* found in one single locality in Italy, and *Meninatherium* Abel, 1910 (type specimen probably destroyed during the Second World War--Kollman, pers. commun.), which is probably synonymous with *Prohyracodon* (Heissig, 1973).

Medium- and large-sized rhinos appeared in Europe during the Upper Oligocene, mainly with *Aceratherium (Mesaceratherium) paulhiacense* and *Diaceratherium lemanense*, predecessors respectively of the Miocene true *Aceratherium* lineage (with a four-digited manus) and the large, semi-aquatic semi-hypsodont brachypotheres of the *D. aginense-D. aurelianensis* lineage (Antunes and Ginsburg, 1983). Rhinocerotids are not known from the Oligocene of Africa (Cooke, 1968; Hamilton, 1973).

In the latest Eocene (Chadronian) of North America, rhinocerotids became larger and more diversified. *Trigonias* is abundant in several Chadronian quarries, and was the last rhinocerotid to retain $I^3/3$ or canines (Figs. 16.1G, 16.3F). More advanced rhinos lost nearly all the anterior teeth except their $I^1/_2$ chisel/tusk combination. The *Subhyracodon-Diceratherium* lineage first appeared in the Duchesnean, and became the dominant American Oligocene large mammal after the extinction of the titanotheres in the early Orellan (Fig. 16.1I,J). In the late Whitneyan, this lineage showed considerable sexual dimorphism in the skull as well as the tusks. Males of *Diceratherium* show paired subterminal nasal rugosities that become well-developed flanges or ridges in the Arikareean taxa. This is the first evidence of rhinos with horns. For almost the entire Whitneyan and Arikareean (from 21 to 31 Ma, almost ten million years), *Diceratherium* reigned unchallenged as the only rhinocerotoid, and the only large mammal, in North America. Throughout in their history, this was the all-time low in rhinocerotoid generic diversity, although there were a number of species of *Diceratherium*, differing chiefly in size. In one quarry, 77 Hill, near Lusk, Wyoming, there is a large sample of both sexes of the large type species, *D. armatum*, and a smaller species, *D. annectens*. In the latest Arikareean (Agate Springs Quarry), the last species of the genus, *D. niobrarense*, came into competition with an immigrant from Europe, *Menoceras*. A few rare specimens of *Diceratherium* are known from the Hemingfordian, and possibly from the Barstovian of Railroad Canyon, Idaho, but the group apparently succumbed to competition from early Miocene immigrant rhinos.

Perhaps the greatest misunderstanding about North American rhinoceroses concerns the "paired horned" rhinos, *Diceratherium* and *Menoceras*. True *Diceratherium* is the end of the endemic North American *Subhyracodon* lineage, and has a long, primitive skull with paired, subterminal nasal ridges (Fig. 16.1J). This differs from *Menoceras*, which has a short, very derived skull with terminal nasal bosses (Fig. 16.1K). A number of derived features show clearly *Menoceras* is more closely related to higher rhinos than it is to *Diceratherium*. These include a shortened basicranium relative to the tooth row, reduced sagittal crests, reduced premaxillae, nasal incision retracted over the posterior part of $P^2$, $I^2$ lost, upper molar cingula weak or absent, strong crochets on the molars, and a shallow anteroventral notch on the atlas. The only similarity between the two genera is the paired nasal horns, and these are not homologous in detail. Instead, they are one of the truly clear examples of evolutionary convergence. If only the hornless female skulls had been known, there would have been no hesitation in putting them in separate genera.

Nevertheless, *Diceratherium* and *Menoceras* have nearly always been confused because of their paired nasal horns. The genus *Menoceras* was separated by Troxell from *Diceratherium* as early as 1921, but virtually all later workers blurred this distinction until Tanner (1969) clarified the differences. *Menoceras* happens to be very abundant in one of the most famous quarries of all, Agate Springs Quarry in Nebraska, and so the misnomer "*Diceratherium cooki*" (actually *Menoceras arikarense*) has been associated with virtually every Agate rhino specimen for over eighty years. Most museum labels and popular and technical books which mention the Agate rhino still have it mislabeled.

The anatomical differences between the two genera are corroborated by their occurrence. *Menoceras* appears abruptly at the very end of the Arikareean (Agate), where it almost completely overwhelms the few *Diceratherium niobrarense* from that locality. *Menoceras* is very similar to the type specimen of *Pleuroceros pleuroceros* Duvernoy, 1853, from the Aquitanian of Europe. It seems clear, then, that *Menoceras* was an immigrant from Europe in the latest Arikareean. In the early Hemingfordian, a slightly larger species, *Menoceras barbouri*, is found in Nebraska, Florida, Texas, Wyoming, and New Mexico. (This name is the senior synonym of *M. marslandensis* and *M. falkenbachi*, and *Moschoedestes delahoensis* Stevens, 1969, from the Castolon l.f., early Hemingfordian of Texas.) By the late Hemingfordian, *Menoceras* was extinct.

**Miocene rhinocerotids of North America**
While North America was experiencing a low diversity of endemic rhinos during the Oligocene, the modern groups of rhinocerotids were developing in Europe, as described earlier. In the Burdigalian, migration reached a peak. A number of endemic groups reached Africa from Europe, and North America received a great many European immigrants as well during the

middle Hemingfordian. Prior to this time, North America had only *Diceratherium* and *Menoceras*, and Asia still sheltered archaic indricotheres and *Cadurcotherium*, the last of the hyracodonts and amynodonts. The great Burdigalian-Hemingfordian interchange completely altered the cast of characters. North America was invaded by two major groups, the aceratheriines and teleoceratines. The primitive aceratheriine *Floridaceras whitei*, the primitive teleoceratine *Brachypotherium americanum*, and an undescribed new genus of primitive aceratheriine occur in one or more of the early and middle Hemingfordian faunas, such as Thomas Farm in Florida, Warm Springs in Oregon, Martin Canyon in Colorado, J. L. Ray Ranch, and Box Butte in Nebraska. By the latest Hemingfordian Sheep Creek Fauna of Nebraska, the long-lived aceratheriines *Peraceras profectum* and *Aphelops megalodus* had appeared, although they are both so primitive that they are hard to distinguish.

From the late Hemingfordian until the late Hemphillian (18-4.5 Ma), the North American rhino fauna consisted of the aceratheriines *Aphelops* and *Peraceras* (until the early Clarendonian), and the teleoceratine *Teleoceras*. Aceratheriines are easy to recognize from a number of derived features, including a lower tusk with a reduced medial flange, a long diastema, a highly retracted nasal incision (to the level of $P^4$), and especially by their loss of the chisel-like upper incisor and reduced premaxilla in most genera (Fig. 16.1L-N). They generally retain the skeletal proportions of the more primitive rhinocerotids, so they were relatively long-legged and adapted for browsing, like the living black rhino.

Along with the relatively primitive aceratheriines came the very derived teleoceratines (Fig. 16.1O-P). From the very beginning, they showed a number of unique features which make nearly every skeletal element distinctive. All teleoceratines had brachycephalic skulls with broad zygomatic arches and flaring lambdoid crests, a nasal incision retracted to anterior $P^3$, nasals shaped like an inverted U in cross-section, hypsodont grazing teeth, strong antecrochets on the molars, and an elongated calcaneal tuber. Their most distinctive feature, however, was their extraordinarily hippo-like skeleton, with a barrel-shaped chest, and short, stumpy legs. Their limb bones were so extremely shortened and compressed that every single element of the tarsus and carpus is immediately recognizable by its flattening. There is a suggestion of an early stage of fusion of carpal elements in later *Teleoceras* (Harrison and Manning, 1983).

The skeleton of *Teleoceras* is so hippo-like that it demands comparison with a living analogue. Despite its low-crowned teeth, *Hippopotamus amphibius* is a grazer that lives in the river by day, but comes out on the plains to graze at night. D. Wright (pers. commun.) has studied the population structure of *Teleoceras* from the late Clarendonian Love Bone Bed, Florida, and found that it matched *Hippopotamus* much better than any browsing rhino. *Teleoceras* is nearly always most abundant in river channel deposits, and the extraordinary Poison Ivy Quarry ash fall assemblage (Clarendonian of Nebraska) appears to have trapped a whole herd of *Teleoceras* in a lake (Voorhies, 1981). Some of these Poison Ivy Quarry rhinos even have grass seeds preserved in their throat regions (Voorhies and Thomasson, 1979).

It is typical for North American Miocene faunas to contain two genera of rhinos, one a browser, the other a grazer. Generally, the grazer can be distinguished from the browser by its hypsodont teeth and other features which permit a diet of abrasive grasses. The browser, on the other hand, often has a prehensile lip, or retracted nasals to support the muscles for a short proboscis for snapping off leaves and twigs. In North America, the browser was often *Aphelops* or some other aceratheriine. The grazer was usually *Teleoceras*, although *Peraceras superciliosum* seemed to mimic *Teleoceras* in many skull and tooth features and was probably also a grazer. This browser-grazer combination was typical throughout the history of rhinos, especially when they occurred in savannah/woodland environments. Significantly, the only such rhino pair still living (the browser *Diceros bicornis* and the grazer *Ceratotherium simum*) are found in East and South Africa, one of the few remaining savannah habitats left on earth.

The long Miocene history of rhinos in North America is fully documented elsewhere (Prothero, in prep.), but its salient features are now becoming clear. *Aphelops* has only three valid species which get progressively larger, more hypsodont, and have more retracted nasals through time. *Peraceras* was less common, and became extinct by the late Clarendonian. There are only three valid species: the primitive, medium-sized *P. profectum*, the large *P. superciliosum*, and the dwarf *P. hessei* (Prothero and Manning, 1987).

The teleoceratine story is more complex. Beginning with the Hemingfordian *Brachypotherium americanum*, *Teleoceras* gets progressively larger until the main lineage reached maximum size with *T. fossiger* in the early Hemphillian. Like many other early Hemphillian forms (*Aepycamelus, Yumaceras, Tapirus, Calippus, Nimravides, Pliohippus, Neohipparion, Epicyon, Leptarctus, Macrogenis, Illingoceras, Barbourofelis, Indarctos,* and *Prosthennops*), the early Hemphillian *T. fossiger* is larger than the late Hemphillian species, *T. hicksi*. An even smaller species, *T. proterum*, is known from the early Hemphillian of Florida (Mixson's Bone Bed), and an unnamed species of *Teleoceras* is known from the latest Hemphillian of Oklahoma (Guymon l.f.). Another dwarf species, *T. meridianum*, occurred in the late Barstovian of the Texas Gulf Coastal Plain along with the dwarf *Peraceras* (Prothero and Sereno, 1982; Prothero and Manning, 1987). Contrary to Matthew (1932), *Teleoceras* does survive into the very late Hemphillian. It is present but very rare in the Upper Bone Valley Formation of Florida, the Eden l.f. of California, the Sawrock l.f. of Kansas, and the Bidahochi Formation of Arizona. It was thought to be extinct because the typical late Hemphillian quarries, such as Coffee Ranch in Texas, and Edson in Kansas, are dominated by *Aphelops mutilus*. A single specimen of *T. hicksi*, however, is present in both quarries.

In addition to the dwarfing and ecological parallelism shown by several rhino species, North American rhino biogeography was very complex and interesting. Dwarf rhinos are particularly diverse in the Texas Gulf Coastal Plain Barstovian, which contains four sympatric species, an all-time high for North America. Prothero and Sereno (1982) suggested that the dwarfs inhabited a more coastal, forested environment, and were comparable to modern dwarf species of hippos, elephants, and Cape buffalo, which prefer browsing in forested habitats. The *Teleoceras* from the Barstovian and Clarendonian Santa Fe Group of New Mexico have peculiarly short nasals without horns and robust premaxillae, and may be an endemic new species. The abundant High Plains quarries of Nebraska, Kansas, Oklahoma, and Texas are typically dominated by the main line species of *Aphelops* and *Teleoceras*, but the northern localities (particularly Montana, South Dakota, and northern Nebraska) sometimes contained *Peraceras superciliosum*, another brachycephalic hippo-like form.

The Miocene record east of the Mississippi is generally poor, except for Florida. The Florida rhino fauna is generally similar to the High Plains fluvial assemblage, except that the early Hemphillian *Teleoceras proterum* from Mixson's Bone Bed is much smaller than High Plains *T. fossiger*. There is a surprising scarcity of rhinos from the western states. Although there are many rich Miocene fossil mammal localities from California, Nevada, Oregon, Arizona, and elsewhere, rhinoceroses are extremely rare compared to their abundance in the High Plains. Horses and camels occur in great numbers in some of these localities, but only a few scraps of rhino are known.

By the latest Hemphillian (earliest Pliocene), rhinoceroses were very scarce in North America. Until recently, there were no rhinos reported from the Blancan, and North American rhinos were assumed to have gone extinct as a result of the Messinian climatic event at the Mio-Pliocene boundary. However, C. Madden and W. Dalquest (pers. commun.) are describing an isolated rhino tooth fragment from the mid-Blancan Beck Ranch locality in Scurry County, Texas. If this specimen is not transported from older deposits (as it appears), then rhinos survived in very small numbers until the mid-Blancan in North America.

## Miocene rhinocerotids of Europe

While North America was dominated by only three genera of rhinos during the most of the Miocene, Eurasia saw far greater diversity. By the middle Miocene (Vindobonian), the menoceratines were extinct, and the fauna was dominated by *Brachypotherium*, primitive aceratheriines, and primitive rhinocerotines. Aceratheriines included *Aceratherium* (*sensu lato*) with related genera or subgenera *Mesaceratherium*, *Alicornops*, *Chilotherium*, and *Dromoaceratherium*. True *Aceratherium* was a medium-sized rhino with a functional fifth metacarpal. Its limbs were long, with proportions like those of the living tapir (Eisenmann and Guérin, 1984). It had brachyodont cheek teeth, and possibly a short proboscis (Hunermann, 1982). It was a browser, and the anatomical similarities with tapirs suggest a similar way of life. The first known species was the Upper Oligocene (Upper Stampian) *Aceratherium* (*Mesaceratherium*) *paulhiacense*, leading to the European lineage composed of the Orleanian/Astaracian *A. platyodon*, the Astaracian *A. lumiarense*, the Astaracian/Early Vallesian *A. tetradactylum*, and the Vallesian/Turolian *A. incisivum* (Guérin, 1980b; Antunes and Ginsburg, 1983). The related genus *Dromoaceratherium* includes *D. mirallesi* of Orelanian/Astaracian age in Spain, and *D. fahlbuschi* from the Astaracian of southern Germany (Heissig, 1972a; Santafé-Llopis, 1978). Another related lineage begins with the little short-legged, three-toed rhino *Alicornops*, first found in the Middle Miocene from Wintershoff West (MN5 zone) and reaching its peak with *A. simorrense* from the Astaracian and Vallesian. *A. simorrense* ranged as far as India (Ginsburg and Guérin, 1979; Guérin, 1979).

A separate lineage of aceratheriines is the genus *Chilotherium*, which mimics the teleoceratines in acquiring hippopotamus-like body proportions and hypsodont cheek teeth. Possibly originating in the Middle Miocene of the Siwaliks, *Chilotherium* immigrated to China, the Middle East (Marageh, Iran) and Europe. It is found in Samos in Greece, Italy (Guérin, 1980b), and *Chilotherium ibericum* migrated to the Iberian Peninsula (Antunes and Ginsburg, 1983). Many species have been described from this huge geographical range, but only four of the Eastern European species (*C. zernowi* from Odessa, Soviet Union, and *C. samium*, *C. schlosseri*, and *C. kowalewskii* from Samos) are considered valid (Heissig, 1975). The derived African genus *Chilotheridium* is endemic to the Miocene of Africa (Hooijer, 1971). These medium-sized grazers may have competed with *Brachypotherium*, although the latter genus had an even larger geographic range.

The teleoceratines first appear in Europe with the Upper Oligocene *Diaceratherium* (not to be confused with the American *Diceratherium*!). The first species was the Stampian form D. lemanense, which was followed by the Aquitanian D. aginense and D. tomerdingense. The lineage culminated with the Burdigalian D. aurelianense. True brachypotheres of the genus *Brachypotherium* replace *Diaceratherium* during the Middle Miocene with the Astaracian B. stehlini and B. brachypus. The last European species was the rare B. goldfussi from the lower Vallesian. The very large East Asian B. perimense and African B. lewisi were more long-lived; the latter species is even found in the Pliocene. *Diaceratherium* and *Brachypotherium* were large to very large rhinos with hypsodont teeth and hippo-like proportions. The reduction in their limbs never reached the extent of American *Teleoceras*, however. Nevertheless, they must have been like *Teleoceras* in their hippo-like aquatic grazing lifestyle.

*Prosantorhinus* was a small teleoceratine with short legs and brachyodont cheek teeth. It is known from *Prosantorhinus* sp. from the Middle Burdigalian, P. douvillei from the Upper Burdigalian and Orleanian, P. germanicus from the Astaracian, and a poorly known species from the Upper Vallesian. *Prosantorhinus* is known only from western Europe and became extinct at the end of the Vallesian (Heissig, 1972a; Guérin, 1980b; Antunes and Ginsburg, 1983).

Another important rhino lineage in the European Miocene was the *Dicerorhinus* group (Fig. 16.1Q, S). The lineage may have originated in the poorly known species "Ceratorhinus" tagicus Roman, a taxon badly in need of revision. Common in the lowermost Aquitanian to upper Burdigalian, some of the material referred to this species should be classified in the genera *Protaceratherium* and *Prosantorhinus*. The first unquestionable *Dicerorhinus* is the medium-sized, cursorial, brachyodont D. (Lartetotherium) sansaniensis from the Orleanian. It was dominant in Astaracian sites of western Europe and Turkey, and survived until the early Vallesian. D. leakeyi from the Middle Miocene of East Africa seems to be anatomically very similar. *Dicerorhinus steinheimensis* from the Astaracian and the lower Vallesian of western Europe may be the smallest rhino ever known in the Neogene, since dwarfing is never observed in Old World rhinos. Possibly derived from the middle Orleanian *Dicerorhinus montesi* (Santafé-Llopis et al., 1987), the large west European *Dicerorhinus schleiermacheri* and its east European relative D. orientalis are among the largest rhinos of the Vallesian and Turolian. All of these Miocene *Dicerorhinus* were tandem-horned, cursorial, brachyodont, and had well-developed tusks. Apparently all were browsers.

There were other rhinos in the European Miocene, but they were rare and apparently had limited success in spreading widely over Europe after immigrating from Asia or Africa. This was probably due to competition from indigenous species. Such rhinos include east Asian *Gaindatherium*, forerunner of the modern one-horned *Rhinoceros*. One isolated species, G. rexmanueli, is known from the Portugese Orleanian. Three species of African tandem-horned rhino *Diceros* (the genus of the living black rhino) occur in the upper Miocene of Spain, Italy, and the Near East (Fig. 16.1U). The best known of these non-African *Diceros* is D. pachygnathus from Greece and Spain. Wagner (1848) originally based the taxon on juvenile material, but Gaudry (1862-1867) fully described it, and all subsequent identifications of D. pachygnathus have been based on these descriptions. Heissig (1975) found that the original material of Wagner belongs to the contemporary species *Dicerorhinus orientalis*, and suggested swapping the original definitions of the two taxa. Until the status of these taxa has been further studied and clarified, we prefer to preserve Gaudry's (1862-1867) concept of the taxon. *Diceros neumayri* from the Near and Middle East, and *D. douariensis* from Tunisia and Italy are the other two species of non-African *Diceros*.

Finally, the immigrant *Hispanotherium* group, a very hypsodont, medium-sized representative of the Iranotheriinae, arrived in Europe in the Miocene. It originated in the lower Miocene of Asia, and is known from Portugal, Spain, Anatolia, the Caucasus, the Siwaliks, Mongolia, and China. The genera *Begertherium*, *Caementodon*, and *Beliajevina* are junior synonyms of *Hispanotherium* (Antunes and Ginsburg, 1983). In Europe, *Hispanotherium* is never found north of the Pyrenees.

## Miocene rhinocerotids of Asia and Africa

East Asian Miocene rhinos are closely related to those of Europe. The first Chinese Neogene rhino is an Agenian *Brachypotherium* sp. from Tibet. In the Orleanian and Astaracian, aceratheriines were represented by *Aceratherium* sp., *Plesiaceratherium* (*P. gracile, P. shanwangensis*), and *Chilotherium* sp. Rhinocerotines are represented by the Orleanian *Dicerorhinus cixianensis* from Shanxi, and iranotheres by the Orleanian *Hispanotherium lingtungensis* (Li, Wu, and Qiu, 1984). Vallesian and Turolian species are well known from the Siwaliks of India and Pakistan (Heissig, 1972b; Guérin, 1979), from Turkey (Heissig, 1972b, 1974, 1975, 1976), and from Iran and China (Li, Wu, and Qiu, 1984). *Gaindatherium browni* from the uppermost Astaracian and lowermost Vallesian of the Siwalik Hills was replaced by *G. vidali* from the Nagri level (Vallesian). This genus was eventually replaced by *Rhinoceros* in the Pliocene and Pleistocene.

The other recent Asian genus *Dicerorhinus* (now represented by the living Sumatran rhino) is known from *D. abeli* from the middle Chinji (uppermost Astaracian/lowermost Vallesian) of India, from *D. ringstroemi* from the Vallesian and Turolian of Turkey and South China, and from *D. orientalis* from the Turolian of the Near East and North China (Fig. 16.1S). *Coelodonta*, the woolly rhino (Fig. 16.1Q), is the most derived of the Dicerorhininae, probably appears in the Ruscinian of North China. Many species of *Chilotherium* are known from the Near and Middle East, including Turkey, India, and China. The last species is *Chilotherium yunnanensis* from the lower Villafranchian. The Iranotheriinae are represented by the large species, *Iranotherium morgani*, from the upper Miocene of Iran, and by several species of *Hispanotherium*. These, in turn, occurred with the first representatives of another very hypsodont rhino group, the Elasmotheriinae (Fig. 16.1R), represented by *Sinotherium* from the Turolian of North China. The very large *Brachypotherium perimense* was a long-lived Indian species which ranged from the Burdigalian to the upper Turolian (Dhok Pathan). Another Asiatic teleoceratine was the small- to medium-sized *Aprotodon fatehjangense*, which lived in India from the Burdigalian to the lower Vallesian.

In Africa, rhinocerotids first appeared in the lower Miocene. The earliest are known from the early Miocene of Libya and Egypt (Hamilton, 1973) with *Aceratherium campbelli* and *Brachypotherium snowi*. In the middle and upper Miocene, the same genera were represented by *A. acutirostratum* from Kenya, Uganda, and Zaïre, and *B. heinzelini* from Kenya, Zaïre, and South Africa. Other African taxa include the teleoceratine *Chilotheridium pattersoni* from Kenya and Uganda, the dicerotines *Paradiceros mukirii* from Kenya and Morocco, and *Diceros douariensis* from Tunisia, and the iranothere *Kenyatherium bishopi* from Nakali, Kenya. The dicerorhinines were represented by *Dicerorhinus leakeyi* from East Africa and *Dicerorhinus primaevus* from Algeria (Guérin, 1980a; Hooijer, 1966, 1968, 1971, 1973; Aguirre and Guérin, 1974).

In general, Miocene Old World rhinoceroses show several interesting features. Like North American *Aphelops* and *Teleoceras*, they show increased hypsodonty associated with the increase in grassland habitats. Hypsodonty arises independently in several rhinocerotid groups, mostly in the aceratheres *Chilotherium* and *Chilotheridium*, the teleoceratines *Teleoceras* and *Brachypotherium*, all the iranotheres (*Hispanotherium*, *Kenyatherium*, *Iranotherium*) and elasmotheres (*Sinotherium* and *Elasmotherium*). The most hypsodont cheek teeth also acquire highly infolded enamel to increase their efficiency. Second, high diversities of rhinos from the same deposit are common in the Miocene. For example, at La Grive Saint Alban (Astaracian of France), *Aceratherium (Alicornops) simorrense*, *Dicerorhinus sansaniensis*, *D. steinheimensis*, and *Brachypotherium* sp. occur together. In the lower Vallesian of Can Ponsic, Spain, there are *Aceratherium incisivum*, *A. simorrense*, *D. sansaniensis*, and *D. steinheimensis* in the same deposit. In the middle Miocene of Kenya, *A. acutirostratum*, *D. leakeyi*, *B. heinzelini*, and *Chilotheridium pattersoni* occur on Rusinga Island. Associations of four rhino species are not rare, as they are in North America, and in some cases, five species occur together. In some cases, one of the associated species is a grazer; the rest are usually browsers.

At the end of the Miocene, there was a worldwide faunal crisis probably associated with the Messinian salinity event in the Mediterranean, and the associated worldwide climatic changes. Many groups of animals went extinct, including all the aceratheriines and most of the teleoceratines. In North America, this meant that the entire rhino fauna was severely decimated, with only one known specimen from the Blancan. In Eurasia, only the rhinocerotines and dicerorhinines survived. In Africa, only dicerotines survived (with two isolated exceptions).

## Plio-Pleistocene rhinocerotids

In Europe and northern Asia, only two rhino lineages are found during the Pliocene and Pleistocene. The first lineage is composed of the genera *Dicerorhinus* and *Coelodonta*. Many Palearctic rhinos are referred to *Dicerorhinus*, the genus of the living Sumatran rhino, but this usage makes the genus a paraphyletic "wastebasket" taxon for a long series of dicerorhinines. The European lineage starts with a Miocene form very near the Sumatran rhino (*Dicerorhinus sumatrensis*), and then the group undergoes many changes. These changes include total loss of incisors, acquisition of an ossified nasal septum with co-ossification of the premaxillae, maxillae, and distal nasal bones, and teeth with increased hypsodonty and complex enamel patterns. Guérin (1980b) proposed the subgenus *Brandtorhinus* for the species without functional incisors or partially ossified nasal septum. The very large *Dicerorhinus megarhinus* from the Ruscinian (lower and middle Pliocene of Europe) possessed visible but non-functional incisors and no bony nasal septum. The larger, but more slender *Dicerorhinus jeanvireti* from the lowermost Villafranchian, and *D. etruscus* from the Villafranchian and early middle Pleistocene, have completely lost their incisors and have an ossified anterior nasal septum. Both of these species were brachyodont browsers.

Other members of the *Dicerorhinus* lineage show an even more completely ossified septum. The very large *D. mercki* (= *D. kirchbergensis*) was an open forest form from the early middle to upper Pleistocene. The medium- to large-sized *D. hemitoechus* was the end of the line, appearing during the end of the middle Pleistocene. Unlike the previous species, *D. hemitoechus* was a semi-hypsodont grazer. All of these Pleistocene species were widespread in Europe, northwest Asia, and the Middle East. *D.*

*etruscus* and *D. mercki* had relatives in the Far East, *D. yunchuchenensis* and *D. choukoutienensis* from China, and *D. japonicus* from Japan.

*Coelodonta* had the same cranial characters of the most derived *Brandtorhinus*, only highly exaggerated. Its teeth were very hypsodont, and the limb skeleton was fully graviportal. The genus seems to have originated in the Upper Villafranchian of northern China and migrated westward. *Coelodonta antiquitatis*, the woolly rhino, arose in China in the Pleistocene and arrived in Europe during the penultimate glaciation. In the Upper Pleistocene, *C. antiquitatis* had the largest range of any known rhino, living or extinct. It extended from South Korea to Scotland to Spain. It was a steppe grazer, well adapted to cold climates, with a broad front lip and a laterally flattened nasal horn well suited for brushing away snow to find grass. Its soft anatomy is well known, since many frozen or mummified carcasses have been found. Many of its anatomical features converge on the white rhino of Africa, even though it belongs to an entirely different lineage. For some reason not yet understood, *C. antiquitatis* never crossed the Bering Land Bridge to North America, even though its frequent companions--such as the woolly mammoth, bison, yak, saiga antelope, and humans--did.

It is not unusual to find three rhino species at the same level in the same site. For example, the cavern of La Fage (Corrèze, France), filled during the Riss glaciation, includes the open forest *Dicerorhinus mercki*, the parkland *Dicerorhinus hemitoechus*, and the steppe *C. antiquitatis* (Guérin, 1973).

The second Eurasian lineage is that of *Elasmotherium*, which originated in China from its sister-taxon *Sinotherium*. *E. caucasicum* occurrred in southeastern Europe and adjacent Asia during the Villafranchian. *E. sibiricum* is known from the middle and upper Pleistocene. *Elasmotherium* was a huge beast, as large as a male Asiatic elephant (*Elephas maximus*), with domed frontal bones and a single frontal horn (Fig. 16.1R). Its cheek teeth were the most specialized of any perissodactyl, and in some ways resemble the specialized teeth of certain rodents. It had only a single premolar, and its three molars were quadrangular in shape, extremely hypsodont with folded enamel and no roots. *Elasmotherium sibiricum* was geographically restricted to the Volga Basin and other tributaries of the Caspian and Black Seas, with possible incursions to central and western Europe. Both the *Dicerorhinus* lineage and the elasmotheres disappear at the end of the Pleistocene with the general extinction of large mammals around 13,000-10,000 years ago.

In southeast Asia, there were two Plio-Pleistocene lineages also, now represented by the genera *Rhinoceros* and *Dicerorhinus*. The one-horned genus *Rhinoceros* (Fig. 16.1T) originated in the Miocene with *Gaindatherium*, and includes the Pliocene *R. sivalensis*, the large Pleistocene *R. paleindicus*, *R. platyrhinus*, *R. sinensis*, and the two living species (*R. unicornis* and *R. sondaicus*). *R. unicornis* appeared in the Middle Pleistocene and includes the peculiar Pleistocene form from Indonesia, *R. unicornis kendengindicus*. *R. sondaicus*, the Javan rhino, can be traced back to the lower Pleistocene with *R. sondaicus sivasondaicus*, and *R. sondaicus guthi*, respectively, as Indonesian and Indochinese Pleistocene sidebranches. *R. unicornis* and *R. sondaicus* still survive in southeast Asia, the first in Assam and Nepal, and the second in Java, Borneo, Malaya, Burma, and Indochina (Groves and Guérin, 1980). *R. unicornis* now occurs in about 19 locations with a total world population (1985 estimate) of less than 2,000 individuals. *R. sondaicus* is mainly restricted to the Udjung Kulon Reserve in western Java, with an estimated population of only 50 individuals left on earth (1987 estimate).

The tandem-horned rhinoceros, *Dicerorhinus sumatrensis*, presently survives in Sumatra, Borneo, Malaya, Burma, and Indochina. It is known from about 18 locations, with a total world population of about 400-900 (1987 estimate). This species is known from the lowermost Pleistocene, and is apparently a relict of the Miocene Dicerorhininae. *R. unicornis* is semi-hypsodont, and is able to graze. *R. sondaicus* and *D. sumatrensis* are brachyodont browsers, inhabiting swamps and dense forests. *R. unicornis* is being bred in several zoos, but *R. sondaicus* and *D. sumatrensis* are both extensively poached and are highly endangered species.

In the Plio-Pleistocene of Africa, there was a different assemblage of rhinos. Besides the dicerotines, which were almost exclusively African, some exotic taxa are also found. The large *Brachypotherium lewisi*, the last of the Teleoceratinae, survived into the Pliocene. The Dicerorhininae made two incursions into North Africa: the Villafranchian *Dicerorhinus africanus*, an African endemic, and the Upper Pleistocene *Dicerorhinus hemitoechus*, a Eurasian species that may have immigrated across Gibraltar. The Dicerotinae, however, were the dominant African group, presently represented by the black rhino, *Diceros bicornis* (Fig. 16.1U), and the white rhino, *Ceratotherium simum* (Fig. 16.1V).

*Diceros* originated from African *Paradiceros* during the Middle Miocene, and was widespread during the Upper Miocene. It occurred from the Middle East (*D. neumayri*) to Italy (*D. douariensis*) to Spain (*D. pachygnathus*). *Diceros bicornis* appeared in the Pliocene and covered all of sub-Saharan Africa, but never reached North Africa or Eurasia. It can be traced back over 4 million years, making it among the most stable and long-lived species on the African savannah. A browser that prefers rugged, hilly, brushy terrain, it once had a clinal distribution of seven subspecies (Groves, 1967), and used to be the most numerous rhino alive, with a population of about 65,000 in 1970. Since that time, however, it has been the most heavily poached, and has been wiped out in all but a few reserves, leaving fewer than 4,000 individuals alive in 1986, and only a few hundred in 1988 (Penny, 1988).

The grazing "white" (or wide-lipped) rhino, *Ceratotherium*, has very hypsodont teeth and a longer skull with an exaggerated occiput that allows it to graze with its head down. The genus appeared in the Pliocene with the long-legged *C. praecox*. The graviportal *C. simum* appeared in the Middle Pleistocene, with two extinct subspecies: *C. simum germanoafricanum* of eastern and southern Africa, and *C. simum mauritanicum* of the Mahgreb (surviving there until the Holocene). There are two living subspecies, *C. simum simum* of South Africa, and *C. simum cottoni* of Central Africa. Only 17 individuals of *C. simum cottoni* survive today in Zaire. On the bright side, however, the South African efforts to save *C. simum simum* are beginning to work. After reaching a low of about 3,000 individuals in 1980, the world population of this subspecies is now up to nearly 4,000 (Penny, 1988).

Next to horses, rhinos have been the most successful group of perissodactyls on this planet. From the Oligocene onward, most terrestrial habitats in the Northern Hemisphere had one or more rhino species as a normal part of the fauna. It is tragic that the rhinoceroses, which have survived so many other crises of environmental change and competition with repeated diversification and migration, may not survive their last crisis--their encounter with humans.

**Acknowledgments**
We thank M. Fortelius and K. Heissig for reviewing the manuscript. Acknowledgment is made to the Donors of the Petroleum Research Fund of the American Chemical Society, and to NSF grant EAR87-08221, for

support of the senior author during this research.

## Bibliography

Abel, O. (1910): Kritische Untersuchungen über die paläogenen Rhinocerotiden Europas. -*Abh. k.k. Reisanst.* 20 (3): 1-52.

Aguirre, E., and Guérin, C. (1974): Première découverte d'un Iranotheriinae (Mammalia, Perissodactyla, Rhinocerotidae) en Afrique: *Kenyatherium bishopi* nov. gen. nov. sp. de la formation vallésienne (Miocène supérieur) de Nakali (Kenya). -*Estudios Geologicos, Madrid*, 30: 229-233.

Antunes, M.T., and Ginsburg, L. (1983): Les Rhinocérotidés du Miocène de Lisbonne--Systématique, écologie, paléobiogéographie, valeur stratigraphique. -*Ciencias da Terra (UNL), Lisboa*, 7: 17-98.

Antunes, M.T., Viret, J., and Zbyszewski, G. (1972): Notes sur la géologie et la paléontologie du Miocène de Lisbonne. -*Bol. Mus. Lab. Miner. Geol. Fac. Ciencias, Lisboa*, 13: 5-23.

Bonis, L. de (1973): Contribution à l'étude des mammifères de l'Aquitanien de l'Agenais: rongeurs, carnivores, perissodactyles. -*Mem. Mus. nat. Hist. natur., Paris*, C, XXVIII: 1-192.

Brunet, M. (1979): Les grands mammifères chefs de file de l'immigration oligocène et le problème de la limite Eocène-Oligocène en Europe. -*Fond. Singer-Polignac édit., Paris*, 1-192.

Cooke, H.B. S. (1968): Evolution of mammals on southern continents. II. The fossil mammal fauna of Africa. -*Quart. Rev. Biol.* 43 (3): 234-264.

Cope, E.D. (1880): The genealogy of American rhinoceroses. -*Amer. Natur.* 14: 540.

Duvernoy, G.-L. (1853): Nouvelles études sur les rhinocéros fossiles. -*C.R. Acad. Sci. Paris*, 36: 117-125.

Eisenmann, V., and Guérin, C. (1984): Morphologie fonctionnelle et environnement chez les Périssodactyles. -*Geobios, Mém. Spéc.* 8: 69-74.

Gaudry, A. (1862-1867): *Animaux fossiles et géologie de l'Attique*. -Paris, 475 pp.

Gaudry, A. (1888): *Les Ancêstres de nos Animaux dans les Temps Géologiques*. -Paris (Bibliothéque scientifique contemporaine), 1-296.

Ginsburg, L., and Guérin, C. (1979): Sur l'origine et l'extension géographique du petit Rhinocérotidé miocène *Aceratherium (Alicornops) simorrense* nov. subgen. -*C.R. Somm. Acad. Sci. Paris, D*, 288: 493-495.

Gregory, W.K., and Cook, H.J. (1928): New material for the study of evolution: a series of primitive rhinoceros skulls (*Trigonias*) from the lower Oligocene of Colorado. -*Proc. Colo. Mus. Nat. Hist.* 8(1): 3-39.

Groves, C. (1967): Geographic variation in the black rhinoceros *Diceros bicornis* (L., 1758). -*Z. f. Säugertierk.* 32 (5): 267-276.

Groves, C. (1983): Phylogeny of the living species of rhinoceros. -*Zeit. Zool. Systematik Evolutionforschung*, 21 (4): 293-313.

Groves, C., and Guérin, C. (1980): Le *Rhinoceros sondaicus annamiticus* (Mammalia, Perissodactyla) d'Indochine: distinction taxonomique et anatomique, relations phylétiques. -*Geobios*, 13(2): 199-208.

Guérin, C. (1973): Les trois espèces de rhinocéros (Mammalia, Perissodactyla) du gisement pléistocène moyen des Abîmes de La Fage à Noailles (Corrèze). -*Nouv. Arch. Mus. Hist. Nat. Lyon*, 11: 55-84.

Guérin, C. (1979): Intérêt biostratigraphique des Rhinocéros du Miocène superieur d'Europe Occidentale. -*Soc. Géol. Fr. édit.*, p. 236.

Guérin, C. (1980a): A propos des rhinocéros (Mammalia, Perissodactyla) néogènes et quaternaires d'Afrique: essai de synthèse sure les espèces et sur les gisements. -*Proc. 8th Panafr. Cong. Prehist. Quat. Stud.*, p. 65-83.

Guérin, C. (1980b): Les rhinocéros

(Mammalia, Perissodactyla) du Miocène terminal au Pléistocène superieur en Europe occidentale. -*Docum. Lab. Géol. Lyon*, 79: 1-1185 (3 vols.).

Guérin, C. (1982): Les Rhinocerotidae (Mammalia, Perissodactyla) du Miocène terminal au Pléistocène superieur en Europe occidentale comparés aux espèces actuelles: tendances évolutives et relations phylogénétiques. -*Geobios*, 15(4): 599-605.

Hanson, C. B. (1989): *Teletaceras radinskyi*, a new primitive rhinocerotid from the late Eocene Clarno Formation, Oregon (this volume, Chapter 20).

Hamilton, W. (1973): North African lower Miocene rhinoceroses. -*Bull. Brit. Mus. Nat. Hist.*, 24(6): 351-395.

Harrison, J.A., and Manning, E.M. (1983): Extreme carpal variability in *Teleoceras* (Rhinocerotidae, Mammalia). -*J. Vert. Paleont.*, 3: 58-64.

Heissig, K. (1969): Die Rhinocerotidae (Mammalia) aus der oberoligozänen Spaltenfüllung von Gaimersheim bei Ingolstadt in Bayern und ihre phylogenetische Stellung. -*Bayer. Akad. Wissens. Abh.*, 138: 1-133.

Heissig, K. (1972a): Die obermiozäne Fossil-Lagerstätte Sandelzhausen. 5. Rhinocerotidae (Mammalia), Systematik und Ökologie. -*Mitt. Bayer. Staatssamml. Paläont. Hist. Geol.*, 12: 57-81.

Heissig, K. (1972b): Paläontologische and geologische Untersuchungen im Tertiär von Pakistan. 5. Rhinocerotidae (Mamm.) aus den unteren und mittleren Siwalik-Schichten. -*Bayer. Akad. Wissens. Abh.*, 152: 1-112.

Heissig, K. (1973): Die Unterfamilien und Tribus der rezenten und fossilen Rhinocerotidae (Mammalia). -*Säugertierk. Mitt.*, 21: 25-30.

Heissig, K. (1974): Neue Elasmotherini (Rhinocerotidae, Mammalia) aus dem Obermiozän Anatoliens. -*Mitt. Bayer. Staatssamml. Paläont. Hist. Geol.*, 14: 21-35.

Heissig, K. (1975): Rhinocerotidae aus dem Jungtertiär Anatoliens. -*Geol. Jahrb.*, 15: 145-151.

Heissig, K. (1976): Rhinocerotidae (Mammalia) aus der Anchitherium-Fauna Anatoliens. -*Geol. Jahrb.*, 19: 3-121.

Heissig, K. (1981): Probleme bei der cladistischen Analyse einer Gruppe mit wenigen eindeutigen Apomorphien: Rhinocerotidae. -*Paläont. Zeit.*, 55: 117-123.

Heissig, K. (1989): The Rhinocerotidae (this volume, Chapter 21).

Hooijer, D.A. (1966): Miocene rhinoceroses of East Africa. -*Bull. Brit. Mus. Nat. Hist.*, 13(2): 119-190.

Hooijer, D.A. (1968): A rhinoceros from the late Miocene of Fort Ternan, Kenya. -*Zool. Mededel.*, 43(6): 77-92.

Hooijer, D.A. (1971): A new rhinoceros from the late Miocene of Loperot, Turkana District, Kenya. -*Bull. Mus. Comp. Zool.*, 142 (3): 339-392.

Hooijer, D.A. (1973): Additional Miocene to Pleistocene rhinoceroses of Africa. -*Zool. Mededel.*, 46 (11): 149-178.

Hugueney, M., and Guérin, C. (1981): Le faune de mammifères de l'Oligocène moyen de Saint-Menoux (Allier). 2° partie: Marsupiaux, Chiroptères, Insectivores, Carnivores, Périssodactyles, Artiodactyles. -*Rev. Sci. Bourbonnais*, p. 52-71.

Hunermann, K.A. (1982): Rekonstruktion des *Aceratherium* (Mammalia, Perissodactyla, Rhinocerotidae) aus dem Jungtertiär vom Höwenegg/Hegau (Baden-Württemberg, BRD). -*Zeit. geol. Wiss.*, 10(7): 929-942.

Li C., Wu W., and Qiu Z. (1984): Chinese Neogene: subdivision and correlation. -*Vert. PalAsiatica*, 22(3): 171-178.

Lucas, S.G., Schoch, R.M., and Manning, E. (1981): The systematics of *Forstercooperia*, a middle to late Eocene hyracodontid (Perissodactyla: Rhinocerotoidea) from Asia and western North America. - *J. Paleont.*, 55: 826-841.

Lucas, S. G., and Sobus, J. (1989): The systematics of indricotheres (this volume, Chapter 19).

Manning, E., Dockery, D.T. III, and Schiebout, J.A. (1986): Preliminary report of a *Metamynodon* skull from the Byram Formation (Lower Oligocene) in Mississippi. -*Miss. Geol.*, 6(2):1-16.

Matthew, W.D. (1931): Critical observations on the phylogeny of the rhinoceroses. -*Univ. Calif. Publ. Geol. Sci.*, 20: 1-8.

Matthew, W.D. (1932): A review of the rhinoceroses with a description of *Aphelops* material from the Pliocene of Texas. -*Univ. Calif. Publ. Geol. Sci.*, 20: 411-480.

McKenna, M.C. (1975): Fossil mammals and early Eocene North American land continuity. -*Ann. Missouri Bot. Garden*, 62: 335-353.

Pavlowa, M. (1892): Études sur l'histoire paléontologiques des Ongules. VI. Les Rhinocéridae de la Russie et le développement des Rhinocéridae en général. -*Bull. Soc. Imp. Natur. Moscou*, 6: 137-221.

Penny, M. (1988): *Rhinos, Endangered Species*. -New York (Facts on File).

Prothero, D. R. (in prep.): The evolution of North American rhinoceroses.

Prothero, D.R., and Manning, E. (1987): Miocene rhinoceroses from the Texas Gulf Coastal Plain. -*J. Paleont.*, 61(2): 388-423.

Prothero, D.R., Manning, E., and Hanson, C.B. (1986): The phylogeny of the Rhinocerotoidea (Mammalia, Perissodactyla). -*Zool. J. Linn. Soc.*, 87: 341-366.

Prothero, D.R., and Sereno, P.C. (1982): Allometry and paleoecology of medial Miocene dwarf rhinoceroses from the Texas Gulf Coastal Plain. -*Paleobiology*, 8: 16-30.

Prothero, D. R., and Schoch, R. M. (1989): The origin and evolution of perissodactyls: summary and synthesis (this volume, Chapter 28).

Radinsky, L.B. (1966): The families of the Rhinocerotoidea (Mammalia, Perissodactyla). -*J. Mamm.*, 47(4): 631-639.

Radinsky, L.B. (1967): A review of the rhinocerotoid family Hyracodontidae (Perissodactyla). -*Bull. Amer. Mus. Nat. Hist.*, 136(1): 1-45.

Romer, A.S. (1966): *Vertebrate Paleontology* (3rd ed.). -Chicago (Univ. Chicago Press).

Santafé-Llopis, J.V. (1978): Rinocerotidos fosiles de España. -Thèse Fac. Sci. Géol. Univ. Barcelone, 1-501.

Santafé-Llopis, J.V., Casanovas-Cladellas, M.L., and Belinchon-Garcia, M. (1987): Una nueva especie de *Dicerorhinus, D. montesi* (Rhinocerotoidea, Perissodactyla) del yacimiento de Bunol (Orleaniense medio) (Valencia, España). -*Paleont. Evol.*, 21: 271-293.

Scott, W.B., and Osborn, H.F. (1898): On the skull of the Eocene rhinoceros, *Orthocynodon*, and the relations of this genus to other members of the group. -*Contrib. Mus. Geol. Archeol. Princeton Coll.*, 3(1): 1-22.

Stevens, M.S., Stevens, J.B., and Dawson, M.R. (1969): New early Miocene formation and vertebrate fauna, Big Bend National Park, Brewster County, Texas. -*Pearce-Sellards Series Texas Mem. Mus.*, 15:1-53.

Tanner, L.G. (1969): A new rhinoceros from the Nebraska Miocene. -*Bull. Univ. Nebraska State Mus.*, 8: 395-412.

Tedford, R.H., Swinehart, J.B., Hunt, R.M., Jr., and Voorhies, M.R. (1985): Uppermost White River and lowermost Arikaree rocks and faunas, White River Valley, northwestern Nebraska, and their correlation with South Dakota. - In: Martin, J.E. (ed.), Fossiliferous Cenozoic deposits of western South Dakota and northwestern Nebraska. -*Dakoterra*, 2(2): 334-352.

Troxell, E.L. (1921): A study of

*Diceratherium* and the diceratheres. - *Amer. J. Sci.*, 202: 197-208.

Viret, J. (1958): Perissodactyla. -In: Piveteau, J. (ed.): *Traité de Paléontologie*, 6 (2): 368-475.

Voorhies, M.R. (1981): Ancient ash fall creates a Pompeii of prehistoric animals. -*Nat. Geog.*, 159(1): 66-75.

Voorhies, M.R., and Thomasson, J.R. (1979): Fossil grass anthoecia within Miocene rhinoceros skeletons: diet in extinct species. -*Science*, 206: 331-333.

Wagner, J.A. (1848): Urweltliche Säugethiere-Überreste aus Griechenland. -*Abh. k. Bayer. Akad. Wiss. II Cl.*, 5(2): 335-378.

Wall, W. P. (1982): Evolution and biogeography of the Amynodontidae (Perissodactyla, Rhinocerotoidea). - *Proc. Third North Amer. Paleon. Conv.*, 2: 563-567.

Wall, W. P. (1989): The phylogenetic history and adaptive radiation of the Amynodontidae (this volume, Chapter 17).

West, R.M., Dawson, M.R., and Hutchinson, J.H. (1977): Fossils from the Paleogene Eureka Sound Formation, N.W.T., Canada: occurrence, climatic and paleogeographic implication. -In: West, R.M. (ed.): Paleontology and plate tectonics. - *Milw. Public Mus. Spec. Publ. Biol. Geol.*, 2: 77-93.

Wood, H.E., II (1927): Some early Tertiary rhinoceroses and hyracodonts. -*Bull. Amer. Paleont.*, 13(50): 5-105.

# 17. THE PHYLOGENETIC HISTORY AND ADAPTIVE RADIATION OF THE AMYNODONTIDAE

## WILLIAM P. WALL

The systematic relationships of amynodontid rhinoceroses are described and diagnoses of taxa to the genus level are given. The Amynodontidae includes two subfamilies, Rostriamynodontinae and Amynodontinae. *Rostriamynodon*, the sole representative of the first subfamily, is the primitive sister taxon to all other amynodontids. The Amynodontinae is divided into three tribes: Amynodontini, Cadurcodontini, and Metamynodontini. The last two tribes represent a dichotomous radiation of amynodontids during the late Eocene and Oligocene. Cadurcodontines evolved a tapir-like proboscis and were presumably terrestial. Metamynodontines show characteristics typical of semi-aquatic animals. Advanced taxa had massive bodies with a very hippo-like appearance. Metamynodontines also display the most extreme dental modifications in the family, having higher-crowned, narrower cheek teeth for improved vertical shear.

## Introduction

The Amynodontidae is an atypical group of Holarctic rhinocerotoids known from the medial Eocene to early Miocene. Research on amynodontids spans more than 100 years, but many questions regarding the origin, systematics, and life habits of this family remain unanswered. Early work on amynodontids was primarily by American and European authors (i.e., Gervais, 1873; Marsh, 1877; Scott and Osborn, 1887; Matthew, 1929; Stock, 1933; and Kretzoi, 1942), but the majority of recent publications are either Russian or Chinese (ie. Gromova, 1954, 1960; Belyayeva, 1971; Chow and Xu, 1965; Xu, 1966; and Xu et al.., 1979). Descriptive taxonomy dominates the literature on amynodontids resulting in 24 genera and 55 species, but many of these taxa are probably not valid (Radinsky, 1969; Wall, 1981). The phylogenetic relationships of amynodontids have recently been reinterpreted by Wall (1982a) and Wall and Manning (1986). The taxonomic arrangement proposed in these last two papers will be followed here.

The mode of life and feeding adaptations of amynodontids have received little attention. Amynodontids are commonly called aquatic rhinoceroses in reference to their presumed amphibious habits. This generalization is the result of superficial comparisons between *Metamynodon* and hippos (Troxell, 1921; Scott, 1941), without recognition of the true diversity of amynodontid body plans. Only one tribe of amynodontids, the Metamynodontini, exhibits skeletal modifications similar to those of hippopotami (Wall, 1981). The postcranial skeleton of cadurcodontines displays features typical of subcursorial ungulates. Cranial modifications culminating in *Cadurcodon* indicate that this tribe developed a tapir-like proboscis (Wall, 1980).

The following review is an attempt to synthesize all available information into a realistic picture of amynodontid systematics.

## Systematic paleontology
Order Perissodactyla
Suborder Ceratomorpha
Superfamily Rhinocerotoidea
Family Amynodontidae Scott and Osborn 1883

*The Evolution of Perissodactyls* (ed. D.R. Prothero & R.M. Schoch) Oxford Univ. Press, New York, 1989

Distribution: medial or late Eocene to late Oligocene of North America and Asia (*Cadurcotherium*, early Miocene of Asia); Oligocene of Europe.

Revised diagnosis: Dental formula = $3\text{-}1/3\text{-}1\ 1/1\ 3/3\text{-}2\ 3/3$; lower canines shear anterior to uppers, both are enlarged; upper and lower P1/1 absent, remaining premolars small; quadratic $M^3$ with well developed post-fossette; $M^3$ metastyle large and deflected labially; upper molars without crista and crochet, generally without antecrochet; anterior rib on upper molar ectolphs positioned far forward; parastyle reduced; lower molars elongate, trigonoid and talonid broadly open lingually; labial groove separating the trigonoid and talonid shallow or absent; preorbital fossa present; large sagittal crest; secondary palate concave; hornless; manus with four digits, pes with three.

Discussion: Amynodontids were placed in a separate family by Scott and Osborn (1883) to differentiate two middle to late Eocene genera, *Amynodon* and *Orthocynodon*, from hyracodontids and rhinocerotids. Scott and Osborn's family diagnosis contained primarily primitive characters and was therefore inadequate to justify a relationship between these two genera. Scott and Osborn (1887) amended the family diagnosis after the discovery of a new genus, *Metamynodon*, in the White River Oligocene. Their revised diagnosis is more accurate and cites several important derived characters: reduction or loss of $P_1$, and enlargement of the sagittal crest.

Osborn (1889) recognized some of the most important characteristics of amynodontids when he added presence of a preorbital fossa, shortening of the nasals, and quadratic shape of the $M^3$ to the family diagnosis. Osborn (1898) placed additional emphasis on the large size of the canines and shortened facial region as means of separating amynodontids from other rhinoceratoids. The amynodontid family diagnosis has been gradually refined since the turn of the century (see Wood, 1927; Gromova, 1954; Radinsky, 1966, 1969; and Wall, 1981).

Dental characters: Enlargement of the canines is a derived character unique within the Rhinocerotoidea, and found in even the earliest representatives of the family. The upper canines are erect and the lowers are slightly procumbent. The lower canines shear anterior to the uppers producing smooth wear surfaces, generally with sharp edges. Most hyracodontids have canines which are only slightly larger than the incisors (a primitive condition), and they lack the characteristic wear pattern of amynodontid canines. Canines in rhinocerotids are reduced or absent. Instead, it is the $I^1/I_2$ shearing complex which becomes dominant in early representatives of the Rhinocerotidae (Radinsky, 1966).

The loss of upper and lower P1/1 is a derived character correlated with reduction in size and importance of the entire premolar series in amynodontids. $P_1$ is absent in all indricotheriines (*Paraceratherium* also lacks $P^1$), and both the upper and lower P1/1 are lost in some rhinocerotids.

The quadratic shape of $M^3$ is an important derived character in amynodontids. The elongate ectoloph and well-developed post-fossette on $M^3$ result from evolution of a dental pattern quite unlike that in other rhinocerotoids. Hyracodontids and rhinocerotids reduce or lose the metastyle, and the ectoloph merges with the metaloph to produce an almost bilophodont $M^3$. Vertical shear along the ectoloph became important early in amynodontid history, and as a result the $M^3$ metastyle is bent labially and enlarged, producing a longer and straighter ectoloph (Wall, 1982a). The relatively simple structure of the protoloph and metaloph on the upper molars is a primitive character.

The importance of vertical shear in amynodontid mastication is also apparent in the lower cheek teeth. The protolophid and hypolophid are oriented in a more oblique direction than in other rhinocerotoids. The reorientation of these two lophs and the reduction in size of the external groove separating the trigonid and talonid create a longer and straighter shearing surface along the labial edge of the tooth.

Amynodontids display several typically rhinocerotoid dental characteristics. First, the upper molars are π-shaped, a pattern in which the protoloph and metaloph are roughly perpendicular to the straight ectoloph. Second, paralophid and metalophid are well developed on the lower molars. And third, $M_3$ lacks a hypoconulid. The loss of the hypoconulid may be an independently derived character. Radinsky (1966) demonstrated that the hypoconulid was also absent in several different groups of tapiroids.

Skull characteristics: The pre-orbital fossa is the most important derived cranial feature in amynodontids. Early representatives of the family have a large, shallow depression in the maxilla. More derived taxa have either a reduced fossa or one which extends medial to the orbit. *Heptodon*, *Hyrachyus*, and *Hyracodon* all lack a preorbital fossae, indicating the absence of such a structure in primitive ceratomorphs. Development of preorbital fossa in several unrelated advanced groups must be the result of independent derivation. *Tapirus* has a groove on either side of the nasal incision which houses nasal diverticula, and two recent rhinocerotids, *Ceratotherium* and *Rhinoceros indicus*, have shallow depressions in the maxillae, but the shape and extent of the fossa is different from that in amynodontids. Since ancestral forms lacked a preorbital fossa, the fossae in advanced ceratomorphs are not homologous, and probably not analogous in function.

Other amynodontid cranial features are not as diagnostic of the entire family, but several evolutionary trends are distinctive. First, amynodontids have an unusually large sagittal crest. Hyracodontids retain a sagittal crest, but it is relatively smaller than in amynodontids. Indricotheriines and rhinocerotids reduce or lose the sagittal crest. The large amynodontid sagittal crest indicates the importance of vertical shear in this family. Second, the amynodontid palate is unusual in being concave. This condition is more highly developed in some of the advanced taxa (Troxell, 1921) and is undoubtedly a derived character. And third, reduction in length of the preorbital region is characteristic of all but the most primitive amynodontids. Snout reduction is an important characteristic in dividing amynodontids into two subfamilies, Rostriamynodontinae and Amynodontinae.

Subfamily Rostriamynodontinae Wall and Manning 1986

Diagnosis: Dental formula = $3/3$ $1/1$ $3/3$ $3/3$; long post-canine diastema; lower molar broad with labial groove separating trigonid from talonid relatively large for the family; preorbital region of the skull long; large, shallow preorbital fossa that does not extend medial to the orbit.

Included genus: *Rostriamynodon*.

Discussion: Wall (1982a) proposed the division of the Amynodontidae into two subfamilies based on snout length. The long snout of *Rostriamynodon* (42% of skull length) compares favorably with a diversity of primitive ceratomorphs such as *Hyrachyus*, *Heptodon*, and *Forstercooperia*, and is therefore a primitive character state (Wall and Manning, 1986). All other amynodontids show a reduced snout, ranging from 32% of skull length in *Amynodon* to 20% in *Cadurcodon* and *Metamynodon*. The lower molars of *Rostriamynodon* are broad compared to those of other amynodontids. *Rostriamynodon* has an $M_3$ width/length ratio of 58% compared to a maximum of 52% in *Amynodon*, the next closest taxon. In this regard, *Rostriamynodon* is intermediate between other amynodontids and primitive ceratomorphs (for which $M_3$ breadth ranges from 60-75% of length). Based on these character polarities, the Rostriamynodontinae is the primitive sister group to the Amynodontinae.

*Rostriamynodon*, the only genus in the

subfamily, comes from the early late Eocene of Inner Mongolia, People's Republic of China. The type specimen, AMNH (American Museum of Natural History, New York) 107635, is a complete skull and jaws which represent the oldest well preserved material of any amynodontid. *Rostriamynodon* is an amynodontid because of the quadratic $M^3$, loss of upper and lower P1/1, enlarged canines, and preorbital fossa. In almost all other characteristics, *Rostriamynodon* is primitive (it is very similar to the contemporaneous *Forstercooperia*). Wall and Manning (1986) provide a detailed description and discussion of the relationships of this taxon.

Subfamily Amynodontinae Scott and Osborn, 1883, *sensu* Wall 1982a

Diagnosis: Dental formula = $^{3-1}/_{3-1}$ $^1/_1$ $^3/_{3-2}$ $^3/_3$; post-canine diastema reduced; lower molars moderately to greatly elongate with labial groove separating trigonid from talonid reduced or lost; preorbital region of skull reduced, never more than 35% of skull length.

Included tribes: Amynodontini, Cadurcodontini, Metamynodontini.

Discussion: Kretzoi (1942) grouped *Amynodon*, *Sharamynodon*, and *Amynodontopsis* into the subfamily Amynodontinae, one of four subfamilies he proposed to subdivide the ten genera of amynodontids recognized at that time. The subfamily Amynodontinae as defined in this paper is very different from Kretzoi's usage of the term. I include all amynodontid genera which exhibit traits derived beyond the condition seen in *Rostriamynodon* in the Amynodontinae. With regard to Kretzoi's remaining subfamilies, my classification reduces his Metamynodontinae to the level of a tribe (Metamynodontini), considers *Paramynodon* to be a primitive member of the Metamynodontini, eliminating the subfamily Paramynodontinae, and regards the Cadurcotheriines as an artificial assemblage of genera and therefore not a valid taxonomic grouping.

The Amynodontinae is a diverse subfamily including 21 named genera. Several genera are represented by fragmentary material, and some are probably not valid taxa. In the following discussion, the majority of genera will be divided into three tribes: Amynodontini, Cadurcodontini, and Metamynodontini. Several genera will be placed into synonymy and the remaining taxa will be listed as Amynodontidae *incertae sedis*, or Rhinocerotoidea *incertae sedis*.

The three tribes in this subfamily can best be distinguished by the relative development of the preorbital fossa (Figure 17.1). In the Amynodontini the preorbital fossa is large but does not extend medial to the orbit. In the Cadurcodontini the preorbital fossa remains large and does extend medial to the orbit, and in the Metamynodontini the preorbital fossa is reduced.

Tribe Amynodontini Scott and Osborn, 1883, *sensu* Wall 1982a

Diagnosis: Dental formula = $^3/_3$ $^1/_1$ $^3/_3$ $^3/_3$; premolar series length approximately 50% of molar series length; labial groove separating talonid and trigonid present; preorbital fossa large but not extending medial to the orbit; preorbital portion of the skull approximately 35% of total skull length; nasal-lacrimal contact present.

Included genus: *Amynodon* (= *Orthocynodon*, Scott and Osborn, 1882)

Discussion: Detailed descriptions and systematic reviews of the genus *Amynodon* have recently been published (Wilson and Schiebout, 1981; and Wall, 1982b). Based on differences in cranial anatomy, I believe that the Asiatic material referred to *Amynodon* is more correctly placed in the genus *Sharamynodon*. The Asiatic specimens show an expansion of the preorbital fossa medial to the orbit, the North American specimens do not. *Amynodon* is therefore limited to the Uintan Eocene of North America. *Amynodon* is advanced compared to *Rostriamynodon* in having a

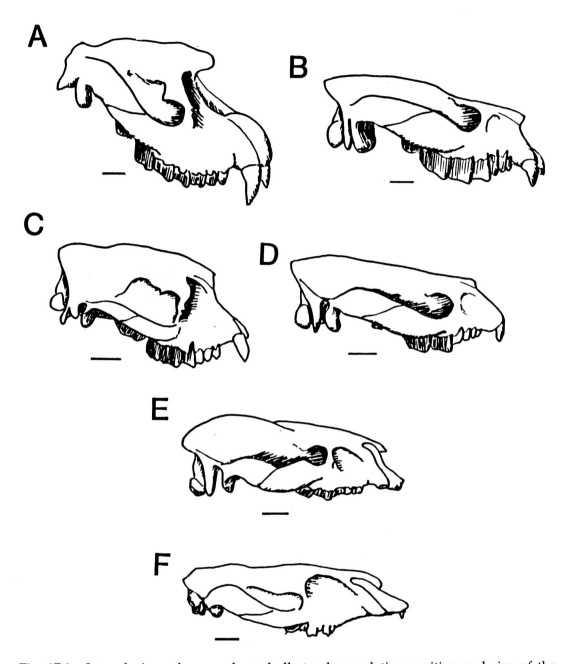

Fig. 17.1. Lateral view of amynodont skulls to show relative position and size of the preorbital fossa. A, *Cadurcodon*; B, *Metamyonodon*; C, *Amynodontopsis*; D, *Zaisanamynodon*; E, *Amynodon*; and F, *Rostriamynodon*. The preorbital fossa in primitive amynodontids (E and F) is large but does not extend medial to the orbit, and in metamynodontines (B and D) the fossa is reduced. Scale line equals 10 cm.

slightly reduced preorbital region and narrower lower molars, but is primitive in almost all other respects. The Amynodontini can be regarded as the primitive sister group to either cadurcodontines or metamynodontines, displaying no derived characteristics which would place them closer to either one of the advanced tribes.

Tribe Cadurcodontini Wall, 1982a

Diagnosis: Dental formula = $3\text{-}1/3\text{-}1$ $1/1$ $3/3\text{-}2$ $3/3$; dentition variable within the tribe, but in none of the genera do the cheek teeth become as high crowned as in some of the more derived taxa of the metamynodontines; labial groove separating trigonid from talonid reduced; premaxilla-nasal contact reduced; preorbital fossa extending medial to the orbit; orbit positioned low on the skull due to large size of frontal sinuses; lacrimal with well-developed rugosites for attachment of strong snout musculature; nasal-lacrimal contact present; zygomatic arch and lower jaw not thickened; skull dolichocephalic.

Included genera: *Sharamynodon, Amynodontopsis, Sianodon*, and *Cadurcodon*.

Discussion: Cadurcodontines are an advanced tribe of amynodontids characterized by the following synapomorphies: large preorbital fossa extending medial to the orbit; enlarged nasal incision; well developed frontal sinuses; and large snout muscle scars. These characteristics suggest a strong evolutionary trend toward the development of a tapir-like proboscis (Wall, 1980).

The four genera within the Cadurcodontini are distinguished by their level of specialization in the snout region and by differences in the dentition. *Sharamynodon* is the most primitive cadurcodontine, retaining three pairs of large incisors and a double-rooted $P_2$. *Amynodontopsis* is a variable genus showing three pairs of small incisors and $P_2$ either single-rooted or lacking. *Sianodon* and *Cadurcodon* are similar in many ways and only one genus may be valid (*Cadurcodon* has priority). Both taxa have reduced the number of incisors and have lost $P_2$, but *Sianodon* is more primitive in skull proportions than *Cadurcodon* (see generic diagnoses).

Cadurcodontines do not exhibit any skeletal traits associated with aquatic habits and therefore post-cranial material can easily be distinguished from the presumably semi-aquatic metamynodontines.

Genus *Sharamynodon* Kretzoi, 1942

Diagnosis: Dental formula = $3/3$ $1/1$ $3/3$ $3/3$; incisors and $P_2$ similar in appearance to *Amynodon* but lower molars relatively narrower; ascending process of premaxilla reduced; maxilla does not form part of the border of the external nares; preorbital fossa extending slightly medial to the orbit; preorbital portion of the skull approximately 25-30% of skull length; premaxilla not fused or thickened; nasal incision extends back to a point above the post-canine diastema.

Discussion: Osborn (1936) described AMNH 20278 (a nearly complete skeleton) and called it a new species of *Amynodon*, *A. mongoliensis*. Kretzoi (1942), however, recognized significant differences between the Asiatic and North American taxa. Kretzoi believed that *A. mongoliensis* was intermediate in position between *Amynodon* and *Amynodontopsis* and therefore placed AMNH 20278 in a new genus, *Sharamynodon*. Chinese paleontologists, apparently unaware of Kretzoi's work, continued to place many of the Eocene Asiatic amynodontids in the genus *Amynodon* (see, for example, Li and Ting, 1983)

A second taxononomic problem with *Sharamynodon* is the extreme splitting of taxa of late Eocene amynodontids. Six species of *Sharamynodon*, including *A. watanabei* (from the Eocene of Japan) are currently recognized (compiled from Savage and Russell, 1983). In addition to these six, Chow and Xu (1965) described a new genus, *Lushiamynodon*, from the Shara Murun Eocene. Five species of *Lushiamynodon* are listed by Li and Ting (1983). *Lushiamynodon* is a primitive cadurcodontine (see Wall, 1981) and similar to *Sharamynodon*. Xu (1965) placed four other species of Eocene cadurcodontines in *Sianodon*. In total there are three genera and fifteen species of primitive cadurcodontines from the Shara Murun Eocene of Asia in the literature. It is very unlikely that all of these taxa are valid. Chow and Xu's (1965) diagnosis of *Lushiamynodon* does not cite any derived characteristics to justify the validity of this taxon. I believe that *Lushiamynodon* and the Eocene species of *Sianodon* are more correctly placed in the genus *Sharamynodon*. The exact number of species of *Sharamynodon* cannot be determined until all of the Asiatic material has been reexamined.

Genus *Amynodontopsis* Stock, 1933

Diagnosis: Dental formula = $3/3$ $1/1$ $3/{2-3}$ $3/3$; incisors small; $P_2$ single-rooted or absent; anticrochet usually present on $M^1$; $M^3$ metastyle strongly deflected labially and with a sharp posterior edge; premaxilla reduced in lateral extent, ending part way up the external nares; right and left premaxilla not fused together or thickened; premaxilla-nasal contact retained due to elongation of the descending process of nasal; nasals reduced but of variable shape; preorbital fossa extending far back medial to the orbit producing a well-defined, narrow anterior rim to the orbit; inner surface of anterior orbital bar concave in continuation with the preorbital fossa; anterior border of orbit located above $M^2$ protoloph; preorbital portion of skull approximately 20% of skull length (based on measurements from the tip of the nasals).

Discussion: *Amynodontopsis* is the most diverse genus of amynodontid. The type species, *A. bodei*, was described by Stock (1933, 1939) from the Duchesnean of California. Wilson and Schiebout (1981) and Bjork (1967) have also recorded the latest Eocene specimens of *A. bodei* from Texas and South Dakota, respectively.

Wall (1980, 1982a) tentatively assigned a large collection of amynodontid material from the Ulan Gochu ( early Oligocene) of Asia to *Amynodontopsis*. The Asiatic material exhibits the following derived characters in common with *A. bodei*: incisors small; $M^3$ metastyle strongly deflected labially with a sharp posterior edge; reduced contact between nasal and premaxilla; and expansion of the preorbital fossa into the medial wall of the anterior border of the orbit. The Asiatic material referred to *Amynodontopsis* is at least specifically different from *A. bodei*. The Asiatic specimens show a reduction in size of the zygomatic arch and shortened nasal bones that are squared off anteriorly. In *A. bodei* the nasals are rounded and overhang the external nares. An alternative taxonomic approach would be to regard the Asiatic material as a sister genus to *Amynodontopsis*.

Genus *Sianodon* Xu, 1965

Diagnosis: Dental formula = $2/2$ $1/1$ $3/2$ $3/3$; lower molars narrow with only a small groove separating talonid and trigonid; $M^3$ metastyle not as strongly deflected labially as in *Amynodontopsis*; premaxilla reduced laterally so that maxilla forms part of the border of the external nares; premaxilla vertically thickened and co-ossified; nasals squared off anteriorly, not overhanging external nares; preorbital fossa extending medial to the orbit but smaller than in *Amynodontopsis*; nasal incision extending back to a point above $P^4$.

Discussion: The diagnosis above is based solely on material referred to Xu's (1965) type species, *S. bahoensis*, from the early Oligocene of Asia. Xu (1966) placed several later Eocene amynodontids into additional species of *Sianodon* with little justification. All of these specimens are similar to *Sharamynodon* and may be referrable to that genus (see prior discussion).

*Sianodon* exhibits a mosaic of characters intermediate between *Sharamynodon* and *Cadurcodon*. *Sianodon* shares with

*Cadurcodon* a reduction in number of incisors, vertical thickening of the premaxilla, and fusion of the right and left premaxillae. *Sianodon* is more primitive than *Cadurcodon* in the retention of two pairs of lower incisors, less extreme posterior expansion of the nasal incision, relatively less massive premaxillae and nasals, and more distinct postorbital constriction. The closest affinities of *Sianodon* are with *Cadurcodon*, but because of the retention of significant primitive character states, it cannot be placed in that genus. It is better to regard *Sianodon* as the primitive sister taxon to *Cadurcodon*.

Genus *Cadurcodon* Kretzoi, 1942

Diagnosis: Dental formula = $^{1-2}/_1$ $^1/_1$ $^3/_2$ $^3/_3$; $M^3$ metastyle not as strongly deflected labially as in *Amynodontopsis*. Premaxilla greatly enlarged with right and left sides fused; premaxilla ends approximately halfway up the narial canal where it contacts a descending process of the nasal; maxilla borders a significant portion of the external nares; nasal incision expanded back to a point above $M^2$; nasals reduced in length but relatively thick; preorbital fossa extending far back medial to the orbit, but not expanded into the orbital wall as in *Amynodontopsis*.

Discussion: Based on a presumed similarity in the lower premolars, Osborn (1923, 1924) assigned amynodontid material from the Ardyn Obo Formation (early Oligocene) to the Eurasian genus *Cadurcotherium* as a separate species, *C. ardynense*. Kretzoi (1942) divided *Cadurcotherium* into three closely related genera: *Cadurcotherium*, *Cadurcamynodon*, and *Cadurcodon* (the last one for the Ardyn Obo amyodontid). Kretzoi based his generic separation of *Cadurcodon* on primitive characters, and therefore his diagnosis did not adequately define this taxon. Kretzoi did, however, recognize the differences between *Cadurcodon* and *Cadurcodontherium*, a taxon I have identified as a metamynodontine (Wall, 1982a). *Cadurcodon* is the most derived cadurcodontine, exhibiting the most extreme specializations in the dentition and cranial characters associated with a proboscis (Wall, 1980). Species-level taxonomy is currently being reviewed by the author.

Tribe Metamynodontini Kretzoi, 1942, *sensu* Wall, 1982a

Diagnosis: Dental formula = $^{1-3}/_{1-3}$ $^1/_1$ $^3/_2$ $^3/_3$; cheek teeth relatively high-crowned, particularly in the more advanced taxa; preorbital portion of skull reduced, but posterior expansion of the nasal incision never as pronounced as in advanced cadurcodontines; skull moderately brachycephalic; zygomatic arch large and massive; preorbital fossa reduced and typically not extending medial to orbit; orbit positioned high on the skull; frontal-maxilla contact present, post-cranial skeleton relatively massive; third trochanter on femur reduced.

Included genera: *Megalamynodon*, *Paramynodon*, *Zaisanamynodon*, *Metamynodon*, and *Cadurcotherium*.

Discussion: The Metamynodontini are a clearly defined group sharing the following derived characteristics: brachcephalic skull; massive zygomatic arches; high crowned cheek teeth; and frontal-maxilla contact (Figure 17.2). Gregory (1920) showed that a nasal-lacrimal contact is primitive for perissodactyls. All non-metamynodontine amynodontids retain the nasal-lacrimal contact; therefore, the presence of a frontal-maxilla contact in metamynodontines is a derived condition.

The Metamynodontini constitute the second major adaptive radiation of amynodontids. Evolutionary trends within the tribe include: a great increase in bulk; increased crown height; narrower cheek teeth; hypertrophy of the canines; thickening of the mandible and zygomatic arch with brachycephaly; and elevation of the orbit on the skull.

Relationships between the five genera of metamynodontines are determined by the following dichotomies. Two relatively

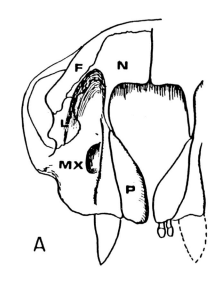

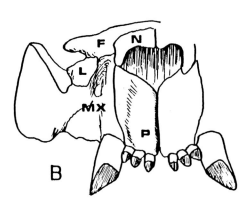

Fig. 17.2. Anterior view of amynodontid skulls to show the primitive perissodactyl nasal-lacrimal contact in *Amynodontopsis* (A) and the derived frontal-maxilla contact of metamynodontines (B is *Metamynodon*).

primitive genera, *Megalamynodon* and *Paramynodon*, are separated from the more derived taxa by their smaller size and less specialized dentition (see generic diagnosis below). *Metamynodon* and *Cadurcotherium* are more derived than *Zaisanamynodon* because of their higher crowned cheek teeth and narrower lower molars. The great reduction in size of the $M^3$ metaloph in *Cadurcotherium* makes it the more derived sister taxon to *Metamynodon*.

Genus *Megalamynodon* Wood 1945

Diagnosis: Dental formula = $3/3$ $1/1$ $3/2$ $3/3$; lower cheek teeth with well-developed labial cingula; $P_3$ relatively larger than in *Paramynodon*; zygomatic arch large and thickened but not increased in height as in *Metamynodon*; glenoid shelf not enlarged as in *Paramynodon*; preorbital region and diastema long for a metamynodontine; body size intermediate between *Amynodon* and *Metamynodon*.

Discussion: *Megalamynodon* is a difficult taxon to diagnose. The single species *M. regalis* from the Duchesnean of North America exibits sufficient derived characteristics to include it in the Metamynodontini, but beyond that, it is essentially primitive. *Megalamynodon* is similar in size and degree of specialization to *Paramynodon* from the late Eocene of Burma. The taxonomic relationship between *Megalamynodon* and *Paramynodon* is uncertain because of their primitive character states. It is possible that the North American and Asiatic forms are the same genus (*Paramynodon* has priority), but the Asiatic species does show the following derived traits; enlarged glenoid shelf and reduced $P_3$.

Wood and Scott (in Scott, 1945) believed that *Megalamynodon* was the "missing link" between *Amynodon intermedius* and *Metamynodon chadronensis*. Although the simplistic monophyletic lineage envisioned by Wood and Scott is probably unrealistic, in general *Megalamynodon* does form a good intermediate between

*Metamynodon* and primitive amynodontids.

Genus *Paramynodon* Matthew 1929

Diagnosis: Dental formula = $3/3?$ $1/1$ $3/2$ $3/3$; a small metamynodontine in the same size range as *Amynodon*; zygomatic arch thick and bowing outward, reaching a dorsal high point midway and then sloping downward posteriorly; extensive glenoid shelf covering the external auditory meatus; orbit positioned moderately high on the skull and far forward over $M_1$; preorbital portion of skull and diastema relatively long.

Discussion: *Paramynodon* is a good example of how failure to recognize individual variation and sexual dimorphism can complicate taxonomic literature. The type species was described by Pilgrim and Cotter (1916) from the Pondaug Formation, late Eocene of Burma, as a new species of *Metamynodon*, *M. burmanicus*. Pilgrim (1925) named a second amynodontid from Burma, *M. cotteri*. Pilgrim based his separation of the two species on differences in shape of the canine cross-section and the larger size of *M. cotteri*. Matthew (1929) recognized differences between the Burma species and North American *Metamynodon* and placed the Asiatic material in a new genus, *Paramynodon*. Matthew did not question the validity of Pilgrim's species-level taxonomy. Colbert (1938) reanalyzed *Paramynodon* and determined that the differences in canine shape were not significant. Colbert also recognized that the size difference between *P. birmanicus* and *P. cotteri* was due to sexual dimorphism.

Genus *Zaisanamynodon* Belyayeva, 1971

Diagnosis: Dental formula = $3/3$ $1/1$ $3/2$ $3/3$; labial groove separating trigonid from talonid greatly reduced; $I^3$ large for an amynodontid; cheek teeth not as high-crowned as in *Metamynodon*; preorbital portion of skull reduced compared to *Paramynodon*; preorbital fossa small; skull highly brachycephalic; a very large amynodontid.

Discussion: *Zaisanamynodon* is similar to *Metamynodon* in many ways, but where differences occur (crown height, width of the lower jaw) *Metamynodon* is more highly derived. Belyayeva (1971) described the partial skeleton of a large amynodontid from the late Eocene of Asiatic Soviet Union as a new genus, *Zaisanamynodon* (type species *Z. borisovi*). However, *Gigantamynodon*, from the late Eocene to the middle Oligocene of Asia, represents a taxonomic problem. Chinese paleontologists (see Li and Ting, 1983) have identified a large Asiatic metamynodontine with morphology similar to *Zaisanamynodon* as *Gigantamynodon*. I believe only one genus of large Asiatic metamynodontine is valid. Normally *Gigantamynodon* would have priority, but since Gromova's (1954) diagnosis is so poor (see Wall and Manning, 1986), I believe *Gigantamynodon* should be considered a *nomen dubium*.

Genus *Metamynodon* Scott and Osborn, 1887
(=*Cadurcopsis*, Kretzoi, 1942)

Diagnosis: Dental formula = $3\text{-}2/?3\text{-}1$ $1/1$ $3/2$ $3/3$; canines large and tusk-like; cheek teeth high-crowned; trigonid and talonid confluent labially; skull brachycephalic; premaxilla and nasal reduced but still contact each other along border of external nares; zygomatic arch very large and massive; preorbital fossa greatly reduced; orbit positioned higher on the skull than in either *Paramynodon* or *Zaisanamynodon*; lower jaw massive.

Discussion: Most of the characteristics cited above are taken from the type species, *M. planifrons*. Both *M. chadronensis* (Wood, 1937) and *M. mckinneyi* (Wilson and Schiebout, 1981) are known from fragmentary material. The three species are easily differentiated by their size and dentition. There are either one or two pairs of lower incisors in *M. planifrons* compared to three pairs of lower incisors in *M. mckinneyi* (the number of incisors in *M. chadronensis* is unknown). *M. planifrons* can also be differentiated by its larger size, more massive lower jaw, and narrower cheek teeth.

Placement of both *M. chadronensis* and *M. mckinneyi* in the genus *Metamynodon* is open to question (particularly for the more primitive *M. mckinneyi* ). A detailed comparison between these two species and *Megalamynodon* is needed to solve this problem. If *M. mckinneyi* is not included in *Metamynodon,* then the genus would be restricted to two species from the early to middle Oligocene of North America.

Genus *Cadurcotherium* Gervais, 1873 (= *Cadurcamynodon,* Kretzoi, 1942)

Diagnosis: Dental formula = $2/1$ $1/1$ $3/2$ $3/3$; very high-crowned cheek teeth; extremely narrow lower molars; confluence of the parastyle and anterior rib on the upper molars; greatly reduced $M^3$ metastyle; a medium to large amynodontid with relatively massive lower jaw and zygomatic arches.

Discussion: *Cadurcotherium* is the most derived stage achieved by the Metamynodontini prior to its extinction at the end of the early Miocene (*Cadurcotherium* is the only amynodontid known from the Miocene). The presence of *Cadurcotherium* in the Oligocene Phosphorite beds of Quercy, France, represents the only successful invasion of an amynodontid into Europe.

Although the cranial anatomy of *Cadurcotherium* is poorly known, two cranial features point to the origin of this genus within the Metamynodontini. These are hypertrophy of the zygomatic arches and increased width of the lower jaw. Dental characteristics also support a metamynodontine relationship for *Cadurcotherium.* The cheek teeth of *Cadurcotherium* can easily be derived through exaggeration of trends already seen in forms like *Metamynodon* but not approached by any cadurcodontine.

**Taxonomic Problems**

Various authors have assigned to the Amynodontidae a large number of genera that are based on fragmentary material or inadequately diagnosed. I have synonymized four of these genera: *Orthocynodon* (= *Amynodon*), *Lushiamynodon* (= *Sharamynodon*); *Cadurcopsis* (= *Metamynodon*); and *Cadurcamynodon* (=*Cadurcotherium*). There are, however, eight genera traditionally placed in the Amynodontidae that I cannot place on the family cladogram (Figure 17.3). *Caenolophus* (Matthew and Granger, 1925), *Teilhardia* (Matthew and Granger, 1926), *Toxotherium* (Wood, 1961), *Huananodon* (You, 1977), and *Euryodon* (Xu et al., 1979) are all known from fragmentary material. Radinsky (1969) placed the first three genera in the Amynodontidae on the basis of large canines. Emry (1979) reexamined *Toxotherium* and decided that it is best to classify this genus as Rhinocerotoidea *incertae sedis*. The four remaining genera are discussed by Wall and Manning (1986). They conclude that a thorough review of these taxa will probably result in lumping *Teilhardia* and *Euryodon* in the genus *Caenolophus* ( as Amynodontidae *incertae sedis*) and regarding *Huananodon* as a *nomen dubium*.

*Hypsamynodon* (Gromova, 1954), *Procadurcodon* (Gromova, 1960), and *Paracadurcodon* (Xu, 1966) are problems because of inadequate diagnoses. *Hypsamynodon* is known only from upper molars from the lower or middle Oligocene of Asia. Gromova described this genus as having high-crowned cheek teeth and near confluence of the anterior rib and parastyle on the molars. Gromova did not mention the greatly reduced $M^3$ metaloph, but it is readily apparent in her figure. All of these characteristics are typical of *Cadurcotherium*, but until better material is available, I prefer not to synonymize the two genera. Gromova's (1960) diagnosis of *Procadurcodon* does not differentiate it from any of several amynodontid genera previously described. The only characters

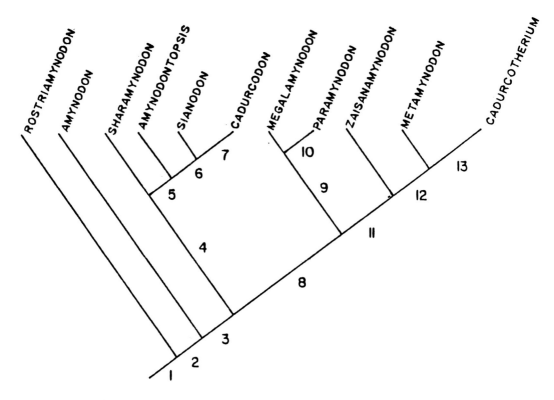

Fig. 17.3. Cladistic relationships of the family Amynodontidae. Character polarities: 1) quadratic $M^3$, loss of upper and lower $P^1$, enlarged canines, elongated talonids, preorbital fossa present; 2) preorbital portion of skull reduced; 3) premaxilla-nasal contact reduced or lost; 4) large preorbital fossa extending medial to orbit; 5) enlarged frontal sinus; 6) reduced number of incisors; 7) reduced number of lower incisors to one pair; 8) preorbital fossa reduced, frontal-maxilla contact present, lower jaw and zygomatic arches thickened, orbit high on skull; 9) zygomatic arches highly convex outward; 10) expanded supraglenoid shelf; 11) reduced diastema, increased size; 12) reduced number of incisors, narrow lower molars; 13) $M^3$ metaloph greatly reduced.

Gromova presented are: moderately high-crowned upper molars; massive canines; $M^3$ metastyle not strongly bent labially; and trigonid and talonid not strongly separated labially. In the absence of any distinguishing characteristics, *Procadurcodon* cannot be clearly defined. Therefore, I believe it should be regarded as Amynodontidae *incertae sedis*. *Paracadurcodon* (Xu, 1966) is based on a description of a lower jaw that is very similar to *Cadurcodon*, however, Xu's photograph is not detailed enough to make an adequate comparison.

## Acknowledgments

For inviting me to participate in this volume, I thank Donald Prothero and Robert Schoch. For useful discussions on amynodontid systematics, I thank Margery Coombs, David Klingener, and Earl Manning. For reviewing the manuscript, I thank Bruce Hanson. For his generosity in allowing me access to specimens at the American Museum, I thank Malcolm McKenna. This research was partially supported by NSF grant DEB-7914783.

## Bibliography

Belyayeva, E.I. (1971): Novye dannye po aminodon tam SSSR. (New data on the amynodonts of the USSR). -*Akad. Nauk. SSSR. Paleo. Tr.* 130: 39-61

Bjork, P.R. (1967): Latest Eocene vertebrates from northwestern South Dakota. -*J. Paleont.*, 41(1): 227-236.

Chow, M., and Y. Xu. (1965): Amynodonts from the upper Eocene of Honan and Shansi. -*Vert. PalAsiatica* 9(2): 190-204

Colbert, E.H. (1938): Fossil mammals from Burma in the American Museum of Natural History. - *Bull. Amer. Mus. Nat. Hist.* 74(6): 255-436.

Emry, R.G. (1979): Review of *Toxotherium* (Perissodactyla: Rhinocerotoidea) with new material from the early Oligocene of Wyoming. -*Proc. Biol. Soc. Wash.*, 92(1): 28-41.

Gervais, J. P. (1873): Sur les fossiles trouves dans les chaux phosphatees du Quercy. -*Bull. Soc. Phil., Paris*, (7) IV: 120-125.

Gregory, W. K. (1920): Studies in comparative myology and osteology, no. IV. A review of the evolution of the lachrymal bone of vertebrates with special reference to that of mammals. -*Bull. Amer. Mus. Nat. Hist.* 42: 95-263.

Gromova, V. (1954): Boltany nosorogi (Amynodontidae) Mongolic. -*Akad Nauk. SSSR . Paleo. Tr.*, 55: 85-189.

Gromova, V. (1960): Pervaya naxodha v sovetshom soyuye Amynodonta. -*Akad. Nauk. SSSR. Paleo. Tr.*, 77: 128-151.

Kretzoi, M. (1942): Auslandische saugetierfossilien der Ungarischen Museum. - *Foldtony, Kozlony.*, 72 (1-3): 139-148.

Li, C., and Ting, S. (1983): The Paleogene mammals of China. - *Bull. Carnegie Mus. Nat. Hist.*, 21: 1-98.

Marsh, O. C. (1877): Notice of some new vertebrate fossils. -*Amer. J. Sci.*, 3rd series,14(81) : 249-256.

Matthew, W. D. (1929): Critical observations upon Siwalik mammals. -*Bull. Amer. Mus. Nat. Hist.*, 56(7): 437-560.

Matthew, W. D., and Granger, W. (1925): New mammals from the Shara Murun Eocene of Mongolia. -*Amer. Mus. Nat. Hist., Novitates*, 196: 1-11.

Matthew, W. D., and Granger, W. (1926): Two new perissodactyls from the Arshanto Eocene of Mongolia. -*Amer. Mus. Nat. Hist., Novitates*, 208: 1-5.

Osborn, H. F. (1889): The Mammalia of the Uinta Formation. Part III, the Perissodactyls. -*Trans. Amer. Phil. Soc.*, 16(3): 505-530.

Osborn, H. F. (1898): The extinct rhinoceroses. -*Mem. Amer. Mus. Nat. Hist.*, 1: 75-164.

Osborn, H. F. (1923): *Cadurcotherium* from Mongolia. -*Amer. Mus. Nat. Hist. Novitates*, 92: 1-2.

Osborn, H. F. (1924): *Cadurcotherium ardynense*, Oligocene, Mongolia. -*Amer. Mus. Nat. Hist. Novitates*, 147:1-4

Osborn, H. F. (1936): *Amynodon mongoliensis* from the upper Eocene of Mongolia. -*Amer. Mus. Nat. Hist. Novitates*, 859: 1-9.

Pilgrim, G. E. (1925): The Perissodactyla of the Eocene of Burma. -*Pal. India.*, 8(3): 1-28.

Pilgrim G. E., and Cotter, G. (1916): Some newly discovered Eocene mammals from Burma. -*Rec. Geol. Surv. India*, 47: 42-77.

Radinsky, L. B. (1966): The families of the Rhinocerotoidea (Mammalia, Perissodactyla). -*J. Mamm.*, 47(4): 631-639.

Radinsky, L. B.(1969): The early evolution of the Perissodactyla. -*Evolution*, 23: 308-323.

Savage, D. E., and Russell, D. E. (1983): *Mammalian Paleofaunas of the World*. Reading, Mass. (Addison-Wesley).

Scott, W. B. (1941): Part 5, Perissodactyla. - In: Scott, W. B. and Jepsen, G. L. (eds.), The mammalian fauna of the White River Oligocene. -*Trans. Amer. Phil. Soc.* 28: 747-980.

Scott, W. B. (1945): The Mammalia of the Duchesne River Oligocene. -*Trans. Amer. Phil. Soc.* 34: 209-252.

Scott, W. B., and Osborn, H. F. (1882): *Orthocynodon*, an animal related to the rhinoceros, from the Bridger Eocene. -*Amer. J. Sci.*, 3(24): 223-225.

Scott, W. B., and Osborn, H. F. (1883): On the skull of the Eocene rhinoceros *Orthocynodon*, and the relation of the genus to other members of the group. - *Contrib. E. M. Mus. Geol. and Arch. Princeton College Bull.*, 3: 1-22.

Scott, W. B., and Osborn, H. F.(1887): Preliminary account of the fossil mammals from the White River Formation contained in the Museum of Comparative Zoology. -*Bull. Mus. Comp. Zool.*, 13:151-171.

Stock, C. (1933): An amynodont skull from the Sespe deposits, California. -*Proc. Nat. Acad. Sci. U. S.*, 19: 762-767.

Stock, C. (1939): Eocene amynodonts from Southern California. -*Proc. Nat. Acad. Sci. U. S.*, 25: 270-275.

Troxell, E. L. (1921): New amynodonts in the Marsh collection. -*Amer. J. Sci.*, 5th series, 2: 21-34.

Wall, W. P. (1980): Cranial evidence for a proboscis in *Cadurcodon* and a review of snout structure in the family Amynodontidae (Perissodactyla, Rhinocerotoidea). -*J. Paleont.*, 54(5): 968-977.

Wall, W. P. (1981): Systematics, phylogeny, and functional morphology of the Amynodontidae (Perissodactyla: Rhinocerotoidea) - Ph.D Thesis, Univ. of Massachusetts, Amherst.

Wall, W. P. (1982a): Evolution and biogeography of the Amynodontidae (Perissodactyla, Rhinocerotoidea. -*3rd. N. Amer. Paleo. Con. Proc.*, 2: 563-567.

Wall, W. P. (1982b): The genus *Amynodon* and its relationship to other members of the Amynodontidae (Perissodactyla, Rhinocerotoidea). -*J. Paleont.*, 56(2): 434-443.

Wall, W. P., and Manning, E. (1986): *Rostriamynodon grangeri* n. gen., n. sp. of amynodontid (Perissodactyla, Rhinocerotoidea) with comments on the phylogenetic history of Eocene Amynodontidae.- *J. Paleont.*, 60(4): 911-919.

Wilson, J. A., and Schiebout, J. A. (1981): Early Tertiary vertebrate faunas, Trans-Pecos Texas: Amynodontidae., -*Texas Mem. Mus. Pearce- Sellards Ser.*, 33: 1-62.

Wood, H. E. (1927): Some early Tertiary rhinoceroses from the Eocene of Mongolia. -*J. Paleont.*, 13: 165-249.

Wood, H. E. (1937): A new, lower Oligocene, amynodont rhinoceros from the Eocene of Mongolia. -*J. Mamm.*, 18(1): 93-94.

Wood, H. E. (1945): Family Amynodontidae. In, Scott, W. B., (ed.): The Mammalia of the Duchesne River Oligocene. -*Trans. Amer. Phil. Soc.*, 34: 209-252.

Wood, H. E. (1961): *Toxotherium hunteri*, a peculiar new Oligocene mammal from Saskatchewan. -*Natural Hist. Papers, Nat. Mus. Canada*, 13: 1-3.

Xu, Y. (1965): A new genus of amynodont from the Eocene of Lantien, Shansi.-*Vert. PalAsiatica*, 9(1): 83-88.

Xu, Y. (1966): Amynodonts of Inner Mongolia. -*Vert. PalAsiatica*, 10(2): 123-190.

Xu, Y., Yan, D., Zhou, S., Han, S., Zhang, Y. (1979): The subdivision of the Red Beds of South China. -In, *The Mesozoic and Cenozoic Red Beds of South China*. Beijing (Science Press), pp. 416-432.

You, Y. (1977): Note on the new genus of early Tertiary Rhinocerotidae from Bose, Guanxi. -*Vert. PalAsiatica*, 15(1): 46-53.

# 18. THE ALLACEROPINE HYRACODONTS

## KURT HEISSIG

The genus *Allacerops* Wood (1932) is a synonym of *Eggysodon* Roman, 1912, but this synonymy does not change the subfamilial name Allaceropinae Wood, 1932. The genera *Eggysodon*, *Prohyracodon* Koch (1897) (=*Meninatherium*, Abel, 1910) and *Ilianodon* Chow and Xu (1961) are included here. The group is characterized by an enlarged canine, only two pairs of incisors in the mandible and somewhat elongated lower premolars. The limbs are tridactyl and less cursorial than in hyracodontines.

### Taxonomic history

*Eggysodon*, *Prohyracodon*, and *Ilianodon* were first considered true rhinoceroses. Stehlin (1930) discussed a symphysis of *Eggysodon*, which was already figured by Osborn (1900), pointing to the differences from the American *Trigonias*. Because of the similarity of the cheek teeth he preferred to retain the genus within the true rhinoceroses, but separated on a subfamilial level. Radinsky (1966) restricted the Rhinocerotidae to the genera with the shearing specialization of $I^1/_2$ and their descendants, referring *Eggysodon* to the Hyracodontidae.

Wood (1932) erected the new genus *Allacerops* on the basis of the eastern form *A. turgaica* (Borissiak, 1915). The confused status of taxonomy of the European forms prevented him from recognizing the valid status of *Eggysodon*, the type species of which he included in his new genus. His subfamily Allaceropinae is based on the same characters used here.

Koch's (1897) name *Prohyracodon* for a small animal from the supposed middle Eocene of Hungary reflects mainly the primitive stage of molarization, the low-crowned teeth and the small size, comparable to the true hyracodonts. Wood (1929), without knowledge of the front teeth, concluded from the fully reduced $M^3$ metastyle that the genus should be placed in the true rhinoceroses.

In 1983 I studied the second specimen of *Meninatherium telleri* Abel in Graz, which consists of a skull fragment and the mandible. The similarity of the upper cheek teeth in size, molarization and $M^3$ structure to *Prohyracodon* was obvious. The only question remaining was the supposed Aquitanian age of its locality, Möttnig. This age, suggested only by some mollusks from the neighboring basin, should be changed to upper Eocene, confirmed by the recent discovery of *Anthracohyus* Pilgrim and Cotter, known from the Upper Eocene of Burma (Heissig, in press). The mandible shows the root of an erect canine and the alveoli of only two pairs of incisors, the second being considerably smaller than the first one. So the front teeth of *Prohyracodon* are known and the genus must be included in the Hyracodontidae again.

Another striking character of this genus, the elongation of the premolars, is present also in *Ilianodon* Chow and Xu from China and some other questionable Asiatic forms. This genus was also thought to be a rhinocerotid, the canine being mistaken as an incisor.

### Taxonomy

Family Hyracodontidae Cope, 1879
Subfamily Allaceropinae Wood, 1932

Diagnosis: Hyracodontidae with strong canine and only two incisor pairs in the mandible. Upper $M^3$ without metastyle, upper premolars slight to fully molarized. Lower premolars lengthened, $P_1$ one- or two-rooted. Manus tridactyl.

*Eggysodon* Roman, 1912
   =*Allacerops* Wood, 1932
Diagnosis (revised): Allaceropinae of medium size, equal to or larger than *Hyracodon*. Upper premolars sub-, semi-, or molariform, with weak or no metacone ribs. $P_1$ single-rooted or lost.
Type species: *E. osborni* (Schlosser, 1902)
Referred species: *E. gaudryi* (Rames, 1886), *E. pomeli* (Roman, 1912), *E. reichenaui* (Deninger, 1903) and *E. turgaicum* (Borissiak, 1915).
Age: Oligocene.

*Prohyracodon* Koch, 1897
   =*Meninatherium* Abel, 1910
Diagnosis: Allaceropinae of small size with rather narrow symphysis. Upper premolars submolariform with medium to weak metacone ribs. $P_1$ double-rooted.
Type species: *P. orientale* Koch, 1897.
Referred species: *P. meridionale* Chow and Xu, 1961, *P. telleri* Abel, 1910.
Age: Middle? to upper Eocene.

*Ilianodon* Chow and Xu, 1961
Diagnosis (revised): Allaceropinae of medium size, but with double-rooted $P_1$. Diastema extremely short.

## Discussion
The status of the different species of *Eggysodon*, as well as *Prohyracodon*, awaits revision. In *Eggysodon*, most species are of similar size, except *E. gaudryi* and *E. turgaicum*, which are somewhat larger. Most specimens in Europe have been found in the fissures of the Quercy Plateau in France, without fixing their stratigraphic position. So the changes of the genus within the Oligocene are not yet known.

The type of *Prohyracodon* disappeared somewhere in the Balkan Penninsula. It could not be discovered in Hungarian collections. In Romania difficulties with cooperation until recently prevented an intensive search. Only one tooth, one of the last upper molars, presented to Wood, is now stored in in the American Musum, representing the type specimen. It is identical in all details with the corresponding tooth of *P. telleri*. The type of this species was also lost in the war, and the second specimen in Graz (Joanneum) should be treated as a neotype.

Both specimens from China, referred to *Prohyracodon*, are nearly of the same size and very similar to the type species.

## Phylogenetic position
The large canine, a plesiomorphic character among Rhinocerotoidea, is further enlarged in later *Eggysodon*. It is not a sign of relationship with *Forstercooperia* and the Indricotheriidae, because of the tridactyl manus observed in *Eggysodon*. The reduction of the third incisor in the mandible, the only clear autapomorphy of the subfamily, already occurs in *Prohyracodon*.

None of the different hyracodonts of Asia show intermediate conditions between Hyracodontinae and Allaceropinae. All have three well-developed lower incisors and a smaller canine. Sometimes there is an elongation of the lower premolars, but it is combined with a long diastema, whereas it is short in the Allaceropinae.

The separation of both subfamilies occurred before both appeared in the upper Eocene. On the other hand, forms bridging the gap between *Prohyracodon* and *Eggysodon* have not yet been found. This question may be answered by paleontological research in China.

## Bibliography
Abel, O. (1910): Kritische Untersuchungen über die paläogenen Rhinocerotiden Europas. -*Abh.k.k. Reisanst.*, 20 (3): 1-52.
Borissiak, A. A. (1915): Ob ostathakh *Epiaceratherium turgaicum* n. sp. -*Bull. Acad. Imp. Sci.*, 1 (3): 781-787.
Chow, M., and Xu, Y. (1961): New primitive

true rhinoceroses from the Eocene of Iliang (Chinese, English summary). -*Vert. PalAsiatica*, 5 (4): 291-304.

Cope, E. D. (1879): On the extinct species of Rhinoceridae of North America and their allies. -*Bull. U.S. Geol. Geogr. Surv. Territ.*, 5(2): 227-237.

Deninger, K. (1903): *Ronzotherium reichenaui* aus dem Oligocän von Weinheim bei Alzey. -*Zeit. deutsch. geol. Ges.*, 55 Abh.: 93-97.

Koch, A. (1897): *Prohyracodon orientalis*, ein neues Ursäugethier aus den mitteleocänen Schichten Siebenbürgens. -*Természetr. Füzetek*, 1897: 481-500.

Osborn, H. F. (1900): Phylogeny of the rhinoceroses of Europe. -*Bull. Amer. Mus. Nat. Hist.*, 13 (19): 229-267.

Pilgrim, G.E., and Cotter, G.P (1916): Some newly discovered Eocene mammals from Burma. -*Rec. Geol. Surv. India*, 47: 42-77.

Radinsky, L. (1966): The families of Rhinocerotoidea (Mammalia, Perissodactyla). -*J. Mamm.*, 47: 631-639.

Rames, J.B. (1886): Note sur l'âge des argiles du Cantal et sur les débris fossiles qu'elles ont fournis. -*Bull. Soc. géol. France* (3) 14: 357-360.

Roman, F. (1912): Les rhinocéridés de l'Oligocène d'Europe. -*Arch. Mus. Sci. Nat. Lyon*, 11 (2): 1-92.

Schlosser, M. (1902): Beiträge zur Kenntnis der Säugethierreste aus den süddeutschen Bohnerzen. -*Geol. Pal. Abh. N. S.*, 5 (9): 117-258.

Stehlin, H. G. (1930): Bemerkungen zur Vordergebiss-formel der Rhinocerotiden. -*Eclog. geol. Hel.*, 23 (2): 644-648.

Wood, H.E. II (1929): *Prohyracodon orientale* Koch, the oldest known true rhinoceros. -*Amer. Mus. Novitates*, 395: 1-7.

Wood, H.E. II (1932): Status of *Epiaceratherium* (Rhinocerotidae). -*J. Mamm.*, 13 (2): 169-171.

# 19. THE SYSTEMATICS OF INDRICOTHERES

## SPENCER G. LUCAS and JAY C. SOBUS

Indricotheriinae Borissiak, 1923, is a subfamily of hyracodontid rhinocerotoids known from the middle-late Eocene of western North America and the middle Eocene - early Miocene of Eurasia. There are four valid genera of indricotheres: *Forstercooperia* Wood, 1939 (= *Pappaceras* Wood, 1963); *Juxia* Chow and Chiu, 1964 (= *Imequincisoria* Wang, 1976); *Urtinotherium* Chow and Chiu, 1963; and *Paraceratherium* Forster Cooper, 1911 (= *Baluchitherium* Forster Cooper, 1913b: = *Indricotherium* Borissiak, 1923; = *Aralotherium* Borissiak, 1939; = *Dzungariotherium* Chiu, 1973). *Benaratherium* Gabuniya, 1955, from the Oligocene of Soviet Georgia probably is an indricothere, but is so poorly known that we consider it a *nomen dubium*. Indricothere monophyly is corroborated by the possession of a functional complex involving the snout and anterior dentition that is an evolutionary novelty not found in other hyracodontids.

## Introduction

The indricotheres are a subfamily of large hyracodontid rhinocerotoids that include the largest land mammals. Indricothere fossils are found in middle Eocene-lower Miocene deposits in Asia, western North America and eastern Europe (Fig. 19.1; Table 19.1). Understanding of indricothere evolution has been clouded by typological taxonomy and the failure of any single worker since Gromova (1959) to review the entire subfamily. In this paper we review the taxonomy (primarily at the generic level) of the indricotheres and present a cladistic hypothesis of their phylogenetic relationships. Here, AMNH = American Museum of Natural History, New York; ANPIN = Paleontological Institute, Academy of Sciences of the USSR, Moscow; IVPP = Institute of Vertebrate Paleontology and Paleoanthropology, Beijing; L = maximum length of a tooth crown; and W = maximum width of a tooth crown. The terms nonmolariform, submolariform, and molariform when referring to upper premolars are used in the sense of Radinsky (1967, p.5).

## Taxonomy

Family Hyracodontidae Cope, 1879
Subfamily Indricotheriinae Borissiak, 1923

1923 Indricotheriinae Borissiak, p. 123.
1923 Baluchitheriinae Osborn, p. 13.
1923 Paraceratheriinae Osborn, p. 13.
1939 Indricotheriidae: Borissiak, p. 271.
1945 Paraceratheriinae: Simpson, p. 142.
1959 Indricotheriidae: Gromova, p. 12.
1981 Indricotheriinae: Lucas, Schoch and Manning, pp. 827, 839.

Included genera: *Forstercooperia* Wood, 1939; *Juxia* Chow and Chiu, 1964; *Urtinotherium* Chow and Chiu, 1963; *Paraceratherium* Forster Cooper, 1911 (for generic synonymies, see the following).

Distribution: Middle-late Eocene of the western United States and middle Eocene - early Miocene of Eurasia (Fig. 19.1; Table 19.1).

Revised diagnosis: Indricotheres are large ($LM_{1-3}$ ranges from 71 to 252 mm) hyracodontid rhinocerotoids distinguished from other hyracodontids by their larger size and derived structure of the snout and anterior dentition (incisors and canines). In the most primitive indricothere, *Forstercooperia*, the cranial rostrum is elongate, and the nasals form the superior margin of a

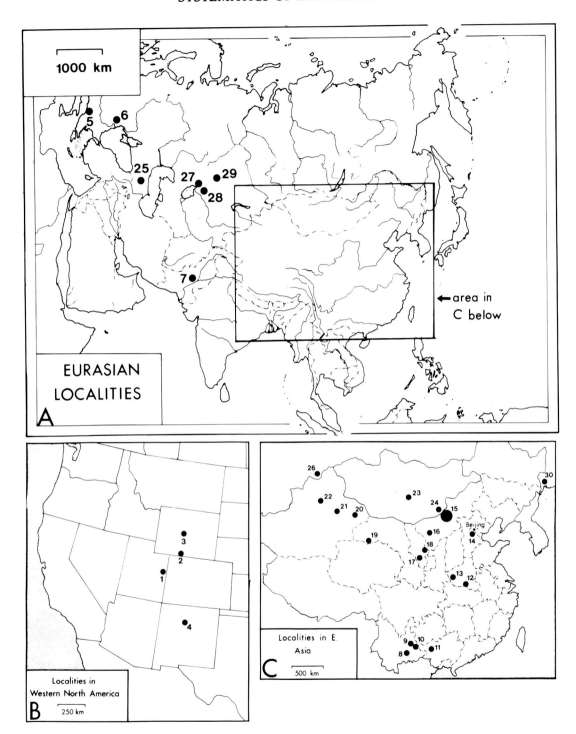

Fig. 19.1. Distribution of indricothere fossils in Eurasia (A), the western United States (B), and eastern Asia (C). See Table 19.1 for data corresponding to the numbered localities.

Table 19.1. Distributional data for the indricotheres (see Fig. 19.1).
Abbreviations are: E = early, F = formation, Gp. = Group, L = late, M = middle.

| Locality (Fig. 19.1) | Location | Unit | Age | Taxa | Source |
|---|---|---|---|---|---|
| United States: | | | | | |
| 1 | Uinta Basin, Utah | Uinta F. | M. Eocene | *F. grandis* | Peterson (1919); Radinsky (1967) |
| 2 | Washakie Basin, Wyoming | Washakie F. | M. Eocene | *F. grandis* | Radinsky (1967) |
| 3 | Absaroka Range, Wyoming | Tepee Trail F. | M. Eocene | *F. minuta*; *F.* cf. *grandis* | Eaton (1985) |
| 4 | Galisteo Basin, New Mexico | Galisteo F. | L. Eocene | *F. minuta* | Lucas (1982) |
| Eastern Europe: | | | | | |
| 5 | Ivangrad, Yugoslavia | ? | ? | *P.* sp. | Petronijevic and Thenius (1957) |
| 6 | Tyurya, Kluj, Romania | ? | ? | *P.* sp. | Gabunia and Iliesku (1960) |
| Pakistan: | | | | | |
| 7 | Bugti Hills, Baluchistan | Chitarwata F. | E. Miocene | *P. bugtiense* | Forster Cooper (1911, 1913a, b, 1923, 1924) |
| People's Republic of China: | | | | | |
| 8 | Lunnan Basin, Yunnan (Lumeiyi-Lunnan area) | Lumeiyi F. | M. Eocene | *F.* sp. | Zhang *et al.* (1978); Zheng *et al.* (1978) |
| 8 | Lunnan Basin, Yunnan (Anyencun-Xiashahe area) | Anyencun F. | M. Eocene-M. Oligocene | *F. totadentata*, *J.* sp., *U.* sp | Chow (1958); Chow *et al.* (1974); Xu and Chiu (1962); Zhang *et al.* (1978) |
| 9 | Qujing Basin, Yunnan | Cajiachong F. | E. Oligocene | *U.* sp. | Tang (1978); Xu (1961) |
| 10 | Luoping Basin, Yunnan | ? | ? | *P.* sp. | Chiu (1962) |
| 11 | Bose Basin, Guangxi | Dongjun F. | M. Eocene | *F.* sp. | Ding *et al.* (1977) |
| 11 | Bose Basin, Guangxi | Gungkang F. | E. Oligocene | Indricotheriinae | Tang *et al.* (1974) |
| 12 | Wucheng Basin, Henan | Lishigou F. | M. Eocene | *F. grandis* | Wang (1976) |
| 12 | Wucheng Basin, Henan | Wulidui F. | L. Eocene | *J. borissiaki* | Wang (1976); Gao (1976) |
| 13 | Lushi Basin, Henan | Lushi F. | M. Eocene | *F. grandis* | Wang (1976) |
| 14 | Changxindian | Changxindian F. | M. Eocene | *F.* sp. | Zhai (1977) |
| 15 | Irdin Manha, Nei Monggol | Arshanto F. | M. Eocene | *F. grandis* | Qi (1979) |
| 15 | Irdin Manha, Nei Monggol | Irdin Manha F. | M. Eocene | *F. totadentata* | Wood (1938) |
| 15 | Huhebolhe Cliff, Nei Monggol | "Irdin Manha" F. | M. Eocene | *F. grandis* | Wood (1963) |
| 15 | Shara Murun, Nei Monggol | "Irdin Manha" F. | M. Eocene | *F. minuta* | Radinsky (1967) |
| 15 | Ula Usu, Baron Sog Mesa, Nei Monggol | Shara Murun F. | L. Eocene | *J. borissiaki* | Chow and Chiu (1964) |
| 15 | Urtyn Obo, Jhama Obo, Nei Monggol | Ulan Gochu/ Urtyn Obo F. | E. Oligocene | *U. incisivum* | Chow and Chiu (1963) |
| 15 | Urtyn Obo, Jhama Obo, Camp Margetts, Nom Khong Obo, Nei Monggol | Houldjin F. | M.-L. Oligocene | *P. transouralicum* | Osborn (1923); Granger and Gregory (1936) |
| 16 | Dengkou (St. Jaques), Nei Monggol | ? | M. Oligocene | *P. transouralicum* | Teilhard de Chardin (1926) |

| | | | | | |
|---|---|---|---|---|---|
| 17 | Tongxin-Guyuan, Ningxia | ? | ? | *P.* sp. | Hu (1962) |
| 18 | Lingwu Basin, Ningxia | Qingshuiying F. | M. Oligocene | *P. transouralicum* | Young and Chow (1956) |
| 19 | Shargaltien Gol, Gansu | ? | ? | *P.* sp. | Bohlin (1946) |
| 20 | Hami Basin, Xinjiang | Yemaquan F. | Oligocene | *P.* sp. | Chow and Xu (1959) |
| 21 | Turpan Basin, Xinjiang | Taoshuyuanzi Gp. (upper part) | M.-L. Oligocene | *P. orgosensis* | Chiu (1962); Xu and Wang (1978) |
| 22 | Junggur Basin, Xingiang | Hesse F. | M.-L. Oligocene | *P. orgosensis* | Chiu (1973) |
| Mongolian People's Republic: | | | | | |
| 23 | Hsanda Gol | Hsanda Gol F. | M. Oligocene | *P. transouralicum* | Osborn (1923); Mellett (1968) |
| 24 | Ergilin-Dzo | Ergilin-Dzo F. | E. Oligocene | *J. borissiaki* | Gabunia and Dashzeveg (1974) |
| Soviet Union: | | | | | |
| 25 | Benara, Georgian S.S.R. | Benara F. | M.-L. Oligocene | "*Benaratherium callistrati*" | Gabunia (1955, 1964) |
| 26 | Zaissan, Kazakhstan | Obailan F. | L. Eocene | *F.* sp. | Gabunia (1977) |
| 27 | Agispe-Petrovskovo | ? | L. Oligocene- E. Miocene | *P. prohorovi* | Gromova (1959) |
| 28 | Chelkar-tenis | Chiliktinskaya F. | L. Oligocene | *P. transouralicum* | Gromova (1959) |
| 29 | Turgai | ? | L. Oligocene | *P. transouralicum* | Gromova (1959) |
| 30 | Artëm | ? | L. Eocene | *J. borissiaki* | Beliajeva (1959) |

fossa that begins above the $C^1$, terminates just anterior to the orbit and arguably was the site of origin of a maxillo-labialis musculature much more developed than in non-indricotheriine hyracodontids. In the most advanced indricothere, *Paraceratherium*, this functional complex culminated in the enlarged $I^1/_1$, loss of $I^{2-3}/_{2-3}$, $C^1/_1$ and $dP^1/_1$, retraction of the nasal incision to above $P^4$ and downturning of the premaxillaries.

Discussion: Borissiak's (1923, p. 123) subfamily name Indricotheriinae has priority over Osborn's (1923, p. 13) names Baluchitheriinae and Paraceratheriinae (imprint date on Borissiak, 1923 is March; on Osborn, 1923, it is May). Thus, Simpson's (1945, p. 142) use of Paraceratheriinae, and the use of Baluchitheriinae by several authors, cannot be upheld. Also, which subfamily name should be used is not now affected by the taxonomic status of *Indricotherium*, which we consider to be a junior subjective synonym of *Paraceratherium* (see Ride *et al.* 1985, Articles 23 and 40). Indeed, Indricotheriinae (or Indricotheriidae) has won general acceptance among those paleontologists who have studied the group (Borissiak, 1939, p. 271; Gromova, 1959, p. 12; Radinsky, 1966, p. 635, 1967; Lucas *et al.*, 1981, p. 839).

Radinsky (1966) first advocated inclusion of the indricotheres in the Hyracodontidae as a subfamily, an allocation upheld in Lucas *et al.* (1981) and in this paper (see later discussion and Fig. 19.8).

Genus *Forstercooperia* Wood, 1939

1919 *Hyrachyus*: Peterson, p. 130, Figs. 18-19.
1938 *Cooperia* Wood, p. 1, Figs. 1-4, 5B.
1939 *Forstercooperia* Wood [dated addendum to Wood, 1938].
1963 *Pappaceras* Wood, p. 2, Figs. 1-2.

Type species: *Forstercooperia totadentata* (Wood, 1938) (=*F. shiwopuensis* Chow, Chang and Ting, 1974).

Included species: the type species, *F. grandis* (Peterson, 1919) (=*Pappaceras confluens* Wood, 1963) and *F. minuta* Lucas, Schoch, and Manning, 1981.

Distribution: Middle to late Eocene of the western United States and Asia (Fig. 19.1; Table 19.1).

Revised diagnosis: *Forstercooperia* is a small indricothere ($LM_{1-3}$=71-140 mm) distinguished from the other indricotheres by the following combination of features: small size; nasal incision above $C^1$; premaxillaries horizontal (not downturned); all incisors present and of essentially equal size; canines and first premolars present, upper premolars usually nonmolariform; and $M^3$ metacone usually present but small.

Discussion: Radinsky (1967) and Lucas *et al.* (1981) justified assignment of *H. grande* to *Forstercooperia* and recognized that *Pappaceras* should be considered a synonym of *Forstercooperia*. On the whole, *Forstercooperia* is smaller than *Juxia* (Figs. 19.2, 19.5) and is readily distinguished from it by features of the snout and anterior dentition (see diagnoses of *Forstercooperia* and *Juxia*).

Lucas *et al.* (1981) revised the species-level taxonomy of *Forstercooperia*, and since that paper appeared, no new material has been discovered that alters their taxonomic conclusions. The three species of *Forstercooperia* recognized by Lucas *et al.* (1981) are primarily distinguished by size differences (smallest to largest: *F. minuta, F. grandis, F. totadentata*) and different degrees of reduction of the $M^3$ metacone (in *F. minuta* a small metacone is present, in *F. grandis* it is smaller or absent, and the $M^3$ is unknown in *F. totadentata*).

Genus *Juxia* Chow and Chiu, 1964

1959 *Eotrigonias*: Beliajeva, p. 83, Figs. 1-4.
1964 *Juxia* Chow and Chiu, p. 264, pl. 1.
1967 *Forstercooperia*: Radinsky, p. 18, Figs. 8D, 9C, 10.
1974 *Juxia*: Chow, Chang, and Ting, p. 268.

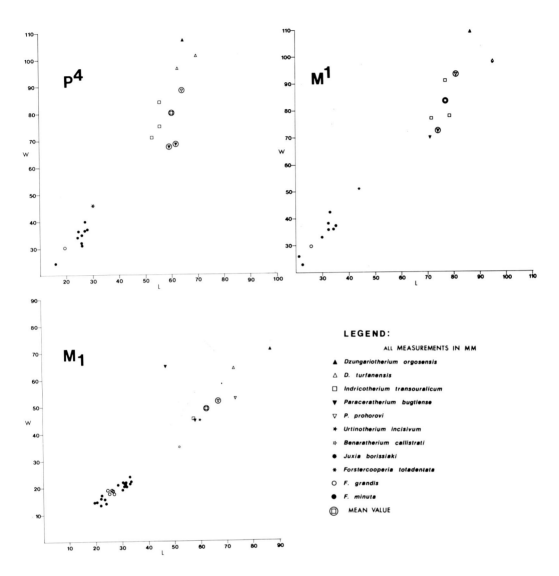

Fig. 19.2. Bivariate plots of selected cheek-tooth data for indricotheres. Data, in part from Gromova (1959) and Lucas *et al.* (1981).

1976 *Imequincisoria*: Wang, p. 104, pls. 1-3.
1981 *Juxia*: Lucas, Schoch, and Manning, p. 828.

Type species: *J. sharamurunense* Chow and Chiu, 1964.

Included species: *J. borissiaki* (Beliajeva, 1959) (= *J. sharamurunense* Chow and Chiu, 1964; = *F. ergiliinensis* Gabuniya and Dashzeveg, 1974).

Distribution: Late Eocene and early Oligocene of Asia (Fig. 19.1; Table 19.1).

Revised diagnosis: *Juxia* is a small indricotheriine ($LM_{1-3}$= 98-116 mm) distinguished from the other indricotheres by the following combination of features: small size; nasal incision above $P^3$; premaxillaries horizontal (not downturned); all incisors present but $I^1/_1$ somewhat larger than the other incisors; canines and first premolars present; upper premolars submolariform; and $M^3$ metacone absent.

Discussion: Radinsky (1967) considered *Juxia* to be a junior subjective synonym of *Forstercooperia*. However, as Chow et al. (1974) and Lucas et al. (1981) argued, the two genera are readily distinguished on the basis of their snout and anterior dental morphology.

Wang (1976) named the new genus *Imequincisoria* (type species = *I. mazhuangensis* Wang, 1976; other included species = *I. micrasis* Wang, 1976) for specimens from the upper Eocene Wulidui Formation in the Wucheng basin, Henan, China. Wang's (1976, p. 109) measurements of the cheek teeth of *I. mazhuangensis* and *I. micrasis* not only fail to distinguish these species from each other, but they fail to differentiate *Imequincisoria* from *Juxia borissiaki* (Fig. 19.2). Furthermore, we are unable to find any meristic differences that justify taxonomic separation, even at the species level, between the two putative species of *Imequincisoria* and between them and *J. borissiaki*. The following features are particularly significant: $I^1/_1$ of *Imequincisoria* are enlarged as in the holotype of *Juxia sharamurunense*, and the incisors of both taxa are essentially equally spaced (compare Fig. 19.3 with Chow and Chiu, 1964, pl. 1, Figs. 1, 3); $P^{2-4}$ of *Imequincisoria* are submolariform as they are in *Juxia* (compare Fig. 19.4A-C with Fig. 19.5D and Chow and Chiu, 1964, pl. 1, Fig. 1); and the $M^3$ metacones of *Imequincisoria* and *Juxia* are equally reduced (compare Fig. 19.4A-C with Chow and Chiu, 1964, pl. 1, Fig. 1). In view of these observations, we consider *Imequincisoria* Wang, 1976, to be a junior subjective synonym of *Juxia* Chow and Chiu, 1964, and *I. mazhuangensis* and *I. micracis* to be synonyms of *J. borissiaki*.

Genus *Urtinotherium* Chow and Chiu, 1963

1958 *Indricotherium*: Chow, p. 264, pl. 1., Figs. 3-4.
1963 *Urtinotherium* Chow and Chiu, p. 230, Fig. 1, pls. 1-2.
1978 *Indricotherium*: Tang, p. 76, pl. 9, Figs. 1-3.

Type and only known species: *U. incisivum* Chow and Chiu, 1963.

Distribution: Early Oligocene of Asia (Fig. 19.1; Table 19.1).

Revised diagnosis: *Urtinotherium* is a large indricotheriine ($LM_{1-3}$ = 209 mm) distinguished from other indricotheres by the following combination of features: large size, $I_1$ much larger than $I_{2-3}$ and procumbent; canines and first premolars present but canines very small; and upper premolars submolariform.

Discussion: *Urtinotherium* is known only from the lower jaw and a few isolated upper cheek-teeth, so the depth of its nasal incision, orientation of the premaxillaries and development of its $M^3$ metacone are unknown. Thus, reference to these structures is omitted from the diagnosis.

Chow (1958) named *I. parvum* for isolated teeth ($P^3$ and $M^2$) from the Lunan Basin, Yunnan, that are somewhat smaller than specimens of *Indricotherium*, although they fall into the size range of *Paraceratherium bugtiense* (compare measurements

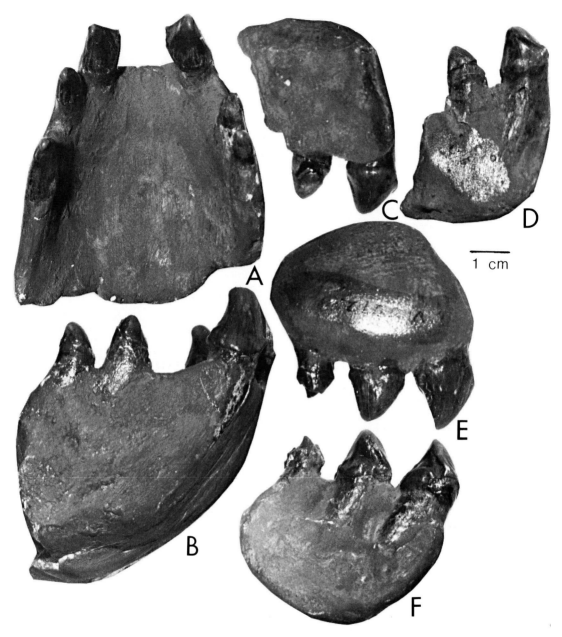

Fig. 19.3. Incisors of "*Imequincisoria*." A-B, "*I. mazhuangensis*," IVPP V 5072, left and right $I_{1-3}$, occlusal (A) and right lateral (B) views. C-D, "*I. micracis*," IVPP V 5073, right $I^{1-2}$ labial (C) and lingual (D) views. E-F, "*I. mazhuangensis*," IVPP V 5072, right $I^{1-3}$, labial (E) and lingual (F) views.

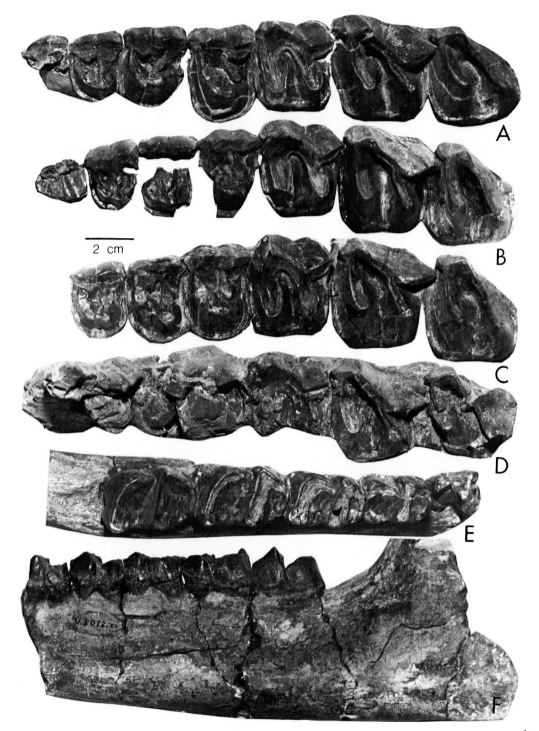

Fig. 19.4. Cheek teeth of "*Imequincisoria*." A, "*I. mazhuangensis*," IVPP V 5072, left $P^1$-$M^3$, occlusal view. B, "*I. micracis*," IVPP V 5073, right $P^1$-$M^3$, occlusal view (photograph reversed). C, "*I. micracis*," IVPP V 5073, left $P^2$-$M^3$, occlusal view. D, "*I. mazhuangensis*," IVPP V 5072, right $P^1$-$M^3$, occlusal view (photograph reversed). E-F, "*I. mazhuangensis*," IVPP V 5072, left dentary fragment with $P_2$-$M_3$, occlusal (E) and labial (F) views.

in Chow, 1958, with Gromova, 1959, Table 5). Similarly, *I. qujingensis* Tang, 1978, from the Qujing basin, Yunnan, is based on isolated upper cheek teeth that differ little in size from those of *I. parvum*. The type specimens of these two taxa are derived from lower Oligocene strata, and their size is what would be expected for *Urtinotherium*, which otherwise is known only from the lower dentition. Therefore, we tentatively asign *I. parvum* and *I. qujingensis* to *Urtinotherium*, believe thay represent the same species, but identify them only as *Urtinotherium* sp., considering them *nomina dubia* at the specific level.

Gabuniya (1955) named *Benaratherium callistrati* and subsequently described and illustrated this taxon in detail (Gabuniya, 1964, pp. 83-106, Figs. 43-53, pls. 3-7). The holotype of *B. callistrati* is a right dentary fragment with $P_2$-$M_3$ (Fig. 6D; Gabuniya, 1964, Figs. 43, 46; pl. 4, Figs. 3-4), and the taxon is otherwise known from a left $M^3$ (Fig. 19.5E), other isolated molars and assorted postcrania, including a metapodial. The length and slenderness of this metapodial and relatively large size ($LM_{1-3}$ = 169 mm) suggest that *Benaratherium* is an indricothere (assuming Gabuniya's association of the metapodial with the dentition in a single taxon is correct). In *Benaratherium*, the $M^3$ metacone is totally suppressed, degree of molarization of the lower premolars is comparable to that of *Urtinotherium*, and size is slightly smaller (about 15%) than known specimens of *Urtinotherium*. However, without the anterior dentition or cranium of *Benaratherium* its validity and relationships with other indricotheres cannot be evaluated. Therefore, pending discovery of additional material of this taxon, we consider *Benaratherium* Gabuniya, 1955 a *nomen dubium*.

Genus *Paraceratherium* Forster Cooper, 1911

1908 *Aceratherium*: Pilgrim, p. 156.
1910 *Aceratherium*: Pilgrim, p. 65.
1911 *Paraceratherium* Forster Cooper, p. 711. pl. 10.
1913a *Thaumastotherium* Forster Cooper, p. 376, Figs. 1-11.
1913b *Baluchitherium* Forster Cooper, p. 504 [proposed as replacement name for *Thaumastotherium* Forster Cooper, 1913; preoccupied by *Thaumastotherium* Kirkaldy, 1908].
1915 *Indricotherium* Borissiak, p. 131, Figs. 1-2.
1939 *Aralotherium* Borissiak, p. 271, Fig. 1.
1973 *Dzungariotherium* Chiu, p. 182, Figs. 1-3, pls. 1-3.

Type species: *P. bugtiense* (Pilgrim, 1908) (="*Baluchitherium*" *osborni* Forster Cooper, 1913a).

Included species: The type species and *P. transouralicum* (Pavlova, 1922) (= *I. asiaticum* Borissiak, 1923; = *I. minus* Borissiak, 1923; = *B. grangeri* Osborn, 1923), *P. prohorovi* (Borissiak, 1939) and *P. orgosensis* (Chiu, 1973) (= *D. turfanensis* Xu and Wang, 1978; = *P. lipidus* Xu and Wang, 1978).

Distribution: Middle Oligocene-early Miocene of Eurasia (Fig. 19.1; Table 19.1).

Revised diagnosis: *Paraceratherium* is a large indricotheriine ($LM_{1-3}$ = 222-252 mm) distinguished from other indricotheres by the following combination of features: large size; nasal incision above $P^4$; premaxillaries downturned; $I^1/_1$ very large and procumbent; $I^{2-3}/_{2-3}$, $C^1$-$C_1$ and $dP^1/_1$ absent; upper premolars submolariform; and $M^3$ metacone absent.

Discussion: Recognition that *Baluchitherium* Forster Cooper, 1913, and *Aralotherium* Borissiak, 1939, are junior subjective synonyms of *Paraceratherium* Forster Cooper, 1911, was already justified by Gromova (1959), and needs little further comment. That Forster Cooper (1923) thought *Baluchitherium* probably is a synonym of *Indricotherium* well reflects the difficulty some authors have had in justifying separation of the genera *Parac*

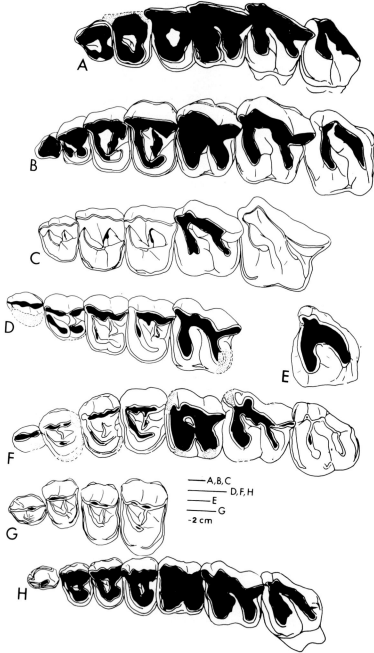

Fig. 19.5. Upper cheek teeth of selected indricotheres. A) *Paraceratherium bugtiense*, left $P^2$-$M^3$ (ater Forster Cooper, 1924, Fig. 11). B, "*Baluchitherium grangeri*," left $P^1$-$M^3$, holotype (after Granger and Gregory, 1936, Fig. 2B). C, "*Indricotherium asiaticum*," left $P^2$-$M^2$ (after Granger and Gregory, 1936, Fig. 2D). D, *Juxia* "*sharamurunense*," left $P^1$-$M^1$ (from Radinsky, 1967, fig. 8D). E, "*Benaratherium callistrati*," left $M^3$ (from Gabuniya, 1964, Fig. 45). F, *Forstercooperia grandis*, left $P^1$-$M^3$ (from Radinsky, 1967, Fig. 11). G, *F. totadentata*, left $P^{1-4}$, holotype (after Radinsky, 1967, Fig. 8A). H. *F.* "*confluens*," left $P^1$-$M^3$, holotype (after Radinsky, 1967, Fig. 8B).

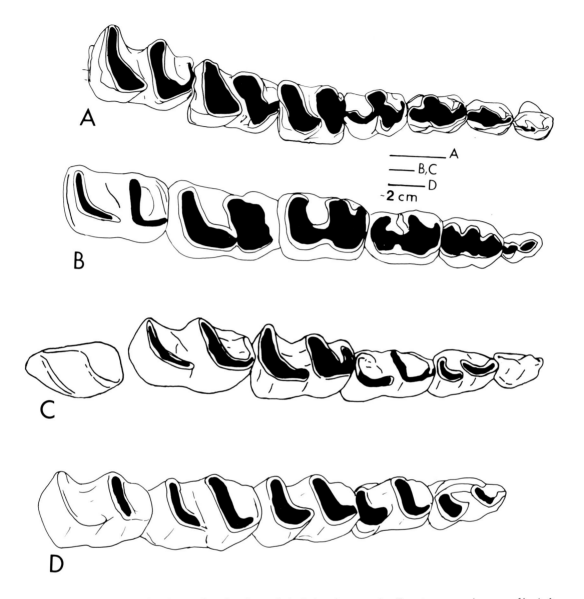

Fig. 19.6. Lower cheek teeth of selected indricotheres. A, *Forstercooperia grandis* (after Radinsky, 1967, Fig. 8C). B, "*Baluchitherium grangeri*" after Granger and Gregory, 1936, Fig. 4A). C, *Paraceratherium prohorovi* (after Gromova, 1959, Fig. 3B) (reversed). D, "*Benaratherium callistrati*" (after Gabuniya, 1964, Fig. 46).

*eratherium, Baluchitherium,* and *Indricotherium.*

Granger and Gregory (1936, pp. 54-62) well argued the case for recognizing only one of these three genera as valid:

> As to generic relationships, our material indicates that both *Baluchitherium* and *Indricotherium* are close to or even synonymous with *Paraceratherium*, although possibly representing slightly different species. In the first place, the type upper molars of *Paraceratherium bugtiense* Pilgrim exhibit no conspicuous differences from those of *Baluchitherium osborni;* secondly, the lower jaw referred to *Paraceratherium bugtiense* by Forster Cooper seems to us to be indistinguishable in generic characters from one of our jaws ([AMNH] No. 26166) that is associated with humerus, radius, ulna and metacarpal III of the general size and characters of Borissiak's *Indricotherium;* thirdly, the cast of the skull referred by Forster Cooper to skulls of *Paraceratherium* reveals essential similarities at all points to our large skulls of *Baluchitherium grangeri* Osborn. . . . Fourthly, the peculiar lower front teeth of *Paraceratherium* are matched precisely in *Baluchitherium osborni* and in Borissiak's *Indricotherium.* Fifthly, we have numerous fully adult limb bones, astragali and metapodials that collectively comprise a closely graded series . . . from the small *Paraceratherium* through *Baluchitherium osborni* to *B. grangeri* and finally to a super-*Indricotherium.* On the other hand, Borissiak has pointed out that in Forster Cooper's *Paraceratherium* the protoloph of the fourth upper molar is higher than in *Indricotherium*, the whole crown is slightly more hypsodont and the cingulum better developed; also the incipient "crochets" of the upper molars are a little more pronounced; assuredly, however, the evidence assembled in Fig. 2 above [see Fig. 19.4 of this paper] is not favorable to the idea that *Paraceratherium, Baluchitherium* and *Indricotherium* are distinct genera, although there are minor and perhaps specific differences in the second upper premolars.

Indeed, the differences between the upper premolars of *Indricotherium* and *Paraceratherium* stressed by Borissiak (1923), Osborn (1923), and Gromova (1959) are arguably intraspecific or, at most, interspecific differences (compare upper premolar variability in *Trigonias:* Gregory and Cook, 1928). Given the virtual identity of the cheek teeth of *Paraceratherium* and *Indricotherium* (compare Fig. 19.5A with 19.5B-C and Fig. 19.6B with 19.6C), the only basis for distinguishing these taxa at the genus-level lies in cranial features. Skulls assigned to these genera (and to *Dzungariotherium* Chiu, 1973) segregate into two groups (Fig. 19.7):

1. Skulls termed *Paraceratherium* and *Dzungariotherium* have relatively slender maxillaries-premaxillaries; nearly horizontal zygomata; shallow skull roofs above the orbits, relatively thin and posteriorly placed mastoid-paroccipital processes (so that the external auditory meati are relatively wide); less posteriorly extended lambdoidal crests and horizontally oriented occipital condyles.

2. The single skull assigned to *Indricotherium* has robust maxillaries-premaxillaries; upturned zygomata; domed frontals above the orbits, thick mastoid-paroccipital processes that are close to the postglenoid processes (so that the external auditory meati are antero-posteriorly constricted); a posteriorly extended lambdoidal crest and vertically oriented occipital condyles.

We advocate the position that these cranial differences, hitherto regarded as indicative of genus-level distinctions, are most reasonably attributed to species-level differences and/or the expression of sexually dimorphic traits. Indeed, virtually all of the differences between skulls termed *Paraceratherium* and *Dzungaroitherium,* and the skull termed *Indricotherium* reflect a single functional complex related to the size of the incisor tusks. The larger tusked form (a male?) has more robust maxillaries-premaxillaries to anchor the upper tusk; a more pronounced mandibular angle indicating a relatively larger masseter muscle to adduct a more anteriorly loaded mandible; an upturned zygomatic arch that

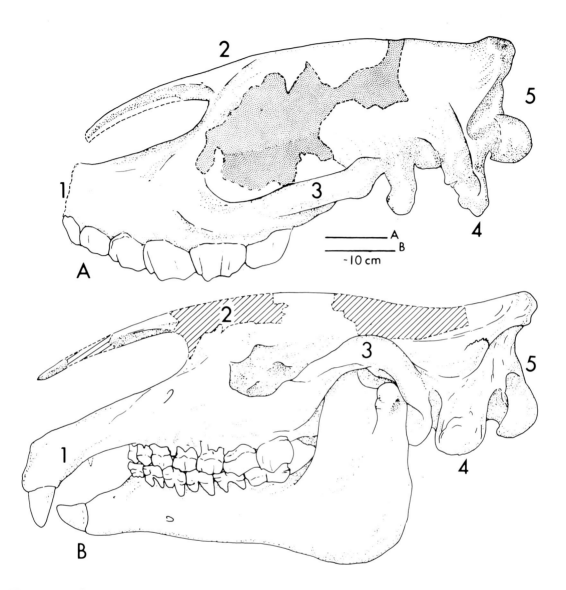

Fig. 19.7. Skulls of (A) *Paraceratherium* and (B) *Indricotherium* showing features traditionally used to distinguish these genera. Numbered features are: 1 = robustness of premaxillaries; 2 = doming of frontals above orbits; 3 = upturned zygomata; 4 = massiveness of mastoid-paroccipital processes and antero-posterior width external auditory meati; 5 = orientation of occipital condyles and lambdoidal crest.

provides a concave surface of origin for this relatively large masseter; a thicker mastoid-paroccipital process for the origin of a larger digastric to abduct the anteriorly loaded lower jaw in a controlled fashion; a more pronounced lambdoidal crest for the insertion of the larger nuchal muscle mass needed to manipulate a more anteriorly loaded skull; and a more vertical occipital condyle that shortens the distance between the occiput and the points of origin of the nuchal musculature so that this musculature is shorter and more powerful. It is not surprising that the skull with the larger tusks and corresponding functional complex also has domed frontals, arguably a male display feature in a sexually dimorphic species (cf. Munthe and Coombs, 1979). Thus, there is a basis for arguing that the cranial differences between *Indricotherium* and *Paraceratherium-Dzungariotherium* may be nothing more than the differences between male (*Indricotherium*) and female (*Paraceratherium*) skulls in a single species, or, at most, different species of a single genus.

Furthermore, there is a small sample of *Paraceratherium-Dzungariotherium* skulls from a limited stratigraphic interval and small geographic area that indicate variability in some of the cranial features used to distinguish *Indricotherium* from *Paraceratherium-Dzungariotherium*. This "population" consists of three skulls from the Turpan basin, Xinjiang, China, described by Xu and Wang (1978). The skull they named *Paraceratherium lipidus* (IVPP V 4322) has relatively robust maxillaries, large mastoid-paroccipital processes, a posteriorly extended lambdoidal crest and relatively vertical occipital condyles (Xu and Wang, 1978, pl. 1). In contrast, the skulls they named *Dzungariotherium turfanensis* have less robust maxillaries, small mastoid-paroccipital processes, lambdoidal crests that do not extend posteriorly and horizontal occipital condyles (Xu and Wang, 1978, pl. 3). We are unable to differentiate these specimens on the basis of tooth size or morphology (note, in particular, that both *P. lipidus* and *D. turfanensis* have $M^{1-2}$ crochets), and thus these skulls suggest significant cranial variation in a single sample. This further strengthens, and in our minds, secures the case for arguing that cranial differences should not be used to distinguish *Indricotherium* from *Paraceratherium* or *Dzungariotherium* from *Paraceratherium*.

We follow Gromova (1959) in recognizing three valid species of *Paraceratherium* named before 1960: *P. transouralicum* (Pavlova, 1922) (= *I. asiaticum* Borissiak, 1923; = *I. minus* Borissiak, 1923; = *B. grangeri* Osborn, 1923), *P. bugtiense* (Pilgrim, 1908) [= *B. osborni* (Forster Cooper, 1913)] and *P. prohorovi* (Borissiak, 1939). Since her work, the following species have been named that are relevant to the species-level taxonomy of *Paraceratherium*: *P. tienshanensis* Chiu, 1962; *D. orgosensis* Chiu, 1973; *P. lipidus* Xu and Wang, 1978; and *D. turfanensis* Xu and Wang, 1978.

Chiu (1962, p.57) based the species *I. intermedium* from Luoping, Yunnan on an isolated "$M^2$", three astragali and a distal end of a metacarpal (IVPP V 2643 1-5). Clearly, these specimens pertain to more that a single individual. Furthermore, there is no way to demonstrate that the "$M^2$" and foot bones belong to a single individual or to a single taxon (material referred by Chiu to *Paraceratherium* is also known from Luoping). In effect, Chiu (1962) only designated syntypes of *I. intermedium*, and we exercise our right (Ride *et al.*, 1985, Article 74b) to designate a lectotype for *I. intermedium*. As lectotype we designate IVPP V 2643-1, the right "$M^2$" (Chiu: pl. 1, Figs. 2a-b). However, we identify this tooth as a $M^1$ because of its relatively weak mesostyle. The size of this tooth (L = 82.5 mm, W = 89.4 mm) is well within the size range of *P. transouralicum* (Gromova, 1959: Table 5). However, we consider Chiu's species is a *nomen dubium*.

Chiu's (1962) *P. tienshanensis*, named for an incomplete right $M^3$ (IVPP V 2370), was originally described as *I.* cf. *I. grangeri* by

Chow and Xu (1959, pl. 1, Fig. 1). This tooth also is within the size range of *P. transouralicum*, although it has a better developed lingual cingulum and slightly higher crown than is typical of *P. transouralicum*. Despite this, we regard *P. tienshanensis* as a *nomen dubium*.

As noted above, we are unable to distinguish *P. lipidus* Xu and Wang, 1978, and *D. turfanensis* Xu and Wang, 1978, from each other, except by cranial differences that are arguably sexually dimorphic. Furthermore, we cannot distinguish these species from *D. orgosensis* Chiu, 1973, the type species of *Dzungariotherium*. We conclude that these three nominal species represent a single species, *P. orgosensis*, distinguished from other *Paraceratherium* by larger tooth size and distinct $M^{1-2}$ crochets. *P. orgosensis* is the largest of known indricotheres (Fig. 19.2).

## Phylogenetic relationships

The phylogenetic hypothesis of the relationships of the Indricotheriinae presented here (Fig. 19.8) recognizes the Triplopodinae (best exemplified by *Triplopus* and *Hyracodon*) as the plesiomorphic sister taxon of the indricotheres. The Hyracodontidae are unified by a cursorial limb structure which decreases the mechanical advantage of muscles exerting ground-reaction force and increases the angular velocity of the limb. This is accomplished by lengthening the distal segments of the leg, principally the metacarpals and metatarsals, and reducing the number of digits contacting the ground to three in the manus (Node 1). *Contra* Lucas *et al.* (1981), conical incisors do not appear to unite this family, since a wide range of incisor morphology from spatulate to conical is found in primitive hyracodontids.

We recognize a unique suite of characters (functional complex) in the anterior facial skeleton that distinguishes the indricotheriines from the triplopodines (Node 2). Although a pronounced nasal incision and some development of a maxillary fossa ocurs in many hyracodontids, in indricotheriines the nasal and maxillary bones are modified for the support of an elaborated muscular snout, which probably supported a short proboscis in the largest and most advanced members of the subfamily.

The initial evolutionary modifications involved in producing this specialized facial anatomy are already present in the most primitive indricotheriine genus, *Forstercooperia*. Thus, in *F. grandis* (AMNH 26660), the nasal bones form a robust shelf which begins above $C^1$, and a high, flattened eminence that terminates above the orbits (Wood, 1963, Fig. 1). This eminence is set at a distinct angle to the rest of the skull in *F. minuta* (Fig. 19.9). A preorbital fossa on the maxillary bones runs parallel to most of the posterior tooth row. This must have provided a wide site of attachment for muscles supporting the snout, notably the dilator naris, and a well developed levator rostri which would have manipulated a muscular rhinarium. The cranium of *F. totadentata*, the largest species of this genus, is known only from an anterior portion of the face preserving complete premaxillae and the anterior nasals and maxillae. This specimen (Wood, 1938, Figs. 1-2), although subjected to lateral crushing, retains a shallow but distinct preorbital fossa. The robust nasal shelf and preorbital fossa thus are present in the three species of *Forstercooperia*. The midfacial region of *Forstercooperia* is long, and the orbits are placed relatively posteriorly, above $M^3$. This contrasts with *Hyracodon*, in which the relatively tall maxillaries contribute to a short, high face, the nasal incision begins at about $P^1$, and the orbits lie above $M^{1-2}$. *Hyrachyus*, arguably the most primitive rhinocerotoid known, is intermediate between *Forstercooperia* and *Hyracodon* in these cranial features. Relative to *Hyrachyus* and *Hyracodon*, the glenoid fossa in *Forstercooperia* more closely approaches the plane of the anterior dentiton.

*Juxia*, *Urtinotherium*, and *Paraceratherium* appear to represent an anagenetic

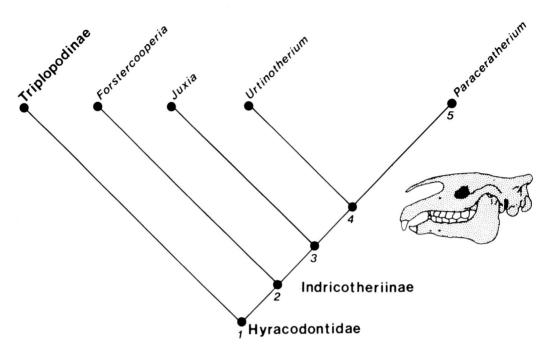

Fig. 19.8. A cladistic hypothesis of the phylogenetic relationships of the indricotheres. See text for discussion of character-states that correspond to the numbered node points.

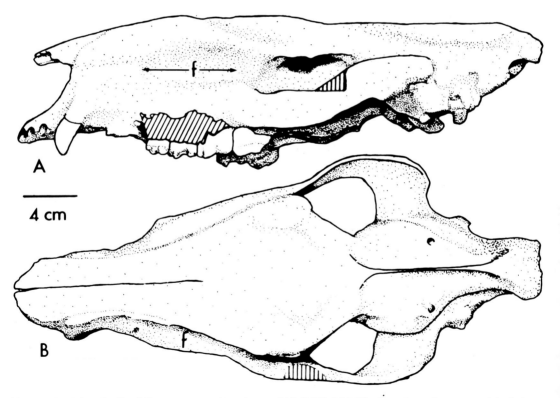

Fig. 19.9. The skull of *Forstercooperia minuta* (AMNH 26643) showing the pre-orbital fossa (f).

evolutionary series (at least at the generic level) showing a consistent increase in body size. This trend is accompanied by the usual allometric changes in linear diminsions (e.g., skull length and length of the nasal shelf). The beginning of the nasal incision in *Juxia* retreats to above $P^3$, which may reflect more than an increase in size alone, since $I^1$ shows a slight relative enlargement. The posterior dentition shows slight molarization of the premolars. These features, although comparatively minor, unite the higher indricotheriine genera (Node 3).

The anterior facial anatomy is further modified in *Urtinotherium* and *Paraceratherium* to emphasize the probable role of the incisors in cropping upper-story vegetation: $I^1/_1$ are enlarged and procumbent, $C^1/_1$ and $dP^1/_1$ are reduced in *Urtinotherium* and lost in *Paraceratherium* (Node 4). Without knowing how they scale with body diminsions, it is difficult to determine whether $I^{2-3}/_{2-3}$ actually undergo reduction (as stated in Lucas *et al.*, 1981) or remain the same relative size. In any case, the dental proportions of *Urtinotherium* and *Paraceratherium* show a feeding pattern which involves extensive reliance on the muscular snout and the incisors.

The largest indricotheriine, *Paraceratherium*, is characterized by an enlarged, procumbent $I^1/_1$ and a set of genuinely discrete character-states: the loss of $I^{2-3}/_{2-3}$, $C^1/_1$ and $dP^1/_1$; the nasal incision begins above $P^4$; the premaxilla is downturned; and $P^4$ is "creased" posterolingually (Node 5).

## Acknowledgments

We thank M. C. McKenna and R. H. Tedford (AMNH) and M. Chow (IVPP) for permission to study specimens in their care. E. Manning and Wang Jingwen influenced some of the ideas present here, and NSF Grant DEB-7919681 supported part of this research.

## Bibliography

Beliajeva, E. (1959): Sur la decouverte de rhinoceros Tertiares anciens dans la province maritime de l'U. R. S. S. -*Vert. PalAsiatica*, 3: 81-91

Bohlin, B. (1946): The fossil mammals from the Tertiary deposit of Taben-buluk, western Kaansu, Pt. II, Simplicidentata.- *Paleont. Sinica* (C), 8b: 1-259.

Borissiak, A. A. (1915): Ob indrikoterii (*Indricotherium* n.g.). -*Geol. Vestnik*, 1(3): 131-134.

Borissiak, A. A. (1923): O rod *Indricotherium* N. G. (sem. Rhinocerotidae). - *Zapiski Ross. Akad. Nauk*, (8) 35 (6): 1-128.

Borissiak, A. A. (1939): O novom predstavitele cem. Indricotheriidae. -In: Anonymous (ed.): *V.A. Obruchevu k 50-letiyu nauchoj i pedagogicheskoj deyatelnosti*. Moscow (Izdatelskovo Akademia Nauk SSSR), 2: 271-276.

Chiu, C. (1962): Giant rhinoceros from Loping, Yunnan, and discussion of the taxonomic characters of *Indricotherium grangeri*. -*Vert. PalAsiatica*, 6: 58-71.

Chiu, C. (1973): A new genus of rhinoceros from Oligocene of Dzungaria, Sinking. -*Vert. PalAsiatica*, 11: 182-191.

Chow, M. (1958): Some Oligocene mammals from Lunnan, Yunnan. -*Vert. PalAsiatica*, 2: 263-268.

Chow, M., and Chiu, C. (1963): A new genus of giant rhinoceros from Oligocene of Inner Mongolia. -*Vert. PalAsiatica*, 7: 230-239.

Chow, M., and Chiu, C. (1964): An Eocene giant rhinoceros. -*Vert. PalAsiatica*, 8: 264-267.

Chow, M., Chang, Y., and Ting, S. (1974): Some early Tertiary Perissodactyla from Lunan basin, E. Yunnan. -*Vert. PalAsiatica*, 12: 262-273.

Chow, M., Li, C., and Chang, Y. (1973): Late Eocene mammalian faunas of Honan and Shansi with notes on some vertebrate fossils collected therefrom. -*Vert. PalAsiatica*, 11: 165-181.

Chow, M., and Xu, Y. (1959): *Indricotherium* from Hami basin, Sinkiang. -*Vert.*

PalAsiatica, 3: 93-98.

Cope, E. D. (1879): On the extinct species of Rhinoceridae of North America and their allies. -Bull. U. S. Geol. Geogr. Surv. Territ., 5 (2): 227-237.

Eaton, J. G. (1985): Paleontology and correlation of the Eocene Tepee Trail and Wiggins Formations in the North Fork of Owl Creek area, southeastern Absaroka Range, Hot Springs County, Wyoming. -J. Vert. Paleont., 5: 345-370.

Forster Cooper, C. (1911): *Paraceratherium bugtiense*, a new genus of Rhinocerotidae from Bugti Hills of Baluchistan - preliminary notice. -Ann. Mag. Nat. Hist., (8)8: 711-716.

Forster Cooper, C. (1913a): *Thaumastotherium osborni*, a new genus of perissodactyls from the Upper Oligocene deposits of the Bugti Hills of Baluchistan - preliminary notice. -Ann. Mag. Nat. Hist., (8)12: 376-381.

Forster Cooper, C. (1913b): Correction of generic name [*Thaumastotherium* to *Baluchitherium* ]. -Ann. Mag. Nat. Hist., (8)12: 504.

Forster Cooper, C. (1923): *Baluchitherium osborni* (?syn. *Indricotherium turgaicum*, Borissiak). -Phil. Trans. Royal Soc. London B, 50: 35-66.

Forster Cooper, C. (1924): On the skull and dentition of *Paraceratherium bugtiense* : a genus of aberrant rhinoceroses from the lower Miocene deposits of Dera Bugti. -Phil. Trans. Royal Soc. London B, 212: 369-394.

Gabuniya, L. (1955): O svoebraznom pred stavitele Indricotheriidae iz Oligotsena Gruzii. -Dokl. Akad. Nauk Arm. SSSR, 21: 177-181.

Gabuniya, L. (1964): *Benarskaya fauna oligotsenovykh pozvonochnykh*. Tbilisi (Metsniereba Press).

Gabuniya, L. (1977): Contribution a la connaissance des mammiferes Paleogenes du bassin de Zaissan (Kazakhstan Central). -Geobios Mem. Spec., 1: 29-37.

Gabuniya, L., and Dashzeveg, D. (1974): Ob Olgotsenovom predstavitele *Forstercooperia*(Hyracodontidae) iz Mongolii. -Soobsh. Akad. Nauk Gruz. SSR, 75: 497-500.

Gabuniya, L., and Iliesku, O. (1960): O pervoij nakhodke ostatkov gigantiskikh nosorogov iz sem. Indricotheriidae v Rumynii. -Dok. Akad. Nauk SSSR, 130: 425-427.

Gao, Y. (1976): Eocene vertebrate localities and horizons of Wucheng and Xichuan basins, Henan. -Vert. PalAsiatica, 14: 26-34.

Granger, W., and Gregory, W. K. (1936): Further notes on the gigantic extinct rhinoceros, *Baluchitherium*, from the Oligocene of Mongolia. -Bull. Amer. Mus. Nat. Hist., 72: 1-73.

Gregory, W. K., and Cook, H. J. (1928): New material for the study of evolution: a series of primitive rhinoceros skulls (*Trigonias*) from the lower Oligocene of Colorado. -Proc. Colorado Mus. Nat. Hist., 8(1): 1-32.

Gromova, V. (1959): Gigantskie nosorogi. -Trudy Paleont. Inst. Akad. Nauk SSSR, 71: 1-164.

Hu, C. (1962): Cenozoic mammalian fossil localities in Kansu and Ningshia. -Vert. PalAsiatica, 6: 162-172.

Kirkaldy, G.W. (1908): Memoir on a few heteropterous Hemiptera from eastern Australia.- Proc. Linn. Soc. New South Wales, 32: 768-788.

Lucas, S.G. (1982): Vertebrate paleontology, stratigraphy, and biostratigraphy of Eocene Galisteo Formation, north-central New Mexico. -New Mex. Bur. Mines Min. Res. Circ., 186: 1-34.

Lucas, S.G., Schoch, R. M., and Manning, E. (1981): The systematics of *Forstercooperia*, a middle to late Eocene hyracodontid (Perissodactyla: Rhinocerotoidea) from Asia and western North America. -J. Paleont., 55: 826-841.

Matthew, W. D., and Granger, W. (1923): The fauna of the Houldjin Gravels. -Amer. Mus. Novit., 97: 1 -6.

Mellett, J. S. (1968): The Oligocene Hsanda Gol Formation, Mongolia: a revised faunal list. -Amer. Mus. Novit., 2318: 1-16.

Munthe, J., and Coombs, M. C. (1979):

Miocene dome-skulled chalicotheres (Mammalia, Perissodactyla) from the western United States: A preliminary discussion of a bizarre structure. -*J. Paleont.*, 53: 77-91.

Osborn, H. F. (1923): *Baluchitherium grangeri*, a giant hornless rhinoceros from Mongolia.-*Amer. Mus. Novit.*, 78: 1-15.

Pavlova, M. (1922): *Indricotherium transouralicum*, n.s. provenant du district de Tourgay. -*Bull. Soc. Natural. Moscou Sec. Geol.*, (2)31: 95-116.

Peterson, O. A. (1919): Report upon the material discovered in the upper Eocene of the Uinta Basin by Earl Douglass in the years 1908-1909, and by O. A. Peterson in 1912. -*Ann. Carnegie Mus.*, 12: 40-168.

Petronijevic, Z., and Thenius, E. (1957): Uber den ersten Nachweis von Indricotherien (Baluchitherien; Rhinocerotidae, Mammalia) im Tertiär von Europa. -*Anz. Math. -Naturw.. Kl. Osterreich Akad. Wiss.*, 9: 153-155.

Pilgrim, G. (1908): The Tertiary and post-Tertiary fresh-water deposits of Baluchistan and Sind, with notices of new vertebrates. -*Rec. Geol. Surv. India*, 37: 139-166.

Pilgrim, G. (1910): Notices of new mammalian genera and species from the tertiaries of India. -*Rec. Geol. Surv. India*, 40: 63-71.

Qi, T. (1979): A general account of the early Tertiary mammalian faunas of Shara Murun area, Inner Mongolia. -*Proc. Second. Congr. Strat. China*: 1-7

Radinsky, L. B. (1964): Notes on Eocene and Oligocene fossil localities in Inner Mongolia. -*Amer. Mus. Novit.*, 2180: 1-11.

Radinsky, L. B. (1966): The families of the Rhinocerotoidea (Mammalia, Perissodactyla). -*J. Mamm.*, 47: 631-639.

Radinsky, L. B. (1967): A review of the rhinocerotoid family Hyracodontidae (Perissodactyla). -*Bull. Amer. Mus. Nat. Hist.*, 136: 1-46.

Ride, W. D. L., Sabrosky, C. W., Bernardi, G., and Melville, R. V. (1985): *International code of zoological nomenclature.* - London (International Trust for Zoological Nomenclature).

Simpson, G. G. (1945): The principles of classification and a classification of mammals. -*Bull. Amer. Mus. Nat. Hist.*, 85: 1-350.

Tang, Y. (1978): New materials of Oligocene mammalian fossils from Qujing Basin, Yunnan. -*Prof. Pap. Strat. Paleont.*, 7: 75-79.

Tang, Y., You, Y., Xu, Q., Qiu, Z., and Hu, Y. (1974): The lower Tertiary of the Baise and Yungle basins, Kwangsi. -*Vert. PalAsiatica*, 12: 279-292.

Teilhard de Chardin, P. (1926): Mammifères Tertiares de Chine et de Mongolie. -*Ann. Paleont.*, 15: 1-51.

Ting, S., Zheng, J., Zhang, Y., and Tong, Y. (1977): The age and characteristics of the Liuniu and the Dongjun faunas, Baise Basin of Guanxi. -*Vert. PalAsiatica*, 15: 35-45.

Tong, Y., and Wang, J. (1980): Subdivision of the Upper Cretaceous and lower Tertiary of the Tantaou Basin, the Lushi Basin and the Lingbao Basin of W. Henan. -*Vert. PalAsiatica*, 18: 21-27.

Wang, J. (1976): A new genus of Forstercooperiinae from the late Eocene of Tongbo, Henan. -*Vert. PalAsiatica*, 14: 104-111.

Wood, H. E., II (1938): *Cooperia totadentata*, a remarkable rhinoceros from the Eocene of Mongolia. -*Amer. Mus. Novit.*, 1012: 1-20 [with addendum dated 23 February 1939].

Wood, H. E., II (1963): A primitive rhinoceros from the late Eocene of Mongolia. -*Amer. Mus. Novit.*, 2146: 1-11.

Xu, Y. (1961): Some Oligocene mammals from Chuching, Yunnan. -*Vert. PalAsiatica*, 5: 315-329.

Xu, Y., and Chiu, C. (1962): Early Tertiary mammalian fossils from Lunan, Yunnan. -*Vert. PalAsiatica*, 6: 313-332.

Xu, Y., and Wang, J. (1978): New materials of giant rhinoceros. -*Mem. Inst. Vert. Paleont. Paleoanthrop.*, 13: 132-140.

Young, C. C., and Chow, M. (1956): Some Oligocene mammals from Lingwu, North

Kansu. -*Acta. Paleont. Sinica*, 4: 447-460.

Zhai, R. (1977): Supplementary remarks on the age of Changxindian Formation. -*Vert. PalAsiatica*, 15: 173-176.

Zhang, Y., You, Y., Ji, H., and Ting, S. (1978): Cenozoic stratigraphy of Yunnan. -*Prof. Pap. Strat. Paleont.*, 7: 1-21.

Zheng, J., Tang, Y., Zhai, R., Ting, S., and Huang, X. (1978): Early Tertiary strata of Lunan Basin, Yunnan. -*Prof. Pap. Strat. Paleont.*, 7: 22-29.

# 20. *TELETACERAS RADINSKYI*, A NEW PRIMITIVE RHINOCEROTID FROM THE LATE EOCENE CLARNO FORMATION OF OREGON

## C. BRUCE HANSON

*Teletaceras*, a new genus of rhinocerotid, includes the most primitive members of the family. A large quarry sample from the uppermost Clarno Formation (Duchesnean) of north-central Oregon provides the hypodigm for the type species, *T. radinskyi*, the least derived and probably earliest representative of the genus and family. It exhibits the derived incisor tusk complex characteristic of the Rhinocerotidae, while retaining an unreduced dental formula, strongly ribbed premolar ectolophs, and many skull characters primitive for the superfamily Rhinocerotoidea. Its tridactyl manus may also be primitive for rhinocerotids and hyracodontids, as are most other carpal and tarsal characters.

Two species originally referred to *Eotrigonias* are transferred to *Teletaceras*, extending its known geographic range to southern California, United States, and Maritime Province, Soviet Union, and its known temporal range to early Chadronian.

### Introduction

The Rhinocerotidae, as now characterized, include previously described forms (*Penetrigonias* and *Trigonias*) as old as Chadronian and possibly latest Duchesnean. *Trigonias* Lucas (1900) has been long recognized as a very primitive rhinocerotid because of its nearly complete incisor formula and presumed primitive tetradactyl manus. Tanner and Martin (1976) described *Penetrigonias* on the basis of an unfortunately incomplete type, but numerous congeneric specimens in existing collections (e.g., the Calf Creek l.f. species "*Subhyracodon*" *sagittatus* Russell, 1982--see discussion following), now document its phyletic position as a rhinocerotid even more primitive than *Trigonias*.

Nonetheless, substantial differences remain between *Penetrigonias* and *Hyrachyus*, a genus historically implicated in rhinocerotid ancestry.

Other "primitive rhinoceroses" of similar or greater age (Wood, 1927, 1929, 1938, 1963) had been assigned to the Rhinocerotidae primarily on the basis of characters now known to have arisen independently in hyracodontids, and they are now placed in the latter family in the classifications of Radinsky (1966) and Prothero, Manning, and Hanson (1986). The Asian genus *Ilianodon* Chow and Xu (1961) bears a tusk thought to be an $I_2$, but its relatively erect orientation suggests that it is a canine and that *Ilianodon* is a hyracodontid as well.

*Teletaceras radinskyi*, the type species of the new genus described here, predates *Penetrigonias* and *Trigonias*. While exhibiting the key rhinocerotid synapomorphies, $I^1$/$I_2$ tusks and reduced premaxilla, it retains many more primitive character states than those genera, bridging much of the remaining "morphologic gap" between *Hyrachyus* and the rhinocerotids. The phyletic relationships of *Teletaceras*, inferred from the type species, have already been reported in Prothero et al. (1986) and in Prothero, Guérin, and Manning (this volume, Chapter 16), where it is identified as the "Clarno rhino," the sister taxon to all other rhinocerotids.

*T. radinskyi* is the most abundantly represented species in its type locality, Hancock Quarry, a prolific bone deposit in Wheeler County, Oregon. Hancock Quarry

*The Evolution of Perissodactyls* (ed. D.R. Prothero & R.M. Schoch) Oxford Univ. Press, New York, 1989

(UCMP locality no. V-75203) is situated within the uppermost subunit of the Clarno Formation, about 10 m below a welded tuff at the base of the overlying John Day Formation. K-Ar dates on this tuff (37.1 and 37.5 Ma; Fiebelkorn *et al.* 1983) are consistent with biostratigraphic correlations of the Hancock Quarry l.f. to the dated interval bracketed by the Porvenir l.f. and Candelaria l.f. of Texas (Wilson, 1977). The combined data places the age of Hancock Quarry in the mid-Duchesnean, about 38 to 39 Ma.

**Abbreviations**

| | |
|---|---|
| AMNH | American Museum of Natural History, New York |
| Fm. | Formation |
| l.f. | local fauna |
| LACM | Los Angeles County Museum |
| LACM (CIT) | California Institute of Technology Collection, now at LACM |
| Ma | million years before present |
| OMSI | Oregon Museum of Science and Industry, Portland |
| SDSM | South Dakota School of Mines and Technology, Rapid City |
| SMNH | Saskatchewan Museum of Natural History, Regina |
| TMM | Texas Memorial Museum, Austin |
| UCMP | University of California Museum of Paleontology, Berkeley |
| UM | University of Minnesota, Minneapolis |
| UOMNH | University of Oregon Museum of Natural History, Eugene |
| UW | University of Wyoming, Laramie |
| WSM | Burke Memorial Washington State Museum, Seattle. |

**Systematic paleontology**

Family Rhinocerotidae Owen (1845)
*Teletaceras* new genus

*Eotrigonias* (?): Stock (1949) (not Wood, 1927)
*Eotrigonias*: Beliaeva (1959) (not Wood, 1927)
*Pappaceras*: Radinsky (1966) (in part) (not Wood, 1963)
*Juxia*: Lucas, Schoch, and Manning (1981) (in part) (not Chow and Chiu, 1964)

*Etymology*.- Greek, *teleta*, initiation, +*a*, without, +*keras*, horn; with reference to the initial phyletic position relative to the family and to the absence of the nasal horn possessed by living members of the family.

*Type species*. - *Teletaceras radinskyi* new species.

*Included species*. - Type species, *T. mortivallis* (Stock, 1949), new combination, and *T. borissiaki* (Beliaeva, 1959), new combination.

*Known distribution*. - Late Eocene (Duchesnean to early Chadronian) of central Oregon and southeastern California, United States, and Maritime Province, Soviet Union.

*Generic diagnosis*. - Dental formula = I3/3, C1/1, P4/4(-3), M3/3. Small rhinocerotids with $I^1$ and $I_2$ tusk complex characteristic of the family, but not as enlarged as in other incisor-bearing rhinocerotids. Differs from all other rhinocerotids in the possession of an unreduced anterior dental series, sharp crease between molar parastyles and paracones, more lingually inflected molar metacone axes, and low connection of molar metalophids to protolophids. Differs additionally from "*Subhyracodon*" *sagittatus* Russell (1982) by the presence of a marked postcanine diastema and single-rooted $P_1$.

*Teletaceras radinskyi* new species
*Etymology*. - Named in honor of the late Dr. Leonard Radinsky in recognition of his insightful contributions to the knowledge of Paleogene ceratomorphs.

*Holotype*. - UCMP 129000, nearly complete skull lacking premaxillae, portion of right zygoma, and occipital crest.

*Type locality*. - UCMP Loc. V75203,

Hancock Quarry, uppermost unit of the Clarno Formation, Wheeler County, Oregon.

*Hypodigm.* - Skulls -- Type and UCMP 129001, 129002; UOMNH 27698; OMSI 616, OMSI (2 unnumb. specimens); maxillae and upper cheek teeth -- UCMP 129003 to 129006, 129012 to 129022; UOMNH 20447, 20452, 20478, 20485, 20539, 20540, 20927, 20928, 21125, 21378, 21381, 21384, 21388, 21390, 21391, 21394, 21401, 21403, 21419, 21426, 27376, 27644, 27647, 27649, 27719, 28311, 28317, 28319; OMSI 829, 830; incisors -- UCMP 129008 to 129010; UOMNH 20483, 28339; canines--UCMP 129011, 129024; UOMNH 20486; dentaries and lower cheek teeth -- UCMP 129026 to 129031, 129033 to 129048, 129051 to 129054; UOMNH 20442, 20445, 20546, 20924, 20937, 20938, 21124, 21383, 21396, 21398, 21406, 21407, 21421, 21432, 21433, 21435, 27645, 27648, 27650 to 27655, 27657 to 27660, 28309, 28312, 28316, 28322, 28325, 28327, 28329, 28333, 28335, 28343, 28347; OMSI 612, WSM 56949, 56952; podials -- UCMP 129055 to 129069; UOMNH 20435, 20443, 20941, 28330, 28338.

*Known distribution.* - Type locality only; Duchesnean (late Eocene) of north-central Oregon.

*Specific diagnosis.* - Paracone and metacone ribs on $P^{2-4}$ ectolophs prominent, subequal, and contiguous (not separated by intervening flat area). Dentition larger and more brachydont than *Teletaceras mortivallis*, smaller than *T. borissiaki*. Mean length $M^{1-3}$, 64 mm; $M_{1-3}$, 65 mm. Crown height index, $0.66 \pm 0.02$.

## Description

In the following description, comparisons are made with *Hyrachyus* (the most primitive member of the Rhinocerotoidea; outgroup for Rhinocerotidae + Hyracodontidae), *Triplopus* and *Hyracodon* (primitive hyracodontids), and *Penetrigonias, Trigonias* and *Subhyracodon* (more derived Paleogene rhinocerotids). As analysis of all characters has already demonstrated the phyletic positions of these taxa relative to *Teletaceras* (Prothero et al., 1986 and Prothero, Guérin, and Manning, this volume), the comparisons will serve to imply the polarities of the described characters.

*Skull.* - The available Hancock Quarry sample of this species includes five nearly complete adult skulls, a sixth lacking the posterior cranium, and the posterior half of a seventh skull. Although none is complete or undistorted, this suite provides enough information for confident graphic reconstruction of all parts of the skull except the premaxilla (Fig. 20.1).

<u>Dorsal aspect</u>: The elongate appearance of the skull in dorsal view is enhanced by the unusually long, gently tapering nasals. Anterior to the orbits, the lateral margins of the dorsal surface converge gradually, describing uniform, laterally concave arcs.

Slightly convex zygomatic arches nearly parallel the midline as in *Hyracodon* and *Hyrachyus*, in contrast with straighter, anteriorly convergent arches in *Subhyracodon occidentalis* and more convex arches in *Penetrigonias sagittatus* and *Trigonias*.

The supratemporal crests converge in smooth arcs from the rear of the supraorbital processes, meeting at the midline to form a low, narrow sagittal crest marked by a medial groove as in *P. sagittatus* and *Subhyracodon copei* but differing from the separate parasagittal crests in *S. occidentalis*. A distinct, rounded notch interrupts the occipital crest at the midline, as in *P. sagittatus*. The crests of *S. occidentalis* and *Trigonias* bear broader indentations. The braincase is less expanded than in other rhinocerotids except *Trigonias*, but comparable to that of *Hyrachyus*.

<u>Lateral aspect</u>: The skull presents an elongate profile, varying little in depth from front to rear. A shallow saddle above the orbits separates the convex dorsal profiles of the muzzle and braincase regions. The convex braincase profile resembles that of most hyracodontids, *Hyrachyus*, and many other perissodactyls, but contrasts with the straight or concave profile of other rhinocerotids.

Strikingly elongate nasals extend

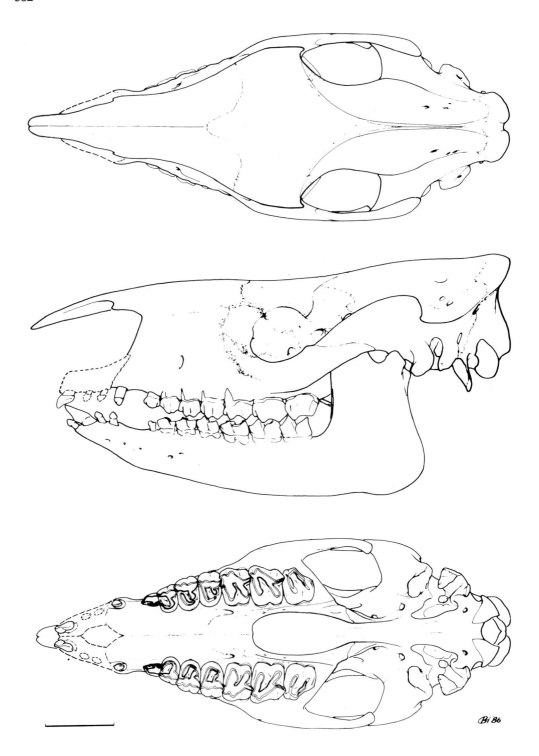

Fig. 20. 1. Composite reconstruction of skull of *Teletaceras radinskyi*. Dorsal (top), lateral (middle), and ventral (bottom) views. Reconstruction based primarily on type specimen (UCMP 129000), supplemented by UCMP 129001 (lateral dimensions), UCMP 129039 (dentary). Scale bar is 5 cm.

further forward than any reasonable reconstruction of the premaxillae. On the ventrolateral side, above the rear of the nasal incision, the nasal bone bears a small but prominent anteriorly directed process similar to that of *P. sagittatus* and *Hyracodon nebrascensis*. The nasal incision extends only to a point above the postcanine diastema, less deep in relation to the dental series than that of other rhinocerotids, but undercutting the nasals to an extent comparable to that of *Trigonias*. The premaxillary suture extends about halfway up the anterior margin of the maxilla, well separated from the nasals.

The center of the orbit lies midway along both the anteroposterior and dorsoventral dimensions. The skull does not exhibit the postorbital extension of *Subhyracodon* and *Trigonias*, but the position of the antorbital rim above anterior $M^3$ is common to all three.

The cheek tooth series is nearly straight in lateral view, and parallel to the dorsal profile of the skull, in contrast with the upward flexure of the anterior cheek-tooth row of later rhinocerotids. The relatively low-crowned teeth and shallow root-bearing portion of the maxilla are largely responsible for the shallow appearance of the mid-portion of the skull compared to that of other rhinocerotids except *P. sagittatus*.

A notch undercuts the posterodorsal end of the zygomatic arch as it does in *Trigonias* and *Subhyracodon* but not *Hyrachyus*.

The postglenoid, posttympanic, and paroccipital processes do not extend as far below the glenoid surface and the external auditory meatus as they do in *Subhyracodon* and *Trigonias*.

Ventral aspect: The narrow palate of *T. radinskyi* bears relatively straight, parallel cheek tooth rows, lacking the slight anterior convergence of the tooth rows of *S. occidentalis*, or the inward curvature of the anterior teeth exhibited by *P. sagittatus* and *Trigonias*. *T. radinskyi* compares more favorably with *Hyrachyus douglassi* (UW 1937; see discussion following) in this respect.

Behind $M^3$, the posterior margin of the maxilla is nearly straight, extending from the root of the zygomatic arch to the pterygoid crest, as in *Hyrachyus* and *P. sagittatus*, but differing from the convex margins in *Trigonias* and *Subhyracodon*. The posterior narial opening incises the palate to the level of the anterior border of $M^2$, farther than any of the compared taxa.

The basicranium of *T. radinskyi* resembles that of *Hyrachyus douglassi* more than that of either *Trigonias* or *Subhyracodon*. The postglenoid processes of the two former species are proportionately shorter with more strongly concave anterior faces, and have thicker (more rounded) posterolateral borders with distinct bulges partly enclosing the external auditory meatus ventrally. In *Teletaceras* and *Trigonias*, a shallow open groove (probably for n. chorda tympani) extends around the medial base of the postglenoid process, whereas in *S. occidentalis* this channel is partly enclosed by a bony bridge. The posttympanic process of *Teletaceras* is triangular in cross section, relatively stout, and has a longer free portion than the compared taxa. As in *Hyrachyus*, the paroccipital processes are quite slender in the lateral dimension. The foramen ovale lies medial to the anterior margin of the postglenoid process, distinctly separated from the middle lacerate foramen.

Dentary. -The dentary of *T. radinskyi* resembles that of *Subhyracodon* in relative proportions, though the former is considerably smaller. The cheek teeth occupy proportionately less of the total length of the jaw, and the ascending ramus is more slender. The symphysis extends back to a point below the anterior end of $P_2$, as in *Subhyracodon* and *Trigonias*, but the symphyseal portion of the jaw is proportionately longer in *T. radinskyi*, as it accomodates a long diastema and two more teeth ($I_3$ and $C_1$) than are usually present in the others. The angle of the jaw has a radius of curvature proportionately smaller

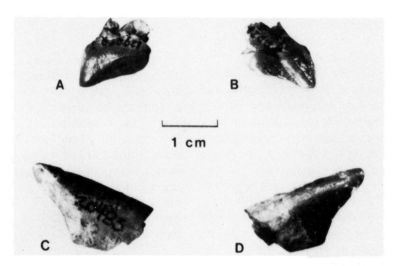

Fig. 20. 2. Incisor tusks referred to *Teletaceras radinskyi*. A. Labial and B. lingual views of left $I^1$, UCMP 129009; C. lingual and D. labial views of right $I_2$, UOMNH 2085. Both specimens from type locality.

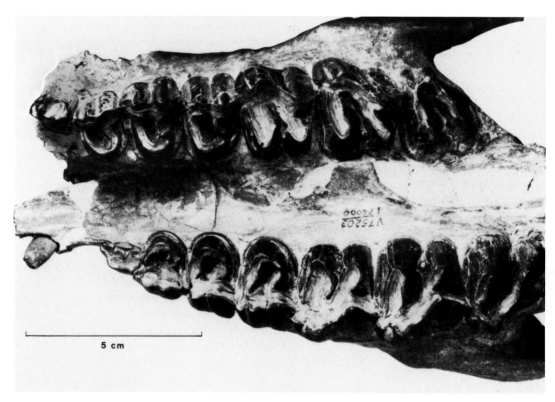

Fig. 20. 3. Ventral view of palate of type specimen of *Teletaceras radinskyi*, UCMP 129000.

than those of *Subhyracodon* and *Trigonias*, and is delimited by more distinct indentations below the prominent postcotyloid process and on the ventral border of the jaw. A pair of mental foramina appears below $P_2$ and anterior $P_3$, and nutrient foramina (usually three on each side) penetrate the ventrolateral sides of the symphysis.

*Dentition.-* <u>Upper teeth</u>: Although the premaxilla is not represented in any of the Hancock Quarry specimens, isolated teeth and wear on incisors in dentaries reveal some aspects of the upper anterior dentition. Three isolated teeth are inferred to be first upper incisors. UCMP 129009 (Figs. 20.2A and B) is a nearly unworn crown, which is relatively small but otherwise similar to the $I^1$ of *Subhyracodon* and *Trigonias*. It has a wear facet compatible with the facets of $I_2$ of referred jaws. The crown is elongate anteroposteriorly and teardrop-shaped in occlusal view, with its greatest transverse dimension near the rear. The medial side is creased near its midline.

The presence of $I^2$ is indicated by a small wear facet on the $I_2$ of UCMP 129039, lateral to the posterior end of the thegosis facet produced by the $I^1$ tusk.

A portion of an $I^3$ root appears to remain in the type specimen (UCMP 129000) just anterior and internal to the left $C^1$. A wear facet on the anterolingual side of $C_1$ in a well-preserved jaw (UCMP 129039) is compatible only with an $I^3$.

$C^1$ is preserved in the type specimen and in UCMP 129003. The crown is small, but the root is proportionately long and massive. Prominent ridges mark the anteromedial and posterior edges of the crown.

The upper cheek teeth (Fig. 20.3) appear quite primitive compared to those of other rhinocerotids. The crowns are low (crown height index = $0.66 \pm 0.02$; see Radinsky, 1967) and irregular, and the morphological difference between the premolars and molars is conspicuous in lateral as well as occlusal view.

$P^1$ (five specimens) has a narrow, rounded triangular outline; its length is nearly twice its width. The large, centrally situated paracone dominates the labial face of the ectoloph, and shallow grooves delimit the small parastyle and metacone. The lingual cusps are quite small and anteroposteriorly elongate, especially the paracone. Cross-lophs are very weak or absent. The lingual cingulum extends only from the tip of the parastyle to the anterior end of the protocone. The size and shape of the hypocone, size of the protocone, and relative tooth width vary within the sample.

$P^2$ through $P^4$ (Fig. 20.4) retain many of the primitive characters of the superfamily. Most of the labial surface on each tooth is occupied by a pair of prominent, subequal, rounded ridges, the paracone and metacone ribs. The curved surfaces of these ribs meet medially, forming a cleft, except near the occlusal edges of unworn teeth. The parastyle is similar in size and shape to the metastyle. The resulting mirror symmetry about the median cleft approximates that seen in *Hyrachyus* and *Triplopus* but differs from the flatter, asymmetrical ectolophs of most rhinocerotids. In occlusal view, the crown outlines of $P^{2-4}$ are also nearly symmetrical fore and aft. The lingual margins range from semicircular on $P^2$ to parabolic on $P^4$.

The protoloph incorporates the protoconule, protocone, and hypocone of $P^{2-4}$ of all specimens except one: the $P^2$ of UOMNH 20447 has a distinct hypocone that appears to have had a metaloph connection. The protoloph of $P^{2-4}$ is isolated from the ectoloph by a deep saddle (most pronounced in $P^2$) in lightly worn teeth. The short metaloph tapers labially from the metaconule to the ectoloph but connects more firmly than does the protoloph. A small flange on the metaconule of most of the specimens may be directed toward the protocone or the small hypocone. The strength of this flange generally decreases from $P^2$ to $P^4$ in a given individual, and the connection is usually

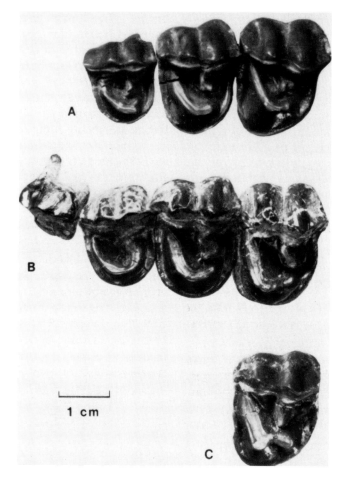

Fig. 20. 4. Occlusal views of upper premolars of *T. radinskyi* showing individual variation. A. P$^{2-4}$, UOMNH 21125; B. P$^{1-4}$, UCMP 129000 (type); C. P$^4$, UOMNH 21391.

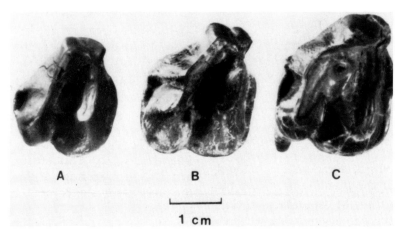

Fig. 20. 5. Occlusal views of third upper molars of *T. radinskyi* showing individual variation. A. UOMNH 21390; B. UCMP 129000 (type); C. UCMP 129019.

with the hypocone in $P^2$ and with the protocone in $P^{3-4}$. The median valleys of $P^{3-4}$ open toward the rear until advanced wear. $P^2$ typically has a closed protoloph-metaloph loop enclosing a central fossette, though one specimen (Fig. 20.4A) has a posterior opening and another a lingual opening. In general, premolar ectoloph features exhibit less intrapopulation variability than do features of the protoloph-metaloph complex.

All specimens lack labial cingula on $P^{2-4}$, but possess anterior and posterior cingula. The lingual cingulum is variable, but any given individual exhibits a progressive decrease in the strength and extent of the lingual cingulum from $P^2$ through $P^4$. It is complete in three of nine $P^2$s, two of thirteen $P^3$s, and none of fourteen $P^4$s.

The roughly square outline of $M^1$ has gently curved anterior and posterior sides, which tend to converge lingually. In unworn teeth, the parastylar fold is smoothly concave near the occlusal edge and narrows to a sharp crease toward the base, but does not continue to the base of the enamel. The ectoloph bears a very prominent dorsoventrally curved paracone rib and a distinct metacone rib. The axis of the metacone cants inward about 45 degrees relative to the sagittal plane (more than in other rhinocerotids) and the ectoloph is relatively short. This forces the metaloph into near alignment with the paracone and midsection of the ectoloph. The protocone bears shallow, rounded grooves delimiting a very weak antecrochet. A metaconule fold is lacking (in most specimens) or very obscure, and cristae are absent. The postfossette is proportionately as large as in *Subhyracodon*. One or two small accessory cuspules or low bulges (sometimes asymmetrical) occur near the lingual end of the median valley in three $M^1$ specimens, including the type, but are absent in eight others. The anterior cingulum extends from the lingual side of the parastyle to the anterolingual side of the protocone. Its edge in occlusal view is a simple arc, convex anteriorly, and in anterior view describes a ventrally concave curve, but with a slight bump where it passes the protoconule in some specimens. This shape is comparable to that of *Hyrachyus*, intermediate between the nearly straight anterior cingulum of *Triplopus* and the more complex curvature in *Subhyracodon*. A short, high posterior cingulum extends straight (in occlusal view) from the ectoloph to the posterolabial side of the hypocone and bears a slight notch behind the postfossette. None of the upper molars has a labial or lingual cingulum.

$M^2$ differs from $M^1$ in its larger size, more labially situated protocone, and more open parastylar fold. The junction of the occlusal edge of the unworn protoloph with the ectoloph is more nearly centered on the parastyle. The cross-lophs diverge a bit more than in $M^1$, and the occlusal surface of the metaloph aligns directly with the paracone in moderate wear stages. The median valleys of both $M^2$s of the type bear single accessory cuspules, not present in the other specimens.

Except for the reduction of the metastyle, $M^3$ differs little from $M^2$. The portion of the ectoloph between the paracone and metacone is a bit more oblique and the parastylar fold is more open and lacks the sharp crease at its basal termination. The metaloph is slightly expanded at the hypocone as it is in the preceding molars, in contrast with the anteroposteriorly flattened $M^3$ hypocone of *Subhyracodon*. The large $M^3$ sample (22 individuals) exhibits a range of variation in the prominence of the metastyle (Fig. 20.5) extending from the condition seen in the type of *Triplopus rhinocerinus* to that in some specimens of *Subhyracodon*. This structure typically arises as a sharp ridge at the labial end of the postfossette and becomes more rounded and flattened toward its occlusal end. In most specimens (including the type; Fig. 20.5B), the ridge apparently extended nearly to the unworn occlusal edge of the ectometaloph, but one specimen (not shown) bears a short ridge extending only 2 mm from the cingulum and has no rounded extension. At the slightly worn $M^3$ occlusal surface of the type, both the inner and outer sides of the ectometaloph have slight parallel flexures at the position of the metastyle. A somewhat variable postfossette extends anterolabially along the ectometaloph farther than in the other rhinocerotids, but less than in *Hyrachyus douglassi*. A small accessory cuspule near the opening of the median valley appears in 3 of 20 specimens. Another has a mure extending from the antecrochet to the metacone.

All of the permanent upper cheek teeth have at least patches of cement on the surface of the ectoloph, in the median valley, and in the postfossette. Cement covers the outer surface of the ectoloph of $M^1$ of the type except at the tip of the metacone, parastyle, and along the paracone. Other teeth have lesser amounts, usually restricted to concave areas.

A nearly unworn $dP^2$ has an asymmetric cordate outline, conspicuously notched between the paracone and metacone and narrowing posterolingually to a rounded point at the hypocone. The ectoloph resembles that of the permanent premolars. The more molariform lingual portion bears a well-developed metaloph and a very oblique protoloph which barely contacts the ectoloph. A small, sharp cingular cuspule blocks the lingual end of the median valley, and a well-defined cingulum extends from here along the protoloph to the ectoloph. The posterior cingulum extends lingually to the posterolingual side of the hypocone. The specimen lacks a crista.

$DP^3$ and $dP^4$ resemble $M^1$ in most respects, except their smaller size, proportionately lower crowns, and the presence (at least in $dP^3$) of a crista. The $dP^4$ specimen has a small cusp near the lingual end of the median valley, and a comparable feature appears to have been broken from the $dP^3$.

<u>Lower teeth:</u> None of the referred jaws has the first lower incisor crown in place, but the three specimens with intact anterior symphyses have roots or alveoli for a small, procumbent $I_1$.

Three jaws include complete $I_2$s, and UOMNH 20483 (Fig. 20.2C and D) is an isolated $I_2$ crown. These teeth are the enlarged, procumbent tusks characteristic of the Rhinocerotidae, but are proportionately smaller than those of other members of the family. The crown cross-section near the base is teardrop-shaped, with the narrow end directed ventromedially. Most of the dorsomedial surface of the crown lacks enamel. A low ridge on the dorsolateral surface of the tooth extends from the base of the enamel toward the tip, decreasing in prominence distally.

All adequately preserved specimens have two alveoli between $I_2$ and the diastema, indicating the presence of both $I_3$ and $C_1$. No $I_3$ crowns are preserved in place, but the alveoli suggest the $I_3$ is smaller than $I_1$ and slightly procumbent. The root of $C_1$ is larger and more erect than that of $I_3$. Its crown resembles that of the upper canine described earlier. The diastema between $C_1$ and $P_1$ averages 23 mm and ranges from 21 to 26 mm (five specimens).

(D)$P_1$ is a small, simple single-rooted tooth which was retained throughout life in all but one of the individuals adequately preserved. The crown is only a bit longer than that of the $C_1$, but not as high. It is lozenge-shaped in occlusal view and has a single anteroposterior ridge along the occlusal surface, terminating at a very faint posterior cusp.

$P_2$ is laterally compressed with distinct paralophid and metalophid, but small, oblique cross-lophids. The labial surface lacks the pronounced fold seen in *Subhyracodon* at the posterior end of the paralophid. The talonid is proportionately smaller and the paraconid less distinct than in other rhinocerotids, though both features vary within the sample. $P_2$ bears faint anterior and posterior cingula, but

Fig. 20. 6. Occlusal view of P$_2$-M$_3$ of *T. radinskyi*. UOMNH 21407.

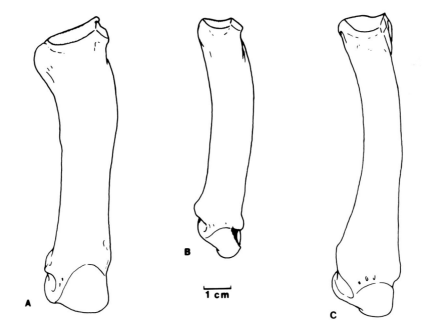

Fig. 20. 7. Right fourth metacarpals of rhinocerotids. A. *Trigonias* sp., a tetradactyl form (UCMP 32011, reversed); B. *T. radinskyi*, (UOMNH 28330); C. *Menoceras arikarense*. a tridactyl form (UCMP 39960). Facet for unciform (shaded) slants laterally in known tetradactyl ceratomorphs and medially in known tridactyl forms, indicating tridactyl manus for *T. radinskyi*.

lacks lingual and labial cingula.

$P_3$ has a nearly molariform trigonid, but the talonid is low and flat with an incomplete hypolophid. A small entoconid appears on most specimens, separated from the abbreviated hypolophid by a shallow cleft. A weak cingulum arches across the anterior face of the crown. The posterior cingulum is restricted to the area of contact with $P_4$ in most specimens, but extends around the talonid in a small percentage of the sample.

$P_4$ is proportionately wider and more molariform than $P_3$, but again the hypolophid is incomplete. The entoconid is even more distinct than in $P_3$ and, in most cases, is situated anterior to the lingual end of the hypolophid. The cingula are similar to those of $P_3$ in distribution and variability.

The lower molars also bear features intermediate between those of *Hyrachyus* and previously recognized rhinocerotids. Indentations on the anterior faces of the transverse lophids accentuate their component cusps so that the lingual and labial ends of the wear surfaces appear expanded, especially at the metaconid. The weak concave paralophid and convex metalophid descend steeply from their posterior connections with the cross-lophids in unworn teeth. The height of the junction between the metalophid and protolophid is little more than one-third the height of the unworn protoconid. The paralophid bends sharply inward near the anterolabial corner of the tooth, much as in *Hyrachyus*, but contrasting with the smoothly arcuate, longer paralophid of *Triplopus*. The anterior cingulum extends only around the transverse portion of the paralophid. The posterior cingulum occupies only the straight medial portion of the posterior molar face, and is weak or absent on $M_3$. No traces of lingual or labial cingula exist. Overall size and the angle between the paralophid and protolophid increase from $M_1$ to $M_3$.

$DP_2$ greatly resembles its permanent successor except in its lower crown, thinner enamel, and long, narrow proportions. The trigonid is especially elongate, and the weak metalophid slants backward more than in the permanent $P_2$.

$DP_3$ likewise differs from $P_3$ in its very long, narrow trigonid, but the talonid is almost fully molariform, differing from $M_1$ only in its narrower dimensions, slightly more oblique cross-lophids, lower crown, and thinner enamel.

*Manus.* -The prepared material from Hancock Quarry includes only one element referred to the manus of *T. radinskyi* -- a complete right fourth metacarpal (UOMNH 28330) -- but it is especially significant as it bears evidence that this species possessed a tridactyl manus. Its size, similarity to the fourth metacarpal of other rhinocerotids, and taphocoenotic association with abundant tooth-bearing specimens of *T. radinskyi* leave little doubt about the taxonomic assignment.

The specimen is 89 mm long and has a maximum transverse diameter of 20 mm across the distal end. In general proportions, it resembles metacarpal 4 of "*Caenopus*" *mitis*, not as slender as in *Menoceras arikarense* or *Subhyracodon occidentalis* (Fig. 20.7). The distal articular surface, however, differs from all of these and from *Trigonias osborni* (but resembles *Hyrachyus*) in its stronger convexity and in the marked step between the lateral and extended medial portions. The latter portion bears a very weak median keel. The triangular proximal end bears subdivided facets for both metacarpals 3 and 5.

Evidence of the lack of a fourth digit is offered by the orientation of the facet for the unciform (Fig. 20.7). This facet is saddle-shaped (transversely concave) in all ceratomorphs, and in *T. radinskyi* the lateral border of the facet is distinctly higher than the medial border. This condition is shared with "*Caenopus*" *mitis*, *Subhyracodon occidentalis*, *Menoceras*, *Hyracodon*, and *Colodon*, all ceratomorphs with a tridactyl manus, whereas in the tetradactyl forms (e.g., *Trigonias*, *Metamynodon*, *Za-*

*isanamynodon, Heptodon,* and *Tapirus*), the unciform facet slants laterally downward and the unciform is subequally shared by metacarpals 4 and 5.

While retaining some of the characters primitive with respect to other rhinocerotids, the metacarpal clearly differs from the long, slender elements which characterize even the least derived hyracodontids.

*Pes.* -The available assemblage from Hancock Quarry includes many more specimens pertaining to the pes of *T. radinskyi* (Fig. 20.8) than to its manus. These specimens were directly compared with elements referred to *Eotrigonias* (?) *mortivallis* Stock (1949) from the Titus Canyon Formation of California, Figured in his description.

Five astragali closely match a specimen [LACM (CIT) 3556] referred to *E.* (?) *mortivallis*. All of the major dimensions of the latter fall within the range of the Hancock Quarry sample. The most apparent difference is the more oblique orientation of the trochlear ridges of *T. radinskyi* with respect to the distal articulation for the navicular. It also has a more pronounced "neck" between the proximal and distal articulations, most noticeable on the lateral side, although in one specimen the medial trochlear ridge encroaches almost as closely on the navicular facet as it does in *E.* (?) *mortivallis*. *T. radinskyi* bears a slight lip on the lateral side of the trochlea, which, as Stock (1949) noted, is absent in *E.* (?) *mortivallis*.

The calcaneum of *T. radinskyi* is represented by four specimens. In addition to its more slender proportions, it differs from the calcanea of both *Subhyracodon occidentalis* (figured in Scott, 1941) and *Trigonias osborni* (UCMP 32011) in the high position and perpendicular orientation of the sustentaculum relative to the main body of the calcaneum. The oval astragalar facet on the sustentaculum compares with that of *Trigonias* but differs from the subquadrate facet in *Subhyracodon*. Just proximolateral to this facet is a

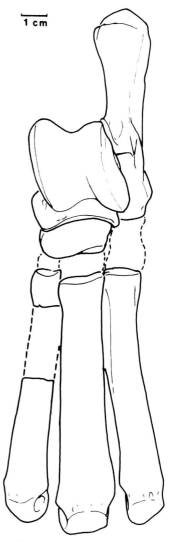

Fig. 20. 8. Composite reconstruction of left tarsus and metatarsus of *T. radinskyi*, dorsal view. Astragalus, UCMP 129061; calcaneum, UCMP 129057; navicular, UCMP 129066; proximal metatarsal 2, UCMP 129066; distal metatarsal 2, UCMP 129067; metatarsal 3, UCMP 129064; metatarsal 4, UOMNH 28338 (reversed).

shallow but distinct depression (absent in both *Trigonias* and *Subhyracodon*), probably homologous to a pit noted by Radinsky (1965) in the calcaneum of *Heptodon* to accomodate the tip of the fibula in extreme tibio-tarsal flexion. Two naviculars from the quarry differ slightly in thickness, and the thinner one resembles that of *S. occidentalis* in overall proportions.

Two incomplete metatarsal 2 specimens resemble the second metatarsal of *E*(?) *mortivallis* except in the more convex facets for metatarsal 3. Metatarsal 3 is proportionately narrower than that of either *Subhyracodon* of *Trigonias* and only slightly wider than the adjacent matatarsals. The proximal end is not expanded as in those genera. The fourth metatarsal of *T. radinskyi* (three specimens) is virtually identical in size and morphology to that of the cotype of *E*. (?) *mortivallis*. These show almost none of the reduction and lateral compression which characterize the medial and lateral metatarsals of the later rhinocerotids.

The distal articulations of all metatarsals are strongly biconvex and weakly keeled, and in metatarsals 2 and 4, the portion nearest the axis of the foot is extended. These are retained primitive characters modified in other rhinocerotids.

An isolated proximal phalanx may have belonged to either the second or fourth tarsal digit. It also closely resembles a comparable element of the *E*. (?) *mortivallis* cotype, but it tapers more toward the distal end. These elements are slightly longer than wide, intermediate in proportions between *Subhyracodon* and the stouter *Trigonias* phalanges.

**Referred Species**

Two species, originally referred to *Eotrigonias* but subsequently transferred to other genera, resemble *Teletaceras radinskyi* closely enough to warrant their inclusion in the same genus.

The genus name *Eotrigonias* is no longer available, as Radinsky (1967) transferred its type species, *E. rhinocerinus*, to *Triplopus*, an assignment with which I agree. Although *E. rhinocerinus* bears a very reduced $M^3$ metastyle, it otherwise resembles other species of *Triplopus* more than it does *Teletaceras*. Compared with the latter, the teeth are more brachydont and the $M^{1-2}$ postfossettes are smaller (primitive characters relative to *Teletaceras*), while the premolar ectolophs bear less prominent ribs and parastyles (derived features).

The lower cheek teeth of *Eotrigonias* (?) *mortivallis*, Stock (1949) from the Titus Canyon Formation, Inyo County, California, closely resemble those of *T. radinskyi*. The only known teeth of the former species are those of the type [LACM (CIT) 3564], with complete $M_2$ and broken $M_1$ and $M_3$) and a subsequently identified jaw fragment representing an older individual (LACM 61303, with incomplete $P_4$-$M_2$ from the same locality--CIT 254). The two species share the steeply descending molar paralophids and metalophids with low, weak connections of the metalophids to protolophids. These contrast with the higher, stronger crests of "*Trigonias* species C" of Russell (1982) (probably referable to *Penetrigonias* ) and other more derived rhinocerotids. Short, sharply bent paralophids and high crowns compare more favorably with *T. radinskyi* than with *Triplopus*. *E.* (?) *mortivallis* differs from *T. radinskyi* in its less obtuse metalophid-hypolophid angle, slightly less elongate trigonid, the absence of a distinct $P_4$ entoconid, and smaller size. The mean length of $M_1$ of *T. radinskyi* exceeds that of *E.* (?) *mortivallis* by 5 mm (4.7 standard deviations), and $M_2$ is even longer in proportion. There is no overlap in the ranges of any tooth measurements. Wood (1963) commented on this species, stating: "As will be shown fully elsewhere (MS), *Eotrigonias* (?) *mortivallis*. . . . is a composite form composed of hyracodont teeth and caenopine [i.e. rhinocerotid] foot bones. The specific name *mortivallis* must go in the genus *Hyracodon*, whatever its validity as a species."

Table 20.1. Statistical summary of upper cheek tooth measurements (in mm) for *Teletaceras*. A-P, anterior-posterior dimension along midline; Tr, transverse dimension from paracone across protocone; N, sample size; SD, standard deviation; CV, coefficient of variation. Measurements for *T. borissiaki* from Beliaeva (1959)

|       |      |    | *T. radinskyi* hypodigm |             |      |      | *T. borissiaki* |
|-------|------|----|------|-------------|------|------|------|
|       |      | N  | Mean | Range       | SD   | CV   |      |
| $P^1$ | A-P  | 5  | 12.2 | 10.9 - 13.7 | 0.95 | 7.83 |      |
|       | Tr   | 5  | 8.7  | 8.0 - 9.4   | 0.50 | 5.71 |      |
| $P^2$ | A-P  | 10 | 12.6 | 11.1 - 13.5 | 0.66 | 5.24 |      |
|       | Tr   | 10 | 16.3 | 15.0 - 18.5 | 1.05 | 6.45 |      |
| $P^3$ | A-P  | 11 | 14.7 | 13.4 - 16.0 | 0.68 | 4.64 |      |
|       | Tr   | 10 | 20.1 | 17.7 - 23.0 | 1.32 | 4.58 |      |
| $P^4$ | A-P  | 14 | 15.6 | 13.8 - 17.2 | 1.80 | 5.15 | --   |
|       | Tr   | 12 | 22.5 | 31.0 - 25.1 | 1.03 | 4.58 | 31   |
| $M^1$ | A-P  | 15 | 20.1 | 18.4 - 21.9 | 0.84 | 4.18 | 27 * |
|       | Tr   | 11 | 24.4 | 23.5 - 26.5 | 0.90 | 3.70 | 33   |
| $M^2$ | A-P  | 18 | 22.6 | 20.9 - 25.4 | 0.96 | 4.23 |      |
|       | Tr   | 14 | 26.7 | 25.0 - 28.8 | 0.98 | 3.68 |      |
| $M^3$ | A-P  | 20 | 22.1 | 20.5 - 25.5 | 1.26 | 5.70 |      |
|       | Tr   | 17 | 25.8 | 23.2 - 27.7 | 1.09 | 4.24 |      |
| $P^{1-4}$ |  | 4  | 55   | 53.0 - 58   | 1.85 | 3.35 |      |
| $M^{1-3}$ |  | 13 | 64   | 59.0 - 70   | 2.67 | 4.17 |      |
| $P^{1-3}$ |  | 3  | 121  | 119.0 - 124 | 2.05 | 1.70 |      |

* Estimated from figure: Length given by Beliaeva (31.0) is along ectoloph.

Table 20.2. Statistical summary of lower cheek tooth measurements (mm) for *Teletaceras*. A-P, anterior-posterior dimension along midline; TrA, transverse dimension across trigonid; TrP, transverse dimension across talonid; N, sample size; SD, standard deviation; CV, coefficient of variation; @, approximate.

| | | *T. radinskyi* hypodigm | | | | | *T. mortivallis* | |
| | | | | | | | LACM (CIT) | |
| LACM | | N | Mean | Range | SD | CV | 3564 | 61303 |
|---|---|---|---|---|---|---|---|---|
| $P_1$ | A-P | 6 | 7.5 | 6.9 - 7.8 | 0.31 | 4.2 | | |
| | TrA | 5 | 4.5 | 4.2 - 4.7 | 0.21 | 4.6 | | |
| $P_2$ | A-P | 16 | 12.9 | 12 - 15.4 | 0.76 | 5.9 | | |
| | TrA | 18 | 7.7 | 7.1 - 8.8 | 0.41 | 5.3 | | |
| $P_3$ | A-P | 25 | 15.9 | 14.6 - 17.5 | 0.66 | 4.1 | | |
| | TrA | 24 | 9.7 | 8.9 - 10.9 | 0.50 | 5.1 | | |
| | TrP | 23 | 9.7 | 8.7 - 10.8 | 0.60 | 6.1 | | |
| $P_4$ | A-P | 35 | 16.6 | 14.7 - 18.6 | 0.85 | 5.1 | 13 @ | |
| | TrA | 31 | 11.4 | 10.2 - 12.4 | 0.49 | 4.3 | | |
| | TrP | 32 | 11.2 | 10.4 - 12.1 | 0.52 | 4.7 | | |
| $M_1$ | A-P | 24 | 19.4 | 18 - 21.7 | 0.99 | 5.1 | 15.2 | 13.6 |
| | TrA | 23 | 12.7 | 11.3 - 14.0 | 0.71 | 5.6 | 10.2 | |
| | TrP | 27 | 13.2 | 11.6 - 14.5 | 0.66 | 5.0 | | |
| $M_2$ | A-P | 38 | 22.6 | 20.5 - 24.8 | 1.03 | 4.6 | 16.7 | 18 @ |
| | TrA | 33 | 14.2 | 13.3 - 15.5 | 0.61 | 4.3 | 11.1 | |
| | TrP | 32 | 14.7 | 12.8 - 17.2 | 0.87 | 6.0 | 11.9 | |
| $M_3$ | A-P | 31 | 23.7 | 22 - 25.7 | 0.88 | 3.7 | 11 @ | |
| | TrA | 28 | 14.7 | 13.6 - 16 @ | 0.56 | 3.8 | | |
| | TrP | 32 | 13.9 | 12.7 - 15.3 | 0.58 | 4.1 | | |
| $P_{1-4}$ | | 13 | 50.2 | 44 @ - 57.2 | 3.54 | 7.1 | | |
| $M_{1-3}$ | | 26 | 64.8 | 60 - 70.5 | 2.38 | 3.7 | 49 | |
| $P_1$-$M_3$ | | 10 | 113.1 | 105.5 - 124.3 | 5.01 | 4.4 | | |
| $dP_2$ | A-P | 1 | 14.5 | | | | | |
| | TrA | 1 | 7.2 | | | | | |
| $dP_3$ | A-P | 1 | 20.5 | | | | | |
| | TrA | 1 | 8.6 | | | | | |
| | TrP | 1 | 10.4 | | | | | |
| $dP_4$ | A-P | 2 | 17.9 | 17.0 - 18.8 | | | | |
| | TrA | 2 | 10.0 | 9.7 - 10.3 | | | | |
| TrP | | 2 | 10.4 | 10 - 10.8 | | | | |

The mentioned manuscript apparently was never published, but Stock's (1949) referral of the teeth and foot bones to a single species now appears justified by their similarity (discussed earlier) to comparable elements of *T. radinskyi*. The resemblances are great enough to dictate assignment of the two populations to the same genus, but differences in the size and morphology of the cheek teeth warrant species-level separation. I therefore transfer *E. (?) mortivallis* to *Teletaceras*.

The fauna associated with *T. mortivallis* indicates an early Chadronian age (Stock, 1949), somewhat younger than the Hancock Quarry l.f.

*Eotrigonias borissiaki* Beliaeva (1959) has suffered a complex nomenclatorial history, at least partly owing to the limitations of the type material. This consists of a maxillary fragment with $M^1$ and incomplete $P^4$ from the Artém coal field, Maritime Province, Soviet Union. Wood (1963) transferred the species to his new genus *Pappaceras*, despite several noted differences and only one positive comparison, "the distinction of the parastyle of $M^1$, much as in *Pappaceras confluens*." Even this similarity is questionable, given the available illustrations. Radinsky (1967) nonetheless agreed with this assignment but synonymized the entire genus *Pappaceras*, as well as *Juxia*, with *Forstercooperia*. Most recently, Lucas, Schoch, and Manning (1981) reassigned *E. borissiaki*, along with *Juxia sharamurenense* Chow and Chiu (1964) to a single monotypic species, *Juxia borissiaki*, citing similar size and "morphologically identical" $P^4$ and $M^1$ of the types of *E. borissiaki* and *J. sharamurenense*. The type of the latter species strongly resembles that of *F. ergilinensis*, and these may well be conspecific. However, the type of *E. borissiaki* bears much greater resemblance to *Teletaceras radinskyi* and differs from *J. sharamurenense* in the following characters: 1) parabolic rather than semicircular lingual margin of $P^4$; 2) incomplete lingual $P^4$ cingulum bearing a bulge at the posterolingual side, as opposed to a complete, uniform lingual cingulum; 3) lingually convergent, rather than subparallel anterior and posterior sides of $M^1$; 4) small, anteriorly directed $M^1$ parastyle, not large and labially deflected; 5) strong paracone rib on $M^1$ ectoloph; 6) narrow connection of $M^1$ protoloph to inner side of parastyle rather than broad connection at paracone ; 7) $M^1$ metaloph short, aligned with paracone at moderate wear stage, and lingually divergent from protoloph--narrow median valley; 8) $M^1$ hypocone smaller with roughly teardrop-shaped (not circular) basal outline; 9) small bridge connecting midpoint of anterior cingulum to protoloph of $M^1$; 10) posterior $M^1$ cingulum short and broad, terminating at posterolabial (vs. posterolingual) side of hypocone; 11) large, anteroposteriorly elongate postfossette; and 12) larger $M^1$ antecrochet. The thickened posterior part of the $M^1$ ectoloph, cited by Beliaeva, may resemble that of *J. sharamurenense* more than *T. radinskyi*, though this is difficult to ascertain from the Fig ures. In size, *E. borissiaki* falls between known specimens of those species, and produces exceptionally large tooth size ranges when considered conspecific with *F. ergiliinensis* .

*Eotrigonias borissiaki* differs somewhat less from the North American genus *Penetrigonias* (discussed later) than from *Juxia*. However, except for its size, it differs from *Penetrigonias* and resembles *T. radinskyi* in its weaker $M^1$ antecrochet, more lingually inflected metacone, more prominent metacone rib, and more directly aligned metaloph, mid-ectoloph, and paracone. Its $P^4$ lingual cingulum is much less complete than in *Penetrigonias*. I therefore recognize *E. borissiaki* as a species of *Teletaceras*.

Notably, the only other published taxon from the Artém l.f., a large amynodontid, *Procadurcodon orientalis*, is congeneric with an undescribed species from

Hancock Quarry. The ages of these faunas therefore appear similar, but it is not clear which is older.

## Discussion

*Teletaceras* and *Penetrigonias* exhibit few autapomorphic characters and fill much of the morphologic gap which formerly separated *Hyrachyus* and known rhinocerotids. Details of this transition can now be re-examined at higher resolution. A number of unpublished or incorrectly referred specimens demonstrate pertinent characters of the related taxa, and deserve mention here.

A well-preserved skull and mandible, UW 1937, from the Washakie B (early Uintan), Sweetwater County, Wyoming, bears a dentition (including unspecialized lower incisors) which closely resembles the type of *Hyrachyus douglassi* Wood (1934). The UW specimen is tentatively referred to that species for purposes of the present discussion. Most of the major skull characters of this advanced *Hyrachyus* are retained in *T. radinskyi*; dorsal skull profile nearly parallel to the tooth row; gently convex, narrow sagittal crest; skull elongate in dorsal view; and occipital crests short transversely and separated by a small medial notch. The basicranium of UW 1937 is almost identical with that of *T. radinskyi* except that the posttympanic and paroccipital processes are slightly shorter in the former. Derived skull characters of *T. radinskyi* relative to *H. douglassi* are narrower free portion of nasals, small process on nasals above nasal incision, straighter nasal-maxillary suture, reduced premaxilla not contacting nasals, slightly reduced sagittal crest, slightly upturned occipital crest, notch at posterior end of zygomatic arch, and the dental features described above. Although the skulls are virtually identical in size, the cheek tooth series of *T. radinskyi* is about 25% longer than that of *H. douglassi*, the postcanine diastema is correspondingly reduced, and the canines are slightly smaller. Overall, the skull of *T. radinskyi* resembles that of *Hyrachyus douglassi* more than it does other rhinocerotids.

Several Chadronian localities in North America (other than Titus Canyon) have yielded specimens of small rhinocerotids which resemble *Teletaceras* in some features but exhibit a number of derived characters. For one of these specimens, Tanner and Martin (1976) named a new genus and species, *Penetrigonias hudsoni*, based on a specimen with $P^{2-4}$ from Sioux County, Nebraska, for which familial assignment was uncertain. These premolars are almost indistinguishable from those included in a nearly complete and clearly rhinocerotid palate (with $P^1$-$M^3$; University of Minnesota, unnumbered specimen) from the Chadron Formation, Pennington County, South Dakota. The two specimens almost certainly represent the same species. Other specimens from the Yoder Formation in Goshen County, Wyoming (SDSM 6353), Porvenir l.f., Texas (TMM 40807-6; Wilson and Schiebout, 1984), the White River Group in Natrona County, Wyoming (AMNH 105019), and the Cypress Hills Formation in southwestern Saskatchewan (SMNH P1635.2, type of *Subhyracodon sagittatus* Russell, 1982, its hypodigm, and SMNH P1635.1, "*Trigonias* species C " in Russell, 1982), all resemble *Penetrigonias hudsoni* to varying degrees. They may represent two or three additional species of that genus, as dental differences are greater than usually seen in a single rhinocerotid species, though all are similar in size ($M^{1-3}$ length 72 to 79.5 mm--about 3-15% larger than the largest *T. radinskyi*). These specimens were assumed congeneric in the analysis presented in Prothero, Manning, and Hanson (1986). Table 1, character set 28 of that paper summarizes the characters which distinguish *Penetrigonias* (and more derived rhinocerotids) from *Teletaceras*.

Of particular interest is the sequence of changes in three characters which have almost invariably entered discussions of the initial evolution of rhinocerotids: the incisor complex, the $M^3$ metacone, and the fourth carpal digit.

*Teletaceras* demonstrates that the lineage ultimately leading to modern rhinos had acquired the uniquely specialized $I^1/I_2$ tusks before the primitive $M^3$ metastyle was completely lost. Specimens here referred to *Penetrigonias* bear either a very faint $M^3$ metastyle ridge or no trace of a metastyle. This further verifies Radinsky's (1966) contention that more than one rhinocerotoid group independently lost the metastyle (it is also completely lost in some hyracodontids), and underscores the limitations of this character in phyletic interpretation.

The evidence for a tridactyl manus in *Teletaceras* is more surprising in light of the generally primitive morphology of the genus and the entrenched assumption that a tetradactyl manus was primitively retained in some of the more advanced rhinocerotids, such as *Trigonias*. Though still open to the interpretation that the loss of metacarpal 5 is an autapomorphic character of *Teletaceras* (independent of its loss in other rhinocerotids), this evidence suggests an alternative hypothesis, that a tridactyl manus is primitive for rhinocerotids plus hyracodontids. A secondarily "revived" fourth digit has already been proposed for the Aceratheriinae (Prothero, et al., 1986). The same may be true for *Trigonias*. The manus is unknown for both *Hyrachyus douglassi* and *Penetrigonias*.

## Acknowledgments

I thank the many individuals who have contributed to the ongoing investigation of the Hancock Quarry l.f., a small part of which is reported here. The Oregon Museum of Science and Industry, through its Hancock Field Station staff, volunteers, and facilities, provided invaluable support during several seasons of field work. Dr. J. Armentrout, former director of the field station, initiated a volunteer excavation program at Hancock Quarry, partly funded by the National Science Foundation, and invited my participation in the project. Most of the tedious excavation was conducted by the student volunteers of the Paleontology Research teams whose dedication and contribution cannot be acknowledged adequately. Specimens collected under this program have been placed on permanent loan from OMSI to UCMP where they were prepared and curated. Very large collections made earlier by UOMNH were generously loaned under the directorship of Dr. J. A. Shotwell for inclusion in this study. I also wish to thank the following individuals for casts and loans of specimens from their institutions: Dr. L. G. Barnes, E. Manning, Dr. R. E. Sloan, and W. Wehr.

Dr. D. E. Savage originally suggested the investigation of the fauna, and has provided advice and support throughout the project. Dr. H. Schorn advised and assisted in photographing the Fig ures.

## Bibliography

Beliaeva, E. I. (1959): Sur le découverte de rhinoceros tertiares anciens dans la Province Maritime de l'U.R.S.S. -*Vert. PalAsiatica*, 3: 81-91.

Chow, M., and Chiu, C. (1964): An Eocene giant rhinoceros. -*Vert. PalAsiatica*, 5: 264-267.

Chow, M., and Xu, Y. (1961): New primitive true rhinoceroses from the Eocene of Iliang, Yunnan. -*Vert. PalAsiatica*, 5: 291-304.

Fiebelkorn, R. B., Walker, G. W., MacLeod, N. S., McKee, E. H., and Smith, J. G. (1983): Index to K-Ar determinations for the state of Oregon. -*Isochron West*, 37: 3-60.

Gabunia, L., and Dashzeveg, D. (1974): Ob Oligotsenovom predstavitele *Forstercooperia* (Hyracodontidae) iz Mongolii [On the Oligocene representatives of *Forstercooperia* (Hyracodontidae) from Mongolia]. -*Soobshcheniya Akad. Nauk Gruz. SSR*, 75: 497-500.

Lucas, F. A. (1900): A new rhinoceros, *Trigonias osborni*, from the Miocene of South Dakota. -*U.S. Nat. Mus. Proc.*, 23: 221-224.

Lucas, S. G., Schoch, R. M., and Manning, E. M. (1981): The systematics of *Forster-*

*cooperia*, a middle to late Eocene hyracodontid (Perissodactyla: Rhinocerotoidea) from Asia and western North America. -*J. Paleont.*, 55: 826-841.

Owen, R. (1845): *Odontography; or a treatise on the comparative anatomy of teeth.*- London (Hippolyte Bailliere).

Prothero, D. R., Guérin, C., and Manning, E. M. (1989): The history of the Rhinocerotoidea (this volume).

Prothero, D. R., Manning, E. M., and Hanson, C. B. (1986): The phylogeny of the Rhinocerotoidea (Mammalia, Perissodactyla). -*Zool. J. Linn. Soc.*, 87: 341-366.

Radinsky, L. B. (1965): Evolution of the tapiroid skeleton from *Heptodon* to *Tapirus*. -*Bull. Mus. Comp. Zool.*, 134: 69-106.

Radinsky, L. B. (1966): The families of the Rhinocerotoidea (Mammalia, Perissodactyla). -*J. Mamm.*, 47: 631-639.

Radinsky, L. B. (1967): A review of the rhinocerotoid family Hyracodontidae (Perissodactyla). -*Bull. Amer. Mus. Nat. Hist.*, 136: 1-46.

Russell, L. S. (1982): Tertiary mammals of Saskatchewan, Part VI: The Oligocene rhinoceroses. -*Royal Ont. Mus. Life Sci. Contrib.*, 133: 1-58.

Scott, W. B. (1941): Perissodactyla. -In: Scott, W. B. and Jepsen, G. L. (eds.): The mammalian fauna of the White River Oligocene. -*Trans. Amer. Phil. Soc.*, n.s., 28: 747-980.

Stock, C. (1949): Mammalian fauna from the Titus Canyon Formation, California. -*Publ. Carnegie Inst. Wash.*, 584: 229-244.

Tanner, L. G., and Martin, L. D. (1976): New rhinocerotoids from the Oligocene of Nebraska. -In: Churcher, C. S. (ed.), *ATHLON: Essays in Paleontology in honor of Loris Shano Russell*: Royal Ont. Mus. Life Sci. Misc. Publ., pp. 210-219.

Wilson, J. A. (1977): Stratigraphic occurrence and correlation of early Tertiary vertebrate faunas, Trans-Pecos, Texas, Part 1: Vieja area. -*Tex. Mem. Mus. Bull.*, 25: 1-42.

Wilson, J. A., and Schiebout, J. (1984): Early tertiary vertebrate faunas, Trans-Pecos Texas: Ceratomorpha less Amynodontidae. -*Texas Mem. Mus. Pearce-Sellards Ser.*, 39: 1-47.

Wood, H. E., II (1927): Some early Tertiary rhinoceroses and hyracodonts. -*Bull. Amer. Paleont.*, 13: 165-264.

Wood, H. E., II (1929): *Prohyracodon orientale* Koch, The oldest known true rhinoceros. -*Amer. Mus. Novit.*, 395: 1-7.

Wood, H. E., II (1934): Revision of the Hyrachyidae. -*Bull. Amer. Mus. Nat. Hist.*, 67: 181-295.

Wood, H. E., II (1938): *Cooperia totadentata*, a remarkable rhinoceros from the Eocene of Mongolia. -*Amer. Mus. Novit.*, 1012: 1-20.

Wood, H. E., II (1963): A primitive rhinoceros from the late Eocene of Mongolia. -*Amer. Mus. Novit.*, 2146: 1-11.

# 21. THE RHINOCEROTIDAE

## KURT HEISSIG

The phylogeny and classification of the Rhinocerotidae are revised on the basis of the newly introduced characters presented by Prothero, Manning, and Hanson (1986) and Groves (1983) and my own observations since the revised classification of Heissig (1973a). The characters used here are discussed in detail, especially to avoid too many parallelisms. The presence of a functional fifth metacarpal in the two fore feet of *Juxia sharamurunense* Chow and Xiu in the American Museum serves the key to the controversy over whether the first true rhinoceroses had a tridactyl or tetradactyl manus. The evidence of a strong relationship of the American *Diceratherium* group with the basic stock of the rhinoceroses, especially the genus *Trigonias*, and on the other hand, the fundamental differences from *Menoceras*, as stated by Prothero, Manning and Hanson (1986), have modified our knowledge of the early history of the family. The hypothesis of a common ancestry of the Teleoceratini and the Rhinocerotinae, brought forward by the same authors, is rejected here. Some characters limiting the adaptational potential of subfamilies and tribes are analyzed in relation to the behavior of the animals.

## Introduction

During the nearly thirty years in the middle of our century, when no specialists in the western world were concerned with detailed study or classification of the rhinoceroses, a huge mass of undescribed material was stored in the museums and collections all over the world. Modern means of transportation led to more extensive digging, and as a result, vast and only partially identified collections of fossil rhinoceroses were waiting for their adequate descriptions or monographs. Since the group was again taken into consideration in the beginning of the 1970s by Guérin, Ginsburg, and Heissig in Europe, Radinsky a little earlier in the United States, and, later, Fortelius in Finland, all authors have felt that a thorough revision of the classification was needed. The basis of the classification was fixed by Radinsky (1966) by restricting the family to members with the chisel-tusk shearing complex of $I_2$ and $I^1$, and their descendants. The later attempts by Heissig (1973a) as well as the phylogenetic and systematic hypotheses presented in this volume, must remain provisional until the materials already collected are described and used as a base of a new classification. We are still far from this goal, but we now need a classification to work with and to arrange our materials. A revised version of my classification (1973a) is presented here, changed by a better understanding of the American species and some strong arguments of my American colleagues.

## Characters and parallelisms in the Rhinocerotidae

Using characters for a phylogenetic analysis means avoiding parallelisms. Most gradually changing characters are an expression of a general tendency among the whole group whereas discrete characters, especially when new structures are formed, may be unique and therefore key characters for the analysis. The loss of an element or a structure may occur very easily and is always suspected to be subject to parallelisms. Nevertheless, we can even use parallel evolved structures, if they follow different ways in different subgroups. In the following list, a lot of single characters currently

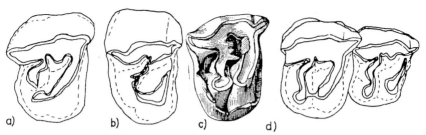

Fig. 21.1. Molarization stages of premolars: a) premolariform b) submolariform c) paramolariform d) semimolariform and molariform

used in the literature, and most of the characters used by Groves (1983) and Prothero, Manning, and Hanson (1986) are arranged in classes of significance. Some missing characters may be ascribed to a supergroup of characters reflecting merely one general tendency. The general significance of others is not sufficiently known to me.

**General trends**

*Skull shortening*-- can be used on the specific, and possibly on the generic level. Single traits of this tendency may have systematic value, as in the shortening of the basicranium. In most cases, we have to consider allometric shortening of face, cranium, and even smaller parts of the skull.

*Skull broadening*-- mainly, but not exclusively, combined with shortening. I have found no means to detect parallelisms.

*Deepening of the nasal notch*-- occurs in most lineages at different times. Forms with strong horns must solve mechanical problems and may be distinguished by the special features of the construction of the nasal bones.

*Molarization of premolars*-- a general tendency in all tribes, but very slow and subject to a high degree of variation in the maxillary. In the mandible, it is completed early. In the advanced forms, it is delayed and possibly even reversed in high-crowned forms. Most, but not all, Rhinocerotini and Dicerotini follow a different way of molarization, passing through a paramolariform instead of a semimolariform stage (Fig. 21.1).

*The complication of the crown pattern of upper teeth by secondary folds as cristae and crochets*-- occurs in most lineages, but may follow different patterns. Parallelisms are extensive, high variability allows reversals. Primitively there is a crochet present in the molars, a crista in the premolars. The crochet arising with molarization in premolars is often split in several short folds in the early stages of molarization, but may unite as a single fold with the crista. The crista of molars is not homologous with the crista of tapiroids, but arises between paracone and metacone.

*Limb shortening*-- is correlated as a general character with the increase of body weight. If exceeding the limits of mediportal conditions it can be used as a character of a group.

*The increase of crown height of the cheek teeth*-- is confined to progressive forms in most groups and may indicate a change of diet, but not necessarily grazing. In most cases it causes a delay of molarization and the formation of cement. In the Elasmotheriini it is a general character, exceeding the degree reached in other groups.

*The reduction of the posterior cingulum in $M^3$*-- is a general character in Rhinocerotidae. It is a good character if reversed. The specific morphology reached by this process differs considerably in several lineages.

*The reduction of the lingual cingula in upper premolars*-- occurs in most, but not all tribes at different times, to a different degree. Reversals of the trend are possible.

*The reduction of the metacone rib of up-*

*per teeth*-- occurs earlier in molars than in premolars. Timing and degree are different in the lineages, but it is going on in nearly all tribes. There is no known reversal.

*The closure of the subaural channel*-- a single trait of skull shortening and therefore occurring in several tribes, but sometimes significant within a lineage. There are no reversals.

*The reduction of the anterior crest of the paralophid in $dP_3$*-- is evolved several times in single genera of different tribes. There are no reversals. In *Coelodonta* it is compensated by a splitting of the paralophid in $dP_2$.

Some characters may be differentiated in divergent directions coming from an intermediate primitive stage. They are listed here, being equally widespread and gradual in change:

*The direction of the premaxillae*-- originally downsloping to enable the contact of the incisors, may be changed to horizontal, when the lower incisors are coming up by curving or by a more upright implantation. In some early genera, such as *Ronzotherium*, there seems to be a slight development in the opposite direction.

*The orientation of the occipital plate*-- reflects the normal skull position and is dependent upon feeding habits. There are deviations from the right angle in both directions, but overhanging of the occipital crest is more widespread and stronger developed, because of its correlation with grazing (Fig. 21.2).

*The articulation of the fibula with the femur*-- may be a primitive feature as indicated by its presence in the tapirs. It is suppressed in different tribes, but the present knowledge of its occurrence is still insufficient. Its occurrence is suppressed in the fossil *Diceros neumayri*, but still developed in both recent Dicerotini. So we must assume the possibility of reversals.

### Trends observed only in single groups

*Further shortening of the limbs, exceeding mediportal conditions*-- is a typical feature in the Teleoceratini. Only the distal limb segments are shortened.

*Shortening of the whole limb length*-- is confined to the *Chilotherium* group, including some *Peraceras* of the Aceratheriini.

*Size increase of the $I_2$*-- is a trend observed in all other subfamilies except Rhinocerotinae.

*The gradual reduction of $P_1$*-- is a character quite different from its early loss in ontogeny. It is observed in most derived genera of the Aceratheriinae and several Diceratheriinae and Elasmotheriini.

*The reduction of protocone constriction from behind and merging of protocone and antecrochet*-- occurs as a general trend among Rhinocerotini and Dicerotini, but is observed in one single species of *Subchilotherium* too.

*The strengthening of the posterior protocone fold*-- the opposite trend, coming from the intermediate primitive condition, occurs as a group tendency independently in Aceratheriinae and Elasmotheriini.

### The loss of characters or elements

*The loss of the distal part of metacarpal V and its digit*-- occurs in all tribes except the stem group Trigoniadini, mostly in the advanced members. In the Rhinocerotinae alone it as a general character of the whole subfamily.

*The loss of all incisors*-- is confined to the Rhinocerotinae, but not as a general character. The reduction is normally rapid, but can be followed in some lineages (Fig. 21.2).

*The loss of $I_1$*-- is correlated with the increase of $I_2$ and a narrow symphysis. It occurs several times in distantly related genera.

*The early loss of $P_1$*-- is a highly variable character and may be subject to reversals easily. The same is true of its prolonged retention.

*The loss of the median lower crest of the mandibular symphysis*-- occurs early in the history of the family, but may be useful to separate lineages in the early evolution.

| | Aceratheriinae (incisors strong, horn weak) | | | | Rhinocerotinae (horn strong, incisors weak or lost) | | | | | |
|---|---|---|---|---|---|---|---|---|---|---|
| | Teleoceratini (short limbs) | | Aceratherini (slender limbs) | | Rhinocerotini and Dicerotini (primitive) | | | Elasmotherini (early derived) | | |
| Grazers | *Teleoceras* (Miocene) | *Aprotodon* (Oligo-Miocene) | *Acerorhinus* (Mio-Pliocene) | *Chilotherium* (Miocene) | *Rhinoceros unicornis* (Recent) | *Ceratotherium* (Recent) | *Coelodonta* (Pleistocene) | *Hispanotherium* (Miocene) | *Iranotherium* (Miocene) | *Elasmotherium* (Pleistocene) |
| Type of adaptation | Limb shortening | Incisors diverging, symphysis broad | Limb shortening | Limb shortening and incisor divergence | Teeth hypsodont Head uptilted | Incisors lost Head lowered | Incisors lost Head lowered | Head lowered Incisors reduced | Head lowered Incisors lost | Head lowered Incisors lost Horn shifted posteriorly |
| Example for browsers | *Brachypotherium* (Miocene) | *Prosantorhinus* (Miocene) | *Aceratherium* (Miocene) | *Mesaceratherium* (Oligo-Miocene) | *Rhinoceros sondaicus* (Recent) | *Diceros* (Recent) | *Dicerorhinus* (Recent) | Unknown | | |

**Stem group**  
**Browsers only**  
Diceratheriinae  
(no known adaptation to grass diet)  
(balance of horn and incisors, or hornless)

Fig. 21.2. The adaptation of the Rhinocerotidae to a grass diet.

*The loss of the horn*-- is confined to the Aceratheriinae, occurs in most, not in all lineages at different times.

## Single characters with some parallelisms

*The formation of a horn*-- occurs three times independently, but may be distinguished by the type of horn, which is unique in each case (Fig. 21.2).

*The reduction and loss of the third articulation between radial and intermedium*-- occurs twice, at the base of the Aceratheriini and the Elasmotheriini, as specialization of the more long-legged slender limbs of these tribes.

*The broadening of the mandibular symphysis*-- occurs twice as an extreme form of specialization in *Aprotodon* and *Chilotherium sensu stricto* but combined with a different position of the incisors.

*The sharpening of the ventral edge of the vomer*-- a character found in *Rhinoceros* only by Groves (1983) is paralleled by *Chilotherium sensu stricto* and may be due to facial shortening.

*A long paralophid in lower molars (and premolars)*-- is a progressive character of Rhinocerotinae, but also present in some unrelated forms. It can be reduced again.

*The closing of the medisinus in upper molars by swollen lingual cusps*-- occurs several times in rather high-crowned forms. It is formed by an antecrochet in *Chilotherium*, by the protocone in *Rhinoceros* and *Ceratotherium*.

*The formation of a postorbital process*-- occurs in several genera, but may be significant in a single lineage.

*A convex lower margin of the mandible*-- was achieved by some unrelated genera, but is of high generic value.

*The inclination of the ramus mandibulae*-- depends upon skull shortening and skull position. Reversals of a tendency are possible.

## Characters with no or rare parallelisms

*The articulation of the ulna with the intermediate*-- is a character uniting all Rhinocerotinae, but never observed in any other tribe.

*The double-rooted $P_1$*-- is an apomorphic character, even if the function is not clear. It is observed in all Rhinocerotinae, where it is retained, and in some specimens of *Subhyracodon*.

*The craniocaudal elongation of the distal facet of the intermedium for the ulnar*-- unites the Rhinocerotini and Dicerotini and is not observed in any other tribe.

*The loss of the naso-lacrimal contact* -- is an autapomorphy of Dicerotini.

*The short trochiter of the caput humeri (tuberculum maius)* - also in the Dicerotini, is autapomorphic.

There are many characters of this class, autapomorphic for single genera. They are not treated here, because the present knowledge does not yet allow a cladogram of all genera. For the Rhinocerotini see Groves (1983).

## Origins

Within the Rhinocerotoidea, the Rhinocerotidae are defined by Radinsky (1966) by the unique shearing complex of $I_2$ against $I^1$. There are many other synapomorphies uniting the family, as shown by Prothero, Manning, and Hanson (1986). The origin of the family was a mystery for a long time because of its retention of a fourth digit in the manus, which was not known in any suspected ancestor, and was definitely lost in all known hyracodonts. The idea of the reappearance of this digit in the Rhinocerotidae is rejected here because it is in the most primitive members of the family where it occurs, being lost in many advanced genera.

Two fore feet of *Juxia sharamurunense*, that I found in the immense treasures of Osborn's Mongolian expeditions, housed in the American Museum of Natural History, show clearly a fully developed fifth metacarpal and its phalanges. So we must now consider the indricotheres as the sister-group of Rhinocerotidae, bound together by a primitive tetradactyl manus, a similar molar and premolar pattern of their earliest members

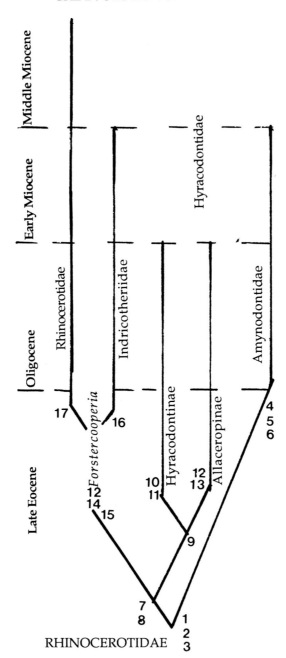

Fig. 21.3. The origin of the Rhinocerotidae. Characters: 1) $M_3$ hypoconulid lost; 2) trigonid lengthened in lower molars; 3) parastyle fused to ectoloph; 4) extreme lengthening of molars; 5) strong canine; 6) limbs massive; 7) lower premolars lengthened; 8) upper premolars broadened; 9) tridactyl manus; 10) diastema shortened; 11) incisiform canine; 12) reduced metastyle of $M^3$; 13) $I_3$ lost; 14) patella broadened asymmetrically; 15) $I^1$ enlarged; 16) $I_1$ enlarged; 17) $I_2$ enlarged.

and the enlarged $I^1$. Both are together the sister-group of the remaining Hyracodontidae *sensu stricto*. It is therefore necessary to exclude the Indricotheriidae again from the hyracodonts and to keep them separate on a family level (Fig. 21.3).

## The earliest Rhinocerotidae

With the beginning of the Oligocene, we find a pair of genera in both North America and Europe. Their common features may represent characters of the common ancestor of the whole family. The skull is narrow and long with tapering hornless nasals and a rather shallow nasal notch. The enlarged lower $I_2$ is straight and nearly horizontally implanted. The upper first incisor is less enlarged and meets the lower one. The upper dentition is still complete. Canine and third incisors are lost in the mandible. The outer wall of the premolars is undulating in the same way as in the indricotheres, with broad paracone and metacone ribs. Sometimes there are sharp metacone ribs in the molars. The premolars are faintly or not molarized. Even in the lower premolars, the entoconid may be isolated or lacking entirely.

The smaller genera of both continents represent the main evolutionary lineages, *Trigonias* Lucas (1900) in America, and *Epiaceratherium* Abel (1910) in Europe. They have larger, more elongate $I^1$, similar to all later Rhinocerotidae, but retain $I^3$ and upper canines. The lower big incisor is rather short with a trigonal outline of the crown and a trigonal cross section. The larger genera *Amphicaenopus* Wood, 1927, in America and *Ronzotherium* Aymard, 1856, in Europe are also similar to each another. The $I^1$ is less elongate, still nearly conical in *Ronzotherium*, but $I^3$ and canine are lost. The first upper incisor in *Amphicaenopus* is intermediate, but more primitive than in *Trigonias*. Both genera have straight, elongate lower tusks with an oval cross sction. They disappear during the Oligocene without descendants.

## The first evolutionary lineages

The record of the early rhinoceroses is poor in the Old World, and rich in America. It is easy to follow the lineage from *Subhyracodon* Brandt, 1878, to *Diceratherium* Marsh, 1875, a sister-group of *Trigonias* with tridactyl manus and without $I^3$ and upper canine. The structure of the upper premolars is very similar to *Trigonias* and the primitive arrangement of the ridges on the outer wall is preserved. In *Diceratherium* the first horns in rhinoceros evolution are formed as lateral protuberances at the sides of the nasals. These groups, mainly known from America, should be united on the subfamiliar level as two tribes: Diceratheriini Dollo, 1885, and Trigoniadini nov. trib. (Fig. 21.4).

The Diceratheriini found their way to the Old World only for a short time. In the upper Oligocene of France, a single skull was found and named "*Rhinoceros pleuroceros*" by Duvernoy (1853). It shows clearly the lateral protuberances of *Diceratherium*, but at the moment it is not clear if it should retain its generic name *Pleuroceros* Roger, 1904, or if it would be better to include it in *Diceratherium*.

Besides the rapidly growing *Ronzotherium* and smaller hyracodonts of the genus *Eggysodon* Roman, 1912, the rhinoceros fauna of the European Oligocene is scarce. The first genus arising from the main stock after *Epiaceratherium* is *Protaceratherium* Abel, 1910, with its more primitive species *albigense* Roman, 1912, from the base of the upper Oligocene. In contrast to the American evolutionary lineage it has not lost its fourth digit in the manus. The metacone rib is flattened even in the premolars. This points to a relationship with the Aceratheriinae. Also in contrast to the American Diceratheriini, the sharp outer protoconid edge is flattened, a character arising already in *Epiaceratherium*. In this species the $I^1$ is still triangular, but reaches the typical bladelike form in the later *Protoceratherium minutum* (Cuvier, 1822), the type species. In this species the premolars are

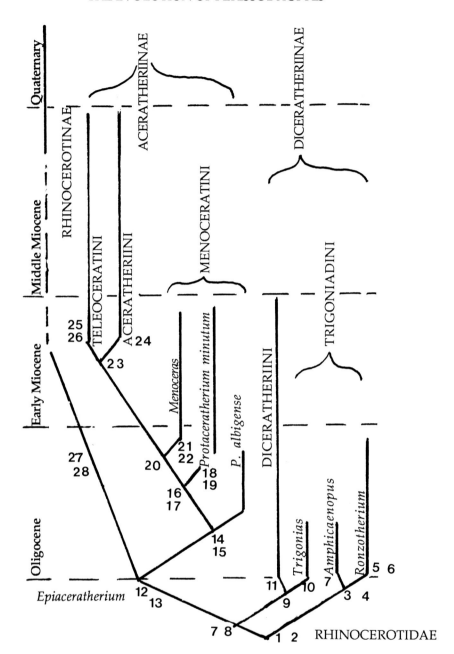

Fig. 21.4. The diversification of the Rhinocerotidae. Characters: 1) incisor shearing complex, $I^1/_2$; 2) general tendency toward deepening of nasal notch; 3) $I_2$ straight and long; 4) $I^3$ and canine lost; 5) symphysis massive and high; 6) size increase; 7) $I^1$ trigonal, elongate; 8) incipient molarization of upper premolars; 9) metacone ribs lost in upper molars; 10) nasals short; 11) tridactyl manus; 12) median lower crest of mandible lost; 13) basicranial axis short; 14) $I^2$ lost; 15) $I_2$ lanceolate, curved; 16) $I^1$ blade-like; 17) metacone rib reduced in premolars; 18) molarized premolars; 19) $I^1$ smaller; 20) split median horn base; 21) tridactyl manus; 22) horn enlarged; 23) nasal notch deepened; 24) posterior articulation of radiale with intermedium lost; 25) limbs massive and short; 26) $I^1$ broad and low; 27) strong subterminal undivided horn base; 28) tridactyl manus, $P_1$ double rooted.

semimolariform to molariform and the elongate lanceolate lower incisors curve upwards, at least in males. The sexual dimorphism is accentuated in the incisors: the females have shorter crowns, whereas the males have long crowns with a loss or thinning of enamel on the lingual side. The limb proportions and a lot of derived characters of the skeleton, especially the form of tarsal 1, are very similar to *Menoceras* Troxell, 1921.

Unfortunately, we cannot follow the steps from this first genus, which shows clear affinities to all modern rhinocerotoid genera, to the later tribes dominating the Palearctic Neogene. The skull is not yet sufficiently investigated to know whether the basicranium is already as short, as it is according to Prothero, Manning, and Hanson (1986), in all higher rhinoceroses.

In the terminal Upper Oligocene, a second immigration wave of rhinoceroses occurred in Europe. Besides the first Aceratheriinae, a *Menoceras*-like animal also occurs, described by others as "*Diceratherium zitteli*" Schlosser, 1902, or "*Diaceratherium florsheimense* " Heller, 1933. Teeth are the majority of remains known, but a well-preserved skull from the Wischberg in Switzerland exhibits the split terminal horn bosses of the *Menoceras*-type. The postcranial skeleton is unknown. Shortened limb bones that may belong to this animal have also been discovered, but whether these are actually part of the same animal remains to be proven.

The evolutionary lineage, from which *Protaceratherium* and later *Menoceras* and the Aceratheriinae split, must have evolved outside of Europe. It must have separated from the ancestral stock of the Rhinocerotinae before curving its lower incisors upwards, before separating the antecrochet of the upper molars by a sharp groove from the protocone. According to Prothero, Manning, and Hanson (1986), this small group should be taken as a separate subfamily, the Menoceratinae. In my opinion, this sister-group of the Aceratheriini and Teleoceratini, without closer relationship to one or the other, should be kept within the Aceratheriinae as a third tribe, Menoceratini.

The different members of the subfamily Aceratheriinae exhibit a different rate of molarization of the premolars. At the end of the Oligocene, all Menoceratini have reached semimolariform to molariform conditions. In the Aceratheriini it is the semimolariform stage that is generally adopted, but the linguodistal edge of the premolars is still rounded and not square. In the lower second premolar there is still an isolated entoconid. The Teleoceratini still exhibit submolariform conditions. In both tribes there are lineages with delayed molarization.

## The Aceratheriini

The less specialized tribe of the advanced Aceratheriinae offers a very complicated phylogeny. Only the morphological characters of the skull, the incisors, and the mandibular symphysis are generally useful, including the presence or absence of upper incisors. The limb bones may be used in more specialized genera.

The earliest known acerathere is *Mesaceratherium* Heissig, 1969, from the upper Oligocene. Its slender limbs and narrow skull are primitive, as are the premolars. A rather peculiar feature is the broad mandibular symphysis with strong upwards curved incisors, shearing against large, chisel-shaped upper ones. These characters are unique in aceratheres and link this genus to *Alicornops* Ginsburg and Guérin, 1979, in the middle Miocene and possibly to *Aceratherium* Kaup, 1834, in the upper Miocene. This may be the only endemic group of aceratheres in Europe.

Another immigration near the end of the lower Miocene brings the third rhinoceros wave to Europe. It comprises another type of acerathere. All forms of this second branch are characterized by the tendency to reduce the upper incisor after the loss of its shearing function, the lower incisor working against a prehensile upper lip. Whereas in *Mesaceratherium* and *Al-*

*icornops* the nasals are still unknown, we know that in at least one line of this second branch a small horn base like those of female *Menoceras* is retained. The first genus to appear is *Plesiaceratherium* Young, 1937, with four species known from China to Spain. It is hornless with a narrow skull and a narrow symphysis. The small upper incisors seldom have traces of wear. The lower ones are flattened and only slightly upturned. The genus disappears during the middle Miocene without descendants.

In the middle Miocene a somewhat later immigration brings *Hoploaceratherium tetradactylum* (Lartet 1837; see Ginsburg and Heissig, this volume, Chapter 22) with a small horn base on unfused nasals, and no upper incisors. The nasals are primitively long and the skull is narrow. The narrow symphysis and the thick, moderately-curved lower incisors, are of the uniform shape uniting *Hoploaceratherium* Ginsburg and Heissig (this volume) with the so called "*Aceratherium*" species of Asia and early *Aphelops* Cope, 1873, and *Peraceras* Cope, 1880. All these forms are somewhat later and therefore have shorter nasals. "*Aceratherium*" *depereti* Borissiak, 1927, also has a small horn that may reflect a real relationship. The American genera seem to be earlier offshoots of this group if they are really homogenous. In Europe the genus terminates with the upper Miocene *H. bavaricum* (Stromer, 1902), with markedly shortened nasals and a broader skull.

In Africa, the genus *Chilotheridium* Hooijer, 1971, with shortened tetradactyl limbs and a small horn far back on the long nasals, seems to be nearer to *Hoploaceratherium* than to *Chilotherium*. The skull form of "*Turkanatherium*" *acutirostratum* Deraniyagala, 1951, points to the same group, but the nasal notch is rather shallow. Both genera occur in the middle and upper Miocene.

The group around *Chilotherium* Ringström, 1924, may have originated from the same group as indicated by the form of the lower incisors. It begins in the early middle Miocene of South Asia with *Subchilotherium* Heissig, 1972, from the lower Siwalik series. Unfortunately, the only known skulls of this genus are of considerably younger age, so that the most primitive condition is unknown. The symphysis is moderately broad as in *Aceratherium*, and the medial flanges of the incisors are not upturned. The structure of the cheek teeth is very near that of the later *Chilotherium* species. The premolars, especially the $P^2$, are shortened in the same way. The limb bones are shortened as in *Chilotherium*, but not in the same degree. The manus is not completely known.

The second genus of this group, *Acerorhinus* Kretzoi, 1942, is better known. It starts in the late middle Miocene with *A. palaeosinensis* (Ringström, 1924) with still rather long nasals and a narrow skull. The general tendencies of broadening of the skull and deepening of the nasal notch can also be seen in this genus. The lower incisors are similar to *Subchilotherium* but less curved. Their medial flanges are upturned as in *Chilotherium* and the mandibular symphysis is hollowed from below. A broadening of the symphysis is not observed. The distance from the nasal notch to the orbit is short, and the facial crista is confluent with the anterior rim of the orbit, forming a nearly vertical straight line. This genus is evolving rapidly during the upper Miocene leading to very complicated structures in the upper cheek teeth (*Sinorhinus brancoi* Schlosser, 1903). In my opinion, the narrow zone of rugosites along the anterior rim of the nasal cannot be interpreted as a horn base. The limbs are shortened, but more massive than in *Chilotherium*, and the manus remains tetradactyl. This genus persists up to the lower Pliocene and comprises the latest Aceratheriini of the Old World.

*Chilotherium sensu stricto* begins in the upper Miocene and exhibits the typical broadened symphysis already in the first species. The medial flanges of the incisors are upturned and the symphysis is hollow below. In contrast to *Acerorhinus*, the skull is broad and the distance between the orbit and the nasal notch is longer. The facial ridge is clearly in front of the orbit and forms a right or obtuse angle. The skull narrows from the frontals to the nasals, not abruptly as in *Acerorhinus*, but gradually. These characters may indicate a long separate history of both genera. The limbs are shorter than in any other genus of the Aceratheriini, but remain slender. Some species show a lateral shifting of tarsal elements. The manus is tridactyl, at least in the type species. In the other species it is not sufficiently known. A general increase in size and a trend to facial shortening is observed during the upper Miocene. The genus disappears, as does *Subchilotherium*, at the end of the Miocene. In spite of the rather high-crowned cheek teeth, there is no sign of neck bending, as in other grazing rhinceroses. I suppose that in all forms fighting with enlarged incisors instead of horns, the head position was horizontal, and that the grazing was possible only by shortening of the limbs to bring the head nearer to the ground. Nevertheless, it remains questionable if *Chilotherium* was a true grazer.

## The Teleoceratini

Whereas the Aceratheriini have always retained rather slender limbs, even while undergoing shortening, the Teleoceratini show the tendency to massive and short distal limb segments throughout their history. In contrast to the Aceratheriini, they never lose their upper first incisor pair and retain their shearing function.

The first representative of this tribe is *Brachydiceratherium* Lavocat, 1951, with the single species *B. lemanense* Pomel, 1853, from the upper Oligocene of France. Its long and narrow skull is quite similar to to the Aceratheriini in the long and tapering nasals with the small, divided horn base at the tips. The nasal incision is deep. The lower incisors are strongly curved and the symphysis is rather broad. The limbs are more massive than in the early Aceratheriini but only slightly shortened. The manus is tetradactyl and shows the primitive third articulation between radial and intermedium. This facet on the radial is replaced in all Aceratheriini by a rough protuberance, indicating the former presence of a facet.

Somewhat later, at the beginning of the lower Miocene, a second genus *Diaceratherium* Dietrich, 1931, came to Europe as an immigrant. The skull is smaller and horned, but the cheek teeth show a divergent morphology and the limbs are more shortened. Both genera coexisted during the lower Miocene in western Europe for some time. The large *Brachydiceratherium* has broad premolars with heavy lingual cingula and broad, low-crowned molars. Also the lower cheek teeth are broad and have strong cingula. The genus disappeared in the upper part of the lower Miocene, probably giving rise to *Brachypotherium* Roger, 1904, in the middle Miocene. This genus is tridactyl in the manus and is widespread in the Old World during the middle and upper Miocene with several species. It persists into the late Pliocene in East Africa. The skull is broad and short, with reduced, hornless nasals and strong premaxillae, bearing a strong upper incisor. The heavy mandible has a curved lower margin and strongly curved, but nearly horizontal implanted incisors. In the upper premolars, the molarization is complete, but the morphology of the other teeth is nearly unchanged. There is a slight shortening of the premolars, especially the first, and the second shows signs of reduction. The limbs are short in the distal segments, the proximal ones being larger. The genus reaches the maximum size of brachydont rhinoceroses in its Indian species, *B. perimense* (Falconer and Cautley, 1847).

The smaller *Diaceratherium* outnumbers *Brachydiceratherium* during the lower Miocene. It shortened its limbs more rapidly, and shows a moderate increase of crown height of the cheek teeth. It retains a nasal horn, and shows only a weak molarization of the premolars and no reduction of the fifth metacarpal. Its evolutionary steps are marked by several species names, *D. as-*

*phaltense* (Depéret and Douxami, 1902), *D. aginense* (Repelin, 1917) and *D. aurelianense* (Nouel, 1866), all in the lower Miocene. The last species, with a moderately shortened skull, extremely shortened limbs, but a tetradactyl manus, tends to develop very high-crowned teeth. It gave rise to the tridactyl genus *Teleoceras* Hatcher, 1894, in America, where it reached hippopotamus-like proportions. In Europe, this lineage disappeared at the end of the lower Miocene.

Contemporaneous to *D. aurelianense*, there existed a smaller offshoot of this lineage, *D. douvillei* (Osborn, 1900). It leads to the genus *Prosantorhinus* Heissig, 1973b, characterized by a saddle-shaped skull with upturned nasals and the fusion of the nasal rugosites to one strong globular horn base. This lineage underwent a slight size reduction combined with the final molarization of the premolars. The crown height is less than than in *D. aurelianense* and does not change. The shortening of the limbs was continuous, but the genus remained tetradactyl. It disappeared in the middle Miocene. There are questionable remains of other species, possibly related to this group. The species *tagicus* Roman, 1907, may be related to *Prosantorhinus* if the similarity of the upper cheek teeth is reliable. It is considerably smaller than all known Teleoceratini.

A third lineage of Teleoceratini remains confined to Asia. It parallels *Chilotherium* in the extreme broadening of the symphysis early in the upper Oligocene, and the early Elasmotheriini in the crown height of the cheek teeth, the low degree of molarization, and the high posterior cingulum of the premolars. This genus *Aprotodon* Forster Cooper, 1915, starts with *A. aralense* (Borissiak, 1944). The skull is entirely primitive with long, slender nasals, but no horn base. The nasal notch is progressive and reached its furthest retraction over the end of the premolar series. These features are contrasted by the specialization of the symphysis. It is very broad and the incisors are curved outwards.

During the history of this genus, known mainly in the Siwalik series, the crown height increased considerably. The premolar molarization is weak but finally reached the semimolariform stage, with a straight metaloph and a simple postfossette. Single teeth and bones are known up to the basal upper Miocene of the Nagri Formation, but the evolution of the skull and mandible are not known.

**The Rhinocerotinae**

All living rhinoceroses have horns as their most conspicuous weapons. Nevertheless, the Asiatic species fight against predators, including man, with their large tusk-like incisors, sharpened by the contact with the upper ones. This behavior was already noticed in the seventeenth century. The horn is used mainly against conspecific rivals. It may have been formed in analogy to the cervid antlers to make fighting less dangerous and to prevent fights by an impressive display.

In the early rhinoceroses, such as *Diceratherium* or *Menoceras*, we find a balance of equally important incisors and horns. Later, the Teleoceratini and Aceratheriini strengthened the incisors, and the Rhinocerotinae strengthened the horn. The alternative structure was lost or reduced, because the skull position for the use of one is incompatible with the use of the other. The Aceratheriini and Teleoceratini retained the primitive horizontal skull position and had to shorten limbs to be able to eat small plants. So we rarely find grazers in these tribes (e.g., *Teleoceras*).

The Rhinocerotinae, comprising also fossil Elasmotheriini, have strengthened their horns. The incisors remained in the primitive stage of early Diceratheriinae or even vanished. They lowered their skull position in order to use their horn, and the muscles once adapted to this position could easily bring the muzzle down to the ground. So the tendency to a grazing diet was followed independently in several lines during this process.

## The Elasmotheriini

The first group undergoing this specialization has reached the highest degree of adaptation to hard and abrasive diet in all large ungulates of the Old World. They finally evolved ever-growing, rootless prismatic cheek teeth. The history of this tribe is dominated by this tendency. The earliest member of this tribe is described as *Caementodon* sp. from the lower Miocene of the Bugti beds. This smallest known member of the Rhinocerotinae showed the first formation of cement in the tooth grooves and an elongation of the molar ectoloph. *Caementodon* Heissig, 1972, from the Siwalk series is a first side branch, characterized by an elongation of molars, heavy cement covers on the cheek teeth, and rather high crowns. The upper and lower incisors are small but present, resembling the incisors of the first rhinoceroses *Trigonias* and *Epiaceratherium*.

The main lineage is represented at the same time by the larger *Beliajevina* Heissig, 1974, from the middle Miocene of Bjelometschetskaja and Anatolia. These animals still have a small incisor, but the crowns of the cheek teeth are higher. In all these genera the molarization of the premolars is delayed and remains on the submolariform stage. From this genus we know the earliest elasmotheriine skull. It was not figured, but in his description Borissiak, 1935, mentions a dome-like horn base at the nasofrontal suture. The limbs are high and show no third articulation of radial with intermedium.

Since the phylogenetic study of Heissig (1976), the number of known elasmotheres has considerably increased. These "newcomers," mainly from China, offer the impression that the position of the horn is rather variable and may have not shifted backward only once to reach the frontals. A second character seems to be more reliable. The zygomatic arches are high and broad, ascending backwards over the level of the skull roof in *Iranotherium* Ringström, 1924, faint and not spreading laterally in *Elasmotherium* Fischer, 1808. In this first character the early genera show an intermediate homogenous type. Most early forms are found in Asia, but there are some offshoots in Europe and Africa. *Tesselodon* Yan, 1979, from the middle Miocene of China is known only by teeth, which are similar to *Beliajevina*. *Shennongtherium* Huang and Yan, 1983, also from the Chinese Miocene, is a high-crowned member of the Rhinocerotini. The better-known genera from the middle Miocene, *Hispanotherium* Crusafont and Villalta, 1947, and *Begertherium* Beliajeva, 1971, both with an intermediate horn position, show only reduced or vestigial incisors and stronger hypsodonty. The zygomatic arch is still unknown in both. *Kenyatherium* Aguirre and Guérin, 1974, known only by teeth from the upper Miocene of Africa, is similar to both, but a relationship to *Iranotherium* cannot be excluded.

With *Iranotherium* from the middle Miocene of the Gobi starts a branch of Elasmotheriini leading to the huge upper Miocene terminal species *I. morgani* Mecquenem, 1929, from Maragheh. It has a strong terminal horn on the nasals and a high ascending zygomatic arch. The incisors are totally lost, and the hypsodonty of the cheek teeth reaches the same degree as the contemporaneous forms of the other branch. These are best known from China and were named *Sinotherium* by Ringström (1924). The only known skull fragment points to a frontal or nasofrontal horn position. Some new species from China may represent the transition from very high-crowned teeth with delayed root formation to the final rootless prismatic stage. At the same time, the fine plication of the enamel is accentuated. A complete skull recently found in the upper Miocene of China shows a terminal horn like *Iranotherium* but faint, straight, zygomatic arches like *Elasmotherium*. It raises the possibility that there was a transition from one genus to the other by reduction of the zygomatic arch and backward shifting of the horn. This genus, *Ningxiatherium* Chen, 1977, may include some of the species known only by teeth.

*Elasmotherium*, the terminal form with

a shortened skull, a domed frontal and an extreme enamel plication, reaches its maximum size in the early Pleistocene and shows some size decrease before dying out. It was confined to the steppes of Asia, and only one specimen is recorded from Europe. It became extinct with the first severe cooling in the middle Pleistocene.

**The Rhinocerotini**
The earliest Rhinocerotini occur in Europe and Asia with the third rhinoceros wave, at the proboscidean datum. That may suggest an African origin, but there are no earlier faunas from Africa containing rhinoceroses. This tribe comprises the more primitive members of the subfamily, but was more diversified and successful than the groups discussed previously. Both tribes are linked by a double-rooted $P_1$, the articulation of the ulna with the intermedium, the presence of one or two strong median horns, and medium-sized to vestigial incisors, horizontally implanted in a long symphysis. Less important similarities are the rather strong metacone ribs, mainly in the upper premolars, and the tridactyl, unspecialized limbs.

In contrast to the Elasmotheriini, most Rhinocerotini remain browsers with a primitive dentition and no marked increase in crown height. The loss of incisors occurs independently in several lines, but not in all. In the manus, the third articulation of the radial with the intermedium is preserved, and the intermedium shows a third facet for the ulna, or at least a lengthening of the distal facet. In this feature, the difference between Aceratheriini and Teleoceratini is paralleled by the difference of the Elasmotheriini and Rhinocerotini and Dicerotini, on the other hand. The origin of this group probably occurs near the *Trigonias-Epiaceratherium* stage of evolution. Most likely the shortening of the basicranium, which it shares with the Aceratheriinae, occurred before the separation of both subfamilies. The upper incisor evolved to its typical blade-like form after the separation of the Elasmotheriini, and the lower incisor is not lengthened in most genera.

The first radiation of the tribe must have occurred before the first record of the group. There are several side branches showing single remaining primitive traits, already lost in the contemporaneous members of the main stock. All these characters are dental, since nothing is known about limb bones and horn formation. In the following, I discuss only the named species.

*Dicerorhinus abeli* (Forster Cooper, 1915) is the best known species of this type. The mandible and lower incisors are of rhinocerotine type, but Forster Cooper was misled by the strong lingual cingulum of the upper premolars and the marked antecrochet constriction in the upper molars, paralleling the aceratheres. In the last premolar there is a faint trace of a bridge uniting the lingual cusps as in a semimolariform stage, unlike the paramolariform condition of most Rhinocerotini, where the cusps are united by their bases. The skull is broad, even if there is a sagittal crest. The same characters as in the early Miocene form from the Bugti beds we find in the smaller *Dicerorhinus steinheimensis* (Jäger, 1835) from the middle Miocene of Europe, except the presence of stronger metacone ribs.

The central stock of the tribe begins with *Lartetotherium* Ginsburg, 1974, in the lower Miocene (Burdigalian) of Europe. The nasals have a conical, subterminal horn base as in *Dicerorhinus*. The frontals show a second horn in the type but not in all referred specimens. The upper incisors are chisel-shaped, the lower ones spatulate and implanted horizontally, even in males. The upper premolars are paramolariform with strong, narrow ribs on the outer wall, widely separated but converging to the ectoloph crest. There is no trace, or only faint traces, of a lingual cingulum. In the molars, the antecrochet and its lingual constriction are faint.

During the middle Miocene, we find rhinoceroses of similar dental type widespread in the Old World. In the Siwalik series of India, *Gaindatherium* Colbert (1934), with its single horn foreshadows the later *Rhinoceros*, but the skull is still long, and the low-crowned teeth are nearly identical to *Lartetotherium*. We can follow this line up to the upper Miocene Nagri beds. From this time onwards, the evolution in South Asia was separate from the rest of the Old World except China. The transition from the *Gaindatherium* lineage to the modern and Pleistocene *Rhinoceros* species is insufficiently known. The upper cheek teeth are high-crowned and block-shaped in the younger genus. There are several side branches, including the huge *Punjabitherium* Khan, 1971, with two horns, showing that the number of median horns is not a reliable character. *Rhinoceros sondaicus* Desmarest, 1822, on the other hand, remains primitive in the dentition, but shares the skull shortening and the upslanting head with the type species, *R. unicornis*. In South Asia, *Dicerorhinus sumatrensis* (Fischer, 1814) persisted nearly unchanged with a dentition like *Lartetotherium*.

In the middle Miocene of Africa, we find *Dicerorhinus leakeyi* Hooijer, 1966, related undoubtedly to the central stock, but a little more primitive in the stronger lingual cingula of the upper premolars and the more primitive stage of molarization, varying between the submolariform and the paramolariform type. The skull is long and low and resembles the other middle Miocene species. The incisors are of the same type. It is possible to trace the second main lineage of the Rhinocerotini back to this species. This lineage is represented mainly by the genus *Stephanorhinus* Kretzoi, 1942, beginning with the species *pachygnathus* Wagner, 1848, from the upper Miocene of Mediterranean area, but also by *Dicerorhinus schleiermacheri* Kaup, 1832, from the same time in Western Europe. While the first species has nearly completely reduced incisors, the second one is larger and more primitive and shows no reduction. Both still have a lingual cingulum in the upper premolars and long, strong nasals with tandem horns. The nasal notch is somewhat retracted. The first species continues during the early Pleistocene with the species *S. etruscus* (Falconer, 1859). It supported its strong horn with an ossified septum to allow further deeping of the nasal notch. The second one continues to the Pliocene *D. megarhinus*, with no septum and reduced, but still relatively large, lower incisors. From *D. megarhinus* De Christol, 1835, we can follow the line to *D. jeanvireti* Guérin, 1972, decreasing in size relative to the large earlier species. A third species in the late upper Miocene, *D. ringstroemi* Arambourg (1959) is a large form without incisors or lingual cingulum. It is possibly a side branch of one of these two species.

During the middle and late Pleistocene, the severe climatic changes caused a permanent shifting, dividing, and reuniting of species areas. There are several well-known species, but their phylogenetic arrangement is still a problem. Both lineages converged in characters with the complete loss of incisors and the development of an ossified nasal septum. So the origins of the steppe-adapted *D. hemitoechus* Falconer, 1868, and the bigger *D. kirchbergensis* Jaeger, 1839, are still unknown. In China *Dicerorhinus choukoutiensis* Wang, 1931, and *D. yunchuchensis* Chow, 1963, are similar and may be related to *D. kirchbergensis* of Europe. The most specialized offshoot of the tribe, *Coelodonta* Bronn, 1831, may be related to this lineage also, as is indicated by the loss of the incisors and the ossified septum. The skull is broad, but in contrast to *Stephanorhinus* and its relatives, very low. There is no form bridging the gap in dental morphology. *Coelodonta* was the only typical grazer of this lineage, with rather high-crowned teeth and a very peculiar morphology, including the newly formed metastyle of the $M^3$ in some specimens. The whole group went extinct during the late Pleistocene without descendants.

### The Dicerotini
The last tribe to appear is the African branch of the tandem-horned rhinoceroses.

Its apomorphic characters set it far apart from the younger Rhinocerotini, but in the middle Miocene of Africa, some poorly-known species seem to bridge the gap.

The earliest known specimens come from the middle Miocene of Anatolia and Chios, a little earlier than the first named species, *Paradiceros mukirii* Hooijer, 1968, from the late middle Miocene of Fort Ternan and Beni Mellal. It is primitive in most dental characters, but in the skull it already shows the characters of the tribe. These include the anteriorly shortened nasals, the outwards-inclined lower border of the orbit, and the lack of functional incisors. The skull resembles the living *Diceros*, but the mandible resembles *Ceratotherium*. The upper premolars already have reduced metacone ribs.

During the upper Miocene, there were two lineages. The more primitive is *Diceros primaevus* (Arambourg, 1959), very similar to *Paradiceros*, with strong lingual cingula in the premolars and low crowns on the cheek teeth. The second species, *Diceros neumayri* (Osborn, 1900) may be its descendant. It expanded its range over western Asia and the southern part of Europe, where it differentiated into ecological types and underwent some evolutionary changes during the upper Miocene. It developed rather high-crowned teeth and an overhanging occipital crest, similar to *Ceratotherium* in its most advanced specimens. Comparable tendencies are observed in *Diceros douariensis* Guérin, 1966, from Tunisia, a rather large animal with high-crowned cheek teeth. The skull morphology is incompletely known.

During the Plio-Pleistocene, the tribe is restricted to Africa again. *Ceratotherium* Gray, 1867, is reported earlier than typical *Diceros* Gray, 1821. It is not yet clear if the splitting up into the living browser-grazer pair occurred in the Pliocene or if the diversification in the upper Miocene led directly into the separate lineages. In this case *D. douariensis* should be suspected to be ancestral to *Ceratotherium*, whereas the living *Diceros bicornis* remained at the evolutionary stage of the more primitive specimens of *D. neumayri*. It is puzzling that in the the living genera the presumed primitive condition of an articulation of the fibula with the femur is retained, whereas it was lost in the upper Miocene species.

**Bibliography**

Abel, O. (1910): Kritische Untersuchungen über die paläogenen Rhinocerotiden Europas. -*Abh.k.k. Reisanst.*, 20 (3): 1-52.

Aguirre, E. & Guérin, C. (1974): Première découverte d'un Iranotheriinae (Mammalia, Perissodactyla, Rhinocerotidae) en Afrique: *Kenyatherium bishopi* nov. gen., nov. sp. de la formation Vallésienne (Miocène supérieur) de Nakali (Kenya). -*Estud. Geol.*, 30 (3) 229-233.

Arambourg, C. (1959): Vertébrés continentaux du Miocène supérieur de l'Afrique du Nord. -*Publ. Serv. carte Géol. Algérie, n.s. Paléont.*, 4: 1-161.

Aymard, A. (1856): Rapport sur la collection de M. Pichot-Dumazel. -*Congr. Sci. France*, 12 (1): 227-245.

Beliajeva, E. I. (1971): On some rhinoceroses, family Rhinocerotidae, from the Neogene of western Mongolia (in Russian). -*Fauna Mesoz. i Kainoz. zap. Mongol*, 3: 78-97.

Borissiak, A. A. (1927): *Aceratherium depereti* n. sp. from the Jilancik-beds. -*Bull. Acad. Sci. St. Petersburg*, (6) 21: 769-786.

Borissiak, A. A. (1935): Neue Materialien zur Phylogenie der Dicerorhinae. -*Dokl. Akad. Nauk SSSR* 38 (8): 381-384.

Borissiak, A. A. (1944): *Aceratherium aralense* n. sp. -*Dokl. Akad. Nauk. SSSR* 43 (1): 30-32.

Brandt, J. F. V. (1878): Tentamen synopseos rhinocerotidum viventium et fossilium. -*Mem. Akad. Sci. St. Petersburg*, (7), 26, 5: ii+ 66 pp.

Bronn, H. G. (1831): Über die fossilen Zähne eines neuen Geschlechtes der Dickhäuter-Ordnung, Coelodonta, Höhlenzahn. -*N. Jahrb. Min. geol. Pal.*, 1831: 51-61.

Chen, G. (1977): A new genus of Iranotheriinae of Ningxia (in Chinese). -*Vert. PalAsiatica*, 15 (2): 143-147.

Chow, M. (1963): A new species of *Dicerorhinus* from Yushe, Shansi, China (Chinese, English summary). -*Vert. PalAsiatica*, 7 (4): 325-329.

Chow, M., and Xiu, C. (1964): An Eocene giant rhinoceros (Chinese, English summary) - *Vert. PalAsiatica*, 8(3): 264-267.

Christol, J. de (1834): *Recherches sur les charactères des grandes espèces de Rhinocéros fossiles*. Montpellier (Martel).

Christol, J. de (1835): Recherches sur les charactères des grandes espèces de Rhinocéros fossiles. -*Ann. sci. Nat. Zool.* 2 sér., 4: 44-112.

Colbert, E. H. (1934): A new rhinoceros from the Siwalik beds of India. -*Amer. Mus. Novitates*, 749: 1-13.

Cope, E. D. (1873): On some new extinct Mammalia from the Tertiary of the plains. -*Paleont. Bull.*, 14: 2 pp.

Cope, E.D. (1880): A new genus of Rhinocerotidae. -*Amer. Natur.*, 14: 540

Crusafont, M., and Villalta, J. F. (1947): Sobre un interesante Rhinoceronte (*Hispanotherium* nov. gen.) del Miocene del Valle del Manzanares. -*Las Ciencias*, 12 (4): 869-883.

Cuvier, G. (1822): *Recherches sur les Ossements Fossiles...* 2d ed. 5 vols. Paris.

Depéret, C. and Douxami, H. (1902): Les Vertébrés Oligocènes de Pyrimont-Challonges (Savoie). -*Abh. Schweiz. Paleont. Ges.*, 29 (1): 1-91.

Deraniyagala, P. E. P. (1951): A new genus and species of hornless Mio-Pliocene rhinoceros. -*Proc. Ceylon Assoc. Sci. Ann. Congr.*, 72:24.

Desmarest, A. G. (1822): *Mammalogie ou description des espèces de mammifères.* Paris (Agasse).

Dietrich, W. O (1931): Neue Nashornreste aus Schwaben (*Diaceratherium tomerdingense* n.g. n. sp.) -*Zeit. Säugetierkde.*, 6 (5): 203-220.

Dollo, L. (1885): Rhinocéros vivants et fossiles. -*Rev. Quest. Sci.*, 17: 293-299.

Duvernoy, G. -L. (1853): Nouvelles études sur les Rhinocéros fossiles. -*C. R. Acad.sci. Paris*, 36: 117-125, 150-154, 159-176.

Falconer, H. (1859): Faunal list. In: Ansted, T. D.: On the geology of Malaga and the southern part of Andalusia. -*Q. J. Geol. Sci. London*, 15: (1): 601-603.

Falconer, H. (1868): *Paleontological memoirs and notes, 2: Mastodon, Elefant, Rhinoceros, ossiferous caves primeval man and his contemporaries.* London (Hardwicke).

Falconer, H., and Cautley, P. (1847): *Fauna antiqua sivalensis, being the fossil zoology of the Sewalik Hills, in the north of India.* London.

Fischer, G. F. v.W., (1808): Notice d'un animal fossile de Sibérie inconnu aux naturalistes. -*Progr. invit. sé. Publ. Soc. Imp. Natural.*: 1-28.

Fischer, G. F. v. W. (1814): *Zoognosia tabulis synopticis illustrata.* -Moscow.

Forster Cooper, C. (1915): New genera and species of mammals from the Miocene deposits of Baluchistan; preliminary notice. -*Ann. Mag. Nat. Hist.*, (8) 16: 404-410.

Ginsburg, L. (1974): Les Rhinocérotidés du Miocène de Sansan (Gers). -*C.R. Acad. Sci. Paris*, 278D: 597-600.

Ginsburg, L., and Guérin, C. (1979): Sur l'origine et l'extension stratigraphique du petit Rhinocérotidé Miocène *Aceratherium* (*Alicornops*) *simorrense* (Lartet, 1851) nov. subgen. -*C.R. somm. Sé. Soc. géol. France*, 3: 114-116.

Ginsburg, L., and Heissig, K. (1989): *Hoploaceratherium* n. gen., a new generic name for "*Aceratherium*" *tetradactylum* (Lartet 1837) (This volume).

Gray, J. (1821): On the natural arrangement of vertebrose animals. -*London Med. Repos.*, 15: 296-310.

Gray, J. (1867): Observations on the preserved specimens and skeletons of the Rhinocerotidae in the collection of the British Museum and Royal College of Surgeons. -*Proc. Zool. Soc. London*, 1865: 1003-1032.

Groves, C. P. (1983) Phylogeny of the living

species of rhinoceros. -*Zeit. Zool. System. Evolutionsf.*, 21 (4): 293-313.

Guérin, C. (1966): *Diceros douariensis* nov. sp., un Rhinocéros du Mio-Pliocène de Tunisie du Nord. -*Doc. Lab. Géol. Fac. Sci., Lyon*, 16: 1-50.

Guérin, C. (1972): Une nouvelle espèce de Rhinocéros (Mammalia, Perissodactyla) à Vialette (Haute Loire, France) et dans d'autres gîsements du Villafranchien inférieur européen: *Dicerorhinus jeanvireti* nov. sp. -*Doc. Lab. Géol. Fac. Sci. Lyon*, 49: 53-150.

Hatcher, J. B. (1894): A median horned rhinoceros from the Loup Fork beds of Nebraska. -*Amer. Geologist*, 13 (3): 149-150.

Heissig, K. (1969): Die Rhinocerotidae aus der oberoligozänen Spaltenfüllung von Gaimersheim bei Ingolstadt in Bayern. -*Abh. Bayer. Akad.Wiss. Math. -Nat. Kl.N.F.* 138: 1-133.

Heissig, K. (1972): Geologische und paläontologische Untersuchungen im Tertiär von Pakistan 5., Rhinocerotidae aus den unteren und mittleren Siwalik-Schichten. -*Abh. Bayer. Akad. Wiss. Math. -Nat. Kl. N. F.* 152: 1-122.

Heissig, K. (1973a): Die Unterfamilien und Tribus der rezenten und fossilen Rhinocerotidae (Mammalia). -*Säugetierkundl. Mitt.*, 21: 25-30.

Heissig, K. (1973b): *Prosantorhinus pro. Brachypodella* Heissig 1972 (Rhinocerotidae, Mammalia) (non *Brachypodella* Beck, 1837 (Gastropoda)). -*Mitt. Bayer. Staatsslg. Paläont. hist. Geol.*, 14: 37.

Heissig, K. (1974): Neue Elasmotheriini aus dem Obermiozän Anatoliens -*Mitt. Bayer. Staatsslg. Paläont. hist. Geol.*,14: 21-35.

Heissig, K. (1976): Rhinocerotidae aus der *Anchitherium*-Fauna Anatoliens. -*Geol. Jb.* B 19: 1-121.

Heller, F. (1933): *Diaceratherium florsheimense* n. sp., ein neuer Rhinocerotidae aus dem Mainzer Becken. -*Senckenb.* 15: 295-302.

Hooijer, D. A. (1966): Fossil Mammals of Africa 21: Miocene rhinoceroses of East Africa. -*Bull. Brit. Mus. (Nat. Hist.) Geol.*, 13: 119-190.

Hooijer, D. A. (1968): A rhinoceros from the late Miocene of Fort Ternan, Kenya. -*Zool. Meded.*, 43: 77-92.

Hooijer, D. A. (1971): A new rhinoceros from the late Miocene of Loperot, Turkana district, Kenya. -*Bull. Mus. Comp. Zool.*, 142 (3): 339 -392.

Huang, G., and Yan, D. (1983): New material of Elasmotheriini from Shennongjia, Hubei (Chinese, English summary) - *Vert. PalAsiatica*, 21 (3): 223-229.

Jäger , G. F. v. (1835-1838): *Über die fossilen Säugethiere, welche in Würtemberg in versciedenen Formationen aufgefunden worden sind...* . Stuttgart.

Kaup, J. J. (1832): Über *Rhinoceros incisivus* Cuvier, und eine neue Art, *Rhinoceros schleirmacheri.* -*Isis von Oken*, 1832: 898-904.

Kaup, J. J. (1834): *Description d'ossemens fossiles de mammifères inconnus jusqu'à présent...*3: 33-64.

Kaup, J. J. (1854): *Beiträge zur näheren Kenntniss der urweltlichen Säugethiere.* Darmstadt.

Khan, E. (1971): *Punjabitherium* gen. nov. - an extinct rhinocerotid of the Siwaliks, Punjab, India. -*Proc. Ind. Nat. Sci. Acad.*, 37A: 105-109.

Kretzoi, M. (1942): Bemerkungen zum System der nachmiozänen Nashorn-Gattungen. -*Földt. Közl.*, 72: 309-318.

Lartet, E. (1837): Sur les débris fossiles trouvés à Sansan et sur les animaux antédiluviens en général. -*C.R. Acad. Sci.*, 5: 158.

Lavocat R. (1951): *Revision de la faune des Mammifères oligocènes d'Auvergne et du Velay.* Paris (Science et Avenir).

Lucas, F. (1900): A new rhinoceros, *Trigonias osborni*, from the Miocene of South Dakota. -*Proc. U.S. Nat. Mus.*, 23: 221-223.

Marsh, O. C. (1875): Notice of new Tertiary mammals IV. -*Amer. Jour. Sci.*, 9: 239-250.

Mecquenem, R. de (1924): Contribution à l' étude des fossiles de Maragha. -*Ann.*

*Paléont.*, 1924: 133-160.

Nouel, E. (1866): Mémoir sur un nouveau rhinocéros fossile. *-Mém. Soc. Agric. Orléans*, 8:241-250.

Osborn, H. F. (1900): Phylogeny of the rhinoceroses of Europe. *-Bull. Amer. Mus. Nat. Hist.*, 13 (19): 229-267.

Pomel, A. (1853): Catalogue méthodique et déscriptif des vertébrés fossiles découverts dans les bassins de la Loire et de l'Allier, part II. *-Ann. Sci. Litt. Indust. Auvergne*, 26: 81-229.

Prothero, D.R., Manning, E., and Hanson, C. B. (1986): The phylogeny of the Rhinocerotoidea (Mammalia, Perissodactyla). *-Zool J. Linn. Soc.*, 87: 341-366.

Radinsky, L. (1966): The families of Rhinocerotoidea (Mammalia, Perissodactyla). *-J. Mamm.*, 47: 631-639.

Repelin, J. (1917): Les rhinocérotidés de l'Aquitanien supérieur de l'Agenais (Laugnac). *-Ann. Mus. Hist. Nat. Marseilles* 16: 1-47.

Ringström, T. (1924): Nashörner der Hipparion-Fauna Nord Chinas. *-Paleont. Sinica* C 1 (4): 1-156.

Roger, O. (1904): Wirbelthierreste aus dem Obermiocän der bayerisch-schwäbischen Hochebene V. *-Ber. Naturw. Ver. Schwab. Neubg. Augsbg.*, 36: 1-22.

Roman, F. (1907): Le Néogène continental dans la basse vallée du Tage (rive droite), 1re partie, paléontologie. *-Mém. Comm. Serv. Géol. Portugal*, 1907: 1-78, 87-88.

Roman, F. (1912): Les rhinocéridés de l'Oligocène d'Europe. *-Arch. Mus. Sci. Nat. Lyon*, 11 (2): 1-92.

Schlosser, M. (1902): Beiträge zur Kenntnis der Säugethierreste aus den süddeutschen Bohnerzen. *-Geol. Pal. Abh. N. S.*, 5 (9): 117-258.

Schlosser, M. (1903): Die fossilen Säugethiere Chinas nebst einer Odontographie der recenten Antilopen. *-Abh. Bayer. Akad. Wiss.* 22: 1-221.

Stromer, E. (1902): Ein Aceratherium-Schädel aus dem Dinotherien-Sand von Niederbayern. *-Geogn. Jahresh.*, 15: 57-64.

Troxell, E.L (1921): A study of *Diceratherium* and the diceratheres. *-Amer. Jour. Sci.*, 202 (10): 197-208.

Wagner, A. (1848): Urweltliche Säugethier-Überreste aus Griechenland. *-Abh. k. Bayer. Akad. Wiss.* II Cl., 5 (2): 335-378.

Wang, K. (1931): Die fossilen Rhinocerotiden von Chou-kou-tien. *-Contr. Nat. Res. Geol. Acad. Sinica*, 1: 69-84.

Wood, H.E., II (1927): Some early Tertiary rhinoceroses and hyracodonts. *-Bull. Amer. Paleont.*, 13 (50): 5-105.

Yan, D. (1979): Einige der fossilen Miozänen Säugetiere der Kreis von Fangxian in der Provinz Hupei (Chin. Germ. summ.). *-Vert. PalAsiatica*, 17 (3): 189-199.

Young, C.C. (1937): On a Miocene mammalian fauna from Shantung. *-Bull. Geol. Soc. China*, 17: 209-243.

# 22. *HOPLOACERATHERIUM*, A NEW GENERIC NAME FOR "*ACERATHERIUM*" *TETRADACTYLUM*

## LEONARD GINSBURG and KURT HEISSIG

The species *tetradactylum* Lartet, 1837 is made the type species of the new genus *Hoploaceratherium*, since its differences from *Aceratherium incisivum* (Kaup, 1832) exceed the species level. Its differences from other primitive Aceratheriini are discussed.

## Introduction

*Rhinoceros tetradactylus* was first described by Lartet in 1837 from Sansan, middle Miocene of France. It was compared to *Aceratherium incisivum* by Kaup (1832), and included in this species by Blainville (1846), Duvernoy (1853), and Kaup (1854). Later authors, such as Osborn (1900) kept it separate as an ancestor of *Aceratherium incisivum* from the upper Miocene of Eppelsheim, Germany. Better knowledge of the Aceratheriini confirmed that the characters common to both species are in fact of tribal rank.

After the separation of the specialized Aceratheriini, such as *Peraceras*, *Aphelops*, *Chilotherium*, and *Chilotheridium* from the main stock by earlier authors, the more primitive species have been distributed among several genera: *Plesiaceratherium* Young, 1937, *Mesaceratherium* Heissig, 1969 and *Alicornops* Ginsburg and Guérin, 1979. In the light of the present knowledge of the Aceratheriini, the differences in fundamental characters of "*A.*" *tetradactylum* and *A. incisivium* far exceed the possible variation within one genus. So we propose a new generic name, *Hoploaceratherium*, for the species *tetradactylum*.

## Taxonomy

Family Rhinocerotidae Owen, 1845
Subfamily Aceratheriinae Dollo, 1885

*Hoploaceratherium*, new genus
*Diagnosis*: Primitive Aceratheriini with a faint horn boss at the tips of the unfused nasals in males. The skull is high with narrow braincase and occiput. There are no upper incisors, but a long premaxillae. The lower $I_2$ is enlarged and moderately curved. There is no space for the $I_1$ in the narrow symphysis. If present, these teeth are displaced labially. The limb bones are moderately slender of a primitive type. Only the centrale in the tarsus shows a specific character in its semilunate outline.
*Type species*: *Rhinoceros tetradactylus* Lartet, 1837.
*Referred species*: *Aceratherium bavaricum* Stromer, 1902.
*Etymology*: Greek *hoplon* = weapon.

## Differences and relationships

The genus is distinguished from all European Aceratheriini by the retention of a primitive type of horn, comparable to the most primitive Teleoceratini. It is furthermore the earliest genus in Europe with a total loss of upper incisors. The size of the lower incisors exceeds that of all other European genera.

Compared with the genera *Mesaceratherium* and *Alicornops* (both with nasals still unknown), it is characterized by the narrower symphysis and the less upturned

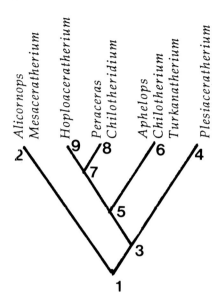

Fig. 22.1. Hypothetical phylogeny of the Aceratheriini. Characters on the nodes as follows: 1) premolars semimolariform; posterior articulation of scaphoid and lunar lost; 2) lower incisors strongly curved; limbs gracile; 3) upper incisors reduced; contact with lower incisors faint; 4) horn lost; lower incisors flattened; lower premolars roughened labially; 5) upper incisor definitely lost; lower incisor strengthened; 6) horn lost; various specializations; 7) mandibular symphysis narrow anteriorly; broad frontals; 8) limbs shorter; other specializations; 9) centrale with semilunate outline.

The position of *Aceratherium* cannot be fixed without knowledge of its premaxillae.

Table 22.1. Measurements of *Hoploaceratherium* and *Aceratherium* (in mm).

| Character | *Hoploaceratherium tetradactylum* (type) | *Aceratherium incisivum* (type) |
|---|---|---|
| Naso-occipital skull length | 532 | 605 |
| Dorsal occiput width | 141 | 150 |
| Zygomatic width | 320 | 370 |
| Frontal width | 193 | 188 |
| Basal width of nasals | 135 | 91 |
| Length of free nasals | 170 | 176 |
| Length, nasal notch to orbit (projected) | 60 | 80 |
| Infraorbital foramen located above: | P2-3 | $P^4$-$M^1$ |
| Nasal notch ends above: | posterior $P^4$ | anterior $P^4$ |
| Anterior rim of orbit above: | anterior $M^2$ | anterior $M^2$ |

incisors. Both small genera have strong upper incisors, lacking in *Hoploaceratherium*.

Compared with *Plesiaceratherium*, the presence of a horn and the lack of upper incisors are the most important features of *Hoploaceratherium*. Some tendencies in *Plesiaceratherium* are similar, but in a more primitive stage. These include the beginning of reduction of the upper incisors, and the weak curving of the lower one. The strong flattening of the lower incisor is in contrast to *Hoploaceratherium*. The narrow symphysis and the narrow skull are also similar. The long, slender limbs of *Plesiaceratherium* are thought to be more primitive than the more massive condition in *Hoploaceratherium*.

Compared with *Aceratherium sensu stricto*, the narrow occiput, but broader frontals and the shorter distance between the orbit and the nasal notch are the distinguishing features of *Hoploaceratherium*. Its symphysis is narrower anteriorly. No trace of a horn ever occurs in *Aceratherium*. The faint frontal rugosities observed by Osborn (1900) have no similarity to the structure of a horn base, as Guérin (1980) has shown.

The presence or absence of upper incisors in *Aceratherium* from the type locality is still an unsolved question. Kaup's (1832) reconstruction provided the type skull with a very large one, belonging undoubtedly to *Brachypotherium goldfussi*. Referred specimens from the Höwenegg, Germany, show the presence of a small $I^1$, but these skulls are too small, compared to the holotype. So we must rely on the cited characters, confirmed by a lot of less important traits in the cheek teeth.

In Asia, there is a nearly contemporaneous species, "*Aceratherium*" *depereti* Borissiak, 1927, that is similar in the lack of upper incisors and the presence of a small horn. It also has a narrow symphysis, but it is more progressive in the broadening of the skull and the shortening of the nasals and premaxillae. Before placing this species in the new genus we need a study of its remains, especially of the skulls in the American Museum. Other Asiatic forms, such as the whole *Chilotherium* group, show different tendencies in the broadening of the mandibular symphysis and the shortening of the limb bones. "*Aceratherium*" *gobiense* has a broad symphysis and is therefore not related. "*Aceratherium*" *aralense* Borissiak is a member of the genus *Aprotodon* and therefore of the tribe Teleoceratini.

In North America we find no related forms, even if in *Peraceras* some forms have a horn and all species of this genus and of *Aphelops* have lost the upper incisors. All are more derived, especially in the shortening of the nasals, and in *Peraceras* the moderate shortening of the limb bones.

*Chilotheridium* in Africa is similar in the long nasals, the presence of a horn and the lack of upper incisors. The backwards-shifted horn position, and the limb shortening are sufficient for a generic separation, but nevertheless both genera seem to be closely related.

The other African genus, *Turkanatherium* Deraniyagala (1951) may be nearer to *Aceratherium* because it lacks a horn, and the nasal notch is shallower. The edentulous premaxillae and the nasals are indeed longer as in this genus and comparable to *Hoploaceratherium*.

## Conclusions

*Hoploaceratherium* is very distinct from all known genera of the Aceratheriini. Some similarities to later forms may indicate phylogenetic relationships. Even some derived forms in other continents may one day be traced back to this genus, the most primitive of all horned Aceratheriini.

## Bibliography

Blainville, H.M.D. de (1846): Des rhinocéros (*G. Rhinoceros* L.). *Ostéographie ou description iconographique comparée*... Paris.

Borissiak, A.A. (1927): *Aceratherium depereti* n.sp. from the Jilancik beds. - *Bull. Acad. Sci. St. Petersburg*, (6) 21:

769-786.

Deraniyagala, P. E. P. (1951): A new genus and species of hornless Mio-Pliocene rhinoceros. -*Proc. Ceylon Assoc. Sci. Ann. Congr.*, 72:24.

Dollo, L. (1885): Rhinocéros vivants et fossiles. -*Rev. Quest. Scientif.*, 17: 293-299.

Duvernoy, G. -L. (1853): Nouvelles études sur les Rhinocéros fossiles. -*C. R. Acad. Sci. Paris*, 36: 117-125, 150-154, 159-176.

Ginsburg, L., and Guérin, C. (1979): Sur l'origine et l'extension stratigraphique du petit Rhinocérotidé Miocène *Aceratherium* (*Alicornops*) *simorrense* (Lartet, 1851) nov. subgen. -*C.R. Somm. Sé. Soc. géol. France*, 3: 114-116.

Guérin, C. (1980): Les rhinocéros (Mammalia, Perissodactyla) du Miocène terminal au Pleistocène supérieur en Europe occidentale. -*Doc. Lab. Géol. Fac. Sci. Lyon*, 79: 1-1185 (3 vols.).

Heissig, K. (1969): Die Rhinocerotidae aus der oberoligozänen Spaltenfüllung von Gaimersheim bei Ingolstadt in Bayern. -*Abh. Bayer. Akad.Wiss. Math. -Nat. Kl.N.F.* 138: 1-133.

Kaup, J. J. (1832): Über *Rhinoceros incisivus* Cuvier und eine neue Art, *Rhinoceros schleiermacheri*. -*Isis*, 1832 (8): 898-904.

Kaup, J. J. (1854): *Beiträge zur näheren Kenntniss der urweltlichen Säugethiere* 1: viii+31 pp.

Lartet, E. (1837): Sur les débris fossiles trouvés à Sansan et sur les animaux antédiluviens en général. -*C.R. Acad. Sci.*, 5: 158.

Osborn, H. F. (1900): Phylogeny of the rhinoceroses of Europe. -*Bull. Amer. Mus. Nat. Hist.*, 13 (19): 229-267.

Owen, R. (1845): *Odontography; or a treatise on the comparative anatomy of teeth.* -London (Hippolyte Bailliere).

Stromer, E. (1902): Ein *Aceratherium*-Schädel aus dem *Dinotherien*-Sand von Niederbayern. -*Geogn. Jahresh.*, 15: 57-64.

Young, C.C. (1937): On a Miocene mammalian fauna from Shantung. -*Bull. Geol. Soc. China*, 17: 209-243.

# 23. TAXONOMY AND BIOCHRONOLOGY OF *EOMOROPUS* AND *GRANGERIA*, EOCENE CHALICOTHERES FROM THE WESTERN UNITED STATES AND CHINA

## SPENCER G. LUCAS and ROBERT M. SCHOCH

A taxonomic revision of the Eocene chalicotheres *Eomoropus* and *Grangeria* recognizes four valid species: *E. amarorum* (Cope, 1881) and *G. anarsius* (Gazin, 1956) from the western United States and *E. quadridentatus* (Zdansky, 1930) and *G. canina* (Zdansky, 1930) from the People's Republic of China. *E. quadridentatus* and *G. canina* localities in China are of middle Eocene age, most probably equivalent to the Uinta "B" of the western United States.

### Introduction

*Eomoropus* Osborn, 1913, is a genus of small, primitive chalicotheres known from the Eocene of China and the western United States. Cope (1881) described the first specimen of *Eomoropus* as a rhinocerotoid, *"Triplopus" amarorum*. Osborn (1913) subsequently recognized that *"T." amarorum* was a chalicothere and proposed the new generic name *Eomoropus*. Since Osborn, two additional species of *Eomoropus*, *E. annectens* Peterson (1919) and *E. anarsius* Gazin (1956), have been proposed for North American specimens.

Zdansky (1930) first recorded *Eomoropus* in Asia and proposed three species of the genus from Shanxi, China: *E. quadridentatus*, *E. major*, and *E. minimus*. Zdansky (1930) also described the new chalicothere genus *Grangeria* from Shandong, China. Later, Chow (1962), proposed another Asian species of *Eomoropus*, *E. ulterior*, for a lower jaw fragment from Yunnan, China.

Radinsky (1964) revised the taxonomy of *Eomoropus*, *Grangeria*, and other Eocene chalicotheres. Some taxonomic problems remained unresolved by Radinsky, primarily because he did not have the opportunity to study firsthand the material described by Zdansky (1930). It is our intention to resolve these problems, basing our taxonomic revision largely on study of the Lagrelius collection of Chinese fossil vertebrates in the Paleontological Museum, Uppsala University, Sweden (Mateer and Lucas, 1985). We also discuss the biochronological significance of *Eomoropus* and *Grangeria*.

### Abbreviations

AMNH   American Museum of Natural History, New York
BGM    Beijing Geological Museum, Beijing
CM     Carnegie Museum of Natural History, Pittsburgh
DNHM   Dinosaur Natural History Museum, Vernal, Utah
FMNH   Field Museum of Natural History, Chicago
IVPP   Institute of Vertebrate Paleontology and Paleoanthropology, Beijing
PMUM   Paleontological Museum, Uppsala University, Uppsala
PU     Princeton University (specimens now in the Yale Peabody Museum, New Haven)
USNM   National Museum of Natural History, Smithsonian Institution, Washington, D.C.

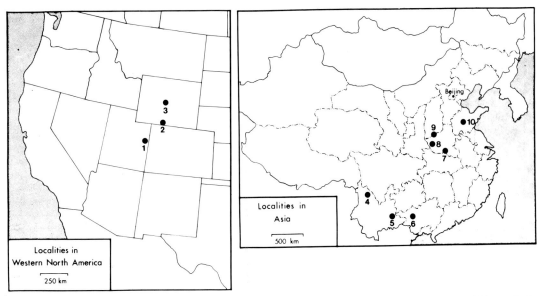

Fig. 23.1. *Eomoropus* and *Grangeria* localities in the western United States (A) and the People's Republic of China (B). Localities are: 1. Uinta Basin, Utah (*E. amarorum, G. anarsius*); 2. Washakie basin, Wyoming (*E. amarorum*); 3. Wind River Basin, Wyoming (*G. anarsius*); 4. Lijiang basin, Yunnan (*E.* sp.); 5. Lunan basin, Yunnan (*E. quadridentatus*); 6. Bose basin, Guangxi (*E. quadridentatus*); 7. Wucheng basin, Henan (*E.* sp.); 8. Lushi basin, Henan (*E.* sp. ); 9. Yuanqu basin, Shanxi (*E. quadridentatus, G. canina*); 10. Xintai basin, Shandong (*G. canina*).

## Systematic paleontology

Order Perissodactyla Owen, 1848
Superfamily Chalicotherioidea Gill, 1872
Family Eomoropidae Matthew, 1929

*Discussion.* Matthew (1929, p. 519) erected the subfamily Eomoropinae (also see Koenigswald, 1932, and Colbert, 1934) as a horizontal taxon (grade) for *Eomoropus*, the most primitive chalicothere. Radinsky (1964) raised the Eomoropinae to family rank to encompass *Paleomoropus, Lophiaspis, Eomoropus, Grangeria,* and *Litolophus*. The paraphyletic nature of the Eomoropidae is clear (Coombs, 1982), and we only use the term here because we have not completed a phylogenetic analysis of early chalicotheres.

In this chapter, we continue to use the term Eomoropidae *sensu* Radinsky (1964) with only three genera—*Eomoropus, Grangeria,* and *Litolophus* included. Thus, we here tentatively agree with Fischer (1977) that *Paleomoropus* Radinsky, 1964, and *Lophiaspis* Depéret, 1910, may not be chalicotheres. *Lunania youngi* Chow, 1957, known only from a diminutive left dentary fragment with $M_{2-3}$ (Chow, 1957, Fig. 2; pl. 1 , Figs. 4-4A), has been provisionally considered a chalicothere (Chow, 1962; Radinsky, 1964), but is so poorly known that its affinities are uncertain. The possibility that *Lunania* is a phenacolophid, not a perissodactyl, should be examined further.

The three eomoropids are readily distinguished from more advanced chalicotheres by a suite of primitive features that include: small size ($M_1$ length 18 mm or less); low-crowned teeth; presence of upper and lower canines of moderate to large size; $P^1$ present; submolariform premolars; premolar rows that are long relative to molar rows; wide and quadrate upper molars with prominent transverse lophs, narrow ectolophs, isolated parastyles, small to absent mesostyles and protocones not in a transverse line with protolophs; lower molar metalophids that terminate labial to the metastylids; $M_3$ hypoconulids; relatively small auditory bullae; tetradactyl manus; tridactyl pes; unspecialized metapodials and long tails (Colbert, 1934, p. 354; Radinsky, 1964, p. 7).

Thus, *Eomoropus, Grangeria,* and *Litolophus* stand apart from all other chalicotheres as the most primitive members of the superfamily. For this reason, generic diagnoses and comparisons in this paper are confined to consideration of only these three genera.

Genus *Eomoropus* Osborn, 1913
1881 *Triplopus*: Cope, p. 389.
1884 *Triplopus*: Cope, p. 687, pl. 55a, Figs. 6-9; pl. 58a, Figs. 2-2a.
1913 *Eomoropus* Osborn, p. 264, Figs. 1, 3A, 4-8.
1930 non *Eomoropus*?: Zdansky, p. 66, pl. 5, Figs. 1-2.
1955 non *Eomoropus*: Gazin, p. 77.
1956 non *Eomoropus*: Gazin, p. 12; pl. 8, Figs. 1-3.
1959 *Eomoropus* (in part): Hu, p. 126, pl. 1, Figs. 1a-c.
1962 non *Eomoropus*: Chow, p. 219, Fig. 1.
1964 *Eomoropus*: Radinsky, p. 9.
1985 *Eomoropus*: Lucas and Schoch, p. 192.

*Type species*: *Eomoropus amarorum* (Cope, 1881) (=*E. annectens* Peterson, 1919).

*Included species*: The type species and *E. quadridentatus* Zdansky, 1930 (= *E. minimus* Zdansky, 1930).

*Distribution*: Middle Eocene of the western United States (Wyoming, Utah) and the People's Republic of China (Henan, Shanxi, Guangxi, Yunnan) (Fig. 23. 1).

*Revised diagnosis*: *Eomoropus* is a small chalicothere (length of $M_1$ = 11-14 mm) distinguished from *Grangeria* by its smaller size, proportionately smaller canines, relatively longer and more shallow cranial rostrum, relatively longer and more shallow mandibular symphysis and mandibular horizontal ramus, relatively smaller and more molariform (less trenchant) $P_2$, and less prominent upper molar metastyles that are not anterolabially extended. *Eomoropus* is distinguished from *Litolophus* by its smaller size, lack of a diastema between $P^1$ and $P^2$, larger and less posteriorly placed $P^2$ protocone, upper molar mesostyles, lack of posterolabial rotation of the $M^3$ metaloph and posterior end of the ectoloph, premolar rows that are long relative to its molar rows, and less oblique lower molar metalophids.

*Discussion*: Radinsky (1964) clearly drew the distinctions between *Eomoropus, Grangeria,* and *Litolophus*. Differences between our diagnosis of *Eomoropus* and Radinsky's reflect two phenomena: 1) our unwillingness and/or inability to repeat some of Radinsky's observations (e.g., that $P^1$ is absent in *Eomoropus*; it is present in *E. quadridentatus* (Fig. 23.2D), and its presence or absence cannot be determined from available specimens of *E. amarorum*); and 2) our assignment of *Eomoropus major* Zdansky, 1930, to *Grangeria*, which clarifies the upper molar morphology of *Grangeria* and justifies assignment of *Eomoropus anarsius* Gazin, 1956, to *Grangeria*.

Most of the differences between *Eomoropus* and *Grangeria* reflect strikingly different functions of the anterior dentition of the two genera. The larger canines and anterior premolars, short face, short mandibular symphysis, and deep mandible of *Grangeria* could be considered male features in a sexually dimorphic taxon whose female members have the *Eomoropus* morphology of small canines and anterior premolars, long face, long mandibular symphysis and shallow mandible. We reject this interpretation because not only is sexual dimorphism along these lines unknown in other chalicotheres (Coombs, 1975), but there is a significant difference in cheektooth size and morphology between *Eomoropus* and *Grangeria* that strongly induces us to consider them separate genera. Similar reasoning justifies recognition of *Litolophus* as a genus distinct from *Eomoropus* and *Grangeria*.

Tentative assignment of *E. major* Zdansky, 1930, and *E. anarsius* Gazin, 1956, to

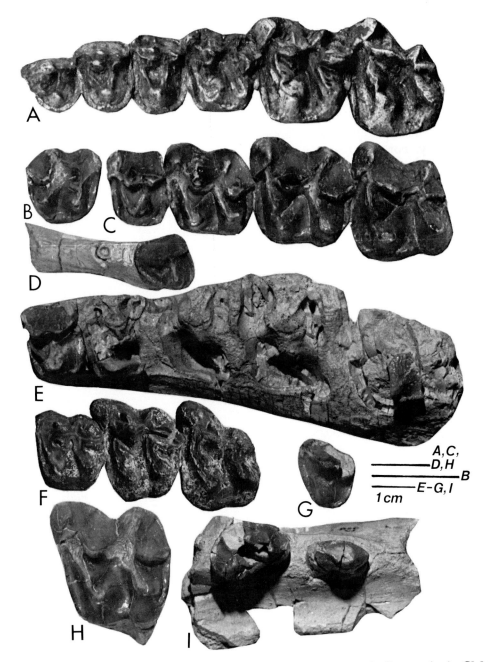

Fig. 23.2. Occlusal views of upper cheek teeth of *Eomoropus* and *Grangeria* A, CM 3109 (holotype of *E. annectens* Peterson, 1919), left $P^2$-$M^3$. B, PMUM 3452 (holotye of *E. minimus* Zdansky, 1930), right $DP^4$. C, PMUM 3451 (lectotype of *E. quadridentatus* Zdansky, 1930) left $P^4$-$M^3$. D, PMUM 3017 (*E. quadridentatus*), left maxillary fragment with $P^1$ root and $P^2$. E, PMUM 3458 (part of holotype of *G. canina* Zdansky, 1930), left maxillary fragment with $P^3$, incomplete $P^4$, roots and alveoli of $M^{1-2}$ and incomplete $M^3$. F, USNM 21097 (holotype of *G. anarsius* Gazin, 1956), left $M^{1-3}$. G, PMUM 3021 (*G. canina*), incomplete left $P^2$. H, PMUM 3453 (holotype of *E. major* Zdansky, 1930), left $M^1$. I, PMUM 3018 (part of holotype of *G. canina* Zdansky, 1930), right maxillary fragment with C alveolus and $P^{1-2}$.

*Grangeria* and *E. ulterior* Chow, 1962, to *Litolophus* was undertaken by Radinsky (1964). The opportunity to study firsthand the Lagrelius Collection in Uppsala, which contains the chalicothere specimens described by Zdansky (1930), supports assignment of *E. major* and *E. anarsius* to *Grangeria* (see later discussion). We uphold Radinsky's (1964) argument that the relatively short lower premolar row and oblique lower molar metalophids of the holotype of *E. ulterior* (Chow, 1962, Fig. 1) justify its transferral to *Litolophus*. In addition, Hu's (1959) isolated $M^3$ referred to *E. major* lacks a mesostyle, so, as pointed out by Chow et al. (1974, p. 271), it should be assigned to *Litolophus*.

*Eomoropus amarorum* (Cope, 1881)
Figs. 23.2A, 23.3B; Tables 23.1, 23.2.
1881 *Triplopus amarorum* Cope, p. 389.
1884 *Triplopus amarorum*: Cope, p. 687, pl. 55a, Figs. 6-9; pl. 58a, Figs. 2-2a
1913 *Eomoropus amarorum*: Osborn, p. 264, Figs. 1, 3A, 4-8.
1919 *Eomoropus annectens* Peterson, p. 139, pl. 36, Fig. 2.
1964 *Eomoropus amarorum*: Radinsky, p. 10, Fig. 2.
1972 *Eomoropus amarorum*: Turnbull, p. 29.
1985 *Eomoropus amarorum*: Lucas and Schoch, p. 192.

*Holotype*: AMNH 5096, incomplete skull with $M^2$ roots and incomplete $M^3$, lower jaw with $I_{1-3}$, C roots and $P_2$-$M_3$, three cervical vertebrae, pelvis, femur, tibia, manus, and pes (Cope, 1884, pl. 55a, Figs. 6-9; pl. 58a, Figs. 2-2a; Osborn, 1913, Figs. 1, 3A, 4-8; Radinsky, 1964, Fig. 2).

*Horizon and locality of holotype*: Uinta "B" interval of the Washakie Formation, Washakie Basin, Wyoming.

*Referred specimens*: From the Uinta "B" interval of the Wagonhound Member, Uinta Formation, Wagonhound Bend, White River, Uinta County, Utah: CM 3109, left maxilla with $P^2$-$M^3$ (holotype of *Eomoropus annectens* Peterson, 1919) (Fig. 23.2A; Peterson, 1919, pl. 36, Fig. 2; Radinsky, 1964, Fig. 2).

From the Uinta "B" interval of the Uinta Formation, gray clays, "upper *D. cornutus* beds," White River, Utah: PU 18067, right maxillary fragment with $M^{2-3}$.

From the Uinta "B" interval of the Adobe Town Member, Washakie Formation, Sweetwater County, Wyoming: FMNH PM 1670, left $P_4$-$M_3$ and tooth fragments (5 km east of Dobeytown Rim) (Fig. 23.3B); FMNH PM 2082, incomplete right pes, distal end of tibia and fibula, proximal half of right femur, phalangeal fragments (1.7 km SW of Wild Horse Spring).

*Revised diagnosis*: *E. amarorum* is distinguished from *E. quadridentatus* by its less molariform premolars. Thus, in *E. amarorum* $P^2$ has a small and relatively indistinct metacone, $P^{3-4}$ are relatively triangular in outline and have relatively small metastyles, $P^{2-4}$ lack complete lingual cingula and $P_2$ lacks a posterolingual ridge.

*Description*: Cope (1881, 1884), Osborn (1913), Peterson (1919), and Radinsky (1964) described most of the specimens we assign to *E. amarorum*, and dental measurements for these specimens are presented here (Tables 23.1-2). The specimens referred here to *E. amarorum* not known to these workers (FMNH PM 1670 and 2082) add no new information to the known morphology of *E. amarorum*.

*Discussion*: The few specimens of *E. amarorum* known from the western United States represent a narrow range of meristic and metric variation, and clearly pertain to one species (see Radinsky's 1964, pp. 10-11 cogent arguments for considiring *E. annectens* a junior subjective synonym of *E. amarorum*).

The meristic difference between North American *E. amarorum* and Asian *E. quadridentatus* (there are no systematic metric differences: Tables 23.1 and 23.2) are minor but consistent differences in the premolars that indicate *E. quadridentatus* has slightly more molariform premolars than *E.*

Table 23.1. Measurements (in mm) of upper cheek teeth of *Eomoropus* and *Grangeria*. L = maximum anteroposterior length of tooth crown; W = maximum transverse width of tooth crown. Asterisks (*) indicate approximate measurements of damaged teeth.

| | $P^2L$ | $P^2W$ | $P^3L$ | $P^3W$ | $P^4L$ | $P^4W$ | $M^1L$ | $M^1W$ | $M^2L$ | $M^2W$ | $M^3L$ | $M^3W$ |
|---|---|---|---|---|---|---|---|---|---|---|---|---|
| *Eomoropus amarorum*: | | | | | | | | | | | | |
| CM 3109 | 10.0 | 9.8 | 10.1 | 12.2 | 10.2 | 13.5 | 13.9 | 14.0 | 17.5 | 17.6 | 17.4 | 20.0 |
| PU 18067 | | | | | | | | | 17.3* | 17.7* | 19.2* | 20.1* |
| *Eomoropus quadridentatus*: | | | | | | | | | | | | |
| BGM V773 | | | 9.7 | 10.8 | 10.0 | 11.8 | | 13.0 | | | | |
| PMUM 3017 | 8.4 | 8.7 | | | | | | | | | | |
| PMUM 3451[a] | | | | | 9.0* | 12.2 | 14.3 | 14.6 | 17.0 | 17.3 | 15.8 | 18.6 |
| PMUM 3451b[a] | | | | | | | 13.5* | 13.5 | 17.3* | 17.6* | 15.9 | 18.7 |
| PMUM 6000 | | | | | 10.3 | 12.4 | 14.8 | 14.5 | 17.9 | 17.4 | 16.2* | 18.2 |
| PMUM 6001 | | | | | | | | | | | 14.9 | 18.2 |
| PMUM 6002 | | | | | 11.1* | 12.5 | | | | | | |
| *Grangeria anarsius* | | | | | | | | | | | | |
| USNM 21097[a] | | | 11.6 | 13.1 | | | 16.3 | 16.4 | 22.4* | 23.8 | 21.2 | 25.0* |
| *Grangeria canina* | | | | | | | | | | | | |
| PMUM 3018 | 12.9* | 14.0* | | | | | | | | | | |
| PMUM 3021 | 13.0 | 14.6 | | | | | | | | | | |
| PMUM 3453 | | | | | | | 20.2 | 19.9 | | | | |
| PMUM 3458[a] | | | 12.9* | 16.9 | 14.4* | 18.7* | 17.3* | 20.2* | 18.0* | 24.4* | 22.6* | 27.9* |

[a] Holotype or lectotype

Table 23.2. Measurements (in mm) of lower cheek teeth of *Eomoropus* and *Grangeria*. AW = trigonid width; L = maximum anteroposterior length of tooth crown; PW = talonid width; W = maximum transverse width of tooth crown. Asterisks (*) indicated approximate measurements of damaged teeth.

| | $P_2L$ | $P_2W$ | $P_3L$ | $P_3W$ | $P_4L$ | $P_4W$ | $M_1L$ | $M_1AW$ | $M_1PW$ | $M_2L$ | $M_2AW$ | $M_2PW$ | $M_3L$ | $M_3AW$ | $M_3PW$ |
|---|---|---|---|---|---|---|---|---|---|---|---|---|---|---|---|
| *Eomoropus amarorum* | | | | | | | | | | | | | | | |
| AMNH 5096[a] | 10.5 | 5.5 | 11.0 | 8.7 | 12.4 | 8.3 | 14.0 | 9.2 | 9.2 | 18.1 | 10.1 | 10.4 | 26.1 | 12.1 | 11.5 |
| FMNH PM 1670 | | | | | 11.4 | 8.1 | 12.9* | | | 16.6 | 8.9 | 9.0 | 21.4* | 10.4 | 10.0* |
| *Eomoropus quadridentatus*: | | | | | | | | | | | | | | | |
| PMUM 3448 | | | 11.5 | 8.6 | | 6.0* | 13.1 | 7.6 | 7.7 | 16.0 | 8.9 | 8.4 | | | 8.1* |
| PMUM 3449-3450 | | 4.1 | | | 10.5* | 7.2* | 11.6* | 7.7* | 8.0 | 14.3 | 8.6 | 8.7 | 19.0 | 9.0 | 8.8 |
| PMUM 3451b[a] | | | | | | | 13.7 | 7.7 | 8.1 | | | | | | |
| PMUM 6004 | | | | | 9.1* | 6.1* | 10.9* | | 7.1* | 14.4 | 8.3 | 8.4 | 19.8 | 9.0 | 8.4 |
| *Grangeria anarsius*: | | | | | | | | | | | | | | | |
| DNHM V-50 | 11.2 | 5.5 | 9.5 | 6.9 | 12.4 | 9.7 | 11.0* | 7.7 | | 17.8 | 9.3 | 9.5 | 25.1 | 10.1 | 9.6 |
| USNM 21097[a] | | | 10.7 | 7.8 | 11.6 | 8.0 | 14.0 | 8.6 | 9.0 | 16.9 | 10.5 | 10.5* | 23.8 | 11.0* | 10.3 |
| *Grangeria canina*: | | | | | | | | | | | | | | | |
| PMUM 3454[a] | | | | | 15.1 | 11.3 | 18.0 | 11.6 | 11.6 | 21.9* | 14.7 | 14.5 | | | |
| PMUM 6097 | | | | | 11.6 | 17.5* | 11.3 | 11.2* | | | | | | | |

[a] Holotype or lectotype

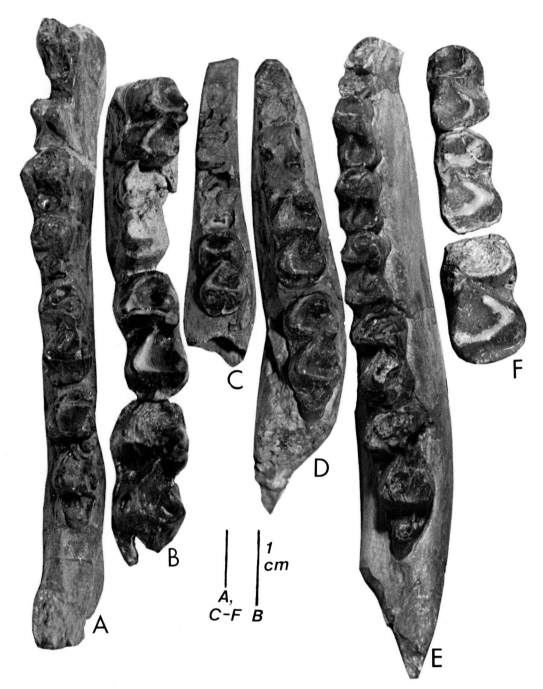

Fig. 23.3. Occlusal views of lower cheek teeth of *Eomoropus* and *Grangeria*. A, DNHM V-50 (*G. anarsius*), left dentary fragment with $P_2$-$M_3$. B, FMNH PM 1670 (*E. amarorum*), left $P_4$-$M_3$. C, PMUM 3451b (same individual as holotype of *E. quadridentatus* Zdansky, 1930), left dentary fragment with $P_{3-4}$ roots and $M_1$. D, PMUM 6004 (*E. quadridentatus*), left dentary fragment with $P_4$-$M_3$. E, USNM 21097 (holotype of *G. anarsius* Gazin, 1956), left dentary fragment with $P_3$-$M_3$. F, PMUM 3454 (part of holotype of *G. canina* Zdansky, 1930), right $P_4$-$M_2$.

*amarorum.* Thus, $P^2$ of *E. quadridentatus* has a more distinct and larger metacone than $P^2$ of *E. amarorum* (compare Fig. 23.2A with Fig. 23.2D); $P_2$ of *E. quadridentatus* has a posterolingual ridge not present on $P_2$ of *E. amarorum* (compare Zdansky, 1930, pl. 4, Fig. 14 and Fig. 23.4B with Radinsky, 1964, Fig. 2); $P^{2-4}$ of *E. quadridentatus* lack the complete lingual cingula present on $P^{2-4}$ of *E. amarorum* (compare Fig. 23.2A with Figs. 23.2C and D); and the $P^{3-4}$ of *E. quadridentatus* have a more quadratic outline and larger parastyles than do those of *E. amarorum* (compare Fig. 23.2A with Fig. 23.2C and Chow et al., 1974, pl. 1, Fig. 4a). Other distinctions drawn between the two species of *Eomoropus* by Radinsky (1964) and Chow et al. (1974) cannot be verified by us: *E. quadridentatus* has a higher $P_3$ paraconid, a shorter $P_4$ trigonid, and a more lingual connection between the cristids obliquae and metalophids on the lower molars. Furthermore, Radinsky's (1964) claim that *E. quadridentatus* has a $P^1$ whereas *E. amarorum* does not cannot be verified from existing specimens.

The conclusion of Chow et al. (1974, p. 271) that *E. quadridentatus* "may be generically distinct" from *E. amarorum* is rejected by us. Instead, we are struck by the close morphological similarity of North American and Asian *Eomoropus*. We are only able to distinguish these taxa on the basis of subtle differences in premolar morphology and suspect that larger samples of *Eomoropus* may even blur these distinctions.

*Eomoropus quadridentatus* Zdansky, 1930
Figs. 23.2B-D, 23.3C-D; 23.4B;
Tables 23.1, 23.2.

1930 *Eomoropus? quadridentatus* Zdansky, p. 62, pl. 3, Figs. 24-25; pl. 4, Figs. 9-14.
1930 *Eomoropus? minus* Zdansky, p. 67, pl. 5, Figs. 3-4.
1964 *Eomoropus quadridentatus*: Radinsky, p. 13.
1964 *Eomoropus? minor (lapsus calami)*: Radinsky, p. 13
1974 *Eomoropus* cf. *quadridentatus (sic)*: Chow, Chang, and Ting, p. 264, pl. 1, Figs. 4-4a.
1985 *Eomoropus minor (lapsus calami)*: Lucas and Schoch, p. 192.
1985 *Eomoropus quadridentatus*: Lucas and Schoch, p. 192.

*Lectotype*: PMUM 3451, left maxillary fragment with $P^4$-$M^3$ (Fig. 23.2C; Zdansky, 1930, pl. 4, Figs. 9-10). Lectotype designated by Radinsky (1964, p. 13).

*Horizon and locality of lectotype*: Zdansky's (1930) "locality 7," Rencun Member of Heti Formation, Erlanggou, Yuanqu Basin, Shanxi, China.

*Referred specimens*: From the same locality and horizon as the holotype: PMUM 3017, left maxillary fragment with $P^1$ root and $P^2$ (Fig. 23.2D; Zdansky, 1930, pl. 4, Figs. 11-12); PMUM 3448, right dentary fragment with $P_3$-$M_2$ and trigonid of erupting $M_3$ (Zdansky, 1930, pl. 3, Figs. 24-25); PMUM 3449-3450, right dentary fragment with $P_2$, $P_4$-$M_3$ (Fig. 23.4B; Zdansky, 1930, pl. 4, Figs. 13-14); PMUM 3451b, right maxillary fragment with $M^{1-3}$, left dentary fragment wit $M_1$ (Fig. 23.3C), left $P_3$ fragment, left $P_1$ and two incisors (apparently representing the same individual as the lectotype PMUM 3451); PMUM 3452, right $DP^4$ (holotype of *Eomoropus minimus* Zdansky, 1930) (Fig. 23.2B; Zdansky, 1930, pl. 5, Figs. 3-4); PMUM 6000, left maxillary fragment with $P^4$-$M^3$; PMUM 6001, left $M^3$; PMUM 6002, right $P^4$; 6003, incomplete left $P^4$; PMUM 6004, left dentary fragment with $P_4$-$M_3$ (Fig. 23.3D).

From the Lumeyi Formation, Lunan Basin, Yunnan, Chaina: BGM V 773, left maxillary fragment with $P^3$-$M^1$ (Chow et al., 1974, pl. 1, Figs. 4-4a).

*Revised diagnosis*: *E. quadridentatus* is distinguished from *E. amarorum* by its more molariform premolars. Thus, in *E. quadri-*

*dentatus* $P^2$ has a relatively large and distinct metacone, $P^{3-4}$ are quadratic in outline and have relatively large metastyles, $P^{2-4}$ lack complete lingual cingula and $P_2$ has a posterolingual ridge.

*Description*: Zdansky (1930, pp. 62-67) provided a detailed and accurate description of specimens in the Lagrelius Collection referred to *E. quadridentatus* that need not be augmented here. Radinsky (1964, p. 14) summarized some salient points of Zdansky's description, and Chow et al. (1974, pp. 264-265) described the specimen from Yunnan assigned to *E. quadridentatus*.

*Discussion*: The only taxonomic problem that need be resolved here is the status of *E. minimus* Zdansky, 1930. Zdansky (1930, p. 67) based this taxon on an isolated upper cheek tooth (Fig. 23.2B) he believed to be an $M^1$ that is from the same locality as the holotype of *E. quadridentatus*. If identified as $M^1$, this tooth is significantly (about 40%) smaller than $M^1$ of *E. quadridentatus*, and this explains Zdansky's motivation in making it the holotype of a distinct species of *Eomoropus*. However, as Radinsky (1964, p. 14) observed, the holotype of *E. minimus* has the proportions of a deciduous tooth (it is longer than wide). Furthermore, our examination of this specimen reveals that it has much thinner and lighter-colored enamel than do the adult cheek teeth of *E. quadridentatus* specimens from Shanxi. These observations strongly support the conclusion that the holotype of *E. minimus* is a deciduous tooth, probably $dP^4$. In light of the fact that it comes from the same locality as the holotype of *E. quadridentatus*, we are reasonably confident in rendering *E. minimus* Zdansky, 1930, a junior subjective synonym of *E. quadridentatus* Zdansky, 1930.

Documented material of *E. quadridentatus* is known only from Shanxi and Yunnan in China. However, there have been some undocumented reports of *Eomoropus* or *E. quadridentatus* from Chinese Eocene localities. Our effort to locate in the IVPP the specimens to substantiate these reports was unsuccessful. For the record (also see Fig. 23.1), these reports are: *Eomoropus* sp. from the Lushi Formation, Lushi basin, Henan (Chow et al., 1973): *Eomoropus* sp. from the Lishigou Formation, Wucheng basin, Henan (Gao, 1976); *Eomoropus* cf. *E. quadridentatus* from the Naduo Formation, Bose basin, Guangxi (Tang and Chow, 1964); and *Eomoropus* sp. from the Xiangshan Formation, Lijiang basin, Yunnan (Zhang et al., 1978; Zheng et al., 1978).

Genus *Grangeria* Zdansky, 1930
1930 *Grangeria* Zdansky, p. 67; pl. 4, Figs. 15-16; pl. 5, Figs. 5-12.
1930 *Eomoropus?*: Zdansky, p. 67, pl. 5, Figs. 1-2.
1930 Unbestimmbare Ungulatenzahne: Zdansky, p. 72; pl. 5, Fig. 13.
1934 *non Grangeria* (in part): Colbert, p. 355, Figs. 1-7.
1935 *non ?Grangeria*: Young and Bien, p. 230, Fig. 5.
1955 *Eomoropus*: Gazin, p. 77.
1956 *Eomoropus*: Gazin, p. 121. pl. 8, Figs. 1-3.
1964 *Grangeria*: Radinsky, p. 14.
1979 *Grangeria?*: Black, p. 395.
1985 *Grangeria*: Lucas and Schoch, p. 192.

*Type species*: *G. canina* Zdansky, 1930 (= *Eomoropus major* Zdansky,1930)

*Included species*: The type species and *G. anarsius* (Gazin, 1956).

*Distribution*: Middle Eocene of the western United States (Utah, Wyoming) and the People's Republic of China (Shanxi, Shandong) (Fig. 23.1).

*Revised Diagnosis*: *Grangeria* is a small chalicothere (length of $M_1$ = 14-18 mm) distinguished from *Eomoropus* by its larger size, proportionately large canines, relatively shorter and deeper cranial rostrum, relatively shorter and deeper mandibular horizontal ramus, relatively larger and less molariform (trenchant) $P_2$, and more prominent upper molar metastyles that are anterolabially extended. *Grangeria* is distinguished from *Litolophus* by its

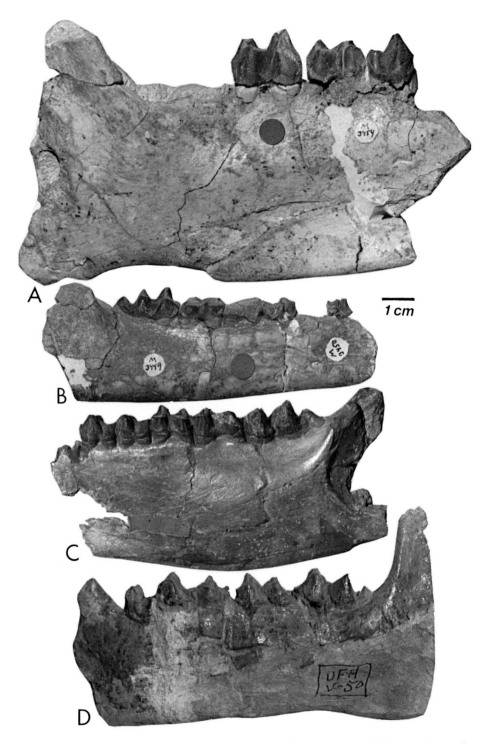

Fig. 23.4. Lateral views of incomplete dentaries of *Eomoropus* and *Grangeria*. A, *Grangeria canina* (holotype), PMUM 3454, right dentary fragment with $P_4$-$M_2$. B, *E. quadridentatus*, PMUM 3449-3450, right dentary fragment with $P_2$ and $P_4$-$M_3$. C, *G. anarsius* (holotype), USNM 21097, left dentary fragment with $P_3$-$M_3$. D, *G. anarsius*, DNMH V-50, left dentary fragment with $P_2$-$M_3$.

proportionately larger canines, relatively shorter and deeper cranial rostrum, relatively shorter and deeper mandibular symphysis and mandibular horizontal ramus, relatively larger and less molariform (trenchant) $P_2$, lack of a diastema between $P^1$ and $P^2$, relatively larger $P^1$, larger and less posteriorly placed $P^2$ protocone, upper molar mesostyles, relatively larger upper molar metastyles, lack of rotation of the $M^3$ metaloph and posterior end of the ectoloph, long premolar row relative to molar row length and less oblique lower molar metalophids.

*Discussion*: Distinction of *Grangeria* from *Eomoropus* and *Litolophus* was discussed earlier in this paper.

*Grangeria canina* Zdansky, 1930
Figures 23.2E, G-I, 23.3F, 23.4A;
Tables 23.1, 23.2.

1930 *Grangeria canina* Zdansky, p. 67; pl. 4, Figs. 15-16; pl. 5, Figs. 5-12.
1930 Unbestimmbare Ungulatenzahne: Zdansky, p. 72; pl. 5, Fig. 13.
1930 *Eomoropus? major* Zdansky, p. 66, pl. 5, Figs. 1-2.
1935 *non ?Grangeria canina*: Young and Bien, p. 230, Fig. 5.
1964 *Grangeria canina*: Radinsky, p. 15.
1964 *Grangeria? major*: Radinsky, p. 17.
1985 *Grangeria canina*: Lucas and Schoch, p. 192.

*Holotype*: PMUM 3018, right maxillary fragment bearing partial C alveolus and $P^{1-2}$ (Fig. 23.2I; Zdansky, 1930, pl. 5, Figs. 5-6); PMUM 3019, right $C^1$ (Zdansky, 1930, pl. 5, Fig. 10); PMUM 3454, right dentary fragment bearing $P_4$-$M_2$ (Figs. 23.3F, 23.4A; Zdansky, 1930, pl. 4, Figs. 15-16); PMUM 3458, left maxillary fragment bearing $P^3$, part of $P^4$, roots and/or alveoli of $M^{1-2}$ and part of $M^3$ (Fig. 23.2E; Zdansky, 1930, pl. 5, Fig. 7); PMUM 6087, left dentary fragment bearing part of $P_4$, complete $M_1$, part of the $M_2$ talonid and a crushed $M_3$ (Zdansky, 1930, pl. 5, Figs. 11-12); PMUM 6091, left dentary fragment bearing $C_1$ alveolus, $P_2$ and part of $P_3$ (Zdansky, 1930, pl. 5, Figs. 8-9; teeth now missing from the specimen). All of these specimens are of the same preservation, from the same locality and arguably belong to the same individual (Zdansky, 1930). However, if this is later shown not to be the case, we believe the holotype should be restricted to the right dentary fragment PMUM 3454.

*Horizon and locality of holotype*: Guanzhuang Formation, locality G2 of Lucas (in prep.), Xintai basin, Shandong, China.

*Referred specimens*: From the same horizon and locality as the holotype: PMUM 3021, incomplete left $P^2$ (Fig. 23.2G; Zdansky, 1930, pl. 5, Fig. 13); PMUM 3455, right dentary fragment bearing posterior root of $P_2$; PMUM 3456, right dentary fragment bearing part of $C_1$ alveolus; PMUM 8146, right $P^1$ and partial right $P^2$; PMUM 8149, fragment of left humerus; PMUM 8150, partial femur and tibia; PMUM 8151 partial left $M_2$; PMUM 8152, fragments of two caudal vertebrae; PMUM 8153, assorted postcrania, including incomplete left humerus, right scapula, ulna, radius and various metapodials and phalanges; PMUM 8154, left tibia and astragalus plus right calcaneum; PMUM 8155, right $P_3$ trigonid and assorted foot bones; PMUM 8156, tooth fragments; PMUM 8157, foot bone fragments; PMUM 8158, incisor; PMUM 8159, five phalanges and two sesamoids; PMUM 8160, assorted foot bones, including two ungual phalanges; PMUM 8161, incomplete right $DP_4$.

From Zdansky's (1930) "locality 7," Rencun Member, Heti Formation, Erlanggou, Yuanqu basin, Shanxi, China: PMUM 3453, left $M^2$ (holotype of *Eomoropus major* Zdansky, 1930) (Fig. 23.2H; Zdansky, 1930, pl. 5, Figs. 1-2).

*Revised diagnosis*: *G. canina* is distinguished from *G. anarsius* by its larger size (length $M_1$ = 18 mm) and more prominent upper molar mesostyles.

*Description*: Zdansky (1930, pp. 67-72)

provided a detailed description and measurements (see Tables 23.1 and 23.2) of the holotype of *G. canina*. Radinsky's (1964, pp. 15-16) brief summary of Zdansky's description stressed morphological features judged to be most important when comparing *G. canina* to other Eocene chalicotheres. Thus, the need for furthur description of the holotype of *G. canina* is obviated.

In addition, a large number of postcranial bones from the same locality as the holotype of *G. canina* pertain to a chalicothere. These bones are of the same color and type of preservation as the holotype of *G. canina*, and may pertain to the same individual as the holotype. These specimens were not mentioned by Zdansky (1930). Here, we note that the manus and pes of *G. canina* are similar to those of *Litolophus gobiensis* (Colbert, 1934, Figs. 5-6). Thus, the metapodials have equal-sized and evenly convex sesamoidal and phalangeal facets, and lack the specializations present on the metapodials of later chalicotheres like *Moropus* (Coombs, 1978; Holland and Peterson, 1914) and *Chalicotherium* (Zapfe, 1979). Radinsky (1964, p. 22) noted that the second phalanges of *L. gobiensis* are relatively short and deep and have long, convex, and grooved distal articular facets, which extend up to the middle of their anterior faces. He thus predicted that the ungual phalanges of *L. gobiensis* are relatively deep, narrow and bore claws. *G. canina* has similar second phalanges and, indeed, its ungual phalanges are relatively deep and narrow and bear shallow dorsal fissures at their distal ends.

*Discussion*: Zdansky (1930, p. 72; pl. 5, Fig. 13) identified an isolated and incomplete upper premolar (PMUM 3021) as "Ungulate, gen. et sp. indet." This tooth (Fig. 23.2G) differs from $P^2$ of PMUM 3018 in having a paraloph that is not completely connected to the ectoloph, a larger parastyle and a weaker metaloph. Nevertheless, the $P^2$ of PMUM 3018 and PMUM 3021 are otherwise identical in size (Table 23.1) and morphology. Therefore, we are certain that identification of PMUM 3021 as a partial left $P^2$ of *G. canina* is justified.

As noted by Schoch and Lucas (1985), the badly damaged, isolated upper molar from the Guanzhuang Formation in Shandong that Young and Bien (1935) tentatively referred to *G. canina* probably belongs to an uintathere.

Assignment of *Eomoropus major* Zdansky, 1930, to *Grangeria* was advocated tentatively by Radinsky (1964, p. 17). The isolated upper molar that is the holotype of *E. major* (Fig. 23.2H) has a large and anterolabially extended parastyle. It is significantly larger than $M^1$ or $M^2$ of *Eomoropus* (Table 23.1). Furthermore, it fits very well onto the $M^1$ alveolus of PMUM 3458 (Fig. 23.2E), part of the holotype of *G. canina*. We thus feel it is both reasonable and parsimonious to identify the holotype of *E. major* as a left $M^1$ of *G. canina*.

*Grangeria anarsius* Gazin, 1956
Figs. 23.2F, 23.3A, E, 23.4C-D;
Tables 23.1, 23.2

1955 *Eomoropus amarorum*: Gazin, p. 77.
1956a *Eomoropus anarsius* Gazin, p. 12; pl. 8, Figs. 1-3.
1964 *Grangeria? anarsius*: Radinsky, p. 16.
1979 *Grangeria? anarsius*: Black, p. 395.
1985 *Grangeria anarsius*: Lucas and Schoch, p. 192.

*Holotype*: USNM 21097, left side of skull with $M^{1-3}$ and left dentary fragment with $P_3$-$M_3$ (Figs. 23.2F, 23.3E, 23.4C; Gazin, 1956, pl. 8, Figs. 1-3).

*Horizon and locality of holotype*: Hendry Ranch Member, Tepee Trail Formation, Dry Creek locality, Wind River Basin, SE 1/4 section 9, T39N, R92W, Fremont County, Wyoming (see Tourtelot, 1955).

*Referred specimens*: From the Uinta "B" interval, Wagonhound Member of Uinta Formation, sec. 24, T8S, R24E, Coyote Basin, Uintah County, Utah: DNHM V-50, left dentary fragment with $P_2$-$M_3$ (Figs. 23.3A, 23.4D).

*Revised diagnosis*: *G. anarsius* is distinguished from *G. canina* by its smaller size (length $M_1$ = 14 mm) and less prominent upper molar mesostyles.

*Description*: Gazin (1956, pp. 12-13) provided a detailed description of the holotype of *G. anarsius*, and Radinsky (1964, pp. 16-17) reviewed some of the salient features of this specimen. The DNHM specimen referred here (Figs. 23.3A, 23.4D) preserves a trenchant $P_2$ not known from USNM 21097 and is otherwise morphologically identical (except for minor size differences: Table 23.2).

*Discussion*: Gazin (1956, pp. 12-13) principally diagnosed *G. anarsius* as a species of *Eomoropus* distinguished by its very large $M^{1-3}$ parastyles and extremely deep lower jaw. Indeed, the relative depth of the mandibular horizontal ramus of USNM 21097 and DNHM V-50 is virtually the same as that of *G. canina*, and readily distinguishes these North American specimens from *Eomoropus* and *Litolophus* (Fig. 23.4). Thus, relevant ratios of the depth of the mandibular ramus below $M_1$, divided by $M_1$ length, are: *E. amarorum* (AMNH 5096, holotype) = 2.1; *E. quadridentatus* (PMUM 3449-3450: Fig. 23.4B) = 1.9; *L. gobiensis* (AMNH 26645, holotype) = 2.2; *G. canina* (PMUM 3454, holotype: Fig. 23.4A) = 3.1; *G. anarsius* USNM 21097, holotype: Fig. 23.4C) = 3.1; and *G. anarsius* (DNHM V-50 (Fig. 23.4D) = 3.0. Moreover, the trenchant $P_2$ of DNHM V-50 (Fig. 23.3A) and the relatively large canines, short and deep cranial rostrum, and large, anterolabially extended parastyles of USNM 21097 (Fig. 23.2F; Gazin, 1956, pl. 2, Figs. 1-2) are features diagnostic of *Grangeria*. We thus feel no hesitation in assigning Gazin's *Eomoropus anarsius* to *Grangeria*. The smaller size (Tables 23.1 and 23.2) and relatively small upper molar mesostyles (compare Figs. 23.2F and G) of North American *Grangeria* justify species-level separation of this material from Asian *G. canina*.

**Biochronology**

The taxonomic revision of *Eomoropus* and *Grangeria* undertaken here clarifies their distribution in Asia and western North America and supports some biochronological conclusions. Specimens of *E. amarorum* from Utah and Wyoming are derived from a limited biochronological interval, Uinta "B." This early Uintan interval is middle Eocene, equivalent to part of the upper Lutetian (Berggren et al., 1985). The type locality of *G. anarsius* in Wyoming is somewhat younger, Uinta "C" (Tourtelot, 1957; Gazin, 1955; Black, 1978, 1979). However, the Utah occurrence of *G. anarsius* is Uinta "B" (Schoch and Lucas, 1981), suggesting a biochronological range for *G. anarsius* of Uinta "B-C."

Asian *E. quadridentatus* is so similar to North American *E. amarorum* that a Uinta "B" correlation for *E. quadridentatus* occurrences in Shanxi (Rencun Member, Heti Formation, Yuanqu basin) and Yunnan (Lumeiyi Formation, Lunnan basin) is a reasonable conclusion. Indeed this correlation does not differ materially from that proposed by Li and Ting (1983) based on a consideration of the entire mammalian fauna from the Chinese *Eomoropus*-producing units.

Asian *G. canina* is more divergent morphologically from North American *G. anarsius* than is *E. quadridentatus* from *E. amarorum*. Nevertheless, *G. canina* occurs together with *E. quadridentatus* in Shanxi. Therefore, its occurrences there and in Shandong (Guanzhuang Formation, Xiantai basin) are most reasonably considered Uinta "B" correlatives. This supports the idea that the type locality of *G. canina* in the Guanzhuang Formation of Shandong is younger than the other fossil-mammal localities in this unit (probable Bridger equivalents: Lucas, in prep.). Zdansky's (1930) original conclusion that the *G. canina* type locality is younger than the other fossil mammal occurences in the Guanzhuang Formation thus is upheld. However, his often cited suggestion that the *G. canina* locality may be as young as early Oligocene cannot be confirmed.

## Acknowledgments

We thank M. C. McKenna and R. H. Tedford (AMNH), Z. Shuonan (BGM), M. Dawson (CM), A. Hamblin (DNHM), W. D. Turnbull (FMNH), M. Chow (IVPP), R. Reyment and S. Stuenes (PMUM), D. Baird (PU) and R. Emry and R. Purdy (USNM) for permission to study specimens in their care. NSF Grant DEB-7919681 and a grant from the Swedish Natural Science Research Council to the senior author supported part of this research.

## Bibliography

Berggren, W. A., Kent, D. V., Flynn, J. J., and Van Couvering, J. A. (1985): Cenozoic geochronology. -Geol. Soc. Amer. Bull., 96: 1407-1418.

Black, C. C. (1978): Paleontology and geology of the Badwater Creek area, central Wyoming. Part 14. The artiodactyls. - Ann. Carnegie Mus., 47: 223-259.

Black, C. C. (1979): Paleontology and geology of the Badwater Creek area, central Wyoming. Part 19. Perissodactyla. -Ann. Carnegie Mus., 48: 391-401.

Chow, M. (1957): On some Eocene and Oligocene mammals from Kwangsi and Yunnan. - Vert. PalAsiatica, 1: 201-214.

Chow, M. (1962): A new species of primitive chalicothere from the Tertiary of Lunan, Yunnan, -Vert. PalAsiatica, 6: 219-224.

Chow, M., Chang, Y., and Ting, S. (1974): Some early Tertiary Perissodactyla from Lunan basin, E. Yunnan. -Vert. PalAsiatica, 12: 262-273.

Chow, M., Li, C., and Chang, Y. (1973): Late Eocene mammalian faunas of Honan and Shansi with notes on some new vertebrate fossils collected therefrom. - Vert. PalAsiatica, 11: 77-85.

Colbert, E. H. (1934): Chalicotheres from Mongolia and China in the American Museum. -Bull. Amer.Mus. Nat. Hist., 67: 353-387.

Coombs, M. C. (1975): Sexual dimorphism in chalicotheres (Mammalia, Perissodactyla) -Syst. Zool., 24: 55-62.

Coombs, M. C. (1978): Reevaluation of early Miocene North American *Moropus* (Perissodactyla, Chalicotheriidae, Schizotheriinae). -Bull. Carnegie Mus., 4: 1-62.

Coombs, M. C. (1982): Chalicotheres (Perissodactyla) as large terrestrial mammals. -Third North Amer. Paleon. Conv. Proc., 1: 99-103.

Cope, E. D. (1881): The systematic arrangement of the Order Perissodactyla. -Proc. Amer. Phil. Soc., 19: 377-403.

Cope, E. D. (1884): The Vertebrata of the Tertiary formations of the West. Book 1. -Rept. U. S. Geog. Surv. Terr., 3: 1-1009.

Depéret, C. (1910): Études sure la famille des lophiodontidés. - Bull. Soc. Geol. France, 10: 558-577.

Fischer, K. (1977): Neue Funde von *Rhinocerolophiodon* (n.gen.), *Lophiodon*, und *Hyrachyus* (Ceratomorpha, Perissodactyla, Mammalia) aus dem Eozän des Geisteltals bei Halle (DDR). -Zeit. Geol. Wiss. Berlin, 5: 909-919.

Gao, Y. (1976): Eocene vertebrate localities and horizons of Wucheng and Xichuan basins, Henan. -Vert. PalAsiatica, 14: 26-34.

Gazin, C. L. (1955): A review of the upper Eocene Artiodactyla of North America. -Smithson. Misc. Coll., 128 (8): 1-96.

Gazin, C. L. (1956): The geology and vertebrate paleontology of the upper Eocene strata in the northeastern part of the Wind River Basin, Wyoming. Part 2. The mammalian fauna of the Badwater area. -Smithson. Misc. Coll., 131 (8): 1-35.

Gill, T. (1872): Arrangement of the families of mammals with analytical table. -Smithson. Misc. Coll., 11: 1-98.

Holland, W. J., and Peterson, O. A. (1914): The osteology of the Chalicotherioidea with special reference to a mounted skeleton of *Moropus elatus* Marsh, now installed in the Carnegie Museum. -Mem. Carnegie Mus., 3: 189-406.

Hu, C. (1959): On some Tertiary chalicotheres of north China. -Palaeovert. et Palaeoanthrop., 1: 125-132.

Koenigswald, G. H. R. von (1932):

*Metaschizotherium fraasi* n. g., n. sp., ein neuer Chalicotheriidae aus dem Obermiocän von Steinheim A. Albuch Bemerkungen zur Systematik der Chalicotheriiden. -*Paleontographica,* Supp. Band, 8(8): 1-24.

Li, C., and Ting, S. (1983): The Paleogene mammals of China. -*Bull. Carnegie Mus.,* 21: 1-93.

Lucas, S. G., and Schoch, R. M. (1985): *Eomoropus* and *Grangeria,* middle Eocene chalicotheres (Mammalia, Perissodactyla) from China and the western United States. -*Abstr. Fourth Int. Ther. Cong.:* 192.

Lucas, S. G. (in prep.): Middle Eocene mammals from the Guanzhuang Formation, Xintai basin, Shandong, China. -*Bull. Geol. Inst. Univ. Uppsala*: in prep.

Mateer, N. J., and Lucas, S. G. (1985): Swedish vertebrate paleontology in China: A history of the Lagrelius Collection. -In: Lucas, S. G. and Mateer, N. J. (eds.): Studies of Chinese Fossil Vertebrates. -*Bull. Geol. Inst. Univ. Uppsala* 11: 1-24.

Matthew, W. D. (1929): Critical observations upon Siwalik mammals. -*Bull. Amer. Mus. Nat. Hist.,* 56: 437-560.

Osborn, H. F. (1913): *Eomoropus* , an American Eocene chalicothere. -*Bull. Amer. Mus. Nat. Hist.,* 32: 261-274.

Owen, R. (1848): Description of teeth and portions of jaws of two extinct anthracotheroid quadrupeds . . ., with an attempt to develop Cuvier's idea of the classification of pachyderms by the number of their toes. -*Q. J. Geol. Soc. London,* 4: 104-141.

Peterson, O. A. (1919): Report upon the material discovered in the upper Eocene of the Uinta Basin by Early Douglass in the years 1908-1909, and by O. A. Peterson in 1912. -*Ann. Carnegie Mus.,* 12: 40-168.

Radinsky, L. B. (1964): *Paleomoropus,* a new early Eocene chalicothere (Mammalia, Perissodactyla), and a revision of Eocene chalicotheres. -*Amer. Mus. Novitates,* 2179: 1-28.

Schoch R. M., and Lucas, S. G. (1981): The sytematics of *Stylinodon,* an Eocene taeniodont (Mammalia) from western North America. -*J. Vert. Paleont.,* 1: 175-183.

Schoch R. M., and Lucas, S. G. (1985): The phylogeny and classification of the Dinocerata (Mammalia, Eutheria). -In Lucas, S.C. and Mateer, N. J. (eds.) Studies of Chinese Fossil Vertebrates. -*Bull. Geol. Inst. Univ. Uppsala,* 11: 31-58.

Tang, X., and Chow, M. (1964): A review of vertebrate-bearing lower Tertiary of south China. -*Vert. PalAsiatica,* 8: 119-133 [reprinted in English in *Int. Geol. Rev.,* 7:1338-1352].

Tourtelot, H. A. (1957): The geology and vertebrate paleontology of upper Eocene strata in the northeastern part of the Wind River Basin, Wyoming. Part 1. Geology. -*Smithson. Misc. Coll.,* 134 (4): 1-27.

Turnbull, W. D. (1972): The Washakie Formation of Bridgerian-Uintan ages, and the related faunas. In: West, R. M. (ed.) *Guidebook Field Conference on Tertiary biostratigraphy of southern and western Wyoming.* Milwaukee, pp. 20-31.

Young, C. C., and Bien, M. N. (1935): Cenozoic geology of the Wenho-Sushui district of central Shantung. -*Bull. Geol. Soc. China,* 14: 221-244.

Zapfe, H. (1979): *Chalicotherium grande* (Blainv.) aus der miozänen Spaltenfullung von Neudorf an der March (Devinská Nová Ves), Tschechoslowakei. -*Neue Denkschriften Naturhist. Mus. Wiens,* 2: 1-282.

Zdansky, O. (1930): Die alttertiaren Saugetiere Chinas nebst stratigraphischen Bemerkungen. -*Paleont. Sin.,* (C) 6 (2): 5-87.

Zhang, Y. (1976): The early Tertiary chalicotheres of the Bose and Yungle basins, Guangxi.-*Vert. Palasiatica,* 14: 128-131.

Zhang, Y., You, Y., Ji, H., and Ding, S. (1978): Cenozoic stratigraphy of Yunnan. -*Prof. Papers Strat. Paleont.* [Geological Publishing House, Beijing], 7: 1-21.

Zheng, J., Tang, Y., Zhai, R., Ding, S., and Huang, X. (1978): Early Tertiary strata of Lunan basin, Yunnan. -*Prof. Papers Strat. Paleont.* [Geological Publishing House, Beijing], 7: 22-29.

# 24. INTERRELATIONSHIPS AND DIVERSITY IN THE CHALICOTHERIIDAE

## MARGERY C. COOMBS

The derived chalicothere family Chalicotheriidae has two subfamilies, the Chalicotheriinae and Schizotheriinae. The Chalicotheriinae consists primarily of the genus *Chalicotherium* and is known from Eurasia and Africa. Chalicotheriinae had robust jaws, conservative cheek teeth, and highly modified postcranials. The Schizotheriinae encompasses at least six genera from Eurasia, Africa, and North America. This group developed more elongated, higher-crowned cheek teeth. Both subfamilies appear to have been browsers and may have erected on their hindlimbs to increase height. Chalicotheriinae appear to have lived in more heavily forested habitats than Schizotheriinae.

## Introduction

The Chalicotherioidea, one of the most unusual perissodactyl groups, is most simply defined on the presence of clawed ungual phalanges. This character proved confusing to early workers, who failed to link clawed postcranials with their clearly perissodactyl dentitions and thus referred the foot elements to the pangolins or edentates. Filhol (1891) was the first to demonstrate that chalicotheres were clawed perissodactyls. Dentally, chalicotheres are distinguished by upper molars having a complete metaloph with no metaconule and an incomplete protoloph in which the paraconule ( = protoconule) is retained. Chalicotheres developed a W-shaped ectoloph early in their history, but this character is not present in the earliest representatives.

Chalicotheres are first known from the Eocene and persisted into the Pleistocene, but they were never a particularly diverse or numerous group. Except for a few cases of unusual concentration, chalicothere remains are uncommon and fragmentary. Following Matthew (1929), Oligocene and later chalicotheres are recognized as a special derived group, the Chalicotheriidae. Eocene chalicotheres have sometimes been placed in the family Eomoropidae, but this group is almost certainly paraphyletic. The present paper concentrates on the Chalicotheriidae. but concludes with observations on Eocene chalicotheres and early evolution in the Chalicotherioidea.

## The Chalicotheriidae

The Chalicotheriidae can be distinguished as a monophyletic derived group of chalicotheres by a series of dental and postcranial characters (Fig. 24.1). The upper molars have shifted from shear emphasizing the transverse lophs to shear emphasizing the ectoloph. Thus the parastyle is smaller, the protoloph is reduced, and the protocone is more posteriorly displaced on the tooth. At the same time the ectoloph becomes stronger and more deeply W-shaped, and the molar as a whole is more elongated. $P^1$ has been lost, leaving three upper, as well as three lower, premolars. On the lower molars the hypoconulid on $M_3$ is reduced or lost. Where adequate postcra-

*The Evolution of Perissodactyls* (ed. D.R. Prothero & R.M. Schoch) Oxford Univ. Press, New York, 1989

nials of a single individual are known, the metatarsals of chalicotheriids are always shorter than the metacarpals, the reverse of the situation in more primitive chalicotheres. The astragalus is shortened by reduction of its neck. Metapodials and phalanges are clearly modified in association with claws.

The Chalicotheriidae is readily divisible into two subfamilies, the Chalicotheriinae and the Schizotheriinae. The Schizotheriinae is first known from the Oligocene of Eurasia (the genus *Schizotherium*); schizotheriine chalicotheres reached North America and Africa in the Miocene. The latest known representatives of the subfamily come from the late Pliocene of Africa. The Chalicotheriinae is first known from the early Miocene of Africa and Eurasia. Chalicotheriines were extinct in Africa by the mid-Miocene, however, and there is no evidence that they ever reached North America. In the late Tertiary, Chalicotheriinae had a broad distribution across Eurasia; the latest known occurrence of this subfamily is in the Pleistocene of China.

## The Chalicotheriinae

This subfamily is defined on the basis of strong postcranial modifications, which are well developed even in the most primitive known representatives (Figs. 24.1, 24.2). These modifications include: strong shortening of the tibia and metatarsals; a change in relative lengths of the metapodials such that Mc (= metacarpal) IV is distinctly longer that Mc III and Mt (= metatarsal) IV longer than Mt III (in perissodactyls the middle digit is usually clearly the longest); loss of Mc V; flattening of the astragalus and eventual loss of its neck; development of a contact between the astragalus and cuboid bones; and strong lateral compression of the ungual phalanges. The result of these changes is a perissodactyl with very odd body proportions, rather like those of a gorilla.

The dentition in the Chalicotheriinae is not at first much modified from the primitive chalicotheriid condition. Chalicotheriine molars remain low-crowned and relatively short throughout their history. Lower canines are retained. The jaws of chalicotheriines are, however, more robust than those of other chalicotheres and the mandibular symphysis is longer. These jaw characters become even more pronounced among derived representatives of the subfamily.

The best reviews of the Chalicotheriinae are by Butler (1965) and Zapfe (1979). The work of Zapfe is particularly valuable, because it describes a large collection of one species (at least 60 individuals of *Chalicotherium grande* from Miocene fissure fills at Neudorf on the March, Czechoslovakia). Although the specimens were mostly disarticulated and unassociated, Zapfe was able to look at individual variation and study the ontogenetic development of certain characters.

Most of the representatives of the Chalicotheriinae are referred to the genus *Chalicotherium*. A second genus, *Nestoritherium*, is also commonly used for material from the Plio-Pleistocene of India and China. As noted later, there is some question whether *Nestoritherium* should remain separate from *Chalicotherium*. The genus *Macrotherium* has sometimes been used for early to mid-Miocene material, but this generic name is a holdover from the early days when chalicothere postcranials and teeth were considered separately, and is a junior synonym of *Chalicotherium* (see Butler, 1965). While it is difficult to unravel the relationships among all species of *Chalicotherium*, major changes within the genus are readily shown by comparisons of the various species (Fig. 24.3).

The most primitive members of the Chalicotheriinae are referred to three species of *Chalicotherium*: *C. rusingense* from the early Miocene (equivalent to Orleanian European land mammal "age") of east Africa, *C. pilgrimi* from the early Miocene (late Agenian or Orleanian) Bugti fauna of Pakistan, and *C. wetzleri* from the early Miocene (Agenian) of Europe. *C.*

## THE EVOLUTION OF PERISSODACTYLS

| Character | *Litolophus* | *Grangeria* | *Eomoropus* | *Schizotherium* | *Borissiakia* | *Moropus* | *Tylocephalonyx* | *Ancylotherium* | *Chalicotherium rusingense/pilgrimi* | *Chalicotherium goldfussi/wuduensis* | *"Nestoritherium"* |
|---|---|---|---|---|---|---|---|---|---|---|---|
| | | | | Schizotheriinae | | | | | Chalicotheriinae | | |
| 1. Mesostyle | □ | ▨ | ▨ | ▨ | ▨ | ▨ | ▨ | ▨ | ▨ | ▨ | ▨ |
| 2. P1 lost | □ | □ | □ | ▨ | ▨ | ▨ | ▨ | ▨ | ▨ | ▨ | ▨ |
| 3. Parastyle reduced | □ | □ | □ | ▨ | ▨ | ▨ | ▨ | ▨ | ▨ | ▨ | ▨ |
| 4. Ectoloph well-developed | □ | □ | □ | ▨ | ▨ | ▨ | ▨ | ■ | ▨ | ▨ | ▨ |
| 5. Protoloph reduced, protocone moved posteriorly | □ | □ | □ | ▨ | ▨ | ▨ | ▨ | ▨ | ▨ | ▨ | ▨ |
| 6. m3 hypoconulid reduced/lost | □ | □ | □ | ▨ | ■ | ■ | ■ | ■ | ■ | ■ | ■ |
| 7. Metatarsals shorter than metacarpals | □ | □ | □ | ▨ | ▨ | ▨ | ▨ | ▨ | ▨ | ▨ | ? |
| 8. Metapodials and phalanges clearly modified for claws | □ | □ | □ | ▨ | ▨ | ▨ | ▨ | ▨ | ▨ | ▨ | ? |
| 9. Astragalus neck reduced/lost | □ | □ | □ | ▨ | ▨ | ▨ | ▨ | ■ | ■ | ■ | ? |
| 10. Molars elongate | □ | □ | □ | ▨ | ■ | ■ | ■ | ■ | □ | □ | □ |
| 11. Claw retraction mechanism | □ | □ | □ | ▨ | ▨ | ▨ | ▨ | ▨ | □ | □ | □ |
| 12. Canines lost | □ | □ | □ | ? | ▨ | ▨ | ▨ | V? | □ | □ | □ |
| 13. M3 posterolingual accessory cusps | □ | □ | □ | □ | ▨ | □ | □ | □ | □ | □ | □ |
| 14. Duplex for digit II manus | □ | □ | □ | □ | □ | ▨ | ▨ | ▨ | □ | □ | □ |
| 15. Mt IV ≥ Mt III | □ | □ | □ | □ | □ | ▨ | □ | □ | ■ | ■ | ■ |
| 16. Lower incisors reduced/lost | □ | □ | □ | □ | □ | □ | □ | ▨ | ▨ | ■ | ■ |
| 17. Domed skull | □ | □ | □ | □ | □ | □ | ▨ | □ | □ | □ | □ |
| 18. M3 with weak hypocone | □ | □ | □ | □ | □ | □ | ▨ | ▨ | □ | □ | □ |
| 19. Crochet | □ | □ | □ | □ | □ | □ | □ | □ | □ | □ | □ |
| 20. McV lost | □ | □ | □ | □ | □ | □ | □ | □ | ▨ | ▨ | ▨ |
| 21. Metacarpals concave dorsally | □ | □ | □ | □ | □ | □ | □ | □ | □ | □ | □ |
| 22. Contact of scaphoid with McII in extreme flexion | □ | □ | □ | □ | □ | □ | □ | ▨ | □ | □ | □ |
| 23. Tibia strongly shortened | □ | □ | □ | □ | □ | □ | □ | □ | ? | ▨ | ? |
| 24. Metatarsals strongly shortened | □ | □ | □ | □ | □ | □ | □ | □ | ▨ | ▨ | ? |
| 25. Ungual phalanges laterally compressed | □ | □ | □ | □ | □ | □ | □ | □ | ▨ | ▨ | ? |
| 26. Astragalus with cuboid contact | □ | □ | □ | □ | ▨ | □ | □ | □ | ? | ■ | ? |
| 27. Protoloph/protoconule lost | □ | □ | □ | □ | □ | □ | □ | □ | □ | ▨ | ■ |
| 28. Robust mandible with well-developed canines | □ | ▨ | □ | □ | □ | □ | □ | □ | ▨ | ■ | ■ |

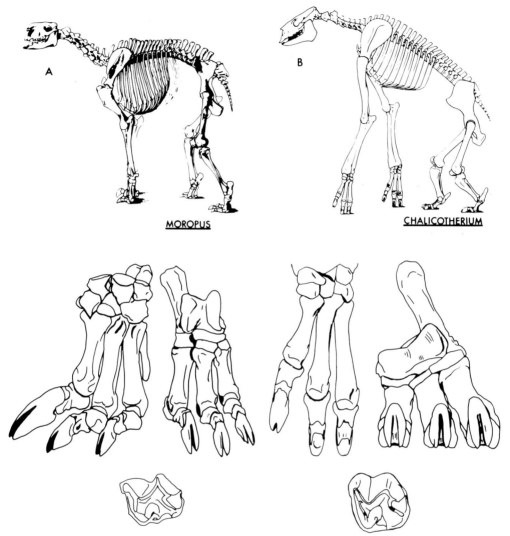

Fig. 24.2. (above) Comparison of morphology of a member of (A) Schizotheriinae (*Moropus elatus*) and (B) Chalicotheriinae (*Chalicotherium grande*). Skeleton of *Moropus* from Osborn (1919); *Chalicotherium* from Zapfe (1979). Manus (left) and pes (right) after Holland and Peterson (1914) and Zapfe (1979). $M^3$ after Munthe and Coombs (1979).

Fig. 24.1. (previous page) Character matrix and cladogram showing distribution of characteristics in the Chalicotherioidea. Hatched boxes indicate a derived state; intermediate and strongly derived states of a character are in some cases indicated respectively by hatched and fully blackened boxes. "?" indicates that a character state is unknown in a particular taxon, but might be expected in view of its distribution among closely related taxa. "V" indicates a variable character within a taxon, "V?" a possibly variable character. Available specimens of *Phyllotillon* place this genus at the same node with *Moropus*, *Tylocephalonyx*, and *Ancylotherium*. See text for further discussion of the characters.

*wetzleri* is known primarily from dental material and is larger than *C. pilgrimi* or *C. rusingense*. The teeth of *C. pilgrimi* are quite primitive (for example, the paracone and metacone are more labial in position than in other species of *Chalicotherium*), but postcranials referred to this species show derived characters which clearly ally them with other species of *Chalicotherium*. *C. rusingense*, described in detail by Butler (1965), is the best known of the three primitive species of *Chalicotherium*. *C. rusingense* has both upper and lower canines, though the upper incisors have evidently been lost. The snout is longer than in more derived species of *Chalicotherium*; the diastema is relatively long compared to the cheek tooth row, and the premolar row is over half the length of the molar row (see Xue and Coombs, 1985, for comparisons of tooth row measurements). A variety of bones of both the manus and pes are known for *C. rusingense*, and all show characters linking them with other species of *Chalicotherium*; nonetheless, the astragalus is less depressed than in more derived species of *Chalicotherium*, and the cuboid facet much smaller (unfortunately, the described astragalus is damaged in this area).

*Chalicotherium grande*, known from the medial Miocene (late Orleanian - Astaracian; see Fig. 24.4) of Europe, is intermediate in many respects between relatively primitive species such as *C. wetzleri* and *C. rusingense* and more derived chalicotheriines such as *Chalicotherium goldfussi*. Known from relatively complete material from Neudorf (Czechoslovakia) and Sansan (France) and fragmentary material from other locales, *C. grande* has lost the upper canines and thus has no teeth in the anterior part of the upper jaw. Canines are consistently present in the lower jaw, but come in two sizes, probably correlated with sexual dimorphism (Zapfe, 1979). Lower incisors are usually present but variably developed. Compared to the molar row, the premolar row is shorter than in *C. wetzleri* or *C. rusingense* but longer than in *C. goldfussi*. The metastylids on the lower molars are reduced. Body reconstructions of *Chalicotherium* (Fig. 24.2) are based primarily on postcranials of *C. grande*. Such postcranial characters as the great length disproportion between forelimb and hindlimb, the pronounced increase in length from Mc II to Mc IV and from Mt II to Mt IV, and the strongly depressed astragalus with large cuboid facet are all clearly shown in *C. grande*.

Another species of *Chalicotherium*, comparable in some respects to *C. grande*, is *C. salinum*, which comes from medial to late Miocene faunas of the Siwalik Group in Pakistan. Pickford (1982), in a recent review of specimens of *C. salinum*, counted a total of 70 fragmentary specimens from a number of localities in the Siwalik region. The temporal range of *C. salinum* is from about 13-6 Ma (see the biostratigraphic zonation by Barry, Lindsay, and Jacobs, 1982). *C. salinum* is, on average, smaller than *C. grande* (Butler, 1965; Pickford, 1982) but is quite similar in a number of respects, such as reduction or loss of the metastylid, proportions of premolar to molar row lengths, and length of the mandibular symphysis, which ends opposite $P_2$ or $P_3$. Pickford distinguished *C. salinum* on the basis of its more distinct paraconule, but this character appears to be somewhat variable within the species. There is some concern that more than one species might be present over the long temporal range of *C. salinum*, but the material is too fragmentary at present to analyze the problem. Several authors, including Badgley *et al.* (1988) and Li, Wu, and Qiu (1984), have listed "*Macrotherium*" (i.e., *Chalicotherium*) *salinum* among the fauna from a hominoid locality at Lufeng, Yunnan Province, China. The material on which this identification is based has not yet been published, but the fauna, an early Turolian equivalent (approximately 8 Ma), does contain elements in common with Siwalik faunas of comparable age.

The most derived species of Chalicotheriinae are *Chalicotherium goldfussi*

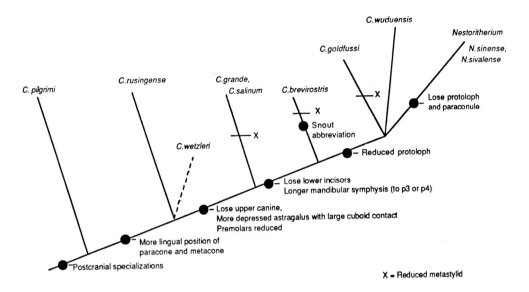

Fig. 24.3. A cladogram showing character distributions in the Chalicotheriinae. See text for further discussion.

Table 24.1. Proportions of length to width of Mt IV of selected Schizotheriinae (Chalicotheriidae).

| Genus and species | Length/ Distal width | Length/ Minimum shaft width | Source |
|---|---|---|---|
| *Schizotherium priscum* | 4.5 | 5.8 | Coombs (1974) |
| *Borissiakia betpakdalensis* | 4.0, 4.2 (2 specimens) | 5.4, 5.9 | Borissiak (1946:94) |
| *Moropus* sp. (Buda, Florida) | 3.6 | 5.2 | Frailey (1979) |
| *Moropus elatus* (8 specimens) | 3.2-3.8 (median = 3.5) | 4.8-5.5 (median = 5.3) | Coombs (1974,1978a) |
| *Moropus* cf. *M. hollandi* | 3.4 | 4.7 (approximate) | Coombs (1978a) |
| *Moropus* sp. (St. -Gérand) | 3.2 | 4.5 | Coombs (1974) |
| ?*Moropus* sp. (Portugal) | 3.4 | 4.5 (*contra* Coombs, 1974) | Figures in Antunes (1966) |
| *Chemositia tugenensis* | 2.9 | 3.8 | Figures in Pickford (1979) |
| *Ancylotherium (A.) pentelicum* | 2.5, 2.8 (2 specimens) | 3.2, 3.2 | Coombs (1974) |
| *Tylocephalonyx skinneri* (5 spec.) | 2.4-3.0 (median = 2.7) | 3.0-3.3 (median = 3.2) | Coombs (1979) |
| *Moropus merriami* | 2.5, 2.9 (2 specimens) | 3.2, 3.6 | Unpubl. measurements |

from the late Miocene of Europe, *Chalicotherium wuduensis* from the late Miocene of China, and *Nestoritherium sivalense* and *Nestoritherium sinense* from the Plio-Pleistocene of Pakistan (Siwaliks) and China. Three of these species (*N. sinense* has no known complete jaws) have an extremely robust lower jaw with a mandibular symphysis extending to opposite $P_3$ or more commonly $P_4$. Lower incisors have been completely lost. In *C. wuduensis* and *C. goldfussi* the premolar row is much shortened compared to the molar row. *C. goldfussi*, *C. wuduensis*, *N. sivalense*, and *N. sinense* all differ from one another in details of the lower dentitions (see Xue and Coombs, 1985). The distinction between *Nestoritherium* and *Chalicotherium* has traditionally rested on the absence of lower incisors and the complete loss of the protoloph and paraconule in *Nestoritherium*. As already noted, *C. goldfussi* and *C. wuduensis* have recently been shown to have lost lower incisors (Garevski and Zapfe, 1983; Xue and Coombs, 1985). *C. goldfussi* has strongly reduced the protoloph, though it still retains a paraconule. Thus *Nestoritherium* actually differs little from derived species of *Chalicotherium*, and a generic distinction is probably not warranted.

*Chalicotherium brevirostris* represents an aberrant medial Miocene (Astaracian equivalent) species of *Chalicotherium* from China. The type is a skull with an oddly abbreviated snout from the Tungur fauna of Inner Mongolia (Colbert, 1934). A lower jaw from Hebei Province subsequently described by Hu (1959) is similarly abbreviated and thus has been tentatively referred to the same species. This jaw has a symphysis extending to $P_4$ as in derived species of *Chalicotherium*, but the whole anterior part of the symphysis has been lost, as apparently have been the lower incisors and canines. As a consequence, *C. brevirostris* appears to be the only speices of *Chalicotherium* to have lost lower canines. In other respects *C. brevirostris* resembles *C. goldfussi* and other relatively derived species of *Chalicotherium*, though the protoloph on the upper molars is still complete as in *C. grande*.

One difficulty in analyzing species of *Chalicotherium* is that so many of them are known only from fragmentary specimens, and the most consistently represented elements are teeth, especially lower teeth. Where adequate samples are available, teeth of *Chalicotherium* can be quite variable in such characters as protoloph and metastylid development. The range of dental variation within a species may in some ways overlap that of another species, as sometimes occurs in comparisons of teeth of *C. grande* and *C. goldfussi*. Thus caution is necessary when using small dental differences in classification, especially when the samples are small.

Most authors believe that the Chalicotheriidae arose in Eurasia and that the early Miocene occurrence of *C. rusingense* in Africa resulted from immigration slightly earlier. Unfortunately the precise time of divergence of the Chalicotheriinae from the Schizotheriinae and the identity of its closest relatives cannot be determined on the basis of present evidence. Chalicotheriines appear rather suddenly as fossils in the early Miocene of several regions. Primitive species of *Chalicotherium* are not too different dentally from *Schizotherium*, an Oligocene representative of the Schizotheriinae. However, no known species of *Schizotherium* shows derived postcranial characters that would ally it with the Chalicotheriinae, and such characters are critical for identifying the closest sister taxon.

### The Schizotheriinae

The Schizotheriinae was rediagnosed by Coombs (1978a) and the relationships among member taxa analyzed by Coombs (1979). The following discussion updates the earlier analyses.

The Schizotheriinae is derived compared to the Chalicotheriinae in having higher-crowned, more elongated molar

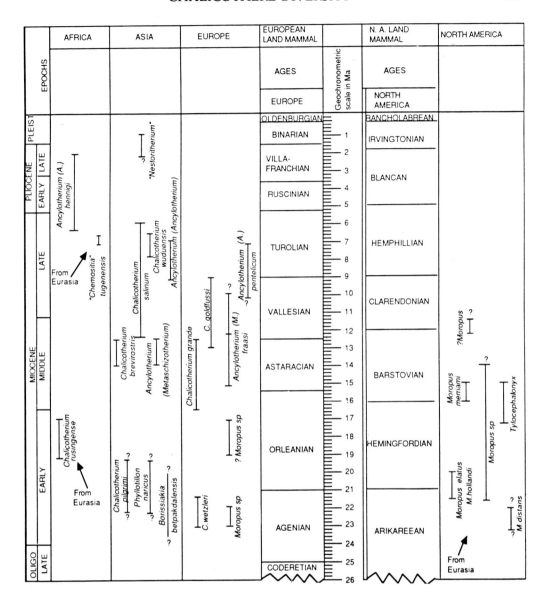

Fig. 24.4. Temporal distribution of Neogene representatives of the Chalicotheriidae. Timescale taken from Berggren *et al.* (1985). The rarity of chalicothere fossils makes it difficult to assess temporal ranges precisely, and exact ages of some deposits from which chalicotheres are known are uncertain. See text for further discussion of localities and ages.

teeth (Figs. 24.1, 24.2). This character is particularly visible in the upper molar ectolophs, which are longer and more vertical than in other chalicotheres. The postcranials in the Schizotheriinae are much less derived than those of even the most primitive known members of the Chalicotheriinae. Nonetheless, a special claw retraction mechanism has evolved in which the claws can be pulled dorsally off the ground during walking; this mechanism affects the morphology of the metapodials and phalanges in particular ways.

The Schizotheriinae contains the following genera: *Schizotherium, Phyllotillon, Borissiakia, Moropus, Tylocephalonyx,* and *Ancylotherium*. Several other schizotheriine genera, including *Metaschizotherium, Chemositia, Gansuodon,* and *Huanghotherium*, have also been erected by various authors. The status of these taxa will be discussed below.

**Schizotherium.** The genus *Schizotherium*, distributed throughout Eurasia during the Oligocene, is generally considered to be the most primitive member of the Schizotheriinae. Although it is the source of the name Schizotheriinae, this genus causes a few problems in the definition of the subfamily. Molars of *Schizotherium* have begun to be elongated, but this character is not so clearly developed as in later schizotheriines. Upper molars of *Schizotherium* can be confused with teeth of primitive species of *Chalicotherium*, such as *C. pilgrimi*, which are unelongated and also have not displaced the paracone and metacone as far lingually as occurs in later species of *Chalicotherium*. Nonetheless, Butler (1965) observed that the length/width ratio of $M^2$ of *C. pilgrimi* (106) was lower than that of *Schizotherium priscum* (110-115) and fell within the range of *C. grande* (104-111). The beginning of tooth elongation in *Schizotherium* is most visible in the increased distance between the parastyle and mesostyle, reflecting the greater length of the anterior limb of the W-shaped ectoloph. *Schizotherium* has merely reduced the hypoconulid on $M_3$, while other Schizotheriinae and all Chalicotheriinae have completely lost this cuspid. Members of this genus seem to have some claw retraction mechanism, but it is weakly developed.

The crista of *Schizotherium*, used by Coombs (1979) as a possible derived character of the genus, is probably in fact primitive. Cristae occur in some Eocene chalicotheres such as *Eomoropus*, primitive species of *Chalicotherium* such as *C. pilgrimi* and *C. rusingense*, and also in the schizotheriine genus *Borissiakia*. Certain critical parts of the skeleton, such as the dentition anterior to the cheek teeth, are unknown for *Schizotherium*.

The most recent review of the species of *Schizotherium* was by Coombs (1978b). Six species have been identified: *S. priscum*, best known from the Phosphorites of Quercy (France), lower to middle Oligocene of Europe; *S. turgaicum* from the *Indricotherium* Beds (middle Oligocene) of Kazakhstan, Soviet Union; *S. chucuae* from the middle to upper Oligocene of the Caucasus region of Georgian S.S.R.; *S. avitum* from the lower Oligocene of Mongolia (Mongolian People's Republic and Inner Mongolia Autonomous Region of China), *S. nabanensis* from the (lower?) Oligocene Gungkang Formation, Guangxi Zhuang Autonomous Region, People's Republic of China; and *S. ordosium* from the Oligocene of the Ho-tau region of Inner Mongolia Autonomous Region, People's Republic of China. "*Chalicotherium*" *modicum*, *Limognitherium ingens*, and "*Ancylotherium*" *gaudryi* are all apparent synonyms of *Schizotherium priscum*. *Kyzylkakhippus orlovi* is synonymous with *Schizotherium turgaicum* (Coombs, 1976).

Unfortunately, the only body part which can be compared among all the species of *Schizotherium* is the lower molars. Species distinctions are made on the basis of size, relative widths of trigonid and talonid, length versus width of $M_2$ or $M_3$, degree of development of $M_3$ hypoconulid, metastylid development, and cingula. Available material is not adequate to give confidence to our understanding of species relationships within this genus. Significant postcranial material is known for *S. priscum* and *S. turgaicum*, and isolated postcranial elements of *Schizotherium* also occur in China. Postcranial elements bespeak in all cases of a primitive chalicotheriid with metatarsals only slightly shorter than the metacarpals, a tall slender astragalus with no cuboid facet on its distal surface, and unfused phalanges. Some postcranial variation appears to exist among the species; for example, *S. priscum* appears to retain a trapezium in the carpus, but *S. turgaicum* does not.

**Borissiakia.** The genus *Borissiakia* was named by Butler (1965) for material originally named *Moropus betpakdalensis* by Flerov (1938) and described in a monograph by Borissiak (1946) as *Phyllotillon betpakdalensis*. *B. betpakdalensis* is one of the better-known chalicotheres, based on abundant but broken material from the Golodnaya Steppe (Betpak-dala), southern Kazakhstan, Soviet Union The fauna is probably Agenian, late Oligocene to earliest Miocene, in age. Neither the genus nor the species has been described from other regions. The naming of *Borissiakia* pays fitting tribute to Borissiak's work on the material from Betpak-dala; his monograph gives detailed descriptions of numerous elements and documents morphological and size variations, all within the context of a single species.

Like *Moropus, Ancylotherium*, and other derived Schizotheriinae, *Borissiakia* is much larger than *Schizotherium* and has much higher-crowned, more elongated molar teeth. It is also derived compared to *Schizotherium* in the complete loss of the $M_3$ hypoconulid. *Borissiakia* is unique among the Schizotheriinae in its modification of $M^3$ in the hypocone region. The posterolingual cingulum around the hypocone is unusually well developed and bears accessory cuspules; the size of these cuspules and their relationship to the hypocone vary in individual specimens. *Borissiakia* is the only member of the Schizotheriinae to have a facet for the cuboid on the distal surface of the astragalus. In this respect *Borissiakia* is convergent to *Chalicotherium*, but in other aspects astragali of the two genera are not at all similar, for the astragalus of *Borissiakia* is much taller, slenderer, and more symmetrical than that of any species of *Chalicotherium*. An additional distinctive character of *Borissiakia* is the large dorsal process on one of the ungual phalanges, probably that belonging to digit II of the manus. This process serves for the insertion of musculature which hyperextends the claw. *Borissiakia* appears to have lost the trapezium from the manus, as have *Schizotherium turgaicum, Moropus hollandi*, and *Ancylotherium*.

Despite its derived characters, *Borissiakia* remains a relatively primitive schizotheriine in a number of respects. Both the tarsals and metatarsals remain long and slender, and thus differ little in proportions from the same elements of *Schizotherium* (Table 24.1). Carpals and proximal ends of metacarpals are slender but deep, while those of more derived members of the Schizotheriinae are shallower in a dorsal to volar direction in association with stronger flexion within the carpus. There is no fusion of phalanges (see later). Upper molars retain a crista, and the protoloph is relatively straight, without strong posterior migration of the protocone.

**Phyllotillon.** The genus *Phyllotillon* was named by Pilgrim (1910) for specimens later described by Pilgrim (1912) and Forster Cooper (1920) from the Bugti Bone Beds of Pakistan. The fauna from these deposits is usually held to be of late Agenian or Orleanian age but needs further study. Pilgrim defined *Phyllotillon* on the basis of elongated molar teeth and the nature of the transverse lophs on the upper premolars. The former character merely identifies *Phyllotillon* as a schizotheriine chalicotheriid; the latter also occurs in *Moropus* and some other genera. The type species of *Phyllotillon*, *P. naricus*, is based on upper and lower cheek teeth and partial jaws, as well as several phalanges. This material differs in some ways from that of other taxa but is inadequate for complete characterization of the genus. When better material is found *Phyllotillon* may prove valid, or *P. naricus* may be better referred to *Ancylotherium* or *Moropus*. Viret (1949) referred early to medial Miocene Schizotheriinae from Europe to *Phyllotillon*, and this practice was followed for many years by workers in France, England, and Portugal. In view of the morphology of some of the European material (see following) and the questionable nature of *Phyllotillon*, this practice is not justified. The name *Phyllotillon* should be restricted for the present to the Bugti material.

*Phyllotillon naricus* is derived relative to *Schizotherium* and *Borissiakia* in having fused the proximal and middle phalanges of digit II of the manus to form what is usually called a duplex bone. Such fusion also occurs in *Moropus*, *Tylocephalonyx*, and *Ancylotherium*. Another character shared by *Phyllotillon* and derived members of the Schizotheriinae is the absence of a crista on upper molars. Most upper molars of *Phyllotillon* lack a crochet, but a few have a very weak one. Even a small crochet might serve as a derived character to link *Phyllotillon* with *Ancylotherium*, which commonly has a large (but also variable) crochet. In other respects, *Phyllotillon* resembles some of the more primitive species of *Moropus*, and it is about medium-sized for that genus. No definitive, comparably primitive material of *Ancylotherium* is known. Based on present knowledge, *Phyllotillon* is not too far removed morphologically from what would be expected in a common ancestor of *Moropus*, *Tylocephalonyx*, and *Ancylotherium*.

**Moropus.** *Moropus* is the best-known genus of the Chalicotheriidae, primarily because of the large quantity of material, including complete skeletons, of *Moropus elatus* excavated from the Agate Spring Quarries of Nebraska, United States This material is the subject of studies by Holland and Peterson (1914), Osborn (1919), Coombs (1975, 1978a), and others. *Moropus* is primarily known from North America, but it probably originated in Eurasia sometime in the late Oligocene.

The most obvious derived character differentiating *Moropus* from other members of the Schizotheriinae is the subequal length relationship of Mt III and Mt IV, with Mt IV commonly slightly longer than Mt III. Most other perissodactyls (including such otherwise derived members of the Schizotheriinae as *Ancylotherium* and *Tylocephalonyx*) have Mt III as the longest metatarsal; *Chalicotherium* is independently derived in having Mt IV much longer than Mt III (and Mc IV longer than Mc III). Correlated with the metatarsal length relationships in *Moropus* is the presence of a small articulation between Mt IV and the ectocuneiform. Taken together, these characters suggest that weight was carried more toward the outside of the hindfoot in *Moropus* than in other taxa. *Moropus* is linked with *Phyllotillon*, *Tylocephalonyx*, and *Ancylotherium* in having proximal and middle phalanges of digit II of the manus fused to form a duplex.

Named species of *Moropus* are as follows: *Moropus distans*, *M. oregonensis*, *M. elatus*, *M. hollandi*, *M. matthewi*, and *M. merriami*. There is in addition a quantity of fragmentary material from various localities which is clearly referable to *Moropus* but inadequate for species designation. The first four of the above species were rediscussed by Coombs (1978a). *M. oregonensis*, based on upper molar teeth, and *M. distans*, based on phalanges and other postcranials, are both from the John Day Formation of Oregon and may be conspecific. Arikareean in age, they are among the earliest chalicotheriids known from North America. *M. elatus* and *M. hollandi* are closely related species from the late Arikareean to earliest Hemingfordian of the North American Great

Plains; *M. elatus* includes the largest known specimens of *Moropus* and is found frequently in the Upper Harrison Formation of northwestern Nebraska (including the famous Agate Spring Quarries) and eastern Wyoming. *M. merriami* is best known from early Barstovian faunas (15 to 16 Ma) of northwestern Nevada. A very similar chalicothere occurs in early Barstovian faunas of the Great Plains. *M. matthewi* is a problematical (but probably valid) species of Hemingfordian or early Barstovian age. A number of fragmentary specimens of *Moropus* are presently insufficient for species designation. Among these are a small Arikareean chalicothere from Florida (Frailey, 1979), early to medial Hemingfordian chalicotheres from the Great Plains (especially those from the Runningwater and Batesland Formations), Hemingfordian material from the West Coast, and a few isolated specimens of medial Barstovian or later age (see, for example, Skinner, 1968).

There is considerable variation within the genus *Moropus* in a number of characters. Size ranges from small (*M. distans* from Oregon and *Moropus* sp. from Florida) to among the larger known chalicotheres (*M. elatus*). Metatarsals rapidly become shorter and wider over time (metatarsals of *M. merriami* are particularly squat), and the labial origin of the metaloph on an unworn $M^3$ becomes increasingly separated from the mesostyle. Fusion of phalanges occurs to yield a duplex for digit II of the manus in all species of *Moropus*, while digit II of the pes has a duplex in all known cases of *M. merriami* but only about 10% of *M. elatus*. Many of the variations appear to have parallels in other schizotheriine lineages.

Coombs (1978c) demonstrated that the premaxilla of *M. elatus* is edentulous. Lower incisors of this species are procumbent and spatulate and may have functioned against an upper horny plate as a ruminant-like vegetation-cropping mechanism. There is no lower canine. Unfortunately the anterior teeth of other species of *Moropus* are not clearly known.

The opinion that *Moropus* originated in Eurasia and migrated to North America is based partly on the absence of North American chalicotheriids prior to the Arikareean and the existence of widespread Oligocene schizotheriine chalicotheriids in Eurasia. It is also probable that *Moropus* can be identified from Eurasia. Coombs (1974) described a juvenile lower jaw and Mt IV from a late Agenian fauna at St.-Gérand-le-Puy (Allier, France) as *Moropus* sp. Assignment of these specimens to *Moropus* was based primarily on the morphology and proportions of Mt IV, which are extremely close to those of *M. elatus*, though the St.-Gérand specimen is much smaller. The Mt IV of the St.-Gérand specimen is clearly articulated with the ectocuneiform, suggesting that metatarsal relationships were similar to those described for *Moropus* above. St.-Gérand may be roughly contemporaneous with the Agate Quarries, from which *M. elatus* comes (see Hunt, 1972; Coombs, 1974).

The precise affinities of somewhat later (Orleanian) schizotheriine chalicotheres from Europe are troublesome. Many of the fragmentary specimens (summarized in Coombs, 1974) are undiagnostic. Others, such as the Mt IV from Charneca do Lumiar, Portugal (Antunes, 1966), may be referable to *Moropus*. The Mt IV from Portugal appears to have articulated with the ectocuneiform and is very close in morphology and proportions to some early Hemingfordian specimens of *Moropus* from North America. Upper molars from Voitsberg (Austria) and Anjou (France) lack the crochet common in *Ancylotherium*, which is the predominant schizotheriine chalicothere in the medial and late Miocene of the Old World. It is not clear whether more than one group of schizotheriine chalicothere migrated from Eurasia to North America (see also *Tylocephalonyx* below), but immigration after the late Hemingfordian seems highly unlikely. There is no evidence, for example, that *Ancylotherium* ever reached North America.

*Tylocephalonyx.* This genus was named by Coombs (1979) for certain specimens of late Hemingfordian and early Barstovian schizotheriine chalicotheres from North America. It is distinctive in having the frontal and parietal bones of the skull dorsally expanded above the braincase to form a hollow, strutted dome. Three skulls showing such doming are known. Postcranial elements of *Tylocephalonyx* are also readily identified, as, for example, by the dorsal to volar compression of the distal end of Mt. III. Unlike *Moropus*, *Tylocephalonyx* still has Mt III as the longest metatarsal and no articulation between the ectocuneiform and Mt IV. *Tylocephalonyx* resembles *Ancylotherium* in having reduced lower incisors and a small hypocone on $M^3$, but it lacks a crochet and the myriad specializations of the manus seen in *Ancylotherium*. No lower canines are present. Like *Moropus*, *Phyllotillon*, and *Ancylotherium*, it has a duplex on digit II of the manus. About 80% of known adults also have fused the proximal and middle phalanges of digit II of the pes into a duplex.

Only one species of *Tylocephalonyx*, *T. skinneri*, has been named. This species occurs in the late Hemingfordian faunas of western Nebraska and Wyoming. Specimens of *Tylocephalonyx* which have not been definitely referred to species are known from Oregon, Montana, Utah, and possibly Colorado.

Comparison of the geographic ranges of *Tylocephalonyx* and *Moropus* in the late Hemingfordian to early Barstovian of North America provides some interesting contrasts, as described by Coombs (1979). *Tylocephalonyx* had a more northern distribution, possibly associated with moister, more forested environments. In the late Hemingfordian, *Tylocephalonyx*, but not *Moropus*, is known from the Sheep Creek fauna of the Great Plains. In the early Barstovian the reverse is true, with chalicotheres close to the Nevada species *Moropus merriami* resident in the lower Snake Creek fauna of the Great Plains, while *Tylocephalonyx* is unknown from that region, though isolated specimens suggest that it persisted farther northwest.

Possible functions of the domed skull of *Tylocephalonyx* were discussed by Munthe and Coombs (1979). The most plausible explanation associates the dome with social behaviors, such as lateral display and low-impact butting. One would expect on this basis that the domes might be more prominent in males, but unfortunately sex cannot be determined for the three animals whose domed skulls are known.

*Ancylotherium.* *Ancylotherium* was the latest surviving genus of the Schizotheriinae, persisting through the Pliocene of east and south Africa. It is confined to the Old World and had a broad range across Asia to Europe and Africa in the late Miocene.

*Ancylotherium* is distinctive in a number of ways. Its upper molars develop the highest crowns and most elongated ectolophs within the Schizotheriinae. A crochet is usually present. The manus has undergone some strong modifications, many of them associated with strong flexion within the carpus and hyperextension of the phalanges. Both Mc V and the trapezium are lost (the trapezium might be fused to Mc II). The carpus has reduced its volar projections; for example, the reduction of the volar process of the lunate and of the disto-volar hook on the magnum. The distal (centrale) process on the scaphoid has been lost. Most significant for extreme flexion of the carpus is evidence that the volar part of the scaphoid contacted Mc II during flexion. The metacarpal shafts are strongly flattened, even concave, on their dorsal surfaces. The dorsal surfaces of the proximal ends of metacarpals and distal carpal row are strongly developed. Many of the above forelimb characters were noted by Schaub (1943) in a detailed discussion of the massive forelimb of a specimen of *Ancylotherium pentelicum* from Pikermi (Greece). *Ancylotherium* resembles *Tylocephalonyx* in having a reduced hypocone on $M^3$ and reduced lower incisors. All known specimens develop a duplex on digit II of the pes as well as on digit II of the manus.

The question whether a lower canine was present in *Ancylotherium* remains uncertain. In agreement with Coombs (1978c), Garevski and Zapfe (1983) noted reduced incisors and the absence of a canine in a specimen of *Ancylotherium* from Greece (Munich AS II 147). However, these authors described a new specimen from Yugoslavia which they believed had a lower canine alveolus but no incisors. They suggested that *Ancylotherium* was sexually dimorphic in having the males retain lower canines. Unfortunately, I have not seen the Yugoslavian specimen, but I am not convinced from the illustrations that a canine was present. Eurasian schizotheriine chalicotheres often have an anterior mental foramen which lies near the canine position, and the Yugoslavian specimen is broken just in that area. The canine question is of some importance because a lower canine has not yet been demonstrated in any other member of the Schizotheriinae.

*Ancylotherium* can be divided into two subgenera: *A. (Ancylotherium)* and *A. (Metaschizotherium). Metaschizotherium* was named as a separate genus by von Koenigswald (1932); Thenius (1953) referred it to *Ancylotherium*, and Zapfe (1967) used it as a subgenus of *Ancylotherium*. Zapfe's usage is, I believe, appropriate, because *Metaschizotherium* is allied to *Ancylotherium* by such characters as the crochet on upper molars, incipient hypocone on $P^4$, and strongly reduced volar process on the lunate. At the same time it is earlier and clearly more primitive than "classic" material of *Ancylotherium* from such localities as Pikermi and Samos. European material clearly referable to *A. (Metaschizotherium)* is of Astaracian age, while that referable to *A. (Ancylotherium)* is Turolian. Schizotheriine material from Vallesian faunas of Eurasia is too rare and fragmentary to clarify how the two subgenera are related across this time gap.

As presently constituted, *Ancylotherium (Metaschizotherium)* has one species, *A. (M.) fraasi*. This species is recognizable from such European Astaracian faunas as Steinheim (Germany), Häder (Germany), Viehhausen (Germany), Sandelzhausen (Germany), La Grive St.-Alban (France), and Krems (Austria). Complete dentitions and limbs have not been described for this species, and it is not yet known how much the manus was specialized (except that the volar process on the lunate was strongly reduced).

*A. (Ancylotherium)* has at present two referred species, *A. (A.) pentelicum* and *A. (A.) hennigi*. The best known of these species is *A. (A.) pentelicum*, whose skull, dentition, and forelimb are relatively completely known and are the basis for identifying definitive characters of *Ancylotherium*. The hindlimb is less well known, but metatarsal proportions and relative metatarsal lengths are available (Mt III > Mt IV). *A. (A.) pentelicum* occurs in a large number of faunas of Turolian (late Miocene) age: Pikermi (Greece), Samos (Greek islands), Halmyropotamos (Greece), Kalimanci and Gorna Sushitsa (Bulgaria), Titov Veles (Yugoslavia), Maragheh (Iran), Novo-Ukrainka (Ukrainian S. S. R.), and Molayan (Afghanistan). It is significantly absent from the Siwalik Group of Pakistan, which includes deposits contemporaneous with those just mentioned.

Several reported specimens almost certainly extend the range of *Ancylotherium* into China. Tung, Huang, and Qiu (1975) named *Huanghotherium anlungense* for some skull material, including $M^2$ and $M^3$, which are very high-crowned and highly elongated; this material comes from the Yushe, Turolian equivalent (see Li, Wu, and Qiu, 1984) fauna at Hoxian, Shanxi Province, China. Wu and Chen (1976) named *Gansuodon pingliangense* for very large elongated upper molar teeth from Pingliang, Gansu Province, China. Both *Huanghotherium* and *Gansuodon* probably belong to *A. (Ancylotherium)*, but it is difficult to evaluate the validity of the species without making a careful comparison with *A. (A.) pentelicum* and *A. (A.)*

*hennigi*. Further evidence of *Ancylotherium* in China is provided by an astragalus from Yushe (Bohlin, 1936) and Mt II and phalanges from an Astaracian equivalent fauna at Tungur (Inner Mongolian Autonomous Region). These specimens also require further study, though it is clear that the earlier Tungur material is more primitive than comparable material of *A. (A.) pentelicum*. The Tungur Mt II (American Museum of Natural History 26581, incorrectly identified as Mt IV of "*Macrotherium*" by Colbert, 1934) resembles specimens of *A. (M.) fraasi* more than those of *A. (A.) pentelicum*.

*A. (Ancylotherium) hennigi* is known from late Miocene (approximately 6.5 Ma) through the late Pliocene (approximately 2 Ma) of east and south Africa. Localities which have yielded *A. (A.) hennigi* include Lukeino, Kaiso, Laetolil, Chemeron, Omo, Olduvai, and Makapansgat (Pickford, 1979; Butler, 1978; Guérin, 1976). Specimens are rare and rather fragmentary, but there is adequate material to affirm reference to *A. (Ancylotherium)*. The worn upper molar from Makapansgat figured by George (1950) has a crochet and resembles in other respects teeth of *Ancylotherium* from Eurasia. Forelimb material from Olduvai Bed I analyzed by Butler (1965) shows forelimb specializations similar to those of *A. (A.) pentelicum* (loss of distal process on scaphoid, very small volar process on lunate, evidence on Mc II for some articulation with the scaphoid during extreme flexion of the carpus, some flattening and dorsal concavity of the shaft of Mc III). On the other hand, *A. (A.) hennigi* is smaller and in some respects more primitive than *A. (A.) pentelicum*. For example, the metacarpals (especially Mc II) are less flattened dorsally. It is possible that *A. (A.) hennigi* is morphologically comparable to an intermediate between *A. (M.) frassi* and *A. (A.) pentelicum*, but the material of all three species is inadequate for complete comparisons.

Pickford (1979) named a new genus and species of schizotheriine chalicothere, *Chemositia tugenensis*, from the Mpesida Beds (6.5-7 Ma) of Kenya. The type includes a femur fragment, calcaneum, astragalus, Mt. IV, and proximal and middle phalanges of a single individual. The primary diagnostic feature is the existence of large facets on the proximal half of the volar surface of the proximal phalanx. As figured by Pickford, this phalanx is indeed peculiar, all the more so in that the rest of the phalanx is little different from that of other schizotheriines. It is difficult to see how the facet could have functioned, and one is tempted to suggest (without actually having seen the specimen) that the proximal volar "facet" is actually a normal volar surface delineated by strong longitudinal volar ridges, as sometimes occurs in schizotheriine phalanges (see, for example, Coombs 1978a, Fig. 19F, for *Moropus elatus*). Whether generic designation for *Chemositia* is warranted thus remains debatable. The Mt IV of the type of *C. tugenensis* is interesting because it can be compared with a number of specimens of *Moropus* and *Ancylotherium* (see Table 24.1). It is shorter and broader than specimens referred to *Moropus* from France and Portugal but less shortened and broad than Mt IV of *A. (A.) pentelicum*. In morphology the Mpesida Mt IV is more similar to the earlier specimens from St.-Gérand (France) and Charneca do Lumiar (Portugal) than to *A. (A.) pentelicum*, but some of these similarities are primitive. It cannot be determined whether the Mpesida Mt IV had a facet for the ectocuneiform, as generally occurs in *Moropus*. Pickford may well be correct in suggesting that the Mpesida material is not referable to *A. (A.) hennigi*, but its affinities are still questionable. Pickford (1979) also suggested that Schizotheriinae might be represented in early Miocene deposits of east Africa, but the evidence for this is equivocal.

## Commentary

It is now appropriate to consider several issues which are relevant to the Chalicotheriidae as a whole, rather than to particular genera or species. Included are questions of sexual dimorphism, diet, habitat, and origins. I have discussed some of these questions elsewhere in different contexts (Coombs, 1975, 1982, 1983).

There is evidence of sexual dimorphism in all chalicotheriid genera, as summarized by Coombs (1975). This dimorphism is reflected primarily in size (and correlated minor allometric) differences within species. Such dimorphism is best documented in the case of *Moropus elatus* from the Agate Spring Quarries of Nebraska (United States), in which lengths of 17 radii and 16 tibiae of different animals yielded a bimodal plot on probability paper (Coombs, 1975). This analysis prompted the synonymy of the smaller sized putative species *Moropus petersoni* with *M. elatus*. Numbers of males and females were approximately equal in the Agate sample. Number equivalence does not occur in all cases, however; in *Borissiakia betpakdalensis* the larger (probably male) size group was more abundant (Borissiak, 1946), while in *Chalicotherium rusingense* smaller sized (female) specimens predominated (Butler, 1965). Unequal numbers are not surprising in view of occasional absolute number disproportions and sexually related habitat and habit differences in various living ungulate species.

The rarity of chalicothere specimens suggests that these animals did not typically congregate in big herds on the plains or near watercourses. Nonetheless, they did sometimes group, as large fossil concentrations at such places as Agate (Nebraska) and Betpak-dala (Kazakhstan) testify. Size sexual dimorphism may suggest polygynous breeding habits, as do the domed skulls of *Tylocephalonyx* interpreted in the context of low-impact butting or wrestling and display by males.

Chalicotheres are clawed herbivores and, as such, bear functional comparison with other large fossil clawed herbivores, such as the ground sloths, homalodotheres, taeniodonts, agriochoeres, and others (Coombs, 1983). While some of the chalicotheres have been interpreted as diggers in search of edible roots, this hypothesis is unlikely. Chalicotheres have few force-increasing modifications of the forelimb, and their relatively low-crowned teeth are poorly suited for a fibrous, gritty diet. Leaves and other browse are much more consistent with the dental and jaw morphology. There are in addition many postcranial characters which suggest that chalicotheriids were bipedal browsers, as has been suggested by such authors as Borissiak (1945), Schaub (1943), Zapfe (1979), and Coombs (1983). Such characters include the hooklike clawed phalanges, the long metacarpals, and extensive side-to-side deviation capability in the wrist. The hindlimb, on the other hand, is adapted for weight bearing; the metatarsals are short relative to metacarpals and become increasingly shorter and wider over time, without regard to the absolute size of the animal. Bipedal browsing seems to have been common to both chalicotheriid subfamilies, but each developed its own specializations. Chalicotheriinae had gorilla-like proportions with very large differences between forelimb and hindlimb lengths (high intermembral index). The short tibia resulted in a very low crural index. Both the manus and pes differed enough from those of Schizotheriinae to suggest some differences in the hook mechanism and stance. Schizotheriinae developed longer limbs and necks and were proportioned somewhat like quadrupedal tree browsers such as *Okapia* (Artiodactyla, Giraffidae). In addition, both forelimb and hindlimb morphology suggest the ability to erect on the hindlimbs while feeding, as do living gerenuks (*Litocranius*, Bovidae) and goats (*Capra*, Bovidae). Such erection would increase browsing height and allow the forelimbs to hook branches within reach. *Ancylotherium* had the most massive forelimb within the Schizotheriinae, and Schaub (1943) suggested that it might have been used in breaking off branches in a manner similar to the trunk and tusks of proboscideans.

Members of the Chalicotheriinae and Schizotheriinae occasionally are known to coexist in the same fauna (Bugti, La Grive, Tungur, Titov Veles, Pikermi), but this is not common. Looking at major range differences, one immediately notes the absence of Chalicotheriinae from the New World and the presence of Chalicotheriinae in the African early Miocene, but not in the late Miocene through Pliocene, when members of the Schizotheriinae prevailed. The presence or predominance of *Chalicotherium* versus *Ancylotherium* in the late Miocene of Eurasia appears to correspond to provincial differences which are environmentally based and reflected in differences in other parts of the fauna. During the Vallesian, *Chalicotherium* is well known from such western European locations as Eppelsheim (Germany) and the Valles Penedes Basin (Spain), while *Ancylotherium* is rare in Europe. Similarly, *Chalicotherium* but not *Ancylotherium* is known from the Siwaliks of Pakistan in both Vallesian and Turolian time. *Ancylotherium* is predominant in the classic Turolian faunas of Pikermi, Samos, Maragheh, and Molayan, while *Chalicotherium* is rare or absent in these faunas. Especially striking are the differences between the Turolian fauna of Molayan (Afghanistan), which includes *Ancylotherium*, and that of the Siwaliks, which includes *Chalicotherium*. The two faunas are quite different, even though they are separated by a distance of only 300 km (Brunet, Heintz, and Battail, 1984). The range of *Ancylotherium* in the Turolian corresponds with the Tethyan Province of Bernor et al., 1979), whose vegetation was dominated by a laurel-pine-oak assemblage with a chaparral undergrowth. Conditions were apparently cooler and drier than in the more closed forests in which *Chalicotherium* predominated. *Chalicotherium* is more often associated with forest primates and suids. The presence of *Ancylotherium* but not *Chalicotherium* in the late Miocene of Africa is not surprising in view of faunal similarities between the Tethyan Province and east Africa in the late Miocene. Forests in east Africa are generally thought to have been more open in the late Miocene than in the early Miocene, when *Chalicotherium rusingense* is found.

Although Eocene chalicotheres are discussed elsewhere (Lucas and Schoch, this volume, Chapter 23), a few observations are appropriate here. There are three genera of mid- to late Eocene chalicotherioids: *Eomoropus* (North America and Asia), *Grangeria* (North America and Asia), and *Litolophus* (Asia). Of these taxa *Litolophus* appears to be the most primitive, because the upper molar mesostyle is less developed and the paracone more labial than in *Grangeria* or *Eomoropus*. *Litolophus* does, however, show greater posterior migration of the protocone, a derived character which is only slightly developed in *Eomoropus*. The metaloph and posterior ectoloph on $M^3$ are oriented somewhat differently in *Litolophus* than in *Eomoropus* and *Gran-geria*. *Grangeria* has disproportionately large canines and a deeper mandible than *Litolophus* or *Eomoropus*. These characters tempt one to suggest a relationship of *Grangeria* to *Chalicotherium*, but postcra-nial characters do not corroborate such affinities (Lucas, pers. comm). The derived canines and mandible of *Grangeria* probably evolved independently.

Chalicotheroidea have often been allied with Equoidea and Brontotherioidea on the basis of the W-shaped ectoloph, but this grouping (Hippomorpha) is not now accepted and the W-shaped ectoloph is considered to have evolved independently in these groups. Borissiak (1946) noted a number of similarities between chalicotheres and brontotheres, but all or most of these characters appear to be primitive or convergent. Radinsky (1964) did not ally chalicotheres to either hippomorphs or ceratomorphs but considered them to be a long-independent perissodactyl lineage.

The early Eocene genus *Paleomoropus* (North America) and the early to medial Eocene genus *Lophiaspis* (Europe) have also been discussed in the context of chalicothere origins (Radinsky, 1964). Fischer (1964, 1977), however, allied both genera with the ceratomorph family Lophiodontidae. Hooker (1984, 1986) joined the Lophiodontidae and Chalicotherioidea under the heading Ancylopoda (previously used for chalicotheres alone), but placed *Paleomoropus* and *Lophiaspis* among the chalicotheres.

In any case, the difficult question of chalicothere affinities rests on the analysis of dental characters among all perissodactyls. Postcranial characters, however useful in discussions of later chalicotheres, are still too poorly known and perhaps too poorly differentiated at early stages to be valuable in such discussions.

**Bibliography**

Antunes, M. T. (1966): Notes sur la géologie et la paléontologie du Miocène de Lisbonne. V. Un schizotheriiné du genre *Phyllotillon* (Chalicotherioidea, Perissodactyla) dans l'Helvétien V-b de Charnaca do Lumiar. Remarques écologies sur la faune de mammifères. - *Bol. Soc. Geol. Portugal*, 16: 159-178.

Badgley, C., Qi, G.Q., Chen, W.Y., and Han, D.F. (1988): Paleoecology of a Miocene tropical upland fauna: Lufeng, China.-*Nat. Geog. Res.*, 4: 178-195.

Barry, J. C., Lindsay, E. C., and Jacobs, L. L. (1982): A biostratigraphic zonation of the middle and upper Siwaliks of the Potwar Plateau of northern Pakistan. - *Palaeogeogr., Palaeoclimat., Palaeoecol.*, 37: 95-130.

Berggren, W. A., Kent, D. V., Flynn, J. J. and Van Couvering, J. A. (1985): Cenozoic geochronology. -*Geol. Soc. Am. Bull.*, 96: 1407-1418.

Bernor, R. L., Andrews, P. J., Solounias, N., and Van Couvering, J. A. H. (1979): The evolution of "Pontian" mammal faunas: some zoogeographic, paleoecologic and chronostratigraphic considerations. - *Ann. Géol. Pays Hellénique, tome hors ser., VIIth Intl. Cong. Mediterranean Neogene, Athens*, 1979: 81-89.

Bohlin, B. (1936): Notes on some remains of fossil mammals from China and Mongolia.- *Bull. Geol. Soc. China*, 15: 321-330.

Borissiak, A. A. (1945): The chalicotheres as a biological type. -*Amer. J. Sci.*, 243: 667-679.

Borissiak, A. A. (1946): [A new chalicothere from the Tertiary of Kazakhstan.] -*Akad. Nauk. SSSR, Trudy Paleont. Inst.*, 13(3): 1-134.

Brunet, M., Heintz, E., and Battail, B. (1984): Molayan (Afghanistan) and the Khaur Siwaliks of Pakistan: an example of biostratigraphic isolation of late Miocene mammalian faunas. - *Geol. en Mijnbouw*, 63: 31-38.

Butler, P. M. (1965): Fossil mammals of Africa No. 18: East African Miocene and Pleistocene chalicotheres. -*Bull. Brit. Mus. (Nat. Hist.) Geol.*, 10: 165-237.

Butler, P. M. (1978): Chalicotheriidae. -In: Maglio, V. J., and Cooke, H. B. S. (eds.): *Evolution of African Mammals*. Cambridge, Mass. (Harvard Univ. Press), pp. 368-370.

Colbert, E. H. (1934): Chalicotheres from Mongolia and China in the American Museum. -*Bull. Amer. Mus. Nat. Hist.*, 67: 353-387.

Coombs, M. C. (1974): Ein Vertreter von *Moropus* aus dem europäischen Aquitanien und eine Zusammenfassung der europäischen postoligozänen Schizotheriinae (Mammalia, Perissodactyla, Chalicotheriidae). -*Sitzungsber. Osterr. Akad. Wiss., Mathem.-naturw. Kl.*, 182: 273-288.

Coombs, M. C. (1975): Sexual dimorphism in chalicotheres (Mammalia, Perissodactyla). -*Syst. Zool.*, 24: 55-62.

Coombs, M. C. (1976): The taxonomic position of the chalicotheriid perissodactyl *Kyzylkakhippus orlovi* from the Oligocene of Kazakhstan. -*Palaeontol.* 19: 191-198.

Coombs, M. C. (1978a): Reevaluation of early Miocene North American *Moropus* (Perissodactyla, Chalicotheriidae,

Schizotheriinae). -*Bull. Carnegie Mus. Nat. Hist.*, 4: 1-62.

Coombs, M. C. (1978b): Additional *Schizotherium* material from China, and a review of *Schizotherium* dentitions (Perissodactyla, Chalicotheriidae).- *Amer. Mus. Novitates*, 2647: 1-18.

Coombs, M. C. (1978c): A premaxilla of *Moropus elatus* Marsh, and evolution of chalicotherioid anterior dentition. -*J. Paleont.*, 52: 118-121.

Coombs, M. C. (1979): *Tylocephalonyx*, a new genus of North American dome-skulled chalicotheres (Mammalia, Perissodactyla). -*Bull. Amer. Mus. Nat. Hist.*,164: 1-64

Coombs, M. C. (1982): Chalicotheres (Perissodactyla) as large terrestrial mammals. -*Third North Amer. Paleont. Conv., Proc.*, 1: 99-103.

Coombs, M. C. (1983): Large mammalian clawed herbivores: a comparative study. -*Trans. Amer. Phil. Soc.*, 73(7): 1-96.

Filhol, H. (1891): *Études sur les Mammifères Fossiles de Sansan*. -Paris (Libraire Acad. Médecine).

Fischer, K. -H. (1964): Die tapiroiden Perissodactylen aus der eozänen Braunkohle des Geiseltales. -*Geologie*, 45: 1-101.

Fischer, K. -H. (1977): Neue Funde von *Rhinocerolophiodon* (n. gen.), *Lophiodon*, und *Hyrachyus* (Ceratomorpha, Perissodactyla, Mammalia) aus dem Eozän des Geiseltals bei Halle (DDR). -*Zeit. Geol. Wiss. Berlin*, 5: 909-919.

Flerov, K. K. (1938): Remains of Ungulata from Betpak-dala. -*C. R. (Doklady) Acad. Sci. URSS.*, 21: 94-96.

Forster Cooper, C. (1920): Chalicotheroidea from Baluchistan. -*Proc. Zool. Soc. London*, 1920: 357-366.

Frailey, D. (1979): The large mammals of the Buda local fauna (Arikareean: Alachua County, Florida).-*Bull. Florida State Mus., Biol. Sci.*, 24: 123-173.

Garevski, R., and Zapfe, H. (1983): Weitere Chalicotheriiden-Funde aus der Pikermi-Fauna von Titov Veles (Mazedonien, Jugoslawien). -*Acta. Mus. Macedonici Sci. Nat.*, 17: 1-20.

George, M. (1950): A chalicothere from the Limeworks Quarry of the Makapan Valley, Potgietersrust District. -*South African J. Sci., Johannesburg*, 46: 241-242.

Guérin, C. (1976): Rhinocerotidae and Chalicotheriidae (Mammalia, Perissodactyla) from the Shungura Formation, Lower Omo Basin.- In: Coppens, Y., Howell, F.C., Isaac, G.L., and Leakey, R.E.F. (eds.): *Earliest Man and Environments in the Lake Rudolf Basin: Stratigraphy, Paleoecology, and Evolution.* Chicago (Univ. Chicago Press), pp. 214-221.

Holland, W. J., and Peterson, O. A. (1914): The osteology of the Chalicotheroidea with special reference to a mounted skeleton of *Moropus elatus* Marsh, now installed in the Carnegie Museum. -*Mem. Carnegie Mus.*, 3: 189-406.

Hooker, J. J. (1984): A primitive ceratomorph (Perissodactyla, Mammalia) from the early Tertiary of Europe. -*Zool. J. Linn. Soc.*, 82: 229-244.

Hooker, J. J. (1986): Mammals from the Bartonian (middle/late Eocene) of the Hampshire Basin, southern England. -*Bull. British Mus. (Nat. Hist.), Geol. ser.*, 39: 191-478.

Hu, C. K. (1959): Chalicotheres from the Tertiary of North China.-*Palaeovert. Palaeoanthrop.*, 1: 125-132.

Hunt, R. M. (1972): Miocene amphicyonids (Mammalia, Carnivora) from the Agate Spring Quarries, Sioux County, Nebraska. -*Amer. Mus. Novitates*, 2506: 1-39.

Koenigswald, G. H. R., von (1932): *Metaschizotherium fraasi* n.g. n.sp., ein neuer chalicotheriide aus dem Obermiocän von Steinheim A. Albuch. -*Palaeontographica*, Suppl. -vol. 8(8): 1-24.

Li, C. K., Wu, W.Y., and Qiu, Z. D. (1984): [Chinese Neogene: subdivision and cor-

relation]. -*Vert. PalAsiatica*, 22: 165-178.
Lucas, S. G., and Schoch, R. M. (1989): Taxonomy and biochronology of *Eomoropus* and *Grangeria*, Eocene chalicotheres from the western United States and China (this volume).
Matthew, W. D. (1929): Critical observations upon Siwalik mammals. -*Bull. Amer. Mus. Nat. Hist.*, 56: 437-560.
Munthe, J., and Coombs, M. C. (1979) Miocene dome-skulled chalicotheres (Mammalia, Perissodactyla) from the western United States: a preliminary discussion of a bizarre structure. -*J. Paleont.*, 53: 77-91.
Osborn, H. F. (1919): Seventeen skeletons of *Moropus*: probable habits of this animal. -*Proc. Nat. Acad. Sci.*, 5: 250-252.
Pickford, M. (1979): New evidence pertaining to the Miocene Chalicotheriidae (Mammalia, Perissodactyla) of Kenya. -*Tertiary Research*, 2: 83-91.
Pickford, M. (1982): Miocene Chalicotheriidae of the Potwar Plateau, Pakistan. -*Tertiary Research*, 4: 13-29.
Pilgrim, G. E. (1910): Notices of new mammalian genera and species from the Tertiaries of India. -*Rec. Geol. Surv. India*, 40: 63-71.
Pilgrim, G. E. (1912): The vertebrate fauna of the Gaj series in the Bugti Hills and the Punjab. -*Palaeont. Indica, Mem. Geol. Surv. India*, new ser., 4(2): 1-83.
Radinsky, L. B. (1964): *Paleomoropus*, a new early Eocene chalicothere, and a revision of Eocene chalicotheres. -*Amer. Mus. Novitates*, 2179: 1-28.
Schaub, S. (1943): Die Vorderextremität von *Ancylotherium pentelicum* Gaudry und Lartet. -*Schweiz. Palaeont. Abh.*, 64: 1-36.
Skinner, M. F. (1968): A Pliocene chalicothere from Nebraska, and the distribution of chalicotheres in the late Tertiary of North America. -*Amer. Mus. Novitates*, 2346: 1-24.
Thenius, E. (1953): Studien über fossile Vertebraten Griechenlands, III. Das Maxillargebiss von *Ancylotherium pentelicum* Gaudry und Lartet. -*Ann. Géol. Pays Helléniques*, 5: 97-106.
Tung, Y. S., Huang, W. P., and Qiu, Z. D. (1975): (*Hipparion* fauna in An-lo, Hohsien, Shansi). -*Vert. PalAsiatica*, 13: 34-47.
Viret, J. (1949): Quelques considerations preliminaires à propos de la revision de la faune des mammifères Miocènes de la Grive St-Alban. -*Bull. Mensuel Soc. Linn. Lyon*, 18: 53-57.
Wu, W. Y., and Chen, G. F. (1976): [A new schizotheriine genus from the Neogene of Pingliang, Gansu]. -*Vert. PalAsiatica*, 14: 194-197.
Xue, X. X., and Coombs, M. C. (1985): A new species of *Chalicotherium* from the upper Miocene of Gansu Province, China. -*J. Vert. Paleont.*, 5: 336-344.
Zapfe, H. (1967): *Ancylotherium* in Obermiozän des Wiener Beckens. -*Ann. Naturhist. Mus. Wien*, 71: 401-411.
Zapfe, H. (1979): *Chalicotherium grande* (Blainv.) aus der miozänen Spaltenfüllung von Neudorf an der March (Devinská Nová Ves), Tschechoslowakei. -*Neue Denkschr. Naturhist. Mus. Wien.*, 2: 1-282.

# 25. THE BRONTOTHERIIDAE: A SYSTEMATIC REVISION AND PRELIMINARY PHYLOGENY OF NORTH AMERICAN GENERA

## BRYN J. MADER

The Brontotheriidae is an extinct perissodactyl family that is defined by the bunoselenodont pattern of the upper molars and in most cases by a shortening of the face and elongation of the skull posterior to the orbit. It is unclear, however, whether these derived cranial proportions occurred in the most primitive brontothere genera. Leidy, Marsh, Cope, and Osborn conducted extensive work on brontothere systematics in the late nineteenth and early twentieth centuries, but, despite their efforts, most recent workers have pointed out the need for a major revision. The present article is the first comprehensive revision of North American brontothere genera in sixty years. In this paper eighteen genera are recognized as valid: *Eotitanops, Palaeosyops, Telmatherium, Mesatirhinus, Metarhinus, Rhadinorhinus, Dolichorhinus, Metatelmatherium, Sthenodectes, Protitanotherium, "Diplacodon" progressum* Peterson (a new generic name for this taxon will be published in the near future), *Eotitanotherium, Notiotitanops, Protitanops, Duchesneodus, Brontops, Menops,* and *Megacerops*. In his now famous brontothere monographs, Osborn (1929) divided the Brontotheriidae into twelve subfamilies. To these, Granger and Gregory (1943) added three new subfamilies, two of which were endemic to Asia and one of which was represented in both Asia and North America. Simpson (1945) essentially adopted the systematic schemes of both Osborn and Granger and Gregory, but reduced the number of North American brontothere subfamilies to eight, primarily by synonymizing ancestral subfamilies with their presumed descendants. In the present paper only two subfamilies are recognized as valid: the Dolichorhininae and the Telmatheriinae. Diplacodonts *sensu lato* and "eubrontotheres" are recognized as valid subgroups of the Telmatheriinae.

### Introduction

The phylogenetic relationships of the family Brontotheriidae to other perissodactyl lineages has never been firmly established. Traditionally brontotheres (= titanotheres) were included in the Hippomorpha (usually assigned subordinal or infraordinal status) along with equoids (Wood, 1937; Scott, 1941; Simpson, 1945; Radinsky, 1964; Romer, 1966; and Schoch 1983, 1984) and occasionally with chalicotherioids (Wood, 1937, and Simpson, 1945). Hooker (1984), however, regarded brontotheres as the primitive sister group to all other perissodactyls.

Within the Perissodactyla, brontotheres are defined by the characteristic bunoselenodont pattern of the upper molars, in which there are a well-developed W-shaped ectoloph and isolated lingual cusps. Paraconules and metaconules are variably developed, being large in primitive forms and reduced or lost in more derived forms. A protoloph or metaloph is never present.

In most brontotheres, the part of the skull that is anterior to the postorbital processes is shorter than the part of the skull that is posterior to the orbit (Figs. 25.1, 25.2). Thus the face is short and the rest of the skull elongated. This is clearly a synapomorphic character that unites the majority of the family. It is uncertain, however, whether the most primitive brontotheres shared this character or were "long faced" like other early perissodactyls.

Brontotheres originated during the early Eocene and are known from early

Eocene localities in both North America and Asia. In North America the group first appeared during the latter part of the Wasatchian land mammal "age" (Lostcabinian). Two genera, *Lambdotherium* and *Eotitanops*, which occurred at this time have generally been regarded as brontotheres. *Lambdotherium* was a somewhat aberrant form and may not belong to the family at all. Steven Wallace, in an unpublished, but rather widely read master's thesis (University of Colorado, 1980), argued that *Lambdotherium* was a palaeothere *sensu lato*, a view that was apparently accepted by Lucas *et al.* (1981). Later, however, Schoch and Lucas (1985) suggested in an abstract that *Lambdotherium* might be a possible sister taxon to the Brontotheriidae. I will not attempt to resolve the question here, but because of the uncertain status of *Lambdotherium*, I will not discuss it further in this paper. Unlike *Lambdotherium*, *Eotitanops* does not exhibit any characters that are atypical of brontotheres. *Eotitanops* is quite similar in its dental morphology to the Bridgerian brontothere genus *Palaeosyops*, but is smaller in size. *Eotitanops* exhibits a number of characters that are primitive for brontotheres, including large canines, long postcanine diastema, and unmolarized premolars.

In the Bridgerian, *Eotitanops* was replaced by the more derived genus *Palaeosyops*, and by late Bridger time two other genera, *Mesatirhinus* and *Telmatherium*, had also appeared. In North America, maximal brontothere diversity occurred in the Uintan when at least eight genera are known. Among these were the first "horned brontotheres," a monophyletic group defined by the presence of true horns and partially molarized premolars.

Most North American brontothere genera disappeared by the end of the Eocene, but one lineage, the horned brontotheres, survived to radiate in the early Oligocene. These Chadronian forms attained very large size and often developed spectacular bifurcated horns and broadly expanded zygomatic arches. All brontotheres became extinct in North America at the end of the Chadronian, but the family survived in Asia until the middle Oligocene.

Most of the original work on brontothere systematics was conducted by Joseph Leidy, O. C. Marsh, and E. D. Cope, roughly between 1870 and 1891. Marsh (1873) was the first to recognize that the Oligocene members of the group were perissodactyls and belonged to a single family that he called the Brontotheridae (emended to Brontotheriidae by Hay, 1902, and all subsequent authors). Marsh (1875) placed the Eocene members of the group, however, in a separate family that he called the Limnohyidae.

Cope (1879) united the Eocene and Oligocene brontotheres into a single family but, unfortunately, also included the genus *Chalicotherium* in this assemblage and called the family Chalicotheriidae. It was not until 1890 that Osborn properly defined the family (which he called the Titanotheriidae) so that it included all of the Eocene and Oligocene genera but omitted extraneous taxa such as *Chalicotherium*.

Leidy, Marsh, and Cope were succeeded in the 1890s by Henry Fairfield Osborn, who remained the primary student of brontothere systematics and evolution until his death in 1935. Osborn culminated his work with a massive, two-volume monograph (1929) that considered aspects of brontothere systematics, evolution, ecology, geographic distribution, and stratigraphy. Unfortunately, despite the wide scope and exhaustive nature of Osborn's work, his systematics are unsatisfactory because they are based on Osborn's own peculiar view of evolution and failed to consider properly such factors as intraspecific variation and taphonomic deformation. Over the last several years there has been a general agreement among perissodactyl systematists that a revision is needed.

In the present revision I will restrict my discussion to the North American genera and will not consider the species level beyond the type species. The determination of valid brontothere species is a vast and complex problem, and treating the subject adequately would require more space than is available here. Brontothere species are, however, part of my ongoing research, and I hope to address the issue in the near future.

In order to provide a complete description of each genus, I have included both primitive and derived characters in my diagnoses. Unique derived characters that can be used to define a genus cladistically are followed in the text by the letter "A" (for autapomorphic) in parentheses. It should be noted that some genera, although clearly distinct, cannot at present be defined in terms of unique derived characters (for example, *Eotitanops, Mesatirhinus,* and *Notiotitanops*). This is usually because these genera are either poorly represented in the fossil record or are primitive within higher brontothere groupings.

Characters that are derived within the Brontotheriidae, but define higher brontothere categories, are identified by an asterisk in parentheses. The asterisk is followed by a number that identifies the node on the cladogram presented in this paper (Fig. 25.3) at which the character is synapomorphic.

As a convenient quantitative indication of size I have included in the generic diagnoses the length of the upper cheek tooth series or the length of the upper cheek tooth series exclusive of the first premolar. I find these measurements to be preferable to the basilar length of the skull, because complete skulls are comparatively rare and do not allow for the sample sizes possible with dental measurements. I also prefer to use the length of the cheek tooth series minus the first premolar, rather than the total length of the series because the first premolar is often missing or damaged.

The size ranges that I report for the Eocene genera are based on my own data, but I have relied on the data reported by Osborn (1929) for all the Oligocene genera except *Protitanops*. Osborn reported the entire length of the cheek tooth series and did not generally provide the length from $P^2$ to $M^3$.

The size ranges reported in this paper should be considered as general indications of size and not as absolute parameters. It should not be surprising, therefore, to find individuals that fall slightly above or below the size ranges reported here, especially for those genera that are represented by comparatively few specimens. I have defined the brontothere genera as I understand them, and my diagnoses are not meant to apply to all of the species that Osborn or others would have included in any given genus. This is especially true of the Chadronian genera in which Osborn (1929) included primitive "horned brontothere" species that he regarded as ancestral to the type species (for example, *Brontops brachycephalus* and *Brontotherium leidyi*). The correct generic assignment of these species remains problematic.

**Abbreviations**

| | |
|---|---|
| ACM | Pratt Museum, Amherst College, Amherst, Mass. |
| AMNH | American Museum of Natural History, New York, N.Y. |
| ANSP | Academy of Natural Sciences of Philadelphia, Philadelphia, Penn. |
| CM | Carnegie Museum of Natural History, Pittsburgh, Penn. |
| FMNH | Field Museum of Natural History, Chicago, Ill. |
| LACM(CIT) | California Institute of Technology collection, Natural History Museum of Los Angeles County, Los Angeles, Calif. |
| SDSM | South Dakota School of Mines and Technology, Rapid City, S.D. |
| TMM | Texas Memorial Museum, University of Texas, Austin, Tex. |
| UCMP | Museum of Paleontology, University of California, Berkeley, Calif. |
| USNM | United States National Museum, Smithsonian Institution, Washington, D.C. |
| YPM | Peabody Museum of Natural History, Yale University, New |

YPM-PU  Haven, Conn.
Princeton University Collection, Peabody Museum of Natural History, New Haven, Conn.

## Revision of North American brontothere genera

### *Eotitanops* Osborn, 1907

*Age*: Wasatchian
*Type Species*: *E. borealis* (Cope, 1880)
*Holotype of type species*: AMNH 4892, a right maxilla with complete $P^4$ to $M^1$ and fragmentary $M^2$ to $M^3$.

*Diagnosis*: Small-sized (average length $P^2$ to $M^3$ is 91 mm in AMNH 14887) brontothere with a well-developed upper canine; long upper and lower diastema; unmolarized premolars; relatively large paraconules; and hypocone or pseudo-hypocone variably present on $M^3$.

*Discussion*: Cope described the holotype maxilla fragment in 1880 and identified it as a new species of *Palaeosyops* that he named *P. borealis*. No other materials were known until 1891 when J. L. Wortman collected a partial skeleton (AMNH 296) that remains the single most complete specimen known. Wortman and Osborn described the skeleton the next year (Osborn and Wortman, 1892) and referred it to *P. borealis* Cope.

In 1897 Osborn placed *P. borealis* in the genus "*Telmatotherium*" (= *Telmatherium*) but later decided that it was an entirely new genus and named it *Eotitanops*. Osborn first used this new generic name in 1907 but did not formally diagnose the genus until 1908.

Between 1909 and 1911, American Museum expeditions led by Walter Granger collected additional *Eotitanops* remains. Among these was a partial skull (AMNH 14887) that, despite its poor condition, is still the most complete skull of *Eotitanops* known. Osborn (1929) designated this skull as the "neotype" for the type species of *Eotitanops*, but because the original type is still preserved, Osborn's specimen has no nomenclatural significance. It is merely a referred specimen.

Osborn (1929) argued that the facial region of his "neotype" skull is longer than the cranial region while in all other brontotheres the face is shorter than the cranium. This was Osborn's main justification for separating *Eotitanops* from middle Eocene brontotheres such as *Palaeosyops*. Wallace (1980) pointed out that Osborn's interpretation of the facial and cranial proportions of the "neotype" skull is open to question. This is because the specimen is very fragmentary and the position of key morphological landmarks is uncertain. Wallace noted that *Eotitanops* is dentally similar to *Palaeosyops* and argued that until a more complete skull of *Eotitanops* is known, cranial proportions cannot be used to separate it from *Palaeosyops*. Wallace, therefore, regarded *Eotitanops* as a junior synonym of *Palaeosyops*.

Despite Wallace's valid objections, I believe that there are enough morphological differences between *Eotitanops* and *Palaeosyops* to justify separation of the two at the generic level. The left zygomatic arch of Osborn's "neotype" skull is complete and shows that this structure was thin and probably relatively straight. It was probably quite similar to the zygomatic arches of *Mesatirhinus* or *Rhadinorhinus*. In contrast, the zygomatic arch of *Palaeosyops* is very robust and is sharply curved. Furthermore, although much of the skull is lacking, the portions of the zygomatic arches and palate that are preserved suggest to me that the skull was dolichocephalic (as stated by Osborn, 1929) or mesaticephalic. In contrast, all skulls of *Palaeosyops* are strongly brachycephalic, and this is a diagnostic character of that genus.

Although the anterior dentition of Osborn's "neotype" skull is incomplete, some differences between it and *Palaeosyops* may also be noted. The left $P^1$ is lacking entirely, and the place on the maxilla to which it had been attached has been covered with plaster. The root of the left upper canine and all of the left $P^2$ are preserved, however. There is a long space between the

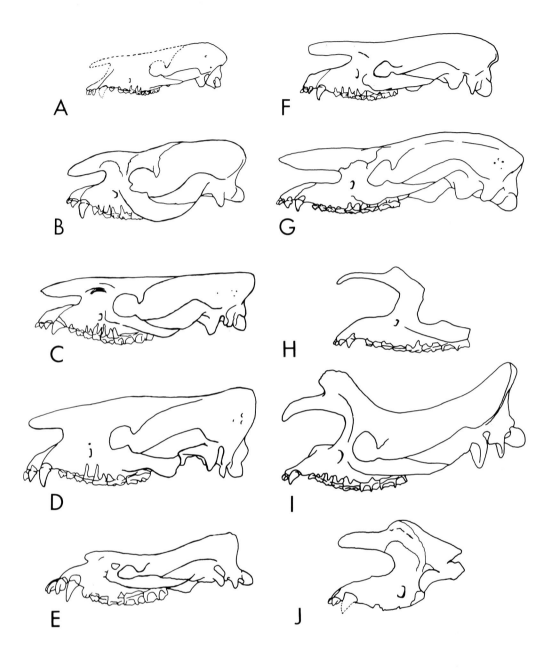

Fig. 25.1 Lateral views of Eocene brontothere skulls. A, *Eotitanops*; B, *Palaeosyops*; C, *Telmatherium*; D, *Metatelmatherium*; E, *Sthenodectes*; F, *Mesatirhinus*; G, *Dolichorhinus*; H, *Eotitanotherium*; I, "*Diplacodon*" *progressum*; and J, *Protitanotherium*. A, B, F, and G after Osborn, 1929; C and D after Osborn, 1908; E after Douglass, 1909; H after Peterson, 1914; I after Peterson, 1934; and J after Hatcher, 1895.

left canine and left $P^2$, so that no matter where the left $P^1$ was originally placed there must have been a considerable diastema present. Skulls of *Palaeosyops* have either an extremely short diastema or no diastema at all.

*Palaeosyops* Leidy, 1870b
(= *Limnohyus* Marsh, 1872; = *Limnohyops* Marsh, 1890; = *Eometarhinus* Osborn, 1919)
*Age*: Bridgerian.
*Type species*: *P. paludosus* Leidy, 1870b
*Lectotype of type species*: USNM 759, an isolated $M_2$.

*Diagnosis*: Medium-sized (length $P^2$ to $M^3$ approximately 129 to 165 mm) brontothere with six upper and lower incisors; large canines; very small or no upper diastema (A) and a moderate lower diastema (mostly between $P_1$ and $P_2$); unmolarized premolars; large paraconules on the molars; hypocone or pseudohypocone variably present on $M^3$; strongly brachycephalic skull (A); robust zygomatic arches that are sharply curved (A); sharply curved nasals that taper distally (A); and a low convexity or dome in the region of the frontoparietal border (A).

*Discussion*: Leidy (1870b) based *Palaeosyops paludosus* on four isolated teeth (Leidy, 1870b). Of these cotypes, Osborn (1929) selected USNM 759, a lower second molar, as the lectotype. Osborn (1929) also designated a lower jaw (AMNH 11680) as a "neotype" for the type species but because the lectotype is still preserved, Osborn's specimen has no nomenclatural significance. Like his "neotype" skull of *Eotitanops borealis*, this jaw is merely a referred specimen.

Based on the length and width of the lectotype molar, I believe that it represents the medium-sized, brachycephalic brontotheres later named *Limnohyus* and *Limnohyops* by Marsh. Because the lectotype appears to be identifiable, I regard the name *Palaeosyops* as valid. If it should later prove that the lectotype is inadequate for diagnostic purposes, the next available name would be *Limnohyus* Marsh (1872; type species *L. robustus*), which is based on a relatively complete skull (YPM 11122).

As early as 1872 Marsh recognized that some of the broad-skulled brontotheres that had been referred to *Palaeosyops* had a hypocone on the $M^3$ and some did not. Marsh proposed that the name *Palaeosyops* be restricted to those animals with a hypocone and that the name *Limnohyus* be given to those without it. Leidy (1872), however, pointed out that the absence of a hypocone was a character originally attributed to *Palaeosyops* (the type series of *Palaeosyops paludosus* included an $M^3$ that lacked a hypocone) and could not be used to define a new genus. Accordingly, Marsh (1890) reversed his previous position and applied the name *Palaeosyops* to specimens without the hypocone on $M^3$ and gave the new generic name *Limnohyops* to those with one.

Osborn (1908 and 1929) adopted Marsh's systematics but recognized a species of *Limnohyops* (*L. monoconus*) that lacked the critical hypocone on $M^3$. Osborn (1929) listed several other characters that he believed distinguished *Palaeosyops* from *Limnohyops*, but most of these characters were trivial. Osborn himself noted (1929, p. 303) that the difference between *Palaeosyops* and *Limnohyops* is less than that observed between species in the modern genus *Cervus*.

My own observations show that the overall similarity between the skulls of individuals with a hypocone on $M^3$ and those without it is too great to justify separating them at the generic level. Most of Osborn's generic distinctions may be attributed to the vagaries of preservation or to individual variation.

Some *Palaeosyops* skulls are clearly more robust than others and have short, wide sagittal crests that form a sharp angle with the temporal crests. Osborn (1929) referred most of these skulls to *Palaeosyops*. Other skulls are more gracile with a longer, thinner sagittal crest that merges more gently with the temporal crests. Osborn regarded these as diagnostic characters of *Limnohyops*. I regard these differences as probable sexual dimorphism, with the more robust skulls representing males and the more gracile skulls representing females. Contrary to a long-held belief, I find no evidence that canine size varies significantly between the sexes.

It is important to point out that Osborn was incorrect in his assertion that skulls of "*Limnohyops*" lack the frontoparietal convexity or dome. Not one of the specimens that Osborn referred to *Limnohyops* has the frontoparietal area well preserved. In fact, in most cases, this area is very badly damaged or missing entirely. I believe that the dome is present in all specimens of *Palaeosyops* but is more prominent in the supposed males.

In 1919 Osborn described a very fragmentary skull (AMNH 17412) from the Huerfano basin and recognized it as a form ancestral to *Metarhinus*. Accordingly, Osborn gave it the name *Eometarhinus* ("dawn *Metarhinus*"). Osborn (1929) upheld this identification, but Robinson (1966) synonymized *Eometarhinus* with *Palaeosyops fontinalis*. The holotype of the type species of *Eometarhinus* is very poorly preserved and therefore difficult to compare adequately with other specimens. Nonetheless, I believe that Robinson was probably correct in his synonymy. Specimens of *Palaeosyops fontinalis* are all rather poorly preserved and in my opinion there is some question as to whether they are truly congeneric with typical *Palaeosyops* material of the Bridger Formation and its equivalents.

*Telmatherium* Marsh, 1872
(= *Leurocephalus* Osborn, Scott, and Speir, 1878; ="*Telmatotherium*" Marsh, 1880; = *Manteoceras* Hatcher, 1895)
*Age*: Bridgerian
*Type species*: *T. validus* Marsh, 1872
*Holotype of type species*: YPM 11120, a partial palate with most of the upper dentition and fragments of the zygomatic arches, nasals, and occipital region preserved.

*Diagnosis*: Moderate to moderately large-sized (length $P^2$ to $M^3$ approximately 160 to 190 mm) brontothere with six upper and lower incisors; large canines; moderate to long upper diastema; unmolarized premolars; paraconules and metaconules usually small and variably present on molars; hypocone or pseudohypocone rarely present on $M^3$; horn-like swelling or prominence developed on frontonasal boundary over facial concavity (*, 4); and a deep pit-like fossa in the middle of the sagittal crest near the back of the skull (A).

*Discussion*: Between 1872 and the mid 1890s, only two fragmentary specimens of *Telmatherium* were known; the type of *T. validus* and YPM-PU 10027, the fragmentary right side of a skull and lower jaw. YPM-PU 10027 was collected in 1877 and was at first recognized as the type of a new genus and species, *Leurocephalus cultridens* (Osborn, Scott, and Speir, 1878). In 1892, however, Earle synonymized *Leurocepahlus* with "*Telmatotherium*" (= *Telmatherium*). (In a list of genera published in 1880, Marsh had invalidly changed the name *Telamatherium* to "*Telmatotherium*." Although "*Telmatotherium*" cannot now be recognized as a valid name, several writers at the time adopted its use).

Complete skulls of *Telmatherium* were not known until 1893 (Osborn 1929, p. 179) or 1894 (Osborn 1929, p. 358), when several specimens were collected by an American Museum expedition led by J. L. Wortman. Wortman wrote from the field that he believed the skulls represented a new genus, for which he suggested the name *Manteoceras*.

Osborn disagreed with Wortman's assessment of the material and (1895) identified the skulls as specimens of *Palaeosyops vallidens* Cope, which Osborn referred to the genus "*Telamatotherium*." Osborn also included under the name "*Telmatotherium*" specimens that would later be placed in the genera *Rhadinorhinus* ("*Telmatotherium*" *diploclonum*) and *Dolichorhinus* ("*Telmatotherium*" *cornutum*).

Hatcher recognized that *"Telmatotherium"* as defined by Osborn was multigeneric. In a postscript to his paper on *Protitanotherium* (1895) he remarked that the type of *"Telmatotherium" vallidens* should be recognized as a new genus which he named *Manteoceras* following the original suggestion of Wortman. In the following sentence he stated that the type of *"Telmatotherium" cornutum* should also be recognized as a new genus and gave it the name *Dolichorhinus*.

In an interesting turn of events, Osborn (1929) concluded that the lectotype lower jaw (AMNH 5098) of Cope's *Palaeosyops vallidens* did not represent the same genus as Wortman's skulls after all, but was congeneric with the holotype skull of *"Telmatotherium" cornutum*, the type species of *Dolichorhinus*. Thus, *Dolichorhinus* would be a synonym of *Manteoceras*.

Osborn (1929) attempted to circumvent this situation by asserting that Hatcher had *"Telmatotherium" vallidens* sensu Osborn in mind when he named *Manteoceras* and not the lectotype of Cope's *Palaeosyops vallidens*. To support this view, Osborn argued that Hatcher's figures of *"Telmatotherium" vallidens* were based on Osborn's original figures of Wortman's skulls and that Hatcher's diagnosis of *Manteoceras* (which was based entirely on cranial characters) was totally inapplicable to the lectotype lower jaw of *Palaeosyops vallidens*. Osborn concluded that *Manteoceras* was the valid generic name for Wortman's skulls because this is what Hatcher had intended, but he could not recognize *Manteoceras vallidens* as the type species.

Osborn thus looked in the subsequent literature for an appropriate type species for *Manteoceras*. In 1899 Matthew had published a faunal list in which the name *Palaeosyops manteoceras* appeared. This was a new species name that Matthew took from an unpublished Osborn manuscript and was meant to apply to Wortman's skulls. Matthew, however, did not designate a type or provide a diagnosis, thus making *Palaeosyops manteoceras* a *nomen nudum* (Osborn, 1929).

In 1902, Hay put the generic name *Manteoceras* and the trivial name *manteoceras* together, and in essence designated a type by specifying that the name was to be applied to Osborn's and Hatcher's figures of Wortman's skulls. These figures illustrated two skulls, AMNH 1569 and AMNH 1570, of which Osborn (1929) chose AMNH 1569 as a lectotype. Osborn (1929) argued that because *"Telmatotherium" vallidens* sensu Osborn was equivalent to *Manteoceras manteoceras* Hay, *M. manteoceras* should be regarded as the type species of the genus *Manteoceras*.

In my opinion, if Osborn was correct in establishing that the lectotype jaw of *Palaeosyops vallidens* is congeneric with Hatcher's *Dolichorhinus*, then the conclusion that *Manteoceras* is a synonym of *Dolichorhinus* is inescapable. Regardless of what Hatcher may have intended he clearly specified that the type specimen of *Palaeosyops vallidens* was to be the type of the genus *Manteoceras*. Thus the type species of *Manteoceras* is Cope's *Palaeosyops vallidens* not *"Telmatotherium" vallidens* sensu Osborn.

After examining the lectotype jaw of *Palaeosyops vallidens*, I am still not certain whether it belongs to *Telmatherium* or *Dolichorhinus*. Although its stratigraphic occurrence (Washakie B) suggests that it might belong to *Dolichorhinus*, the specimen is fragmentary and resembles both genera. In the present paper, I have provisionally recognized *Manteoceras* as a junior synonym of *Telmatherium* and have accepted *Dolichorhinus* as a distinct and valid genus. If later work should prove that the jaw is referable to *Dolichorhinus*, however, then the name *Dolichorhinus* could be recognized as a junior synonym of *Manteoceras*. Because this would generate a great deal of confusion and create instability in the literature, I would strongly recommend that any author formally synonymizing *Dolichorhinus* and *Manteoceras* recognize *Dolichorhinus* as having priority over *Manteoceras* (see discussion of *Dolichorhinus*).

It is very clear that *Manteoceras, sensu* Hay (1902) and Osborn (1929), is congeneric with *Leurocephalus*. Skulls referred to *Manteoceras* by Osborn have distinct hornlike prominences anterior to the orbit near the frontonasal suture. The holotype of *Leurocephalus cultridens* includes a fragment of the right nasal from this region that clearly shows the characteristic prominence. Osborn (1929) recognized this fact but regarded *Leurocephalus* as generically distinct from *Manteoceras*, arguing that the similar horn morphology was the result of convergent evolution. Osborn cited several minor differences between the type of *Leurocephalus cultridens* and skulls of *Manteoceras* to justify placing them in separate genera. My own examination of these materials, however, shows that the differences are trivial and easily attributed to intraspecific variation and differences in preservation. Gregory arrived at a similar conclusion (see Osborn 1929, footnote on page 339; and Granger and Gregory, 1943, p. 357) and noted that certain specimens were morphologically intermediate between *Leurocephalus* (*Telmatherium cultridens*) and *Manteoceras*. Granger and Gregory (1943, Fig. 4), however, continued to regard *Manteoceras* and *Telmatherium* as valid genera that were closely related but ancestral to two different lineages.

Earle (1891) and Osborn (1929) both recognized *Leurocephalus* as a junior synonym of *Telmatherium*. Although the type of *Telmatherium validus* is fragmentary, it appears to conform closely in both size and morphology to the type of *Leurocephalus cultridens* and Wortman's skulls of *Manteoceras*. Unfortunately, the diagnostic frontonasal region and sagittal crest are not preserved, making identification somewhat difficult. The type of *Telmatherium validus* is distinctly different, however, from the skulls of the other two Bridger genera that I recognize as valid, *Palaeosyops* and *Mesatirhinus*. The type of *Telmatherium validus* lacks the thick, curved zygomatic arches and large paraconules that characterize *Palaeosyops* and is much larger than specimens of *Mesatirhinus*.

*Mesatirhinus* Osborn, 1908

*Age*: Bridgerian
*Type species*: *M. megarhinus* (Earle, 1891)
*Holotype of type species*: YPM-PU 10008, a partial skull.

*Diagnosis*: Medium-sized (length $P^2$ to $M^3$ approximately 128 to 145 mm) brontothere with six upper incisors; reduced canine (*, 3); upper diastema usually short but occasionally long; unmolarized premolars; small paraconules and metaconules variably present on the molars; no hypocone on $M^3$; well-developed suborbital protuberance (*, 3); and nasals that are moderately flared distally.

*Discussion*: In 1891 Earle described the skull of a small brontothere that he named *Palaeosyops megarhinus*. In 1908 Osborn recognized that this skull was generically distinct from *Palaeosyops* and gave it the new generic name *Mesatirhinus*. *Mesatirhinus* is allied to *Rhadinorhinus* and *Dolicorhinus* and shares with these genera a distinct suborbital protuberance. In *Mesatirhinus* the suborbital protuberance is always well developed. Osborn (1929) asserted that *Mesatirhinus* had incipient horns, but I have not found any horn or hornlike prominence present in any of the several specimens that I have examined.

*Metarhinus* Osborn, 1908

*Age*: Uintan
*Type species*: *M. fluviatalis* Osborn, 1908
*Holotype of type species*: AMNH 1500, a partial skull lacking the nasals.

*Diagnosis*: Medium-sized (length $P^2$ to $M^3$ approximately 138 to 145 mm) brontothere with six upper incisors; reduced canine (*, 3); no upper diastema, unmolarized premolars; hypocone occasionally present on $M^3$; small suborbital protuberance (*, 3); prominent orbits; very deep lateral nasal incision; and spoon-shaped nasals that are flared distally and constricted at the base

(A).

*Discussion*: The holotype of the type species of *Metarhinus* was described by Osborn in 1908. Later, Riggs (1912) described a fine collection of Uintan brontotheres collected in 1910 by the Field Museum in the Uinta Basin of Utah. Among these specimens were skulls with deep nasal incisions and prominent orbits, most of which Riggs referred to *Metarhinus*. One of the skulls, however, he referred to a new genus and species, *Rhadinorhinus abbotti*. The type skull of *R. abbotti* had short, distally tapered nasals, while the skulls he referred to *Metarhinus* had broad, spoon-shaped nasals.

Both skull morphologies are easily distinguished, although I am not convinced that they warrant distinction at the generic level. It might be more appropriate to recognize them as separate species of the same genus, or even as sexual variants of a single species.

A major nomenclatural difficulty arises from the fact that the holotype of *Metarhinus fluviatilis* lacks the critical nasals and is in generally poor condition. After examining the specimen, I am not certain whether it represents the spoon-nosed or short-nosed variety of brontothere. It is possible, therefore, that *Metarhinus* may be a *nomen dubium*, or a senior synonym of *Rhadinorhinus*.

For the purposes of this paper, I have provisionally accepted Riggs' (1912) and Osborn's (1929) conclusion that the type of *Metarhinus fluviatilis* represents the spoon-nosed variety of brontothere, and that this variety is generically distinct from the short-nosed variety. I would like to emphasize, however, that in my opinion, neither of these conclusions is certain.

Skulls referred to *Metarhinus* are approximately the same size as larger skulls of *Mesatirhinus* and smaller skulls of *Rhadinorhinus*. The teeth of all three genera are virtually indistinguishable. *Metarhinus* may be distinguished from *Mesatirhinus* primarily by the morphology of the nasal and orbital regions. The lateral nasal incision of *Metarhinus* extends very far posteriorly so that it reaches the proximity of the orbit, and the orbit itself is very prominent. Both of these characteristics occur in *Rhadinorhinus*, also. In contrast, the lateral nasal incision of *Mesatirhinus* is typical of brontotheres. The primary diagnostic character of *Metarhinus* that distinguishes it from both *Mesatirhinus* and *Rhadinorhinus* is the shape of the nasals, which are constricted proximally, giving them their distinctive spoon-shaped appearance. The nasals of *Mesatirhinus* are not constricted at the base, and are in many respects similar to those of *Dolichorhinus*, while those of *Rhadinorhinus* are short and taper to a point.

*Rhadinorhinus* Riggs, 1912
*Age*: Uintan
*Type species*: *R. abbotti* Riggs, 1912
*Holotype of type species*: FMNH 12179, a skull.

*Diagnosis*: Medium-sized (length $P^2$ to $M^3$ approximately 143 to 148 mm) brontothere with six upper incisors; reduced canine (*, 3); no upper diastema; unmolarized premolars; hypocone occasionally present on $M^3$; small to large suborbital protuberance (*, 3); prominent orbits; very deep lateral nasal incision; and nasals that are short and strongly tapered distally (A).

*Discussion*: The holotype skull of *Rhadinorhinus abbotti* was described by Riggs (1912) and was part of the collection of Uintan brontotheres obtained by the Field Museum in 1910. The specimen differed from skulls in the collection that Riggs referred to *Metarhinus* primarily in the shape of the nasals, but also in the small size of the suborbital protuberance (process). In the same paper, Riggs referred the type skull of "*Telmatotherium*" *diploclonum* Osborn (AMNH 1560) to the genus *Rhadinorhinus* as well. In 1908, Osborn had referred this specimen to *Metarhinus*, but in 1929 he accepted Riggs' conclusion and referred the specimen to *R. abbotti* instead. (The 1908 paper does not clearly state that

Osborn meant to refer AMNH 1560 to the genus *Metarhinus*, but in 1929 he claimed that this was his intention). Unfortunately, AMNH 1560 lacks the nasals, and in view of this fact, I believe that its correct generic assignment is uncertain.

It is possible that *Rhadinorhinus* is a junior synonym of *Metarhinus* and that the short-nosed variety of brontotheres, here referred to *Rhadinorhinus*, is not generically distinct from the spoon-nosed variety referred to *Metarhinus* (see discussion earlier). I have only provisionally accepted Riggs' (1912) and Osborn's (1929) conclusion that both names are valid and that both varieties are distinct at the generic level.

Skulls referred to *Rhadinorhinus* are approximately the same size as larger skulls of *Mesatirhinus* and *Metarhinus*. The primary diagnostic character that distinguishes *Rhadinorhinus* from both *Mesatirhinus* and *Metarhinus* is the shape of the nasals, which are short and tapered to a point. *Rhadinorhinus* may be further distinguished from *Mesatirhinus* by the extreme posterior extent of the lateral nasal incision, and by the prominence of the orbit, characteristics that it shares with *Metarhinus*.

Both Riggs (1912) and Osborn (1929) reported that the suborbital protuberance is small in *Rhadinorhinus*, but actually the size of the structure varies. In CM 3098 (incorrectly identified as *Metarhinus* in the Carnegie Museum catalogue), the protuberance is small, while in CM 3510 (also misidentified as *Metarhinus*), it is large. It is not clear whether this difference in morphology has any taxonomic significance, or merely represents intraspecific variation. I suspect that it may represent sexual dimorphism.

*Dolichorhinus* Hatcher, 1895
*Age*: Uintan
*Type species*: *D. cornutum* (Osborn, 1895) [?= *D. hyognathus* (Osborn, 1890)]
*Holotype of type species*: AMNH 1851, a complete skull (*D. cornutum*); YPM-PU 10273, a partial lower jaw (*D. hyognathus*).

*Diagnosis*: Moderately large-sized (length $P^2$ to $M^3$ approximately 164 to 193 mm) brontothere with six upper and lower incisors; large to medium-sized canines; long upper and lower diastema (lower diastema between $P_1$ and $P_2$ and between $P_1$ and the canine); unmolarized premolars; small paraconules and metaconules variably present on the molars; hypocone or pseudohypocone often present on $M^3$; hyperdolichocephalic skull (A); large suborbital protuberance (*, 3) with a small anterior flange and a larger posterior flange (A); nasals broad and flared distally; small angular horn at frontonasal boundary over orbit (A); and a wide, rounded cranial vertex with no sagittal crest (A).

*Discussion*: In 1890 Osborn described an extremely elongate lower jaw (YPM-PU 10273) that he provisionally referred to the genus *Palaeosyops* and named *P. hyognathus*. Six years later, Osborn (1895) described a hyperdolichocephalic skull (AMNH 1851) that he referred to the genus "*Telmatotherium*" (= *Telmatherium*) and named *T. cornutum*. Both of these specimens clearly belong to to the genus that would later be named *Dolichorhinus*.

Hatcher (1895) recognized that "*Telmatotherium*" as diagnosed by Osborn consisted of three distinct forms (see discussion of *Telmatherium*). Hatcher proposed that the long-skulled brontotheres that Osborn had named "*Telmatotherium*" *cornutum* be placed in their own genus for which he suggested the name *Dolichorhinus*. Thus the type species of *Dolichorhinus* would be *D. cornutum* (Osborn), the type specimen of which is AMNH 1851, the complete skull.

Osborn (1929), however, argued that the lower jaw (YPM-PU 10273) that he had named *Palaeosyops hyognathus* represented the same species as *Dolichorhinus cornutum*. Thus *Dolichorhinus cornutum* would be a junior synonym of *D. hyognathus*. I have not yet decided whether Osborn was correct in synonymizing the two species and have noted both names and types here.

Osborn (1929) provisionally referred to

the genus *Dolichorhinus* the lectotype lower jaw of *Palaeosyops vallidens* Cope, which Hatcher (1895) had made the type species of the genus *Manteoceras*. *Manteoceras* was named in the same paper as *Dolichorhinus*, and was mentioned first, although priority has never been formally established.

I have not yet determined if Osborn was correct in regarding *Palaeosyops vallidens* as congeneric with *Dolichorhinus*. In the current paper, I provisionally regard *P. vallidens* as being a specimen of *Telmatherium*, while accepting *Dolichorhinus* as a distinct and valid genus (see discussion of *Telmatherium*). If later work proves Osborn to have been correct, however, the name *Dolichorhinus* could be regarded as a junior synonym of *Manteoceras*. Because this would result in severe nomenclatural instability and confusion, I strongly recommend that any author attempting to formally synonymize *Dolichorhinus* and *Manteoceras* choose *Dolichorhinus* as the name having priority (see Recommendation 24A of the International Code of Zoological Nomenclature [Ride et al., 1985]).

*Dolichorhinus* was one of the most highly derived brontothere genera. The hyperdolichocephalic skull, rounded cranial vertex, and small horn are all unique derived characters that define the genus. The horn of *Dolichorhinus*, although similar to that of the so-called "horned brontotheres," was apparently independently evolved. The sister genera of *Dolichorhinus* (*Mesatirhinus* and *Rhadinorhinus*) are hornless.

As in *Mesatirhinus* and *Rhadinorhinus*, a suborbital protuberance is present in *Dolichorhinus*. In *Dolichorhinus*, this protuberance is always large and is divided into a smaller anterior flange and a larger posterior flange.

*Metatelmatherium* Granger and Gregory, 1938
(includes *Telmatherium ultimum* Osborn, 1908, and *Manteoceras uintensis* Douglass, 1909)

*Age*: Uintan

*Type species*: *M. cristatum* Granger and Gregory, 1938

*Holotype of type species*: AMNH 26411, a skull and lower jaw.

*Diagnosis*: Large-sized (in *Metatelmatherium ultimum* length of $P^2$ to $M^3$ is approximately 200 to 208 mm) brontothere with six upper incisors; large canines; long upper and lower diastema (in AMNH 2060 the lower diastema is between both the canine and $P_1$ and between $P_1$ and $P_2$); unmolarized premolars; paraconules and metaconules not present in the few skulls known; pseudohypocone occasionally present on $M^3$; zygomatic arch often with a prominent flange on the ventral surface of the jugal near to where it borders on the squamosal (A); and the lateral incision of the external nares shifted forward so as to lie over the upper diastema or $P^1$ in an uncrushed specimen (A).

*Discussion*: The type species of *Metatelmatherium* is from Asia and is slightly larger than the American species, *M. ultimum* (Granger and Gregory, 1938, 1943). Osborn described the holotype skull and lower jaw of *M. ultimum* (AMNH 2060) in 1908 and recognized it as a new species of *Telmatherium*. After examining this specimen, however, Granger and Gregory (1938) concluded that *Telmatherium ultimum* Osborn was congeneric with one of the Asian skulls in the American Museum collection and that this genus was distinct from *Telmatherium*. They applied the new generic name *Metatelmatherium* to these materials.

In 1909, Douglass described the front part of a skull from the Uinta Basin in Utah which he identified as a new species of *Manteoceras*, *M. uintensis*. In 1929, Osborn upheld this identification. My own observations, however, suggest that this skull (CM 2388) is not a specimen of *Telmatherium* (= *Manteoceras*) but instead has the large canine, long diastema, and forwardly placed lateral incision of the external nares that are diagnostic of *Metatelmatherium*. The zygomatic arches are imperfectly preserved but may have had the diagnostic flange on the underside of the jugal as well.

Osborn (1929) gave several characters that he believed allied Douglass's specimen to *Manteoceras* (i.e., *Telamatherium*) but these are all quite trivial. I believe that the similarities to *Metatelmatherium* heavily outweigh any similarities to *Telmatherium*.

Most specimens of *Metatelmatherium ultimum* have the distinct flange on the base of the jugal that is apomorphic for the genus. One skull (CM 11380), however, lacks the characteristic flange but resembles specimens of *Metatelmatherium ultimum* in all other respects. This skull is from a slightly lower horizon (Sand Wash) than the other materials and so may represent another species. I am more inclined, however, to regard the differences as intraspecific variation and probably sexual dimorphism.

*Metatelmatherium ultimum* is known with certainty from only five skulls: AMNH 2060 (holotype), AMNH 2004, CM 2388 (holotype of "*Manteoceras*" *uintensis*), CM 2339 (identified in the Carnegie Museum catalog as "*Manteoceras*" *uintensis*), and CM 11380 (discussed earlier, and also earlier identified as "*M.*" *uintensis*). Another skull (AMNH 2029), which is very badly crushed, also appears to represent this genus. Osborn (1929) identified this skull as "*Manteoceras*" *uintensis* (?).

*Sthenodectes* Gregory, 1912

*Age*: Uintan

*Type species*: *S. incisivum* (Douglass, 1909)

*Holotype of type species*: CM 2398, a skull crushed dorsoventrally.

*Diagnosis*: Large-sized (average length $P^2$ to $M^3$ in CM 2398 is 195 mm) brontothere with six very large, spatulate upper incisors (A); very long, pointed canine (A); no upper diastema (A); and unmolarized premolars.

*Discussion*: The type skull of *Sthenodectes incisivum* was described by Douglass in 1909 who provisionally identified it as a new species of *Telmatherium*, *T. incisivum*. Despite this identification, Douglass believed that the skull probably represented a genus distinct from *Telmatherium* but preferred not to create a new generic name at that time.

A short time later, W. K. Gregory of the American Museum obtained permission to study the specimen and like Douglass concluded that it was generically distinct from *Telmatherium*. Gregory (1912) gave it the new generic name *Sthenodectes* (literally "strong biter"), in allusion to the large size of the canines and incisors.

*Sthenodectes* is unquestionably a valid genus and its immense incisors are unique among brontotheres. *Sthenodectes* is an extremely rare genus and is known at present from only two or three skulls: the type of *S. incisivum* in the Carnegie Museum, a partial skull in the Field Museum (FMNH 12165, erroneously identified as FMNH 12168 in Osborn, 1929) that was collected from the same locality (Riggs, 1912), and another partial skull and jaw in the Carnegie Museum (CM 11437) that Peterson (1934) referred to the genus.

Skulls from the Pruett Formation of Texas that Wilson (1977) named *Sthenodectes australis* (holotype: TMM 41723-3) are not referable to the genus *Sthenodectes* at all. Instead, they should be regarded as belonging to the group I call diplacodonts *sensu lato* (see section on higher categories of brontotheres). In many respects, the holotype of "*Sthenodectes*" *australis* is similar to *Protitanotherium*, but its exact generic assignment remains unclear. It is possible that it represents an entirely new genus.

*Protitanotherium* Hatcher, 1895

*Age*: Uintan

*Type species*: *P. emarginatum* (Hatcher, 1895)

*Holotype of type species*: YPM-PU 11242, anterior part of skull with lower jaws.

*Diagnosis* (based on holotype of type species only): Large-sized brontothere with six upper and lower incisors; large canines; long upper and lower diastema; short, broad nasals; and short but prominent frontonasal horns that are laterally directed and elliptical in cross-section.

*Discussion*: In 1895 Hatcher described an unusual brontothere skull with small horns that he provisionally referred to the genus *Diplacodon* Marsh, 1875 (type species *D. elatus*). Although Marsh (1875) had asserted that *Diplacodon* lacked horns, Hatcher pointed out that Marsh's claim was purely conjectural, because the type of *Diplacodon elatus* (YPM 11180) lacked the entire horn and nasal region. Hatcher argued that in all other respects his specimen appeared to be identical to *Diplacodon elatus* but larger in size. He therefore recognized the specimen as a new species of *Diplacodon* which he named *Diplacodon emarginatum*. Hatcher proposed that the generic name *Protitanotherium* be applied to the specimen if there should prove to be hornless forms with the same dental characters.

Osborn (1929) recognized the genus *Protitanotherium* as valid and argued that it was distinct from *Diplacodon* regardless of whether *Diplacodon* had horns or not. Most subsequent authors have also recognized *Protitanotherium* as valid, though in an abstract Schoch and Lucas (1985) synonymized *Protitanotherium* with *Diplacodon*.

The holotype of *Diplacodon elatus* (YPM 11180) is a badly crushed skull and is difficult to compare to the type of *Protitanotherium emarginatum*. The two specimens have only the upper canines, diastema, and part of the first upper premolars in common. The canine of *Protitanotherium* is large and robust while that of *Diplacodon* is much smaller in size. The canine of *Diplacodon* is much more similar in size to that of *Eotitanotherium osborni* or "*Diplacodon*" *progressum*. Contrary to a long held belief, canine size does not appear to be a strongly sexual dimorphic character in brontotheres (see discussions of *Palaeosyops* and *Duchesneodus*). The difference in canine size between *Diplacodon* and *Protitanotherium* is, therefore, probably a valid taxonomic character. This, and my suspicion that *Diplacodon* may be a senior synonym of *Eotitanotherium*, causes me to regard *Protitanotherium* as a valid genus distinct from *Diplacodon*.

"*Diplacodon*" *progressum* Peterson, 1934
*Age*: Uintan
*Holotype*: CM 11879A, a skull and lower jaw.

*Diagnosis*: Large-sized (length $P^2$ to $M^3$ approximately 207 to 221 mm) brontothere with six upper incisors; reduced canine (*, 5); long upper diastema; one lingual cusp on $P^1$, one or two lingual cusps on $P^2$, and two poorly separated lingual cusps on $P^3$ and $P^4$ (*, 5); paraconules and metaconules occasionally present on molars; hypocone and pseudohypocone occasionally present on $M^3$; robust nasals that are rounded distally and curved downward; relatively large, bulbous horns that are roughly circular in cross-section (A); and a widened cranial vertex with no sagittal crest (*, 5).

*Discussion*: The holotype (YPM 11180) of the type species of *Diplacodon, D. elatus*, is a skull that is almost crushed flat but with the grinding dentition and one canine well preserved. *Diplacodon elatus* has two lingual cusps on the third and fourth upper premolars while skulls of "*Diplacodon*" *progressum* tend to have a single elongate lingual cusp or two lingual cusps that are poorly separated on these teeth. The premolars of *Diplacodon elatus* are much more similar to those of *Eotitanotherium osborni* than they are to "*Diplacodon*" *progressum*.

The largest specimen (based on the length of the upper cheek tooth series exclusive of the first premolar and the length of the upper molar series) of "*Diplacodon*" *progressum* is 6 to 7% smaller than the type of *Diplacodon elatus*. *Diplacodon elatus* falls, however, in the size range of *Eotitanotherium*.

In my opinion, "*Diplacodon*" *progressum* is generically distinct from the type species of *Diplacodon*, and a new name will have to be proposed for it. I am currently working on a more thorough treatment of the subject and intend to propose a new name in the near future. *"Diplacodon" progressum* is currently known from eight specimens, two of which have previously been misidentified as specimens of *Eotitanotherium*. These specimens are AMNH 21887 (identified as *Eotitanotherium* in the American Museum catalog), CM 11879A (holotype of *"Diplacodon" progressum*), CM 2858 (paratype of *Eotitanotherium osborni*), CM 11881 (paratype of *"Diplacodon" progressum*), FMNH 14632, FMNH 14633, FMNH 14799, and FMNH 15446.

*Eotitanotherium* Peterson, 1914b
(= *Diploceras* Peterson, 1914a)
*Age*: Uintan
*Type species*: *E. osborni* (Peterson, 1914a)
*Holotype of type species*: CM 2859, the anterior part of a skull with jaws, atlas, axis, scapula, pelvis, and various foot bones.

*Diagnosis*: Large-sized (length $P^2$ to $M^3$ approximately 221 to 236 mm) brontothere with six upper incisors; reduced canine (*, 5); long upper and lower diastema; one lingual cusp on $P^1$, one or two lingual cusps on $P^2$, and two well separated lingual cusps on $P^3$ and $P^4$ (*, 5); no hypocone on $M^3$ in any of the three specimens known; gracile nasals that are distally rounded and down-turned; and short, angular horns that are elliptical in cross-section (A).

*Discussion*: Peterson first applied the name *Diploceras* (1914a) to this brontothere but changed it to *Eotitanotherium* (1914b) when it came to his attention that the name *Diploceras* was occupied by a mollusk. *Eotitanotherium* may be a junior synonym of *Diplacodon* Marsh (1875), which was based on a crushed skull. Dental measurements of the holotype of the type species of *Diplacodon* (*D. elatus*) fall within the size range of *Eotitanotherium*; like *Eotitanotherium*, *Diplacodon* has well-separated lingual cusps on the third and fourth upper premolars. In contrast, the contemporaneous brontothere *"Diplacodon" progressum* is about 10% smaller in size and has very poor separation of the lingual cusps on the posterior premolars (the primitive condition).

I have refrained from formally synonymizing *Eotitanotherium* with *Diplacodon* here because of the poor condition of the type of *Diplacodon elatus*, the relatively few known specimens of *Eotitanotherium*, and our poor understanding of Uintan horned brontotheres in general. *Eotitanotherium* is currently known from three specimens in the Carnegie Museum. These specimens are CM 2859 (the holotype of *Eotitanotherium osborni*), CM 11828 (identified in the Carnegie Museum catalog as *Diplacodon* sp.), and CM 11895 (identified in the catalog simply as "Brontotheriid"). No complete skulls of *Eotitanotherium* are known. Peterson's paratype of *Eotitanotherium osborni* (CM 2858) and a specimen referred to *Eotitanotherium* in the American Museum (AMNH 21887) do not represent *Eotitanotherium* but are specimens of *"Diplacodon" progressum*.

*Notiotitanops* Gazin and Sullivan, 1942
*Age*: Probably Uintan or Duchesnean
*Type species*: *N. mississippiensis* Gazin and Sullivan, 1942
*Holotype of type species*: USNM 16646, basal part of a skull with dentition and portions of both rami.

*Diagnosis* (based on holotype of type species only): Large-sized (average length $P^2$ to $M^3$ in USNM 16646 is 210.5 mm) brontothere with four reduced, globular upper incisors (*, 6); reduced canine (*, 5); long upper diastema; single lingual cusp on $P^1$ and two poorly separated lingual cusps on $P^2$ to $P^4$ (*, 5); no paraconules or metaconules on $M^2$ and $M^3$ ($M^1$ too worn to tell); and a rudimentary hypocone on $M^3$.

*Discussion*: The type and only known specimen of *Notiotitanops mississippiensis* was collected in 1940 in marine sediments on the Covington Farm, approximately two and one-half miles south of Quitman, in Clarke County, Mississippi. It was the first brontothere to be found "at a place in the southern states remote from the recorded distribution of titanotheres" in the Rocky Mountain and Great Plains regions (Gazin and Sullivan, 1942).

The holotype of *Notiotitanops mississippiensis* represents a brontothere similar to *Protitanotherium*, "*Diplacodon*" *progressum*, *Eotitanotherium*, and *Duchesneodus*, but exhibits an unusual combination of primitive and derived dental characters. Like "*Diplacodon*" *progressum* and *Eotitanotherium osborni* it has small canines and a long diastema, but it is more similar to *Duchesneodus* and the Chadronian brontotheres in the reduced number and globular morphology of the upper incisors.

Given its dental characters, *Notiotitanops mississippiensis* almost certainly had horns, although the frontonasal region was not preserved in the type specimen.

*Duchesneodus* Lucas and Schoch, 1982
*Age*: Duchesnean
*Type species*: *D. uintensis* (Peterson, 1931)
*Holotype of type species*: CM 11809, a lower jaw.

*Diagnosis*: Large-sized (length $P^2$ to $M^3$ approximately 210 to 248 mm) brontothere with four upper (*, 6) and six lower incisors that are reduced and globular in shape; reduced canines (*, 5); no diastema (*, 7); first premolars occasionally lacking (*, 7); single lingual cusp on $P^1$ and two lingual cusps on $P^2$ to $P^4$ (*, 5); hypocone consistently present on $M^3$ (*, 7); zygomatic arches moderately expanded posteriorly and apparently developing a wing-shape in males (A); small frontonasal horns that are roughly circular in cross-section; small to large convexity or dome in the region of the frontoparietal border (A); and a broad, "saddle-shaped" cranial vertex (*, 7).

*Discussion*: Between the years 1929 and 1930, Carnegie expeditions led by J. L. Kay collected several brontothere skulls, lower jaws, and postcranial elements from a single quarry in the Duchesne River Formation, eleven miles west of Vernal in northeastern Utah. Although the general location of this important site was indicated on a map in Peterson and Kay (1931), the exact map location has never been published. Its exact location is in the SE 1/4 SW 1/4 NW 1/4 Sec. 33, T.4S, R.20E, Uintah County, Utah. The quarry is located on the southeastern face of a hill. Peterson made a preliminary description of this material (1931) and recognized it as a new species of *Teleodus* Marsh (1890), which he named *T. uintensis*.

The holotype lower jaw (YPM 10321) of the type species of *Teleodus*, *T. avus*, was first described by Marsh in 1890. It was from Oligocene deposits and was unusual in its retention of six lower incisors. Marsh regarded this as its primary diagnostic character. In all other respects the jaw is similar to that of typical Chadronian horned brontotheres.

Lower jaws from the Carnegie quarry also have six lower incisors, and like the holotype of *Teleodus avus*, have partially molarized premolars, short canines with rounded cross-sections, short diastemas, and frequently lack $P_1$. Peterson, therefore, had a rather convincing set of characters to justify inclusion of his material in *Teleodus*.

Upon re-examining Marsh's type of *Teleodus avus*, however, Lucas and Schoch (1982) came to a surprising conclusion. They determined that the lateral incisors were either added by a technician during the initial preparation of the specimen or are retained decidous incisors. Thus the presence of six lower incisors is not the natural condition of this lower jaw. Lucas and Schoch (1982) regarded *Teleodus* as a probable junior synonym of *Brontops* or *Brontotherium*.

Peterson's "*Teleodus*" *uintensis*, which naturally possesses six lower incisors, is a distinct genus that Lucas and Schoch (1982) named *Duchesneodus*. The *Duchesneodus* material in the Carnegie Museum is of considerable value because it provides the rare opportunity to examine what is likely to be a single brontothere population. It is possible, therefore, to determine the extent of such critical factors as individual variation and sexual dimorphism.

Peterson (1931) observed that some of the skulls in the collection, which he presumed to be males, were more brachycephalic than others and had especially well-developed parietal convexities. Scott (1945) added to these observations that presumed male skulls had larger horns and more robust zygomatic arches than presumed females. My own observations and those independently made by Lucas and Schoch (this volume, Chapter 27) confirm the findings of Peterson and Scott. Two cranial morphologies occur in the collection. Most skulls have low dorsal convexities in the frontoparietal region, relatively short horns, and moderately expanded zygomatic arches. One skull (CM 11767), however, has a very large convexity in the frontoparietal region that forms a huge dome similar to that of the dome-skulled chalicothere *Tylocephalonyx* (Coombs, 1979). The horns of this specimen are slightly larger than most of the other skulls in the collection and the zygomatic arches are much more expanded. Along with Peterson and Scott, I regard these morphological differences as sexual dimorphism with most of the skulls representing females, and CM 11767 and at least one other skull representing males. Peterson (1931) believed that female skulls in the collection had smaller canines than males. I, however, have not found this to be the case. As in *Palaeosyops*, there does not appear to be any correlation between canine size and the sex of the individual.

For a more detailed discussion of intraspecific variation exhibited by specimens from the *Duchesneodus* quarry, see Lucas and Schoch (this volume, Chapter 27).

*Protitanops* Stock, 1936

*Age*: Chadronian

*Type species*: *P. curryi* Stock, 1936

*Holotype of type species*: LACM (CIT) 1854, a skull and lower jaws.

*Diagnosis* (based on holotype of type species only): Large-sized (length $P^2$ to $M^3$ is 254.5 mm in LACM (CIT) 1854) brontothere with four reduced, globular incisors (*, 6) (the morphology and number of lower incisors is unknown); reduced canine (*, 5); long upper diastema; single lingual cusp on $P^1$ and two lingual cusps on $P^{2-4}$ (*, 5); paraconules and metaconules lacking on $M^{2-3}$ of LACM (CIT) 1854 ($M^1$ is too worn in this specimen to tell whether these structures are present); hypocone on $M^3$ (*, 7); zygomatic arches that are moderately wing-shaped (?A); robust nasals that are rounded distally and curve downward; relatively large horns that are roughly elliptical in cross-section; and a widened cranial vertex with no sagittal crest (*, 5).

*Discussion*: The type specimen of *Protitanops curryi* was collected in 1934 by H. Donald Curry of the California Institute of Technology during an expedition to the northern Death Valley region. The specimen was found in a canyon east of Thimble Peak in the Grapevine Mountains and was preserved in the lower red beds of the Titus Canyon Formation. These sediments are considered early Oligocene (Chadronian) in age (Stock, 1936; Savage and Downs, 1954).

Despite its Oligocene occurrence, *Protitanops* is a relatively primitive horned brontothere. In general appearance, the skull of *Protitanops* is somewhat similar to that of "*Diplacodon*" *progressum*. In both genera the canine is small, the diastema long, and the cranial vertex wide but not attaining the "saddle shape" of *Duchesneodus*, *Brontops*, *Menops*, and *Megacerops*. The nasals, horns, and zygomatic arches of "*Diplacodon*" *progressum* and *Protitanops* are also somewhat similar, but the horns of *Protitanops* are more elliptical in shape and the zygomatic arches more wing-like in their development than those of "*Diplacodon*" *progressum*. *Protitanops* is further distinguished from "*Diplacodon*" *progressum* by its larger size and by the more derived condition of the upper incisors, which are reduced to four in number, and are globular in shape.

The dentition of *Protitanops* is quite similar to that of *Notiotitanops*, and like that genus exhibits a combination of primitive and derived characters. *Protitanops* and *Notiotitanops* are similar to the primitive horned brontothere genera *"Diplacodon" progressum* and *Eotitanotherium osborni* in the structure of the canine and long upper diastema, but are more similar to the derived horned genera *Duchesneodus*, *Brontops*, *Menops*, and *Megacerops* in the reduced number and globular morphology of the upper incisors and possibly the presence of a hypocone on $M^3$. Because the type and only known specimen of *Notiotitanops mississippiensis* is so fragmentary, its exact phyletic relationship to *Protitanops* cannot be easily determined.

*Protitanops* is known with certainty from the type specimen of *P. curryi* only. Two specimens (UCMP 126100 and UCMP 126101) from the Clarno Formation of Oregon have been identified as *Protitanops*, but in my opinion this identification is doubtful.

### *Brontops* Marsh, 1887
(= *Diploclonus* Marsh, 1890)
*Age*: Chadronian
*Type species*: *B. robustus* Marsh, 1887
*Holotype of type species*: YPM 12048, a skull and skeleton.

*Diagnosis*: Very large-sized (length $P^1$ to $M^3$ approximately 300 to 375 mm according to data published in Osborn, 1929) brontothere with four (or fewer) upper (*, 6) and lower incisors (*, 8) that are reduced and globular in shape; no diastema (*, 7); partially molarized premolars (*, 5); hypocone on $M^3$ (*, 7); massive zygomatic arches that are thickened or swelled posteriorly (A); short, anteriorly directed horns that are rounded to elliptical in cross-section (A); and a wide, "saddle-shaped" cranial vertex (*, 7).

*Discussion*: *Brontops* is the only Chadronian brontothere genus whose type species is based upon a large part of the skeleton. In fact, there are probably more mounted skeletons of *Brontops* in United States museums than of any other brontothere genus. These skeletons include YPM 12048 (holotype of *B. robustus*), AMNH 518 (skull, forelimbs, and thorax only), and SDSM 4912.

In 1890, Marsh named a new genus, *Diploclonus*, whose primary diagnostic feature was the presence of a small knob or hornlet on the medial edge of the horn. In all other respects, most specimens referred to *Diploclonus* are very similar to specimens of *Brontops*. Osborn (1929) provisionally accepted *Diploclonus* as valid, but was doubtful as to the significance of the hornlet and questioned its use as a diagnostic character at the generic level. Osborn also noted that some of the skulls that he referred to *Brontotherium* and *Menodus* exhibited this character as well.

In my opinion, the presence or absence of a hornlet alone is not an adequate diagnostic character for separating specimens at the generic level. Most of the specimens that have been referred to *Diploclonus* are, in fact, specimens of *Brontops*. This includes another almost complete skeleton in the Pratt Museum of Amherst College (ACM 327).

### *Menops* Marsh, 1887
(= *Allops* Marsh, 1887; = *Menodus, sensu* Osborn, 1929)
*Age*: Chadronian
*Type species*: *M. varians* Marsh, 1887
*Holotype of type species*: YPM 12060, a skull.

*Diagnosis*: Very large-sized (length $P^1$ to $M^3$ approximately 330 to 385 mm according to data published in Osborn, 1929) brontothere with four (or fewer) upper (*, 6) and lower incisors (*, 8) that are reduced and globular in shape; reduced canine (*, 5); no upper or lower diastema (*, 7); partially molarized premolars (*, 5); hypocone on $M^3$ (*, 7); massive zygomatic arch that is not thickened or expanded laterally; horns relatively large, laterally directed, and strongly trihedral in cross-section (A); and wide "saddle-shaped" cranial vertex (*, 7).

*Discussion*: The generic names *Menops* and *Allops* (type species: *A. serotinus*) were both proposed by Marsh (1887) in the same paper. *Menops* is discussed first, and as first

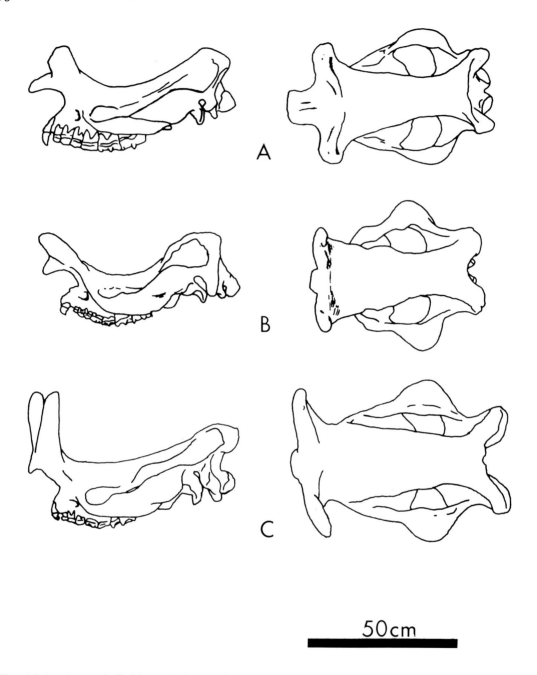

Fig. 25.2. Lateral (left) and dorsal (right) views of Oligocene brontothere skulls. A, *Menops*; B, *Brontops*; C, *Megacerops*. All illustrations after Osborn, 1929.

reviser, I assign it nomenclatural priority. Both genera are based on well-preserved skulls, and the type of *Allops serotinus* (USNM 4251) includes some postcranial material as well.

Osborn (1929) considered *Menops* to be a junior synonym of *Menodus* Pomel, 1849, but recognized *Allops* as a valid genus. He regarded *Menodus* and *Allops* as close relatives, however, and combined them into the subfamily Menodontinae. Because the holotype of the type species of *Menodus* (*M. giganteus*) had been lost (Osborn speculated that it was destroyed in the "great fire" of St. Louis), Osborn (1929) designated a skull (AMNH 505) as the neotype of the type species.

Osborn never actually saw the holotype of *Menodus giganteus*. The "great fire" in which Osborn presumed the specimen had been lost occurred in 1849, eight years before his birth. Osborn concluded from Leidy's published descriptions, measurements, and illustrations that the type jaw fragment of *Menodus giganteus* represented the same genus as YPM 2010, a skull with trihedral horns that Marsh (1874) had named *Brontotherium ingens* and that Osborn (1902) had referred to *Titanotherium*. Under Osborn's direction, a model of the lower jaw fragment of *Menodus giganteus* was sculpted using Leidy's data and figures. This model was then compared to the skulls of brontotheres with trihedral horns until one was found that "fit very well the lower teeth of the type" (Osborn, 1929). This skull (AMNH 505) was then designated the neotype of *Menodus giganteus*.

There is little question that the type of *Allops serotimus* represents the same genus as Osborn's neotype of *Menodus giganteus*. Figures 184, 375, and 399 in Osborn (1929) clearly show that the type of *A. serotinus* has the same trihedral horns as the neotype of *M. giganteus*. I regard this as an apomorphic condition that is diagnostic at the generic level. Osborn himself appeared doubtful concerning the validity of *Allops* as a genus distinct from *Menodus*, for although he specifically referred to *Allops* as a genus in some places (for example, 1929, p. 506), he referred to *Allops* as a subgenus elsewhere (1929, p. 537). Simpson, in his classification of mammals (1945), appears to have regarded the genera as synonymous.

The type of *Menops varians* also appears to have the same diagnostic horn morphology as *Allops serotinus*. I have not had the opportunity to examine the type skull myself, but Osborn clearly stated that the horn is trihedral in cross-section although it is crushed downward so as to "decrease the acuteness of the trihedral section" (Osborn, 1929). I therefore regard *Menops* and *Allops* as synonymous, both representing the same genus as Osborn's neotype of *Menodus*.

*Menops* and *Allops* were clearly based upon diagnostic material. I am uncertain, however, whether the type jaw fragment of *Menodus giganteus* was diagnosable. *Menodus* may have been a *nomen dubium*. Although Osborn's neotype of *M. giganteus* is clearly diagnostic, given the dubious methods by which he determined that it represented the type species of *Menodus*, I am inclined to ignore Osborn's neotype in favor of the holotype of *Menops varians*. I believe that *Menops* is a more satisfactory nomenclatural term than *Menodus* because *Menops* was based on more complete type material, the holotype of *Menops varians* is still available for study, and the name *Menops* was established in the literature long before Osborn designated a neotype for *Menodus*.

Clark and Beerbower (1967) contended that the brontotheres recognized here as *Brontops*, *Menops*, and *Megacerops* were in fact members of a single, highly variable species, *Menodus giganteus*. This conclusion was based largely on a consideration of four horn cores found within thirty feet of one another in a single "titanothere graveyard" deposit one foot thick. Clark and Beerbower argued that it is difficult to imagine that each horn core represents a distinct species, and felt it was more appropriate to regard them all as pertaining to a single species population, perhaps a local interbreeding unit. This conclusion somewhat parallels that of Osborn (1896), who suggested that all Chadronian brontotheres were members of a single genus, *Titanotherium*. Six years later, however, Osborn (1902) reversed this opinion and recognized four valid genera.

My own observations of numerous complete skulls suggests that there is too much variability in Chadronian brontotheres to be encompassed by a single genus, much less a single species. Skulls, with few exceptions, seem to fall into three distinct categories defined primarily by the morphology of the horns and zygomatic arches. In the present paper, I have recognized each of these categories as distinct genera.

Although I have not examined the materials described by Clark and Beerbower, two of the horn cores with triangular cross-sections appear to represent the genus *Menops*. One, which is long and flattened, probably represents *Megacerops*. The fourth, which is very short and ovoid in cross-section could represent *Brontops* or *Megacerops*, depending upon how much of the horn core is missing.

*Megacerops* Leidy, 1870a

(= "*Megaceratops*" Cope, 1873; = *Brontotherium* Marsh, 1873; = *Titanops* Marsh, 1887)

*Age*: Chadronian

*Type species*: *M. coloradensis* Leidy, 1870a

*Holotype of type species*: ANSP 13362, fragmentary horns.

*Diagnosis*: Very large-sized (length $P^1$ to $M^3$ approximately 300 to 355 mm according to data published in Osborn, 1929) brontothere with four (or fewer) upper (*, 6) and lower incisors (*, 8) that are reduced and have a globular shape; reduced canine (*, 5); no diastema (*, 7); partially molarized premolars (*, 5); hypocone on $M^3$ (*, 7); zygomatic arches that are broadly expanded and wing-shaped (A); horns that are long, laterally directed and roughly circular in cross-section (A); and a wide, "saddle-shaped" cranial vertex (*, 7).

*Discussion*: In 1870 Leidy described the fragmentary horns and nasals that comprise the holotype of *Megacerops coloradensis* (Leidy, 1870a). These were the first non-dental brontothere remains to be collected and at first there was some question as to what kind of animal the holotype represented. Leidy suspected that it might pertain to the *Titanotherium* remains that were being discovered at the time in the White River badlands, but also compared the holotype to the artiodactyls *Megalomeryx* and *Sivatherium*.

In 1873, Cope described a partial skull (AMNH 6348) that he believed to be a new species of Leidy's *Megacerops*. Objecting to the name *Megacerops* on etymologic grounds, Cope emended it to "*Megaceratops*" and named the new species "*Megaceratops*" *acer*. As Osborn (1929) pointed out, however, according to the rules of zoological nomenclature, the name "*Megaceratops*" cannot be accepted as valid and *Mergacerops* must be used instead.

In 1873, Marsh named a new genus, *Brontotherium* (type species *B. gigas*) based upon the remains of three individuals. In 1929 Osborn selected from these co-types a lower jaw (YPM 12009) as a lectotype. Fourteen years later, Marsh (1877) named yet another genus, *Titanops* (type species *T. curtus*) the holotype of which was a complete skull (YPM 12013). In addition to the type species Marsh also recognized a second species that he named *Titanops elatus* (holotype YPM 12061, a skull and jaw). In 1902 Osborn observed that the jaw of *Titanops elatus* was identical to the lectotype jaw of *Brontotherium gigas* and synonymized the genera. Osborn upheld this synonymy in 1929.

Comparing typical skulls that Osborn (1929) referred to *Megacerops* (for example, Cope's "*Megacerartops*" *acer*) and *Brontotherium* (Marsh's *Titanops curtus* and *Titanops elatus*) I find only one significant difference between them. Specimens referred to *Brontotherium* have a crest that connects the horns medially, while specimens referred to *Megacerops* do not. I do not believe that this single character provides sufficient grounds for separating the specimens on the generic level, and so regard these materials as congeneric. I am not certain at present whether the crest has any diagnostic value at the species level or whether it varies intraspecifically.

Osborn (1929) reported that he could not locate the holotype of *Megacerops coloradensis* and believed that the specimen was lost. In fact, the holotype is preserved in the Academy of Natural Sciences of Philadelphia and now has the catalog number ANSP 13362. The horns of the holotype of *Megacerops coloradensis* are relatively thick and short and are similar in size and morphology to the horns of some skulls referred to *Megacerops*, but also to AMNH 1476 (holotype of *Brontops bicornutus*) and AMNH 2251 (referred to *Brontops dispar*). It is uncertain, therefore, that the holotype of *Megacerops coloradensis* represents the same genus as Cope's "*Megaceratops*" *acer* and Marsh's *Titanops curtus*.

If *Megacerops coloradensis* is not congeneric with "*Megaceratops*" *acer* and *Titanops curtus*, then the name *Brontotherium* might be applied to those specimens instead. I am not certain, however, that *Brontotherium* will prove to be a valid taxonomic term. Although I have not yet examined the lectotype jaw of *Brontotherium gigas*, I suspect that it may be difficult to distinguish from jaws of *Brontops*. *Brontotherium* may, therefore, be a *nomen dubium*. If both *Megacerops* and *Brontotherium* should prove to be invalid names, then the next available name for these specimens would be *Titanops* Marsh, 1887.

## Higher categories of brontotheres and brontothere phylogeny

Figure 25.3 presents a hypothesis of relationships for selected brontothere genera. This is a preliminary cladogram I am presenting in the hope that others will find it useful and test its conclusions.

The first comprehensive treatment of higher brontothere categories was presented by Osborn in 1914. In this paper Osborn recognized eleven subfamilies that he defined as follows: Lambdotheriinae (*Lambdotherium*), Eotitanopinae (*Eotitanops*), Palaeosyopinae (*Palaeosyops* and *Limnohyops*), Telmatheriinae (*Telmatherium* and *Sthenodectes*), Manteoceratinae (*Manteoceras* and *Protitanotherium*), Dolichorhinae (*Dolichorhinus, Mesatirhinus, Metarhinus, Rhadinorhinus,* and *Sphenocoelus*), Diplacodontinae (*Diplacodon*), Menodontinae (*Menodus* and *Allops*), Brontopinae (*Brontops* and *Diploclonus*), Megaceropinae (*Megacerops*) and Brontotheriinae (*Brontotherium*). Osborn admitted to a liberal use of subfamily divisions and stated that a "more conservative usage would divide the titanotheres into four subfamilies only" (Osborn, 1914). Unfortunately, Osborn did not specify which four subfamily names would be appropriate should a more conservative approach be adopted.

In 1929 Osborn presented essentially the same system that he had published in 1914, but placed *Rhadinorhinus* in its own subfamily (the Rhadinorhininae) and added *Brachydiastematherium, Eometarhinus, Eotitanotherium,* and *Teleodus* to the Manteoceratinae, Dolichorhininae (emended from Dolichorhinae), Diplacodontinae, and Brontopinae, respectively. Osborn regarded the Manteoceratinae as ancestral to the Brontopinae and suggested, but with less certainty, that the Eotitanopinae and Rhadinorhininae were ancestral to the Palaeosyopinae and Megaceropinae respectively. Osborn also suggested that the Diplacodontinae might have given rise to both the Menodontinae and Brontotheriinae.

In their revision of the Mongolian brontotheres, Granger and Gregory (1943) created three new subfamily names, two of which were endemic to Asia (although they regarded these Asian subfamilies as descendants of North American subfamilies). Granger and Gregory placed *Metatelmatherium* and two purely Asian genera, which they regarded as possible

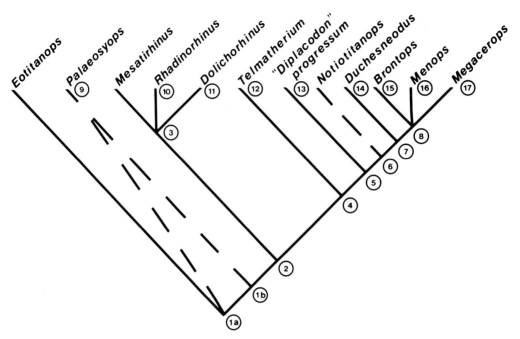

Fig. 25.3. Hypothesis of relationship for selected brontothere genera. Key to characters: 1a (family Brontotheriidae), bunoselenodont molars, ?face relatively short and skull posterior to orbit elongated (see 1b); 1b, ?face relatively short and skull posterior to orbit elongated; 2, shear more emphasized on ectoloph of molars; 3, (subfamily Dolichorhininae), presence of suborbital protuberance, reduced canine; 4 (subfamily Telmatheriinae), frontonasal horn-like swelling or prominence over facial concavity, ? sagittal crest reduced; 5 (Diplacodonts *sensu lato*), frontonasal swelling developed into a true horn, cranial vertex widened, canines reduced, premolars partially molarized; 6, four upper incisors that are reduced and globular in shape (still six lower incisors?); 7 ("Eubrontotheres"), cranial vertex wide and "saddle-shaped", diastema lost, hypocone well developed and consistently present on $M^3$; 8, four (or fewer) lower incisors; 9 (genus *Palaeosyops*), strongly brachycephalic skull, pronounced convexity or dome in region of frontoparietal boundary, massive zygomatic arches that are strongly curved, strongly curved nasals that are tapered distally, extreme reduction or loss of upper diastema; 10 (genus *Rhadinorhinus*), nasals short and strongly tapered distally, pronounced rim around anterior edge of orbit; 11 (genus *Dolichorhinus*), hyperdolichocephalic skull, very large suborbital protuberance that is divided into a small anterior flange and a larger posterior flange, short and angular frontonasal horn over orbit, wide and rounded cranial vertex with loss of saggital crest; 12 (genus *Telmatherium*), prominent pit-like fossa in center of saggital crest; 13 (*"Diplacodon" progressum*), relatively large bulbous horns that are roughly circular in cross-section; 14 (genus *Duchesneodus*), prominent convexity or dome in region of frontoparietal boundary, moderately expanded zygomatic arches that are apparently wing-shaped in males; 15 (genus *Brontops*), massive zygomatic arches that are swelled posteriorly, relatively large horns that are directed anteriorly and are circular to elliptical in cross-section; 16 (genus *Menops*), moderate sized horns that are trihedral in cross-section; 17 (genus *Megacerops*), broadly expanded zygomatic arches with a distinct wing-shape, long to very long horns that are directed laterally and are roughly circular to elliptical in cross-section. The genera *Eotitanops*, *Mesatirhinus*, and *Notiotitanops* cannot at present be defined in terms of unique derived characters because they are primitive within higher brontothere groupings. In addition, *Eotitanops* and *Notiotitanops* are poorly represented in the fossil record. Valid genera not included in the cladogram: *Metarhinus*, *Metatelmatherium*, *Sthenodectes*, *Protitanotherium*, *Eotitanotherium* (? =*Diplacodon*) and *Protitanops*.

descendants of the North American species *Telmatherium cultridens*, in the new subfamily Metatelmatheriinae.

In his classification of mammals, Simpson (1945) essentially adopted the systematics of both Osborn, and Granger and Gregory, but reduced the number of subfamilies by synonymizing ancestral subfamilies with their descendants. Thus Simpson combined the Eotitanopinae and Palaeosyopinae into the Palaeosyopinae, and the Manteoceratinae and Brontopinae into the Brontopinae. Although Simpson largely accepted the systematic schemes of previous authors, he did on occasion exercise his own judgment. For example, although Osborn thought that both the Menodontinae and Brontotheriinae might be derived from the Diplacodontinae, Simpson synonymized only the Menodontinae with the Diplacodontinae. Simpson regarded the Brontotheriinae as a valid subfamily separate from the Menodontinae and synonymized the Megaceropinae with it. Simpson also did not accept Osborn's view that the Rhadinorhininae were ancestral to the Megaceropinae. Instead, Simpson placed *Rhadinorhinus* back in the Dolichorhininae.

With one exception (the Embolotheriinae), Simpson regarded all Asian subfamilies recognized by Granger and Gregory as junior synonyms of North American subfamilies. The Metatelmatheriinae were recognized as a junior synonym of the Telmatheriinae, apparently because Granger and Gregory regarded *Metatelmatherium* as a descendant of *Telmatherium* (see Granger and Gregory, 1943, Fig. 4).

Osborn's systematics were obviously flawed by an overabundant use of the subfamily level of classification. Five out of the twelve subfamilies that he recognized in 1929 were monogeneric. With the synonymies introduced in the present paper, another three of these subfamilies (Palaeosyopinae, Menodontinae, and Brontopinae) must also be regarded as monogeneric and a fourth (the Diplacodontinae) as probably monogeneric. Only three (Telmatheriinae, Manteoceratinae, and Dolichorhininae) of Osborn's twelve subfamilies are reliably known to include more than a single genus.

To Simpson's credit he reduced the number of subfamilies from twelve to eight, but because he followed the general systematic schemes of Osborn and Granger and Gregory, most of his subfamilies are not valid. The Palaeosyopinae, Telmatheriinae, and Dolichorhininae are probably not monophyletic as he defined them and the Brontopinae, Menodontinae, and Brontotheriinae are essentially synonymous (at least with regard to the North American genera that Simpson included under these names).

In this paper I recognize only two North American subfamilies as valid: the Telmatheriinae (Fig. 25.3, node 4) and Dolichorhininae (Fig. 25.3, node 3). Another subfamily, the Metatelmatheriinae, might also be recognized as valid if my suspicion is correct that *Metatelmatherium* is related to *Sthenodectes*. The Metatelmatheriinae would then join the Telmatheriinae and Dolichorhininae at node 2 in an unresolved trichotomy. *Eotitanops* and *Palaeosyops* are primitive sister genera to these subfamilies and in my opinion should not be assigned individual subfamily names.

The Telmatheriinae as defined by Osborn and Simpson would now include the genera *Telmatherium*, *Metatelmatherium*, and *Sthenodectes*, and as such it is probably not a monophyletic group. The Manteoceratinae as defined by Osborn would be monophyletic if it were extended to include all of the "horned brontotheres" of the Uintan, Duchesnean, and Chadronian. Unfortunately, *Manteoceras* is regarded here as a junior synonym of *Telmatherium*, so the name Manteoceratinae is not appropriate for the subfamily. I have instead adopted the name Telmatheriinae, which was named in the same paper as the Manteoceratinae (Osborn, 1914) and is equally available as a nomenclatural term. I now define the Telmatheriinae, however, so as to exclude *Metatelmatherium* and

*Sthenodectes* and to include *Telmatherium* and all "horned" brontotheres.

The Diplacodontinae (*sensu* Osborn, 1914) would be valid if defined to include all brontotheres with true horns and partially molarized premolars, thus making it a subset of the Telmatheriinae (node 5, diplacodonts *sensu lato*). The Dolichorhininae are monophyletic as defined by Riggs (1912) and Osborn (1914) but not by Osborn (1929) and Simpson (1945), who incorporated *Eometarhinus* (=*Palaeosyops*) into the family.

In an abstract, Schoch and Lucas (1985) divided the Brontotheriidae into two main subgroups; the paleobrontotheres and eubrontotheres. Paleobrontotheres were defined as the typical brontotheres of the middle Eocene and included all genera from *Palaeosyops* (presumably including *Eotitanops*) to *Diplacodon*. The eubrontotheres were meant to include *Duchesneodus* and the typical Oligocene horned genera. The paleobrontotheres are a paraphyletic assemblage of genera and not a natural group. The eubrontotheres are monophyletic (node 7) and may be regarded as a subset of the diplacodonts *sensu lato*.

According to Osborn (1929), large horns evolved independently in three different brontothere lineages. Over the past few decades this view has become firmly entrenched in both the primary and secondary literature. My own belief is that relatively large horns evolved only once in brontothere phylogeny (node 5), although the genus *Dolichorhinus* did evolve a short horn independently.

## Acknowledgments

I would like to thank Margery Coombs and Spencer Lucas for their thoughtful comments on this manuscript. I thank W. Alden Hamblin and Dee Wallace for information about the location of the *Duchesneodus* quarry. Partial funding for the research presented here was provided by a grant from the Theodore Roosevelt Memorial Fund of the American Museum of Natural History and the Rose M. Louer Fellowship Fund of the Field Museum of Natural History.

## Bibliography

Clark, J., and Beerbower, J.R. (1967): Geology, paleontology, and paleoecology of the Chadron Formation. -In: Clark, J., Beerbower, J.R., and Kietzke, K.K.: Oligocene sedimentation, stratigraphy, paleoecology, and paleoclimatology in the Big Badlands of South Dakota. - *Fieldiana Geol. Mem.*, 5: 21-74.

Coombs, M. C. (1979): *Tylocephalonyx*, a new genus of North American dome-skulled chalicotheres (Mammalia, Perissodactyla). -*Bull. Amer. Mus. Nat. Hist.*, 164 (1): 1-64

Cope, E.D. (1873): Second notice of extinct Vertebrata from the Tertiary of the plains. -*Paleont. Bull.* 15:1-6

Cope, E.D. (1879): On the extinct species of Rhinoceridae of North America and their allies. -*Bull. U.S. Geol. and Geog. Surv. Terr.*, 5: 227-237

Cope, E.D. (1880): The badlands of the Wind River and their fauna. -*Amer. Naturalist*, 14 (10): 745-748.

Douglass, E. (1909): Preliminary descriptions of some new titanotheres from the Uinta deposits. -*Ann. Carn. Mus.*, 6(2): 304-313.

Earle, C. (1891): On a new species of *Palaeosyops, Palaeosyops megarhinus* sp. nov. -*Amer. Naturalist,* 25(289): 45-47.

Earle, C. (1892): A memoir upon the genus *Palaeosyops* Leidy and its allies. -*J. Acad. Nat. Sci. Phil.*, 2d ser., 9(6): 267-388.

Gazin, C.L., and Sullivan, J.M. (1942): A new titanothere from the Eocene of Mississippi, with notes on the correlation between the marine Eocene of the Gulf Coastal Plain and continental Eocene of the Rocky Mountain region. -*Smithson. Misc. Coll.*, 101(13):1-13

Granger, W., and Gregory, W.K. (1938): Addendum. A new titanothere genus from the upper Eocene of Mongolia and North America. -In: Colbert, E.H. (ed.): Fossil mammals from Burma in the American

Museum of Natural History. -*Bull. Amer. Mus. Nat. Hist.*, 74(6): 435-436.

Granger, W., and Gregory, W.K. (1943): A revision of the Mongolian titanotheres. -*Bull. Amer. Mus. Nat. Hist*, 80(10): 349-389.

Gregory, W. K. (1912): Note on the upper Eocene titanotheroid *Telmatherium? incisivum* Douglass, from the Uinta Basin. -*Science*, new ser., 35(901): 545

Hatcher, J. B. (1895): On a new species of *Diplacodon*, with a discussion of the relations of that genus to *Telmatotherium*. -*Amer. Naturalist*, 29:1084-1090

Hay, O. P. (1902): Bibliography and catalogue of the fossil Vertebrata of North America. -*Bull. U.S. Geol. Surv.* 179:1-868.

Hooker, J. J. (1984): A primitive ceratomorph (Perissodactyla, Mammalia) from the early Tertiary of Europe. -*Zool. J. Linn. Soc.*, 82:229-244.

Leidy, J. (1870a): Remarks on *Megacerops coloradensis*. -*Proc. Acad. Nat. Sci. Phila.*, 22:1-2.

Leidy, J. (1870b): On fossils from Church Buttes, Wyoming Territory. -*Proc. Acad. Nat. Sci. Phila.*, 22: 113-114.

Leidy, J. (1872): Remarks on fossil mammals of Wyoming. -*Proc. Acad. Nat. Sci. Phila.*, 24: 240-242.

Lucas, S.G., and Schoch, R.M. (1982): *Duchesneodus*, a new name for some titanotheres (Perissodactyla, Brontotheriidae) from the Late Eocene of western North America. -*J. Paleont.*, 56(4): 1018-1023.

Lucas, S.G., Schoch, R.M., Manning, E., and Tsentas, C. (1981): The Eocene biostratigraphy of New Mexico. -*Bull. Geol. Soc. Amer.*, Part I, 92: 951-967.

Marsh, O.C. (1872): Preliminary description of new Tertiary mammals, Part I. - *Amer. J. Sci.*, 3d ser., 4: 122-130.

Marsh, O.C. (1873): Notice of new Tertiary mammals. -*Amer. J. Sci.*, 3d ser., 5: 407-411, 485-488.

Marsh, O.C. (1874): On the structure and affinities of the Brontotheridae. -*Amer. J. Sci.*, 3d ser., 7:81-88.

Marsh, O.C. (1875): Notice of new Tertiary mammals, IV. -*Amer. J. Sci.*, 3d ser., 9:239-250.

Marsh, O.C. (1880): *List of genera established by Professor O.C. Marsh, 1862-1879.* New Haven, Conn.

Marsh, O.C. (1887): Notice of new fossil mammals. -*Amer. J. Sci.*, 3d ser., 34(35): 323-331.

Marsh, O.C. (1890): Notice of new Tertiary mammals. -*Amer. J. Sci.*, 3d ser., 39:523-525.

Matthew, W.D. (1899): A provisional classification of the fresh-water Tertiary of the west. -*Bull. Amer. Mus. Nat. Hist.*, 12(3): 19-75.

Osborn, H.F. (1890): The Mammalia of the Uinta formation, Part 3, the Perissodactyla. -*Trans. Amer. Phil. Soc.*, new ser., 16(9):461-572.

Osborn, H.F. (1895): Fossil mammals of the Uinta Basin, Expedition of 1894. -*Bull. Amer. Mus. Nat. Hist.*, 7(2): 71-105.

Osborn, H.F. (1896): The cranial evolution of *Titanotherium*. -*Bull. Amer. Mus. Nat. Hist.*, 8(9): 157-197.

Osborn, H.F. (1897): The Huerfano lake basin, southern Colorado, and its Wind River and Bridger fauna. -*Bull. Amer. Mus. Nat. Hist.*, 9(21): 247-258.

Osborn, H.F. (1902): The four phyla of Oligocene titanotheres. -*Bull. Amer. Mus. Nat. Hist.*, 16(8):91-109.

Osborn, H.F. (1907): Tertiary mammal horizons of North America. -*Bull. Amer. Mus. Nat. Hist.*, 23(11): 237-253.

Osborn, H.F. (1908): New or little known titanotheres from the Eocene and Oligocene. -*Bull. Amer. Mus. Nat. Hist.*, 24(32): 599-617.

Osborn, H.F. (1914): Recent results in the phylogeny of the titanotheres. -*Bull. Geol. Soc. Amer.*, 25: 403-405.

Osborn, H.F. (1919): New titanotheres from the Huerfano. -*Bull. Amer. Mus. Nat. Hist.*, 41(15): 557-569.

Osborn, H.F. (1929): The titanotheres of ancient Wyoming, Dakota, and Nebraska. -*U.S. Geol. Surv., Monograph* 55: 1-953.

Osborn, H.F., Scott, W.B., and Speir, F.

(1878): Paleontological report of the Princeton scientific expedition of 1887. - *E.M. Mus. Geol. and Arch. Princeton Coll. Contr.* (1): 1-106.

Osborn, H.F., and Wortman, J.L. (1892): Fossil mammals of the Wasatch and Wind River beds, collection of 1891. - *Bull. Amer. Mus. Nat. Hist.*, 4(1) art. 11: 81-147.

Peterson, O.A. (1914a): A new titanothere from the Uinta Eocene. -*Ann. Carn. Mus.*, 9(1-2):29-52.

Peterson, O.A. (1914b): Correction of generic name. -*Ann. Carn. Mus.*, 9: 220.

Peterson, O.A. (1931): A new species of the genus *Teleodus* from the upper Uinta of northeastern Utah. -*Ann. Carn. Mus.*, 20(3-4): 307-312.

Peterson, O.A. (1934): New titanotheres from the Uinta Eocene in Utah. -*Ann. Carn. Mus.*, 22 (3-4): 351- 361.

Peterson, O.A., and Kay, J. L. (1931): The upper Uinta Formation of northeastern Utah. -*Ann. Carn. Mus.*, 20 (3-4): 293-306.

Pomel, A. (1849): Description d'un os maxillaire fossile de *Palaeotherium* par Hiram Prout, Am. J. Sci. and Arts by Silliman's [sic] and J. Dana, 2$^e$ série, vol 3, no. 8, p. 248. -*Bibliotèque Univ. Genève (suppl.), Arch. Sci. Phys. Nat.*, 10:73-75.

Radinsky, L.B. (1964): *Paleomoropus*, a new early Eocene chalicothere (Mammalia, Perissodactyla), and a revision of Eocene chalicotheres. -*Amer. Mus. Nov.*, 2179: 1-28.

Ride, W. D. L., Sabrosky, C. W., Bernardi, G., and Melville, R. V. (1985): *International Code of Zoological Nomenclature* (3rd ed.).- London (International Trust for Zoological Nomenclature).

Riggs, E.S. (1912): New or little known titanotheres from the lower Uinta formations. -*Field Mus. Pub. 159, Geol. Ser.*, 4(2): 17-41.

Robinson, P. (1966): Fossil Mammalia of the Huerfano formation, Eocene, of Colorado. -*Bull. Peabody Mus. Nat. Hist.*, 21: 1-95.

Romer, A. S. (1966): *Vertebrate Paleontology*, 3d ed. - Chicago (Univ. of Chicago Press).

Savage, D.E., and Downs, T. (1954): Cenozoic land and life of southern California. -*Calif. Div. Mines Bull.*, 170: 43-58.

Schoch, R. M. (1983): Relationships of the earliest perissodactyls (Mammalia, Eutheria). -*Geol. Soc. Amer., Abst. Prog.*, 15: 144 (abstract).

Schoch, R. M. (1984): Two unusual specimens of *Helaletes* in the Yale Peabody Museum collections, and some comments on the ancestry of the Tapiridae (Perissodactyla, Mammalia). -*Postilla, Yale Peabody Mus.*,193:1-20.

Schoch, R.M., and Lucas, S.G. (1985): The Brontotheriidae, a group of Eocene and Oligocene perissodactyls from North America, Asia, and Eastern Europe. -In: Fuller, W.A., Nietfeld, M.T., and Harris, M.A. (eds.): *Abstracts of papers and posters, Fourth International Theriological Congress, Edmonton, Alberta, Canada*: abstract 0558.

Scott, W.B. (1941): The mammalian fauna of the White River Oligocene. Part 5, Perissodactyla. -*Trans. Amer. Phil. Soc.*, 28: 747-980.

Scott, W.B. (1945): The Mammalia of the Duchesne River Oligocene. -*Trans. Amer. Phil. Soc.*, (new series) 34(3): 209-253.

Simpson, G.G. (1945): The principles of classification and a classification of mammals. -*Bull. Amer. Mus. Nat. Hist.*, 85: 1-350.

Stock, C. (1936): Titanotheres from the Titus Canyon Formation, California. -*Proc. Nat. Acad. Sci.*, 229 (11): 656-661.

Wallace, S.M. (1980): A revision of North American early Eocene Brontotheriidae (Mammalia, Perissodactyla). -Univ. of Colorado (Master's thesis), 1-157.

Wilson, J.A. (1977): Early Tertiary vertebrate faunas, Big Bend area, Trans-Pecos Texas: Brontotheriidae. -*Pearce-Sellards Series, Texas Mem. Mus.*, 25: 1-17.

Wood, H.E. (1937): Perissodactyl suborders. -*J. Mamm.*, 18(1): 106.

# 26. EUROPEAN BRONTOTHERES

## SPENCER G. LUCAS and ROBERT M. SCHOCH

Three taxa of brontotheres have been named from Europe, but only two of these, *Diplacodon* (=*Brachydiastematherium*) *transilvanicum* Böckh and Maty, 1876, from the Eocene of Romania, and ?*Menodus rumelicus* Toula, 1892, from the Eocene of Bulgaria, are apparently based on specimens actually from Europe. *Titanotherium bohemicum* Kiernik, 1913, is a *nomem dubium* and is arguably based on North American specimens imported to Europe.

## Discussion

Apparently only two *bona fide* brontothere occurrences are known from Europe, although three taxa have been described. *Brachydiastematherium transilvanicum* Böckh and Maty, 1876 (in Böckh, 1876), from Andrashaza, Romania is known only from its holotype, an incomplete lower jaw (Figs. 26.1A-B; Böckh, 1876, pls. 17, 18; Osborn, 1929, Figs. 100, 315C). Although Osborn (1929) noted the close similarity of *Brachydiastematherium* and North American "*Protitanotherium*" and considered the European genus valid, Lucas (1983) argued that *Brachydiastematherium* is a junior subjective synonym of *Diplacodon* (= *Protitanotherium*). We uphold this synonymy and note that the only significant difference between the holotype of *B. transilvanicum* and specimens of *Diplacodon* is the peculiar cuspidate cingula on the lower incisors and canine of the European form (Fig. 26.1A). Recognizing *Brachydiastematherium* as a synonym of *Diplacodon* supports the notion that the Andrashaza locality, which is also the type locality of the rhinocerotoid *Prohyracodon orientalis* Koch, 1897, is middle Eocene (Uintan equivalent: Osborn, 1929, p. 382).

The two other putative occurrences of brontotheres in Europe have received the names *Menodus? rumelicus* Toula, 1892, and *Titanotherium bohemicum* Kiernik, 1913. As elaborated upon later, there is reason to believe that the latter of these taxa is based on specimens imported from North America and mistakenly assumed to have come from Europe (Osborn, 1929). However, workers subsequent to Osborn (1929) have ignored both of these taxa (e.g., Heissig, 1979; Yanovskaya, 1980; Russell et al., 1982).

The lectotype of *Menodus? rumelicus* designated by Osborn (1929, p. 230) is a right dentary fragment with M3 (Fig. 26.1C-D; Toula, 1892, pl. 1; Osborn, 1929, Figs. 193, 795B) reputedly found near Burgas, Bulgaria. Additional material is confined to lower teeth and jaw fragments (Fig. 26.1E, 2C-D; Toula, 1892, 1896; Osborn, 1929, Figs. 463, 464A, 795A). Relative to the supposed locality from which these specimens originated, Osborn (1929, p. 230) wrote:

> On account of the extreme rarity of titanotheres in Europe it seems important to note the published evidence concerning the provenience of the type and referred specimens of this species. According to Toula the specimens were received from his friend G. N. Zlatarski in Sofia. Toula does not state that Zlatarski himself collected the specimens. He states only that they must have come from near the railroad at Kajali, from the great heaps of material which had been dug up in search of usable rubble

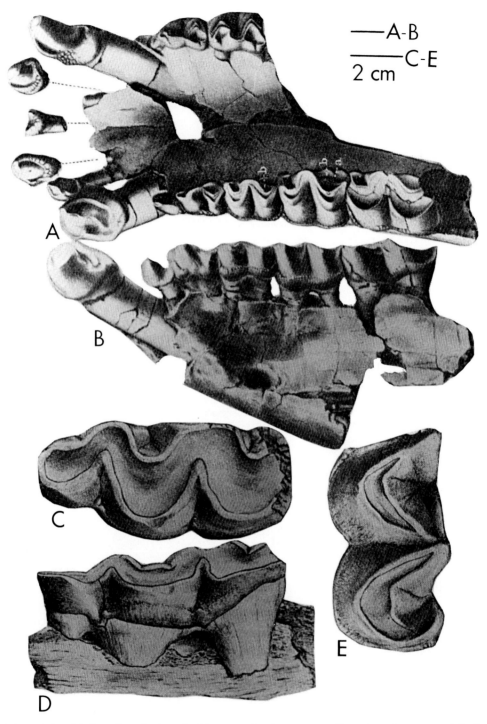

Fig. 26.1. A-B. *Diplacodon* (=*Brachydiastematherium transilvanicum*), part of holotype incomplete lower jaw, occlusal (A) and left lateral (B) views (from Böckh, 1876, pl. 17, Fig. 1, and pl. 18, Fig. 1). C-E, ?*Menodus rumelicus*, lectotype right dentary fragment with $M_3$, occlusal (C) and right lateral (D) views, and lectoparatype right $M_2$ in occlusal view (E) (from Osborn, 1929, Fig. 193).

("tauglichem Schotter"), and that these "Schottermassen" should correspond at best with that isolated remnant of a formation at Lidscha, northwest of Burgas, of which he had already spoken in his first report on the geology of the eastern Balkans. He writes: "I have referred to these "Schotter" as Belvedereschotter, and I believe, from the condition of preservation of the specimens from Kajali, and especially from the rusty sand grains still adhering to them, that they must be referred to the same kind of rock." Besides the specimens of titanotheres Toula records a lower molar and a canine of a "middle-sized rhinoceros" from the same locality. Later he received from the same locality, this also from Zlatarski, a fragment of the lower jaw of a titanothere that included the symphyseal region (Toula, 1896.1, pp. 922-924). But Toula has not dispoved the possibility that these specimens may have been imported from America, perhaps by laborers returning home from the western United States.

At this point it seems impossible to make a definitive statement as to the provenience of the material referred to *Menodus? rumelicus* without further study of the original material, its adhering matrix, and further examination of the outcrops in the Burgas region.

Osborn (1929) tentatively assigned *Menodus? rumelicus* to *Brontotherium*. Due to the undiagnostic nature of the material upon which this taxon is based, we suggest that perhaps the name *Menodus? rumelicus* is not certainly applicable to, or diagnostic of, any particular species-level taxon and is perhaps best considered a *nomen dubium*.

Nikolov and Heissig (1985) described and illustrated brontothere teeth found in the upper Eocene strata near the Black Sea coast of Bulgaria. They reassigned Toula's (1892) *Menodus (?) rumelicus* as *Sivatitanops (?) rumelicus*, and referred these teeth to this taxon. Clearly, there are autochthonous brontothere fossils in Bulgaria, but we are skeptical of the identification by Nikolov and Heissig (1985), since we regard the taxon *Menodus (?) rumelicus* as a *nomen dubium*.

Kiernik's (1913) *Titanotherium bohemicum* is based on a right dentary fragment with $M_3$ (Fig. 26.2A-B; Osborn, 1929, Figs. 206, 795C) reputedly collected from alluvium near Prague, Czechoslovakia. Concerning the supposed locality from which this specimen originated, Osborn (1929, p. 240) noted:

The specimen, a fragment of the lower jaw containing the third right lower molar, was received with a lot of fossils from the diluvium near Prague. It was supposed to have come from the lime pits of Podbaba, near Prague, and to have been sold by one of the workers in the lime pits to Herr Baumeister Kuchta (died 1910). He gave it, along with other prehistoric specimens, to Herr Rozanek, who in turn gave it to Herr Jira, who presented it to the Institute for Comparative Anatomy at Prague. After carefully considering the possibility that the specimen might have been of American provenience the author, Herr Kiernik, inclines rather to the view that it really came from Bohemia, although not from Prodbaba, but from the fresh-water Tertiary deposits of Tuchoritz (northwestern Bohemia). The well-known fauna of Tuchoritz is, however, of lower Miocene facies.

Osborn (1929, p. 241) continues:

According to Dr. W. K. Gregory, who has compared a cast of the type of *Titanotherium bohemicum* with various American titanotheres, the type specimen is closely similar to one in the American Museum of Natural History referred to *Menodus*

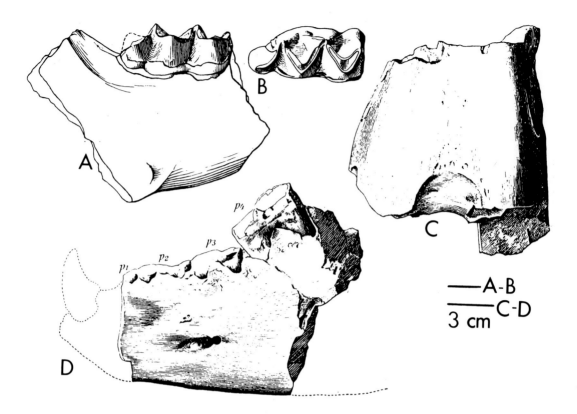

Fig. 26.2. A-B. *Titanotherium bohemicum*, holotype right dentary fragment with M3, right lateral (A) and occlusal (B) views (from Osborn, 1929, Fig. 795C). C-D. ?*Menodus rumelicus*, symphyseal fragment with incomplete P4 in ventral (C) and left lateral (D) views (from Osborn, 1929, Fig. 463).

*giganteus* (Am. Mus. 1007). It differs chiefly in the greater width of the anterior lobe of M3. It appears indeed to be specifically referable to *Menodus giganteus*, and it seems possible that it is in reality an American specimen which became mixed with the collection of fossils from Podbaba, near Prague.

In his monograph Osborn (1929) alternatively indicated that *Titanotherium bohemicum* is a synonym of *Menodus giganteus* or that it is a distinct species, *Menodus bohemicus*. We suggest that perhaps *Titanotherium bohemicum* is best considered a *nomen dubium* at the specific level, pending further study.

**Bibliography**

Bakalov, P., and Nikolov, I. (1962): *Fosilite na Bulgariya Tertsiern; Bozainitsi*. -Sofia (Bulg. Akad. na Naurite).

Böckh, J. (1876): *Brachydiastematherium transilvanicum* Böckh et Maty, ein neues Pachydermen-Genus aus den eocanen Schichten Siebenburgens. -*K. Geol. Anstalt Mitt. aus Jahrb.* 4: 125-150.

Heissig, K. (1979): Die hypothetische Rolle Sudosteuropas bei den Saugetierwanderungen im Eozän und Oligozän. -*N. Jb. Geol. Paläont. Mh.* 1979 (2): 83-96.

Kiernik, E. (1913): O nowym gatunku *Titanotherium*, ein neuer Titanotheriumfund in Europa. -*Internat. Acad. Sci. Carcovie Bull.* (B) 10: 1211-1225.

Koch, A. (1897): *Prohyracodon orientalis,* ein neues Ursaugethier aus den mitteleocänen Schichten Siebenburgens. - *Termesz. Fuz.* 20: 481-500.

Lucas, S. G. (1983): *Protitanotherium* (Mammalia, Perissodactyla) from the Eocene Baca Formation, west-central New Mexico. - *New Mex.J. Sci.,* 23: 39-47.

Nikolov, I., and Heissig, K. (1985): Fossile Säugetiere aus dem Obereozän und Unteroligozän Bulgariens und ihre Bedeutung für die Palaeogeographie. -*Mitt. Bayer. Staatsslg. Paläont. Hist. Geol.,* 25: 61-79.

Osborn, H.F. (1929): The titanotheres of ancient Wyoming, Dakota, and Nebraska. - *U. S. Geol. Surv. Monograph* 55: 1-953.

Russell, D. E., Hartenberger, J.-L., Pomerol, C., Sen, S., Schmidt-Kittler, N., and Vianey-Liaud, M. (1982): Mammals and stratigraphy: The Paleogene of Europe. -*Palaeovert. Mem. Extraordinaire,* 77 pp.

Toula, F. (1892): Zwie neuer Saugethierfundorte auf der Balkanhalbinsel. - *Akad. Wiss. Wien Sitzungsber.* 101 (1): 608-615.

Toula, F. (1896): Über einen neuen Rest von *Leptodon? (Titanotherium?) rumelicus* Toula spec. -*Deutsch. Geol. Gessell. Zeit.,* 48: 922-924.

Yanovskaya, N. M. (1980): Brontoterii Mongolii [The brontotheres of Mongolia]. - *Sovmestnaya Sovetsko-Mongol'skaya Paleontologicheskaya Ekspeditsaya, Trudy [The Joint Soviet-Mongolian Paleontological Expedition, Transactions]* 12: 1-219.

# 27. TAXONOMY OF *DUCHESNEODUS* (BRONTOTHERIIDAE) FROM THE LATE EOCENE OF NORTH AMERICA

## SPENCER G. LUCAS and ROBERT M. SCHOCH

Two valid species of the brontothere *Duchesneodus*--*D. uintensis* (Peterson, 1931) (=*Teleodus californicus* Stock, 1935; = *Teleodus thyboi* Bjork, 1967) and *D. primitivus* (Lambe, 1908) -- are known from the late Eocene (Duchesnean) of western North America. A quarry sample of 11 individuals of *D. uintensis* from northeastern Utah documents marked sexual dimorphism characterized by a neotenic female and a male with a dome-shaped cranial convexity, broad and robust zygomata, and large, blunt horns. *D. primitivus* from the Cypress Hills Formation, Saskatchewan, suggests the presence of a Duchesnean horizon slightly older than the Southfork local fauna.

### Introduction
The taxonomy of brontotheres has remained in chaos since Osborn (1929) last revised the family. Only now (Mader, this volume, Chapter 25) have significant inroads been made into clarifying the generic taxonomy of the Brontotheriidae. Here, we present a species-level revision of *Duchesneodus*, a distinctive brontothere genus from the late Eocene of western North America. Critical to this revision is a quarry sample of *D. uintensis* from northeastern Utah, which reveals significant sexual dimorphism and other intraspecific variation in this species. In this paper, BMNH = British Museum of Natural History, London; CM = Carnegie Museum of Natural History, Pittsburgh; F:AM = Frick Collection, American Museum of Natural History, New York; FMNH = Field Museum of Natural History, Chicago; L = maximum length of a tooth crown; LACM = Los Angeles County Museum of Natural History (mostly specimens formerly in the California Institute of Technology); NMC = National Museum of Canada, Ottawa; NMMNH = New Mexico Museum of Natural History, Albuquerque; SDSM = South Dakota School of Mines, Rapid City; TMM = Texas Memorial Museum, Austin; UNM = University of New Mexico, Albuquerque; UUVP = University of Utah, Salt Lake City; and W = maximum width of a tooth crown. Terminology of the landmarks on brontothere cheek-teeth follows Osborn (1929). Zygomacephalic index (Osborn, 1929) = (width across zygomatic arches/premaxillary-occipital condyle length) x 100. Cranial breadth index = (maximum width of skull roof above orbits/nasal-occiput length) x 100.

### Previous studies
The history of the taxonomy of *Duchesneodus* begins with Marsh's (1890, p. 524) introduction of the name *Teleodus* (type species = *T. avus*) for an incomplete lower jaw from Chadronian (early Oligocene) strata in South Dakota. Although Hatcher (1893, p. 217) and Osborn (1896, p. 194) considered *Teleodus* Marsh, 1890 synonymous with *Titanotherium* Leidy, 1852, and Osborn (1902, p. 99) considered it synonymous with *Megacerops* Leidy, 1870, Osborn (1929) upheld the validity of *Teleodus*. Osborn (1929, p. 482) also expanded the taxonomic concept *Teleodus* to include a lower jaw from Saskatchewan that Lambe (1908, p. 49) named *Megacerops primitivus*. Russell (1934, p. 58) upheld this expanded concept, and also assigned a partial cranium

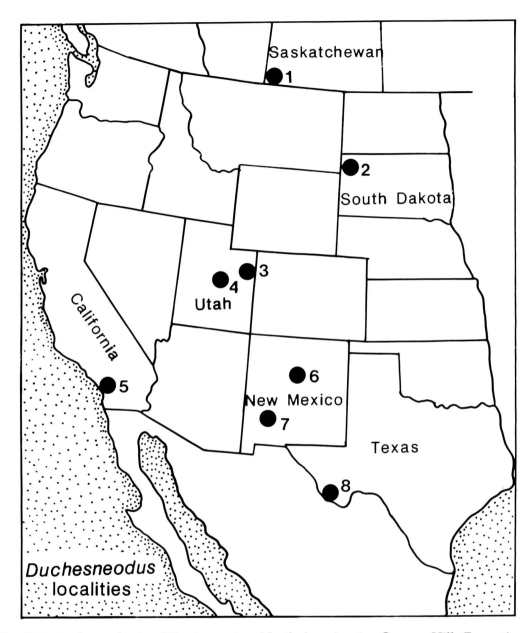

Fig. 27.1. *Duchesneodus* localities in western North America. 1 = Cypress Hills Formation; 2 = Slim Buttes Formation; 3 = Duchesne River Formation; 4 = Green River Formation; 5 = Sespe Formation; 6 = Galisteo Formation; 7 = Rubio Peak Formation; 8 = Devil's Graveyard Formation and Chambers Tuff.

from Saskatchewan to *T. primitivus*. Peterson (1931) further extended usage of *Teleodus* to specimens from the Duchesne River Formation in Utah for which he proposed the new species, *T. uintensis*. Scott (1945) provided additional documentation of *T. uintensis* from Utah, and Stock (1935, 1938) and Bjork (1967) named two more species of *Teleodus*: *T. californicus* from the Sespe Formation in California and *T. thyboi* from the Slim Buttes Formation in South Dakota. Specimens of *Teleodus* were also reported from the Green River Formation in Utah (Nelson et al., 1980), the Galisteo Formation in New Mexico (Lucas and Kues, 1979; Lucas, 1982) and the Chambers Tuff in Texas (Clark et al., 1967; Wilson, 1978).

Recently we (Lucas and Schoch, 1982) presented evidence that the holotype lower jaw of *T. avus* represents a different genus than the genus represented by all other specimens assigned to *Teleodus*. Thus, we coined the name *Duchesneodus* [type species = *D. uintensis* (Peterson, 1931)] for the genus that encompasses the species named by Lambe (1908), Peterson (1931), Stock (1935) and Bjork (1967). Indeed, as we (1982) indicated, all specimens that have been assigned to *Teleodus*, except the holotype of *T. avus*, pertain to *Duchesneodus*. Since 1982, specimens of *Duchesneodus* have also been reported from the Rubio Peak Formation of southwestern New Mexico (Lucas, 1983) and the Devil's Graveyard Formation in Texas (Wilson and Stevens, 1986).

**Taxonomy**

Order Perissodactyla Owen, 1848
Family Brontotheriidae Marsh, 1873
Genus *Duchesneodus* Lucas and Schoch, 1982

1890 non *Teleodus* Marsh, p. 524.
1908 *Megacerops* (in part): Lambe, p. 49, pl. 6, Figs. 4-5.
1929 *Teleodus* (in part): Osborn, p. 481, Figs. 204, 413A.
1931 *Teleodus* (in part): Peterson, p. 307, Figs. 1-2, pls. 12-15.
1935 *Teleodus* (in part): Stock, p. 458, pls. 1-2.
1967 *Teleodus* (in part): Bjork, p. 231, pl. 27, Figs. 2-3, Table 2.
1982 *Duchesneodus* Lucas and Schoch, p. 1022, Text-fig. 1A-B.

*Type species*: *D. uintensis* (Peterson, 1931) (= *Teleodus californicus* Stock, 1935; = *Teleodus thyboi* Bjork, 1967).

*Included species*: The type species and *D. primitivus* (Lambe, 1908).

*Distribution*: Latest Eocene (Duchesnean) of the western United States (South Dakota, Utah, New Mexico, Texas and California) and western Canada (Saskatchewan) (Fig. 27.1).

*Revised diagnosis*: Moderately large brontotheres ($LM^{1-3}$ is 150-215 mm); dental formula $I^2/_3 C^1/_1 P^{4-3}/_{4-3} M^3/_3$; incisors small and round (globular); canines short with round cross-sections; premolars molariform (upper premolars with distinct tetartacones, lower premolars with distinct entoconids); $M^3$ with distinct hypocone; labial faces of lower cheek teeth vertical or nearly vertical; labial cingula absent or very poorly developed on cheek teeth; nasals bear two small, blunt horns that are round to trihedral in cross-section; skull roof with distinct convexity ("dome") between occiput and orbits; skull dolicocephalic to mesaticephalic (zygomacephalic index of adult skulls is 57-80).

*Discussion*: The diagnosis of *Duchesneodus* is not stated in differential terms because of the current vagaries of brontothere taxonomy. We consider *Duchesneodus* to be the first "eubrontothere" (cf. Schoch and Lucas, 1985; Mader, this volume) in that it has globular incisors, a concave-upward skull roof, distinct horns, small canines and little or no postcanine diastemata, and $M^3$ hypocones. These features and large size distinguish *Duchesneodus* and other eubrontotheres from earlier, North American Eocene "paleobrontotheres:" *Palaeosyops*, *Telmatherium*, *Mesatirhinus*, and allied taxa. Relatively small size, the presence of three lower incisors, relatively small horns and a

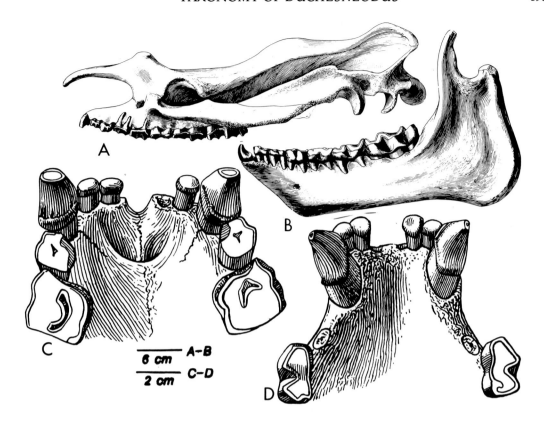

Fig. 27.2. Some craniodental morphology of *Duchesneodus uintensis* (from Peterson, 1931). A, Left lateral view of skull, CM 11759. B, Left lateral view of lower jaw, CM 11809. C, Occlusal view of upper anterior dentition, CM 11759. D, Occlusal view of lower anterior dentition, CM 11761.

convexity on the skull roof distinguish *Duchesneodus* from the more derived, North American early Oligocene eubrontotheres which many authors, following Scott (1940) and Clark *et al.* (1967), have termed *Menodus* but which Mader (1989) allocates to *Brontops*, *Menops*, and *Megacerops*. *Duchesneodus* thus strikes us as a very distinctive genus of brontotheres that lies at or near the base of the eubrontothere evolutionary radiation of the Oligocene.

*Duchesneodus uintensis* (Peterson, 1931)
1931 *Teleodus uintensis* Peterson, p. 308, Figs. 1-2, pls. 12-15.
1935 *Teleodus californicus* Stock, p. 458, pls. 1-2.
1938 *Teleodus* cf. *californicus*: Stock, p. 508, pl. 2.
1945 *Teleodus uintensis*: Scott, p. 240, pl. 7.
1967 *Teleodus thyboi* Bjork, p. 231, pl. 27, Figs. 2-3, Table 2.
1977 *Menodus bakeri* (in part): Wilson, p. 8, Table 4.
1979 *Teleodus uintensis*: Lucas and Kues, p. 227, Fig. 5.
1980 *Teleodus uintensis*: Nelson, Madsen, and Stokes, p. 130, Figs. 2-3, Table 1.
1982 *Teleodus* cf. *T uintensis*: Lucas, p. 24, Fig. 14, Table 2.
1982 *Duchesneodus uintensis*: Lucas and Schoch, p. 1022, Text-Fig. 1A-B.
1983 *Duchesneodus* sp.: Lucas, p. 190, Fig. 4, Table 1.
1986 ?*Duchesneodus* cf. *uintensis* (in part): Wilson and Stevens, p. 226, Figs. 2-3, Tables 3-4.

*Holotype*: CM 11809, lower jaw with

left C, $P_2$-$M_3$ and right $I_{1-2}$, C, $P_2$-$M_3$ (Peterson, 1931, pl. 13; Scott, 1945, pl. 7, Fig. 2; Lucas and Schoch, 1982, Text-Fig. 1A-B; this paper, Fig. 27.2B).

*Horizon and locality of holotype*: LaPoint Member of Duchesne River Formation, *Duchesneodus* quarry approximately 20 km west of Vernal, Uintah County, Utah (Peterson, 1931). Hereafter, this quarry is referred to as the *Duchesneodus* quarry (see Mader, this volume, Chapter 25, for additional information regarding this quarry).

*Referred specimens*: From the Slim Buttes Formation, Harding County, South Dakota, SDSM locality V582: SDSM 63689, skull with left $I^3$, C, $P^{1-2}$ roots, $P^3$-$M^3$ and right C, $P^4$-$M^2$ (holotype of *Teleodus thyboi* Bjork, 1967; illustrated by Bjork, 1967, pl. 27, Fig. 3); SDSM 59164, crushed right $M^2$ (identified as $M^3$ by Bjork, 1967, p. 231); SDSM 59167, right $dP^{3-4}$; SDSM 59170, right $dP^3$; SDSM 59171, left $P^4$; SDSM 59178, tooth fragments; SDSM 59180, tooth fragments; SDSM 59181, left dentary fragment with $M_1$.

From the Slim Buttes Formation, Harding County, South Dakota, SDSM locality V6244: SDSM 63690, maxillary-premaxillary fragment with left $I^{2-3}$, right $I^{2-3}$, C, $P^1$ alveolus and $P^{2-3}$ (Bjork, 1967, pl. 27, Fig. 2); SDSM 63691, left maxillary fragment with $P^4$-$M^2$; SDSM 63692, left $P^3$ and tooth fragments; SDSM 63695, left dentary fragment with incomplete $M_3$; SDSM 63715, canine fragment.

From the Slim Buttes Formation, NE1/4 SW1/4, sec. 11, T16N, R8E, Harding County, South Dakota: FMNH PM 26425, right $M^2$.

From the Duchesne River Formation at the same locality (i.e., *Duchesneodus* quarry) as the holotype of *D. uintensis*: CM 9958, anterior part of skull; CM 9959, posterior portion of skull; CM 9960, right maxillary with $M^{1-3}$; CM 11754, skull (Nelson *et al.*, 1980, Fig. 3); CM 11757, skull; CM 11758, juvenile skull (Fig. 27.4); CM 11759, skull (Peterson, 1931, Fig. 1, pl. 12; Scott, 1945, pl. 7, Fig. 1; this paper, Figs. 27.2A, C, 27.3D-F); CM 11760, juvenile skull; CM 11761, lower jaw (Peterson, 1931, Fig. 2, pls. 14-15; Scott, 1945, pl. 7, Fig. 2; this paper, Fig. 27.2D); CM 11766, skull, vertebrae and humerus; CM 11767, skull (Fig. 27.3A-C); CM 11815, skull; CM 11816, skull; CM 11821, lower jaw; CM 11822, lower jaw; CM 11866, skull (Scott, 1945, pl. 7, Fig. 1); postcrania from the quarry are catalogued as CM 9459, 9501, 11756, 11831, 11832, 11834 and 12106.

From the LaPoint Member of the Duchesne River Formation at the east wall of Halfway Hollow: FMNH PM 2240, palate.

From the Green River Formation at Spring City Cuesta, Sanpete County, Utah: UUVP 9501, incomplete skull with left $P^2$-$M^3$ and right $P^2$-$M^2$ plus impression of anterior part of palate with cast of $P^1$ (Nelson *et al.*, 1980).

From the Sespe Formation at Pearson Ranch, California Institute of Technology (CIT) locality 150: LACM 973, left maxillary with $P^{1-3}$; LACM 974, symphysis with right $P_{3-4}$ and $M_2$; LACM 975, right dentary fragment with $M_{1-2}$; LACM 1118, left dentary fragment with $P_4$; LACM 1119, symphysis fragment; LACM 1120, symphysis with right and left C's and left $P_{2-4}$ (paratype of *Teleodus californicus* Stock, 1935); LACM 1126, symphysis and right dentary fragment with C, $P_2$-$M_3$ and left $P_4$; LACM 1388, right dentary with $P_4$-$M_3$; 1398, symphysis and left dentary with $P_3$-$M_2$ and part of $M_3$, isolated left $C_1$, $P_{1-2}$ and two lower incisor fragments (holotype of *Teleodus californicus* Stock, 1935; illustrated by Stock, 1935, pl. 1, Fig. 1); LACM 1834, right maxillary with $C^1$ and $P^{1-2}$ (Stock, 1935, pl. 1, Fig. 2); LACM 4793, right dentary with $P_4$-$M_3$; LACM 4794, left dentary with $M_{1-2}$; LACM 53183, right dentary with $P_4$-$M_1$. A large number of edentulous jaw fragments, postcrania and isolated teeth from CIT 150 catalogued in the LACM pertain to *D. uintensis* and are grouped here for brevity. Edentulous jaw fragments: LACM 53215-53217, 53219. Postcrania: LACM 53218, 53220-53229. Isolated canines: LACM 1028, 1030, 1097, 1098, 1129, 1136, 1831 (Stock, 1935, pl. 2, Fig. 8); 53109-53112, 53184-53190; Isolated incisors: LACM 1027, 1029, 1837 (Stock, 1935, pl. 2, Fig. 9), 1838 (Stock, 1935, pl. 2, Fig. 10), 53193-53214. Isolated lower premolars: 1005, 1012, 1024, 1080, 1083, 1085, 1088, 1110-1113, 1135, 53117, 53148-53160, 53192. Isolated lower molars: LACM 1014, 1078, 1082, 1084, 1085, 1835 (Stock, 1935, pl. 2, Fig. 5), 1836 (Stock, 1935, pl. 2, Fig. 6), 53161-53178, 53180-53182. Isolated upper premolars: LACM 1011 (Stock, 1935, pl. 2, Fig. 2), 1013, 1022, 1023, 1090, 1091 (Stock, 1935, pl. 2, Fig. 11), 1092, 1115, 1116, 1132, 53113-53116, 53118-53120, 53122. Isolated upper molars: LACM 1004 (Stock, 1935, pl. 2, Fig. 1), 1006-1008, 1009 and 1094 (Stock, 1935, pl. 2, Fig. 7), 1025, 1081, 1093, 1095 (Stock, pl. 2, Fig. 4), 1096, 1127, 1128, 1832, 1833 (Stock, 1935, pl. 2, Fig. 3), 53124-53139, 53191.

From the Sespe Formation at CIT locality 292: LACM 2143, palate (Stock, 1938, pl. 2).

From the Galisteo Formation, north-central New Mexico at locality W7 of Lucas (1982): F:AM 108510, left $P^2$ (Lucas, 1982, Fig. 14C); 108521, left $P^2$; 108524, skull (Lucas, 1982, Fig. 14A-B; Kues, 1982, photograph on p. 191).

From the Galisteo Formation, north-central New Mexico at locality C6 of Lucas (1982): F:AM 108529, nasals (Lucas, 1982, Fig. 14E-F); 108530, left $I_{2-3}$ and C (Lucas, 1982, Fig. 14D).

From the Galisteo Formation, north-central New Mexico at locality T2 of Lucas (1982): UNM GE-070, right dentary fragment with $P_2$-$M_3$ (Lucas and Kues, 1979, Fig. 5).

From the Rubio Peak Formation along Turkey Creek near Winston, Sierra County, New Mexico: NMMNH P-3051, lower jaw bearing left $P_2$-$M_3$, right $P_4$-$M_2$, incomplete left C and right $M_3$ and roots of the right C, $P_2$ and $P_3$ (Lucas, 1983, Fig. 4, illustrated a cast of this specimen, UNM-MSB-A3369).

From the Chambers Tuff, Presidio County, Texas: FMNH PM 136, skull and lower jaw (blue cliff horizon, Rifle Range Hollow); FMNH PM 482, skull (north of big cliff, big red horizon); TMM 40062-1, right dentary fragment with $P_2$-$M_1$.

From the Skyline channels of the Devil's Graveyard Formation, Brewster County, Texas: TMM 41580-30, premaxillary with left and right $I^{2-3}$ and left C (Wilson and Stevens, 1986, Fig. 2); TMM 41715-6, left dentary fragment with $P_2$-$M_3$ (Wilson and Stevens, 1986, Fig. 3); TMM 41853-17, palate (Needle Creek).

*Revised diagnosis*: *D. uintensis* is distinguished from *D. primitivus* by the less vertical labial faces of its cheek teeth (brachydonty) and virtual lack or lack of postcanine diastemata.

*Description*: The references listed above in the synonymy provide extensive descriptions and illustrations of many specimens of *D. uintensis* and thus obviate the need for a detailed description here (see especially Scott, 1945). However, there are two aspects of the morphology of *D. uintensis* -- sexual dimorphism and juvenile structure -- that warrant brief description and interpretation.

There are eleven skulls of *D. uintensis* from the *Duchesneodus* quarry, of which two are juvenile. Of the nine adult skulls, we judge three to be male and six female. Skulls we judge to be male (e.g., Fig. 27.3A-C) have a very prominent, dome-shaped convexity on the skull roof, relatively large, blunt horns and broadly flaring zygomata with massive roots and robust, flattened arches. Furthermore, male skulls are mesaticephalic (zygomacephalic index = 75-80) but have a relatively narrow skull roof (cranial breadth index = approximately 30).

In contrast, skulls we judge to represent females (e.g., Fig. 27.3D-F) have a low, broad, cranial convexity, small, blunt horns and narrow zygoma with relatively weak roots and thin arches. These skulls are more dolicocephalic than are those of the males

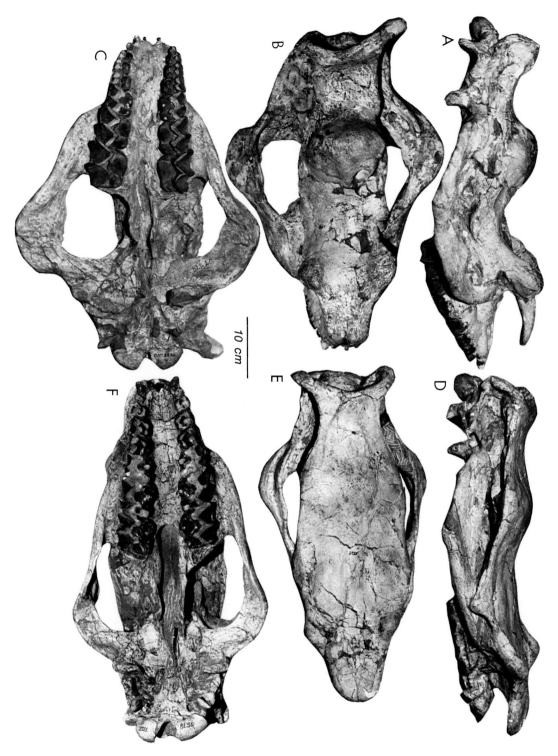

Fig. 27.3. Male and female skulls of *Duchesneodus uintensis* from the *Duchesneodus* quarry in the Duchesne River Formation near Vernal, Utah. A-C, CM 11767, male skull, right lateral (A), dorsal (B) and ventral (C) views. D-F, CM 11759, female skull, right lateral (D), dorsal (E) and ventral (F) views.

(zygomacephalic index = 57-65) but have a skull roof that is somewhat wider relative to its length (cranial breadth index = approximately 35). The male and female specimens of *D. uintensis* do not differ in cheek-tooth size or in the structure and size of the canines and incisors. Sexual dimorphism along these lines has already been noted in eubrontotheres (Osborn, 1929, p. 793), but to our knowledge the *Duchesneodus* quarry represents the most adequate documentation yet available.

The two juvenile skulls of *D. uintensis* known from the *Duchesneodus* quarry are CM 11758 and CM 11760. Of these, CM 11758 (Fig. 27.4) is the better preserved, although it is somewhat distorted and in need of further preparation. CM 11758 has a low, broad convexity on its skull roof even though the nasal horns are little more than low swellings near the front of the snout (Fig. 27.4A). The zygomacephalic index is 67, intermediate between the indices for adult male and female skulls from the quarry. The skull roof of the juvenile, however, is very broad (cranial breadth index = 52).

The $dP^{3-4}$ of the juvenile skull are worn and damaged but do not appear to differ significantly, except in size, from those of *Menodus giganteus* illustrated by Osborn (1929, pl. 25). Indeed, CM 11758 is somewhat older ontogenetically than the specimen Osborn illustrated, and thus is older than his "stage 7," the youngest stage of brontothere ontogeny Osborn (1929, p. 454) discussed. Although CM 11758 is about two-thirds of adult size (judged by cranial length), its permanent cheek teeth are already being emplaced.

CM 11758 more resembles female skulls than male skulls in the quarry sample by having small horns, a low and broad cranial convexity, and relatively narrow, thin, and weakly rooted zygomatic arches. We do not believe that this is because CM 11758 is a juvenile female; in fact, we are unable to sex CM 11758. Instead, we believe the similarities between the juvenile and female skulls of *D. uintensis* indicate that the adult female of *D. uintensis* is neotenic (*sensu* Gould, 1977); i. e., the female retains formerly juvenile characters by retardation of somatic development. This type of sexual dimorphism is common among large, living ungulates (e.g., *Bison*: McDonald, 1981) and thus is not surprising in a large brontothere perissodactyl.

*Discussion*: Specimens we assign to *D. uintensis* represent a relatively narrow range of metric and morphological variation readily subsumed under one species. Metric variation (note the coefficients of variation in Table 27.1) of *D. uintensis* is comparable to that of other single species of large fossil mammals, including a quarry sample of the rhinoceros *Ceratotherium praecox* from the Pliocene of South Africa (Hooijer, 1972), a quarry sample of the pantodont *Coryphodon molestus* from the Eocene of New Mexico (Lucas, 1984) and specimens assigned to a single species of late Eocene amynodontid rhinoceros, *Amynodon advenus* (Wall, 1982). Most of the nonmetric variation among specimens of *D. uintensis* was described above in the discussion of sexual dimorphism and juvenile structure.

Stock (1935, p. 460) only distinguished *T. californicus* from *T. uintensis* by the former's "precocious reduction of the lower incisors from three to two in some individuals." However, it is clear that the incisors of *D. uintensis* are vestigial teeth. As Scott (1945) noted, even adult individuals with well worn cheek teeth have essentially unworn incisors. Moreover, there is a great range in incisor number - from zero to two - in specimens that represent a single species of Chadronian eubrontothere (Osborn, 1929). Thus, we think the presence of one specimen with two lower incisors in the Sespe sample of *Duchesneodus* does not warrant recognition of *D. californicus* as a species separate from *D. uintensis*.

Bjork (1967, p. 231) listed three features in his diagnosis of *T. thyboi*: "primitive, stout, curved, canines; short diastema between C and $P^1$; mesostyle strongly developed to absent on $P^4$." The $P^4$

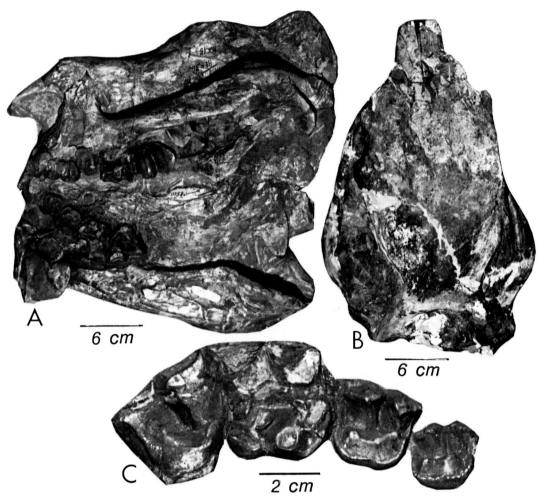

Fig. 27.4. CM 11758, juvenile skull and upper dentition of *Duchesneodus uintensis* from the *Duchesneodus* quarry in the Duchesne River Formation near Vernal, Utah. A-B, Left lateral (A) and dorsal (B) views of skull. C, Occlusal view of right $P^{1-2}$ and $dP^{3-4}$.

Table 27.1. Tooth measurements (in cm) and summary statistics of selected specmens of *D. uintensis*, and the holotype of *D. primitivus*. CV = coefficient of variation; n = sample size; s = standard deviation; x = mean; and * = approximate measurement of a damaged tooth.

*Duchesneodus uintensis*  LOWER TEETH

| SPECIMEN | $P_2L$ | $P_2W$ | $P_3L$ | $P_3W$ | $P_4L$ | $P_4W$ | $M_1L$ | $M_1W$ | $M_2L$ | $M_2W$ | $M_3L$ | $M_3W$ |
|---|---|---|---|---|---|---|---|---|---|---|---|---|
| CM 11761 (d) | 1.9 | 1.2* | 2.4 | 1.6* | 2.7 | 1.9 | 3.5 | 2.3 | 4.8 | 2.8 | 6.6 | 2.9 |
| CM 11809 (a) | 2.2 | 1.6 | 2.6 | 2.0 | 3.2 | 2.4 | 3.6 | 2.4 | 4.7 | 3.1 | 7.8 | 3.1 |
| FMNH PM136 | | | | | | 2.4 | 4.4* | 2.8 | 6.1 | 3.3 | 8.7 | |
| LACM 1120 | 2.3 | 1.5 | 2.8 | 2.1 | 3.1 | 2.4 | | | | | | |
| LACM 1126 | | | 2.3 | 1.4 | 3.1 | 2.2 | 3.5 | 2.4 | 4.9 | 2.8 | - | 3.0 |
| LACM 1387 | 2.1 | 1.3 | 2.5 | 1.7 | 3.0 | 1.8 | 4.6 | 2.3 | 5.7 | 2.9 | | |
| LACM 1388 | | | | | 2.9 | 2.1 | 4.0 | 2.5 | 5.3 | 3.0 | 7.4 | 2.9 |
| LACM 1389 | | | 2.6 | 1.7 | 2.8 | 2.3 | 3.9 | 2.4 | 5.2* | 2.9 | 7.0* | 2.9 |
| LACM 1398 (b) | 2.2 | 1.3 | 2.4 | 1.9 | 2.8 | 2.3 | 3.9 | 2.7 | 4.8 | 3.2 | - | 3.1* |
| NMMNH P3051 | 2.2 | 1.7 | 3.1 | 2.2 | 3.4 | 2.6 | 4.6 | 2.9 | 5.6 | 3.4 | 7.9 | 3.3 |
| TMM 41715-6 | | | 2.4 | - | 2.8 | 2.0 | 3.9 | 2.5 | 5.0 | 3.1 | 7.5 | 3.0 |
| UNM GE-070 | | | 3.0 | 2.3 | 3.6 | 2.7 | 4.9 | 3.0 | 5.8 | 3.6 | 8.5 | 3.6 |
| n | 6 | 6 | 10 | 9 | 11 | 12 | 11 | 11 | 11 | 11 | 8 | 9 |
| x | 2.15 | 1.43 | 2.61 | 1.88 | 3.04 | 2.56 | 4.16 | 2.56 | 5.26 | 3.10 | 7.68 | 3.09 |
| s | 0.23 | 0.18 | 0.26 | 0.28 | 0.27 | 0.26 | 0.45 | 0.23 | 0.45 | 0.24 | 0.66 | 0.22 |
| CV | 5.85 | 12.5 | 9.91 | 15.0 | 8.80 | 11.50 | 10.7 | 9.15 | 8.58 | 7.90 | 8.63 | 7.07 |

*Duchesneodus primitivum* (holotype of *Megacerops primitivus* Lambe, 1908)

| | $P_2L$ | $P_2W$ | $P_3L$ | $P_3W$ | $P_4L$ | $P_4W$ | $M_1L$ | $M_1W$ | $M_2L$ | $M_2W$ | $M_3L$ | $M_3W$ |
|---|---|---|---|---|---|---|---|---|---|---|---|---|
| NMC 6421 | 2.6 | 1.8 | 3.2 | 2.3 | 3.5 | 2.7 | 4.8 | 2.9 | 5.8 | 3.3 | 8.2 | 3.3 |

*Duchesneodus uintensis*  UPPER TEETH

| SPECIMEN | $P^2L$ | $P^2W$ | $P^3L$ | $P^3W$ | $P^4L$ | $P^4W$ | $M^1L$ | $M^1W$ | $M^2L$ | $M^2W$ | $M^3L$ | $M^3W$ |
|---|---|---|---|---|---|---|---|---|---|---|---|---|
| CM 11754 | 2.1 | 2.7 | 2.6 | 3.2 | 3.5 | 3.9 | 4.6 | 4.5 | 6.1 | 5.3 | 6.5 | 6.1 |
| CM 11759 (d) | 2.0 | 2.4 | 2.5 | 3.0 | 2.8 | 3.5 | 4.4 | 3.8 | 5.0 | 5.2 | 5.6 | 5.6 |
| F:AM 108521 | 2.0* | 2.7 | 2.5 | 2.6 | 3.1 | 4.4 | 4.5* | 5.0 | 6.0 | 6.5 | 7.2 | 7.3 |
| FMNH PM136 | 2.3 | 2.7 | 2.6 | 3.4 | 3.2 | 4.4 | 5.0 | 5.2 | 6.8 | 6.3 | 7.4 | 6.6 |
| FMNH PM482 | 2.6* | 3.3* | 3.0* | 4.4* | 3.6* | 4.9* | 4.6* | 5.8 | 5.4* | 6.8 | 6.6 | 7.0 |
| FMNH PM22410 | 2.8 | 3.3 | 3.5 | 4.0 | 4.3 | 4.8 | 5.7 | 5.6 | 7.6 | 6.8 | 8.3 | 8.0 |
| LACM 2143 | 2.2* | 2.7* | 2.7* | 3.2 | 3.0 | 4.4 | 4.3 | 4.6 | 5.5 | 5.5 | 6.7 | 6.4 |
| SDSM 63689 (c) | | | 2.9 | 3.2 | 3.5 | 4.1 | 4.6 | 4.6 | 5.8 | 5.7 | 6.0* | 5.7* |
| TMM 41853-17 | 2.1 | 2.4 | 2.8 | 3.1 | 3.1* | 3.9 | 4.5 | 4.4 | 5.7 | 5.4 | 6.2 | 5.7 |
| n | 8 | 8 | 9 | 9 | 9 | 9 | 9 | 9 | 9 | 9 | 9 | 9 |
| x | 2.26 | 2.78 | 2.79 | 3.46 | 3.34 | 4.26 | 4.69 | 4.83 | 5.98 | 5.94 | 6.72 | 6.49 |
| s | 0.27 | 0.33 | 0.30 | 0.44 | 0.42 | 0.42 | 0.40 | 0.59 | 0.72 | 0.62 | 0.77 | 0.78 |
| SD | 12.1 | 11.8 | 10.7 | 12.7 | 12.5 | 10.0 | 8.6 | 12.3 | 12.0 | 10.4 | 11.4 | 11.9 |

(a) Holotype of *Teleodus uintensis* Peterson, 1931
(b) Holotype of *Teleodus californicus* Stock, 1935
(c) Holotype of *Teleodus thyboi* Bjork, 1967
(d) Paratypes of *Teleodus uintensis* Peterson, 1931

mesostyle is, as Bjork (1967) noted, variable among specimens he assigned to *T. thyboi*. Also, we do not judge the canines or diastemata of specimens of *Duchesneodus* from South Dakota to differ significantly from those of some specimens in the *Duchesneodus* quarry sample. Therefore, we do not recognize *D. thyboi* as a species distinct from *D. uintensis*. Furthermore, we believe all other specimens previously assigned to *Teleodus* or *Duchesneodus*, except for the holotypes of *T. avus* Marsh, 1890 and *Megacerops primitivus* Lambe, 1908 (see following), pertain to *D. uintensis*. Metrically and morphologically these specimens of *D. uintensis*, from South Dakota, Utah, California, New Mexico, and Texas, do not significantly extend the range of metric and morpological variation present in the *Duchesneodus* quarry sample.

*Duchesneodus primitivus* (Lambe, 1908)
1908 *Megacerops primitivus* Lambe, p. 49, pl. 6, Figs. 4-5.
1929 *Teleodus primitivus*: Osborn, 1929, p. 482, Figs. 204, 413A.
1934 *Teleodus primitivus* (in part): Russell, p. 58, Fig. 3B.
1982 *Duchesneodus primitivus*: Lucas and Schoch, p. 1022.

*Holotype and only known specimen*: NMC 6421, incomplete lower jaw with left and right $I_{1-3}$ and C, right $P_{2-3}$, and left $P_1$-$M_3$ (Lambe, 1908, pl. 6, Fig s. 4-5; Osborn, 1929, Fig s. 204, 413A; Russell, 1934, Fig. 3B; this paper, Fig. 27.5A-C).

*Horizon and locality of holotype*: Cypress Hills Formation, Saskatchewan.

*Revised diagnosis*: A species of *Duchesneodus* distinguished from *D. uintensis* by its relatively large postcanine diastemata and vertical labial faces of the lower cheek teeth.

*Description*: Lambe (1908, pp. 49-50) provided an accurate and detailed description of the holotype and only known specimen of *D. primitivus* obviating the need for further description here.

*Discussion*: The holotype and only known specimen of *D. primitivus*, NMC 6421, can be distinguished from specimens of *D. uintensis* by its relatively long postcanine diastemata and the essentially vertical labial aspects of its cheek teeth. Peterson (1931) first drew attention to these distinctions, and we follow him in separating *D. primitivus* and *D. uintensis*. It is unfortunate that only one specimen of *D. primitivus* is known and that its provenance within the Cypress Hills Formation of Sakatchewan is not known with precision.

Russell (1934, p. 58) tentatively referred a skull roof in the Cutler collection of the British Museum (BMNH M 14103: Figs. 27.5D-F) to *Megacerops primitivus* "based on the small size and the relatively primitive structure." Other specimens in the Cutler collection that are within the size range of *D. primitivus* are lower molars catalogued under BMNH M 14098 and BMNH M 14099 from "Calf Creek, Saskatchewan."

Unlike Russell (1934), we do not refer BMNH M 14103 to *D. primitivus*. This dolicocephalic, large-horned skull does not fit into our concept of the cranial morphology of *Duchesneodus*, despite the fact that it has a very small cranial convexity. Instead, the BMNH skull more readily fits into Osborn's (1929) concept of the cranial morphology of *Megacerops*. If the cranial morphology of BMNH M 14103 could be demonstrated to correspond to the lower jaw and dental morphology of NMC 6421 by discovery of associated material, then Lambe's (1908) original concept of *Megacerops primitivus* for a taxon with a dentition convergent on that of *Duchesneodus* could be upheld.

Here we adopt the conservative course of assigning NMC 6421 to *Duchesneodus* and not including BMNH M 14103 with it in *D. primitivus*. This strongly suggests that a Duchesnean-age horizon, perhaps slightly older than (or equivalent to?) the Southfork local fauna (Storer, 1984), is present in the Cypress Hills Formation.

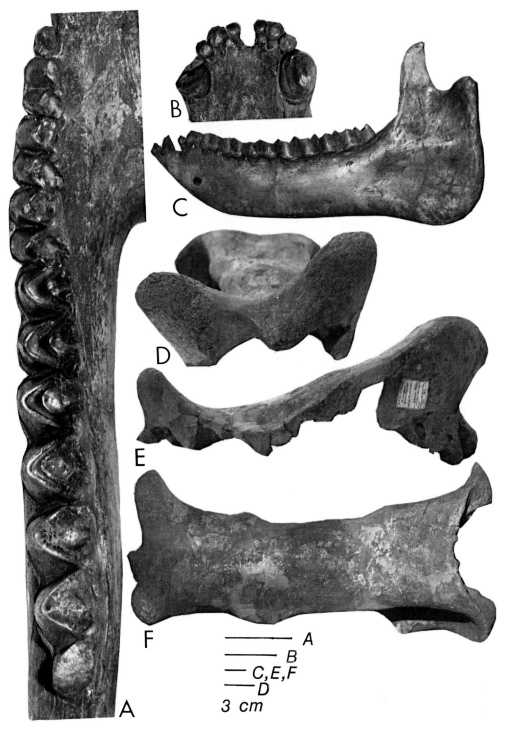

Fig. 27.5. Holotype lower jaw of *Duchesneodus primitivus* and a skull roof that Russell (1934) referred to that taxon. A-C, NMC 6421, holotype lower jaw of *D. primitivus*, occlusal view of left $P_1$-$M_3$ (A), occlusal view of left and right $I_{1-3}$ and C (B) and left lateral view of lower jaw (C). D-F, BMNH M 14103, skull roof that Russell (1934) assigned to *Megacerops primitivus*, anterior (D), left lateral (E) and dorsal (F) views.

## Acknowledgments

We thank L. Barnes, P. Bjork, M. Dawson, J. Hooker, D. Russell, W. Turnbull, and J. Wilson for access to specimens of *Duchesneodus* in their care; and B. Mader for comments on an earlier version of this paper. NSF grant DEB-7919681 to the senior author supported part of this research.

## Bibliography

Bjork, P. R. (1967): Latest Eocene vertebrates from northwestern South Dakota. - *J. Paleont*, 41: 227-236.

Clark, J., Beerbower, J. R., and Kietzke, K. K. (1967): Oligocene sedimentation, stratigraphy, paleoecology and paleoclimatology in the Big Badlands of South Dakota. - *Fieldiana: Geology Mem.*, 5: 1-158.

Gould, S. J. (1977): *Ontogeny and Phylogeny*.- Cambridge, Mass. (Belknap Press).

Hatcher, J. B. (1893): The *Titanotherium* beds. - *Am. Nat.* 27: 204-221.

Hooijer, D. A. (1972): A late Pliocene rhinoceros from Langebaanweg, Cape Province. - *Ann. S. Afr. Mus.* 59: 151-191.

Kues, B. S. (1982): *Fossils of New Mexico.* - Albuquerque (University of New Mexico Press).

Lambe, L. M. (1908): The Vertebrata of the Oligocene of the Cypress Hills, Saskatchewan. - *Contribs. Canadian Paleont.*, 3: 1-64.

Leidy, J. (1852): Description of the remains of extinct Mammalia and Chelonia from the Nebraska Territory.-In: Owen, D.D., *Report of a geological survey of Wisconsin, Iowa, and Minnesota, and incidentally a portion of Nebraska Territory.* -Philadelphia, pp. 553-572.

Leidy, J. (1870): [Description of a new genus and species, *Megacerops coloradensis*]. - *Proc. Acad. Nat. Sci. Phila.*, 1870: 1-2.

Lucas, S. G. (1982): Vertebrate paleontology, stratigraphy, and biostratigraphy of Eocene Galisteo Formation, north-central New Mexico. - *New Mex. Bur. Mines Min. Res. Circular*, 186: 1-34.

Lucas, S. G. (1983): The Baca Formation and the Eocene-Oligocene boundary in New Mexico. - *New Mex. Geol. Soc. Guidebook Field Conf.*, 34: 187-192.

Lucas, S. G. (1984): Systematics, biostratigraphy and evolution of early Cenozoic *Coryphodon* (Mammalia, Pantodonta). - New Haven, Conn.: Ph.D. Diss., Yale Univ., 649 pp.

Lucas, S. G. and Kues, B. S. (1979): Vertebrate biostratigraphy of the Eocene Galisteo Formation, north-central New Mexico. - *New Mex. Geol. Soc. Guidebook Field Conf.*, 30: 225-229

Lucas, S. G. and Scoch, R. M. (1982): *Duchesneodus*, a new name for some titanotheres (Perissodactyla, Brontotheriidae) from the late Eocene of western North America. - *J. Paleont.*, 56: 1018-1023.

Mader, B. (1989): The Brontotheriidae: a systematic revision and preliminary phylogeny of North American genera (this volume).

Marsh, O. C. (1873): Notice of new Tertiary mammals. - *Amer. J. Sci.*, 5: 407-410, 485-488.

Marsh, O. C. (1890): Notice of new Tertiary mammals. - *Amer. J. Sci.*, 39: 523-525.

McDonald, J.N. (1981): *North American Bison: their Classification and Evolution.* - Berkeley (Univ. Calif. Press,).

Nelson, M.E., Madsen, J.H., and Stokes, W. L. (1980): A titanothere from the Green River Formation, central Utah: *Teleodus uintensis* (Perissodactyla: Brontotheriidae). - *Univ. Wyo. Contribs. Geol.*, 18: 127-134.

Osborn, H.F. (1896): The cranial evolution of *Titanotherium*. - *Bull. Am. Mus. Nat. Hist.*, 8: 157-197.

Osborn, H. F. (1902): The four phyla of Oligocene titanotheres. - *Bull. Amer. Mus. Nat. Hist.*, 16: 91-109.

Osborn, H.F. (1929): The titanotheres of ancient Wyoming, Dakota, and Nebraska. - *U.S. Geol. Surv. Monograph* 55: 1-953.

Owen, R. (1848): Description of teeth and portions of jaws of two extinct anthracotheroid quadrupeds . . . with an attempt to develop Cuvier's idea of the classification of pachyderms by the number of their toes. -*Q. J. Geol. Soc. London*, 4: 103-141.

Peterson, O. A. (1931): New species of the genus *Teleodus* from the upper Uinta of northeastern Utah. - *Ann. Carnegie Mus.*, 20: 307-312.

Russell, L. S. (1934): Revision of the lower Oligocene vertebrate fauna of the Cypress Hills, Saskatchewan. - *Royal Can. Inst. Trans.*, 20: 49-67.

Schoch, R. M., and S. G. Lucas (1985): The Brontotheriidae, a group of Eocene and Oligocene perissodactyls from North America, Asia and eastern Europe. - In: Fuller, W.A., Nietfeld, M.T., and Harris, M.A. (eds.): *Abstracts of Papers and Posters, Fourth Int. Ther. Congr., Edmonton, Alberta, Canada*: abstract 0558.

Scott, W. B. (1940): The mammalian fauna of the White River Oligocene. Part V - Perissodactyla. - *Trans. Amer. Phil. Soc.*, 28: 747-980.

Scott, W. B. (1945): The Mammalia of the Duchesne River Oligocene. - *Trans. Amer. Phil. Soc.*, 34: 209-253.

Stock, C. (1935): Titanothere remains from the type Sespe of California. - *Proc. Nat. Acad. Sci.*, 21: 456-462.

Stock, C. (1938): A titanothere from the type Sespe of California. - *Proc. Nat. Acad. Sci.*, 24: 507-512.

Storer, J. E. (1984): Fossil mammals of the Southfork local fauna (early Chadronian) of Saskatchewan. -*Can. J. Earth Sci.* 21: 1400-1405.

Wall, W.P. (1982): The genus *Amynodon* and its relationship to other members of the Amynodontidae (Perissodactyla, Rhinocerotoidea). - *J. Paleont.* 56: 434-443.

Wilson, J. A. (1977): Early Tertiary vertebrate faunas Big Bend area Trans-Pecos Texas: Brontotheriidae. - *Texas Mem. Mus., Pearce-Sellards Series*, 25: 1-17.

Wilson, J. A. (1978): Stratigraphic occurrence and correlation of early Tertiary vertebrate faunas, Trans-Pecos Texas, Part 1: Vieja area. - *Texas Mem. Mus. Bull.*, 25: 1-42.

Wilson, J. A., and Stevens, M. S. (1986): Fossil vertebrates from the latest Eocene, Skyline channels, Trans-Pecos Texas. - *Contrib. Geol., Univ. Wyo., Spec. Pap.*, 3: 221-235.

# 28. ORIGIN AND EVOLUTION OF THE PERISSODACTYLA: SUMMARY AND SYNTHESIS

## DONALD R. PROTHERO and ROBERT M. SCHOCH

The Perissodactyla appear to have originated in Asia and/or Africa during the late Paleocene, where they diverged from their close relatives, the tethytheres (proboscideans, sirenians, desmostylians) and arsinoitheres. Contrary to many accounts, they are not as closely related to phenacodontids as they are to tethytheres, and perissodactyls did not come from Central America. Many unique synapomorphies strongly suggest that hyraxes (long considered "subungulates" related to elephants) are in fact perissodactyls. Since the Order Perissodactyla included the hyrax when it was created by Owen in 1848, we return the hyraxes to the Perissodactyla. The hyraxes were apparently the first group to split from the rest, and became isolated in Africa (along with tethytheres and arsinoitheres). There they underwent an endemic radiation, converging in some ways with bovids, pigs, tapirs, and chalicotheres; eventually they spread to Eurasia.

The non-hyracoid perissodactyls (Mesaxonia of Marsh, 1884) split into three major infraorders. The first group, the Titanotheriomorpha, was dominant in both Asia and North America during most of the later Eocene, migrating back and forth over the Bering Strait before finally becoming extinct during the Oligocene. The other two infraorders were widespread over Holarctica in the early Eocene, where they began to diverge into Hippomorpha (pachynolophids, equids, and palaeotheres) and Moropomorpha (isectolophids, lophiodonts plus chalicotheres, tapiroids *sensu lato*, and rhinocerotoids).

The hippomorph radiation began in the latest Paleocene and earliest Eocene with the "wastebasket" taxon *Hyracotherium*, which includes the most primitive pachynolophids, equids, and palaeotheres. If Hooker (this volume, Chapter 6) is correct, then the type species of *Hyracotherium*, *H. leporinum*, is most closely related to palaeotheres, and the most primitive equids must be referred to another genus (possibly *Protorohippus*). After the early Eocene, equids became endemic to North America (with several back-migrations to Eurasia), while pachynolophids and palaeotheres became endemics which dominated Europe in the later Eocene. Both European groups were decimated during the early Oligocene ("la Grande Coupure") and disappeared by the mid-Oligocene.

The moropomorph radiation began with *Homogalax* and a number of "tapiroid" forms which were widespread over Holarctica in the early Eocene. By the mid-Eocene, they had diverged into a number of groups dominant in Asia and North America (isectolophids, ancylopods, lophialetids and other tapiroids *sensu lato*, and rhinocerotoids). Most of these groups reached their maximum diversity in the later Eocene of Asia, and then were decimated in the Oligocene. Chalicotheres, on the other hand, diversified in the Miocene of Eurasia, where they became large clawed herbivores adapted for pulling down branches on trees. Lophiodonts were closely related to chalicotheres, and were important tapir-like animals in Europe before their late Eocene extinction.

In the later Eocene of Asia and North America, rhinocerotoids diversified into amynodonts (some of which were large, aquatic forms with a proboscis), hyracodonts (long-legged forms which reached gigantic size in Asia), and rhinocerotids (true rhinoceroses). Amynodonts and hyracodonts were both severely affected by the late Eocene extinctions event, and were almost completely gone by the Oligocene. Rhinocerotids, on the other hand, began to dominate and radiate in the latest Eocene, filling the niches previously occupied by titanotheres and amynodonts in North America, and by

*The Evolution of Perissodactyls* (ed. D.R. Prothero & R.M. Schoch) Oxford Univ. Press, New York, 1989

palaeotheres, tapiroids, and lophiodonts in Eurasia.

In the Miocene, equids and rhinocerotids were very diverse, and among the dominant large herbivores in Holarctica. Both were severely affected by the Messinian event at the end of the Miocene. During the Pleistocene, tapirs and equids crossed to South America via Panama. After the Pleistocene extinctions, only relicts of hyraxes in Africa, equids in Africa and Eurasia, rhinocerotids in Africa and southeast Asia, and tapirs in South America and southeast Asia were left from previously worldwide distributions of all of these groups. The wild populations of all of these groups (especially tapirs and rhinos) are now facing extinction from human population pressure and poaching.

## Introduction

The Perissodactyl Workshop at the Fourth International Theriological Congress in 1985, and this resulting volume, brought together much new information about the Perissodactyla. Some of it radically changes the prevailing orthodoxies about the group which are still appearing in popular articles and books (e.g., Monroe, 1985; Savage and Long, 1986), textbooks (e.g., Carroll, 1988) and review articles (e.g., MacFadden, 1988). Many of these ideas have radical implications for mammalogy (e.g., the relationships of hyraxes, and our new perissodactyl classification) and evolutionary biology (e.g., revised ideas about horse evolution). In this chapter, we review the contributions to this volume, and place them in the larger context of much recent research about ungulate evolution. We have also compiled generic range charts of all the valid perissodactyl genera (Fig. 28.1), which necessitate a discussion of the diversification, biogeography and evolutionary patterns in the Perissodactyla through the entire Cenozoic. We hope this will help bring the next generation of textbooks and review articles up to date.

As this book went to press, new argon-argon dates from some of the classic North American terrestrial vertebrate-bearing sequences indicate that our later Eocene and Oligocene correlations will have to be radically revised. These dates place the Duchesnean/Chadronian boundary around 36-37 Ma, the Chadronian/Orellan boundary around 33 Ma, the Orellan/Whitneyan boundary around 31.8 Ma, and the Whitneyan/Arikareean boundary around 29 Ma. At the same time, revised magnetostratigraphic correlations are in agreement with the increasing number of analyses that place the Eocene/Oligocene boundary at about 34 Ma. Although the revision of the timescale is not yet complete, it appears from these correlations that the Bridgerian, Uintan, and Duchesnean are middle Eocene, the Chadronian (classically considered early Oligocene) is actually late Eocene, and the Orellan, Whitneyan, and early Arikareean are early Oligocene. The Terminal Eocene Event is actually the Chadronian/Orellan extinction, although most authors (e.g., Prothero, 1985) have labeled this the "mid-Oligocene event." Similarly, the Duchesnean/Chadronian faunal turnover is actually the middle/late Eocene (Bartonian/Priabonian) event.

These radical changes in the timescale came too late to incorporate into most of the book. Except for this chapter and Chapter 10, it was impossible to rewrite the other chapters to reflect these new dates. Thus, for example, Mader (this volume, Chapter 25) refers to North American Oligocene brontotheres throughout his chapter, although it appears now that the extinction of brontotheres in North America occurred at the end of the Eocene. The range charts in this chapter (Fig. 28.1) were also finished long ago, and could not be completely redrawn at such a late stage in preparation of this book. Therefore, the reader is cautioned to read the terms "Eocene" and "Oligocene" in this book with the above revisions in mind, and substitute "late Eocene" whenever the author uses "early Oligocene" to mean Chadronian.

## Origin of the Perissodactyla

Prior to this volume, the prevailing orthodoxy (e.g., Radinsky, 1966a; Sloan, 1970, 1987; Van Valen, 1978; Gingerich, 1976)

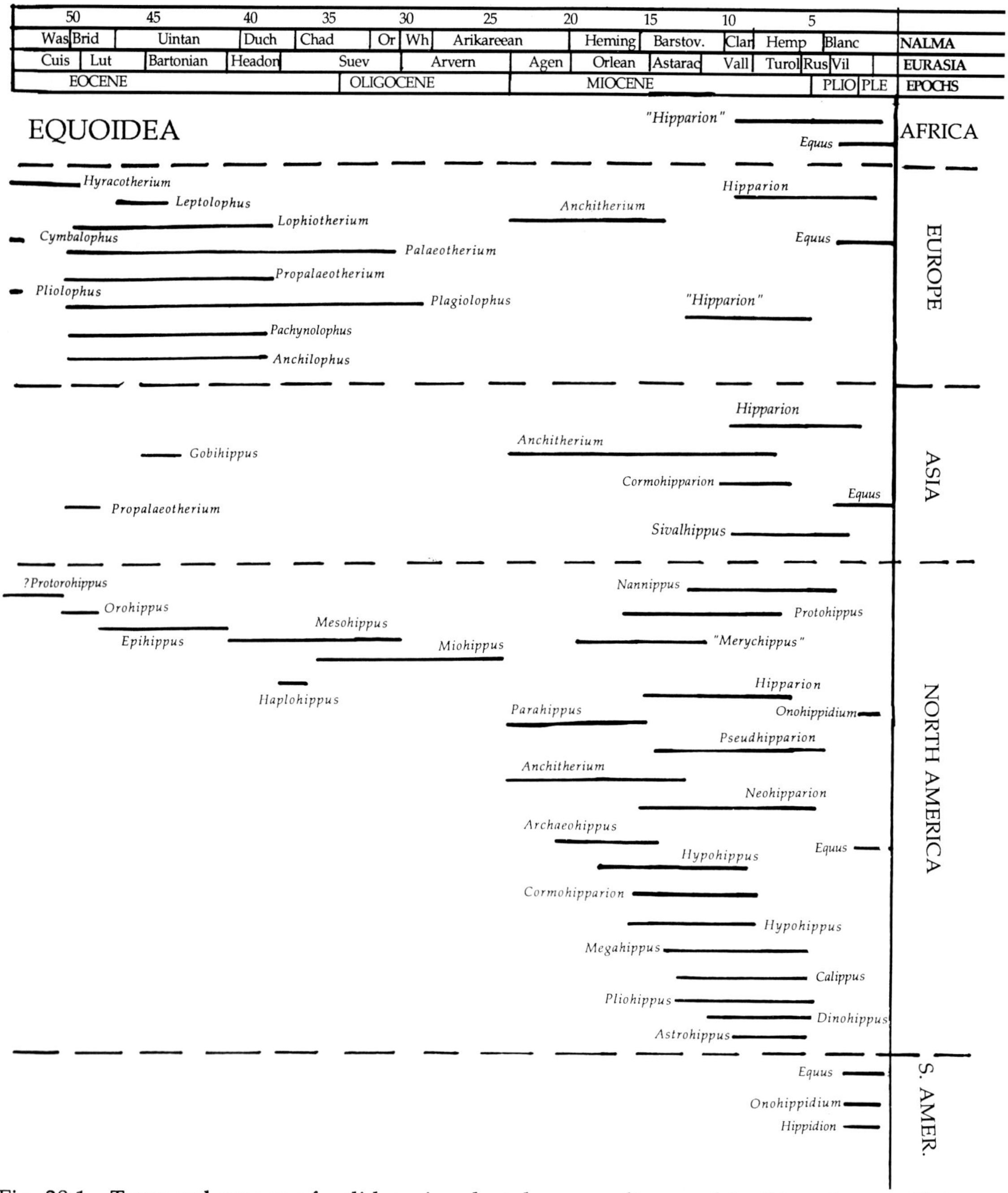

Fig. 28.1. Temporal ranges of valid perissodactyl genera discussed in this volume. Note that neither the timescale nor the ranges have been adjusted for the recent recalibration of the North American later Eocene-Oligocene (see p. 505). Timescale abbreviations: NALMA= North American land mammal "ages," as follows: Wasatchian, Bridgerian, Uintan, Duchesnean, Chadronian, Orellan, Whitneyan, Arikareean, Hemingfordian, Barstovian, Clarendonian, Hemphillian, Blancan. Eurasian chronology abbreviations for the following: Cuisian, Lutetian, Bartonian, Headonian, Suevian, Arvernian, Agenian, Orleanian, Astaracian, Vallesian, Turolian, Ruscinian, Villafranchian.

# SUMMARY OF PERISSODACTYL EVOLUTION

| 50 | 45 | 40 | 35 | 30 | 25 | 20 | 15 | 10 | 5 | |
|---|---|---|---|---|---|---|---|---|---|---|
| Was | Brid | Uintan | Duch | Chad | Or | Wh | Arikareean | Heming | Barstov. | Cla | Hemp | Blanc | **NALMA** |
| Cuis | Lut | Bartonian | Headon | Suev | Arvern | Agen | Orlean | Astarac | Vall | Turol | Rus | Vil | **EURASIA** |
| EOCENE | | | | OLIGOCENE | | | MIOCENE | | | PLIO | PLE | **EPOCHS** |

## HYRACOIDEA

*Seggeurius*, *Microhyrax*, *Geniohyus*, *Bunohyrax*, *Pachyhyrax*, *Megalohyrax*, *Saghatherium*, *Selenohyrax*, *Titanohyrax*, *Meroehyrax*, *Prohyrax*, *Parapliohyrax*, *Procavia*, *Gigantohyrax*, *Heterohyrax*, *Dendrohyrax* — AFRICA

*Geniohyus*, *Sogdohyrax*, *Pliohyrax*, *Kvabebihyrax*, *Postschizotherium* — ASIA

## TITANOTHERIOMORPHA

*Eotitanops*, *Palaeosyops*, *Sthenodectes*, *Duchesneodus*, *Metarhinus*, *Telmatherium*, *Brontops*, *Rhadinorhinus*, *Menops*, *Mesatirhinus*, *Megacerops*, *Dolichorhinus*, *Lambdotherium*, *Protitanotherium*, *"Diplacodon"*, *Xenicohippus*, *Eotitanotherium*, *Metatelmatherium* — NORTH AMERICA

*Metatelmatherium*, *Desmatotitan*, *Epimanteoceras*, *Protitan*, *Microtitan*, *Dolichorhinoides*, *Gnathotitan*, *Rhinotitan*, *Parabrontops*, *Titanodectes*, *Embolotherium*, *Metatitan*, *Hyotitan* — ASIA

507

## "TAPIROIDS" AND ANCYLOPODA

| Ma | 50 | | 45 | | 40 | | 35 | | 30 | | 25 | | 20 | | 15 | | 10 | | 5 | | |
|---|---|---|---|---|---|---|---|---|---|---|---|---|---|---|---|---|---|---|---|---|---|
| NALMA | Was | Brid | | Uintan | | Duch | | Chad | | Or | Wh | Arikareean | | Heming | Barstov. | | Clar | Hemp | Blanc | | |
| EURASIA | Cuis | Lut | | Bartonian | | Headon | | Suev | | | Arvern | | Agen | Orlean | Astarac | | Vall | Turol | Rus | Vil | |
| EPOCHS | EOCENE | | | | | | | | OLIGOCENE | | | | | MIOCENE | | | | | PLIO | PLE | |

Tapirus — S. AM.

NORTH AMERICA
- Dilophodon
- Desmatotherium
- Helaletes
- Heptodon
- Plesiocolopirus
- Selenaletes
- Colodon
- Paleomoropus
- Homogalax
- Tapiravus
- Isectolophus
- Protapirus
- Miotapirus
- Schizotheriodes
- Toxotherium
- Tapirus

ASIA
- Deperetella
- Chasmotherium
- Helaletes
- Colodon
- Heptodon
- Indolophus
- Homogalax
- Lunania
- Isectolophus
- Sastrilophus
- Kalakotia
- Aulaxolophus
- Palaeotapirus
- Breviodon
- Parabreviodon
- Megatapirus
- Teleolophus
- Rhodopagus
- Tapirus
- Pataecops
- Eoletes
- Schlosseria
- Simplaletes
- Lophialetes

EUROPE
- Lophidochoerus
- Tapiriscus
- Lophiodon
- Eotapirus
- Paralophiodon
- Lophiaspis
- Tapirus
- Atalonodon
- Protapirus
- Chasmotherium
- Palaeotapirus-Paratapirus

AFRICA
- Chalicotherium
- Ancylotherium

EUROPE
- Moropus
- Chalicotherium
- Schizotherium
- Ancylotherium
- Phyllotillon

ASIA
- Nestoritherium
- Litolophus
- Chalicotherium
- Eomoropus
- Ancylotherium
- Grangeria
- Phyllotillon
- Borissiakia

N. AMER.
- Moropus
- Tylocephalonyx

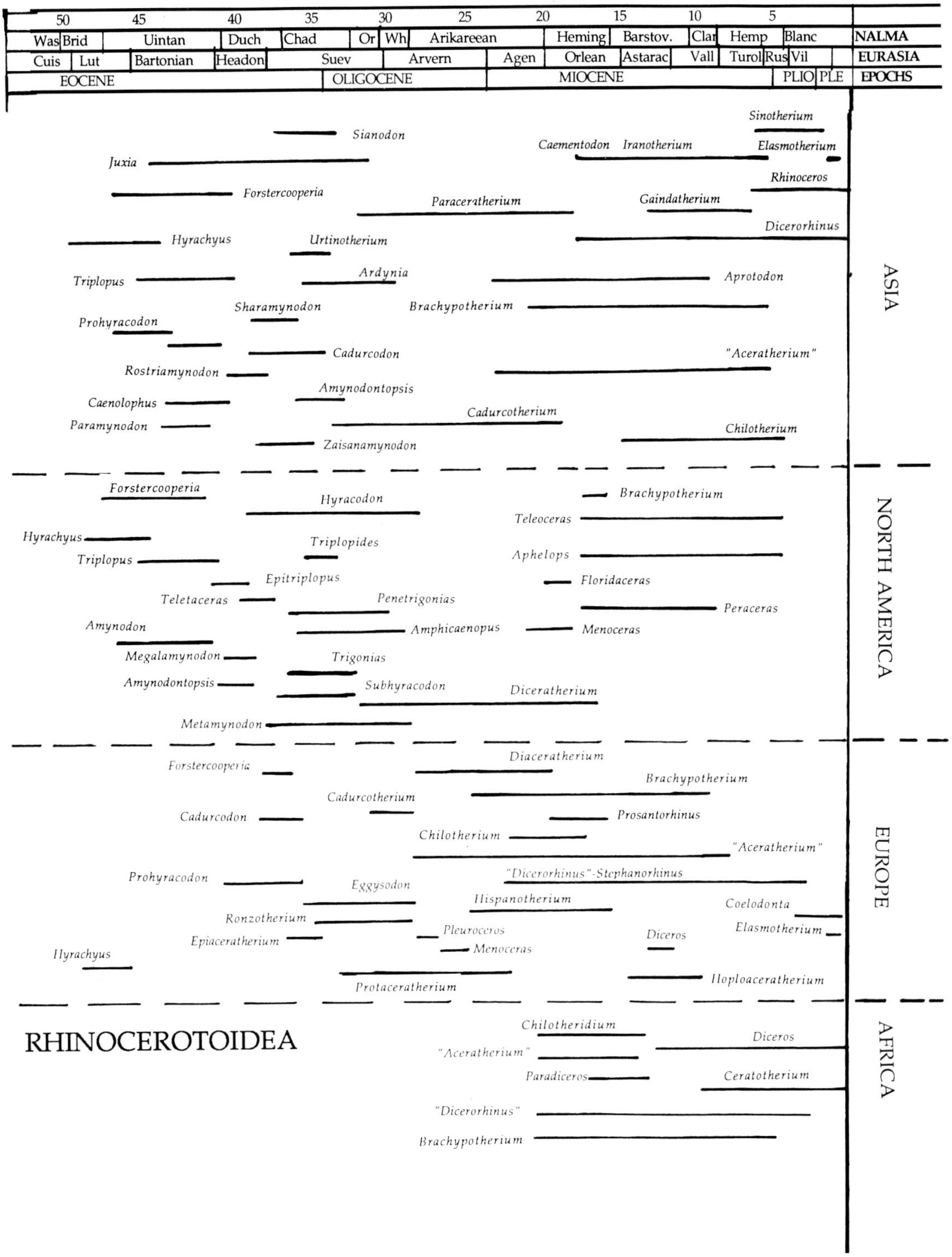
50 45 40 35 30 25 20 15 10 5
Was Brid Uintan Duch Chad Or Wh Arikareean Heming Barstov. Clar Hemp Blanc NALMA
Cuis Lut Bartonian Headon Suev Arvern Agen Orlean Astarac Vall Turol Rus Vil EURASIA
EOCENE OLIGOCENE MIOCENE PLIO PLE EPOCHS
ASIA
Sinotherium
Sianodon
Caementodon
Iranotherium
Elasmotherium
Juxia
Rhinoceros
Forstercooperia
Paraceratherium
Gaindatherium
Dicerorhinus
Hyrachyus
Urtinotherium
Triplopus
Ardynia
Aprotodon
Sharamynodon
Brachypotherium
Prohyracodon
Cadurcodon
"Aceratherium"
Rostriamynodon
Amynodontopsis
Caenolophus
Cadurcotherium
Paramynodon
Chilotherium
Zaisanamynodon
NORTH AMERICA
Forstercooperia
Hyracodon
Brachypotherium
Teleoceras
Hyrachyus
Triplopides
Aphelops
Triplopus
Epitriplopus
Floridaceras
Teletaceras
Penetrigonias
Peraceras
Amynodon
Amphicaenopus
Menoceras
Megalamynodon
Trigonias
Amynodontopsis
Subhyracodon
Diceratherium
Metamynodon
EUROPE
Diaceratherium
Forstercooperia
Brachypotherium
Cadurcotherium
Cadurcodon
Prosantorhinus
Chilotherium
"Aceratherium"
"Dicerorhinus"-Stephanorhinus
Prohyracodon
Eggysodon
Hispanotherium
Coelodonta
Ronzotherium
Elasmotherium
Pleuroceros
Epiaceratherium
Diceros
Menoceras
Hyrachyus
Protaceratherium
Hoploaceratherium
AFRICA
Chilotheridium
Diceros
"Aceratherium"
Paradiceros
Ceratotherium
"Dicerorhinus"
Brachypotherium
RHINOCEROTOIDEA

derived the Perissodactyla from the phenacodontids. Some went so far as to derive *Hyracotherium* from a known species (e.g., Sloan, 1987, implicated *Desmatoclaenus mearae*). Sloan (1970) and Gingerich (1976, pp. 86-88) speculated on paleoclimatological grounds that perissodactyls came from unknown phenacodontids living in the late Paleocene of Central America. Many of these "derivationist" scenarios were based on comparison of shared primitive characters predicated by the search for ancestral forms. They also tended to look only at dental evidence, and to neglect the possibility of migration from other regions. This practice has been likened by McKenna *et al.* (1977) to "connect-the-dots" art, and in the case of the perissodactyls, it led to erroneous conclusions.

Two important contributions have led to new conclusions concerning the origin of the perissodactyls. The first is the description of *Radinskya* (McKenna *et al.*, this volume, Chapter 3). Although McKenna *et al.* ultimately assigned it to the phenacolophid arsinoitheres, they pointed out that *Radinskya* shares some derived similarities with the perissodactyls. There is also a strong resemblance to the most primitive tethythere, the Chinese Paleocene form *Minchenella*, which was also once considered a phenacolophid (Domning *et al.*, 1986). The Chinese Paleocene fauna strongly suggests that arsinoitheres, tethytheres, and perissodactyls are very closely related.

This strikingly confirms the second line of evidence derived from cladistic analysis of the ungulates (McKenna and Manning, 1977; Prothero *et al.*, 1988). Prothero *et al.* (1988) considered all ungulate taxa, looking only at derived characters, and especially at the non-dental characters neglected in the phylogenies of Sloan (1970, 1987) and Van Valen (1978). They concluded that perissodactyls, arsinoitheres, and tethytheres were much more closely related to each other than they were to phenacodonts. A number of shared derived characters support this contention (Prothero *et al.*, 1988, Table 8.1). Prothero *et al.* (1988) placed tethytheres and perissodactyls as closest sister-taxa, with arsinoitheres as the next outgroup. The evidence of *Radinskya* may place arsinoitheres nearer to perissodactyls.

This conclusion has been tested by recent detailed work on the petrosals of arsinoitheres. According to N. Court (pers. commun.), *Arsinoitherium* shares the most derived similarities in its petrosal with elephants, and does not resemble other tethytheres, hyracoids, or mesaxonians as closely. If this work is substantiated, then arsinoitheres might become another group of tethytheres. In addition, Court's work clearly shows that perissodactyls (including hyraxes), arsinoitheres, and tethytheres are a monophyletic group.

Putting this all together, it is clear that a group consisting of tethytheres, arsinoitheres, and perissodactyls (including hyraxes) was diversifying in the late Paleocene of eastern Asia. Prothero *et al.* (1988) labeled this group of higher ungulates the "Pantomesaxonia" of Franz (1924), following M. Fischer (1986). Unfortunately, we have since learned that Franz's usage of Pantomesaxonia included a heterogeneous, paraphyletic assemblage of non-artiodactyl ungulates, and so it does not seem appropriate to resurrect it as a name for the higher ungulates (node 47 in Fig. 8.1, Prothero *et al.*, 1988). Thus, we create the Grandorder Altungulata (new taxon) to include the higher ungulates: tethytheres, arsinoitheres, hyraxes, and mesaxonians (this volume, Chapter 29).

Since tethytheres, arsinoitheres, and hyracoids (the latter representing the perissodactyls) became African endemics in the Eocene and Oligocene, it seems likely that Africa may enter into this as well. Africa was separated from Eurasia by a narrow Tethyan seaway during the Paleocene (Savage and Russell, 1983). The fossil evidence from the Paleocene and Eocene of Africa is very poor, and presently does little to test this hypothesis. A small late

Paleocene fauna from Morocco (Cappetta et al., 1978) contains no ungulates. The next-youngest assemblages are middle Eocene and already contain hyracoids, sirenians, and *Moeritherium* (Sudre, 1979; Savage, 1969).

At the present, two hypotheses seem plausible. One postulates that the Altungulata diverged in the late Paleocene of Asia (possibly closer to Tethys, but including China), and that their three earliest offshoots (tethytheres, arsinoitheres, hyracoids) crossed the Tethys and became isolated in Africa during the Eocene. However, it is not unreasonable to suggest that there may have been more connection between Africa and Eurasia across the Tethys than previously supposed. The Paleocene fauna reported by Cappetta et al. (1978) is not particularly endemic, but has similarities to the Chinese Paleocene (Sloan, 1987). In this case, there may have been common faunal elements on both sides of the Tethys during the Paleocene, and the tethytheres, arsinoitheres, and hyracoids became endemic to Africa in the Eocene when Tethys became a more difficult barrier to cross.

All of this does not deny the fact that phenacodonts have many derived similarities with perissodactyls and other altungulates. Some of these characters were discussed by Prothero et al. (1988). However, the evidence clearly shows that the phenacodont-altungulate split predates the diversification of altungulates in the late Paleocene. This is not surprising, since phenacodonts were found all over Holarctica by the early Paleocene, and were a very diverse group throughout that epoch. It is clear, however, that no known phenacodont was "ancestral" to the Perissodactyla or any other altungulate group, contrary to Sloan (1987), Van Valen (1978), and Radinsky (1969).

## Hyracoids as Perissodactyla

Another striking new development has been the renewed evidence for the perissodactyl affinities of hyracoids. As reviewed by M. Fischer (this volume, Chapter 4), the idea goes back a long way, and Owen included the hyrax when he coined the term "Perissodactyla" in 1848. For various reasons, the idea was less popular in this century, and most authors treated hyracoids as a separate order with no known affinities, or suggested that they might be related to elephants and other "subungulates." As more studies have been done on mammalian phylogeny in the last ten years, however, the case for hyrax affinities has undergone much more scrutiny. Some authors (e.g., Novacek, 1982, 1986; Novacek and Wyss, 1986; McKenna, 1987; Novacek et al., 1988) have found morphological data, and others have found molecular data (discussed by M. Fischer, this volume, Chapter 4), which suggest that hyraxes are "paenungulates" related to elephants. Others, however, have suggested that hyracoids belong with other perissodactyls (McKenna, 1975b; McKenna and Manning, 1977; M. Fischer, 1986; this volume, Chapter 4; Prothero et al., 1988). Although there are some morphological characters that seem to support tethythere-hyracoid affinities (Novacek, 1982, 1986; Novacek and Wyss, 1986; Novacek et al., 1988), M. Fischer (1986; this volume, Chapter 4) argues that many of them are invalid or of doubtful taxonomic importance.

Similarly, the molecular evidence is not particularly strong, since it is based on very few amino acid substitutions and only sampled for a few proteins (Prothero et al.,1988). The molecular data matrices of Wyss et al. (1987) reveal just how weak this molecular evidence is. Only one shared derived amino acid substitution in the $\alpha$-lens crystalline, one substitution in $\alpha$-hemoglobin, and possibly two in $\beta$-hemoglobin could be used to support hyrax-proboscidean affinities in the most parsimonious arrangements of the data. There are no data yet for hyracoids or proboscideans in several other key proteins: pancreatic ribonucleases, cytochrome c, fibrinopeptides A and B. There are no hyrax data for myoglobin. Thus, the molecular evidence has a long way to go before it provides a strong case for hyrax-proboscidean affinities.

By contrast, M. Fischer has detailed an impressive array of striking and unique shared derived characters that occur only in hyracoids and mesaxonians. The most bizarre of these is the inflated Eustachian sac, a feature with uncertain functional significance and therefore little reason to suspect as a parallelism. M. Fischer showed that the detailed morphology of the hooves, the shoulder musculature, the appendages to the iris in the eye, and the retention of the tuber maxillaris after tooth eruption are among the many unique synapomorphies found only in hyraxes and mesaxonians. In addition, there are many skeletal features that can also be seen in the fossils, such as the strong dental similarities, and the feet with reduced first and fifth metapodials and enlarged third metapodials. Many features of the basicranium and cranial circulation (Cifelli, 1982; Wible, 1986, 1987), including the extrabullar internal carotid artery, the loss of the promontory and stapedial sulci, and the large, bridged tympanohyal, support hyracoid-mesaxonian affinities.

Novacek et al. (1988) found that some of M. Fischer's characters (such as the reduced acromion) do not occur in *Hyracotherium* (but this does not establish that they are absent from other primitive mesaxonians as well). In the latest incarnation of their "paenungulate" hypothesis, Novacek et al. (1988) support the monophyly of hyraxes and tethytheres with five characters (Novacek et al., 1988, Table 3.1). Most of these characters are open to question, however. For example, as M. Fischer (this volume, Chapter 4) points out, amastoidy also occurs in pangolins, whales, and dermopterans, and in some suids and rhinoceroses. Novacek et al. (1988) stand by their use of the serial carpus character, questioning M. Fischer's contention that it developed secondarily in hyraxes due to rotatory midcarpal joint. Even if M. Fischer's interpretation is wrong, however, the serial carpus is still not a very strong character. As Gregory (1910, p. 452) pointed out, it also occurs in some rodents and insectivores, *Hyaenodon*, and some phenacodonts and meniscotheres. Indeed, within a genus, it is not even consistent. Radinsky (1966a) pointed out that *Phenacodus primaevus* has a serial carpus, but that of *P. copei* is alternating. And the zonary placentation character, as M. Fischer points out (this volume, Chapter 4), also occurs in aardvarks and carnivores.

This leaves only the posterior extension of the jugal, and the bifurcate M. styloglossus, as characters supporting hyrax-tethythere affinities. By contrast, most of the characters supporting hyrax-mesaxonian monophyly discussed previously (especially the Eustachian sac, iris appendages, detailed hoof morphology, and extrabullar internal carotid artery) are not only unique among the Eutheria, but truly bizarre and hard to imagine evolving in parallel.

Certainly, the next important step in testing this hypothesis is to find whether the hyracoid-mesaxonian features are found in their extinct sister-taxa. At the present, however, we feel that there is a sufficiently strong case for hyracoid-mesaxonian affinities to include the hyraxes in the perissodactyls. Consequently, we gave hyraxes full coverage in this volume, and placed hyraxes in the formal classification as a suborder of the Perissodactyla (Prothero and Schoch, this volume, Chapter 29). After all, hyraxes were in Owen's (1848) original definition of the Perissodactyla.

Some zoologists may object to the demotion of the long-established Order Hyracoidea and the confusion generated by redefining the Perissodactyla. However, this is one place where we feel a phylogenetic classification will do much more good than retaining the separate orders, with the implication that nothing is known of their relationships. With the new definition of the Perissodactyla, it becomes necessary to use a different term for the non-hyracoid perissodactyls, and Marsh's (1884) long-established term Mesaxonia becomes available. Simpson (1945, p. 136) used this term as a monotypic superorder synonymous with Perissodactyla. However, it is certainly appropriate to use Mesaxonia in the sense Marsh (1884) intended: for horses, rhinos, tapirs, and their extinct relatives, but not hyraxes.

Hyraxes have a long and successful history that is only now beginning to be appreciated (Meyer, 1978; Rasmussen, this volume, Chapter 5). Most of their evolution took place in isolation in Africa, where they had no competition from other groups of ungulates. Consequently, they developed the ecological equivalents of pigs, bovids, anthracotheres, chalicotheres, tapirs, and some equids. They also ranged in body size from the housecat-sized *Microhyrax* to the rhino-sized *Titanohyrax*. This size range is already present in the earliest known hyraxes from the middle Eocene of Algeria. They reached the known peak of their success in the Oligocene of Africa, with as many as eight genera in the Fayum deposits of Egypt. In the early Miocene, artiodactyls and mesaxonians invaded from Eurasia and many hyraxes went extinct. By the late Miocene and Pliocene, a second radiation of large, hypsodont hyraxes spread widely over Eurasia, where they competed with chalicotheres and horses. Some developed eyes and nares on the top of the skull, possibly for an aquatic habitat, or enlarged tusk-like incisors. These hyraxes lasted until the Pleistocene in China, after which the group survived only in Africa. The Quaternary African forms include the huge *Gigantohyrax*, but the three living genera are all much smaller in body size. Even so, they are specialized in their ecological habits, even though they may be sympatric on the same rock outcropping, or *kopje. Procavia* and *Heterohyrax* live mainly in areas of rocky scrub, whereas *Dendrohyrax* is arboreal. Because they live in rocky areas, they are not so severely threatened by human populations that have driven most other wild perissodactyls to the brink of extinction.

### Infraordinal relationships within the Mesaxonia

Schoch (this volume, Chapter 2) reviewed some of the early ideas about the relationships of the families of the Mesaxonia. The two most widely accepted subdivisions, Hippomorpha and Ceratomorpha of Wood (1937), have undergone many changes in meaning and acceptance. The Hippomorpha, originally consisting of horses, palaeotheres, titanotheres, and chalicotheres, has been discredited since there are no shared derived characters to support the monophyly of this grouping. The Ceratomorpha (tapiroids and rhinocerotoids), on the other hand, has been increasingly supported by shared derived characters as a good monophyletic group.

The most thorough and exhaustive effort to analyze all the shared derived characters of both the dentition and the rest of the skeleton in all the major infraordinal groups within the Mesaxonia is presented by Hooker (this volume, Chapter 6). He found three major divisions of the Mesaxonia: hippomorphs (horses, palaeotheres, and pachynolophids), "tapiromorphs" (= moropomorphs: isectolophids, chalicotheres, lophiodonts, and ceratomorphs), and titanotheres. Two rather weak characters appeared to unite the titanotheres and equoids, but Hooker chose not to use them, nor to create a group for titanotheres plus equoids. Thus, all three groups are here treated as infraorders in an unresolved trichotomy within the Suborder Mesaxonia. The Ceratomorpha are clearly supported as a monophyletic group in Hooker's analysis, but are a sister-group to the ancylopods (chalicotheres plus lophiodonts) within a larger group, the Moropomorpha (= Tapiromorpha *sensu* Hooker, this volume, Chapter 6). The Hippomorpha cannot be used in the old sense to include chalicotheres and titanotheres, but Hooker revised its contents to include pachynolophids, palaeotheres, and equids. In general, we find his

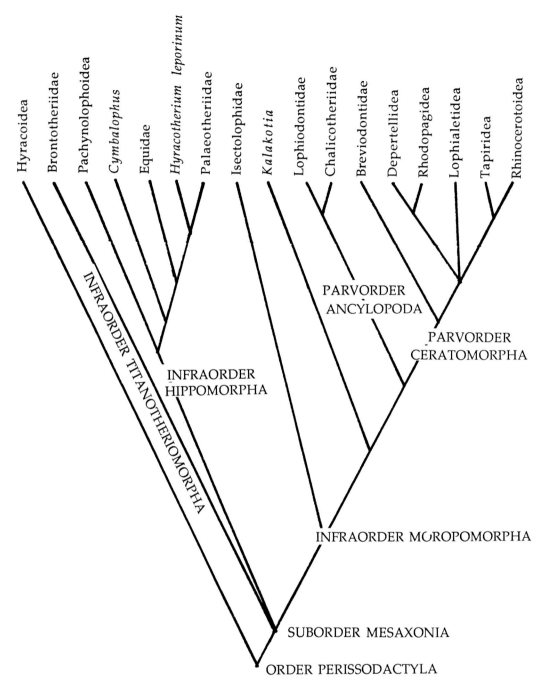

Fig. 28.2. Phylogenetic relationships of the major groups of perissodactyls. For discussion and character states at the nodes, see text and also Hooker (this volume, Chapter 6).

conclusions well supported and convincing, and so have adopted his scheme in our classification. Our primary reservation is that there are a number of unique postcranial characters (cited by Borissiak, 1945, 1946) that seem to unite chalicotheres and titanotheres. The most striking of these is the fusion of the centrale to the scaphoid in the carpus. This character would have to be a parallelism in Hooker's scheme.

**Infraorder Hippomorpha**
The first infraorder covered in this book includes the diverse and successful equids, plus their sister taxa, the pachynolophids and palaeotheres. According to Hooker, they are united by the presence of a $P^3$ paraconule and by the proximity of the optic foramen to the posteroventral orbital foramina (MacFadden, 1976, Fig. 6). The pachynolophids (including "*Hyracotherium*" sp. from Rians, France) split off first, leaving the main group of hippomorphs, the Equoidea (i.e., equids and palaeotheriids; Hooker, this volume, Fig. 6.5; see also Fig. 29.2). The Equoidea (except for *Cymbalophus*) are united by several synapomorphies: a notched preparaconule crista and preparacrista junction on the upper molars and notched protocristid on the lower molars; and foramen ovale and medial lacerate foramen separated by a narrow bony bridge (MacFadden, 1976, Fig. 5). All equoids (including *Cymbalophus*) have broad, less tapered $P_3$ trigonids. The foramen ovale bridge is not known in *Cymbalophus*, so this character might apply to the entire Equoidea.

One of the striking results of Hooker's analysis is that the primitive hippomorph *Hyracotherium* (as presently constituted) appears to be a wastebasket taxon for all primitive pachynolophids, equids, and palaeotheres. The last revision of this genus (Kitts, 1956) did not consider the stratigraphic separation of the samples and is now thought to have lumped too many species together (Gingerich, 1980; Hooker, 1980). Hooker concludes that some species should be placed in the Pachynolophidae ("*Hyracotherium*" sp. from Rians), some as sister-taxon to the Palaeotheriidae (*Hyracotherium leporinum*, the type species), and some as independent genera ("*H.*" *cuniculum*, now *Cymbalophus* ; "*H.*" *tapirinum*, which could be resurrected as *Systemodon* Cope, 1881; "*H.*" *vulpiceps*, which could be resurrected as *Pliolophus* Owen, 1858). Franzen (this volume, Chapter 7), on the other hand, views the character polarities differently. As a consequence, he restricts the content of the Palaeotheriidae considerably, and places pachynolophids, *Propalaeotherium*, and *Lophiotherium* in the equids.

It is unclear what generic name should be applied to the most primitive equids, such as "*Hyracotherium*" *vasacciense* and synonymous North American species. According to Bakker, Cooke, and Schain (unpublished manuscript, accepted but never resubmitted to this volume), the type species of *Eohippus*, *E. validus*, is related to chalicotheres, and is not a horse. Cope (1872a) originally named the common North American horse *Lophiotherium vasacciense*, but that genus refers to a European palaeothere. This animal was subsequently referred to the primate genus *Notharctus* (Cope, 1872b), and then to *Orotherium*, a Bridgerian genus that may be a synonym of *Orohippus* (Cope, 1873). The next available generic name that was applied to a Wasatchian equid is *Protorohippus* Wortman (1896), based on *Hyracotherium venticolum* Cope, 1881, from the Lost Cabin Member of the Wind River Formation (late Wasatchian). To our knowledge, this may be the first valid generic name for early Eocene equids from North America. Since the differences between these species are very slight and they share much symplesiomorphic similarity, we would not be surprised if there is resistance to breaking up the *Hyracotherium* wastebasket along cladistic lines. If the dental distinctions made by Hooker (this volume, Chapter 6) are valid, however, then we must conclude that *Hyracotherium* is more closely related to the palaeotheres, and that the North American forms must be placed in a different genus. We leave it to the next reviser of Eocene horses to decide whether the first true horse should be called *Hyracotherium*, *Systemodon*, *Orotherium*, *Protorohippus*, or some other genus.

Once the hippomorphs began to radiate, they were widespread all over Holarctica in the early Eocene. *"Hyracotherium"* is among the commonest taxa in early Eocene deposits of North America and Europe. *"Hyracotherium" gabuniai* Dashzeveg, 1979 (regarded by Hooker, 1984, as a ceratomorph) and *Propachynolophus* are reported from the early Eocene of Mongolia and China. The entire early Eocene fauna was very cosmopolitan because of several Holarctic dispersal routes along Beringia, the Greenland-Barents Shelf, and via the Wyville Thompson Ridge through Greenland, Iceland, the Faeroes, and Scotland (McKenna, 1975a; 1983a, b). By the middle Eocene, however, there was increasing endemism. Europe began to be isolated, and its perissodactyl fauna came to be dominated by pachyno-lophids and palaeotheres (along with lophiodonts). These were the only significant perissodactyls in Europe until the Grande Coupure in the Oligocene brought in rhinos and other ungulate competitors. The later Eocene of Europe was the heyday for the non-equid hippomorphs, as described by Franzen (this volume, Chapter 7). Palaeotheres got to be quite large, and some *Palaeotherium* were very similar in size and morphology to the modern tapir, complete with retracted nasals (indicating a short proboscis) and selenolophodont molars. Specimens of *Propalaeotherium* from Messel, Germany, preserve soft anatomy and stomach contents indicating that palaeotheres browsed on leaves and fruits (Sturm, 1978). Although the lophiodonts were extinct by the middle Eocene, and pachynolophids and palaeotheres were severely decimated by the Eocene/Oligocene event, some taxa (*Palaeotherium*, *Pseudopalaeotherium*, *Plagiolophus*) managed to persist into the early Oligocene. By the mid-Oligocene, however, all the non-equid hippomorphs were extinct.

## Horse evolution

While Europe was the domain of endemic palaeotheres, pachynolophids, and lophiodonts during the middle and late Eocene, horses were found in the rest of Holarctica. Dashzeveg (1979) reported a horse he called *Gobihippus menneri* from the late Eocene of Mongolia, and Zdansky (1930) reported *Propalaeotherium sinense* from the middle Eocene of China. Most of the perissodactyls in the middle and late Eocene of Asia were not equoids, however, but tapiroids, amynodonts, hyracodonts, and chalicotheres. In North America, horses formed a fairly continuous lineage from *"Hyracotherium"* (= ?*Protorohippus*) to *Orohippus* (and the doubtfully distinct *Haplohippus*) to *Epihippus* in the middle and late Eocene. By the Oligocene, however, horses became much more diverse. Contrary to the popular myth of a single lineage of horses passing gradually through *Mesohippus* and *Miohippus*, Prothero and Shubin (this volume, Chapter 10) found that both of these horses were highly speciose, with many sympatric species spanning millions of years. Nor do the two genera intergrade. *Miohippus* is a distinctly larger horse with numerous distinguishing characters, and overlaps *Mesohippus* in temporal range by almost five million years.

By the late Oligocene, *Miohippus* split into two well-established groups, the persistently primitive, browsing anchitheriine horses, and the higher-crowned, more cursorial equines. There has been little recent work on the anchitheriines, yet they were a very successful group. They persisted in small numbers in North America, often living sympatrically with many species of equines. Since they were browsing horses, they subdivided the environment with the more grazing equines. Anchitheriines got to be quite large. One species of *Hypohippus* was as big as a modern horse, and almost twice the size of its Miocene contemporaries. Contrary to popular notions that anchitheriines were slowly evolving, MacFadden (1986) has shown that the groups increased the occlusal surface area of their teeth, and corresponding body size, quite rapidly. Anchitheriines were also very successful at spreading around Holarctica. *Anchitherium* was the first post-Eocene horse to leave North America, occurring widely in the early Miocene of Europe and Asia. *Hypohippus* also traveled across the Bering Land Bridge to China, where it gave rise to the closely related *Sinohippus*. The small horse *Archaeohippus*, long thought to be an anchitheriine, is now considered to be an equine (Evander, this volume, Chapter 8; Hulbert, this volume, Chapter 11).

The beginning of the equine lineage through *Kalobatippus*, *Parahippus*, and "*Merychippus*," is becoming better known (Evander, this volume, Chapter 8). One of the biggest problems is another taxonomic wastebasket, the mid-Miocene horse "*Merychippus*." According to Evander (1986), the type species of the genus, *M. insignis*, is based on two deciduous premolars and only a few specimens can be referred to this species with any confidence. Most of the Barstovian horses referred to this genus may have different generic allocations. This is even more critical when some species of "*Merychippus*" are sister taxa to various hipparionines, and others to some equinines (Hulbert, this volume, Fig. 11.1). According to Hulbert, for example, "*Merychippus*" *carrizoensis* and "*M.*" *stylodontus* are sister-taxa of, and could be referred to, *Pliohippus*, "*M.*" *coloradense* to *Pseudhipparion* or *Neohipparion*, and "*M.*" *goorisi* to *Cormohipparion* or *Nannippus*.

Whatever nomenclature is finally adopted for these horses, it is clear that there was an enormous radiation of horses by the mid-Miocene. MacFadden (1985, 1986, 1988) reviewed much of the recent literature on evolutionary trends in Miocene horses, so there is no need to do so again here. The biggest single area of controversy is over the systematics of hipparionine horses, which were extremely diverse and migrated repeatedly to the Old World during the Miocene (Woodburne, this volume, Chapter 12; Alberdi, this volume, Chapter 13). The primary argument concerns the use of morphological characters in hipparionine systematics. Prior to the work of Skinner, MacFadden, Woodburne, and Bernor, hipparionine systematics emphasized dental characters. Since there is a tremendous amount of parallelism in dental characters, and there was much oversplitting based on trivial differences in teeth, hipparionine systematics were a mess. The work of the scientists named above (summarized by Woodburne, this volume, Chapter 12) has used additional characters, particularly the facial fossa, as evidence of hipparionine relationships.

The controversy is far from settled. Many European workers are skeptical of the facial fossa as a character (discussed by Alberdi, this volume, Chapter 13), although many (but not all) North American workers use it. MacFadden (1980, 1984) demonstrated that the facial fossa was consistent within several quarry samples of horses (for example, *Hipparion tehonense* from Frick MacAdams Quarry, Clarendonian of Texas, or *Cormohipparion occidentale* from Hans Johnson Quarry, Clarendonian of Nebraska). Nevertheless, the controversy continues (e.g., Forstén, 1982, and reply by MacFadden and Skinner, 1982; Eisenmann *et al.*, 1987, and reply by MacFadden, 1987). MacFadden (this volume, Chapter 9) discusses the issue of character variability, and shows that equids (fossil and living) are no more variable than any other group of mammals. Thus, he supports his argument that the facial fossa is not overly variable within a single population. In our opinion, any analysis that is based on more characters should be preferred to those based on a single suite of characters, unless those additional characters can clearly be shown to be due to individual variation within populations.

Regardless of the taxonomy one adopts for hipparionines, it has become clear that there were several migrations of hipparionine horses to the Old World (Woodburne, this volume, Chapter 12). According to Woodburne, there were at least two different migration events. The first took place about 12 Ma (= million years ago) with the appearance of *Hippotherium primigenium* in the Vallesian of western Europe (derived from *Cormohipparion* in North America). It was followed by the migration of *Hipparion sensu stricto* (derived from North American *Hipparion sensu stricto*) about 9.5 Ma. This clearly demolishes the old notion that a single migration of hipparions from North America marked the beginning of the Vallesian (once thought to be the Mio-Pliocene boundary, but now considered late middle Miocene) all over the Old World (Berggren and Van Couvering, 1978). Surprisingly, references to the "*Hipparion* datum" are still widely found in the literature.

Horse diversity reached an all-time peak worldwide in the late Miocene (Clarendonian-Vallesian). By the beginning of the Pliocene, all of the anchitheriines, most of the hipparionines (except for *Calippus* in North America and several Old World hipparions), and many of the archaic equinines in North America (*Protohippus, Calippus, Astrohippus,* and *Pliohippus*) were extinct. The main lineage of Pliocene to Recent horse evolution took place in the Tribe Equini (the equinines), beginning with *Dinohippus*. One group, the hippidions (*Onohippidium* and *Hippidion*), had a highly retracted narial incision, and presumably some sort of snout. They evolved in South America after the late Pliocene reconnection of the Panamanian land bridge.

Another late Pliocene immigrant to South America was the living genus *Equus*, which spread widely all over the world after its origin in North America in the early Pliocene. *Equus* also spread to the Old World around 2.6 Ma (Lindsay et al., 1980), whereupon it became common in almost all faunas. It entered Africa in the late Pliocene, where the zebras became diversified (Churcher and Richardson, 1978). *Equus* also spread widely over Asia in the Pleistocene. Because *Equus* is very abundantly represented all over the world in the Pleistocene, it has been subject to the same confusion in taxonomy as the hipparionines. This is due to oversplitting of taxa based on inadequate samples, usually isolated teeth. The conundrum of *Equus* systematics has not yet been completely resolved, but Winans (this volume, Chapter 14) attempts to resolve some of the problems of North American *Equus* by using multivariate morphometrics. Of the 59 named species, she reduces the complexity to just five distinct subgeneric groups, even fewer than recognized by Kurtén and Anderson (1980).

In terms of numbers of individuals, number of species, or ability to spread geographically, *Equus* is undoubtedly the most successful perissodactyl that ever lived. Ironically, it became extinct in its homeland, North America, during the megafaunal extinctions at the beginning of the Holocene. It also became extinct in South America, and greatly reduced in Eurasia. But as domesticated descendants of the Asian *E. przewalskii*, it has been reintroduced to these areas, as well as to places like Australia that have never had perissodactyls. Thanks to domestication, *Equus* is the only living perissodactyl that has increased in numbers and range, rather than having been diminished by the growth of human populations.

## Infraorder Moropomorpha

The next great infraorder of mesaxonians is the Moropomorpha. This group includes not only the Ceratomorpha (tapiroids and rhinocerotoids), but also the ancylopods (chalicotheres and lophiodonts), as recognized by Hooker (1984; this volume, Chapter 6). Hooker defines the Moropomorpha (= Tapiromorpha *sensu* Hooker) by the loss of the lower molar lingual postcristid branch, and the development of the lower molar hypolophid and upper molar metaloph. Both of these characters represent a precocious development of metaloph-hypolophid bilophodonty, a feature that is characteristic of nearly all moropomorphs.

Moropomorph systematics has long been in a very confused state, because of the tremendous amount of shared primitive similarity of most forms (as reviewed by Schoch, this volume, Chapters 2, 15). Much of this was cleared up by the monographs of Radinsky (1963, 1965), but his work still included some paraphyletic groups. In particular, the Family Helaletidae was long used as a wastebasket family to include all the non-isectolophid, non-tapirid "tapiroids." As is apparent from the phylogenies of Hooker (this volume, Chapter 6) and Schoch (this volume, Chapter 15), the various "tapiroids" are not a monophyletic group. Isectolophids are the most primitive sister-taxon of all other moropomorphs, but most of the "tapiroids" (*Breviodon*, deperetellids, rhodopagids, lophialetids *sensu stricto*, *Heptodon*) and Tapiroidea *sensu stricto* are united with the rhinocerotoids as the Ceratomorpha.

The Moropomorpha began with the early Eocene form *Homogalax*, which was very abundant in North American faunas, and also found in Asia (Chow and Li, 1965). By the late early Eocene, moropomorphs had begun to diversify into a variety of taxa, including the North American "tapiroids" *Heptodon*, *Helalates*, *Selenaletes*, *Plesiocolopirus*, *Desmatotherium*, *Dilophodon*, and *Isectolophus*, the ancylopod *Paleomoropus*, and the rhinocerotoid *Hyrachyus*. In the middle Eocene of Europe, there was a similar fauna which included the ancylopods *Lophiodon*, *Paralophiodon*, and *Lophiaspis*, the rhinocerotoids *Hyrachyus* and *Chasmotherium*, but no "tapiroids." Asian Middle Eocene faunas, on the other hand, contained few lophiodonts or equoids, but a great abundance of "tapiroids" (*Colodon*, *Helaletes*, *Deperetella*, *Teleolophus*, *Rhodopagus*, *Pataecops*, *Eoletes*, *Lophialetes*, *Schlosseria*, *Breviodon*), chalicotheres (*Grangeria*), and a diversity of rhinocerotoids, including *Hyrachyus*, amynodonts (*Lushiamynodon*, *Caenolophus*) and hyracodonts (*Triplopus*, *Urtinotherium*, *Forstercooperia*). The Moropomorpha reached their maximum diversity in the later Eocene, especially in Asia and North America. By the early Oligocene, their diversity had declined greatly. Of the "tapiroids," only *Colodon* and *Protapirus* survived in North America, and *Colodon* and *Teleolophus* in Asia.

Each of these groups of moropomorphs developed different specializations. Many of the "tapiroid" families developed more and more strongly bilophodont molars, presumably for browsing. The true tapiroids, in addition, began to develop a deeply incised narial notch, presumably for support of a prehensile lip or proboscis. This tendency was carried to an extreme in the Family Tapiridae, which retract the nasals nearly to the top of the head, and greatly reduce the nasal bones. Tapirs changed very little after the Oligocene, remaining at a low diversity throughout the Tertiary of North America, Europe, and Asia. In the late Pliocene, they migrated across the Panamanian Isthmus along with many other North American forms, and became established in South America. They even reached the size of a rhino with the giant form *Megatapirus* from the Pleistocene of China. In the late Pleistocene, tapirs went extinct over most of their range except for one species in southeast Asia (the Malayan tapir, *Tapirus indicus*) and three species in South America. All four species are greatly endangered, primarily due to the destruction of their tropical rain forest habitat.

## The Ancylopoda

Even more surprising is the conclusion that chalicotheres and lophiodonts were also primitive moropomorphs, unrelated to the palaeotheres or other equoids. Hooker (this volume, Chapter 6) modified Cope's (1889) taxon Ancylopoda for this group (originally constructed for the chalicotheres alone). According to Hooker, the

Ancylopoda is united by the shared possession of labially expanded $M^3$ parastyles, distal recurving of the upper molar protocone and hypocone, and lower molar protoconid and hypoconid. They also share many derived characters in the feet, first noticed by Osborn (1913).

The chalicotheres and lophiodonts typically occupied "no man's land" in perissodactyl classification. As reviewed by Schoch (this volume, Chapter 2), they were often allied with the hippomorphs (e.g., Simpson, 1945), with the titanotheres (e.g., Borissiak, 1945, 1946), or placed in their own suborder with no implication of relationships (e.g., Radinsky, 1964). Our classification (this volume, Chapter 29) reflects Hooker's conclusion that they are the sister-taxon of the Ceratomorpha, which includes most of the "tapiroids" (except isectolophids, *Kalakotia*, and *Aulaxolophus*) and rhinocerotoids. This is supported by derived characters, such as the distolingual position of the upper premolar metacone relative to the paracone, and the slight convergence of the upper molar metacone and hypocone, causing labial bending of the pre- and postmetacristae.

The close affinity of the chalicotheres and lophiodonts might also explain why there have been so many controversial forms that have been switched from one group to another. For example, *Paleomoropus* and *Lophiaspis* were assigned to the chalicotheres by Radinsky (1964), but were placed in the lophiodonts by K.-H. Fischer (1964, 1977). Similarly, *Toxotherium* and *Schizotheriodes* were placed in the tapiroids (Radinsky, 1964; Schiebout, 1977), amynodonts (Emry, 1979), or hyracodonts (Wilson and Schiebout, 1984), but Prothero et al. (1986) gave evidence to suggest that they, too, were lophiodonts. For the present, we place *Paleomoropus, Lophiaspis, Toxotherium,* and *Schizotheriodes* with the lophiodonts (Schoch, this volume, Chapter 15).

Even though they were more closely related to chalicotheres, lophiodonts converged on tapirids in many features. They reached their acme during the middle Eocene in Europe, where they were endemic forms, along with palaeotheres and pachynolophids. Some species of *Lophiodon* were rhino-sized, with huge bilophodont teeth and a tapir-like proboscis. Lophiodonts were reviewed by K.-H. Fischer (1964, 1977), although there has been little recent work on the group. Unlike the palaeotheres, pachynolophids, and "tapiroids," lophiodonts were extinct by the late Eocene, when the climate had begun to change worldwide.

Chalicotheres, on the other hand, had a unique ecological niche. They developed hook-like claws and long forelimbs, presumably for pulling down branches and browsing (Coombs, 1982, 1983). *Chalicotherium* itself had proportions much like a gorilla, and knuckle-walked with its claws held inward, like a ground sloth (Zapfe, 1979; see Fig. 24.2). Their first *bona fide* representatives are the "Eomoropidae," a paraphyletic group (discussed by Lucas and Schoch, this volume, Chapter 23) which is found in the late Eocene of both China and western North America. In the Oligocene, only *Schizotherium* is known, and it is restricted to Eurasia. Chalicotheres diversified and spread out in the early Miocene (Coombs, 1982; this volume, Chapter 24). They were a predominantly Eurasian group, although they were never particularly common, probably because they lived in a restricted forest habitat. Two subfamilies are recognized: the Schizotheriinae and the Chalicotheriinae. The Schizotheriinae had more hypsodont, elongated molars, and a special claw-retraction mechanism, but were not as gorilla-like in body proportions as the Chalicotheriinae. The chalico-theriine genus *Chalicotherium* spread to Africa in the early Miocene, and *Nestoritherium* was the last surviving member of the family, persisting until the Pleistocene in China. The schizotheriines spread to North America in the early Miocene (Hemingfordian), where well-known forms such as *Moropus* and the bizarre dome-skulled *Tylocephalonyx* occurred (Coombs, 1978, 1979). By the Pliocene, schizotheriines were extinct in both North America and Eurasia, but *Ancylotherium* persisted in the Plio-Pleistocene of Africa. Our early hominid ancestors must have known the last of the chalicotheres in both Africa and China, but sadly, chalicotheres did not survive to join the horses, hyraxes, rhinos, and tapirs as members of the living fauna.

## The Rhinocerotoidea

The largest and most ecologically diverse group of perissodactyls is the Rhinocerotoidea. Rhinocerotoids occupied an enormous range of ecological niches, from gigantic tree-top browsers (the indricotheres) to small, dog-sized running forms (the hyracodontines), to hippo-like river-dwelling grazers (some amynodonts, aceratherines, and teleoceratines), to forms with tapir-like proboscises (cadurcodontines). Although we associate rhinos with horns, most fossil rhinos were hornless. Indeed, the first horned rhinos had paired horns near the tip of their nasals, a feature that evolved twice independently. Surprisingly, there has been little detailed work on fossil rhinoceroses in over fifty years. Recently, however, there has been renewed interest in the group. Most of the recent research is reviewed by Prothero, Manning, and Hanson (1986) and Prothero, Guérin, and Manning (this volume, Chapter 16), so it is unnecessary to go over the details here.

The most primitive rhinocerotoid was *Hyrachyus*, which was widespread over Eurasia and North America in the middle Eocene. Although Radinsky (1966b, 1967, 1969) placed this taxon in the Tapiroidea based on shared primitive characters (see Hopson, this volume, Chapter 1), most authors have since placed it in the Rhinocerotoidea (Prothero *et al.*, this volume, Chapter 16). By the late middle Eocene, the three major families of rhinocerotoids had begun to diversify in North America and Asia. However, Europe was apparently cut off from Asia by the Turgai Straits in the late Eocene, allowing an endemic fauna of palaeotheres, lophiodonts, and pachynolophids (discussed earlier) to evolve. When immigrant rhinos and other ungulates entered Europe in the early Oligocene, these endemic perissodactyls went into decline.

The first family of rhinocerotoids was the Amynodontidae, reviewed by Wall (this volume, Chapter 17). Amynodonts were particularly common in late Eocene faunas of Asia, and slightly less common in North America. Beginning with primitive, long-faced forms like *Rostriamynodon* (Wall and Manning, 1986), they diverged into two subfamilies: the tapir-like cadurcodontines, which had a well-developed proboscis, and the more hippo-like, aquatic metamynodontines. Both subfamilies were reduced in diversity by the early Oligocene. In North America, only the hippo-like form *Metamynodon* survived to the mid-Oligocene, when it went extinct. In Asia, however, amynodonts persisted until the middle Miocene of Pakistan, where *Cadurcotherium* was the last survivor of this once diverse group.

The Family Hyracodontidae had a similar history of diversification and geographic dispersal. All hyracodontids, regardless of size, can be recognized by their long, slender metapodials. Beginning in the late Eocene with *Triplopus*, they were common in the middle and late Eocene of both Asia and North America. Three subfamilies are recognized. The hyracodontines were all small, cursorial forms, known primarily from North America. By the Oligocene, only *Hyracodon* was common in North America. It persisted until the end of the Whitneyan, the last of its group to go extinct. After the Grande Coupure, hyracodonts also migrated into Europe, where the small, tusked allaceropines were found (Heissig, this volume, Chapter 18).

The most spectacular hyracodonts were the indricotheres, which reached gigantic sizes. They are reviewed by Lucas and Sobus (this volume, Chapter 19). Heissig (this volume, Chapter 21) argued that the indricotheres are rhinocerotids, because one specimen of *Forstercooperia* has a primitive tetradactyl manus, rather than the derived tridactyl condition. However, the hyracodontid affinities of indricotheres are clearly supported by their metapodial elongation, which persists even in gigantic forms that by all rights should have become graviportal. In addition, the enlarged incisors of indricotheres do not resemble the chisel-tusk incisor combination seen in rhinocerotids. Beginning with the small form, *Forstercooperia*, from the late Eocene of both Asia and North America (Lucas et al., 1981), indricotheres became the largest land mammals to have ever lived, and were restricted to the Oligocene of Asia. The largest of them all was *Paraceratherium* (= *Baluchitherium, Indricotherium*, according to Lucas and Sobus, this volume, Chapter 19), which reached 18 feet (6 meters) at the shoulder, and could browse on the tops of trees. By the Miocene, indricotheres had vanished from Asia, the last of their family.

The most successful rhinocerotoids, however, were the true rhinoceroses (Family Rhinocerotidae), which include all five living species. According to Radinsky (1966b), the family is restricted to those forms with a chisel-like $I^1$ and a tusk-like $I_2$. The oldest known member of the family is *Telataceras* from the middle Eocene (Duchesnean) of Oregon, described by Hanson (this volume, Chapter 20). Hanson also refers specimens from the middle Eocene of California and Asia to this genus. By the early Oligocene, rhinocerotids had spread over Holarctica and begun to diversify, replacing groups that had been dominant in the middle Eocene, such as amynodonts, hyracodonts, palaeotheres, lophiodonts, and pachynolophids. As discussed by Heissig (this volume, Chapter 21), they included a variety of forms in Europe, such as *Ronzotherium* and *Epiaceratherium*. In North America, there were several genera, but the most successful was the *Subhyracodon-Diceratherium* lineage, which persisted for almost 20 million years.

In the late Oligocene of Europe, a number of distinct subfamilies and tribes of rhinocerotids began to diverge. There were the prehensile-lipped aceratheriines, the hippo-like teleoceratines, the paired-horned menoceratines, and the primitive members of the dicerorhinine lineage, which includes the living Sumatran rhino. In the early Miocene (Hemingfordian-Orleanian), several of these groups migrated to Asia and North America. Teleoceratines, aceratheriines, and dicerorhinines all became established in Asia in the early Miocene, where they were common elements of the fauna. North America first saw migration of the menoceratines from Europe in the latest Arikareean, followed by immigration of the aceratheriines and teleoceratines in the late Hemingfordian. The browsing aceratheriines *Aphelops* and *Peraceras*, and the grazing teleoceratine *Teleoceras* became important elements of nearly every North American Miocene fauna. Africa acquired teleoceratines, aceratheriines, and dicerorhinines in the mid-Miocene, and the endemic dicerotines (including the living African black and white rhinos) developed on that continent.

At the end of the Miocene, rhinos, like horses and many other land mammal groups, suffered greatly from the Messinian crisis and the terminal Miocene extinctions. With the exception of one Blancan rhino specimen from Beck Ranch in Texas, all of the aceratheriines and nearly all of the teleoceratines went extinct, thus wiping out the rhino fauna of North America. In Eurasia, only the rhinocerotines and dicerorhinines survived. In Africa, only the dicerotines and the last of the teleoceratines persisted until the Pliocene. The gap left by this extinction event was filled by a renewed radiation of Plio-Pleistocene rhinos, mostly from the dicerorhinines. These were widespread across Eurasia, culminating in the woolly rhinoceros, *Coelodonta*. Also characteristic of the Asian Pleistocene were the elephantine elasmotheres, which had a huge, single horn on their forehead.

They originated in China, but were restricted to Siberia and the Volga Basin and Poland in the Pleistocene. By the terminal Pleistocene extinction, most of these Eurasian rhinos became extinct.

Today, only relicts of this originally worldwide distribution of rhinos survive. Southeast Asia has two rhinocerotinines, the Indian rhino (*Rhinoceros unicornis*) and the Javan rhino (*R. sondaicus*). The last member of the long-lived dicerorhinine lineage survives in the Sumatran rhino, *Dicerorhinus sumatrensis*. In Africa, two dicerotines remain: the white rhino, *Ceratotherium simum*, and the black rhino, *Diceros bicornis*. All five of these species are being hunted to extinction by poachers, since their horns are extremely valuable (Penny, 1988). Ironically, although rhinos have long been the most diverse group of perissodactyls, they are now so threatened by humans that they may not outlast the less diverse horses, hyraxes, or tapirs.

**Infraorder Titanotheriomorpha**
The third and final major infraordinal group of mesaxonians includes the brontotheres ( = titanotheres) and their primitive sister-taxon, *Lambdotherium*. Hooker (this volume, Chapter 6) created a new suborder, Titanotheriomorpha, for this group, although we have lowered it to infraordinal rank to coordinate it with the rest of the classification (Prothero and Schoch, this volume, Chapter 29). Hooker defines this group on the basis of the following shared derived characters: an overhanging occiput, the loss of the $P^4$ metaconule, the convergence of the upper molar paracone-protocone and metacone-hypocone, the flexion of the centrocrista, a centrocristal mesostyle, the notching of the preparaconule crista-paracrista and lower protocristid, and the lingual migration of the upper metaconule and the lower preultimate molar hypoconulid.

The affinities of titanotheres have long been controversial, although typically they were clustered with the equoids in the Hippomorpha, or with the chalicotheres (see Schoch, this volume, Chapter 2). As discussed above, Hooker found two rather weak derived characters that appeared to support a relationship between titanotheres and hippomorphs (loss of $I_3$ distal cusp, upper molar cingular metastyle), but chose not to use this evidence to unite these two groups. Thus, we follow Hooker in classifying mesaxonians in an unresolved trichotomy of three infraorders: Hippomorpha, Moropomorpha, and Titanotheriomorpha (Fig. 28.2).

Of all the perissodactyl groups neglected over the last few decades, titanotheres have been the least studied. Although a few isolated papers have been published, there had been no significant reviews of the group since Osborn's 1929 monograph. Perhaps because this work was so intimidating in its size and the magnitude of its errors, and probably also because titanotheres are big and difficult to work with, it took sixty years before another scientist would critically evaluate Osborn's monograph *in toto*. Fortunately, we are able to include a revision of North American brontotheres by Mader (this volume, Chapter 25), the first and only significant review of the entire group since 1929.

Titanotheres apparently began with *Lambdotherium*, a fairly common taxon in the late early Eocene (late Wasatchian) of North America. There is some question as to whether *Lambdotherium* is really a titanothere. In an unpublished study, Wallace (1980) argued that *Lambdotherium* was really a palaeothere, but Hooker (this volume, Chapter 6) and Schoch and Lucas (1985) suggested that it had derived characters of the brontotheres. Whatever its affinities, Mader (this volume, Chapter 25) has shown that the rest of the Brontotheriidae, beginning with *Eotitanops* and *Palaeosyops*, are a good monophyletic group. *Eotitanops* is also found in the late Wasatchian, and *Palaeosyops* replaces it in the early middle Eocene (Bridgerian). Brontotheres then undergo a big radiation, diverging into several genera and spreading back and forth between North America and

Eurasia. By the late middle Eocene, they had become the largest land mammals in Eurasia and North America, sharing that niche with the uintatheres and amynodonts.

Osborn (1929) greatly oversplit the group, creating dozens of invalid species and genera, and dubious subfamilies he called "phyla." Mader (this volume, Chapter 25) reduces that mess to only seventeen valid North American genera, most from the Uintan (late middle Eocene). Lucas and Schoch (this volume, Chapter 27) show that a single quarry sample of *Duchesneodus* provides a good index of intrapopulation variability, which will be essential in future systematic studies of titanotheres. Unfortunately, there has been no similar revision of the Asian titanotheres, which were greatly oversplit by Granger and Gregory (1943) and by recent Chinese workers. We have not attempted to synonymize these invalid Asian genera in the classification adopted in this volume (Chapter 29), but the diversity of the group must surely be exaggerated. Whatever taxonomy is adopted, however, titanothere diversity was considerably reduced by the late Eocene (Chadronian). At that point, titanotheres reached their maximum size, and most had well-developed, paired blunt horns on their noses. Much work remains to be done on Chadronian titanotheres, since the best collections are still in their field wrappings in the Frick Collection of the American Museum of Natural History (New York). In Asia, the bizarre embolotheres, with their single blunt horn, were the culmination of the group in the ?Oligocene. A specimen of *Brachydiastematherium* is known from the Oligocene of Romania, but generally titanotheres are not found in European Eocene or Oligocene faunas (Lucas and Schoch, this volume, Chapter 26).

At the peak of their size and horn development, titanotheres became extinct. Earlier workers attributed their extinction to factors such as "racial senescence," but recent work has shown that the extinction of titanotheres coincides with the extinction of a number of archaic forms at the end of the Eocene (labeled the "mid-Oligocene event" by Prothero, 1985). Titanotheres were among the many victims of the terminal Eocene climatic event that resulted in global cooling and glaciation, lowered sea level, and resulting changes in vegetation (reviewed by Prothero, 1985). Once they became extinct, titanotheres were never truly replaced in North America. Oligocene and Miocene rhinos never reached their size. In Asia titanotheres competed with, and were succeeded by, the giant indricotheres, although both groups were extinct by the Miocene.

## Acknowledgments

We thank J. J. Hooker, S. G. Lucas, and M. C. McKenna for helpful reviews of this chapter and the classification which follows. M.C. Coombs, C. Guérin, K. Heissig, B. J. Mader, E. Manning, T. Rasmussen, and M.O. Woodburne graciously contributed or checked much of the stratigraphic range data for Fig. 28.1. The senior author was partially supported by a Guggenheim Fellowship, and by NSF grant EAR87-08221 during the preparation of this paper.

## Bibliography

Alberdi, M.-T. (1989): A review of Old World hipparionine horses (this volume, Chapter 13).

Bakker, R.A., Cooke, J.C., and Schain, J. (in prep.): The dawn horses revisited and the basal bushiness of the Perissodactyla. -*Hunteria* (in press).

Berggren, W.A. and Van Couvering, J.A. (1978): Biochronology. -*Amer. Assoc. Petrol. Geol. Stud. Geol.*, 6: 39-56.

Borissiak, A.A. (1945): The chalicotheres as a biological type. -*Amer. J. Sci.*, 243: 667-679.

Borissiak, A.A.(1946): [A new chalicothere from the Tertiary of Kasakhstan]. -*Akad. Nauk. SSSR, Trudy Paleont. Inst.*, 13 (3): 1-134.

Cappetta, H., Jaeger, J.-J., Sabatier, M., Sudre, J., and Vieney-Liaud, M. (1978): Découverte dans le Paléocène du Maroc des plus anciens mammifères euthériens d'Afrique. -*Geobios*, 11(2): 257-263.

Carroll, R.L. (1988): *Vertebrate Paleontology and Evolution*. -New York (W.H. Freeman).

Chow, M., and Li, C. (1965): *Homogalax* and *Heptodon* of Shantung. -*Vert. PalAsiatica*, 9: 19-21.

Churcher, C.S., and Richardson, M.L. (1978): Equidae. -In: Maglio, V.J., and Cooke, H.B.S. (eds): *Evolution of African mammals*. - Cambridge, Mass. (Harvard Univ. Press), pp. 379-422.

Cifelli, R.L. (1982): The petrosal structure of *Hyopsodus* with respect to that of some other ungulates, and its phylogenetic implications. -*J. Paleont.*, 56: 795-805.

Coombs, M.C. (1978): Reevaluation of early Miocene North American *Moropus* (Perissodactyla, Chalicotheriidae, Schizotheriinae). -*Bull. Carnegie Mus. Nat. Hist.*, 4: 1-62.

Coombs, M.C. (1979): *Tylocephalonyx*, a new genus of North American dome-skulled chalicothere (Mammalia, Perissodactyla). -*Bull. Amer. Mus. Nat. Hist.*, 164: 1-64.

Coombs, M.C. (1982): Chalicotheres (Perissodactyla) as large terrestrial mammals. -*Proc. Third North Amer. Paleo. Conv.*, 1: 99-103.

Coombs, M.C. (1983): Large mammalian clawed herbivores: a comparative study. -*Trans. Amer. Phil. Soc.*, 73(7): 1-96.

Coombs, M. C. (1989): Interrelationships and diversity in the Chalicotheriidae (this volume, Chapter 24).

Cope, E.D. (1872a): On a new genus of Pleurodira from the Eocene of Wyoming. -*Proc. Amer. Phil. Soc.*, 12: 1-6.

Cope, E.D. (1872b): Third account of new Vertebrata from the Bridger Eocene of Wyoming Valley. -*Paleont. Bull.*, 3: 1-4.

Cope, E.D. (1873): On the extinct Vertebrata of the Eocene of Wyoming, observed by the expedition of 1872, with notes on the geology. -*Sixth Ann. Rept. U.S. Geol. Surv. Terr.*, pp. 545-649.

Cope, E.D. (1881): On the Vertebrata of the Wind River Eocene beds of Wyoming. - *Bull. U.S. Geol. Geogr. Surv. Terr.*, 6: 183-202.

Cope, E.D. (1889): The Vertebrata of the Swift Current River, II. -*Amer. Nat.*, 23: 151-155.

Dashzeveg, D. (1979): On an archaic representative of the equoids (Mammalia, Perissodactyla) from the Eocene of central Asia. -*Trans. Joint Soviet-Mongolian Paleont. Exped.*, 8: 10-22.

Domning, D.P., Ray, C.E., and McKenna, M.C. (1986): Two new Oligocene desmostylians and a discussion of tethytherian systematics. -*Smithson. Contrib. Paleobiol.*, 59: 1-56.

Eisenmann, V., Sondaar, P., Alberdi, M.-T., and De Giuli, C. (1987): Is horse phylogeny becoming a playfield in the game of theoretical evolution? -*J. Vert. Paleont.*, 7(2): 224-229.

Emry, R.J. (1979): Review of *Toxotherium* (Perissodactyla: Rhinocerotoidea) with new material from the early Oligocene of Wyoming. -*Proc. Biol. Soc. Washington*, 92: 28-41.

Evander, R. (1986): The taxonomic status of *Merychippus insignis* Leidy. -*J. Paleont.*, 60: 1277-1279.

Evander, R. (1989): Phylogeny of the family Equidae (this volume, Chapter 8).

Fischer, K.-H. (1964): Die tapiroiden Perissodactylen aus der eozänen Braunkohle des Geiseltales. -*Geologie,* Berlin, 43: 1-101.

Fischer, K.-H. (1977): Neue funde von *Rhinocerolophiodon* (n. gen.), *Lophiodon*, und *Hyrachyus* (Ceratomorpha, Perissodactyla, Mammalia) aus dem Eozän des Geiseltals bei Halle (DDR). - *Zeit. Geol. Wiss.*, 5(7): 909-919.

Fischer, M.S. (1986): Die Stellung der Schliefer (Hyracoidea) im phylogenetischen System der Eutheria. -*Cour. Forsch. -Inst. Senckenberg*, 84: 1-132.

Fischer, M. S. (1989): Hyracoids, the sister-group of perissodactyls (this volume, Chapter 4).

Forstén, A.M. (1982): The status of the genus *Cormohipparion* Skinner and Mac-

Fadden (Mammalia, Equidae). -*J. Paleont.*, 56: 1332-1335.

Franz, V. (1924): *Die Geschichte der Organismen.* -Jena (G. Fischer).

Franzen, J. L. (1989): Origin and systematic position of the Palaeotheriidae (this volume, Chapter 7).

Gingerich, P.D. (1976): Cranial anatomy and evolution of early Tertiary Plesiadapidae (Mammalia, Primates). - *Univ. Mich. Pap. Paleont.* 15: 1-141.

Gingerich, P.D. (1980): Evolutionary patterns in early Cenozoic mammals. -*Ann. Rev. Earth Planet. Sci.*, 8: 407-424.

Gingerich, P.D. (1981): Variation, sexual dimorphism, and social structure in the early Eocene horse *Hyracotherium* (Mammalia, Perissodactyla). -*Paleobiology*, 7(4): 443-455.

Granger, W., and Gregory, W.K. (1943): A revision of Mongolian titanotheres. - *Bull. Amer. Mus. Nat. Hist.*, 80(10): 349-389.

Gregory, W. K. (1910): The orders of mammals. -*Bull. Amer. Mus. Nat. Hist.*, 27: 1-524.

Hanson, C. B. (1989): *Teletaceras radinskyi*, a new primitive rhinocerotid from the late Eocene Clarno Formation, Oregon (this volume, Chapter 20).

Heissig, K. (1989a): The allaceropine hyracodonts (this volume, Chapter 18).

Heissig, K. (1989b): The Rhinocerotidae (this volume, Chapter 21).

Hooker, J. J. (1980): The succession of *Hyracotherium* (Perissodactyla, Mammalia) in the English early Eocene. -*Bull. Brit. Mus. Nat. Hist. (Geol.)*, 33(2): 101-114.

Hooker, J. J. (1984): A primitive ceratomorph (Perissodactyla, Mammalia) from the early Tertiary of Europe. -*Zool. J. Linn. Soc. London*, 82: 229-244.

Hooker, J. J. (1989): Character polarities in early perissodactyls and their significance for *Hyracotherium* and infraordinal relationships (this volume, Chapter 6).

Hopson, J. A. (1989): Leonard Burton Radinsky (1937-1985) (this volume, Chapter 1).

Hulbert, R. C., Jr. (1989): Phylogenetic interrelationships and evolution of North American late Neogene Equidae (this volume, Chapter 11).

Kitts, D.B. (1956): American *Hyracotherium* (Perissodactyla, Equidae). - *Bull. Amer. Mus. Nat. Hist.*, 110(1): 7-60.

Kurtén, B., and Anderson, E. (1980): *Pleistocene Mammals of North America.* - New York (Columbia Univ. Press).

Lindsay, E.H., Opdyke, N.D., and Johnson, N.M. (1980): Pliocene dispersal of the horse *Equus* and late Cenozoic mammalian dispersal events. -*Nature*, 87: 135-138.

Lucas, S. G., and Schoch, R. M. (1989a): European brontotheres (this volume, Chapter 26).

Lucas, S. G., and Schoch, R. M. (1989b): Taxonomy of *Duchesneodus* (Brontotheriidae) from the late Eocene of North America (this volume, Chapter 27).

Lucas, S.G., Schoch, R.M., and Manning, E. (1981): The systematics of *Forstercooperia*, a middle to late Eocene hyracodontid (Perissodactyla: Rhinocerotoidea) from Asia and western North America. - *J. Paleont.*, 55: 826-841.

Lucas, S. G., and Sobus, J. C. (1989): The systematics of indricotheres (this volume, Chapter 19).

MacFadden, B.J. (1976): Cladistic analysis of primitive equids, with notes on other perissodactyls. -*Syst. Zool.*, 24: 1-14.

MacFadden, B.J. (1980): The Miocene horse *Hipparion* from North America and from the type locality in southern France. - *Palaeont.*, 23: 617-635.

MacFadden, B.J. (1984): Systematics and phylogeny of *Hipparion, Neohipparion, Nannippus*, and *Cormohipparion* (Mammalia, Equidae) from the Miocene and Pliocene of the New World. -*Bull. Amer. Mus. Nat. Hist.*, 179(1): 1-195.

MacFadden, B.J. (1985): Patterns of phylogeny and rates of evolution in fossil horses: hipparions from the Miocene and Pliocene of North America. -*Paleobiology*, 11: 245-257.

MacFadden, B.J. (1986): Fossil horses from

"Eohippus" (*Hyracotherium*) to *Equus*: scaling, Cope's Law, and the evolution of body size. -*Paleobiology*, 12(4): 355-369.

MacFadden, B.J. (1987): Systematics, phylogeny, and evolution of fossil horses: a rational alternative to Eisenmann et al. (1987). -*J. Vert. Paleont.*, 7(2): 230-235.

MacFadden, B.J. (1988): Horses, the fossil record, and evolution, a current perspective. -*Evol. Biol.*, 22: 131-158.

MacFadden, B. J. (1989): Dental character variation in paleopopulations and morphospecies of fossil horses (this volume, Chapter 9).

MacFadden, B.J., and Skinner, M.F. (1982): Hipparion horses and modern phylogenetic interpretation: comments on Forstén's view of *Cormohipparion*. -*J. Paleont.*, 56: 1336-1342.

Mader, B. J. (1989): The Brontotheriidae: a systematic revision and preliminary phylogeny of North American genera (this volume, Chapter 25).

Marsh, O.C. (1884): Dinocerata. A monograph of an extinct order of gigantic mammals. -*Monogr. U.S. Geol. Surv.*, 10: 1-237.

McKenna, M.C. (1975a): Fossil mammals and early Eocene North American land continuity. -*Ann. Missouri Bot. Garden*, 62: 335-353.

McKenna, M.C. (1975b): Toward a phylogenetic classification of the Mammalia. - In: Luckett, W.P., and Szalay, F.S. (eds.): *Phylogeny of the Primates.* - New York (Plenum), pp. 21-46.

McKenna, M. C. (1983a): Cenozoic paleogeography of North Atlantic land bridges.- In: Bott, M. H., Saxov, S., Talwani, M., and Thiede, J. (eds.): *Structure and Development of the Greenland-Scotland Ridge.*- New York (Plenum), pp. 351-399.

McKenna, M. C. (1983b): Holarctic land mass rearrangement, cosmic events, and Cenozoic terrestrial organisms. - *Ann. Missouri Bot. Garden*, 70: 459-489.

McKenna, M. C. (1987): Molecular and morphological analysis of high-level mammalian interrelationships. - In: Patterson, C. (ed.): *Molecules and Morphology: Conflict or Compromise?* - Cambridge (Cambridge Univ. Press), pp. 55-92.

McKenna, M. C., Chow, M. C., Ting, S. Y., and Luo, Z. (1989): *Radinskya yupingae*, a perissodactyl-like mammal from the late Paleocene of southern China (this volume, Chapter 3).

McKenna, M.C., Engelmann, G.F., and Barghoorn, S.F. (1977): Review of "Cranial anatomy and evolution of early Tertiary Plesiadapidae (Mammalia, Primates)" by Philip D. Gingerich. - *Syst. Zool.*, 26(2): 233-238.

McKenna, M.C., and Manning, E. (1977): Affinities and palaeobiogeographic significance of the Mongolian Paleogene genus *Phenacolophus*. -*Geobios, Mém. Spec.*, 1: 61-85.

Meyer, G. (1978): Hyracoidea. -In: Maglio, V.J., and Cooke, H.B.S. (eds): *Evolution of African Mammals*. Cambridge, Mass. (Harvard Univ. Press), pp. 284-314.

Monroe, J.S. (1985): Basic created kinds and the fossil record of perissodactyls. - *Creation/Evol.*, 5(2): 4-30.

Novacek, M.J. (1982): Information for molecular studies from anatomical and fossil evidence on higher eutherian phylogeny. -In: Goodman, M. (ed): *Macromolecular Sequences in Systematic and Evolutionary Biology*. New York (Plenum Press), pp. 2-41

Novacek, M.J. (1986): The skull of leptictid insectivorans and the higher-level classification of eutherian mammals. -*Bull. Amer. Mus. Nat. Hist.*, 183: 1-112.

Novacek, M.J., and Wyss, A. (1986): Higher-level relationships of the recent eutherian orders: morphological evidence. -*Cladistics*, 2: 257-287.

Novacek, M. J., Wyss, A. R., and McKenna, M.C. (1988): The major groups of eutherian mammals.- In: Benton, M.J. (ed.): *The Phylogeny and Classification of the Tetrapods*. Oxford (Clarendon Press), 2: 31-71.

Osborn, H.F. (1913): *Eomoropus*, an American Eocene chalicothere. -*Bull. Amer.*

*Mus. Nat. Hist.*, 23: 261-274.

Osborn, H.F. (1929): The titanotheres of ancient Wyoming, Dakota, and Nebraska. -*Monogr. U.S. Geol. Surv.*, 55: 1-953 (2 vols.).

Owen, R. (1848): Description of the teeth and portions of jaws of two extinct anthracotheroid quadrupeds (*Hyopotamus vectianus* and *Hyop. bovinus*) discovered by the Marchioness of Hastings in the Eocene deposits of the N.W. coast of the Isle of Wight: with an attempt to develop Cuvier's idea of the classification of Pachyderms by the number of their toes. -*Q. J. Geol. Soc. London*, 4: 103-141.

Owen, R. (1858): Description of a small lophiodont mammal (*Pliolophus vulpiceps* Owen) from the London Clay near Harwich. -*Quart. J. Geol. Soc. London*, 14: 54-71.

Penny, M. (1988): *Rhinos, Endangered Species.* -New York (Facts on File).

Prothero, D.R. (1985): North American mammalian diversity and Eocene-Oligocene extinctions. -*Paleobiology*, 11(4): 389-405.

Prothero, D. R., Guérin, C., and Manning, E.M. (1989): The history of the Rhinocerotoidea (this volume, Chapter 16).

Prothero, D.R., Manning, E.M., and Fischer, M. (1988): The phylogeny of the ungulates. -In: Benton, M.J. (ed.): *The Phylogeny and Classification of the Tetrapods.* Oxford (Clarendon Press), 2: 201-234.

Prothero, D.R., Manning, E.M. and Hanson, C.B. (1986): The phylogeny of the Rhinocerotoidea (Mammalia, Perissodactyla). -*Zool. J. Linn. Soc.*, 87: 341-366.

Prothero, D. R., and Schoch, R. M. (1989): Classification of the Perissodactyla (this volume, Chapter 29).

Prothero, D. R., and Shubin, N. (1989): The evolution of Oligocene horses (this volume, Chapter 10).

Radinsky, L. B. (1963): Origin and evolution of North American Tapiroidea. -*Bull. Yale Peabody Mus.*, 17: 1-106.

Radinsky, L. B. (1964): *Paleomoropus,* a new early Eocene chalicothere (Mammalia, Perissodactyla), and a revision of Eocene chalicotheres. -*Amer. Mus. Novit.*, 2179: 1-28.

Radinsky, L. B. (1965): Early Eocene Tapiroidea of Asia. -*Bull. Amer. Mus. Nat. Hist.*, 129: 181-262.

Radinsky, L. B. (1966a): The adaptive radiation of the phenacodontid condylarths and the origin of the Perissodactyla. -*Evolution,* 20: 408-417.

Radinsky, L.B. (1966b): The families of the Rhinocerotoidea (Mammalia, Perissodactyla). -*J. Mamm.*, 47: 631-639.

Radinsky, L.B. (1967): *Hyrachyus, Chasmotherium,* and the early evolution of helaletid tapiroids. -*Amer. Mus. Novit.*, 2313: 1-23.

Radinsky, L. B. (1969): The early evolution of the Perissodactyla. -*Evolution*, 23: 308-328.

Rasmussen, D. T. (1989): The evolution of the Hyracoidea: a review of the fossil evidence (this volume, Chapter 5).

Savage, D. E., and Russell, D. E. (1983) *Mammalian Paleofaunas of the World.* -Reading, Mass. (Addison-Wesley).

Savage, R.J.G. (1969): Early Tertiary mammal locality in southern Libya. -*Proc. Geol. Soc. London*, 1648: 98-101.

Savage, R.J.G., and Long , M.R. (1986): *Mammal Evolution, an Illustrated Guide.* -New York (Facts on File).

Schiebout, J.A. (1977): *Schizotheroides* [sic] (Mammalia, Perissodactyla) from the Oligocene of Trans-Pecos Texas. -*J. Paleont.*, 51: 455-458.

Schoch, R. M. (1989a): A brief historical review of perissodactyl classification (this volume, Chapter 2).

Schoch, R. M. (1989b): A review of the tapiroids (this volume, Chapter 15).

Schoch, R.M., and Lucas, S.G. (1985): The Brontotheriidae, a group of Eocene and Oligocene perissodactyls from North America, Asia and Eastern Europe. -In: Fuller, W.A., Nietfeld, M.T., and Harris, M.A. (eds.): *Abstracts of papers and posters, Fourth Int. Theriological Congress, Edmonton, Alberta, Canada,* abstract number 0558.

Simpson, G.G. (1945): The principles of classification and a classification of mammals. -*Bull. Amer. Mus. Nat. Hist.*, 85: 1-350.

Sloan, R.E. (1970): Cretaceous and Paleocene terrestrial communities of western North America. -*Proc. North Amer. Paleon. Conv.* E: 427-453.

Sloan, R.E. (1987): Paleocene and latest Cretaceous mammal ages, biozones, magnetozones, rates of sedimentation and evolution. -*Geol. Soc. Amer. Spec. Paper*, 209: 165-200.

Sturm, M. (1978): Maw contents of an Eocene horse (*Propalaeotherium*) out of the oil shale of Messel near Darmstadt. -*Cour. Forsch. -Inst. Senckenberg*, 30: 120-122.

Sudre, J. (1979): Nouveaux Mammifères éocènes du Sahara occidental. -*Palaeovert.*, 9: 83-115.

Van Valen, L. (1978): The beginning of the age of mammals. - *Evol. Theory*, 4: 45-80.

Wall, W. P. (1989): The phylogenetic history and adaptive radiation of the Amynodontidae (this volume, Chapter 17).

Wall, W., and Manning, E. (1986): *Rostriamynodon grangeri*, n.gen., n. sp. of amynodontid (Perissodactyla, Rhinocerotoidea) with comments on the phylogenetic history of Eocene Amynodontidae. -*J. Paleont.*, 60(4): 911-919.

Wallace, S.M. (1980): A revision of North American early Eocene Brontotheriidae (Mammalia, Perissodactyla). -Univ. Colorado (Master's thesis), 1-157.

Wible, J. R. (1986): Transformation in the extracranial course of the internal carotid artery in mammalian phylogeny. - *J. Vert. Paleont.*, 6: 313-325.

Wible, J.R. (1987): The eutherian stapedial artery: character analysis and implications for superordinal relationships. -*Zool. J. Linn. Soc.*, 91: 107-135.

Wilson, J.A., and Schiebout, J.A. (1984): Early Tertiary vertebrate faunas, Trans-Pecos Texas: Ceratomorpha less Amynodontidae. -*Pearce-Sellards Series, Tex. Mem. Mus.*, 39: 1-47.

Winans, M. C. (1989): A quantitative study of North American fossil species of the genus *Equus* (this volume, Chapter 14).

Wood, H.E. II (1937): Perissodactyl suborders. -*J. Mamm.*, 18: 106.

Woodburne, M. O. (1989): Hipparion horses: a pattern of worldwide dispersal and endemic evolution (this volume, Chapter 12).

Wortman, J. L. (1896): Species of *Hyracotherium* and allied Perissodactyla from the Wahsatch and Wind River beds of North America. -*Bull. Amer. Mus. Nat. Hist.*, 8: 81-110.

Wyss, A.R., Novacek, M.J., and McKenna, M.C. (1987): Amino acid sequence versus morphological data and the interordinal relationships of mammals. -*Mol. Biol. Evol.*, 4(2): 99-116.

Zapfe, H. (1979): *Chalicotherium grande* (Blainv.) aus der miozänen Spaltenfüllung von Neudorf an der March (Devinská Nová Ves), Tschechoslowakei. -*Neue Denkschriften Naturhist. Mus. Wien*, 2: 1-282.

Zdansky, O. (1930): Die alttertiären Säugetiere Chinas nebst Stratigraphischen Bemerkungen. -*Paleontol. Sinica*, C (2): 1-87.

# 29. CLASSIFICATION OF THE PERISSODACTYLA

## DONALD R. PROTHERO and ROBERT M. SCHOCH

The classification below includes all the valid genera of perissodactyls that we currently recognize. The classification attempts to be phylogenetic, as far as possible. We have created the minimum number of new taxa necessary to establish the phylogenetic relationships, but have changed the rank of many previously established suprageneric taxa to maintain their correct place in the hierarchy. The phylogenetic relationships in this classification are based on the phylogenies found in the summary chapter (Prothero and Schoch, this volume, Chapter 28). Original references to the names cited below can be found in Simpson (1945), the *Zoological Record* (published annually by the Zoological Society of London), and the series, *Bibliography of Fossil Vertebrates* (see Gregory et al., 1988).

Class Mammalia Linnaeus, 1758
Subclass Theria Parker and Haswell, 1897
Infraclass Eutheria Gill, 1872
Superorder Ungulata Linnaeus, 1766
  Grandorder Altungulata  Prothero and Schoch, 1989 (this volume, Chapter 28)
    Order Tethytheria McKenna, 1975 (new rank) [contents not subdivided here]
    Order Embrithopoda Andrews, 1906 [contents not subdivided here, except those relevant to perissodactyls]
        *Radinskya* McKenna, Chow, Ting, and Luo, 1989 (this volume, Chapter 3)
        *Heptaconodon*  Zdansky, 1930
        [?*Lunania*  Chow, 1957]
    Order Perissodactyla Owen, 1848
      Suborder Hyracoidea  Huxley, 1869
        Family Pliohyracidae Osborn, 1899
          Subfamily Geniohyinae Andrews, 1906
            *Seggeurius* Crochet, 1986
            *Geniohyus*  Andrews, 1904
          Subfamily Saghatheriinae Andrews, 1906
            *Microhyrax*  Sudre, 1979
            *Bunohyrax*  Schlosser, 1910
            *Pachyhyrax*  Schlosser, 1910
            *Megalohyrax*  Andrews, 1903

             *Saghatherium* Andrews and Beadnell, 1902
             *Titanohyrax* Matsumoto, 1922
             *Selenohyrax* Rasmussen and Simons, 1988
             *Thyrohyrax* Meyer, 1973
             *Meroehyrax* Whitworth, 1954
          Subfamily Pliohyracinae Osborn, 1899
             *Prohyrax* Stromer, 1926
             *Parapliohyrax* Lavocat, 1961
             *Pliohyrax* Osborn, 1899
             *Soqdohyrax* Dubrovo, 1978
             *Kvabebihyrax* Gabunia and Vekua, 1966
             *Postschizotherium* von Koenigswald, 1932
       Family Procaviidae Thomas, 1892
             *Gigantohyrax* Kitching, 1965
             *Procavia* Storr, 1780
             *Heterohyrax* Gray, 1868
             *Dendrohyrax* Gray, 1868

Suborder Mesaxonia Marsh, 1884
  Infraorder Hippomorpha Wood, 1937
    Superfamily Pachynolophoidea Pavlow, 1888
      Family Pachynolophidae Pavlow, 1888
          *Pachynolophus* Pomel, 1847
          *Anchilophus* Gervais, 1852
    Superfamily Equoidea Gray, 1821
          *Cymbalophus* Hooker, 1984 (= "*Hyracotherium*" *cuniculum*)
          *Systemodon* (= "*Hyracotherium*" *tapirinum*) Cope, 1881
          *Pliolophus* (= "*Hyracotherium*" *vulpiceps*) Owen, 1858
          *Hyracotherium sensu stricto* Owen, 1841   (not Owen, 1840, *contra* Simpson, 1945)
      Family Equidae Gray, 1821
          *Protorohippus* Wortman, 1896
          *Gobihippus* Dashzeveg, 1979
          *Orohippus* Marsh, 1872
          *Haplohippus* McGrew, 1953
          *Epihippus* Marsh, 1877 (= *Duchesnehippus*)
          *Mesohippus* Marsh, 1875
          *Miohippus* Marsh, 1874 (= *Pediohippus*)
        Subfamily Anchitheriinae Leidy, 1869
          *Anchitherium* Meyer, 1844
          *Hypohippus* Leidy, 1858
          *Megahippus* McGrew, 1938
          *Sinohippus* Zhai, 1962
        Subfamily Equinae Gray, 1821
          *Kalobatippus* Osborn, 1915
          *Archaeohippus* Gidley, 1906
          *Parahippus* Leidy, 1858
          *Merychippus sensu stricto* Leidy, 1857

Tribe Hippotheriini Bonaparte, 1850 (new rank) [= Hipparionini Quinn, 1955]
*Cormohipparion* Skinner and MacFadden, 1977
*Nannippus* Matthew, 1926
*Hipparion* de Christol, 1832 (= *Proboscidipparion, Notohipparion*)
*Neohipparion* Gidley, 1903
*Pseudhipparion* Ameghino, 1904 (= *Griphippus*)
*Stylohipparion* van Hoepen, 1932
*Hippotherium* Kaup, 1833
Tribe Equini Gray, 1821
*Protohippus* Leidy, 1858
*Calippus* Matthew and Stirton, 1930
*Pliohippus* Marsh, 1874
*Astrohippus* Stirton, 1940
*Hippidion* Owen, 1869 (= *Parahipparion, Hyperhippidium*)
*Onohippidium* Moreno, 1891
*Dinohippus* Quinn, 1955
*Equus* Linnaeus, 1858 (= *Plesippus, Asinus, Hippotigris, Dolichohippus, Onager, Hemionus, Kraterohippus, Kolpohippus, Sterrohippus, Neohippus*)
Family Palaeotheriidae Bonaparte, 1850
*Propachynolophus* Lemoine, 1891
*Propalaeotherium* Gervais, 1849
*Lophiotherium* Gervais, 1849
*Palaeotherium* Cuvier, 1804
*Plagiolophus* Pomel, 1847 (=*Paloplotherium*)
*Pseudopalaeotherium* Franzen, 1972
*Paraplagiolophus* Depéret, 1917
*Leptolophus* Remy, 1965
*Cantabrotherium* Casanova-Cladellas and Santafé-Llopis, 1987
?Hippomorpha *incertae sedis*
Family Indolophidae Schoch, 1984
*Indolophus* Pilgrim, 1925

Infraorder Moropomorpha Schoch, 1984 (new rank, =Tapiromorpha *sensu* Hooker, 1984)
Subinfraorder Isectolophomorpha, new taxon
Family Isectolophidae Peterson, 1919
*Homogalax* Hay, 1899
*Isectolophus* Scott and Osborn, 1887 (= *Parisectolophus, Schizolophodon*)
*Sastrilophus* Sahni and Khare, 1971
Subinfraorder Rhinotapiromorpha, new taxon
*Kalakotia* Ranga Rao, 1972
*Aulaxolophus* Ranga Rao, 1972
Parvorder Ancylopoda Cope, 1889 (new rank)
Superfamily Lophiodontoidea Gill, 1872
Family Lophiodontidae Gill, 1872
*Lophiodon* Cuvier, 1822 (= *Pernatherium,* ?*Hypsolophiodon,* ?*Leptolophiodon*)

*Paralophiodon* Dedieu, 1977 (= *Rhinocerolophiodon*)
*Schizotheriodes* Hough, 1955
*Toxotherium* Wood, 1961
*Paleomoropus* Radinsky, 1964
*Lophiaspis* Depéret, 1910
*Atalonodon* Dal Piaz, 1929
*Lophiodochoerus* Lemoine, 1880
Superfamily Chalicotherioidea Gill, 1872
Family Chalicotheriidae Gill, 1872
*Litolophus* Radinsky, 1964
*Olsenia* Matthew and Granger, 1925
*Eomoropus* Osborn, 1913
*Grangeria* Zdansky, 1930
[?*Lunania* Chow, 1957]
Subfamily Schizotheriinae Holland and Peterson, 1913
*Borissiakia* Butler, 1965
*Phylotillon* Pilgrim, 1910
*Macrotherium* Lartet, 1837
*Moropus* Marsh, 1877
*Tylocephalonyx* Coombs, 1979
*Ancylotherium* Gaudry, 1863 (= *Metaschizotherium, Postschizotherium, Hoangkotherium, Gansuodon, Chemositia*)
Subfamily Chalicotheriinae Gill, 1872
*Chalicotherium* Kaup, 1833
*Nestoritherium* Kaup, 1859

Parvorder Ceratomorpha Wood, 1937 (new rank)
*Breviodon* Radinsky, 1965
*Parabreviodon* Reshetov, 1975
Magnafamily Deperetellidea Radinsky, 1965
Family Deperetellidae Radinsky, 1965
*Deperetella* Matthew and Granger, 1925 (= *Cristindentinus, Diplolophodon*)
*Teleolophus* Matthew and Granger, 1925
Family Rhodopagidae Reshetov, 1975
*Rhodopagus* Radinsky, 1965
*Pataecops* Radinsky, 1966 (= *Pataecus*)
Magnafamily Lophialetidea Matthew and Granger, 1925
Family Eoletidae Schoch, 1989 (this volume, Chapter 2)
*Eoletes* Birjukov, 1975
Family Lophialetidae Matthew and Granger, 1925
*Schlosseria* Matthew and Granger, 1926
*Simplaletes* Qi, 1980
*Lophialetes* Matthew and Granger, 1925
Magnafamily Tapiridea Burnett, 1830
*Heptodon* Cope, 1882
Superfamily Tapiroidea Burnett, 1830
*Helaletes* Marsh, 1872 (=*Veragromovia, Chasmotheroides*)

      *Desmatotherium* Scott, 1883
      *Selenaletes* Radinsky, 1966
      *Dilophodon* Scott, 1883 (= *Heteraletes*)
      *Haagella* Heissig, 1978
      *Plesiocolopirus* Schoch, 1989 (this volume, Chapter 15)
      *Colodon* Marsh, 1890 (= *Paracolodon, Mesotapirus*)
   Family Tapiridae Burnett, 1830
      *Protapirus* Filhol, 1877 (= *Tanyops*)
      *Eotapirus* Cerdeno and Ginsburg, 1988
      *Miotapirus* Schlaikjer, 1937
      *Palaeotapirus* Filhol, 1888
      *Paratapirus* Depéret, 1902
      *Tapiriscus* Kretzoi, 1951
      *Tapiravus* Marsh, 1877
      *Megatapirus* Matthew and Granger, 1923
      *Tapirus* Brünnich, 1772 (= *Pinchacus, Cinchacus, Tapirella,*
         *Elasmognathus, Acrocodia, Tapir, Syspotamus, Rhinochoerus, Tapyra*)

Superfamily Rhinocerotoidea Owen, 1845
      *Hyrachyus* Leidy, 1871 (= *Colonoceras, Metahyrachyus, Ephyrachyus*)
      *Chasmotherium* Rütimeyer, 1862
   Family Amynodontidae Scott and Osborn, 1883
      *Caenolophus* Matthew and Granger, 1925 (= *Teilhardina, Euryodon*)
      *Rostriamynodon* Wall and Manning, 1986
      *Amynodon* Marsh, 1877 (= *Orthocynodon*)
     Subfamily Cadurcodontinae Wall, 1982 (new rank)
      *Sharamynodon* Kretzoi, 1942 (= *Lushiamynodon*)
      *Amynodontopsis* Stock, 1933
      *Sianodon* Xu, 1965
      *Cadurcodon* Kretzoi, 1942 (= *Paracadurcodon*)
     Subfamily Metamynodontinae Kretzoi, 1942
      *Megalamynodon* Wood, 1945
      *Paramynodon* Matthew, 1929
      *Zaisanamynodon* Beliaeva, 1971
      *Metamynodon* Scott and Osborn, 1887 (= *Cadurcopsis*)
      *Cadurcotherium* Gervais, 1873 (= *Cadurcamynodon, Hypsamynodon*)
   Family Amynodontidae *incertae sedis*:
     *Gigantamynodon* Gromova, 1954
     *Huananodon* You, 1977
     *Procadurcodon* Gromova, 1960
     *Mesamynodon* Peterson, 1931

Grandfamily Rhinocerotida Owen, 1845 (new rank)
   Family Hyracodontidae Cope, 1879
     Subfamily Indricotheriinae Borissiak, 1923
      *Forstercooperia* Wood, 1939 (=*Pappaceras*)
      *Juxia* Chow and Chiu, 1963 (=*Imequincisoria, Guixia*)
      *Urtinotherium* Chow and Chiu, 1963

*Paracheratherium* Forster Cooper, 1911 (= *Baluchitherium, Thaumastotherium, Aralotherium, Dzungariotherium*)
*Indricotherium* Borissiak, 1915
Subfamily Indricotheriinae *incertae sedis*
*Benaratherium* Gabuniya, 1955
Subfamily Hyracodontinae Cope, 1879
*Triplopus* Cope, 1880 (= *Prothyracodon, Eotrigonias, Ephyrachyus*)
*Hyracodon* Leidy, 1856
*Epitriplopus* Wood, 1927
*Ardynia* Matthew and Granger, 1923 (= *Ergilia*)
*Triplopides* Radinsky, 1965
Subfamily Allaceropinae Wood, 1932
*Prohyracodon* Koch, 1897 (= *Meninatherium*)
*Ilianodon* Chow and Xu, 1961
*Eggysodon* Roman, 1911 (= *Allacerops* Wood, 1932)
Family Rhinocerotidae Owen, 1845
*Teletaceras* Hanson, 1989 (this volume, Chapter 20)
*Ronzotherium* Aymard, 1886
*Epiaceratherium* Abel, 1910
*Penetrigonias* Tanner and Martin, 1976
*Trigonias* Lucas, 1900
*Amphicaenopus* Wood, 1927
Subfamily Diceratheriinae Dollo, 1885
*Subhyracodon* Brandt, 1878 (= *Caenopus, Leptaceratherium*)
*Diceratherium* Marsh, 1875
Subfamily Menoceratinae Prothero, Manning, and Hanson, 1986
*Menoceras* Troxell, 1921 (= *Moschoedestes*)
*Pleuroceros* Roger, 1898
*Protaceratherium* Abel, 1910
Subfamily Aceratheriinae Dollo, 1885
*Mesaceratherium* Heissig, 1969
*Proaceratherium* Ginsburg and Hugueney, 1980
*Floridaceras* Wood, 1966
*Aphelops* Cope, 1873
*Peraceras* Cope, 1880
*Aceratherium* Kaup, 1832
*Hoploaceratherium* Ginsburg and Heissig, 1989 (this volume, Chapter 22)
*Dromoaceratherium* Crusafont and Villalta, 1955
*Plesiaceratherium* Young, 1937
*Alicornops* Ginsburg and Guérin, 1979
*Subchilotherium* Heissig, 1972
*Chilotherium* Ringström, 1924
*Chilotheridium* Hooijer, 1971
*Acerorhinus* Kretzoi, 1942
*Sinorhinus* Schlosser, 1903
Subfamily Rhinocerotinae Owen, 1845
Tribe Teleoceratini Hay, 1902
*Diaceratherium* Dietrich, 1931

    *Brachydiceratherium* Lavocat, 1951
    *Prosantorhinus* Heissig, 1973
    *Teleoceras* Hatcher, 1894
    *Brachypotherium* Roger, 1904
    *Aprotodon* Forster Cooper, 1915
    *Brachypodella* Heissig, 1973
  Tribe Rhinocerotini Owen, 1845
   Subtribe Dicerorhinina Ringström, 1924 (new rank)
    *Dicerorhinus* Gloger, 1841 (=*Didermocerus, Ceratorhinus*)
    *Stephanorhinus* Kretzoi, 1942
    *Lartetotherium* Ginsburg, 1974
    *Coelodonta* Bronn, 1831 (= *Tichorhinus*)
   Subtribe Elasmotheriina Bonaparte, 1845 (new rank)
    *Hispanotherium* Crusafont and Villalta, 1947 (=*Begertherium,*
     *Caementodon, Beliajevina, Tesseledon*)
    *Iranotherium* Ringström, 1924
    *Kenyatherium* Aguirre and Guérin, 1974
    *Sinotherium* Ringström, 1922 (=*Parelasmotherium* )
    *Elasmotherium* Fischer, 1808
   Subtribe Rhinocerotina Owen, 1845 (new rank)
    *Gaindatherium* Colbert, 1934
    *Rhinoceros* Linnaeus, 1758
   Subtribe Dicerotina Ringström, 1924 (new rank)
    *Paradiceros* Hooijer, 1968
    *Diceros* Gray, 1821
    *Ceratotherium* Gray, 1867
 Rhinocerotoidea *incertae sedis*:
   *Symphyssorrhachis* Belyajeva, 1954
   *Meschotherium* Gabunia, 1964
   *Ninxiatherium* Chen, 1977
   *Punjabitherium* Kahn, 1971

Infraorder Titanotheriomorpha Hooker, 1989 (new rank)
 Superfamily Brontotherioidea Marsh, 1873
  Family Lambdotheriidae Cope, 1889
    *Lambdotherium* Cope, 1880
    *Xenicohippus* Bown and Kihm, 1981
  Family Brontotheriidae Marsh, 1873
    *Eotitanops* Osborn, 1907
    *Pakititanops* West, 1980
    *Palaeosyops* Leidy, 1870 (= *Limnohyops, Limnohyus, Eometarhinus*)
   Subfamily Dolichorhininae Riggs, 1912
    *Mesatirhinus* Osborn, 1908
    *Rhadinorhinus* Riggs, 1912
    *Dolichorhinus* Hatcher, 1895
   Subfamily Telmatheriinae Osborn, 1914
    *Telmatherium* Marsh, 1872 (= *Leurocephalus, Manteoceras*)
    *Metatelmatherium* Granger and Gregory, 1938

*Sthenodectes* Gregory, 1912
*Protitanotherium* Hatcher, 1895
*Eotitanotherium* Peterson, 1914
*Notiotitanops* Gazin and Sullivan, 1942
*Duchesneodus* Lucas and Schoch, 1982
*Teleodus* Marsh, 1890
*Brontops* Marsh, 1887 (= *Diploclonus*)
*Menops* Marsh, 1887 (= *Allops*)
*Megacerops* Leidy, 1870 (= *Brontotherium, Titanops*)
Asian brontotheres [synonymies and relationships not recently revised; many are invalid]
  *Sivatitanops* Pilgrim, 1925
  *Desmatotitan* Granger and Gregory, 1943
  *Hyotitan* Granger and Gregory, 1943
  *Epimanteoceras* Granger and Gregory, 1943
  *Protitan* Granger and Gregory, 1943
  *Microtitan* Granger and Gregory, 1943
  *Dolichorhinoides* Granger and Gregory, 1943
  *Gnathotitan* Granger and Gregory, 1943
  *Rhinotitan* Granger and Gregory, 1943
  *Pachytitan* Granger and Gregory, 1943
  *Parabrontops* Granger and Gregory, 1943
  *Metatitan* Granger and Gregory, 1943
  *Titanodectes* Granger and Gregory, 1943
  *Embolotherium* Osborn, 1929
  *Protembolotherium* Tang, You, Xu, Qiu, and Hu, 1974
  *Arctotitan* Chow, 1978
  *Dianotitan* Chow and Hu, 1959
Brontotheriidae *incertae sedis*:
  *Menodus* Pomel, 1849 (*nomen dubium*)
  *Symborodon* Cope, 1873
  *Sphenocoelus* Osborn, 1895
  *Protitanops* Stock, 1936
  *Diplacodon* Marsh, 1875
  *Brachydiastematherium* Böckh and Maty, 1876
  *Titanotherium* Leidy, 1852

**Bibliography**

Gregory, J. T., Backskai, J.A., Shkurkin, G.V., Winans, M.C., and Rauscher, B.H. (1988): *Bibliography of Fossil Vertebrates, 1985.* -Los Angeles (Society of Vertebrate Paleontology).

Simpson, G.G. (1945): The principles of classification and a classification of mammals.- *Bull. Amer. Mus. Nat. Hist.*, 85: 1-350.